# AN INTRODUCTION TO
# ANALYSIS

# AN INTRODUCTION TO
# ANALYSIS

## SECOND EDITION

## WILLIAM R. WADE
UNIVERSITY OF TENNESSEE

Prentice Hall
Upper Saddle River, NJ 07458

Library of Congress Cataloging-in-Publication Data

Wade, W. R.
    An introduction to analysis / William R. Wade. --2nd ed.
      p. cm.
    Includes bibliographical references and index.
    ISBN 0-13-014409-6
    1. Mathematical analysis. I. Title.
  QA300.W25    1999
  515–dc21                    99-28230
                                  CIP

Acquisition editor: George Lobell
Editor-in chief: Jerome Grant
Executive managing editor: Kathleen Schiaparelli
Managing editor: Linda Mihatov Behrens
Editorial assistants: Nancy Bauer/Gale Epps
Assistant VP production/manufacturing: David W. Riccardi
Manufacturing manager: Trudy Pisciotti
Manufacturing buyer: Alan Fischer
Art director: Jayne Conte
Cover Designer: Bruce Kenselaar
Marketing manager: Melody Marcus

Printed in the United States of America
10 9 8 7 6 5 4 3 2 1

ISBN 0-13-014409-6

PRENTICE-HALL OF AUSTRALIA PTY. LIMITED, SYDNEY
PRENTICE-HALL INTERNATIONAL (UK) LIMITED, LONDON
PRENTICE-HALL CANADA INC., TORONTO
PRENTICE-HALL HISPANOAMERICA, S.A., MEXICO
PRENTICE-HALL OF INDIA PRIVATE LIMITED, NEW DELHI
PRENTICE-HALL OF JAPAN, INC., TOKYO
SIMON & SCHUSTER ASIA PTE. LTD., SINGAPORE
EDITORA PRENTICE-HALL DO BRASIL, LTDA., RIO DE JANEIRO

# Contents

# Preface

This text provides a bridge from "sophomore" calculus to graduate courses which use analytic ideas, e.g., real and complex analysis, partial and ordinary differential equations, numerical analysis, fluid mechanics, and differential geometry. For a two-semester course, the first semester should end with Chapter 8. For a three-quarter course, the second quarter should begin in Chapter 6 and end somewhere in the middle of Chapter 11.

Our presentation is divided into two parts. The first half, Chapters 1 through 7 together with Appendices A and B, gradually introduces the central ideas of analysis in a one-dimensional setting. The second half, Chapters 8 through 15 together with Appendices C through F, covers multidimensional theory. More specifically, Chapter 1 introduces the real number system as a complete, ordered field, Chapters 2 through 5 covers calculus on the real line, and Chapters 6 and 7 discusses infinite series, including uniform and absolute convergence. Chapter 8 is a short bridge where we explore convergence of sequences and functions in $\mathbf{R}^n$, and the connection between linear functions and matrices. After these foundations have been laid, the material becomes more abstract with the introduction of certain topological concepts. These can be covered (according to the instructor's preference) in the concrete Euclidean space setting (Chapter 9), or in the abstract metric space setting (Chapter 10). (Since either of these chapters provides the necessary foundation for the rest of the book, the instructors are free to choose the approach they feel best suits their aims. There is no need to cover both chapters in the same course.) Chapters 11 through 13 develops the machinery and theory of vector calculus. Chapter 14 gives a short introduction to Fourier series, including summability and convergence of Fourier series, growth of Fourier coefficients, and uniqueness of trigonometric series. And Chapter 15 gives a short introduction to differentiable manifolds which culminates in a proof of Stokes's Theorem on differentiable manifolds.

Separating the one-dimensional from the $n$-dimensional material is not the most efficient way to present the material, but it does have two advantages. Introduction of the more abstract, geometric concepts can be postponed until students have been given a thorough introduction to analysis on the real line. And, students have two chances to master some of the deeper ideas of analysis (e.g., convergence of sequences, limits of functions, and uniform continuity).

A course based on this text is suitable not only for mathematics majors (late undergraduate or first-year graduate), but also for science and engineering majors who need a careful introduction to the concepts and methods of analytic proofs (e.g., those planning to pursue Ph.D.'s in theoretical physics, electrical engineering, or operations research). Such students (especially those specializing in applied mathematics) must be prepared to work in higher dimensions, able to change variables in multiple integrals, optimize functions of several variables, apply the fundamental theorems of vector calculus, and understand the curl, divergence, and gradient operators from both a mechanical and a geometric standpoint. For this reason, we have avoided the temptation to emphasize the one variable material at the expense of the multidimensional case.

Since for many students this is the last (for some the only) place to see a rigorous development of vector calculus, we focus our attention on classical, nitty-gritty analysis. By avoiding abstract concepts such as vector spaces and the Lebesgue integral, we have room for a thorough, comprehensive introduction. We include sections on improper integration, the gamma function, Lagrange multipliers, the Inverse and Implicit Function Theorem, Green's Theorem, Gauss's Theorem, Stokes's Theorem, and a full account of the change of variables formula for multiple Riemann integrals.

We assume the reader has completed a three-semester or four-quarter sequence in elementary calculus. Because many of our students now take their elementary calculus in high school (where theory may be almost nonexistent), we assume that the reader is familiar only with the mechanics of calculus, i.e., can differentiate, integrate, and graph simple functions of the form $y = f(x)$ or $z = f(x, y)$. We also assume the reader has had an introductory course in linear algebra, i.e., can add, multiply, take determinants of matrices with real entries, and is familiar with Cramer's Rule. (Appendix C, which contains an exposition of all definitions and theorems from linear algebra used in the text, can be used as review if the instructor deems it necessary.)

Always we emphasize the fact that the concepts and results of analysis are based on simple geometric considerations and on analogies with material already known to the student. The aim is to keep the course from looking like a collection of tricks and to share enough of the motivation behind the mathematics so that students are prepared to construct their own proofs when asked. We begin complicated proofs with a short paragraph (marked STRATEGY:) which shows why the proof works, e.g., the Archimedean Principle (Theorem 1.22), Density of Rationals (Theorem 1.24), Cauchy's Theorem (Theorem 2.29), Change of Variables in **R** (Theorem 5.34), Riemann's Theorem about rearrangements (Theorem 6.29), the Implicit Function Theorem (Theorem 11.47), the arc length formula (Theorem 13.5), and Stokes's Theorem on Manifolds (Theorem 15.45). We precede abstruse definitions or theorems with a short paragraph which describes, in simple terms, what behavior we are examining, and why, e.g., Cauchy sequences, one-sided limits, upper and lower Riemann sums, the Integral Test, Abel's Formula, uniform convergence, the total derivative, compact sets, differentiable curves and surfaces, smooth curves, and orientation equivalence. And we include examples to show why each hypothesis of a major theorem is necessary, e.g., the Nested Interval Property, the Bolzano–Weierstrass Theorem, the Mean Value Theorem, the Heine–Borel Theorem, the Inverse Function Theorem, the existence of exact differentials, and Fubini's Theorem.

We have made this text very flexible by including, but distinguishing, core material from enrichment material. The core material, occupying fewer than 389 pages, can be covered easily in a one-year course. The enrichment material is included for two reasons: curious students can use it to delve deeper into the core material or as a jumping off place to pursue more general topics, and instructors can use it to supplement their course or to add variety from year to year.

Enrichment material appears in enrichment sections, marked with a superscript $e$, or in core sections, where it is marked with an asterisk. Exercises which use enrichment material are also marked with an asterisk, and the material needed to solve them is cited in the Answers and Hints section. To make course planning easier, each enrichment section begins with a statement which indicates whether that section uses material from any other enrichment section. Since no core material depends on enrichment material, any of the latter can be skipped without loss in the integrity of the course.

Each section contains a collection of exercises designed to allow students the luxury and thrill of discovering mathematics on their own. These exercises range from very elementary (to be sure the student understands the concepts introduced in that section) to more challenging (to give the student practice in using these concepts to expand the theory). To minimize frustration, some of the more difficult exercises have several parts which serve as an outline to a solution of the problem. To keep from producing students who know theory but cannot apply it, each set of exercises contains a mix of computational and theoretical assignments. (Exercises which play a prominent role later in the text are marked with a box. These exercises are an integral part of the course and all of them should be assigned.)

Since many students have difficulty reading and understanding mathematics, we have paid close attention to style and organization. We have consciously limited the vocabulary, kept notation consistent from chapter to chapter, and presented the proofs in a unified style. Individual sections are determined by subject matter, not by length of lecture, so that students can comprehend related results in a larger context. Examples and important remarks are numbered and labeled so that students can read the text in small chunks. (Many of these, included for the student's benefit, need not be covered in class.) Paragraphs are short and focused so that students are not overwhelmed by long-winded explanations. To help students discern between central results and peripheral ones, the word "Theorem" is used relatively sparingly; preliminary results and results which are only used in one section are called Remarks, Lemmas, and Examples.

How does the second edition differ from the first? We have corrected a number of misprints which appeared in the first edition. We have simplified the presentation in several ways: We have avoided "limits through a set", sequential compactness, made the section on limits supremum and infimum optional, and completely re-worked the proof of the change of variables formula for multivariable functions. We have numbered definitions, theorems, examples, etc. consecutively so one can find a cited result more easily. We have divided some of the longer chapters into smaller pieces. We have reorganized the presentation of Euclidean spaces, separating the computational material from the topological material. We have included an optional chapter on metric spaces (which can be covered instead of the chapter on the topology of Euclidean spaces, for those who prefer a more abstract approach). We have reorganized the presentation of series, gathering the more esoteric material into separate enrichment sections. And, we have added more straightforward exercises, especially in the chapters on limits, integration, infinite series, and differentiation.

I wish to thank Mr. P.W. Wade and Professors S. Fridli, G.S. Jordan, Mefharet Kocatepe, J. Long, M.E. Mays, M.S. Osborne, P.W. Schaefer, F.E. Schroeck, and Ali Sinan Sertoz, who carefully read parts of the first edition and made many valuable suggestions and corrections. Also, I wish to express my gratitude to Ms. C.K. Wade for several lively discussions of a pedagogical nature, which helped shape the organization and presentation of this material.

If there remain any typographical errors, I plan to keep an up-to-date list at my web site (http://www.math.utk.edu/~wade). If you find errors which are not listed at that site, I would appreciate your contacting me at the e-mail address below.

iwade@utk.edu

# Chapter 1

# The Real Number System

You have already had several calculus courses in which you evaluated limits, differentiated functions, and computed integrals. You may even remember some of the major results of calculus, such as the Chain Rule, the Mean Value Theorem, and the Fundamental Theorem of Calculus. Although you are probably less familiar with multivariable calculus, you have taken partial derivatives, computed gradients, and evaluated certain line and surface integrals.

In view of all this, you must be asking: Why another course in calculus? The answer to this question is twofold. Although some proofs may have been presented in earlier courses, it is unlikely that the subtler points (e.g., completeness of the real numbers, uniform continuity, and uniform convergence) were covered. Moreover, the skills you acquired were mostly computational; you were rarely asked to prove anything yourself. This course develops the theory of calculus carefully and rigorously from basic principles and gives you a chance to learn how to construct your own proofs. It also serves as an introduction to analysis, an important branch of mathematics which provides a foundation for numerical analysis, functional analysis, harmonic analysis, differential equations, differential geometry, real analysis, complex analysis, and many other areas of specialization within mathematics.

Every rigorous study of mathematics begins with certain undefined concepts, primitive notions on which the theory is based, and certain postulates, properties which are assumed to be true and need no proof. Our study will be based on the primitive notions of set and real numbers and on four postulates (containing a total of 18 different properties), which will be introduced in the first three sections of this chapter.

## 1.1 ORDERED FIELD AXIOMS

In this section, we explore the algebraic structure of the real number system. We shall use standard set theoretic notation. For example, $\emptyset$ represents the *empty set* (the set with no elements), $a \in A$ means that $a$ is an *element of* $A$, and $a \notin A$ means that $a$ is *not* an element of $A$. We can represent a given finite set in two ways. We

1

can list its elements directly, or we can describe it using sentences or equations. For example, the set of solutions to the equation $x^2 = 1$ can be written as

$$\{1, -1\} \quad \text{or} \quad \{x : x^2 = 1\}.$$

A set $A$ is said to be a *subset* of a set $B$ (notation: $A \subseteq B$) if every element of $A$ is also an element of $B$. If $A$ is a subset of $B$ but there is at least one element $b \in B$ which does not belong to $A$, we shall call $A$ a *proper subset* of $B$ (notation: $A \subset B$). Two sets $A$ and $B$ are said to be *equal* (notation: $A = B$) if $A \subseteq B$ and $B \subseteq A$. If $A$ and $B$ are not equal, we write $A \neq B$. A set $A$ is said to be *nonempty* if $A \neq \emptyset$.

The *union* of two sets $A$ and $B$ (notation: $A \cup B$) is the set of elements $x$ such that $x$ belongs to $A$ or $B$ or both. The *intersection* of two sets $A$ and $B$ (notation: $A \cap B$) is the set of elements $x$ such that $x$ belongs to both $A$ and $B$. The *complement* of $B$ *relative to* $A$ (notation: $A \setminus B$, sometimes $B^c$ if $A$ is understood) is the set of elements $x$ such that $x$ belongs to $A$ but does not belong to $B$. For example,

$$\{-1, 0, 1\} \cup \{1, 2\} = \{-1, 0, 1, 2\}, \qquad \{-1, 0, 1\} \cap \{1, 2\} = \{1\},$$

$$\{1, 2\} \setminus \{-1, 0, 1\} = \{2\} \quad \text{and} \quad \{-1, 0, 1\} \setminus \{1, 2\} = \{-1, 0\}.$$

Let $X$ and $Y$ be sets. The *Cartesian product* of $X$ and $Y$ is the set of *ordered pairs* defined by

$$X \times Y := \{(x, y) : x \in X, y \in Y\}.$$

(The symbol := means "equal by definition" or "is defined to be.") Two points $(x, y), (z, w) \in X \times Y$ are said to be *equal* if $x = z$ and $y = w$.

Let $X$ and $Y$ be sets. A *relation* on $X \times Y$ is any subset of $X \times Y$. Let $\mathcal{R}$ be a relation on $X \times Y$. The *domain* of $\mathcal{R}$ is the collection of $x \in X$ such that $(x, y)$ belongs to $\mathcal{R}$. When $(x, y) \in \mathcal{R}$, we shall frequently write $x \mathcal{R} y$.

A *function* $f$ from $X$ into $Y$ (notation: $f : X \to Y$) is a relation on $X \times Y$ whose domain is $X$ (notation: $\text{Dom}\,(f) := X$) such that for each $x \in X$ there is one and only one $y \in Y$ which satisfies $(x, y) \in f$. In this case we say that $f$ is *defined* on $X$, and call $y$ the *value* of $f$ at $x$ (notation: $y = f(x)$ or $f : x \longmapsto y$). Notice that by the definition of equality of ordered pairs, two functions $f$ and $g$ are equal if and only if they have the same domain and values, i.e., $\text{Dom}\,(f) = \text{Dom}\,(g)$ and $f(x) = g(x)$ for all $x \in \text{Dom}\,(f)$. (For the definition of the *domain* of a function defined by a formula, see Section 1.4.)

We shall denote the set of real numbers by $\mathbf{R}$. We shall assume that $\mathbf{R}$ is a field, i.e., that $\mathbf{R}$ satisfies the following postulate.

**POSTULATE 1.** [THE FIELD AXIOMS]. There are functions $+$ and $\cdot$, defined on $\mathbf{R}^2 := \mathbf{R} \times \mathbf{R}$, which satisfy the following properties for every $a, b, c \in \mathbf{R}$:

**The Closure Properties.** $a + b$ and $a \cdot b$ belong to $\mathbf{R}$.

**The Associative Properties.** $a + (b + c) = (a + b) + c$ and $a \cdot (b \cdot c) = (a \cdot b) \cdot c$.

**The Commutative Properties.** $a + b = b + a$ and $a \cdot b = b \cdot a$.

**The Distributive Law.** $a \cdot (b + c) = a \cdot b + a \cdot c.$

**Existence of the Additive Identity.** There is a unique element $0 \in \mathbf{R}$ such that $0 + a = a$ for all $a \in \mathbf{R}$.

**Existence of the Multiplicative Identity.** There is a unique element $1 \in \mathbf{R}$ such that $1 \neq 0$ and $1 \cdot a = a$ for all $a \in \mathbf{R}$.

**Existence of Additive Inverses.** For every $x \in \mathbf{R}$ there is a unique element $-x \in \mathbf{R}$ such that

$$x + (-x) = 0.$$

**Existence of Multiplicative Inverses.** For every $x \in \mathbf{R} \setminus \{0\}$ there is a unique element $x^{-1} \in \mathbf{R}$ such that

$$x \cdot (x^{-1}) = 1.$$

We note in passing that the word "unique" can be dropped from the statement of Postulate 1 (see Appendix A).

We shall frequently denote $a + (-b)$ by $a - b$, $a \cdot b$ by $ab$, $a^{-1}$ by $\dfrac{1}{a}$ or $1/a$, and $a \cdot b^{-1}$ by $\dfrac{a}{b}$ or $a/b$. Notice that by the existence of additive and multiplicative inverses, the equation $x + a = 0$ can be solved for each $a \in \mathbf{R}$ and the equation $ax = 1$ can be solved for each $a \in \mathbf{R}$ provided $a \neq 0$.

From these few properties (i.e., from Postulate 1), one can derive all the usual algebraic laws of real numbers, including the following:

$$(1) \qquad\qquad\qquad (-1)^2 = 1,$$

$$(2) \qquad 0 \cdot a = 0, \quad -a = (-1) \cdot a, \quad -(-a) = a, \qquad a \in \mathbf{R},$$

$$(3) \qquad\qquad -(a - b) = b - a, \qquad a, b \in \mathbf{R},$$

and

$$(4) \qquad a, b \in \mathbf{R} \quad \text{and} \quad ab = 0 \quad \text{imply} \quad a = 0 \quad \text{or} \quad b = 0.$$

We want to keep our attention sharply focused on analysis. Since the proofs of algebraic laws like these lie more in algebra than analysis (see Appendix A), we will not present them here. In fact, with the exception of the absolute value and the Binomial Formula, we will use all material usually presented in a high school algebra course (including the quadratic formula and graphs of the conic sections) without further explanation as the need arises.

Postulate 1 is sufficient to derive all algebraic laws of $\mathbf{R}$ but it does not completely describe the real number system. The set of real numbers also has an order relation, i.e., a concept of "less than."

**POSTULATE 2.** [THE ORDER AXIOMS]. There is a relation $<$ on $\mathbf{R} \times \mathbf{R}$ which has the following properties:

**The Trichotomy Property.** Given $a, b \in \mathbf{R}$, one and only one of the following statements holds:

$$a < b, \quad b < a, \quad \text{or} \quad a = b.$$

**The Transitive Property.**

$$a < b \quad \text{and} \quad b < c \quad \text{imply} \quad a < c.$$

**The Additive Property.**

$$a < b \quad \text{and} \quad c \in \mathbf{R} \quad \text{imply} \quad a + c < b + c.$$

**The Multiplicative Properties.**

$$a < b \quad \text{and} \quad c > 0 \quad \text{imply} \quad ac < bc,$$

and

$$a < b \quad \text{and} \quad c < 0 \quad \text{imply} \quad bc < ac.$$

By $b > a$ we shall mean $a < b$. By $a \le b$ and $b \ge a$ we shall mean $a < b$ or $a = b$. If $a < b$ and $b < c$, we shall write $a < b < c$.

We shall call a number $a \in \mathbf{R}$ *nonnegative* if $a \ge 0$ and *positive* if $a > 0$. Postulate 2 has a slightly simpler formulation using the set of positive elements as a primitive concept (see Exercise 11). We have introduced Postulate 2 as above because these are the properties we use most often.

The real number system $\mathbf{R}$ contains certain special subsets: the set of *natural numbers*

$$\mathbf{N} := \{1, 2, \dots\},$$

obtained by beginning with 1 and successively adding 1's to form $2 := 1 + 1$, $3 := 2 + 1$, etc.; the set of *integers*

$$\mathbf{Z} := \{\dots - 2, -1, 0, 1, 2, \dots\}$$

(*Zahlen* is German for number); the set of *rationals* (or fractions or quotients)

$$\mathbf{Q} := \left\{ \frac{m}{n} : m, n \in \mathbf{Z} \text{ and } n \neq 0 \right\};$$

and the set of *irrationals*

$$\mathbf{Q}^c := \mathbf{R} \setminus \mathbf{Q}.$$

Equality in $\mathbf{Q}$ is defined by

$$\frac{m}{n} = \frac{p}{q} \quad \text{if and only if} \quad mq = np.$$

Recall that each of the sets $\mathbf{N}, \mathbf{Z}, \mathbf{Q}$, and $\mathbf{R}$ is a proper subset of the next, i.e.,

$$\mathbf{N} \subset \mathbf{Z} \subset \mathbf{Q} \subset \mathbf{R}.$$

For example, every rational is a real number (because $m/n := mn^{-1}$ is a real number by Postulate 1), but $\sqrt{2}$ is an irrational.

Since we did not really define $\mathbf{N}$ and $\mathbf{Z}$, we must make certain assumptions about them. If you are interested in the definitions and proofs, see Appendix A.

**1.1 Remark.** *We assume that* **N** *and* **Z** *satisfy the following properties.*

    i) *Given $n \in \mathbf{Z}$, one and only one of the following statements holds:*
$$n \in \mathbf{N}, \quad -n \in \mathbf{N}, \ or \quad n = 0.$$

    ii) *If $n \in \mathbf{N}$ then $n + 1 \in \mathbf{N}$ and $n \geq 1$.*

    iii) *If $n \in \mathbf{N}$ and $n \neq 1$, then $n - 1 \in \mathbf{N}$.*

    iv) *If $n \in \mathbf{Z}$ and $n > 0$, then $n \in \mathbf{N}$.*

Using these properties and induction, we can prove that **N** and **Z** are closed under addition and multiplication (see Remark 1.13). We can also prove that **Q** satisfies Postulate 1 (see Exercise 6 below).

We notice in passing that none of the other special subsets of **R** satisfies Postulate 1. **N** satisfies all but three of the properties in Postulate 1: **N** has no additive identity (since $0 \notin \mathbf{N}$), **N** has no additive inverses (e.g., $-1 \notin \mathbf{N}$), and only one of the nonzero elements of **N** (namely, 1) has a multiplicative inverse. **Z** satisfies all but one of the properties in Postulate 1: only two nonzero elements of **Z** have multiplicative inverses (namely, 1 and $-1$). $\mathbf{Q}^c$ satisfies all but four of the properties in Postulate 1: $\mathbf{Q}^c$ does not have an additive identity (since $0 \notin \mathbf{R} \setminus \mathbf{Q}$), does not have a multiplicative identity (since $1 \notin \mathbf{R} \setminus \mathbf{Q}$), and does not satisfy either closure property. Indeed, since $\sqrt{2}$ is irrational, the sum of irrationals may be rational ($\sqrt{2} + (-\sqrt{2}) = 0$) and the product of irrationals may be rational ($\sqrt{2} \cdot \sqrt{2} = 2$).

Notice that any subset of **R** satisfies Postulate 2. Thus **Q** satisfies both Postulates 1 and 2. The remaining two postulates, introduced in Sections 1.2 and 1.3 below, identify properties which **Q** does not possess. In particular, these four postulates distinguish **R** from each of its special subsets **N**, **Z**, **Q**, and $\mathbf{Q}^c$. These postulates actually characterize **R**; i.e., **R** is the only set which satisfies Postulates 1 through 4. [Such a set is called a *complete Archimedean ordered field*. We may as well admit a certain arbitrariness in choosing this approach. **R** has been axiomized in at least five other ways (e.g., as a one-dimensional continuum or as a set of binary decimals with certain arithmetic operations). The decision to present **R** using Postulates 1 through 4 is partly based on economy and partly on personal taste.]

Using all four postulates, one can define a function $f : x \longmapsto x^\alpha$ for any $x > 0$ and $\alpha \in \mathbf{R}$ (see Exercise 5 in Section 5.3) so that the following properties hold: $x^0 = 1$, $x^\alpha > 0$, $x^\alpha \cdot x^\beta = x^{\alpha+\beta}$, and $(x^\alpha)^\beta = x^{\alpha \cdot \beta}$ for all $\alpha, \beta \in \mathbf{R}$ and all $x > 0$, and if $\alpha = n \in \mathbf{N}$ then $x^n = x \cdot \ldots \cdot x$ (there are $n$ factors here). We also define $0^\alpha := 0$ for $\alpha > 0$. [The symbol $0^0$ is left undefined because it is indeterminate (see Example 4.22).]

Because it would be impractical to wait until Chapter 5 to use $x^\alpha$ for examples, we shall accept these properties as given and use them as the need arises. We shall also accept, as given, the *trigonometric functions* (whose formulas are) represented by $\sin x$, $\cos x$, $\tan x$, $\cot x$, $\sec x$, $\csc x$, the *exponential function* $e^x$ and its inverse, the *natural logarithm*

$$\log x := \int_1^x \frac{dt}{t},$$

defined and real-valued for each $x \in (0, \infty)$. [Although this last function is denoted by $\ln x$ in elementary calculus texts, most analysts denote it, as we did just now, by $\log x$. We will follow this practice throughout this text.]

Notice that $x^\alpha \cdot x^{-\alpha} = x^{\alpha - \alpha} = x^0 = 1$. By the uniqueness of multiplicative inverses, it follows that $x^{-\alpha} = (x^\alpha)^{-1} = 1/x^\alpha$ for all $\alpha \in \mathbf{R}$ and $x > 0$.

If $\alpha = 1/m$ for some $m \in \mathbf{N}$, we shall denote $x^\alpha$ by $\sqrt[m]{x}$. (We shall also write $\sqrt{x}$ for $\sqrt[2]{x}$.) Hence

$$( \sqrt[m]{x})^n = (x^{1/m})^n = x^{n/m} = (x^n)^{1/m} = \sqrt[m]{x^n}$$

for all $m \in \mathbf{N}$, $n \in \mathbf{Z}$, and $x > 0$. In particular, $( \sqrt[m]{x})^m = \sqrt[m]{x^m} = x$ for all $x > 0$ and $m \in \mathbf{N}$. Notice that since $x^\alpha$ is positive when $x > 0$, $\sqrt[m]{x} \geq 0$ for all $x \geq 0$ and $m \in \mathbf{N}$.

Postulates 1 and 2 can be used to derive all the usual algebraic laws regarding real numbers and inequalities (e.g., see implications (5) through (9) below). Since arguments based on inequalities are of fundamental importance to analysis, we begin to supply details of proofs at this stage.

What is a proof? Every mathematical result (for us this includes examples, remarks, lemmas, and theorems) has hypotheses and a conclusion. There are three main methods of proof: mathematical induction, direct deduction, and contradiction.

*Mathematical induction*, a special method for proving statements that depend on positive integers, will be covered in Section 1.2.

To construct a *deductive proof* we assume the hypotheses to be true and proceed step by step to the conclusion. Each step is justified by a hypothesis, a definition, a postulate, or a mathematical result which has already been proved. (Actually, this is usually the way we write a proof. When constructing your own proofs, you may find it helpful to work forward from the hypotheses as far as you can and then work backward from the conclusion, trying to meet in the middle.)

To construct a *proof by contradiction* we assume the hypotheses to be true, the conclusion to be false, and work step by step deductively until a *contradiction* occurs; i.e., a statement which is obviously false or which is contrary to the assumptions made. At this point the proof by contradiction is complete. The phrase "suppose to the contrary" always indicates a proof by contradiction (e.g., see the proof of Theorem 1.9 below).

Here are some examples of deductive proofs. (Note: The symbol ∎ indicates that the proof or solution is complete.)

**1.2 Example.** If $a \in \mathbf{R}$, prove

$$(5) \qquad\qquad a \neq 0 \quad \text{implies} \quad a^2 > 0.$$

In particular, $-1 < 0 < 1$.

PROOF. Suppose $a \neq 0$. By the Trichotomy Property, either $a > 0$ or $a < 0$.

*Case 1.* $a > 0$. Multiply both sides of this inequality by $a$. By the first Multiplicative Property, we obtain $a^2 = a \cdot a > 0 \cdot a = 0$.

*Case 2.* $a < 0$. Multiply both sides of this inequality by $a$. Since $a < 0$, it follows from the second Multiplicative Property that $a^2 = a \cdot a > 0 \cdot a = 0$. This proves $a^2 > 0$ when $a \neq 0$.

Since $1 \neq 0$, it follows that $1 = 1^2 > 0$. Adding $-1$ to both sides of this inequality, we conclude that $0 = 1 - 1 > 0 - 1 = -1$. ∎

$$0 < \sqrt[n]{a} < 1 \qquad \qquad \sqrt[n]{a} > 1$$

**1.3 Example.** If $a \in \mathbf{R}$, prove

$$0 < a < 1 \text{ implies } 0 < a^n < a \text{ and } a > 1 \text{ implies } a^n > a$$
$$\text{for } n \in \mathbf{N}.$$

(6)         $0 < a < 1$   implies   $0 < a^2 < a$,   and   $a > 1$   implies   $a^2 > a$.

PROOF. Suppose $0 < a < 1$. Multiply both sides of this inequality by $a$. By the first Multiplicative Property,

$$0 = 0 \cdot a < a \cdot a = a^2 < 1 \cdot a = a.$$

On the other hand, if $a > 1$, then $a > 0$ by Example 1.2 and the Transitive Property. Multiplying $a > 1$ by $a$, we conclude that $a^2 = a \cdot a > 1 \cdot a = a$. ∎

Similarly (see Exercise 4), we can prove

$$0 \leq a < b \text{ imply } 0 \leq a^n < b^n \text{ for } n \in \mathbf{N}$$
$$0 \leq a < b \text{ imply } 0 < \sqrt[n]{a} < \sqrt[n]{b} \text{ for } n \in$$

(7)              $0 \leq a < b$   and   $0 < c < d$   imply   $ac < bd$,

(8)              $0 \leq a < b$   implies   $0 \leq a^2 < b^2$   and   $0 \leq \sqrt{a} < \sqrt{b}$,

and

(9)              $0 < a < b$   implies   $\dfrac{1}{a} > \dfrac{1}{b} > 0$.

Although it may seem both pedantic and unnecessary to include proofs of such well known (yes, perhaps even obvious) laws, we include them here for several reasons. We want this book to be reasonably self-contained, because this will make it easier for you to begin to construct your own proofs. We want the first proofs you see in this book to be easily understood, because they deal with familiar properties which are unobscured by new concepts. And we want to form a habit of proving all statements, even "obvious" statements like these. The reason for this hard headed approach is that some "obvious" statements are false. For example, some students think it obvious that any continuous function must be differentiable at some point. Others think it obvious that if every rational in $[0, 1]$ is covered by a small interval, then the sum of the lengths of those intervals must exceed 1. We shall see that both these statements, and many others equally "obvious," are false (e.g., see Theorem 7.62 and Remark 9.43). In particular, we harbor a skepticism that demands proofs of all statements, even the "obvious" ones.

What, then, are you allowed to use when solving the exercises? You may use any property of real numbers (e.g., $2 + 3 = 5$, $2 < 7$, or $\sqrt{2}$ is irrational) without reference or proof. You may use any algebraic property of real numbers involving equal signs (e.g., $(x + y)^2 = x^2 + 2xy + y^2$ or $(x + y)(x - y) = x^2 - y^2$) and the techniques of calculus to find local maxima or minima of a given function without reference or proof. After completing the exercises in Section 1.2, you may also use any algebraic property of real numbers involving inequalities (e.g., $0 < a < b$ implies $0 < a^x < b^x$ for all $x > 0$) without reference or proof. (To illustrate how to use the Well-Ordering Principle and the Completeness Axiom, we have included some proofs of properties like these, e.g., see Remarks 1.13, 1.25 and 1.26.)

Much of analysis deals with estimation (of error, of growth, of volume, etc.), in which inequalities and the following concept play a central role.

**1.4 DEFINITION.** The *absolute value* of a number $a \in \mathbf{R}$ is the number

$$|a| = \begin{cases} a & a \geq 0 \\ -a & a < 0. \end{cases}$$

When proving results about the absolute value, we can always break the proof up into several cases, depending on when the parameters are positive, negative, or zero. Here is a typical example.

**1.5 Remark.** *The absolute value is multiplicative; i.e., $|ab| = |a|\,|b|$ for all $a, b \in \mathbf{R}$.*

PROOF. We consider four cases.

*Case 1.* $a = 0$ or $b = 0$. Then $ab = 0$, so by definition, $|ab| = 0 = |a|\,|b|$.

*Case 2.* $a > 0$ and $b > 0$. By the first Multiplicative Property, $ab > 0^2 = 0$. Hence by definition $|ab| = ab = |a|\,|b|$.

*Case 3.* $a > 0$ and $b < 0$, or, $b > 0$ and $a < 0$. By symmetry, we may suppose $a > 0$ and $b < 0$. (That is, if we can prove it for $a > 0$ and $b < 0$, then by reversing the roles of $a$ and $b$, we can prove it for $a < 0$ and $b > 0$.) By the second Multiplicative Property, $ab < 0$. Hence by Definition 1.4,

$$|ab| = -(ab) = (-1)(ab) = a((-1)b) = a(-b) = |a|\,|b|.$$

*Case 4.* $a < 0$ and $b < 0$. By the second Multiplicative Property, $ab > 0$. Hence by Definition 1.4,

$$|ab| = ab = (-1)^2(ab) = (-a)(-b) = |a|\,|b|. \quad \blacksquare$$

We shall soon see that there are more efficient ways to prove results about absolute values than breaking the argument into cases.

The following result is useful when solving inequalities involving absolute value signs.

**1.6 THEOREM.** Let $a \in \mathbf{R}$ and $M \geq 0$. Then $|a| \leq M$ if and only if $-M \leq a \leq M$.

PROOF. Suppose first that $|a| \leq M$. Multiplying by $-1$, we also have $-|a| \geq -M$.

*Case 1.* $a \geq 0$. Then by Definition 1.4 and hypothesis,

$$-M \leq 0 \leq a = |a| \leq M.$$

*Case 2.* $a < 0$. Then

$$-M \leq -|a| = a < 0 \leq M.$$

This proves $-M \leq a \leq M$ in either case.

Conversely, if $-M \leq a \leq M$ then $a \leq M$ and $-M \leq a$. Multiplying the second inequality by $-1$, we have $-a \leq M$. Consequently, $|a| = a \leq M$ if $a \geq 0$, and $|a| = -a \leq M$ if $a < 0$.  ∎

Note: In a similar way we can prove that $|a| < M$ if and only if $-M < a < M$. Here is another useful result about absolute values.

**1.7 THEOREM.** *The absolute value satisfies the following three properties.*

    i) [POSITIVE DEFINITE] *For all* $a \in \mathbf{R}$, $|a| \geq 0$, *with* $|a| = 0$ *if and only if* $a = 0$.
    ii) [SYMMETRIC] *For all* $a, b \in \mathbf{R}$, $|a - b| = |b - a|$.
    iii) [TRIANGLE INEQUALITIES] *For all* $a, b \in \mathbf{R}$,

$$|a + b| \leq |a| + |b|, \qquad |a - b| \geq |a| - |b|, \quad and \quad \big| |a| - |b| \big| \leq |a - b|.$$

PROOF. i) If $a \geq 0$ then $|a| = a \geq 0$. If $a < 0$ then by Definition 1.4 and the second Multiplicative Property, $|a| = -a = (-1)a > 0$. Thus $|a| \geq 0$ for all $a \in \mathbf{R}$.

If $|a| = 0$ then by definition $\pm a = 0$. Hence $a = 1 \cdot a = (\pm 1)^2 a = (\pm 1)(\pm a) = (\pm 1) \cdot 0 = 0$. Thus $|a| = 0$ implies $a = 0$. Conversely, $|0| = 0$ by definition.

ii) By Remark 1.5, $|a - b| = |-1| \, |b - a| = |b - a|$.

iii) To prove the first inequality, notice that $|x| \leq |x|$ holds for any $x \in \mathbf{R}$. Thus Theorem 1.6 implies $-|a| \leq a \leq |a|$ and $-|b| \leq b \leq |b|$. Adding these inequalities we obtain

$$-(|a| + |b|) \leq a + b \leq |a| + |b|.$$

Hence by Theorem 1.6 again, $|a + b| \leq |a| + |b|$.

The second inequality follows immediately from the first, since

$$|a| - |b| = |a - b + b| - |b| \leq |a - b| + |b| - |b| = |a - b|.$$

To prove the third inequality, notice that by Theorem 1.6 we need to show

$$-|a - b| \leq |a| - |b| \leq |a - b|.$$

The right-hand inequality has already been proved. Hence by Remark 1.5,

$$|b| - |a| \leq |b - a| = |-1| \, |a - b| = |a - b|.$$

Multiplying this last inequality by $-1$ we conclude that

$$-|a - b| \leq |a| - |b|.  \quad ∎$$

Some students mistakenly mix absolute values and the Additive Property to conclude that $b < c$ implies $|a + b| < |a + c|$. It is important from the beginning to recognize that this implication is false unless both $a + b$ and $a + c$ are nonnegative. For example, if $a = 1$, $b = -5$, and $c = -1$, then $b < c$ but $|a + b| = 4$ is not less than $|a + c| = 0$. A correct way to estimate using absolute value signs usually involves one of the triangle inequalities.

**1.8 Example.** Prove that if $-2 < x < 1$ then $|x^2 + x| < 6$.

PROOF. By hypothesis, $|x| < 2$. Hence by the triangle inequality and Remark 1.5,
$$|x^2 + x| \leq |x|^2 + |x| < 4 + 2 = 6. \quad \blacksquare$$

The following result (which is equivalent to the Trichotomy Property) will be used many times in this and subsequent chapters.

**1.9 THEOREM.** *Let $x, y, a \in \mathbf{R}$.*

i) *$x < y + \varepsilon$ for all $\varepsilon > 0$ if and only if $x \leq y$.*
ii) *$x > y - \varepsilon$ for all $\varepsilon > 0$ if and only if $x \geq y$.*
iii) *$|a| < \varepsilon$ for all $\varepsilon > 0$ if and only if $a = 0$.*

PROOF. i) Suppose to the contrary that $x < y + \varepsilon$ for all $\varepsilon > 0$ but $x > y$. Set $\varepsilon_0 = x - y > 0$ and observe that $y + \varepsilon_0 = x$. Hence by the Trichotomy Property, $y + \varepsilon_0$ cannot be greater than $x$. This contradicts the hypothesis for $\varepsilon = \varepsilon_0$. Thus $x \leq y$.

Conversely, suppose $x \leq y$ and $\varepsilon > 0$ is given. Either $x < y$ or $x = y$. If $x < y$ then $x + 0 < y + 0 < y + \varepsilon$ by the Additive and Transitive Properties. If $x = y$ then $x < y + \varepsilon$ by the Additive Property. Thus $x < y + \varepsilon$ for all $\varepsilon > 0$ in either case. This completes the proof of part i).

ii) Suppose $x > y - \varepsilon$ for all $\varepsilon > 0$. By the second Multiplicative Property, this is equivalent to $-x < -y + \varepsilon$, hence by part i), equivalent to $-x \leq -y$. Multiplying this inequality by $-1$, we conclude that $x \geq y$.

iii) Suppose $|a| < \varepsilon$ for all $\varepsilon > 0$. By Theorem 1.6, this is equivalent to $-\varepsilon < a < \varepsilon$. It follows from parts i) and ii) that $0 \leq a \leq 0$. We conclude by the Trichotomy Property that $a = 0$. $\blacksquare$

Let $a$ and $b$ be real numbers. A *closed interval* is a set of the form

$$[a, b] := \{x \in \mathbf{R} : a \leq x \leq b\}, \qquad [a, \infty) := \{x \in \mathbf{R} : a \leq x\},$$

$$(-\infty, b] := \{x \in \mathbf{R} : x \leq b\}, \quad \text{or} \quad (-\infty, \infty) := \mathbf{R},$$

and an *open interval* is a set of the form

$$(a, b) := \{x \in \mathbf{R} : a < x < b\}, \quad (a, \infty) := \{x \in \mathbf{R} : a < x\},$$

$$(-\infty, b) := \{x \in \mathbf{R} : x < b\}, \quad \text{or} \quad (-\infty, \infty) := \mathbf{R}.$$

By an *interval* we mean a closed interval, an open interval, or a set of the form

$$[a, b) := \{x \in \mathbf{R} : a \leq x < b\} \quad \text{or} \quad (a, b] := \{x \in \mathbf{R} : a < x \leq b\}.$$

Notice, then, that when $a < b$, the intervals $[a, b]$, $[a, b)$, $(a, b]$, and $(a, b)$ correspond to line segments on the real line, but when $b < a$, these "intervals" are all the empty set.

A nonempty interval $I$ is said to be *bounded* if it has the form $[a, b]$, $(a, b)$, $[a, b)$, or $(a, b]$ for some $-\infty < a \leq b < \infty$, in which case the numbers $a, b$ will be called the *endpoints* of $I$. All other nonempty intervals will be called *unbounded*. An interval with endpoints $a, b$ is called *degenerate* if $a = b$ and *nondegenerate* if $a < b$. Thus a degenerate open interval is the empty set, and a degenerate closed interval is a point.

Analysis has a strong geometric flavor. Geometry enters the picture because the real number system can be identified with the real line in such a way that $a < b$ if and only if $a$ lies to the left of $b$ (see Figures 1.1, 2.1, and 2.2 below). This gives us a way of translating analytic results on $\mathbf{R}$ into geometric results on the number line, and vice versa. We close with several examples.

The absolute value is closely linked to the idea of length. The *length* of a bounded interval $I$ with endpoints $a, b$ is defined to be $|I| := |b - a|$. And the *distance* between any two points $a, b \in \mathbf{R}$ is defined by $|a - b|$.

Inequalities can be interpreted as statements about intervals. By Theorem 1.6, $|a| \leq M$ if and only if $a$ belongs to the closed interval $[-M, M]$. And, by Theorem 1.9, $a$ belongs to the open intervals $(-\varepsilon, \varepsilon)$ for all $\varepsilon > 0$ if and only if $a = 0$.

We will use this point of view in Chapters 2 through 5 to give geometric interpretations to the calculus of functions defined on $\mathbf{R}$, and in Chapters 11 through 13 to extend this calculus to functions defined on the Euclidean spaces $\mathbf{R}^n$.

## EXERCISES

*In each of the following exercises, verify the given statement carefully, proceeding step by step. Validate each step which involves an inequality by using some statement found in this section.*

**1**. **This exercise is used in Section 6.3.** The *positive part* of an $a \in \mathbf{R}$ is defined by

$$a^+ := \frac{|a| + a}{2}$$

and the *negative part* by

$$a^- := \frac{|a| - a}{2}.$$

a) Prove that $a = a^+ - a^-$ and $|a| = a^+ + a^-$.
b) Prove that

$$a^+ = \begin{cases} a & a \geq 0 \\ 0 & a \leq 0 \end{cases} \quad \text{and} \quad a^- = \begin{cases} 0 & a \geq 0 \\ -a & a \leq 0. \end{cases}$$

**2.** Solve each of the following inequalities for $x \in \mathbf{R}$.
a) $|x - 2| < 5$.   b) $|1 - x| < 4$.   c) $|x^2 - x - 1| < x^2$.   d) $|x^2 + x| < 2$.

**3.** Suppose $a, b, c \in \mathbf{R}$ and $a \leq b$.
a) Prove that $a + c \leq b + c$.
b) If $c \geq 0$, prove $a \cdot c \leq b \cdot c$.

**4.** Prove (7), (8), and (9). Show that each of these statements is false if the hypothesis $a \geq 0$ or $a > 0$ is removed.

**5. a)** Prove if $0 < a < 1$ and $b = 1 - \sqrt{1-a}$, then $0 < b < a$.
  **b)** Prove if $a > 2$ and $b = 1 + \sqrt{a-1}$, then $2 < b < a$.
  **c)** The *arithmetic mean* of $a, b \in \mathbf{R}$ is $A(a,b) = (a+b)/2$ and the *geometric mean* of $a, b \in [0, \infty)$ is $G(a,b) = \sqrt{ab}$. If $0 \leq a \leq b$, prove $a \leq G(a,b) \leq A(a,b) \leq b$. Prove that $G(a,b) = A(a,b)$ if and only if $a = b$.

**6. a)** Interpreting a rational $m/n$ as $m \cdot n^{-1} \in \mathbf{R}$ and assuming that $\mathbf{Z}$ is closed under addition and multiplication (see Remark 1.13 below), use Postulate 1 to prove

$$\frac{m}{n} + \frac{p}{q} = \frac{mq + np}{nq}, \quad \frac{m}{n} \cdot \frac{p}{q} = \frac{mp}{nq}, \quad -\frac{m}{n} = \frac{-m}{n}, \quad \text{and} \quad \left(\frac{\ell}{n}\right)^{-1} = \frac{n}{\ell}$$

  for $m, n, p, q, \ell \in \mathbf{Z}$ and $n, q, \ell \neq 0$.
  **b)** Prove that Postulate 1 holds with $\mathbf{Q}$ in place of $\mathbf{R}$.
  **c)** Prove that the sum of a rational and an irrational is always irrational. What can you say about the product of a rational and an irrational?
  **d)** Let $m/n, p/q \in \mathbf{R}$ with $n, q > 0$. Prove

$$\frac{m}{n} < \frac{p}{q} \quad \text{if and only if} \quad mq < np.$$

  (Restricting this observation to $\mathbf{Q}$ gives a definition of "$<$" on $\mathbf{Q}$.)

**7. a)** Prove that $|x| \leq 1$ implies $|x^2 - 1| \leq 2|x - 1|$.
  **b)** Prove that $-1 \leq x \leq 2$ implies $|x^2 + x - 2| \leq 4|x - 1|$.
  **c)** Prove that $|x| \leq 1$ implies $|x^2 - x - 2| \leq 3|x + 1|$.
  **d)** Prove that $|x - 1| < 1$ implies $|x^3 + x - 2| < 8|x - 1|$.

**8.** For each of the following, find all values of $n \in \mathbf{N}$ which satisfy the given inequality.

  a) $$\frac{1 - n}{1 - n^2} < 0.01.$$

  b) $$\frac{n^2 + 2n + 3}{2n^3 + 5n^2 + 8n + 3} < 0.025.$$

  c) $$\frac{n - 1}{n^3 - n^2 + n - 1} < 0.002.$$

**9.** Prove that
$$(a_1 b_1 + a_2 b_2)^2 \leq (a_1^2 + a_2^2)(b_1^2 + b_2^2)$$
  for all $a_1, a_2, b_1, b_2 \in \mathbf{R}$.

**10.** Suppose $x, a, y, b \in \mathbf{R}$, $|x - a| < \varepsilon$, and $|y - b| < \varepsilon$ for some $\varepsilon > 0$.
  **a)** Prove $|xy - ab| < (|a| + |b|)\varepsilon + \varepsilon^2$.
  **b)** Prove $|x^2 y - a^2 b| < \varepsilon(|a|^2 + 2|ab|) + \varepsilon^2(|b| + 2|a|) + \varepsilon^3$.

**11. a)** Let **P** represent the collection of positive real numbers. Prove that **P** satisfies the following two properties.

i) For each $x \in \mathbf{R}$, one and only one of the following hold:

$$x \in \mathbf{P}, \quad -x \in \mathbf{P}, \quad \text{or} \quad x = 0.$$

ii) Given $x, y \in \mathbf{P}$, both $x + y$ and $x \cdot y$ belong to **P**.

**b)** Suppose **R** contains a subset **P** (not necessarily the set of positive numbers) which satisfies properties i) and ii). Define $x \prec y$ by $y - x \in \mathbf{P}$. Prove that Postulate 2 holds with $\prec$ in place of $<$.

## 1.2 THE WELL-ORDERING PRINCIPLE

In this section we introduce the Well-Ordering Principle, a postulate which distinguishes the set **N** from the sets **Z**, **Q**, and **R**. We use it to establish the Principle of Induction and prove the Binomial Formula, a result which shows how to expand powers of a binomial expression, i.e., an expression of the form $a + b$.

The Well-Ordering Principle is different from the preceding postulates in a fundamental way. Postulates 1 and 2 were statements about the algebraic structure of **R**, namely, about finite sums and products of elements of **R**. Postulate 3 is a statement about the "direction" of **N** under the order relation $<$, namely, about existence of least elements of subsets of **N**. Before we state the Well-Ordering Principle, we make precise what we mean by a least element.

**1.10 DEFINITION.** A number $x$ is a *least element* of a set $E \subset \mathbf{R}$ if $x \in E$ and $x \leq a$ for all $a \in E$.

Note: Because French mathematicians (e.g., Borel, Jordan, and Lebesgue) did fundamental work on the connection between analysis and set theory, and *ensemble* is French for set, analysts frequently use $E$ to represent a general set.

**POSTULATE 3.** [THE WELL-ORDERING PRINCIPLE]. Every nonempty subset of **N** has a least element.

Notice that the Well-Ordering Principle is not satisfied by the number systems **Z**, **Q**, and **R** since none of these systems contains a least element.

Our first application of the Well-Ordering Principle is the *Principle of Mathematical Induction* (which, under mild assumptions, is equivalent to the Well-Ordering Principle–see Appendix A).

**1.11 THEOREM.** *Suppose for each $n \in \mathbf{N}$ that $A(n)$ is a proposition (i.e., a verbal statement or formula) which satisfies the following two properties:*

i) *$A(1)$ is true.*

ii) *For every $k \in \mathbf{N}$ for which $A(k)$ is true, $A(k+1)$ is also true.*

*Then $A(n)$ is true for all $n \in \mathbf{N}$.*

PROOF. Suppose the theorem is false. Then the set $E = \{n \in \mathbf{N} : A(n) \text{ is false}\}$ is nonempty. Hence by Postulate 3, $E$ has a least element, say $x$.

By hypothesis i), $x \neq 1$. Since $x \in E \subseteq \mathbf{N}$, it follows from Remark 1.1iii) that $x - 1 \in \mathbf{N}$. But $x - 1 < x$ and $x$ is a least element of $E$. Consequently, $A(x - 1)$ is true. Applying hypothesis ii) to $k = x - 1$, we see that $A(x) = A(k + 1)$ must also be true, i.e, $x \notin E$, a contradiction. $\blacksquare$

Recall that if $x_0, x_1, \ldots, x_n$ are real numbers and $0 \leq j \leq n$, then

$$\sum_{k=j}^{n} x_k := x_j + x_{j+1} + \cdots + x_n$$

denotes the sum of the $x_k$'s as $k$ ranges from $j$ to $n$. The following examples illustrate the fact that the Principle of Mathematical Induction can be used to prove a variety of statements involving integers.

**1.12 Example.** Prove

$$\sum_{k=1}^{n} (3k - 1)(3k + 2) = 3n^3 + 6n^2 + n$$

for $n \in \mathbf{N}$.

Proof. Let $A(n)$ represent the statement

$$\sum_{k=1}^{n} (3k - 1)(3k + 2) = 3n^3 + 6n^2 + n.$$

For $n = 1$ the left side of this equation is $2 \cdot 5$ and the right side is $3 + 6 + 1$. Therefore, $A(1)$ is true. Suppose $A(n)$ is true for some $n \geq 1$. Then

$$\sum_{k=1}^{n+1} (3k - 1)(3k + 2) = (3n + 2)(3n + 5) + \sum_{k=1}^{n} (3k - 1)(3k + 2)$$

$$= (3n + 2)(3n + 5) + 3n^3 + 6n^2 + n = 3n^3 + 15n^2 + 22n + 10.$$

On the other hand, a direct calculation reveals

$$3(n + 1)^3 + 6(n + 1)^2 + (n + 1) = 3n^3 + 15n^2 + 22n + 10.$$

Therefore, $A(n+1)$ is true when $A(n)$ is. We conclude by induction that $A(n)$ holds for all $n \in \mathbf{N}$. $\blacksquare$

Next, we show that $\mathbf{N}$ and $\mathbf{Z}$ satisfy the Closure Properties.

**1.13 Remark.** Prove that if $n, m \in \mathbf{N}$ (respectively, $\in \mathbf{Z}$), then $n + m$ and $nm$ belong to $\mathbf{N}$ (respectively, to $\mathbf{Z}$).

Proof. Since $n \in \mathbf{Z}$ if and only if $n = 0$, $n \in \mathbf{N}$, or $-n \in \mathbf{N}$, it suffices to prove the closure properties for $\mathbf{N}$, namely, to show that given $n \in \mathbf{N}$, both $n + m$ and $nm$ belong to $\mathbf{N}$ for all $m \in \mathbf{N}$. We shall prove these by induction on $m$.

Fix $n \in \mathbf{N}$ and consider the set $A := \{m \in \mathbf{N} : n + m \in \mathbf{N}\}$. Recall from Remark 1.1ii) that $n \in \mathbf{N}$ implies $n+1 \in \mathbf{N}$, i.e., $1 \in A$. Suppose $m \in A$ for some $m \geq 1$, i.e., $n+m \in \mathbf{N}$. Then by Remark 1.1ii) and associativity, $n + (m+1) = (n+m)+1 \in \mathbf{N}$, i.e., $m + 1 \in A$. Thus, by induction, $A = \mathbf{N}$ and closure holds for addition.

Now consider the set $B := \{m \in \mathbf{N} : n \cdot m \in \mathbf{N}\}$. Clearly, $n \cdot 1 = n \in \mathbf{N}$, i.e., $1 \in B$. Next, if some $m \geq 1$ belongs to $B$, then $n(m + 1) = nm + n \in \mathbf{N}$ by the closure of addition. Thus $m + 1 \in B$ and the proof is complete by induction. ∎

Two formulas encountered early in an algebra course are the perfect square and cube formulas:

$$(a + b)^2 = a^2 + 2ab + b^2 \quad \text{and} \quad (a + b)^3 = a^3 + 3a^2 b + 3ab^2 + b^3.$$

Our next application of the Principle of Mathematical Induction is a generalization of these formulas from $n = 2$ or $3$ to arbitrary $n \in \mathbf{N}$.

Recall that Pascal's Triangle is the triangular array of integers whose rows begin and end with 1's with the property that an interior entry on any row is obtained by adding the two numbers in the preceding row immediately above that entry. Thus the first few rows of Pascal's Triangle are as below.

$$
\begin{array}{ccccccccccccc}
 &  &  &  &  &  & 1 &  &  &  &  &  & \\
 &  &  &  &  & 1 &  & 1 &  &  &  &  & \\
 &  &  &  & 1 &  & 2 &  & 1 &  &  &  & \\
 &  &  & 1 &  & 3 &  & 3 &  & 1 &  &  & \\
 &  & 1 &  & 4 &  & 6 &  & 4 &  & 1 &  & \\
 & 1 &  & 5 &  & 10 &  & 10 &  & 5 &  & 1 & \\
1 &  & 6 &  & 15 &  & 20 &  & 15 &  & 6 &  & 1
\end{array}
$$

Notice that the third and fourth rows are precisely the coefficients which appeared in the perfect square and cube formulas above.

One can write down a formula for each entry in each row of the Pascal Triangle. The first (and only) entry in the first row is

$$\binom{0}{0} := 1.$$

Using the notation $0! := 1$ and $n! := 1 \cdot 2 \cdots (n-1) \cdot n$ for $n \in \mathbf{N}$, define the *binomial coefficient n over k* by

$$\binom{n}{k} := \frac{n!}{(n - k)! k!}$$

for $0 \leq k \leq n$ and $n = 0, 1, \ldots$.

The following result shows that the binomial coefficient $n$ over $k$ does produce the $(k + 1)$–st entry in the $(n + 1)$–st row of Pascal's Triangle.

**1.14 Lemma.** *If $n, k \in \mathbf{N}$ and $1 \leq k \leq n$, then*

$$\binom{n+1}{k} = \binom{n}{k-1} + \binom{n}{k}.$$

PROOF. By definition,

$$\binom{n}{k-1} + \binom{n}{k} = \frac{n! \, k}{(n-k+1)!k!} + \frac{n!(n-k+1)}{(n-k+1)!k!}$$

$$= \frac{n!(n+1)}{(n-k+1)!k!} = \binom{n+1}{k}. \ \blacksquare$$

Binomial coefficients can be used to expand the $n$th power of a sum of two terms.

**1.15 THEOREM** [BINOMIAL FORMULA]. *If $a, b \in \mathbf{R}$ and $n \in \mathbf{N}$, then*

$$(a+b)^n = \sum_{k=0}^{n} \binom{n}{k} a^k b^{n-k}.$$

PROOF. The proof is by induction on $n$. The formula is obvious for $n = 1$. Suppose the formula is true for some $n \in \mathbf{N}$. Then by the inductive hypothesis and Postulate 1,

$$(a+b)^{n+1} = (a+b)(a+b)^n$$

$$= (a+b) \left( \sum_{k=0}^{n} \binom{n}{k} a^k b^{n-k} \right)$$

$$= \left( \sum_{k=0}^{n} \binom{n}{k} a^{k+1} b^{n-k} \right) + \left( \sum_{k=0}^{n} \binom{n}{k} a^k b^{n-k+1} \right)$$

$$= \left( a^{n+1} + \sum_{k=1}^{n} \binom{n}{k-1} a^k b^{n-k+1} \right) + \left( b^{n+1} + \sum_{k=1}^{n} \binom{n}{k} a^k b^{n-k+1} \right)$$

$$= a^{n+1} + \sum_{k=1}^{n} \left( \binom{n}{k-1} + \binom{n}{k} \right) a^k b^{n-k+1} + b^{n+1}.$$

Hence it follows from Lemma 1.14 that

$$(a+b)^{n+1} = a^{n+1} + \sum_{k=1}^{n} \binom{n+1}{k} a^k b^{n-k+1} + b^{n+1} = \sum_{k=0}^{n+1} \binom{n+1}{k} a^k b^{n-k+1};$$

i.e., the formula is true for $n+1$. We conclude by induction that the formula holds for all $n \in \mathbf{N}$. $\blacksquare$

## EXERCISES

**1**. **This exercise is used in Sections 1.4, 2.4, and 5.1.** Prove the following formulas hold for all $n \in \mathbf{N}$.

a) $\displaystyle\sum_{k=1}^{n} k = \frac{n(n+1)}{2}$,

b) $\displaystyle\sum_{k=1}^{n} k^2 = \frac{n(n+1)(2n+1)}{6}$,

c) $\displaystyle\sum_{k=1}^{n} \frac{a-1}{a^k} = 1 - \frac{1}{a^n}$,   $a \neq 0$,

d) $\displaystyle\sum_{k=1}^{n}(2k-1)^2 = \frac{n(4n^2-1)}{3}$.

**2.** Use the Binomial Formula to prove each of the following.

a) $2^n = \sum_{k=0}^{n} \binom{n}{k}$ for all $n \in \mathbf{N}$.

b) $(a+b)^n \geq a^n + na^{n-1}b$ for all $n \in \mathbf{N}$ and $a, b \geq 0$.

c) $(1+1/n)^n \geq 2$ for all $n \in \mathbf{N}$.     $\lim_{n \to \infty} \left(1 + \frac{1}{n}\right)^n = e.$

**3.** Let $n \in \mathbf{N}$. Write

$$\frac{(x+h)^n - x^n}{h}$$

as a sum, none of whose terms has an $h$ in the denominator.

**4.** a) Suppose $0 < x_1 < 1$ and $x_{n+1} = 1 - \sqrt{1 - x_n}$ for $n \in \mathbf{N}$. Prove $0 < x_{n+1} < x_n < 1$ holds for all $n \in \mathbf{N}$.

b) Suppose $x_1 \geq 2$ and $x_{n+1} = 1 + \sqrt{x_n - 1}$ for $n \in \mathbf{N}$. Prove $2 \leq x_{n+1} \leq x_n \leq x_1$ holds for all $n \in \mathbf{N}$.

**5.** Suppose $0 < x_1 < 2$ and $x_{n+1} = \sqrt{2 + x_n}$ for $n \in \mathbf{N}$. Prove $0 < x_n < x_{n+1} < 2$ holds for all $n \in \mathbf{N}$.

**6.** Prove that each of the following inequalities holds for all $n \in \mathbf{N}$.

$$n < 2^n, \qquad n^2 \leq 2^n + 1, \quad \text{and} \quad n^3 \leq 3^n.$$

use   use of $a^n < b^n$

**7**. **This exercise is used in Section 2.3.** Prove that $0 \leq a < b$ implies $0 \leq a^n < b^n$ and $0 \leq \sqrt[n]{a} < \sqrt[n]{b}$ for all $n \in \mathbf{N}$.

**8.** In the next section we shall prove that the square root of an integer $m$ is rational if and only if $m = k^2$ for some $k \in \mathbf{N}$. Assume this result is true.

a) Prove that $\sqrt{n+3} + \sqrt{n}$ is rational for some $n \in \mathbf{N}$ if and only if $n = 1$.

b) Find all $n \in \mathbf{N}$ such that $\sqrt{n+7} + \sqrt{n}$ is rational.

**9.** Prove that $2^n + 3^n$ is a multiple of 5 for all odd $n \in \mathbf{N}$.

**10.** Let $a_0 = 3$, $b_0 = 4$, and $c_0 = 5$.

a) Let $a_k = a_{k-1} + 2$, $b_k = 2a_{k-1} + b_{k-1} + 2$, and $c_k = 2a_{k-1} + c_{k-1} + 2$ for $k \in \mathbf{N}$. Prove that $c_k - b_k$ is constant for all $k \in \mathbf{N}$.

b) Prove that the numbers defined in part a) satisfy

$$a_k^2 + b_k^2 = c_k^2$$

for all $k \in \mathbf{N}$.

## 1.3   THE COMPLETENESS AXIOM

In this section we introduce the last of four postulates which describe $\mathbf{R}$. To formulate this postulate, which distinguishes $\mathbf{Q}$ from $\mathbf{R}$, we need the following concepts.

**1.16 DEFINITION.**   Let $E \subset \mathbf{R}$ be nonempty.

i) The set $E$ is said to be *bounded above* if there is an $M \in \mathbf{R}$ such that $a \le M$ for all $a \in E$.

ii) A number $M$ is called an *upper bound* of the set $E$ if $a \le M$ for all $a \in E$.

iii) A number $s$ is called a *supremum* of the set $E$ if $s$ is an upper bound of $E$ and if $s \le M$ for all upper bounds $M$ of $E$. (In this case we shall say that $E$ *has a supremum* $s$ and shall write $s = \sup E$.)

Notice by iii) that a supremum of a set $E$ (when it exists) is the smallest (or least) upper bound of $E$. By definition, then, in order to prove that $s = \sup E$ for some set $E \subset \mathbf{R}$ we must show two things: $s$ is an upper bound, AND $s$ is the smallest upper bound.

**1.17 Example.**   If $E = [0, 1]$, prove that $\sup E = 1$.

PROOF.   By the definition of interval, 1 is an upper bound of $E$. Let $M$ be any upper bound of $E$; i.e., $M \ge x$ for all $x \in E$. Since $1 \in E$, it follows that $M \ge 1$. Thus 1 is the smallest upper bound of $E$. ∎

The following two remarks answer the questions: how many upper bounds and suprema can a given set have?

**1.18 Remark.**   *If a set has one upper bound, it has infinitely many upper bounds.*

PROOF.   If $M_0$ is an upper bound for a set $E$, then so is $M$ for any $M > M_0$. ∎

**1.19 Remark.**   *If a set has a supremum, then it has only one supremum.*

PROOF.   Let $s_1$ and $s_2$ be suprema of the same set $E$. Then both $s_1$ and $s_2$ are upper bounds of $E$, whence by Definition 1.16iii), $s_1 \le s_2$ and $s_2 \le s_1$. We conclude by the Trichotomy Property that $s_1 = s_2$. ∎

(Note: This proof illustrates a general principle. When asked to prove $a = b$, it is often easier to verify $a \le b$ and $b \le a$.)

The next result, a fundamental property of suprema, shows that the supremum of a set $E$ always lies near points of $E$ (see Figure 1.1 for an illustration).

**1.20 THEOREM** [APPROXIMATION PROPERTY FOR SUPREMA].   *If $E$ has a supremum and $\varepsilon > 0$ is any positive number, then there is a point $a \in E$ such that*

$$\sup E - \varepsilon < a \le \sup E.$$

PROOF.   Suppose the theorem is false. Then there is an $\varepsilon_0 > 0$ such that no element of $E$ lies between $s_0 := \sup E - \varepsilon_0$ and $\sup E$. It follows that $a \le s_0$ for all $a \in E$; i.e., $s_0$ is an upper bound of $E$. Thus, by Definition 1.16iii), $\sup E \le s_0 =$

$\sup E - \varepsilon_0$. Subtracting $\sup E$ from both sides of this inequality, we conclude that $\varepsilon_0 \leq 0$, a contradiction. ∎

The Approximation Property can be used to show that the supremum of any subset of integers is itself an integer.

**1.21 Remark.** *If $E \subset \mathbf{N}$ has a supremum, then $\sup E \in E$.*

PROOF. Suppose $s := \sup E$ and apply the Approximation Property to choose an $x_0 \in E$ such that $s - 1 < x_0 \leq s$. If $s = x_0$, then $s \in E$ as promised. Otherwise, $s - 1 < x_0 < s$ and we can apply the Approximation Property again to choose $x_1 \in E$ such that $x_0 < x_1 < s$.

Subtract $x_0$ from this last inequality to obtain $0 < x_1 - x_0 < s - x_0$. Using the leftmost inequality, we have by Remarks 1.1iv) and ii) that $x_1 - x_0 \geq 1$. On the other hand, since $x_0 > s - 1$ the right-most inequality implies $x_1 - x_0 < s - (s - 1) = 1$, a contradiction. ∎ *1.1 iv) If $x \in \mathbb{Z}$ and $n > 0$, then $n \in \mathbb{N}$ ii) If $n \in \mathbb{N}$, then $n + 1 \in \mathbb{N}$ and $n \geq 1$*

*$x_0, x_1 \in E \subset \mathbb{N}$*

The existence of suprema is the last major assumption about **R** we make.

**POSTULATE 4.** [THE COMPLETENESS AXIOM]. If $E$ is a nonempty subset of **R** which is bounded above, then $E$ has a (finite) supremum.

We shall use this property many times. Our first four applications deal with the distribution of integers and rationals among real numbers.

**1.22 THEOREM** [ARCHIMEDEAN PRINCIPLE]. *Given positive real numbers $a$ and $b$, there is an integer $n \in \mathbf{N}$ such that $b < na$.*

STRATEGY: The idea behind the proof is simple. By the Completeness Axiom and Remark 1.21, any nonempty subset of integers which is bounded above has a "largest" integer. If $k_0$ is the largest integer which satisfies $k_0 a \leq b$, then $n = (k_0 + 1)$ (which is larger than $k_0$) must satisfy $na > b$. In order to justify this application of the Completeness Axiom, we have two details to attend to: 1) Is the set $E := \{k \in \mathbf{N} : ka \leq b\}$ bounded above? 2) Is $E$ nonempty? The answer to the second question depends on whether $b < a$ or not. Here are the details.

PROOF. If $b < a$ set $n = 1$. If $a \leq b$, consider the set $E = \{k \in \mathbf{N} : ka \leq b\}$. $E$ is nonempty since $1 \in E$. Since $ka \leq b$ for all $k \in E$ and $a > 0$, it follows from the Multiplicative Property that $k \leq b/a$ for all $k \in E$; i.e., $E$ is bounded above by $b/a$. Thus, by the Completeness Axiom and Remark 1.21, $E$ has a supremum $s$ which belongs to $E$, in particular, $s \in \mathbf{N}$. *$E \subset \mathbb{N}$ has a sup, then $\sup E \in E$.*

Set $n = s + 1$. Then $n \in \mathbf{N}$ and (since $n$ is larger than $s$), $n$ cannot belong to $E$. Thus $na > b$. ∎

Notice in Example 1.17 and Remark 1.21 that the supremum of $E$ belonged to $E$. The following result shows that this is not always the case.

**1.23 Example.** Let $A = \{1, \frac{1}{2}, \frac{1}{4}, \frac{1}{8}, \dots\}$ and $B = \{\frac{1}{2}, \frac{3}{4}, \frac{7}{8}, \dots\}$. Prove that $\sup A = \sup B = 1$.

PROOF. It is clear that 1 is an upper bound of both sets. It remains to see that 1 is the smallest upper bound of both sets.

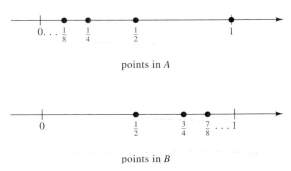

points in $A$

points in $B$

**Figure 1.1**

For $A$, this is trivial. Indeed, if $M$ is any upper bound of $A$ then $M \geq 1$ (since $1 \in A$). On the other hand, if $M$ is an upper bound for $B$, but $M < 1$, then $1 - M > 0$. Since (9) implies $1/(1 - M) > 0$, we can choose (by the Archimedean Principle) an $n \in \mathbf{N}$ such that $n > 1/(1 - M)$. It follows (do the algebra) that $x_0 := 1 - 1/n > M$. Since $x_0 \in B$, this contradicts the assumption that $M$ is an upper bound of $B$ (see Figure 1.1). ∎

The next proof shows how the Archimedean Principle is used to establish scale.

**1.24 THEOREM** [DENSITY OF RATIONALS]. *If $a, b \in \mathbf{R}$ satisfy $a < b$, then there is a $q \in \mathbf{Q}$ such that $a < q < b$.*

STRATEGY: To motivate the proof, consider the special case $a = 1/4$ and $b = 1/3$. We want to find a fraction $q = m/n$ such that $1/4 < m/n < 1/3$. No such $m$ exists if $1 \leq n \leq 6$ because the fractions $p/n$ are spaced too far apart; $1/6$ is too small and $2/6$ is too large. If $n$ is large enough, however, so that some of the fractions $p/n$ belong to the interval $(1/4, 1/3)$ (e.g., $n = 7$), then an acceptable value for $m$ is $m = k_0 - 1$, where $k_0$ is the smallest integer satisfying $b \leq k_0/n$. How large should $n$ be? In order for $p/n$ to belong to $(a, b)$ we need an $n$ which satisfies

$$\frac{1}{n} < b - a.$$

Such an $n$ can be chosen by the Archimedean Principle. We begin our formal proof at this point.

PROOF. Since $b - a > 0$, use the Archimedean Principle to choose an $n \in \mathbf{N}$ which satisfies $n(b - a) > 1$.

*Case 1.* $b > 0$. Consider the set $E = \{k \in \mathbf{N} : b \leq k/n\}$. By the Archimedean Principle, $E$ is nonempty. Hence, by the Well-Ordering Principle, $E$ has a least element, say $k_0$. Set $m = k_0 - 1$ and $q = m/n$. Since $m < k_0$ and $k_0$ is a least element of $E$, $m \notin E$. This can happen two ways. Either $m \leq 0$ or $b > m/n = q$. In either case we obtain $q < b$. On the other hand, since $k_0 \in E$ implies $b \leq k_0/n$, it follows from the choice of $n$ that

$$a = b - (b - a) < \frac{k_0}{n} - \frac{1}{n} = \frac{k_0 - 1}{n} = q.$$

*Case 2.*  $b \leq 0$.  Choose (by the Archimedean Principle) a $k \in \mathbf{N}$ such that $k + b > 0$. By Case 1, there is an $r \in \mathbf{Q}$ such that $k + a < r < k + b$. Therefore, $q := r - k$ belongs to $\mathbf{Q}$ and satisfies the inequality $a < q < b$. ∎

Here is another application of the Archimedean Principle to the distribution of numbers.

**1.25 Remark.** *If $x > 1$ and $x \notin \mathbf{N}$, then there is an $n \in \mathbf{N}$ such that $n < x < n+1$.*

PROOF.  By the Archimedean Principle, the set $E = \{m \in \mathbf{N} : x < m\}$ is nonempty. Hence by the Well-Ordering Principle, $E$ has a least element, say $m_0$.

Set $n = m_0 - 1$. Since $m_0 \in E$, $n + 1 = m_0 > x$. Since $m_0$ is least, $n = m_0 - 1 \leq x$. Since $x \notin \mathbf{N}$, we also have $n \neq x$. Therefore, $n < x < n + 1$. ∎

Using this last result, we can prove that the set of irrationals is nonempty.

**1.26 Remark.** *If $n \in \mathbf{N}$ is not a perfect square (i.e., if there is no $m \in \mathbf{N}$ such that $n = m^2$), then $\sqrt{n}$ is irrational.*

PROOF.  Suppose to the contrary that $n \in \mathbf{N}$ is not a perfect square but $\sqrt{n} \in \mathbf{Q}$; i.e., $\sqrt{n} = p/q$ for some $p, q \in \mathbf{N}$. Choose by Remark 1.25 an integer $m_0 \in \mathbf{N}$ such that

$$(10) \qquad\qquad m_0 < \sqrt{n} < m_0 + 1.$$

Consider the set $E := \{k \in \mathbf{N} : k\sqrt{n} \in \mathbf{Z}\}$. Since $q\sqrt{n} = p$ we know that $E$ is nonempty. Thus by the Well-Ordering Principle, $E$ has a least element, say $n_0$.

Set $x = n_0(\sqrt{n} - m_0)$. By (10), $0 < \sqrt{n} - m_0 < 1$. Multiplying this inequality by $n_0$, we find that

$$(11) \qquad\qquad 0 < x < n_0.$$

Since $n_0$ is a least element of $E$, it follows from (11) that $x \notin E$. On the other hand,

$$x\sqrt{n} = n_0(\sqrt{n} - m_0)\sqrt{n} = n_0 n - m_0 n_0 \sqrt{n} \in \mathbf{Z}$$

since $n_0 \in E$. Moreover, since $x > 0$ and $x = n_0\sqrt{n} - n_0 m_0$ is the difference of two integers, $x \in \mathbf{N}$. Thus $x \in E$, a contradiction. ∎

For some applications, we also need the following concepts.

**1.27 DEFINITION.** Let $E \subset \mathbf{R}$ be nonempty.

i) The set $E$ is said to be *bounded below* if there is an $m \in \mathbf{R}$ such that $a \geq m$ for all $a \in E$.

ii) A number $m$ is called a *lower bound* of the set $E$ if $a \geq m$ for all $a \in E$.

iii) A number $t$ is called an *infimum* of the set $E$ if $t$ is a lower bound of $E$ and if $t \geq m$ for all lower bounds $m$ of $E$. In this case we shall say that *$E$ has an infimum $t$* and write $t = \inf E$.

iv) $E$ is said to be *bounded* if it is bounded above and below.

When a set $E$ contains its supremum (respectively, its infimum) we shall frequently write max $E$ for sup $E$ (respectively, min $E$ for inf $E$).

[Some authors call the supremum the *least upper bound* and the infimum the *greatest lower bound*. We will not use this terminology because it is somewhat old-fashioned and because it confuses some students, since the **least** upper bound of a given set is always greater than or equal to the **greatest** lower bound.]

To relate suprema to infima, we define the *reflection* of a set $E \subseteq \mathbf{R}$ by

$$-E := \{x : x = -a \text{ for some } a \in E \}.$$

For example, $-(1, 2] = [-2, -1)$.

The following result shows that the supremum of a set is the same as the negative of its reflection's infimum. This can be used to prove a completeness axiom for infima (see Exercise 6b).

**1.28 THEOREM.** *Let $E \subseteq \mathbf{R}$ be nonempty.*

i) *$E$ has a supremum if and only if $-E$ has an infimum, in which case*

$$\inf(-E) = -\sup E.$$

ii) *$E$ has an infimum if and only if $-E$ has a supremum, in which case*

$$\sup(-E) = -\inf E.$$

PROOF. The proofs of these statements are similar. We prove only the first statement.

Suppose $E$ has a supremum $s$ and set $t = -s$. Since $s$ is an upper bound for $E$, $s \geq a$ for all $a \in E$, so $-s \leq -a$ for all $a \in E$. Therefore, $t$ is a lower bound of $-E$. Suppose that $m$ is any lower bound of $-E$. Then $m \leq -a$ for all $a \in E$, so $-m$ is an upper bound of $E$. Since $s$ is the supremum of $E$, it follows that $s \leq -m$, i.e., $t = -s \geq m$. Thus $t$ is the infimum of $-E$ and $\sup E = s = -t = -\inf(-E)$.

Conversely, suppose $-E$ has an infimum $t$. By definition, $t \leq -a$ for all $a \in E$. Thus $-t$ is an upper bound for $E$. Since $E$ is nonempty, $E$ has a supremum by the Completeness Axiom. ∎

Theorem 1.28 allows us to obtain information about infima from results about suprema and vice versa (see the proof of the next theorem and Exercises 5 and 6).

We shall use the following result many times.

**1.29 THEOREM** [MONOTONE PROPERTY]. *Suppose that $A \subseteq B$ are nonempty subsets of $\mathbf{R}$.*

i) *If $B$ has a supremum, then $\sup A \leq \sup B$.*
ii) *If $B$ has an infimum, then $\inf A \geq \inf B$.*

PROOF. i) Since $A \subseteq B$, any upper bound of $B$ is an upper bound of $A$. Therefore, $\sup B$ is an upper bound of $A$. It follows from the Completeness Axiom that $\sup A$ exists, and from Definition 1.16iii) that $\sup A \leq \sup B$.

Sup $B$ is an upper bound of $A$.
So   $\sup A \leq \sup B$.   by def. 1.16

ii) Clearly, $-A \subseteq -B$. Thus by part i), Theorem 1.28, and the second Multiplicative Property,

$$\inf A = -\sup(-A) \geq -\sup(-B) = \inf B. \quad \blacksquare$$

It is convenient to extend the definition of suprema and infima to all subsets of **R**. To do this we expand the definition of **R** as follows. By an *extended real number* $x$ we mean either $x \in \mathbf{R}$, $x = \infty$, or $x = -\infty$. Let $E \subseteq \mathbf{R}$ be nonempty. We shall define $\sup E = +\infty$ if $E$ is unbounded above and $\inf E = -\infty$ if $E$ is unbounded below. Finally, we define $\sup \emptyset = -\infty$ and $\inf \emptyset = +\infty$. Notice, then, that the supremum of a subset $E$ of **R** (respectively, the infimum of $E$) is finite if and only if $E$ is bounded above (respectively, bounded below). Moreover, under the convention $-\infty \leq a$ and $a \leq \infty$ for all $a \in \mathbf{R}$, the Monotone Property still holds for this extended definition; i.e., if $A$ and $B$ are subsets of **R** and $A \subseteq B$, then $\sup A \leq \sup B$ and $\inf A \geq \inf B$.

## EXERCISES

**1.** Find the infimum and supremum of each of the following sets.
   a) $E = \{4, 3, 2, 1, 8, 7, 6, 5\}$.
   b) $E = \{x \in \mathbf{R} : x^2 - 3x - 5 = 0\}$.
   c) $E = [a, b)$, where $a < b$ are real numbers.
   d) $E = \{p/q \in \mathbf{Q} : p^2 < 2q^2 \text{ and } p, q > 0\}$.
   e) $E = \{x \in \mathbf{R} : x = 1 + (-1)^n \text{ for } n \in \mathbf{N}\}$.
   f) $E = \{x \in \mathbf{R} : x = 1/n - (-1)^n \text{ for } n \in \mathbf{N}\}$.
   g) $E = \{1 + (-1)^n/n : n \in \mathbf{N}\}$.

**2.** Show that if $E$ is a nonempty bounded subset of **Z**, then both $\sup E$ and $\inf E$ exist and belong to $E$.

**3.** [DENSITY OF IRRATIONALS] **This exercise is used in Section 3.3.**
   Prove that if $a < b$ are real numbers, then there is an irrational $\xi \in \mathbf{R}$ such that $a < \xi < b$.

**4.** Prove that for each $a \in \mathbf{R}$ and each $n \in \mathbf{N}$ there exists a rational $r_n$ such that $|a - r_n| < 1/n$. *For fixed a,* $a - \frac{1}{n}, a + \frac{1}{n}$   $a - 1, a + 1$   $a - \frac{1}{n}, a + \frac{1}{n}$.

**5.** [APPROXIMATION PROPERTY FOR INFIMA] **This exercise is used in many sections including 2.2 and 5.1.**

   a) By modifying the proof of Theorem 1.20, prove that if a set $E \subset \mathbf{R}$ has a finite infimum and $\varepsilon > 0$ is any positive number, then there is a point $a \in E$ such that $\inf E + \varepsilon > a \geq \inf E$.

   b) Give a second proof of the Approximation Property for Infima by using Theorem 1.28.

**6.** a) Prove that a lower bound of a set need not be unique but the infimum of a given set $E$ is unique.

   b) Prove that if $E$ is a nonempty subset of **R** which is bounded below, then $E$ has a finite infimum.

The product of any nonzero rational number and any irrational is irrational.
The sum of two positive irrational number is not always irrational
e.g.   $2 - \sqrt{2} > 0, \sqrt{2} > 0,$   $(2 - \sqrt{2}) + \sqrt{2} = 2 \in \mathbf{Q}$.

**7.** a) Prove that if $x$ is an upper bound of a set $E \subseteq \mathbf{R}$ and $x \in E$, then $x$ is the supremum of $E$.

b) Make and prove an analogous statement for the infimum of $E$.

c) Show by example that the converse of each of these statements is false.

**8.** Let $x_n \in \mathbf{R}$ and suppose that there is an $M \in \mathbf{R}$ such that $|x_n| \leq M$ for $n \in \mathbf{N}$. Prove that $s_n = \sup\{x_n, x_{n+1}, \dots\}$ defines a real number for each $n \in \mathbf{N}$ and $s_1 \geq s_2 \geq \dots$. Prove an analogous result about $t_n = \inf\{x_n, x_{n+1}, \dots\}$.

**9.** Prove that if $a$ and $b$ are real numbers and $0 \leq a < b$, then there exist $n, m \in \mathbf{N}$ such that $a < m/10^n < b$.

**10.** Suppose $E, A, B \subset \mathbf{R}$ and $E = A \cup B$. Prove that if $E$ has a supremum and both $A$ and $B$ are nonempty, then $\sup A$ and $\sup B$ both exist, and $\sup E$ is one of the numbers $\sup A$ or $\sup B$.

$$\text{Sup}(A \cup B) = \max\{\text{Sup}(A), \text{Sup}(B)\}$$

## 1.4 FUNCTIONS, COUNTABILITY, AND THE ALGEBRA OF SETS

In this section we examine the role functions play in distinguishing one kind of infinite set from another and use this point of view to obtain more information about the special subsets of $\mathbf{R}$ introduced in Section 1.1. We also introduce "transfinite" unions and intersections of sets and examine what happens to them under images and inverse images by functions.

We begin with some preliminary remarks. For the first half of this course, most of the concrete functions we consider will be *real-valued functions of a real variable*, i.e., functions $f : E \to \mathbf{R}$ where $E \subseteq \mathbf{R}$. We shall often call such functions simply *real functions*.

When a real function $f$ is given by a formula, the *domain* of $f$ is defined to be the largest subset of $\mathbf{R}$ on which $f$ is defined and real-valued. For example, if $f(x) = 1/x$ and $g(x) = \sqrt{x}$, then $\text{Dom}(f) = \{x \in \mathbf{R} : x \neq 0\}$ and $\text{Dom}(g) = \{x \in \mathbf{R} : x \geq 0\}$. We will also use the usual abuse of notation by representing such functions by their formulas, e.g., calling $1/x$ a function.

We assume you are familiar with the trigonometric functions (whose formulas are) represented by $\sin x$, $\cos x$, $\tan x$, $\cot x$, $\sec x$, $\csc x$, the natural logarithm $\log x$ and its inverse $e^x$, and the power functions $x^\alpha$ which are defined by

$$x^\alpha := e^{\alpha \log x}, \qquad x > 0, \quad \alpha \in \mathbf{R}.$$

We also assume you can differentiate algebraic combinations of these functions using the basic formulas

$$(\sin x)' = \cos x, \quad (\cos x)' = -\sin x, \quad (\tan x)' = \sec^2 x, \quad (e^x)' = e^x, \quad x \in \mathbf{R},$$

$$(\log x)' = \frac{1}{x}, \quad \text{and} \quad (x^\alpha)' = \alpha x^{\alpha-1}, \quad x > 0, \quad \alpha \in \mathbf{R}.$$

(For a derivation of these identities based on fundamental properties, see Exercise 4 in Section 4.3 and Exercises 4 and 5 in Section 5.3.) Even with these assumptions, we shall repeat some material from elementary calculus.

$$(u^x)' = (e^{\ln u \cdot x})' = \ln u \, e^{\ln u \cdot x} = \ln u \cdot u^x$$

$$\left| \int_a^b f(x)\,dx \right| \leq \int_a^b |f(x)|\,dx$$

Let $f : X \to Y$. Although, by the definition of a function, each $x \in X$ is assigned a *unique* (meaning one and only one) $y = f(x) \in Y$, there is nothing which keeps two $x$'s from being assigned to the same $y$, and nothing which says every $y \in Y$ corresponds to some $x \in X$. Functions which satisfy these additional properties are important enough to warrant separate terminology.

**1.30 DEFINITION.** Let $f$ be a function from a set $X$ into a set $Y$.

i) $f$ is said to be *one–to–one* (1–1) on $X$ if

$$x_1, x_2 \in X \quad \text{and} \quad f(x_1) = f(x_2) \quad \text{imply} \quad x_1 = x_2.$$

ii) $f$ is said to take $X$ *onto* $Y$ if for each $y \in Y$ there is an $x \in X$ such that $y = f(x)$.
    *if $Y \subseteq \operatorname{Rng} f$, $Y = \operatorname{Rng} f$.*

For example, the function $f(x) = x^2$ is 1–1 from $[0, \infty)$ onto $[0, \infty)$ but not 1–1 on any open interval containing 0.

Some authors call 1–1 functions *injections*, onto functions *surjections*, and 1–1 onto functions *bijections*. Here is a simple, useful characterization of bijections from one set $X$ to another $Y$.

**1.31 THEOREM.** *Let $X$ and $Y$ be sets and $f : X \to Y$. Then $f$ is 1–1 from $X$ onto $Y$ if and only if there is a unique function $g$ from $Y$ onto $X$ which satisfies*

$$(12) \qquad f(g(y)) = y, \qquad y \in Y,$$

*and*

$$(13) \qquad g(f(x)) = x, \qquad x \in X.$$

PROOF. Suppose $f$ is 1–1 and onto. Let $y \in Y$ and choose a unique $x \in X$ such that $f(x) = y$. Define $g(y) = x$. By construction of $g$, (12) and (13) are satisfied. Moreover, it is clear that $g$ takes $Y$ onto $X$.

Conversely, suppose there is a function $g$ from $Y$ onto $X$ which satisfies (12) and (13). If $x_1, x_2 \in X$ and $f(x_1) = f(x_2)$, then it follows from (13) that $x_1 = g(f(x_1)) = g(f(x_2)) = x_2$. Thus $f$ is 1–1 on $X$. If $y \in Y$ and $x = g(y)$, then (12) implies that $f(x) = f(g(y)) = y$. Thus $f$ takes $X$ onto $Y$.

Finally, suppose $h$ is another function which satisfies (12) and (13), and $y \in Y$. Choose $x \in X$ such that $f(x) = y$. Then by (13),

$$h(y) = h(f(x)) = x = g(f(x)) = g(y),$$

i.e., $h = g$ on $Y$. It follows that the function $g$ is unique. ∎

If $f$ is 1–1 from a set $X$ onto a set $Y$, we shall say that $f$ *has an inverse function*. We shall call the function $g$ given in Theorem 1.31 the *inverse* of $f$, and denote it by $f^{-1}$. [Note: This is different from the function $(f(x))^{-1} := 1/f(x)$.] Notice by (12) and (13) that

$$f(f^{-1}(y)) = y \quad \text{and} \quad f^{-1}((f(x)) = x$$

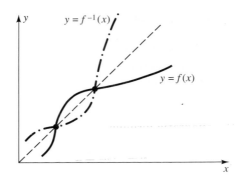

**Figure 1.2**

for all $y \in Y$ and $x \in X$.

Let $f$ be a real-valued function of a real variable. If $f$ has an inverse function $f^{-1}$ and $y = f(x)$, we have by definition that $(x, f(x)) = (f^{-1}(y), y)$. Hence, the graph of $y = f^{-1}(x)$ is a reflection of the graph of $y = f(x)$ about the line $y = x$ (see Figure 1.2).

How can we prove that a given function $f$ is 1–1 on a set $E$? By definition, we must show that if $a, b \in E$ and $f(a) = f(b)$, then $a = b$. One way to accomplish this is to solve $f(a) = f(b)$ for $a$, hoping to get $b$ as an answer. For example, to show $f(x) = e^x + 1$ is 1–1 on $\mathbf{R}$, suppose $e^a + 1 = e^b + 1$, subtract 1 from both sides, and take the logarithm of the resulting expression. We obtain $a = b$, so $f$ is 1–1 on $\mathbf{R}$. This simplistic approach will not work if $f$ is suitably complicated, because it is not possible to solve all algebraic expressions, e.g., general polynomials of degree $\geq 5$. Fortunately, if $f$ is differentiable, there is a simple sufficient condition to prove that $f$ is 1–1 on a given interval.

**1.32 Remark.** *If $f$ is differentiable on an open interval $I$ and $f'(x) \neq 0$ for all $x \in I$, then $f$ is 1–1 on $I$.*

PROOF. You may remember the Mean Value Theorem (see Theorem 4.15 below) from elementary calculus: if $a < b$ and the derivative $f'$ of a function $f$ exists at every point in an interval $[a, b]$, then there is a $c \in (a, b)$ such that $f(b) - f(a) = (b - a)f'(c)$. Suppose $f$ is not 1–1 on $I$. Then there exist points $a \neq b$ in $I$ such that $f(a) = f(b)$. Hence by the Mean Value Theorem,

$$0 = \frac{f(b) - f(a)}{b - a} = f'(c)$$

for some $c$ between $a$ and $b$, hence some $c \in I$. This contradicts the fact that $f'$ is never zero on $I$. ∎

The following example shows that the inverse function $y = f^{-1}(x)$ can sometimes be found by treating $y = f(x)$ as a relation which implicitly defines $x = f^{-1}(y)$ and solving for $x$.

**1.33 Example.** Prove $f(x) = e^x - e^{-x}$ is 1–1 on $\mathbf{R}$ and find a formula for $f^{-1}$.

SOLUTION. Since $f'(x) = e^x + e^{-x} > 0$, it is never zero on $\mathbf{R}$. Hence by Remark 1.32, $f$ is 1–1 on $\mathbf{R}$.

To find $f^{-1}$, let $y = e^x - e^{-x}$. Multiplying this equation by $e^x$ and collecting all nonzero terms on one side of the equation, we have

$$e^{2x} - ye^x - 1 = 0,$$

a quadratic in $e^x$. By the quadratic formula,

$$e^x = \frac{y \pm \sqrt{y^2 + 4}}{2}.$$

Since $e^x$ is always positive, the minus sign must be discarded. Taking the logarithm of this last identity, we obtain $x = \log(y + \sqrt{y^2 + 4}) - \log 2$. Therefore,

$$f^{-1}(x) = \log(x + \sqrt{x^2 + 4}) - \log 2. \quad \blacksquare$$

*The following material will not be used until Chapters* 9 *and* 10.

Functions which have inverses can be used to "count" infinite sets. Before we make a formal definition, let us examine what it means to count a finite set of objects $E$. When we count $E$ we assign a number $n \in \mathbf{N}$ to each object in $E$; i.e., we construct a function $f$ from a subset of $\mathbf{N}$ to $E$. For example, if $E$ has three objects, then the "counting" function takes $\{1, 2, 3\}$ to $E$. Now in order to count $E$ properly, we must be careful to avoid two pitfalls. We must not count any element of $E$ more than once (i.e., $f$ must be 1–1), and we cannot miss any element of $E$ (i.e., $f$ must take $\{1, 2, 3\}$ onto $E$). Accordingly, we make the following definition.

**1.34 DEFINITION.** Let $E$ be a set.

   i) $E$ is said to be *finite* if $E = \emptyset$ or if there is an $n \in \mathbf{N}$ and a 1–1 function from $\{1, 2, \dots, n\}$ onto $E$.
   ii) $E$ is said to be *countably infinite* if there is a 1–1 function from $\mathbf{N}$ onto $E$.
   iii) $E$ is said to be *countable* if $E$ is either finite or countably infinite.
   iv) $E$ is said to be *uncountable* if $E$ is not countable.

Loosely speaking, a set is countably infinite if it has the same number of elements as $\mathbf{N}$, finite if it has less, uncountable if it has more.

To show a set $E$ is countably infinite, we must prove that there is a 1–1 function from $\mathbf{N}$ onto $E$. For example, the set of even integers $E = \{2, 4, \dots\}$ is countably infinite because $f(k) := 2k$ is 1–1 and takes $\mathbf{N}$ onto $E$. Thus, two infinite sets can have the same number of elements even though one is a proper subset of the other. (In fact, this property can be used as a definition of "infinite set.")

The following result shows that not every infinite set is countably infinite.

**1.35 Remark** [CANTOR'S DIAGONALIZATION ARGUMENT]. *The open interval* $(0, 1)$ *is uncountable.*

STRATEGY. We shall suppose to the contrary that $(0, 1)$ is countable, shall write a list of "all" its elements, and will construct a new number $x$ which is different from each number in that list. How can we determine whether two numbers are different? For example, why is $0.1234 \neq 0.1254$? Because they have different decimal expansions. Thus, we shall construct $x$ so that its decimal expansion differs by at least one digit from the decimal expansion of EVERY number in our list.

But there is an added complexity which we must take into account before we have a complete proof. Every finite decimal has an alternate expansion which terminates in 9's, e.g., $0.5 = 0.4999\ldots$ and $0.24 = 0.23999\ldots$ (see Exercise 10 in Section 2.2). Hence, when specifying the decimal expansion of $x$ we must avoid decimals which terminate in 9's.

PROOF. Suppose there is a 1–1 function $f$ which takes $\mathbf{N}$ onto the interval $(0, 1)$. Write the numbers $f(j)$, $j \in \mathbf{N}$, in decimal notation, using the finite expansion when possible, i.e.,

$$f(1) = 0.\alpha_{11}\alpha_{12}\ldots,$$

$$f(2) = 0.\alpha_{21}\alpha_{22}\ldots,$$

$$f(3) = 0.\alpha_{31}\alpha_{32}\ldots,$$

$$\ldots,$$

where $\alpha_{ij}$ represents the $j$th digit in the decimal expansion of $f(i)$ and none of these expansions terminates in 9's. Let $x$ be the number whose decimal expansion is given by $0.\beta_1\beta_2\ldots$, where

$$\beta_k := \begin{cases} \alpha_{kk} + 1 & \text{if } \alpha_{kk} \leq 5 \\ \alpha_{kk} - 1 & \text{if } \alpha_{kk} > 6. \end{cases}$$

Clearly, $x$ is a number in $(0, 1)$ whose decimal expansion does not even contain a 9, much less terminate in 9's. Since $f$ is onto, there is a $j \in \mathbf{N}$ such that $f(j) = x$. But then the decimal expansions of $f(j)$ and $x$ must be identical, e.g., $\alpha_{jj} = \beta_j := \alpha_{jj} \pm 1$. It follows that $0 = \pm 1$, a contradiction. ∎

It is natural to ask about the countability of the sets $\mathbf{Z}$, $\mathbf{Q}$, and $\mathbf{R}$. To answer these questions, we prove the following result.

**1.36 THEOREM.** *If $A \subseteq B$ and $B$ is countable, then $A$ is countable.*

PROOF. If $A$ is finite, then $A$ is countable by definition. If $A$ is infinite, then $B$ is infinite. Thus by hypothesis, there is a 1–1 function $f$ from $\mathbf{N}$ onto $B$. Since $A \subseteq B$ and $f$ is onto, the set $\{k \in \mathbf{N} : f(k) \in A\}$ is nonempty. Let $k_1$ be the least element of this set. Since $A$ is infinite, the set $\{k \in \mathbf{N} : f(k) \in A \setminus \{f(k_1)\}\}$ is nonempty, so it has a least element, say $k_2$. By induction, there exist integers $k_1 < k_2 < \ldots$ such that $k_j$ is the least element of the set

$$\{k \in \mathbf{N} : f(k) \in A \setminus \{f(k_1), f(k_2), \ldots f(k_{j-1})\}\}.$$

Since each $k_j$ is least, it is clear that

(14)
$$A = \{f(k_1), f(k_2), \dots\}.$$

Consider the function $g : \mathbf{N} \to B$ defined by $g(j) = f(k_j)$, $j \in \mathbf{N}$. Since $f$ is 1–1, $g$ is also 1–1. By (14), $g$ takes $\mathbf{N}$ onto $A$. Hence, $A$ is countably infinite by Definition 1.34. ∎

**1.37 Remark.** *If $A$ is uncountable and $A \subseteq B$, then $B$ is uncountable. In particular, $\mathbf{R}$ is uncountable.*

PROOF. If $B$ were countable, then $A$ would also be countable by Theorem 1.36. This proves the first statement. Since the interval $(0,1)$ is uncountable (by Remark 1.35) and a subset of $\mathbf{R}$, it follows that $\mathbf{R}$ is uncountable. ∎

We now prove that a countable union of countable sets is countable.

**1.38 THEOREM.** *Let $A_1, A_2, \dots$ be sets and*

$$E = \bigcup_{j \in \mathbf{N}} A_j := \bigcup_{j=1}^{\infty} A_j := \{x : x \in A_j \quad \text{for some } j \in \mathbf{N}\}.$$

*If each $A_j$ is countable, then so is $E$.*

PROOF. By Theorem 1.36, we may suppose that each $A_j$ is countably infinite and $A_j \bigcap A_k = \emptyset$ for $k \neq j$. Thus, for each $j \in \mathbf{N}$ we can write

$$A_j = \{a_{jk} : k \in \mathbf{N}\}$$

with $a_{jk} = a_{pq}$ if and only if $j = p$ and $k = q$.

List the elements of $E$ as an infinite matrix with elements of $A_j$ forming the $j$th row. This suggests that $E$ can be "counted" in rising diagonals (follow the arrows in Figure 1.3).

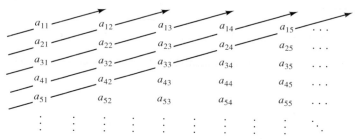

**Figure 1.3**

To explicitly construct a 1–1 function from $E$ onto $\mathbf{N}$ which embodies this suggestion, set $f(a_{11}) = 1$, $f(a_{21}) = 2$, and $f(a_{12}) = 3$. To describe what happens to the $N$th rising diagonal, notice that its elements are given by

$$a_{N,1}, a_{N-1,2}, a_{N-2,3}, \dots, a_{2,N-1}, a_{1,N};$$

i.e., it contains $N$ elements $a_{jk}$ whose indices satisfy $j + k = N + 1$. Since the sum of integers $1 + 2 + \cdots + (N - 1)$ is given by $(N - 1)N/2$ (see Exercise 1a in Section 1.2), there are $(N - 1)N/2$ elements in the first $N - 1$ rising diagonals of $E$. Thus, define $f$ on the $N$th rising diagonal by

$$f(a_{jk}) = \frac{(N - 1)N}{2} + k = \frac{(j + k - 2)(j + k - 1)}{2} + k.$$

By construction, $f$ is a 1–1 function from $E$ onto $\mathbf{N}$; i.e., $f^{-1}$ is a 1–1 function from $\mathbf{N}$ onto $E$. In particular, $E$ is countably infinite. ∎

A similar proof establishes that if $A$ and $B$ are countable, then $A \times B$ is countable. You may use this to solve Exercise 11 below.

**1.39 Remark.** *The sets $\mathbf{Z}$ and $\mathbf{Q}$ are countably infinite, but the set of irrationals is uncountable.*

PROOF. $\mathbf{Z} = \mathbf{N} \cup (-\mathbf{N}) \cup \{0\}$ and $\mathbf{Q} = \bigcup_{n=1}^{\infty} \{p/n : p \in \mathbf{Z}\}$ are both countable by Theorem 1.38.

If $\mathbf{R} \setminus \mathbf{Q}$ were countable, then $\mathbf{R} = (\mathbf{R} \setminus \mathbf{Q}) \cup \mathbf{Q}$ would also be countable, a contradiction of Remark 1.37. ∎

Theorem 1.38 had something to say about a countable union of sets. In Chapter 9 and 10, we must use uncountable unions and intersections. Here is some notation which will prove useful in this regard. A collection of sets $\mathcal{E}$ is said to be *indexed by* a set $A$ if there is a function $F$ from $A$ onto $\mathcal{E}$. In this case $A$ is called the *index set* of $\mathcal{E}$, and we shall represent $F(\alpha)$ by $E_{\alpha}$. In particular, we shall represent a collection of sets indexed by $A$ as

$$\mathcal{E} = \{E_{\alpha}\}_{\alpha \in A}.$$

**1.40 DEFINITION.** Let $\mathcal{E} = \{E_{\alpha}\}_{\alpha \in A}$ be a collection of sets.
   i) The *union* of the collection $\mathcal{E}$ is the set

$$\bigcup_{\alpha \in A} E_{\alpha} := \{x : x \in E_{\alpha} \quad \text{for some } \alpha \in A\}.$$

ii) The *intersection* of the collection $\mathcal{E}$ is the set

$$\bigcap_{\alpha \in A} E_{\alpha} := \{x : x \in E_{\alpha} \quad \text{for all } \alpha \in A\}.$$

There is an easy way to get from unions to intersections and vice versa.

**1.41 THEOREM** [DEMORGAN'S LAWS]. *Let $X$ be a set and $\{E_{\alpha}\}_{\alpha \in A}$ be a collection of subsets of $X$. If for each $E \subseteq X$ the symbol $E^c$ represents the set $X \setminus E$, then*

$$(15) \qquad \left( \bigcup_{\alpha \in A} E_{\alpha} \right)^c = \bigcap_{\alpha \in A} E_{\alpha}^c$$

*and*

(16)
$$\left( \bigcap_{\alpha \in A} E_\alpha \right)^c = \bigcup_{\alpha \in A} E_\alpha^c.$$

PROOF. Suppose $x$ belongs to the left side of (15), i.e., $x \in X$ and $x \notin \bigcup_{\alpha \in A} E_\alpha$. By definition, $x \in X$ and $x \notin E_\alpha$ for all $\alpha \in A$. Hence, $x \in E_\alpha^c$ for all $\alpha \in A$; i.e., $x$ belongs to the right side of (15). These steps are reversible. This verifies (15). A similar argument verifies (16). ∎

The following concepts will be used frequently in subsequent chapters.

**1.42 DEFINITION.** Let $X$ and $Y$ be sets and $f : X \to Y$. The *image* of a set $E \subseteq X$ under $f$ is the set

$$f(E) := \{y \in Y : y = f(x) \text{ for some } x \in E\}.$$

*(handwritten annotation: $x \in E \Rightarrow f(x) \in f(E)$  $\Rightarrow y \in f(E)$  since $y = f(x)$)*

The *inverse image* of a set $E \subseteq Y$ under $f$ is the set

(17)
$$f^{-1}(E) := \{x \in X : f(x) = y \text{ for some } y \in E\}.$$

*(handwritten annotation: $f(x) \in E$.  $x \in f^{-1}(E) \Leftrightarrow f(x) \in E$.)*

Notice that equation (17) makes sense whether $f$ is 1–1 or not; i.e., $f$ need not be 1–1 for $f^{-1}(E)$ to be defined. In particular, $f^{-1}(E)$ is the inverse image of $E$ under $f$, not the image of $E$ under the inverse function $f^{-1}$ (unless $f$ is 1–1). In fact, the inverse function $f^{-1}$ exists on $f(X)$ if and only if the inverse image $f^{-1}(\{y\})$ contains at most one point for all $y \in Y$.

Another indication that the inverse image of a set is different from its image under the inverse function can be seen by examining $f^{-1}(f(E))$. If $f$ were 1–1, this set would be $E$ (see Exercise 6 below). In general, however, the best one can say is $f^{-1}(f(E)) \supseteq E$ for any $E \subseteq \operatorname{Dom} f$ (see Theorem 1.43v) below). If $f$ is not 1–1, $E$ can be a PROPER subset of $f^{-1}(f(E))$. For example, let $f : \mathbf{R} \to \mathbf{R}$ be defined by $f(x) = x^2$ and set $E = [0, 1)$. Then $f(E) = [0, 1)$, so $f^{-1}(f(E)) = (-1, 1) \supset E$.

The following result, which plays a prominent role in Chapters 9 and 12, describes images and inverse images of unions and intersections of sets.

**1.43 THEOREM.** Let $X$ and $Y$ be sets and $f : X \to Y$.

i) If $\{E_\alpha\}_{\alpha \in A}$ is a collection of subsets of $X$, then

$$f\left( \bigcup_{\alpha \in A} E_\alpha \right) = \bigcup_{\alpha \in A} f(E_\alpha) \quad \text{and} \quad f\left( \bigcap_{\alpha \in A} E_\alpha \right) \subseteq \bigcap_{\alpha \in A} f(E_\alpha).$$

ii) If $B$ and $C$ are subsets of $X$, then

$$f(C \setminus B) \supseteq f(C) \setminus f(B).$$

iii) If $\{E_\alpha\}_{\alpha \in A}$ is a collection of subsets of $Y$, then

$$f^{-1}\left( \bigcup_{\alpha \in A} E_\alpha \right) = \bigcup_{\alpha \in A} f^{-1}(E_\alpha) \quad \text{and} \quad f^{-1}\left( \bigcap_{\alpha \in A} E_\alpha \right) = \bigcap_{\alpha \in A} f^{-1}(E_\alpha).$$

*iv) If $B$ and $C$ are subsets of $Y$, then*

$$f^{-1}(C \setminus B) = f^{-1}(C) \setminus f^{-1}(B).$$

*v) If $E \subseteq f(X)$, then $f(f^{-1}(E)) = E$, but if $E \subseteq \text{Dom } f$, then $f^{-1}(f(E)) \supseteq E$.*

PROOF. i) By definition, $y \in f(\cup_{\alpha \in A} E_\alpha)$ if and only if $y = f(x)$ for some $x \in E_\alpha$ and $\alpha \in A$. This is equivalent to $y \in \cup_{\alpha \in A} f(E_\alpha)$. Similarly, $y \in f(\cap_{\alpha \in A} E_\alpha)$ if and only if $y = f(x)$ for some $x \in \cap_{\alpha \in A} E_\alpha$. This implies that for all $\alpha \in A$ there is an $x_\alpha \in E_\alpha$ such that $y = f(x_\alpha)$. Therefore, $y \in \cap_{\alpha \in A} f(E_\alpha)$.

ii) If $y \in f(C) \setminus f(B)$, then $y = f(c)$ for some $c \in C$ but $y \neq f(b)$ for any $b \in B$. It follows that $y \in f(C \setminus B)$. Similar arguments prove parts iii), iv), and v). ∎

The set inequalities in parts i), ii), and v) are equalities when $f$ is 1–1 (see Exercise 6 below). When $f$ is NOT 1–1, they can be strict. For example, if $f : \mathbf{R} \to \mathbf{R}$ is defined by $f(x) = x^2$, $E_1 = \{1\}$, and $E_2 = \{-1\}$, then $f(E_1 \cap E_2) = \emptyset$ is a proper subset of $f(E_1) \cap f(E_2) = \{1\}$.

## EXERCISES

*[handwritten: graph is always increasing, or always decreasing, then 1-1]*

**1.** For each of the following, prove $f$ is 1–1 on $E$ and find a formula for $f^{-1}$.

    a) $f(x) = 3x - 7$, $E = \mathbf{R}$.
    b) $f(x) = e^{1/x}$, $E = (0, \infty)$.
    c) $f(x) = \tan x$, $E = (-\pi/2, \pi/2)$.
    d) $f(x) = x^2 + 3x - 6$, $E = [-3/2, \infty)$.
    e) $f(x) = 3x - |x| + |x - 2|$, $E = \mathbf{R}$.
    f) $f(x) = x/(x^2 + 1)$, $E = [-1, 1]$.

**2.** Suppose $A$ is finite and $f$ is 1–1 from $A$ onto $B$. Prove that $B$ is finite.

**3.** Prove that the set of odd integers $\{1, 3, \dots\}$ is countably infinite.

**4.** Find $f(E)$ and $f^{-1}(E)$ for each of the following.

    a) $f(x) = 1 - 5x$, $E = (-3, 1)$.
    b) $f(x) = x^2$, $E = [-1, 4]$.
    c) $f(x) = x^2 + x$, $E = [-2, 1)$.  *[handwritten: $y = x^2 + x$  $x^2 + x - y = 0$  $x = \dfrac{-1 \pm \sqrt{1 + 4y}}{2}$]*
    d) $f(x) = \log(x^2 + x + 1)$, $E = (1/2, 5]$.
    e) $f(x) = \sin x$, $E = [0, \infty)$.

**5.** Give a simple description of each of the following sets.

$$\text{a) } \bigcup_{x \in [0,1]} [x - 1, x + 1]. \qquad \text{b) } \bigcap_{x \in [0,1]} [x - 1, x + 1].$$

$$\text{c) } \bigcup_{k \in \mathbf{N}} [0, 1/k]. \qquad \text{d) } \bigcap_{k \in \mathbf{N}} [0, 1/k].$$

**6.** Let $X$, $Y$ be sets and $f : X \to Y$. Prove that the following are equivalent.

    a) $f$ is 1–1 on $X$.

b) $f(A \setminus B) = f(A) \setminus f(B)$ for all subsets $A$ and $B$ of $X$.    $f(A) \setminus f(B) \subseteq f(A \setminus B)$

c) $f^{-1}(f(E)) = E$ for all subsets $E$ of $X$.    $E \subseteq f^{-1}(f(E))$

d) $f(A \cap B) = f(A) \cap f(B)$ for all subsets $A$ and $B$ of $X$.    $f(A \cap B) \subseteq f(A) \cap f(B)$

**7.** Prove (16).

**8.** Prove Theorem 1.43iii), iv), and v).

**9.** Let $f : A \to B$ and $g : B \to C$ and define $g \circ f : A \to C$ by $(g \circ f)(x) := g(f(x))$.

    a) Show that if $f, g$ are 1–1 (respectively, onto), then $g \circ f$ is 1–1 (respectively, onto).

    b) [Pigeonhole Principle] Prove that if $f$ is 1–1 from $A$ into $B$, then $f^{-1}$ is 1–1 from $f(A)$ onto $A$.

    c) Suppose $g$ is 1–1 from $B$ onto $C$. Prove that $f$ is 1–1 on $A$ (respectively, onto $B$) if and only if $g \circ f$ is 1–1 on $A$ (respectively, onto $C$).

**10.** Suppose $n \in \mathbf{N}$ and $\phi : \{1, 2, \ldots, n\} \to \{1, 2, \ldots, n\}$.

    a) Prove that $\phi$ is 1–1 if and only if $\phi$ is onto.

    b) Suppose $E$ is a finite set and $f : E \to E$. Prove that $f$ is 1–1 on $E$ if and only if $f$ takes $E$ onto $E$.

**11.** A number $x_0 \in \mathbf{R}$ is called *algebraic of degree $n$* if it is the root of a polynomial of the form $P(x) = a_n x^n + a_{n-1} x^{n-1} + \cdots + a_1 x + a_0$, where $a_j \in \mathbf{Z}$ and $a_n \neq 0$. A number $x_0$ which is not algebraic is called *transcendental*.

    a) Prove that if $n \in \mathbf{N}$ and $q \in \mathbf{Q}$, then $n^q$ is algebraic.

    b) Prove that for each $n \in \mathbf{N}$ the collection of algebraic numbers of degree $n$ is countable.

    c) Prove that the collection of transcendental numbers is uncountable. (Two famous transcendental numbers are $\pi$ and $e$. For more information on transcendental numbers and their history, see Kline [5].)

If $f : A \to B$, $D \subseteq A$, $E \subseteq B$, and $a \in A$, then:

$a \in D \Rightarrow f(a) \in f(D)$

$f(a) \in f(D) \Rightarrow a \in D$, if $f$ is one-to-one.

$a \in f^{-1}(E) \Leftrightarrow f(a) \in E$.

$f = \{(x, f(x)) : x \in A\}$

9.c) If $g$ is 1–1 and onto, and $g \circ f$ is onto, then $f$ is onto.

    if $g$ is not 1–1, this is false.

    If $g \circ f$ is onto, then $g$ is onto.

    If $g \circ f$ is 1–1, then $f$ is 1–1.

# Chapter 2

# Sequences in **R**

## 2.1 LIMITS OF SEQUENCES

An *infinite sequence* (more briefly, a *sequence*) is a function whose domain is **N**. A sequence $f$ whose *terms* are $x_n := f(n)$ will be denoted by $x_1, x_2, \ldots, \{x_n\}_{n \in \mathbf{N}}$, $\{x_n\}_{n=1}^{\infty}$, or $\{x_n\}$. Thus $1, 1/2, 1/4, 1/8, \ldots$ represents the sequence $\{1/2^{n-1}\}_{n \in \mathbf{N}}$, $-1, 1, -1, 1, \ldots$ represents the sequence $\{(-1)^n\}_{n \in \mathbf{N}}$, and $1, 2, 3, 4, \ldots$ represents the sequence $\{n\}_{n \in \mathbf{N}}$.

It is important not to confuse a sequence $\{x_n\}_{n \in \mathbf{N}}$ with the set $\{x_n : n \in \mathbf{N}\}$; these are two entirely different concepts. For example, as sequences, $1, 2, 3, 4, \ldots$ is different from $2, 1, 3, 4, \ldots$, but as sets, $\{1, 2, 3, 4, \ldots\}$ is identical with $\{2, 1, 3, 4, \ldots\}$. Again, the sequence $1, -1, 1, -1, \ldots$ is infinite, but the set $\{(-1)^n : n \in \mathbf{N}\}$ has only two points.

The limit concept is one of the fundamental building blocks of analysis. Recall from elementary calculus that a sequence of real numbers $\{x_n\}$ converges to a number $a$ if $x_n$ gets near $a$ (i.e., the distance between $a$ and $x_n$ gets small) as $n$ gets large. Thus, given $\varepsilon > 0$ (no matter how small), $|x_n - a|$ gets smaller than $\varepsilon$ as $n$ gets large. This leads us to a formal definition of the limit of a sequence.

**2.1 DEFINITION.** A sequence of real numbers $\{x_n\}$ is said to *converge* to a real number $a \in \mathbf{R}$ if for every $\varepsilon > 0$ there is an $N \in \mathbf{N}$ (which in general depends on $\varepsilon$) such that

$$n \geq N \quad \text{implies} \quad |x_n - a| < \varepsilon.$$

We shall use the following phrases and notation interchangeably: a) $\{x_n\}$ converges to $a$; b) $x_n$ converges to $a$; c) $a = \lim_{n \to \infty} x_n$; d) $x_n \to a$ as $n \to \infty$; e) the *limit* of $\{x_n\}$ exists and equals $a$.

When $x_n \to a$ as $n \to \infty$, you can think of $x_n$ as a sequence of approximations to $a$, and $\varepsilon$ as an upper bound for the error of these approximations. The number $N$ in Definition 2.1 is chosen so that the error is less than $\varepsilon$. In general, the smaller $\varepsilon$ gets, the larger $N$ must be. (See, for example, Figure 2.1 below.)

Notice by definition that $x_n$ converges to $a$ if and only if $|x_n - a| \to 0$ as $n \to \infty$. In particular, $x_n \to 0$ if and only if $|x_n| \to 0$ as $n \to \infty$.

**Figure 2.1**

When using Definition 2.1 to prove a particular limit exists, we must show that for any $\varepsilon > 0$, no matter how small, we can find an $N$ such that $n \geq N$ implies $|x_n - a| < \varepsilon$. In particular, $\varepsilon$ is usually introduced before $N$ is specified.

**2.2 Example.** Prove that $1/n \to 0$ as $n \to \infty$.

PROOF. Let $\varepsilon > 0$. Use the Archimedean Principle to choose $N \in \mathbf{N}$ such that $N > 1/\varepsilon$. By taking the reciprocal of this inequality, we see that $n \geq N$ implies $1/n \leq 1/N < \varepsilon$. Since $1/n$ are all positive, it follows that $|1/n| < \varepsilon$ for all $n \geq N$. ∎

Let $\mathcal{P}(n)$ be a property indexed by $\mathbf{N}$. We shall say that "$\mathcal{P}(n)$ holds for large $n$" if there is an $N_0 \in \mathbf{N}$ such that $\mathcal{P}(n)$ is true for all $n \geq N_0$. Hence by definition, $x_n$ converges to $a$ if and only if $|x_n - a|$ is small for large $n$. What we mean by this is that by choosing $N_0$ large enough we can be sure that $|x_n - a|$ is less than any prescribed positive quantity $\varepsilon$ for all $n \geq N_0$.

The following two results show that a given sequence can have no limits or one limit, but no more.

**2.3 Example.** The sequence $\{(-1)^n\}_{n \in \mathbf{N}}$ has no limit.

PROOF. Suppose $(-1)^n \to a$ as $n \to \infty$ for some $a \in \mathbf{R}$. Given $\varepsilon = 1/2$, there is an $N \in \mathbf{N}$ such that $n \geq N$ implies $|(-1)^n - a| < \varepsilon$. For $n$ odd this implies $|1 + a| = |-1 - a| < 1/2$, and for $n$ even this implies $|1 - a| < 1/2$. Hence,

$$2 = |1 + 1| \leq |1 - a| + |1 + a| < \frac{1}{2} + \frac{1}{2} = 1,$$

a contradiction. ∎

**2.4 Remark.** *A sequence can have at most one limit.*

PROOF. Suppose $x_n$ converges to both $a$ and $b$. By definition, given $\varepsilon > 0$, there are integers $N_1$ and $N_2$ such that $n \geq N_1$ implies $|x_n - a| < \varepsilon/2$, and $n \geq N_2$ implies $|x_n - b| < \varepsilon/2$. Let $N = \max\{N_1, N_2\}$. By the choice of $N_1$ and $N_2$, $n \geq N$ implies both $|x_n - a| < \varepsilon/2$ and $|x_n - b| < \varepsilon/2$. Thus it follows from the triangle inequality that

$$|a - b| \leq |a - x_n| + |x_n - b| < \varepsilon;$$

i.e., $|a - b| < \varepsilon$ for all $\varepsilon > 0$. We conclude, by Theorem 1.9 that $a = b$. ∎

Notice that in the proof of Remark 2.4 we forced two properties which held for $n \geq N_j$, $j = 1, 2$, to hold for $n \geq N$ by setting $N$ equal to the maximum of $N_1$ and $N_2$. It is clear that by this same process, if $N_1, \ldots, N_q$ have been chosen so that for each $j$ a property $\mathcal{P}_j$ holds when $n > N_j$ and if $N = \max\{N_1, \ldots, N_q\}$, then all $q$ properties $\mathcal{P}_1, \ldots, \mathcal{P}_q$ hold simultaneously when $n > N$. We shall use this device frequently below, but rarely write $N$ explicitly as a maximum of integers $N_j$ again.

Sometimes the following device is used to correct a sequence which behaves badly or to speed up convergence of another which converges slowly.

**2.5 DEFINITION.** By a *subsequence* of a sequence $\{x_n\}_{n \in \mathbf{N}}$, we shall mean a sequence of the form $\{x_{n_k}\}_{k \in \mathbf{N}}$, where each $n_k \in \mathbf{N}$ and $n_1 < n_2 < \cdots$.

For example, $\{1/2^n\}$ is a subsequence of $\{1/n\}$ (since $n_k := 2^k$ satisfies $n_1 < n_2 < \ldots$) which converges much faster to zero than the original; $\{1^n\}_{n \in \mathbf{N}}$ is a subsequence of $\{(-1)^n\}_{n \in \mathbf{N}}$ (since $n_k := 2k$ satisfies $n_1 < n_2 < \ldots$) which converges whereas the original sequence $(-1)^n$ does not converge at all (see Example 2.3 above).

If $x_n \to a$ as $n \to \infty$, then the $x_n$'s get near $a$ as $n$ gets large. Since $n_k$ gets large as $k$ does, it comes as no surprise that any subsequence of a convergent sequence also converges. $\{1, \frac{1}{2}, 2, \frac{1}{3}, 3, \cdots\}$ not bounded, but subsequ. $\{\frac{1}{2}, \frac{1}{3}, \frac{1}{4}, \cdots\}$ converges to 0. (infinitly many subseq. converges to 0)

**2.6 THEOREM.** *If $\{x_n\}_{n \in \mathbf{N}}$ converges to $a$ and $\{x_{n_k}\}_{k \in \mathbf{N}}$ is any subsequence of $\{x_n\}_{n \in \mathbf{N}}$, then $x_{n_k}$ converges to $a$ as $k \to \infty$.*

PROOF. Let $\varepsilon > 0$ and choose $N \in \mathbf{N}$ such that $n \geq N$ implies $|x_n - a| < \varepsilon$. Since $n_k \in \mathbf{N}$ and $n_1 < n_2 < \ldots$, it is clear that $n_k \geq k$ for all $k \in \mathbf{N}$. Hence, $k \geq N$ implies $|x_{n_k} - a| < \varepsilon$, i.e., $x_{n_k} \to a$ as $k \to \infty$. ∎

The following concepts also play an important role for the theory of sequences.

**2.7 DEFINITION.** Let $\{x_n\}$ be a sequence of real numbers.

   i) $\{x_n\}$ is said to be *bounded above* if there is an $M \in \mathbf{R}$ such that $x_n \leq M$ for all $n \in \mathbf{N}$.

   ii) $\{x_n\}$ is said to be *bounded below* if there is an $m \in \mathbf{R}$ such that $x_n \geq m$ for all $n \in \mathbf{N}$.

   iii) $\{x_n\}$ is said to be *bounded* if it is bounded both above and below.

It is easy to check (see Exercise 5 below) that $\{x_n\}$ is bounded if and only if there is a $C > 0$ such that $|x_n| \leq C$ for all $n \in \mathbf{N}$. In this case we shall say that $\{x_n\}$ is *bounded*, or *dominated*, by $C$.

Is there a relationship between convergent sequences and bounded sequences?

**2.8 THEOREM.** *Every convergent sequence is bounded.* (converse is false).

STRATEGY: The idea behind the proof is simple (see Figure 2.1). Suppose $x_n \to a$ as $n \to \infty$. By definition, for large $N$ the sequence $x_N, x_{N+1}, \ldots$ must be close to $a$, hence bounded. Since the finite sequence $x_1, \ldots, x_{N-1}$ is also bounded, it should follow that the whole sequence is bounded. We now make this precise.

PROOF. Given $\varepsilon = 1$ there is an $N \in \mathbf{N}$ such that $n \geq N$ implies $|x_n - a| \leq 1$. Hence by the triangle inequality, $|x_n| \leq 1 + |a|$ for all $n \geq N$. On the other hand, if $1 \leq n \leq N$, then

$$|(x_n - a) + a| \leq |x_n - a| + |a|$$
$$\leq 1 + |a|$$

$$|x_n| \leq M := \max\{|x_1|, |x_2|, \ldots, |x_N|\}.$$

Therefore, $\{x_n\}$ is dominated by $1 + |a| + M$. ∎

Notice that by Example 2.3, the converse of Theorem 2.8 is false.

## EXERCISES

**1.** Using the method of Example 2.2, prove each of the following limits exist.

    a) $3 + 1/n \to 3$ as $n \to \infty$.
    b) $2(1 - 1/n) \to 2$ as $n \to \infty$.
    c) $(5 + n)/n^2 \to 0$ as $n \to \infty$.
    d) $\pi - 3/\sqrt{n} \to \pi$ as $n \to \infty$.

**2.** Suppose $x_n$ is a sequence of real numbers which converges to 1 as $n \to \infty$. Using Definition 2.1, prove each of the following limits exist.

    a) $1 - x_n \to 0$ as $n \to \infty$.
    b) $3x_n + 1 \to 4$ as $n \to \infty$.
    c) $(2 + x_n^2)/x_n \to 3$ as $n \to \infty$.

**3.** a) Prove that $\{(-1)^n\}$ has some subsequences which converge and others which do not converge.
    b) Find a convergent subsequence of $n + (-1)^{3n}n$.

**4.** a) Suppose $\{b_n\}$ is a sequence of nonnegative numbers which converges to 0, and $\{x_n\}$ is a real sequence which satisfies $|x_n - a| \le b_n$ for large $n$. Prove that $x_n$ converges to $a$.
    b) What happens to part a) if "$\le b_n$" is replaced by "$\le Cb_n$" for some fixed positive constant $C$?

**5.** Suppose $x_n \in \mathbf{R}$.

    a) Prove that $\{x_n\}$ is bounded if and only if there is a $C > 0$ such that $|x_n| \le C$ for all $n \in \mathbf{N}$.
    b) Suppose that $\{x_n\}$ is bounded. Prove that if $\delta_n \to 0$ as $n \to \infty$, then $\delta_n x_n \to 0$ as $n \to \infty$.

**6.** a) Suppose $\{x_n\}$ and $\{y_n\}$ converge to the same point. Prove that $x_n - y_n \to 0$ as $n \to \infty$.
    b) Prove that the sequence $\{n\}$ does not converge.
    c) Show that the converse of part a) is false.

**7.** a) Let $a$ be a fixed real number and define $x_n := a$ for $n \in \mathbf{N}$. Prove that the "constant" sequence $x_n$ converges.
    b) What does $\{x_n\}$ converge to?

**8.** Suppose $\{x_n\}$ is a sequence in $\mathbf{R}$. Prove that $x_n$ converges to $a$ if and only if EVERY subsequence of $x_n$ also converges to $a$.

## 2.2  LIMIT THEOREMS

One of the biggest challenges we face (both for theory and applications) is deciding whether a given sequence converges or not. Once we know it converges, we can often use other techniques to approximate or evaluate its limit.

One way to identify convergent sequences is by comparing a sequence whose convergence is in doubt with another whose convergence property is already known (see Examples 2.10 and 2.21 below). The following result is the first of many theorems which addresses this issue.

**2.9 THEOREM** [SQUEEZE THEOREM]. *Suppose $\{x_n\}$, $\{y_n\}$, and $\{w_n\}$ are real sequences.*

i) *If $x_n \to x$ and $y_n \to x$ as $n \to \infty$, and if there is an $N_0 \in \mathbf{N}$ such that*

$$x_n \leq w_n \leq y_n \quad \text{for} \quad n \geq N_0,$$

*then $w_n \to x$ as $n \to \infty$.*

ii) *If $x_n \to 0$ as $n \to \infty$ and $\{y_n\}$ is bounded, then $x_n y_n \to 0$ as $n \to \infty$.*

PROOF. i) Let $\varepsilon > 0$. Since $x_n$ and $y_n$ converge, use Definition 2.1 and Theorem 1.6 to choose $N_1, N_2 \in \mathbf{N}$ such that $n \geq N_1$ implies $-\varepsilon \leq x_n - x \leq \varepsilon$ and $n \geq N_2$ implies $-\varepsilon \leq y_n - x \leq \varepsilon$. Set $N = \max\{N_0, N_1, N_2\}$. If $n \geq N$ we have by hypothesis and the choice of $N_1$ and $N_2$ that

$$x - \varepsilon \leq x_n \leq w_n \leq y_n \leq x + \varepsilon;$$

i.e., $|w_n - x| \leq \varepsilon$ for $n \geq N$. We conclude that $w_n \to x$ as $n \to \infty$.

ii) Suppose $x_n \to 0$ and there is an $M > 0$ such that $|y_n| \leq M$ for $n \in \mathbf{N}$. Let $\varepsilon > 0$ and choose an $N \in \mathbf{N}$ such that $n \geq N$ implies $|x_n| \leq \varepsilon/M$. Then $n \geq N$ implies

$$|x_n y_n| \leq M \frac{\varepsilon}{M} = \varepsilon.$$

We conclude that $x_n y_n \to 0$ as $n \to \infty$. ∎

The following example shows how the Squeeze Theorem can be used to find the limit of a complicated sequence by ignoring its "less important" factors.

**2.10 Example.** Find $\lim_{n \to \infty} 2^{-n} \cos(n^3 - n^2 + n - 13)$.

SOLUTION. The factor $\cos(n^3 - n^2 + n - 13)$ looks intimidating, but it is superfluous for finding the limit of this sequence. Indeed, since $|\cos x| \leq 1$ for all $x \in \mathbf{R}$ the sequence $\{2^{-n} \cos(n^3 - n^2 + n - 13)\}$ is dominated by $2^{-n}$. Since $2^n > n$ it is clear by Example 2.2 and the Squeeze Theorem that both $2^{-n} \to 0$ and $2^{-n} \cos(n^3 - n^2 + n - 13) \to 0$ as $n \to \infty$. ∎

The Squeeze Theorem can also be used to construct convergent sequences with certain properties. To illustrate how this works, we now prove a result which connects suprema and infima with convergent sequences.

**2.11 THEOREM.** *Let $E \subset \mathbf{R}$. If $E$ has a finite supremum (respectively, a finite infimum), then there is a sequence $x_n \in E$ such that $x_n \to \sup E$ (respectively, $x_n \to \inf E$) as $n \to \infty$.*

$$\lim_{n \to \infty} \sup E - \tfrac{1}{x} = \sup E$$

PROOF. Suppose $E$ has a finite supremum. For each $n \in \mathbf{N}$, choose (by the Approximation Property for Suprema) an $x_n \in E$ such that $\sup E - 1/n < x_n \le \sup E$. Then by the Squeeze Theorem and Example 2.2, $x_n \to \sup E$ as $n \to \infty$. Similarly, there is a sequence $y_n \in E$ such that $y_n \to \inf E$. ∎

Here is another result which helps to evaluate limits of specific sequences. This one works by viewing complicated sequences in terms of simpler components.

**2.12 THEOREM.** *Suppose $\{x_n\}$ and $\{y_n\}$ are real sequences and $\alpha \in \mathbf{R}$. If $\{x_n\}$ and $\{y_n\}$ are convergent, then*

i)
$$\lim_{n \to \infty} (x_n + y_n) = \lim_{n \to \infty} x_n + \lim_{n \to \infty} y_n,$$

ii)
$$\lim_{n \to \infty} (\alpha x_n) = \alpha \lim_{n \to \infty} x_n,$$

*and*

iii)
$$\lim_{n \to \infty} (x_n y_n) = (\lim_{n \to \infty} x_n)(\lim_{n \to \infty} y_n).$$

*If, in addition, $\lim_{n \to \infty} y_n \neq 0$, then*

iv)
$$\lim_{n \to \infty} \frac{x_n}{y_n} = \frac{\lim_{n \to \infty} x_n}{\lim_{n \to \infty} y_n}.$$

*(In particular, all these limits exist.)*

PROOF. Suppose $x_n \to x$ and $y_n \to y$ as $n \to \infty$.

i) Let $\varepsilon > 0$ and choose $N \in \mathbf{N}$ such that $n \ge N$ implies $|x_n - x| < \varepsilon/2$ and $|y_n - y| < \varepsilon/2$. Thus $n \ge N$ implies

$$|(x_n + y_n) - (x + y)| \le |x_n - x| + |y_n - y| < \frac{\varepsilon}{2} + \frac{\varepsilon}{2} = \varepsilon.$$

ii) It suffices to show $\alpha x_n - \alpha x \to 0$ as $n \to \infty$. But $x_n - x \to 0$ as $n \to \infty$, hence, by the Squeeze Theorem, $\alpha(x_n - x) \to 0$ as $n \to \infty$.

iii) By Theorem 2.8, the sequence $\{x_n\}$ is bounded. Hence, by the Squeeze Theorem the sequences $\{x_n(y_n - y)\}$ and $\{(x_n - x)y\}$ both converge to 0. Since

$$x_n y_n - xy = x_n(y_n - y) + (x_n - x)y,$$

it follows from part i) that $x_n y_n \to xy$ as $n \to \infty$. A similar argument establishes part iv) (see Exercise 3). ∎

$$\lim (x_n y_n - xy) = \lim x_n y_n - \lim xy = \lim x_n y_n - xy = 0$$
$$\lim x_n y_n = xy.$$

Theorem 2.12 can be used to evaluate limits of sums, products, and quotients. Here is a typical example.

**2.13 Example.** Find $\lim_{n\to\infty}(n^3 + n^2 - 1)/(1 - 3n^3)$.

SOLUTION. Multiplying the numerator and denominator by $1/n^3$, we find

$$\frac{n^3 + n^2 - 1}{1 - 3n^3} = \frac{1 + (1/n) - (1/n^3)}{(1/n^3) - 3}.$$

By Example 2.2 and Theorem 2.12iii), $1/n^k = (1/n)^k \to 0$, as $n \to \infty$, for any $k \in \mathbf{N}$. Thus by Theorem 2.12i), ii), and iv),

$$\lim_{n\to\infty} \frac{n^3 + n^2 - 1}{1 - 3n^3} = \frac{1 + 0 - 0}{0 - 3} = -\frac{1}{3}. \quad \blacksquare$$

The sequence $\{\log n\}_{n\in\mathbf{N}}$ fails to converge in a different way than $\{n(-1)^n\}_{n\in\mathbf{N}}$ does. (Indeed, the terms $\log n$ get steadily larger as $n \to \infty$, but the terms $n(-1)^n$ bounce back and forth between large positive values and large negative values.) It is sometimes convenient to emphasize this difference by generalizing limits to include extended real numbers.

**2.14 DEFINITION.** Let $\{x_n\}$ be a sequence of real numbers.

i) $\{x_n\}$ is said to *diverge* to $+\infty$ (notation: $x_n \to +\infty$ as $n \to \infty$ or $\lim_{n\to\infty} x_n = +\infty$) if for each $M \in \mathbf{R}$ there is an $N \in \mathbf{N}$ such that

$$n \geq N \quad \text{implies} \quad x_n > M.$$

ii) $\{x_n\}$ is said to *diverge* to $-\infty$ (notation: $x_n \to -\infty$ as $n \to \infty$ or $\lim_{n\to\infty} x_n = -\infty$) if for each $M \in \mathbf{R}$ there is an $N \in \mathbf{N}$ such that

$$n \geq N \quad \text{implies} \quad x_n < M.$$

Notice by Definition 2.14i) that $x_n \to +\infty$ if given $M \in \mathbf{R}$, $x_n$ is greater than $M$ for sufficiently large $n$; i.e., eventually $x_n$ exceeds every number $M$ (no matter how large and positive $M$ is). Similarly, $x_n \to -\infty$ if $x_n$ eventually is less than every number $M$ (no matter how large and negative $M$ is).

It is easy to see that the Squeeze Theorem can be extended to infinite limits (see Exercise 6). The following is an extension of Theorem 2.12.

**2.15 THEOREM.** *Suppose $\{x_n\}$ and $\{y_n\}$ are real sequences such that $x_n \to +\infty$ (respectively, $x_n \to -\infty$) as $n \to \infty$.*

i) *If $y_n$ is bounded below (respectively, $y_n$ is bounded above), then*

$$\lim_{n\to\infty} (x_n + y_n) = +\infty \quad (\text{respectively}, \lim_{n\to\infty} (x_n + y_n) = -\infty).$$

ii) *If $\alpha > 0$, then*

$$\lim_{n\to\infty} (\alpha x_n) = +\infty \quad (\text{respectively}, \lim_{n\to\infty} (\alpha x_n) = -\infty).$$

iii) *If* $y_n > M_0$ *for some* $M_0 > 0$ *and all* $n \in \mathbf{N}$, *then*

$$\lim_{n \to \infty} (x_n y_n) = +\infty \quad (\text{respectively,} \quad \lim_{n \to \infty} (x_n y_n) = -\infty).$$

iv) *If* $\{y_n\}$ *is bounded and* $x_n \neq 0$, *then*

$$\lim_{n \to \infty} \frac{y_n}{x_n} = 0.$$

PROOF. We suppose for simplicity that $x_n \to +\infty$ as $n \to \infty$.

i) By hypothesis, $y_n \geq M_0$ for some $M_0 \in \mathbf{R}$. Let $M \in \mathbf{R}$ and set $M_1 = M - M_0$. Since $x_n \to +\infty$, choose $N \in \mathbf{N}$ such that $n \geq N$ implies $x_n > M_1$. Then $n \geq N$ implies $x_n + y_n > M_1 + M_0 = M$.

ii) Let $M \in \mathbf{R}$ and set $M_1 = M/\alpha$. Choose $N \in \mathbf{N}$ such that $n \geq N$ implies $x_n > M_1$. Since $\alpha > 0$, we conclude that $\alpha x_n > \alpha M_1 = M$ for all $n \geq N$.

iii) Let $M \in \mathbf{R}$ and set $M_1 = M/M_0$. Choose $N \in \mathbf{N}$ such that $n \geq N$ implies $x_n > M_1$. Then $n \geq N$ implies $x_n y_n > M_1 M_0 = M$.

iv) Let $\varepsilon > 0$. Choose $M_0 \in \mathbf{R}$ such that $|y_n| \leq M_0$ and $M_1 > 0$ so large that $M_0/M_1 < \varepsilon$. Choose $N \in \mathbf{N}$ such that $n \geq N$ implies $x_n > M_1$. Then $n \geq N$ implies

$$\left| \frac{y_n}{x_n} \right| = \frac{|y_n|}{x_n} < \frac{M_0}{M_1} < \varepsilon. \quad \blacksquare$$

If we adopt the conventions

$$x + \infty = \infty, \quad x - \infty = -\infty, \qquad x \in \mathbf{R},$$

$$\infty + \infty = \infty, \quad -\infty - \infty = -\infty,$$

$$x \cdot \infty = \infty, \quad x \cdot (-\infty) = -\infty, \qquad x > 0,$$

$$x \cdot \infty = -\infty, \quad x \cdot (-\infty) = \infty, \qquad x < 0,$$

$$\infty \cdot \infty = (-\infty) \cdot (-\infty) = \infty, \quad \text{and} \quad \infty \cdot (-\infty) = (-\infty) \cdot \infty = -\infty,$$

then Theorem 2.15 contains the following corollary.

**2.16 COROLLARY.** *Let* $\{x_n\}$, $\{y_n\}$ *be real sequences, and* $\alpha$, $x$, $y$ *be extended real numbers. If* $x_n \to x$ *and* $y_n \to y$, *as* $n \to \infty$, *then*

$$\lim_{n \to \infty} (x_n + y_n) = x + y$$

*(provided the right side is not of the form* $\infty - \infty$*), and*

$$\lim_{n \to \infty} (\alpha x_n) = \alpha x, \qquad \lim_{n \to \infty} (x_n y_n) = xy$$

*(provided none of these products is of the form* $0 \cdot \pm\infty$*).*

We have avoided the cases $\infty - \infty$ and $0 \cdot \pm\infty$. These and other "indeterminate forms" will be covered by l'Hôpital's Rule in Section 4.3.

Theorems 2.12 and **2.15** show how the limit sign interacts with the algebraic structure of $\mathbf{R}$. (Namely, they say that the limit of a sum (product, quotient) is the sum (product, quotient) of the limits.) The following theorem shows how the limit sign interacts with the order structure of $\mathbf{R}$.

**2.17 THEOREM** [COMPARISON THEOREM]. *Suppose $\{x_n\}$ and $\{y_n\}$ are convergent sequences. If there is an $N_0 \in \mathbf{N}$ such that*

(1)  $$x_n \leq y_n \quad \text{for} \quad n \geq N_0,$$

*then*

$$\lim_{n\to\infty} x_n \leq \lim_{n\to\infty} y_n.$$

*In particular, if $x_n \in [a, b]$ converges to some point $c$, then $c$ must belong to $[a, b]$.*

PROOF.    Suppose the first statement is false, i.e., that (1) holds but $x := \lim_{n\to\infty} x_n$ is greater than $y := \lim_{n\to\infty} y_n$. Set $\varepsilon = (x - y)/2$. Choose $N_1 > N_0$ such that $|x_n - x| < \varepsilon$ and $|y_n - y| < \varepsilon$ for $n \geq N_1$. Then for such an $n$,

$$x_n > x - \varepsilon = x - \left(\frac{x-y}{2}\right) = y + \left(\frac{x-y}{2}\right) = y + \varepsilon > y_n,$$

which contradicts (1). This proves the first statement.

We conclude by noting that the second statement follows from the first, since $a \leq x_n \leq b$ implies $a \leq c \leq b$. ∎

One way to remember this result is that it says the limit of an inequality is the inequality of the limits, provided these limits exist. We shall call this process "taking the limit of an inequality." Since $x_n < y_n$ implies $x_n \leq y_n$, the Comparison Theorem contains the following corollary: if $\{x_n\}$ and $\{y_n\}$ are convergent real sequences, then

$$x_n < y_n, \quad n \geq N_0, \quad \text{implies} \quad \lim_{n\to\infty} x_n \leq \lim_{n\to\infty} y_n.$$

In particular, if $x_n < M$ for $n$ large and $\{x_n\}$ converges, then $\lim_{n\to\infty} x_n \leq M$. It is important to notice that these results are false if, in the conclusion, $\leq$ is replaced by $<$. For example,

$$\frac{1}{n^2} < \frac{1}{n} \quad \text{but} \quad \lim_{n\to\infty} \frac{1}{n^2} = \lim_{n\to\infty} \frac{1}{n} = 0.$$

**EXERCISES**

**1.** Prove that each of the following sequences converges to zero.
  a) $x_n = \sin((n^4 + n + 1)/(n^2 + 1))/n$.
  b) $x_n = n/(n^2 + 1)$.
  c) $x_n = (\sqrt{2n} + 1)/(n + 1)$.
  d) $x_n = n/2^n$.

**2.** Find the limit (if it exists) of each of the following sequences.
  a) $x_n = (1 + n - 3n^2)/(3 - 2n + n^2)$.
  b) $x_n = (n^3 + n - 5)/(5n^3 + n - 1)$.
  c) $x_n = \sqrt{2n^2 - 1}/(n + 1)$.
  d) $x_n = \sqrt{n+1} - \sqrt{n}$.

$\lim \sqrt{x_n} \to \sqrt{\lim x_n}.$

**3.** Prove Theorem 2.12iv).

**4.** Suppose that $x \in \mathbf{R}$, $x_n \geq 0$, and $x_n \to x$ as $n \to \infty$. Prove $\sqrt{x_n} \to \sqrt{x}$ as $n \to \infty$. (For the case $x = 0$ you may wish to use (8) in Section 1.1.)

**5.** Prove that given $x \in \mathbf{R}$ there is a sequence $r_n \in \mathbf{Q}$ such that $r_n \to x$ as $n \to \infty$.

**6.** Suppose $x$ and $y$ are extended real numbers and $\{x_n\}$, $\{y_n\}$, and $\{w_n\}$ are real sequences.

   a) If $x_n \to x$ and $y_n \to x$, as $n \to \infty$, and $x_n \leq w_n \leq y_n$ for $n \in \mathbf{N}$, prove that $w_n \to x$ as $n \to \infty$.

   b) If $x_n \to x$ and $y_n \to y$, as $n \to \infty$, and $x_n \leq y_n$ for $n \in \mathbf{N}$, prove that $x \leq y$.

**7.** Using the result in Exercise 4, show the following.

   a) Suppose $x_1 \geq 0$ and $x_{n+1} = \sqrt{2 + x_n}$ for $n \in \mathbf{N}$. If $x_n \to x$ as $n \to \infty$, prove $x = 2$.

   b) Suppose $0 \leq x_1 \leq 1$ and $x_{n+1} = 1 - \sqrt{1 - x_n}$ for $n \in \mathbf{N}$. If $x_n \to x$ as $n \to \infty$, prove $x = 0$ or 1.

**8.** Prove Corollary 2.16.

**9.** Interpret a decimal expansion $0.a_1a_2\ldots$ as

$$0.a_1a_2\cdots = \lim_{n \to \infty} \sum_{k=1}^{n} \frac{a_k}{10^k}.$$

Prove $0.5 = .4999\ldots$ and $1 = .999\ldots$

**10.** **This exercise was used in Section 1.4.**

   a) Suppose $0 \leq y < 1/10^n$ for some integer $n \geq 0$. Prove there is an integer $0 \leq w \leq 9$ such that

$$\frac{w}{10^{n+1}} \leq y < \frac{w}{10^{n+1}} + \frac{1}{10^{n+1}}.$$

   b) Prove that given $x \in [0, 1)$ there exist integers $0 \leq x_k \leq 9$ such that for all $n \in \mathbf{N}$,

$$\sum_{k=1}^{n} \frac{x_k}{10^k} \leq x < \sum_{k=1}^{n} \frac{x_k}{10^k} + \frac{1}{10^n}.$$

   c) Prove that given $x \in [0, 1)$ there exist integers $0 \leq x_k \leq 9$, $k \in \mathbf{N}$, such that

$$x = \lim_{n \to \infty} \sum_{k=1}^{n} \frac{x_k}{10^k}.$$

(Note: The numbers $x_k$ are called *digits* of $x$, and $0.x_1x_2\ldots$ is called a *decimal expansion* of $x$. Unless $x$ is a rational number whose denominator is of the form $2^i5^j$ for some integers $i \geq 0$, $j \geq 0$, this expansion is unique; i.e., there is only one sequence of integers $\{x_k\}$ which satisfies part c). On the other hand, if $x$ is a rational number whose denominator is of the form $2^i5^j$, then there are two sequences $\{x_k\}$ which satisfy part c), one which satisfies $x_k = 0$ for large $k$ and one which satisfies $x_k = 9$ for large $k$ (see Exercise 9 above). We shall identify the second sequence by saying it *terminates* in 9's.)

## 2.3   THE BOLZANO–WEIERSTRASS THEOREM

Notice that although the sequence $\{(-1)^n\}$ does not converge, it has convergent subsequences. In this section we shall prove that this is a general principle. Namely, we shall establish the Bolzano–Weierstrass Theorem, which states that every bounded sequence has a convergent sequence.

We begin with a special case (monotone sequences) for which the Bolzano–Weierstrass Theorem is especially transparent. Afterward, we shall use this special case to obtain the general result.

**2.18 DEFINITION.** Let $\{x_n\}_{n \in \mathbf{N}}$ be a sequence of real numbers.

i) $\{x_n\}$ is said to be *increasing* (respectively, *strictly increasing*) if $x_1 \leq x_2 \leq \dots$ (respectively, $x_1 < x_2 < \dots$).

ii) $\{x_n\}$ is said to be *decreasing* (respectively, *strictly decreasing*) if $x_1 \geq x_2 \geq \dots$ (respectively, $x_1 > x_2 > \dots$).

iii) $\{x_n\}$ is said to be *monotone* if it is either increasing or decreasing.
[Some authors call decreasing sequences *nonincreasing* and increasing sequences *nondecreasing*.]

If $\{x_n\}$ is increasing (respectively, decreasing) and converges to $a$, we shall write $x_n \uparrow a$ (respectively, $x_n \downarrow a$), as $n \to \infty$. Clearly, every strictly increasing sequence is increasing, and every strictly decreasing sequence is decreasing. Also, $\{x_n\}$ is increasing if and only if the sequence $\{-x_n\}$ is decreasing.

By Theorem 2.8, any convergent sequence is bounded. We now establish the converse of this result for monotone sequences. (For an extension to extended real numbers, see Exercise 3 below.)

**2.19 THEOREM** [Monotone Convergence Theorem]. *If $\{x_n\}$ is monotone and bounded, then $\{x_n\}$ has a finite limit.*

PROOF. We shall actually prove that an increasing sequence converges to its supremum, and a decreasing sequence converges to its infimum.

i) Suppose $\{x_n\}$ is increasing and bounded above. By the Completeness Axiom, the supremum $a := \sup\{x_n : n \in \mathbf{N}\}$ exists and is finite. Let $\varepsilon > 0$. By the Approximation Property for Suprema, choose $N \in \mathbf{N}$ such that

$$a - \varepsilon < x_N \leq a.$$

Since $x_N \leq x_n$ for $n \geq N$ and $x_n \leq a$ for all $n \in \mathbf{N}$, it follows that $a - \varepsilon < x_n \leq a$ for all $n \geq N$. In particular, $x_n \uparrow a$ as $n \to \infty$.

ii) If $\{x_n\}$ is decreasing with infimum $b := \inf\{x_n : n \in \mathbf{N}\}$, then $\{-x_n\}$ is increasing with supremum $-a$ (see Theorem 1.28). Hence, by part i) and Theorem 2.12ii),

$$b = -(-b) = -\lim_{n\to\infty}(-x_n) = \lim_{n\to\infty} x_n. \quad \blacksquare$$

The Monotone Convergence Theorem is used most often to show that a limit exists. Once existence has been established, it is often easy to find the value of that limit by using Theorems 2.9 and 2.12. The following examples illustrate this fact.

**2.20 Example.** If $|a| < 1$, then $a^n \to 0$ as $n \to \infty$.

PROOF. It suffices to prove that $|a|^n \to 0$ as $n \to \infty$. First, we notice that $|a|^n$ is monotone decreasing since by the Multiplicative Property, $|a| < 1$ implies $|a|^{n+1} < |a|^n$ for all $n \in \mathbf{N}$. Next, we observe that $|a|^n$ is bounded below (by 0). Hence by the Monotone Convergence Theorem, $L := \lim_{n \to \infty} |a|^n$ exists.

Suppose $L \neq 0$. Taking the limit of the algebraic identity $|a|^{n+1} = |a| \cdot |a|^n$, as $n \to \infty$, we see by Theorem 2.12 that $L = |a| \cdot L$. Since $L$ is not zero, it follows that $|a| = 1$, a contradiction. ∎

**2.21 Example.** If $a > 0$, then $a^{1/n} \to 1$ as $n \to \infty$.

PROOF. We consider three cases.

*Case 1.* $a = 1$. Then $a^{1/n} = 1$ for all $n \in \mathbf{N}$, and it follows that $a^{1/n} \to 1$ as $n \to \infty$.

*Case 2.* $0 < a < 1$. We first show that $\{a^{1/n}\}$ is increasing and bounded above. Indeed, $\{a^{1/n}\}$ is bounded above by 1 since by Exercise 7 in Section 1.2, $0 < a^{1/n} < 1$ for all $n \in \mathbf{N}$. If $\{a^{1/n}\}$ is not increasing, then there is an $n \in \mathbf{N}$ such that $a^{1/n} > a^{1/(n+1)}$. Taking the $n(n+1)$-st power of this inequality, we obtain $a^{n+1} > a^n$, i.e., $a > 1$, a contradiction. Thus $\{a^{1/n}\}$ is increasing and bounded above. Hence, by Theorems 2.19 and 2.17, $L := \lim_{n \to \infty} a^{1/n}$ exists and satisfies $L \leq 1$, and $0 < a^{1/n} \leq L$; i.e., $0 < a \leq L^n$ holds for all $n \in \mathbf{N}$. To show $L = 1$, suppose to the contrary that $L < 1$. Since $L^n \to 0$ as $n \to \infty$ (see Example 2.20), it follows from $0 < a \leq L^n$ and the Squeeze Theorem that $a = 0$, a contradiction.

*Case 3.* $a > 1$. Then $0 < 1/a < 1$. Hence, by Theorem 2.12 and Case 2,

$$\lim_{n \to \infty} a^{1/n} = \lim_{n \to \infty} \frac{1}{1/a^{1/n}} = \frac{1}{\lim_{n \to \infty}(1/a)^{1/n}} = 1. \quad ∎$$

Next, we introduce a monotone property for sequences of sets.

**2.22 DEFINITION.** A sequence of sets $\{I_n\}_{n \in \mathbf{N}}$ is said to be *nested* if

$$I_1 \supseteq I_2 \supseteq \cdots.$$

In Chapters 3, 8, and 9, we shall use this concept to study continuous functions. Here, we use it to prove the Bolzano–Weierstrass Theorem. All of these applications depend in a fundamental way on the following result.

**2.23 THEOREM** [NESTED INTERVAL PROPERTY]. *If $\{I_n\}_{n \in \mathbf{N}}$ is a nested sequence of nonempty closed bounded intervals, then*

$$E = \bigcap_{n \in \mathbf{N}} I_n := \{x : x \in I_n \text{ for all } n \in \mathbf{N}\} \qquad a_n \leq x \leq b_n \quad \text{for all } n.$$

*contains at least one number. Moreover, if the lengths of these intervals satisfy* $|I_n| \to 0$ *as* $n \to \infty$, *then $E$ contains exactly one number.*

PROOF. Let $I_n = [a_n, b_n]$. Since $\{I_n\}$ is nested, the real sequence $\{a_n\}$ is increasing and bounded above by $b_1$, and $\{b_n\}$ is decreasing and bounded below by $a_1$

$$a_n \leq b_n \quad \text{for all } n.$$

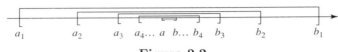

**Figure 2.2**

(see Figure 2.2). Thus by Theorem 2.19, there exist $a, b \in \mathbf{R}$ such that $a_n \uparrow a$ and $b_n \downarrow b$ as $n \to \infty$. Since $a_n \leq b_n$ for all $n \in \mathbf{N}$, it also follows from the Comparison Theorem that $a_n \leq a \leq b \leq b_n$. Hence, a number $x$ belongs to $I_n$ for all $n \in \mathbf{N}$ if and only if $a \leq x \leq b$. This proves that $E = [a, b]$.

Suppose now that $|I_n| \to 0$ as $n \to \infty$. Then $b_n - a_n \to 0$ as $n \to \infty$, and we have by Theorem 2.12 that $b - a = 0$. In particular, $E = [a, a] = \{a\}$ contains exactly one number. ∎

The next two results show that neither of the hypotheses of Theorem 2.23 can be relaxed.   $\bigcap_{n \in \mathbf{N}} [0, \frac{1}{n}] = 0$     $\bigcap_{n \in \mathbf{N}} (0, \frac{1}{n}] = \emptyset$ .

**2.24 Remark.** *The Nested Interval Property is not true if "closed" is omitted.*

PROOF. The intervals $I_n = (0, 1/n)$, $n \in \mathbf{N}$, are bounded and nested but not closed. If there were an $x \in I_n$ for all $n \in \mathbf{N}$, then $0 < x < 1/n$; i.e., $n < 1/x$ for all $n \in \mathbf{N}$. Since this contradicts the Archimedean Principle, it follows that the intervals $I_n$ have no point in common. ∎

**2.25 Remark.** *The Nested Interval Property is not true if "bounded" is omitted.*

PROOF. The intervals $I_n = [n, \infty)$, $n \in \mathbf{N}$ are closed and nested but not bounded. Again, they have no point in common. ∎

We are now prepared to prove the main result of this section.

**2.26 THEOREM** [THE BOLZANO–WEIERSTRASS THEOREM]. *Every bounded sequence of real numbers has a convergent subsequence.*

PROOF. We begin with a general observation. Let $\{x_n\}$ be any sequence. If $E = A \cup B$ are sets and $E$ contains $x_n$ for infinitely many values of $n$, then at least one of the sets $A$ or $B$ also contains $x_n$ for infinitely many values of $n$. (If not, then $E$ contains $x_n$ for only finitely many $n$, a contradiction.)

Let $\{x_n\}$ be a bounded sequence. Choose $a, b \in \mathbf{R}$ such that $x_n \in [a, b]$ for all $n \in \mathbf{N}$, and set $I_0 = [a, b]$. Divide $I_0$ into two halves, say, $I' = [a, (a + b)/2]$ and $I'' = [(a + b)/2, b]$. Since $I_0 = I' \cup I''$, at least one of these half intervals contains $x_n$ for infinitely many $n$. Call it $I_1$, and choose $n_1 > 1$ such that $x_{n_1} \in I_1$. Notice that $|I_1| = |I_0|/2 = (b - a)/2$.

Suppose closed intervals $I_0 \supset I_1 \supset \cdots \supset I_m$ and natural numbers $n_1 < n_2 < \cdots < n_m$ have been chosen such that for each $0 \leq k \leq m$,

$$(2) \qquad |I_k| = \frac{b - a}{2^k}, \quad x_{n_k} \in I_k, \quad \text{and} \quad x_n \in I_k \quad \text{for infinitely many } n.$$

To choose $I_{m+1}$, divide $I_m = [a_m, b_m]$ into two halves, say $I' = [a_m, (a_m + b_m)/2]$ and $I'' = [(a_m + b_m)/2, b_m]$. Since $I_m = I' \cup I''$, at least one of these half intervals contains $x_n$ for infinitely many $n$. Call it $I_{m+1}$, and choose $n_{m+1} > n_m$ such that $x_{n_{m+1}} \in I_{m+1}$. Since

$$|I_{m+1}| = \frac{|I_m|}{2} = \frac{b-a}{2^{m+1}},$$

it follows by induction that there is a nested sequence $\{I_k\}_{k \in \mathbf{N}}$ of nonempty closed bounded intervals which satisfy (2) for all $k \in \mathbf{N}$.

By the Nested Interval Property, there is an $x \in \mathbf{R}$ which belongs to $I_k$ for all $k \in \mathbf{N}$. Since $x \in I_k$, we have by (2) that

$$|x_{n_k} - x| \leq |I_k| \leq \frac{b-a}{2^k}$$

for all $k \in \mathbf{N}$. Hence, by the Squeeze Theorem, $x_{n_k} \to x$ as $k \to \infty$. ∎

### EXERCISES

**1.** Prove that

$$x_n = \frac{(n^2 + 20n + 35)\sin(n^3)}{n^2 + n + 1}$$

has a convergent subsequence.

**2.** Suppose $E \subset \mathbf{R}$ is a nonempty bounded set and $\sup E \notin E$. Prove there exists a strictly increasing sequence $\{x_n\}$ which converges to $\sup E$ such that $x_n \in E$ for all $n \in \mathbf{N}$.

**3.** a) Suppose $\{x_n\}$ is a monotone increasing sequence in $\mathbf{R}$ (not necessarily bounded above). Prove that there is an extended real number $x$ such that $x_n \to x$ as $n \to \infty$.

b) State and prove an analogous result for decreasing sequences.

**4.** Suppose $0 < x_1 < 1$ and $x_{n+1} = 1 - \sqrt{1 - x_n}$ for $n \in \mathbf{N}$. Prove $x_n \downarrow 0$ as $n \to \infty$ and $x_{n+1}/x_n \to 1/2$, as $n \to \infty$. (This is Exercise 4.3 in Apostol [1].)

**5.** Let $0 < x_1 \leq 3$ and $x_{n+1} = \sqrt{2x_n + 3}$ for $n \in \mathbf{N}$. Prove that $x_n \uparrow 3$ as $n \to \infty$.

**6.** Suppose $x_1 \geq 2$ and $x_{n+1} = 1 + \sqrt{x_n - 1}$ for $n \in \mathbf{N}$. Prove that $x_n \downarrow 2$ as $n \to \infty$. What happens when $1 \leq x_1 < 2$?   $x_n \uparrow 2$ as $n \to \infty$, or $x_n \uparrow 1$ as $n \to \infty$.

**7.** Prove

$$\lim_{n \to \infty} x^{1/(2n-1)} = \begin{cases} 1 & x > 0 \\ 0 & x = 0 \\ -1 & x < 0. \end{cases}$$

**8.** Suppose $x_0 \in \mathbf{R}$ and

$$x_n = \frac{1 + x_{n-1}}{2}$$

for $n \in \mathbf{N}$. Prove $x_n \to 1$ as $n \to \infty$.

**9.** Let $0 < y_1 < x_1$ and set

$$x_{n+1} = \frac{x_n + y_n}{2} \quad \text{and} \quad y_{n+1} = \sqrt{x_n y_n} \qquad n \in \mathbf{N}.$$

a) Prove $0 < y_n < x_n$ for all $n \in \mathbf{N}$.
b) Prove $y_n$ is increasing and bounded above, and $x_n$ is decreasing and bounded below.
c) Prove that $0 < x_{n+1} - y_{n+1} < (x_1 - y_1)/2^n$ for $n \in \mathbf{N}$.
d) Prove that $\lim_{n \to \infty} x_n = \lim_{n \to \infty} y_n$. (This common value is called the *arithmetic-geometric* mean of $x_1$ and $y_1$.)

**10.** Suppose $x_0 = 1$, $y_0 = 0$,

$$x_n = x_{n-1} + 2y_{n-1},$$

and

$$y_n = x_{n-1} + y_{n-1}$$

for $n \in \mathbf{N}$. Prove that $x_n^2 - 2y_n^2 = \pm 1$ for $n \in \mathbf{N}$ and

$$\frac{x_n}{y_n} \to \sqrt{2} \quad \text{as } n \to \infty.$$

**11.** [ARCHIMEDES] Suppose $x_0 = 2\sqrt{3}$, $y_0 = 3$,

$$x_n = \frac{2x_{n-1}y_{n-1}}{x_{n-1} + y_{n-1}},$$

and

$$y_n = \sqrt{x_n y_{n-1}}$$

for $n \in \mathbf{N}$.

a) Prove that $x_n \downarrow x$ and $y_n \uparrow y$, as $n \to \infty$, for some $x, y \in \mathbf{R}$.
b) Prove $x = y$ and

$$3.14155 < x < 3.14161.$$

(The actual value of $x$ is $\pi$.)

## 2.4 CAUCHY SEQUENCES

In this section we introduce an extremely powerful and widely used concept.

By definition, if $\{x_n\}$ is a convergent sequence, then there is a point $a \in \mathbf{R}$ such that $x_n$ is near $a$ for large $n$. If the $x_n$'s are near $a$, they are certainly near each other. This leads us to the following concept.

**2.27 DEFINITION.** A sequence of points $x_n \in \mathbf{R}$ is said to be *Cauchy* if for every $\varepsilon > 0$ there is an $N \in \mathbf{N}$ such that

(3) $$n, m \geq N \quad \text{imply} \quad |x_n - x_m| < \varepsilon.$$

The next two results show how this concept is related to convergence.

**2.28 Remark.** *If $\{x_n\}$ is convergent, then $\{x_n\}$ is Cauchy.*

PROOF. Suppose $x_n \to a$ as $n \to \infty$. Then by definition, given $\varepsilon > 0$ there is an $N \in \mathbf{N}$ such that $|x_n - a| < \varepsilon/2$ for all $n \geq N$. Hence, if $n, m \geq N$, it follows from the triangle inequality that

$$|x_n - x_m| \leq |x_n - a| + |x_m - a| < \frac{\varepsilon}{2} + \frac{\varepsilon}{2} = \varepsilon. \quad \blacksquare$$

The following result shows that the converse of Remark 2.28 is also true (for real sequences).

**2.29 THEOREM** [CAUCHY]. *Let $\{x_n\}$ be a sequence of real numbers. Then $\{x_n\}$ is Cauchy if and only if $\{x_n\}$ converges (to some point $a$ in $\mathbf{R}$).*

STRATEGY. By Remark 2.28, we need only show that every Cauchy sequence converges. Suppose $\{x_n\}$ is Cauchy. Since the $x_n$'s are near each other, the sequence $\{x_n\}$ should be bounded. Hence by the Bolzano–Weierstrass Theorem, $\{x_n\}$ has a convergent subsequence, say $x_{n_k}$. This means that for large $k$, the $x_{n_k}$'s are near some point $a \in \mathbf{R}$. But since $\{x_n\}$ is Cauchy, the $x_n$'s should be near the $x_{n_k}$'s for large $n$, hence also near $a$. Thus the full sequence must converge to that same point $a$. Here are the details.

PROOF. Suppose $\{x_n\}$ is Cauchy. Choose $N \in \mathbf{N}$ such that $|x_N - x_m| < 1$ for all $m \geq N$. By the triangle inequality

$$|x_m| < 1 + |x_N| \quad \text{for} \quad m \geq N.$$

Therefore, $\{x_n\}$ is bounded by $M = \max\{|x_1|, |x_2|, \ldots, |x_{N-1}|, 1 + |x_N|\}$.

By the Bolzano–Weierstrass Theorem, $\{x_n\}$ has a convergent subsequence, say $x_{n_k} \to a$ as $k \to \infty$. Let $\varepsilon > 0$. Since $x_n$ is Cauchy, choose $N_1 \in \mathbf{N}$ such that

$$n, m \geq N_1 \quad \text{implies} \quad |x_n - x_m| < \frac{\varepsilon}{2}.$$

Since $x_{n_k} \to a$ as $k \to \infty$, choose $N_2 \in \mathbf{N}$ such that

$$k \geq N_2 \quad \text{implies} \quad |x_{n_k} - a| < \frac{\varepsilon}{2}.$$

Fix $k \geq N_2$ such that $n_k \geq N_1$. Then

$$|x_n - a| \leq |x_n - x_{n_k}| + |x_{n_k} - a| < \varepsilon$$

for all $n \geq N_1$. Thus $x_n \to a$ as $n \to \infty$. $\quad \blacksquare$

This result is extremely useful because it is often easier to show a sequence is Cauchy than to show it converges. The reason for this, as the following example shows, is that we can prove that a sequence is Cauchy even when we have no idea what its limit is.

**2.30 Example.** Prove that any real sequence $\{x_n\}$ which satisfies

$$|x_n - x_{n+1}| \leq \frac{1}{2^n}, \qquad n \in \mathbf{N},$$

is convergent.

Proof. If $m > n$, then

$$\begin{aligned}
|x_n - x_m| &= |x_n - x_{n+1} + x_{n+1} - x_{n+2} + \cdots + x_{m-1} - x_m| \\
&\leq |x_n - x_{n+1}| + |x_{n+1} - x_{n+2}| + \cdots + |x_{m-1} - x_m| \\
&\leq \frac{1}{2^n} + \cdots + \frac{1}{2^{m-1}} \\
&= \frac{1}{2^{n-1}} \sum_{k=1}^{m-n} \frac{1}{2^k} = \frac{1}{2^{n-1}}\left(1 - \frac{1}{2^{m-n}}\right).
\end{aligned}$$

(This last step uses Exercise 1c) in Section 1.2 for $a = 2$.) It follows that $|x_n - x_m| < 1/2^{n-1}$ for all integers $m > n \geq 1$. But given $\varepsilon > 0$ we can choose $N \in \mathbf{N}$ so large that $n \geq N$ implies $1/2^{n-1} < \varepsilon$. We have proved that $\{x_n\}$ is Cauchy. By Theorem 2.29, therefore, it converges to some real number. ∎

The following result shows that a sequence is not necessarily Cauchy just because $x_n$ is near $x_{n+1}$ for large $n$.

**2.31 Remark.** *A sequence which satisfies $x_{n+1} - x_n \to 0$ is not necessarily Cauchy.*

Proof. Consider the sequence $x_n := \log n$. By basic properties of logarithms (see Exercise 4 in Section 5.3),

$$x_{n+1} - x_n = \log(n+1) - \log n = \log((n+1)/n) \to \log 1 = 0$$

as $n \to \infty$. $\{x_n\}$ cannot be Cauchy, however, because it does not converge; in fact, it diverges to $+\infty$ as $n \to \infty$. ∎

$$\left|\int_a^b f(s)\,ds\right| \leq \int_a^b |f(x)|\,dx.$$

$$n! \leq 2^{n-1}$$

Geometric with $r = \frac{1}{2}$.

**EXERCISES**

$$\frac{1}{2^n} + \frac{1}{2^{n+1}} + \cdots + \frac{1}{2^m} = \frac{1}{2^n}\left(1 + \frac{1}{2} + \frac{1}{2^2} + \cdots + \frac{1}{2^{m-n}}\right) \leq \frac{1}{2}\left(1 + \frac{1}{2} + \frac{1}{2^2} + \cdots\right)$$
$$= \frac{1}{2}\left(\frac{1}{1-\frac{1}{2}}\right)$$

**1.** Prove (without using Theorem 2.29) that the sum of two Cauchy sequences is Cauchy.

**2.** Prove that if $\{x_n\}$ is a sequence which satisfies

$$|x_n| \leq \frac{1+n}{1+n+2n^2}$$

for all $n \in \mathbf{N}$, then $\{x_n\}$ is Cauchy.

**3.** Suppose that $x_n \in \mathbf{N}$ for $n \in \mathbf{N}$. If $\{x_n\}$ is Cauchy, prove that there are numbers $a$ and $N$ such that $x_n = a$ for all $n \geq N$.

**4.** Let $\{x_n\}$ be a sequence of real numbers. Suppose for each $\varepsilon > 0$ there is an $N \in \mathbf{N}$ such that $m \geq n \geq N$ implies $\left| \sum_{k=n}^{m} x_k \right| < \varepsilon$. Prove that

$$\lim_{n \to \infty} \sum_{k=1}^{n} x_k$$

exists and is finite.

**5.** Let $\{x_n\}$ be Cauchy. Prove that $\{x_n\}$ converges if and only if at least one of its subsequences converges (compare with Exercise 8 in Section 2.1).

**6.** Prove that $\lim_{n \to \infty} \sum_{k=1}^{n} (-1)^k / k$ exists and is finite.

**7.** Let $\{x_n\}$ be a sequence. Suppose there is an $a > 1$ such that

$$|x_{k+1} - x_k| \leq a^{-k}$$

for all $k \in \mathbf{N}$. Prove that $x_n \to x$ for some $x \in \mathbf{R}$.

**8.** a) A subset $E$ of $\mathbf{R}$ is said to be *sequentially compact* if every sequence $x_n \in E$ has a convergent subsequence whose limit belongs to $E$. Prove that every closed bounded interval is sequentially compact.

b) Prove that there exist bounded intervals in $\mathbf{R}$ which are not sequentially compact.

c) Prove that there exist closed intervals in $\mathbf{R}$ which are not sequentially compact.

**9.** a) Let $E$ be a subset of $\mathbf{R}$. A point $a \in \mathbf{R}$ is called a *cluster point* of $E$ if $E \cap (a - r, a + r)$ contains infinitely many points for every $r > 0$. Prove that $a$ is a cluster point of $E$ if and only if for each $r > 0$, $E \cap (a - r, a + r) \setminus \{a\}$ is nonempty.

b) Prove that every bounded infinite subset of $\mathbf{R}$ has at least one cluster point.

## $^e$2.5 LIMITS SUPREMUM AND INFIMUM  *This section uses no material from any other enrichment section.*

In some situations (for example, the Root Test in Section 6.3), we shall use the following generalization of limits.

**2.32 DEFINITION.** Let $\{x_n\}$ be a real sequence. The *limit supremum* of $\{x_n\}$ is the extended real number

(4)
$$\limsup_{n \to \infty} x_n := \lim_{n \to \infty} (\sup_{k \geq n} x_k),$$

and the *limit infimum* of $\{x_n\}$ is the extended real number

$$\liminf_{n \to \infty} x_n := \lim_{n \to \infty} (\inf_{k \geq n} x_k).$$

Before we proceed, we must show that the limits in Definition 2.32 exist as extended real numbers. To this end, let $\{x_n\}$ be a sequence of real numbers and consider the sequences

$$s_n = \sup_{k \geq n} x_k := \sup\{x_k : k \geq n\} \quad \text{and} \quad t_n = \inf_{k \geq n} x_k := \inf\{x_k : k \geq n\}.$$

Each $s_n$ and $t_n$ is an extended real number and, by the Monotone Property, $s_n$ is a decreasing sequence and $t_n$ an increasing sequence of extended real numbers. In particular, there exist extended real numbers $s$ and $t$ such that $s_n \downarrow s$ and $t_n \uparrow t$ as $n \to \infty$ (see Exercise 3 in Section 2.3). These extended real numbers are, by Definition 2.32, the limit infimum and limit supremum of the sequence $\{x_n\}$.

Here are two examples of how to compute limits supremum and limits infimum.

**2.33 Example.** Find $\limsup_{n\to\infty} x_n$ and $\liminf_{n\to\infty} x_n$ if $x_n = (-1)^n$.

SOLUTION. Since $\sup_{k\geq n}(-1)^k = 1$ for all $n \in$ **N**, it follows from Definition 2.32 that $\limsup_{n\to\infty} x_n = 1$. Similarly, $\liminf_{n\to\infty} x_n = -1$. ∎

**2.34 Example.** Find $\limsup_{n\to\infty} x_n$ and $\liminf_{n\to\infty} x_n$ if $x_n = 1 + 1/n$.

SOLUTION. Since $\sup_{k\geq n}(1 + 1/k) = 1 + 1/n$ for all $n \in$ **N**, $\limsup_{n\to\infty} x_n = 1$. Since $\inf_{k\geq n}(1 + 1/k) = 1$ for all $n \in$ **N**, $\liminf_{n\to\infty} x_n = 1$. ∎

These examples suggest that there is a connection between limits supremum, limits infimum, and convergent subsequences. The next several results make this connection clear.

**2.35 THEOREM.** *Let $\{x_n\}$ be a sequence of real numbers, $s = \limsup_{n\to\infty} x_n$, and $t = \liminf_{n\to\infty} x_n$. Then there are subsequences $\{x_{n_k}\}_{k\in\mathbf{N}}$ and $\{x_{\ell_j}\}_{j\in\mathbf{N}}$ such that $x_{n_k} \to s$ as $k \to \infty$ and $x_{\ell_j} \to t$ as $j \to \infty$.*

PROOF. We will prove the result for the limit supremum. A similar argument establishes the result for the limit infimum. Let $s_n = \sup_{k\geq n} x_k$ and observe that $s_n \downarrow s$ as $n \to \infty$.

*Case 1.* $s = \infty$. Then by definition $s_n = \infty$ for all $n \in$ **N**. Since $s_1 = \infty$, there is an $n_1 \in$ **N** such that $x_{n_1} > 1$. Since $s_{n_1+1} = \infty$, there is an $n_2 \geq n_1 + 1 > n_1$ such that $x_{n_2} > 2$. Continuing in this manner, we can choose a subsequence $\{x_{n_k}\}$ such that $x_{n_k} > k$ for all $k \in$ **N**. Hence, it follows from the Squeeze Theorem that $x_{n_k} \to \infty = s$ as $k \to \infty$.

*Case 2.* $s = -\infty$. Since $s_n \geq x_n$ for all $n \in$ **N**, it follows from the Squeeze Theorem that $x_n \to -\infty = s$ as $n \to \infty$.

*Case 3.* $-\infty < s < \infty$. Set $n_0 = 0$. By Theorem 1.20 (the Approximation Property for Suprema), there is an integer $n_1 \in$ **N** such that $s_{n_0+1} - 1 < x_{n_1} \leq s_{n_0+1}$. Similarly, there is an integer $n_2 \geq n_1 + 1 > n_1$ such that $s_{n_1+1} - 1/2 < x_{n_2} \leq s_{n_1+1}$. Continuing in this manner, we can choose integers $n_1 < n_2 < \ldots$ such that

(5)
$$s_{n_{k-1}+1} - \frac{1}{k} < x_{n_k} \leq s_{n_{k-1}+1}$$

for $k \in \mathbf{N}$. Since $s_{n_{k-1}+1} \to s$ as $k \to \infty$, we conclude by the Squeeze Theorem that $x_{n_k} \to s$ as $k \to \infty$. ∎

This observation leads directly to a characterization of limits in terms of limits infimum and limits supremum.

**2.36 THEOREM.** *Let $\{x_n\}$ be a real sequence and $x$ be an extended real number. Then $x_n \to x$ as $n \to \infty$ if and only if*

$$(6) \qquad \limsup_{n \to \infty} x_n = \liminf_{n \to \infty} x_n = x.$$

PROOF. Suppose $x_n \to x$ as $n \to \infty$. Then $x_{n_k} \to x$ as $k \to \infty$ for all subsequences $\{x_{n_k}\}$. Hence, by Theorem 2.35, $\limsup_{n \to \infty} x_n = x$ and $\liminf_{n \to \infty} x_n = x$, i.e., (6) holds.

Conversely, suppose (6) holds.

*Case 1.* $x = \pm\infty$. By considering $\pm x_n$ we may suppose that $x = \infty$. Thus given $M \in \mathbf{R}$ there is an $N \in \mathbf{N}$ such that $\inf_{k \geq N} x_k > M$. It follows that $x_n > M$ for all $n \geq N$, i.e., $x_n \to \infty$ as $n \to \infty$.

*Case 2.* $-\infty < x < \infty$. Let $\varepsilon > 0$. Choose $N \in \mathbf{N}$ such that

$$\sup_{k \geq N} x_k - x < \frac{\varepsilon}{2} \quad \text{and} \quad x - \inf_{k \geq N} x_k < \frac{\varepsilon}{2}.$$

Let $n, m \geq N$ and suppose for simplicity that $x_n > x_m$. Then

$$|x_n - x_m| = x_n - x_m \leq \sup_{k \geq N} x_k - x + x - \inf_{k \geq N} x_k < \frac{\varepsilon}{2} + \frac{\varepsilon}{2} = \varepsilon.$$

Thus $\{x_n\}$ is Cauchy and converges to some finite real number. But by Theorem 2.35, some subsequence of $\{x_n\}$ converges to $x$. We conclude that $x_n \to x$ as $n \to \infty$. ∎

Theorem 2.35 also leads to the following geometric interpretation of limits supremum and limits infimum.

**2.37 Remark.** *Let $\{x_n\}$ be a sequence of real numbers. Then $\limsup_{n \to \infty} x_n$ (respectively, $\liminf_{n \to \infty} x_n$) is the largest value (respectively, the smallest value) to which some subsequence of $\{x_n\}$ converges. Namely, if $x_{n_k} \to x$ as $k \to \infty$, then*

$$(7) \qquad \liminf_{n \to \infty} x_n \leq x \leq \limsup_{n \to \infty} x_n.$$

PROOF. Suppose $x_{n_k} \to x$ as $k \to \infty$. Fix $N \in \mathbf{N}$ and choose $K$ so large that $k \geq K$ implies $n_k \geq N$. Clearly,

$$\inf_{j \geq N} x_j \leq x_{n_k} \leq \sup_{j \geq N} x_j$$

for all $k \geq K$. Taking the limit of this inequality as $k \to \infty$, we obtain

$$\inf_{j \geq N} x_j \leq x \leq \sup_{j \geq N} x_j.$$

Taking the limit of this last inequality as $N \to \infty$ and applying Definition 2.32, we obtain (7). ∎

We close this section with several other properties of limits supremum and limits infimum.

**2.38 Remark.** *If $\{x_n\}$ is any sequence of real numbers, then*

$$\liminf_{n\to\infty} x_n \leq \limsup_{n\to\infty} x_n.$$

PROOF. Since $\inf_{k\geq n} x_k \leq \sup_{k\geq n} x_k$ for all $n \in \mathbf{N}$, this inequality follows from Theorem 2.17 (the Comparison Theorem). ∎

The following result is an immediate consequence of Definition 2.32, the Comparison Theorem, and the Monotone Convergence Theorem.

**2.39 Remark.** *A real sequence $\{x_n\}$ is bounded above if and only if $\limsup_{n\to\infty} x_n < \infty$, and is bounded below if and only if $\liminf_{n\to\infty} x_n > -\infty$.*

The following result shows we can take limits supremum and limits infimum of inequalities.

**2.40 THEOREM.** *If $x_n \leq y_n$ for $n$ large, then*

$$(8) \qquad \limsup_{n\to\infty} x_n \leq \limsup_{n\to\infty} y_n \quad and \quad \liminf_{n\to\infty} x_n \leq \liminf_{n\to\infty} y_n.$$

PROOF. If $x_k \leq y_k$ for $k \geq N$, then $\sup_{k\geq n} x_k \leq \sup_{k\geq n} y_k$ and $\inf_{k\geq n} x_k \leq \inf_{k\geq n} y_k$ for any $n \geq N$. Taking the limit of these inequalities as $n \to \infty$, we obtain (8). ∎

## EXERCISES

**1.** Find the limit infimum and the limit supremum of each of the following sequences.

    a) $x_n = 3 - (-1)^n$.
    b) $x_n = \cos(n\pi/2)$.
    c) $x_n = (-1)^{n+1} + (-1)^n/n$.
    d) $x_n = \sqrt{1 + n^2}/(2n - 5)$.
    e) $x_n = y_n/n$, where $\{y_n\}$ is any bounded sequence.
    f) $x_n = n(1 + (-1)^n) + n^{-1}((-1)^n - 1)$.
    g) $x_n = (n^3 + n^2 - n + 1)/(n^2 + 2n + 5)$.

**2.** Suppose $\{x_n\}$ is a real sequence. Prove that

$$-\limsup_{n\to\infty} x_n = \liminf_{n\to\infty}(-x_n)$$

and

$$-\liminf_{n\to\infty} x_n = \limsup_{n\to\infty}(-x_n).$$

**3.** Let $\{x_n\}$ be a real sequence and $r \in \mathbf{R}$.

    a) Prove that

$$\limsup_{n\to\infty} x_n < r \quad \text{implies} \quad x_n < r$$

for $n$ large.

b) Prove that
$$\limsup_{n\to\infty} x_n > r \quad \text{implies} \quad x_n > r$$

for infinitely many $n \in \mathbf{N}$.

**4.** Suppose $\{x_n\}$ and $\{y_n\}$ are real sequences.

a) Prove that
$$\liminf_{n\to\infty} x_n + \liminf_{n\to\infty} y_n \leq \liminf_{n\to\infty}(x_n + y_n)$$
$$\leq \limsup_{n\to\infty} x_n + \liminf_{n\to\infty} y_n$$
$$\leq \limsup_{n\to\infty}(x_n + y_n) \leq \limsup_{n\to\infty} x_n + \limsup_{n\to\infty} y_n,$$

provided none of these sums is of the form $\infty - \infty$.

b) Show that if $\lim_{n\to\infty} x_n$ exists, then
$$\liminf_{n\to\infty}(x_n + y_n) = \lim_{n\to\infty} x_n + \liminf_{n\to\infty} y_n$$

and
$$\limsup_{n\to\infty}(x_n + y_n) = \lim_{n\to\infty} x_n + \limsup_{n\to\infty} y_n.$$

c) Show by examples that each of the inequalities in part a) can be strict.

**5.** Let $\{x_n\}$ and $\{y_n\}$ be real sequences.

a) Suppose that $x_n \geq 0$ and $y_n \geq 0$ for each $n \in \mathbf{N}$. Prove
$$\limsup_{n\to\infty}(x_n y_n) \leq \left(\limsup_{n\to\infty} x_n\right)\left(\limsup_{n\to\infty} y_n\right),$$

provided the product on the right is not of the form $0 \cdot \infty$. Show by example that this inequality can be strict.

b) Suppose $x_n \leq 0 \leq y_n$ for $n \in \mathbf{N}$. Prove
$$\left(\liminf_{n\to\infty} x_n\right)\left(\limsup_{n\to\infty} y_n\right) \leq \liminf_{n\to\infty}(x_n y_n),$$

provided none of these products is of the form $0 \cdot \infty$.

**6.** Suppose that $x_n \geq 0$ and $y_n \geq 0$ for all $n \in \mathbf{N}$. Prove that if $x_n \to x$ as $n \to \infty$ ($x$ may be an extended real number), then
$$\limsup_{n\to\infty}(x_n y_n) = x \limsup_{n\to\infty} y_n,$$

provided none of these products is of the form $0 \cdot \infty$.

**7.** Prove that
$$\limsup_{n\to\infty} x_n = \inf_{n\in\mathbf{N}}\left(\sup_{k\geq n} x_k\right) \quad \text{and} \quad \liminf_{n\to\infty} x_n = \sup_{n\in\mathbf{N}}\left(\inf_{k\geq n} x_k\right)$$

for any real sequence $\{x_n\}$.

**8.** Suppose $x_n \geq 0$ for $n \in \mathbf{N}$. Under the interpretation $1/0 = \infty$ and $1/\infty = 0$, prove

$$\limsup_{n \to \infty} \left( \frac{1}{x_n} \right) = \frac{1}{\liminf_{n \to \infty} x_n} \quad \text{and} \quad \liminf_{n \to \infty} \left( \frac{1}{x_n} \right) = \frac{1}{\limsup_{n \to \infty} x_n}.$$

**9.** Let $x_n \in \mathbf{R}$. Prove that $x_n \to 0$ as $n \to \infty$ if and only if

$$\limsup_{n \to \infty} |x_n| \leq 0.$$

# Chapter 3

# Continuity on $\mathbf{R}$

## 3.1 TWO-SIDED LIMITS

In the previous chapter we studied limits of real sequences. In this chapter we examine limits of *real functions*, i.e., functions whose domains and ranges are subsets of $\mathbf{R}$.

Recall from elementary calculus that a function $f(x)$ converges to a limit $L$, as $x$ approaches $a$, if $f(x)$ is near $L$ when $x$ is near $a$. Here is a precise definition of this concept.

**3.1 DEFINITION.** Let $a \in \mathbf{R}$, $I$ be an open interval which contains $a$, and $f$ be a real function defined everywhere on $I$ except possibly at $a$. Then $f(x)$ is said to *converge to $L$, as $x$ approaches $a$*, if for every $\varepsilon > 0$ there is a $\delta > 0$ (which in general depends on $\varepsilon$, $f$, and $a$) such that

(1) $$0 < |x - a| < \delta \quad \text{implies} \quad |f(x) - L| < \varepsilon.$$

In this case we write

$$L = \lim_{x \to a} f(x)$$

and call $L$ the *limit* of $f(x)$ as $x$ approaches $a$.

As was the case for sequences, the $\varepsilon$ represents the maximal error allowed in the approximation $f(x)$ to $f(a)$. The number $\delta$ represents the tolerance allowed in the measurement $x$ of $a$ which will produce an approximation $f(x)$ which is acceptably close to the value $f(a)$.

According Definition 3.1, to show a function has a limit we must begin with a general $\varepsilon > 0$ and describe how to choose a $\delta$ which satisfies (1).

**3.2 Example.** Suppose $f(x) = mx + b$ for some $m, b \in \mathbf{R}$. Prove

$$f(a) = \lim_{x \to a} f(x)$$

for all $a \in \mathbf{R}$.

PROOF. If $m = 0$ there is nothing to prove. Otherwise, given $\varepsilon > 0$, set $\delta = \varepsilon/|m|$. If $|x - a| < \delta$, then

$$|f(x) - ma - b| = |m|\,|x - a| < |m|\delta = \varepsilon.$$

Thus by definition, $f(x) \to f(a)$ as $x \to a$. ∎

Sometimes, in order to determine $\delta$ one must break $f(x) - L$ into two factors, replacing the less important factor by an upper bound.

**3.3 Example.** If $f(x) = x^2 + x - 3$, prove that $f(x) \to -1$ as $x \to 1$.

PROOF. Let $\varepsilon > 0$ and set $L = -1$. Notice that

$$f(x) - L = x^2 + x - 2 = (x - 1)(x + 2).$$

If $0 < \delta \le 1$, then $|x - 1| < \delta$ implies $0 < x < 2$, so by the triangle inequality, $|x + 2| \le |x| + 2 < 4$. Set $\delta = \min\{1, \varepsilon/4\}$. It follows that if $|x - 1| < \delta$, then

$$|f(x) - L| = |x - 1|\,|x + 2| < 4\delta \le \varepsilon.$$

Thus by definition, $f(x) \to L$ as $x \to 1$. ∎

Before continuing, we would like to draw your attention to two features of Definition 3.1: the assumption that $f$ be defined on an open interval $I$, and the assumption that $0 < |x - a|$. First, notice that if $I = (c, d) \supset \{a\}$ and $\delta_0 := \min\{a - c, d - a\}$, then $|x - a| < \delta_0$ implies $x \in I$. Hence, the assumption that $f$ be defined on some open interval containing $a$ is made so that $f(x)$ is defined for all $x$ satisfying $|x - a| < \delta$ when $\delta$ sufficiently small. Next, notice that the assumption $|x - a| > 0$ is equivalent to $x \ne a$. Thus the function $f$ need not be defined at $a$ in order for $f$ to have a limit at $a$.

The next result shows that even when $f$ is defined at $a$, the value of the limit of a function at $a$ is, in general, independent of the value of the function at $a$.

**3.4 Remark.** *Let $a \in$ **R**, $I$ be an open interval which contains $a$, and $f, g$ be real functions defined everywhere on $I$ except possibly at $a$. If $f(x) = g(x)$ for all $x \in I \setminus \{a\}$ and $f(x) \to L$ as $x \to a$, then $g(x)$ also has a limit as $x \to a$, and*

$$\lim_{x \to a} g(x) = \lim_{x \to a} f(x).$$

PROOF. Let $\varepsilon > 0$ and choose $\delta > 0$ by Definition 3.1 so that (1) holds and $|x - a| < \delta$ implies $x \in I$. Suppose $0 < |x - a| < \delta$. We have $f(x) = g(x)$ by hypothesis and $|f(x) - L| < \varepsilon$ by (1). It follows that $|g(x) - L| < \varepsilon$. ∎

Thus to prove that a function $f$ has a limit, we may begin by simplifying $f$ algebraically.

**3.5 Example.** Prove that

$$g(x) = \frac{x^3 + x^2 - x - 1}{x^2 - 1}$$

has a limit as $x \to 1$.

PROOF. Set $f(x) = x + 1$ and observe by Example 3.2 that $f(x) \to 2$ as $x \to 1$. Since

$$g(x) = \frac{x^3 + x^2 - x - 1}{x^2 - 1} = \frac{(x + 1)(x^2 - 1)}{x^2 - 1} = f(x)$$

for $x \neq \pm 1$, it follows from Remark 3.4 that $g(x)$ has a limit as $x \to 1$ (and that limit is 2). ∎

There is a close connection between limits of functions and limits of sequences.

**3.6 THEOREM** [SEQUENTIAL CHARACTERIZATION OF LIMITS]. *Let $a \in \mathbf{R}$, $I$ be an open interval which contains $a$, and $f$ be a real function defined everywhere on $I$ except possibly at $a$. Then*

$$L = \lim_{x \to a} f(x)$$

*exists if and only if $f(x_n) \to L$ as $n \to \infty$ for every sequence $x_n \in I \setminus \{a\}$ which converges to $a$ as $n \to \infty$.*

PROOF. Suppose $f$ converges to $L$ as $x$ approaches $a$. Then given $\varepsilon > 0$ there is a $\delta > 0$ such that (1) holds. If $x_n \in I \setminus \{a\}$ converges to $a$ as $n \to \infty$, then choose an $N \in \mathbf{N}$ such that $n \geq N$ implies $|x_n - a| < \delta$. Since $x_n \neq a$, it follows from (1) that $|f(x_n) - L| < \varepsilon$ for all $n \geq N$. Therefore, $f(x_n) \to L$ as $n \to \infty$.

Conversely, suppose $f(x_n) \to L$ as $n \to \infty$ for every sequence $x_n \in I \setminus \{a\}$ which converges to $a$. If $f$ does not converge to $L$ as $x$ approaches $a$, then there is an $\varepsilon > 0$ (call it $\varepsilon_0$) such that the implication "$0 < |x - a| < \delta$ implies $|f(x) - L| < \varepsilon_0$" does not hold for any $\delta > 0$. Thus, for each $\delta = 1/n$, $n \in \mathbf{N}$, there is a point $x_n \in I$ which satisfies two conditions: $0 < |x_n - a| < 1/n$ and $|f(x_n) - L| \geq \varepsilon_0$. Now the first condition and the Squeeze Theorem (Theorem 2.9) imply that $x_n \neq a$ and $x_n \to a$ so by hypothesis, $f(x_n) \to L$, as $n \to \infty$. In particular, $|f(x_n) - L| < \varepsilon_0$ for $n$ large, which contradicts the second condition. ∎

Thus to show the limit of a function $f$ does not exist as $x \to a$, we need only find two sequences converging to $a$ whose images under $f$ have different limits.

**3.7 Example.** Prove that

$$f(x) = \begin{cases} \sin \dfrac{1}{x} & x \neq 0 \\ 0 & x = 0 \end{cases}$$

has no limit as $x \to 0$.

PROOF. By examining the graph of $y = f(x)$ (see Figure 3.1), we are lead to consider two extremes:

$$a_n := \frac{2}{(4n + 1)\pi} \quad \text{and} \quad b_n := \frac{2}{(4n + 3)\pi}, \qquad n \in \mathbf{N}.$$

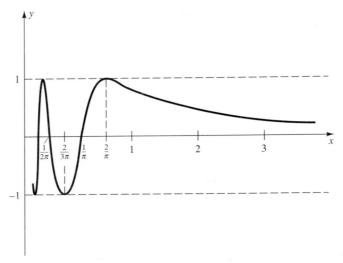

**Figure 3.1**

Clearly, both $a_n$ and $b_n$ converge to 0 as $n \to \infty$. On the other hand, since $f(a_n) = 1$ and $f(b_n) = -1$ for all $n \in \mathbf{N}$, $f(a_n) \to 1$ and $f(b_n) \to -1$ as $n \to \infty$. Thus by Theorem 3.6, the limit of $f(x)$, as $x \to 0$, cannot exist. ∎

Theorem 3.6 also allows us to translate results about limits of sequences to results about limits of functions. The next three theorems illustrate this principle.

Before stating these results, we need additional notation. Suppose $f$ and $g$ are real functions. The *pointwise sum*, $f + g$, of $f$ and $g$ is defined by

$$(f + g)(x) := f(x) + g(x), \qquad \mathrm{Dom}\,(f + g) = \mathrm{Dom}\,(f) \cap \mathrm{Dom}\,(g);$$

the *scalar product*, $\alpha f$, of a scalar $\alpha \in \mathbf{R}$ with $f$ by

$$(\alpha f)(x) := \alpha f(x), \qquad \mathrm{Dom}\,(\alpha f) = \mathrm{Dom}\,(f);$$

the *pointwise product*, $fg$, of $f$ and $g$ by

$$(fg)(x) := f(x)g(x), \qquad \mathrm{Dom}\,(fg) = \mathrm{Dom}\,(f) \cap \mathrm{Dom}\,(g);$$

and the *pointwise quotient*, $f/g$, of $f$ and $g$ by

$$\left(\frac{f}{g}\right)(x) := \frac{f(x)}{g(x)}, \qquad \mathrm{Dom}\,(f/g) = \{x \in \mathrm{Dom}\,(f) \cap \mathrm{Dom}\,(g) : g(x) \neq 0\}.$$

The following result is a function analogue of Theorem 2.12.

**3.8 THEOREM.** *Suppose $a \in \mathbf{R}$, $I$ is an open interval which contains $a$, and $f$, $g$ are real functions defined everywhere on $I$ except possibly at $a$. If $f(x)$ and $g(x)$*

converge as $x$ approaches $a$, then so do $(f + g)(x)$, $(fg)(x)$, $(\alpha f)(x)$, and $(f/g)(x)$ (when the limit of $g(x)$ is nonzero). In fact,

$$\lim_{x \to a} (f + g)(x) = \lim_{x \to a} f(x) + \lim_{x \to a} g(x),$$

$$\lim_{x \to a} (\alpha f)(x) = \alpha \lim_{x \to a} f(x),$$

$$\lim_{x \to a} (fg)(x) = \lim_{x \to a} f(x) \lim_{x \to a} g(x),$$

and (when the limit of $g(x)$ is nonzero)

$$\lim_{x \to a} \left( \frac{f}{g} \right)(x) = \frac{\lim_{x \to a} f(x)}{\lim_{x \to a} g(x)}.$$

PROOF. Let

$$L := \lim_{x \to a} f(x) \quad \text{and} \quad M := \lim_{x \to a} g(x).$$

If $x_n \in I \setminus \{a\}$ converges to $a$, then by Theorem 3.6, $f(x_n) \to L$ and $g(x_n) \to M$ as $n \to \infty$. By Theorem 2.12i), $f(x_n) + g(x_n) \to L + M$ as $n \to \infty$. Since this holds for any sequence $x_n \in I \setminus \{a\}$ which converges to $a$, we conclude by Theorem 3.6 that

$$\lim_{x \to a} (f + g)(x) = L + M = \lim_{x \to a} f(x) + \lim_{x \to a} g(x).$$

The other rules follow in an analogous way from Theorem 2.12ii) through iv). ∎

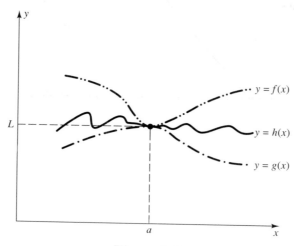

**Figure 3.2**

Similarly, the Sequential Characterization of Limits can be combined with Theorems 2.9 and 2.17 to prove the following results.

**3.9 THEOREM** [SQUEEZE THEOREM FOR FUNCTIONS]. *Suppose $a \in$* **R**, *I is an open interval which contains $a$, and $f, g, h$ are real functions defined everywhere on $I$ except possibly at $a$.*

i) *If $g(x) \leq h(x) \leq f(x)$ for all $x \in I \setminus \{a\}$, and*

$$\lim_{x \to a} f(x) = \lim_{x \to a} g(x) = L,$$

*then the limit of $h$ exists, as $x \to a$, and*

$$\lim_{x \to a} h(x) = L.$$

ii) *If $|g(x)| \leq M$ for all $x \in I \setminus \{a\}$ and $f(x) \to 0$ as $x \to a$, then*

$$\lim_{x \to a} f(x)g(x) = 0.$$

The preceding result is illustrated in Figure 3.2.

**3.10 THEOREM** [COMPARISON THEOREM FOR FUNCTIONS]. *Suppose $a \in$* **R**, *I is an open interval which contains $a$, and $f, g$ are real functions defined everywhere on $I$ except possibly at $a$. If $f$ and $g$ have a limit as $x$ approaches $a$ and*

$$f(x) \leq g(x), \qquad x \in I \setminus \{a\},$$

*then*

$$\lim_{x \to a} f(x) \leq \lim_{x \to a} g(x).$$

We shall refer to this last result as taking the limit of an inequality.

The limit theorems (Theorems 3.8, 3.9, and 3.10) allow us to prove limits exist without resorting to $\varepsilon$'s and $\delta$'s.

**3.11 Example.** Prove

$$\lim_{x \to 1} \frac{x - 1}{3x + 1} = 0.$$

PROOF. By Example 3.2, $x - 1 \to 0$ and $3x + 1 \to 4$ as $x \to 1$. Hence, by Theorem 3.8,

$$\lim_{x \to 1} \frac{x - 1}{3x + 1} = \frac{0}{4} = 0. \quad \blacksquare$$

**EXERCISES**

**1.** Using Definition 3.1, prove that each of the following limits exist.

a)
$$\lim_{x \to 2} x^2 - x + 1 = 3.$$

b)
$$\lim_{x \to 1} \frac{x^2 - 1}{x - 1} = 2.$$

c)
$$\lim_{x \to 1} x^3 + x + 1 = 3.$$

**2.** Decide which of the following limits exist and which do not. Prove that your answer is correct.

a)
$$\lim_{x \to 0} \cos\left(\frac{1}{x}\right).$$

b)
$$\lim_{x \to 0} x \sin\left(\frac{1}{x}\right).$$

c)
$$\lim_{x \to 1} \frac{1}{\log x}.$$

**3.** Evaluate the following limits using results from this section.
[You may assume that $\sin x$, $1 - \cos x$, and $\sqrt[3]{x}$ converge to 0 as $x \to 0$.]

a)
$$\lim_{x \to 0} \frac{x^2 + \cos x}{2 - \tan x}.$$

b)
$$\lim_{x \to 1} \frac{x^2 + x - 2}{x^3 - x}.$$

c)
$$\lim_{x \to \sqrt{\pi}} \frac{\sqrt[3]{\pi - x^2}}{x + \pi}.$$

d)
$$\lim_{x \to 1} \frac{x^n - 1}{x - 1}, \qquad n \in \mathbf{N}.$$

e)
$$\lim_{x \to 0} x \sin\left(\frac{1}{x^2}\right).$$

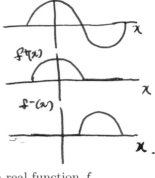

**4.** Using Definition 3.1, prove that

$$\lim_{x \to 0} x^n \sin\left(\frac{1}{x}\right)$$

exists for all $n \in \mathbf{N}$.

**5.** Prove Theorem 3.9.

**6.** Prove Theorem 3.10.

$\boxed{7}$. **This exercise is used in Sections 3.2 and 5.2.** For each real function $f$ define the *positive part* of $f$ by

$$f^+(x) = \frac{|f(x)| + f(x)}{2}, \qquad x \in \text{Dom}\,(f)$$

and the *negative part* of $f$ by

$$f^-(x) = \frac{|f(x)| - f(x)}{2}, \qquad x \in \text{Dom}\,(f).$$

a) Prove $f^+(x) \geq 0$, $f^-(x) \geq 0$, $f(x) = f^+(x) - f^-(x)$ and $|f(x)| = f^+(x) + f^-(x)$ hold for all $x \in \text{Dom}\,(f)$. (Compare with Exercise 1 in Section 1.1.)

b) Prove that if

$$L = \lim_{x \to a} f(x)$$

exists, then $f^+(x) \to L^+$ and $f^-(x) \to L^-$ as $x \to a$ .

**8.** Suppose $f$ is a real function.

a) Prove that if

$$L = \lim_{x \to a} f(x)$$

exists, then $|f(x)| \to |L|$ as $x \to a$ .

b) Show that the converse of a) is false.

$\boxed{9}$. **This exercise is used in Sections 3.2 and 5.2.**

Let $f, g$ be real function and for each $x \in \text{Dom}\,(f) \cap \text{Dom}\,(g)$ define

$$\underline{(f \vee g)(x) := \max\{f(x), g(x)\}} \quad \text{and} \quad \underline{(f \wedge g)(x) := \min\{f(x), g(x)\}}.$$

a) Prove

$$(f \vee g)(x) = \frac{(f + g)(x) + |(f - g)(x)|}{2}$$

and

$$(f \wedge g)(x) = \frac{(f + g)(x) - |(f - g)(x)|}{2}$$

for all $x \in \text{Dom}\,(f) \cap \text{Dom}\,(g)$.

b) Prove that if

$$L = \lim_{x \to a} f(x) \quad \text{and} \quad M = \lim_{x \to a} g(x)$$

exist, then $(f \vee g)(x) \to L \vee M$ and $(f \wedge g)(x) \to L \wedge M$ as $x \to a$.

## 3.2   ONE-SIDED LIMITS AND LIMITS AT INFINITY

In the previous section we defined the limit of a real function. In this section we expand that definition to handle more general situations.

What is the limit of $f(x) := \sqrt{x - 1}$ as $x \to 1$? A reasonable answer is that the limit is zero. This function, however, does not satisfy Definition 3.1 because it is not defined on an OPEN interval containing $a = 1$. Indeed, $f$ is only defined for $x \geq 1$. To handle such situations, we introduce "one-sided" limits.

To this end, let $I$ be a nonempty open interval with $a$ as its left endpoint. A function $f : I \to \mathbf{R}$ is said to *converge to L as x approaches a from the right* if for every $\varepsilon > 0$ there is a $\delta > 0$ (which in general depends on $\varepsilon$, $f$, and $a$) such that

(2)                    $a < x < a + \delta$   implies   $|f(x) - L| < \varepsilon.$

In this case we call $L$ the *right-hand limit* of $f$ at $a$, and denote it by

$$f(a+) := L =: \lim_{x \to a+} f(x).$$

Similarly, if $I$ is a nonempty open interval with $a$ as its right endpoint, then a function $f : I \to \mathbf{R}$ is said to *converge to L as x approaches a from the left* if for every $\varepsilon > 0$ there is a $\delta > 0$ (which in general depends on $\varepsilon$, $f$, and $a$) such that

$$a - \delta < x < a   \text{ implies }   |f(x) - L| < \varepsilon,$$

in which case we call $L$ the *left-hand limit* of $f$ at $a$ and denote it by

$$f(a-) := L =: \lim_{x \to a-} f(x).$$

It is easy to check that when two-sided limits are replaced with one-sided limits, all the limit theorems from the previous section hold. We shall use them as the need arises without further comment.

Existence of a one-sided limit can be established by these limit theorems or, as in the following example, by appealing directly to the definition.

**3.12 Example.** If $f(x) = \sqrt{x}$ prove

$$\lim_{x \to 0+} f(x) = 0.$$

PROOF. Let $\varepsilon > 0$ and set $\delta = \varepsilon^2$. If $0 < x < \delta$, then $|f(x)| = \sqrt{x} < \sqrt{\delta} = \varepsilon.$ ∎

Not every function has one-sided limits (see Example 3.7). Moreover, even when a function has one-sided limits, it may not have a two-sided limit.

**3.13 Example.** Show that

$$f(x) = \begin{cases} x + 1 & x \geq 0 \\ x - 1 & x < 0 \end{cases}$$

has one-sided limits at $a = 0$ but that $\lim_{x \to 0} f(x)$ does not exist.

PROOF. Let $\varepsilon > 0$ and set $\delta = \varepsilon$. If $0 < x < \delta$, then $|f(x) - 1| = |x| < \delta = \varepsilon$. Hence $\lim_{x \to 0+} f(x)$ exists and equals 1. Similarly, $\lim_{x \to 0-} f(x)$ exists and equals $-1$. However, if $x_n = (-1)^n/n$, then $f(x_n) = (-1)^n(1 + 1/n)$ does not converge as $n \to \infty$. Hence by the Sequential Characterization of Limits, $\lim_{x \to 0} f(x)$ does not exist. ∎

The following result shows that if both one-sided limits at a point $a$ exist and are equal, then the two-sided limit at $a$ exists.

**3.14 THEOREM.** *Let $f$ be a real function. Then the limit*

$$\lim_{x \to a} f(x)$$

*exists and equals $L$ if and only if*

$$(3) \qquad L = \lim_{x \to a+} f(x) = \lim_{x \to a-} f(x).$$

PROOF. If the limit $L$ of $f(x)$ exists as $x \to a$, then given $\varepsilon > 0$ choose $\delta > 0$ such that $0 < |x - a| < \delta$ implies $|f(x) - L| < \varepsilon$. Since any $x$ which satisfies $a < x < a + \delta$ or $a - \delta < x < a$ also satisfies $0 < |x - a| < \delta$, it is clear that both the left and right limits of $f(x)$ exist as $x \to a$ and satisfy (3).

Conversely, suppose (3) holds. Then given $\varepsilon > 0$ there exist $\delta_j > 0$, $j = 1$ and 2, such that $a < x < a + \delta_1$ (respectively, $a - \delta_2 < x < a$) implies

$$|f(x) - L| < \varepsilon.$$

Set $\delta = \min\{\delta_1, \delta_2\}$. Then $0 < |x - a| < \delta$ implies both $a < x < a + \delta_1$ and $a - \delta_2 < x < a$. Hence (1) holds, i.e., $f(x) \to L$ as $x \to a$. ∎

The definition of limits of real functions can be expanded to include extended real numbers. We say that $f(x) \to L$ as $x \to \infty$ (respectively, as $x \to -\infty$) if there exists a $c > 0$ such that $f : (c, \infty) \to \mathbf{R}$ (respectively, $f : (-\infty, -c) \to \mathbf{R}$) and given $\varepsilon > 0$ there is an $M \in \mathbf{R}$ such that $x > M$ (respectively, $x < M$) implies $|f(x) - L| < \varepsilon$. In this case we shall write

$$\lim_{x \to \infty} f(x) = L \qquad (\text{respectively}, \ \lim_{x \to -\infty} f(x) = L).$$

We say that $f(x) \to +\infty$ (respectively, $f(x) \to -\infty$) as $x \to a$ if there is an open interval $I$ containing $a$ such that $f : I \to \mathbf{R}$ and given $M \in \mathbf{R}$ there is a $\delta > 0$ such that $0 < |x - a| < \delta$ implies $f(x) > M$ (respectively, $f(x) < M$). In this case we shall write

$$\lim_{x \to a} f(x) = +\infty \qquad (\text{respectively}, \ \lim_{x \to a} f(x) = -\infty).$$

Obvious modifications define $f(x) \to \pm\infty$ as $x \to a+$ and $x \to a-$, and $f(x) \to \pm\infty$ as $x \to \pm\infty$.

**3.15 Example.** Prove $1/x \to 0$ as $x \to \infty$.

PROOF. Given $\varepsilon > 0$, set $M = 1/\varepsilon$. If $x > M$, then $1/x < 1/M = \varepsilon$. Thus $1/x \to 0$ as $x \to \infty$. ∎

**3.16 Example.** Prove

$$\lim_{x \to 1-} f(x) := \lim_{x \to 1-} \frac{x+2}{2x^2 - 3x + 1} = -\infty.$$

PROOF. Let $M \in \mathbf{R}$. We must show that $f(x) < M$ for $x$ near but to the left of 1 (no matter how large and negative $M$ is). Without loss of generality, assume that $M < 0$. As $x$ converges to 1 from the left, $2x^2 - 3x + 1$ is negative and converges to 0. (Observe $2x^2 - 3x + 1$ is a parabola opening upward with roots $1/2$ and 1.) Therefore, choose $\delta \in (0, 1)$ such that $1 - \delta < x < 1$ implies $3/M < 2x^2 - 3x + 1 < 0$, i.e., $1/(2x^2 - 3x + 1) < M/3$. Notice $0 < x < 1$ also implies $2 < x + 2 < 3$. It follows that

$$f(x) = \frac{x+2}{2x^2 - 3x + 1} < M$$

for all $1 - \delta < x < 1$. ∎

In order to unify the presentation of one-sided, two-sided, and infinite limits, we introduce the following notation. Let $a$ be an extended real number, and $I$ be a nondegenerate open interval which either contains $a$ or has $a$ as one of its endpoints. Suppose further that $f$ is a real function defined on $I$ except possibly at $a$. If $a$ is finite and $I$ contains $a$, then

$$(4) \qquad\qquad\qquad \lim_{\substack{x \to a \\ x \in I}} f(x)$$

will denote $\lim_{x \to a} f(x)$ (when it exists); if $a$ is a finite left endpoint of $I$, then (4) will denote $\lim_{x \to a+} f(x)$ (when it exists); if $a$ is a finite right endpoint of $I$, then (4) will denote $\lim_{x \to a-} f(x)$ (when it exists); if $a = \pm\infty$ is an endpoint of $I$, then (4) will denote $\lim_{x \to \pm\infty} f(x)$ (when each exists).

Using this notation, we can state a Sequential Characterization of Limits valid for two-sided, one-sided, and infinite limits.

**3.17 THEOREM.** *Let $a$ be an extended real number, and $I$ be a nondegenerate open interval which either contains $a$ or has $a$ as one of its endpoints. Suppose further that $f$ is a real function defined on $I$ except possibly at $a$. Then*

$$\lim_{\substack{x \to a \\ x \in I}} f(x)$$

*exists and equals $L$ if and only if $f(x_n) \to L$ for all sequences $x_n \in I$ which satisfy $x_n \neq a$ and $x_n \to a$ as $n \to \infty$.*

PROOF. Since we have already proved this for two-sided limits, we must show it for the remaining eight cases which notation (4) represents. Since the proofs are similar, we shall give the details for only one of these cases, namely the case when $a$ belongs to $I$ and $L = \infty$. Thus we must prove that $\underline{f(x) \to \infty \text{ as } x \to a \text{ if and only}}$ $\underline{\text{if } f(x_n) \to \infty \text{ for any sequence } x_n \in I \text{ which converges to } a \text{ and satisfies } x_n \neq a}$ $\overline{\text{for } n \in \mathbf{N}}$.

Suppose first that $f(x) \to \infty$ as $x \to a$. If $x_n \in I$, $x_n \to a$ as $n \to \infty$, and $x_n \neq a$, then given $M \in \mathbf{R}$ there is a $\delta > 0$ such that $0 < |x - a| < \delta$ implies $f(x) > M$, and there is an $N \in \mathbf{N}$ such that $n \geq N$ implies $|x_n - a| < \delta$. Consequently, $n \geq N$ implies $f(x_n) > M$, i.e., $f(x_n) \to \infty$ as $n \to \infty$ as required.

Conversely, suppose to the contrary that $f(x_n) \to \infty$ for any sequence $x_n \in I$ which converges to $a$ and satisfies $x_n \neq a$ but $f(x)$ does NOT converge to $\infty$ as $x \to a$. By the definition of "convergence" to $\infty$ there are numbers $M_0 \in \mathbf{R}$ and $x_n \in I$ such that $|x_n - a| < 1/n$ and $f(x_n) \leq M_0$ for all $n \in \mathbf{N}$. The first condition implies $x_n \to a$ but the second condition implies that $f(x_n)$ does not converge to $\infty$ as $n \to \infty$. This contradiction proves 3.17 in the case $a \in I$ and $L = \infty$. ∎

Using Theorem 3.17, we can prove limit theorems which are function analogues of Theorem 2.15 and Corollary 2.16. We leave this to the reader and will use these results as the need arises.

These limit theorems can be used to evaluate infinite limits.

**3.18 Example.** Prove

$$\lim_{x \to \infty} \frac{2x^2 - 1}{1 - x^2} = -2.$$

PROOF.  Since the limit of a product is the product of the limits, we have by Example 3.15 that $1/x^m \to 0$ as $x \to \infty$ for any $m \in \mathbf{N}$. Multiplying numerator and denominator of the expression above by $1/x^2$ we have

$$\lim_{x \to \infty} \frac{2x^2 - 1}{1 - x^2} = \lim_{x \to \infty} \frac{2 - 1/x^2}{-1 + 1/x^2} = \frac{\lim_{x \to \infty}(2 - 1/x^2)}{\lim_{x \to \infty}(-1 + 1/x^2)} = \frac{2}{-1} = -2. \quad ∎$$

**EXERCISES**

$|x| = \sqrt{x^2}.$    so    if $x \geq 0$, $x = \sqrt{x^2}$
if $x < 0$, $-x = \sqrt{x^2}$

**1.** Using definitions (rather than limit theorems) prove

$$\lim_{x \to a+} f(x)$$

exists and equals $L$ in each of the following cases.

   a) $f(x) = |x|/x$, $a = 0$, and $L = 1$.
   b) $f(x) = -1/x$, $a = 0$, and $L = -\infty$.
   c) $f(x) = (x - 1)/(x^2 + x - 2)$, $a = -2$, and $L = \infty$.
   d) $f(x) = 1/(x^2 - 1)$, $a = 1$, and $L = \infty$.

**2.** Evaluate the following limits when they exist.

a)
$$\lim_{x \to 0+} \frac{x+1}{x^2 - 2x}.$$

b)
$$\lim_{x \to 1-} \frac{x^3 - 3x + 2}{x^3 - 1}.$$

c)
$$\lim_{x \to \pi+} (x^2 + 1) \sin x.$$

d)
$$\lim_{x \to 0+} \frac{x}{|x|}.$$

e)
$$\lim_{x \to \pi/2-} \frac{\tan x}{x}.$$

**3.** Evaluate the following limits when they exist.

a)
$$\lim_{x \to \infty} \frac{3x^2 - 13x + 4}{1 - x - x^2}.$$

b)
$$\lim_{x \to \infty} \frac{x^2 + x + 2}{x^3 - x - 2}.$$

c)
$$\lim_{x \to -\infty} \frac{x^3 - 1}{x^2 + 2}.$$

d)
$$\lim_{x \to \infty} \arctan x.$$

[You may assume that $\tan x \to L$ as $x \to a$, $x \in (-\pi/2, \pi/2)$, if and only if $\arctan x \to a$ as $x \to L$.]

e)
$$\lim_{x \to \infty} \frac{\sin x}{x^2}.$$

f)
$$\lim_{x \to -\infty} x^2 \sin x.$$

**4**. **This exercise is used many places**. Recall that a *polynomial of degree n* is a function of the form

$$P(x) = a_n x^n + a_{n-1} x^{n-1} + \cdots + a_1 x + a_0$$

where $a_j \in \mathbf{R}$ for $j = 0, 1, \ldots, n$ and $a_n \neq 0$.

a) Prove $\lim_{x \to a} x^n = c^n$ for $n = 0, 1, \cdots$ .

b) Prove that if $P$ is a polynomial, then

$$\lim_{x \to a} P(x) = P(a)$$

for every $a \in \mathbf{R}$.

c) Suppose that $P$ is a polynomial and $P(a) > 0$. Prove $P(x)/(x-a) \to \infty$ as $x \to a+$, $P(x)/(x-a) \to -\infty$ as $x \to a-$, but

$$\lim_{x \to a} \frac{P(x)}{x-a}$$

does not exist.

**5.** Prove that $(\sin(x+3) - \sin 3)/x$ converges to 0 as $x \to \infty$.

**6.** Prove that $\sqrt{1 - \cos x}/\sin x \to \sqrt{2}/2$ as $x \to 0+$.

**7.** Prove the following comparison theorems for real functions.

a) If $f(x) \geq g(x)$ and $g(x) \to \infty$ as $x \to a$, then $f(x) \to \infty$ as $x \to a$.

b) If $f(x) \leq g(x) \leq h(x)$ and

$$L := \lim_{x \to \infty} f(x) = \lim_{x \to \infty} h(x),$$

then $g(x) \to L$ as $x \to \infty$. [Sequential characterization of limit tend to ∞]

**8.** Suppose $f : [a, \infty) \to \mathbf{R}$ for some $a \in \mathbf{R}$. Prove that $f(x) \to L$ as $x \to \infty$ if and only if $f(x_n) \to L$ for any sequence $x_n \in (a, \infty)$ which converges to $\infty$ as $n \to \infty$.

**9.** Suppose that $f : [0, 1] \to \mathbf{R}$ and $f(a) = \lim_{x \to a} f(x)$ for all $a \in [0, 1]$. Prove that $f(q) = 0$ for all $q \in \mathbf{Q} \cap [0, 1]$ if and only if $f(x) = 0$ for all $x \in [0, 1]$.

## 3.3  CONTINUITY

In elementary calculus, a function is called *continuous* at $a$ if $a \in \text{Dom } f$ and $f(x) \to f(a)$ as $x \to a$. In particular, it is tacitly assumed that $f$ is defined on BOTH sides of $a$. Here, we introduce a more general concept of continuity which includes functions, like $\sqrt{x}$ at $a = 0$, which are defined on only one side of some points in their domain.

**3.19 DEFINITION.** Let $E$ be nonempty subset of $\mathbf{R}$ and $f : E \to \mathbf{R}$.

    i) $f$ is said to be *continuous at a point* $a \in E$ if given $\varepsilon > 0$ there is a $\delta > 0$ (which in general depends on $\varepsilon$, $f$, and $a$) such that

(5)
$$|x - a| < \delta \quad \text{and} \quad x \in E \quad \text{imply} \quad |f(x) - f(a)| < \varepsilon. \qquad \lim_{x \to a} f(x) = f(a).$$

    ii) $f$ is said to be *continuous on $E$* (notation: $f : E \to \mathbf{R}$ is continuous) if $f$ is continuous at every $x \in E$.

    iii) $f$ is said to be *continuous* if it is continuous on its domain Dom $(f)$.

The following result shows that if $E$ is an open interval which contains $a$, then "$f$ is continuous at $a \in E$" means "$f(x) \to f(a)$ as $x \to a$." (Therefore, we shall abbreviate "$f$ is continuous at $a \in E$" by "$f$ is continuous at $a$" when $E$ is an open interval.)

**3.20 Remark.** *Let $I$ be an open interval which contains a point $a$ and $f : I \to \mathbf{R}$. Then $f$ is continuous at $a \in I$ if and only if*

$$f(a) = \lim_{x \to a} f(x).$$

1. $f(a)$ is defined $(a \in \text{Dom } f)$
2. $\lim_{x \to a} f(x)$ exist
3. $\lim_{x \to a} f(x) = f(a)$

PROOF. Suppose $I = (c, d)$ and set $\delta_0 := \min\{|c - a|, |d - a|\}$. If $\delta < \delta_0$, then $|x - a| < \delta$ implies $x \in I$. Therefore, condition (5) is identical to (1) when $f(a) = L$, $E = I$, and $\delta < \delta_0$. It follows that $f$ is continuous at $a \in I$ if and only if $f(x) \to f(a)$ as $x \to a$. ∎

By repeating the proof of Theorem 3.6, we can establish a sequential characterization of continuity which is valid on any nonempty set.

**3.21 THEOREM.** *Suppose $E$ is a nonempty subset of $\mathbf{R}$ and $f : E \to \mathbf{R}$. Then the following statements are equivalent:*

    i) *$f$ is continuous on $E$.*

    ii) *If $x_n \in E$ converges to some $x \in E$, then $f(x_n) \to f(x)$ as $n \to \infty$.*

If $x_n \to a$ then $\lim_{n \to \infty} f(x_n) = f(a)$    if $\lim_{x \to a} f(x) = f(a)$.

In particular, $\sqrt{x}$ is continuous on $I = [0, \infty)$ by Exercise 4 in Section 2.2.

By combining Theorem 3.21 with Theorem 2.12, we obtain the following result.

**3.22 THEOREM.** *Let $E$ be a nonempty subset of $\mathbf{R}$ and $f, g : E \to \mathbf{R}$. If $f, g$ are continuous at a point $a \in E$ (respectively, continuous on the set $E$), then so are $f \pm g$, $fg$, and $\alpha f$ (for any $\alpha \in \mathbf{R}$). Moreover, $f/g$ is continuous at $a \in E$ when $g(a) \neq 0$ (respectively, on $E$ when $g(x) \neq 0$ for all $x \in E$).*

It follows from Exercises 7, 8, and 9 in Section 3.1 that if $f, g$ are continuous at a point $a$ or on a set $E$, then so are $|f|$, $f^+$, $f^-$, $f \vee g$, and $f \wedge g$. We also notice by Exercise 4 in Section 3.2, that every polynomial is continuous on $\mathbf{R}$.

Many complicated functions can be broken into simpler pieces, using sums, products, quotients, and the following operation.

**3.23 DEFINITION.** Let $A$ and $B$ be subsets of **R**. Suppose further that $f$ is a real function defined on $A$ and $g$ is a real function defined on $B$. If $f(A) \subseteq B$, then the *composition* of $g$ with $f$ is the function $g \circ f : A \to \mathbf{R}$ defined by

$$(g \circ f)(x) := g(f(x)), \qquad x \in A.$$

The following result contains information about when a limit sign and something else (in this case, the computation of a function) can be interchanged. We shall return to this theme many times, identifying conditions under which one can interchange any two of the following objects: limits, integrals, derivatives, infinite summations, and computation of a function (see especially Sections 7.1, 7.2, and 11.1, and the entry "interchange the order of" in the Index).

**3.24 THEOREM.** *Suppose $A$ and $B$ are subsets of* **R**, *$f$ is a real function defined on $A$, $g$ is a real function defined on $B$, and $f(A) \subseteq B$.*

i) *If $A := I$ is a nondegenerate interval which contains $a$, if*

$$L := \lim_{\substack{x \to a \\ x \in I}} f(x)$$

*exists and belongs to $B$, and if $g$ is continuous at $L \in B$, then*

$$\lim_{\substack{x \to a \\ x \in I}} (g \circ f)(x) = g\left( \lim_{\substack{x \to a \\ x \in I}} f(x) \right).$$

ii) *If $f$ is continuous at $a \in A$ and $g$ is continuous at $f(a) \in B$, then $g \circ f$ is continuous at $a \in E$.*

PROOF. Let $\varepsilon > 0$ and choose (by Definition 3.19) a $\delta > 0$ such that $|y - L| < \delta$ and $y \in B$ imply $|g(y) - g(L)| < \varepsilon$.

To prove part i), let $x_n \in I \setminus \{a\}$ converge to $a$ as $n \to \infty$. By the Sequential Characterization of Limits (Theorem 3.17), it suffices to prove that $(g \circ f)(x_n) \to g(L)$ as $n \to \infty$. This is easy. Since $L$ exists, $f(x_n) \to L$ as $n \to \infty$ by Theorem 3.17. Hence there is an $N \in \mathbf{N}$ so large that $n \geq N$ implies $|f(x_n) - L| < \delta$. Since $y := f(x_n)$ and $L$ belong to $B$, it follows that $|g \circ f(x_n) - g(L)| := |g(y) - L| < \varepsilon$ for all $n \geq N$. By definition, then, $\lim_{n \to \infty} (g \circ f)(x_n) = g(L)$, as required. A similar proof, using the Sequential Characterization of Continuity (Theorem 3.21) proves part ii). ∎

For many applications, it is important to be able to find the maximum or minimum of a given function. As a first step in this direction, we introduce the following concept.

**3.25 DEFINITION.** Let $E$ be a nonempty subset of **R**. A function $f : E \to \mathbf{R}$ is said to be *bounded* on $E$ if there is an $M \in \mathbf{R}$ such that $|f(x)| \leq M$ for all $x \in E$.

We shall refer to this last inequality by saying $f$ is *dominated* by $M$ on $E$.

Notice that whether a function $f$ is bounded or not on a set $E$ depends on $E$ as well as on $f$. For example, $f(x) = 1/x$ is bounded on $[1, \infty)$ (by 1) but not on $(0, 2)$. Again, the function $f(x) = x^2$ is bounded on $(-2, 2)$ (by 4) but not on $[0, \infty)$.

The following result, which we shall use often, shows that a continuous function on $[a, b]$ is always bounded.

**3.26 THEOREM** [EXTREME VALUE THEOREM]. *If $I$ is a closed, bounded interval and $f : I \to \mathbf{R}$ is continuous on $I$, then $f$ is bounded on $I$. Moreover, if*

$$M = \sup_{x \in I} f(x) \quad and \quad m = \inf_{x \in I} f(x),$$

*then there exist points $x_m, x_M \in I$ such that*

(6) $$f(x_M) = M \quad and \quad f(x_m) = m.$$

PROOF. Suppose first that $f$ is not bounded on $I$. Then there exist $x_n \in I$ such that

(7) $$|f(x_n)| > n, \quad n \in \mathbf{N}.$$

Since $I$ is bounded, we know (by the Bolzano–Weierstrass Theorem) that $\{x_n\}$ has a convergent subsequence, say $x_{n_k} \to a$ as $k \to \infty$. Since $I$ is closed, we also know (by the Comparison Theorem) that $a \in I$. In particular, $f(a) \in \mathbf{R}$. On the other hand, substituting $n_k$ for $n$ in (7) and taking the limit of this inequality as $k \to \infty$, we have $|f(a)| = \infty$, a contradiction. Hence, the function $f$ is bounded on $I$.

We have proved that both $M$ and $m$ are finite real numbers. To show there is an $x_M \in I$ such that $f(x_M) = M$, suppose to the contrary that $f(x) < M$ for all $x \in I$. Then the function

$$g(x) = \frac{1}{M - f(x)}$$

is continuous, hence, bounded on $I$. In particular, there is a $C > 0$ such that $|g(x)| = g(x) \le C$. It follows that

(8) $$f(x) \le M - \frac{1}{C}$$

for all $x \in I$. Taking the supremum of (8) over all $x \in I$, we obtain $M \le M - 1/C < M$, a contradiction. Hence, there is an $x_M \in I$ such that $f(x_M) = M$. A similar argument proves that there is an $x_m \in I$ such that $f(x_m) = m$. ∎

We shall sometimes refer to (6) by saying that the supremum and infimum of $f$ are *attained* on $I$. We shall also call the value $M$ (respectively, $m$) the *maximum* (respectively, the *minimum*) of $f$ on $I$.

Neither of the hypotheses on the interval $I$ in Theorem 3.26 can be relaxed.

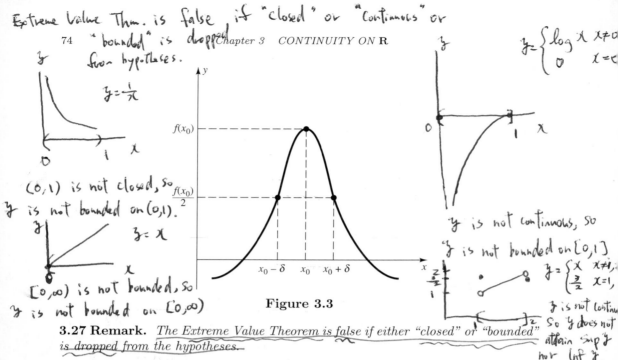

*(Handwritten annotations around the figures:)*

Extreme Value Thm. is false if "closed" or "Continuous" or "bounded" is dropped from hypotheses.

$y = \frac{1}{x}$

$(0,1)$ is not closed, so $y$ is not bounded on $(0,1)$.

$y = x$

$[0,\infty)$ is not bounded, so $y$ is not bounded on $[0,\infty)$

$y = \begin{cases} \log x & x \neq 0 \\ 0 & x = 0 \end{cases}$

$y$ is not continuous, so $y$ is not bounded on $[0,1]$

$y = \begin{cases} x & x \neq 1, \\ \frac{3}{2} & x = 1, \end{cases}$

$y$ is not continuous so $y$ does not attain sup $y$ nor inf $y$.

**Figure 3.3**

*(Figure labels:)* $f(x_0)$, $\frac{f(x_0)}{2}$, $x_0 - \delta$, $x_0$, $x_0 + \delta$

**3.27 Remark.** *The Extreme Value Theorem is false if either "closed" or "bounded" is dropped from the hypotheses.*

PROOF. The interval $(0,1)$ is bounded but not closed, and the function $f(x) = 1/x$ is continuous and unbounded on $(0,1)$. The interval $[0,\infty)$ is closed but not bounded, and the function $f(x) = x$ is continuous and unbounded on $[0,\infty)$. ∎

What more can be said about continuous functions? One useful conceptualization of functions which are continuous on an interval is that their graphs have no holes or jumps (see Theorem 3.29 below). Our proof of this fact is based on the following elementary observation.

**3.28 Lemma** [SIGN PRESERVING PROPERTY]. *Let $f$ be a real function. If $f$ is continuous at a point $x_0$ and $f(x_0) > 0$, then there are positive numbers $\varepsilon$ and $\delta$ such that*

$$|x - x_0| < \delta \quad \text{implies} \quad f(x) > \varepsilon.$$

STRATEGY: The idea behind this proof is simple. If $f(x_0) > 0$, then $f(x) > f(x_0)/2$ for $x$ near $x_0$ (see Figure 3.3). Here are the details.

PROOF. By (1), given $\varepsilon = f(x_0)/2$ choose $\delta > 0$ such that $|x - x_0| < \delta$ implies $|f(x) - f(x_0)| < \varepsilon$. It follows that

$$-\frac{f(x_0)}{2} < f(x) - f(x_0) < \frac{f(x_0)}{2}.$$

Solving the left-hand inequality, we see that $f(x) > f(x_0)/2 = \varepsilon$ holds for all $|x - x_0| < \delta$. ∎

A real number $y_0$ is said to *lie between* two numbers $c$ and $d$ if $c < y_0 < d$ or $d < y_0 < c$.

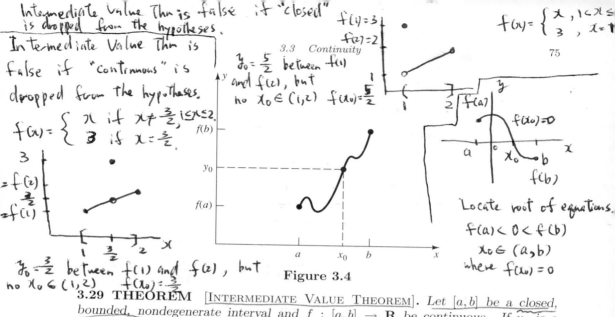

**Figure 3.4**

**3.29 THEOREM** [INTERMEDIATE VALUE THEOREM]. *Let $[a,b]$ be a closed, bounded, nondegenerate interval and $f : [a,b] \to \mathbf{R}$ be continuous. If $y_0$ is a real number which lies between $f(a)$ and $f(b)$, then there is an $x_0 \in (a,b)$ such that $f(x_0) = y_0$.*

PROOF. We may suppose that $f(a) < y_0 < f(b)$. Extend $f$ to $\mathbf{R}$ by $f(x) := f(a)$ if $x < a$ and $f(x) := f(b)$ if $x > b$, and consider the set $E = \{x \in [a,b] : f(x) < y_0\}$. Since $a \in E$ and $E \subseteq [a,b]$, $E$ is a nonempty bounded subset of $\mathbf{R}$. Hence, by the Completeness Axiom, $x_0 := \sup E$ is a finite real number. We guess that $f(x_0) = y_0$ (see Figure 3.4).

To prove this guess is correct, use Theorem 2.11 to choose $x_n \in E$ such that $x_n \to x_0$ as $n \to \infty$. Since $E \subseteq [a,b]$, $x_0 \in [a,b]$ (see Theorem 2.17). Hence, by continuity of $f$ and the definition of $E$, we have $f(x_0) = \lim_{n\to\infty} f(x_n) \le y_0$.

To show $f(x_0) = y_0$, suppose to the contrary that $f(x_0) < y_0$. Then $y_0 - f(x)$ is a continuous function whose value at $x = x_0$ is positive. Hence, by Lemma 3.28, we can choose positive numbers $\varepsilon$ and $\delta$ such that $y_0 - f(x) > \varepsilon > 0$ for $|x - x_0| < \delta$. In particular, any $x$ which satisfies $x_0 < x < x_0 + \delta$ also satisfies $f(x) < y_0$, a contradiction of the fact that $x_0 = \sup E$.

Finally, since $y_0$ equals neither $f(a)$ nor $f(b)$, $x_0$ cannot equal $a$ or $b$. Thus $x_0 \in (a,b)$ as required. ∎

If $f$ fails to be continuous at a point $a$, we say that $f$ is *discontinuous* at $a$ and call $a$ a *point of discontinuity* of $f$. How badly can a function behave near a point of discontinuity? The following examples can be interpreted as answers to this question. (See also Exercise 9 in Section 9.5.)

**3.30 Example.** Prove that the function

$$f(x) = \begin{cases} \dfrac{|x|}{x} & x \neq 0 \\ 1 & x = 0 \end{cases}$$

is continuous on $(-\infty, 0)$ and $[0, \infty)$, discontinuous at 0, and that both $f(0+)$ and $f(0-)$ exist.

PROOF. Since $f(x) = 1$ for $x \geq 0$, it is clear that $f(0+) = 1$ exists and $f(x) \to f(a)$ as $x \to a$ for any $a > 0$. In particular, $f$ is continuous on $[0, \infty)$. Similarly, $f(0-) = -1$ and $f$ is continuous on $(-\infty, 0)$. Finally, since $f(0+) \neq f(0-)$, the limit of $f(x)$ as $x \to 0$ does not exist by Theorem 3.14. Therefore, $f$ is not continuous at $0$. ∎

**3.31 Example.** Assuming $\sin x$ is continuous on **R**, prove that the function

$$
f(x) = \begin{cases} \sin \dfrac{1}{x} & x \neq 0 \\ 1 & x = 0 \end{cases}
$$

is continuous on $(-\infty, 0)$ and $(0, \infty)$, discontinuous at $0$, and neither $f(0+)$ nor $f(0-)$ exists. (See Figure 3.2.)  $\lim\limits_{x \to a} \left(\frac{f}{g}\right)(x) = \frac{\lim\limits_{x \to a} f(x)}{\lim\limits_{x \to a} g(x)} \quad g(x) \neq 0.$

PROOF. The function $1/x$ is continuous for $x \neq 0$ by Theorem 3.8. Hence, by Theorem 3.24, $f(x) = \sin(1/x)$ is continuous on $(-\infty, 0)$ and $(0, \infty)$. To prove $f(0+)$ does not exist, let $x_n = 2/((2n+1)\pi)$, and observe (see Appendix B) that $\sin(x_n) = (-1)^n$, $n \in \mathbf{N}$. Since $x_n \downarrow 0$ but $(-1)^n$ does not converge, it follows from Theorem 3.21 (the Sequential Characterization of Continuity) that $f(0+)$ does not exist. A similar argument proves $f(0-)$ does not exist. ∎

**3.32 Example.** The *Dirichlet function* is defined on **R** by

$$
f(x) := \begin{cases} 1 & x \in \mathbf{Q} \\ 0 & x \notin \mathbf{Q}. \end{cases}
$$

Prove that every point $x \in \mathbf{R}$ is a point of discontinuity of $f$. (Such functions are called *nowhere continuous*.)

PROOF. By Theorem 1.24 and Exercise 3 in Section 1.3 (Density of Rationals and Irrationals), given any $a \in \mathbf{R}$ and $\delta > 0$ we can choose $x_1 \in \mathbf{Q}$ and $x_2 \in \mathbf{R} \setminus \mathbf{Q}$ such that $|x_i - a| < \delta$ for $i = 1, 2$. Since $f(x_1) = 1$ and $f(x_2) = 0$, $f$ cannot be continuous at $a$. ∎

**3.33 Example.** Prove that the function

$$
f(x) = \begin{cases} \dfrac{1}{q} & x = \dfrac{p}{q} \in \mathbf{Q} \quad \text{(in reduced form)} \\ 0 & x \notin \mathbf{Q}. \end{cases}
$$

is continuous at every irrational in the interval $(0, 1)$ but discontinuous at every rational in $(0, 1)$.

PROOF. Let $a$ be a rational in $(0, 1)$ and suppose that $f$ is continuous at $a$. If $x_n$ is a sequence of irrationals which converges to $a$, then $f(x_n) \to f(a)$; i.e., $f(a) = 0$. But $f(a) \neq 0$ by definition. Hence, $f$ is discontinuous at every rational in $(0, 1)$.

Let $a$ be an irrational in $(0, 1)$. We must show $f(x_n) \to f(a)$ for every sequence $x_n \in (0, 1)$ which satisfies $x_n \to a$ as $n \to \infty$. We may suppose that $x_n \in \mathbf{Q}$. For each $n \in \mathbf{N}$, write $x_n = p_n/q_n$ in reduced form. Since $f(a) = 0$, it suffices to show $q_n \to \infty$ as $n \to \infty$. Suppose to the contrary that there exist integers $n_1 < n_2 < \ldots$ such that $|q_{n_k}| \le M < \infty$ for $k \in \mathbf{N}$. Since $x_{n_k} \in (0,1)$, it follows that the set

$$E := \left\{ x_{n_k} = \frac{p_{n_k}}{q_{n_k}} : k \in \mathbf{N} \right\}$$

contains only a finite number of points. Hence, the limit of any sequence in $E$ must belong to $E$, a contradiction since $a$ is such a limit and is irrational. ∎

To see how counterintuitive Example 3.33 is, try to draw a graph of $y = f(x)$. Stranger things can happen.

**3.34 Remark.** *The composition of two functions $g \circ f$ can be nowhere continuous, even though $f$ is discontinuous only on $\mathbf{Q}$ and $g$ is discontinuous at only one point.*

PROOF. Let $f$ be the function given in Example 3.33 and set

$$g(x) = \begin{cases} 1 & x \neq 0 \\ 0 & x = 0. \end{cases}$$

Clearly,

$$(g \circ f)(x) = \begin{cases} 1 & x \in \mathbf{Q} \\ 0 & x \notin \mathbf{Q}. \end{cases}$$

Hence, $g \circ f$ is the Dirichlet function, nowhere continuous by Example 3.32. ∎

In view of Example 3.33 and Remark 3.34, we must be skeptical of proofs which rely exclusively on geometric intuition. And although we shall use geometric intuition to suggest methods of proof for many results in subsequent chapters, these suggestions will always be followed by a careful rigorous proof which contains no fuzzy reasoning based on pictures or sketches no matter how plausible they seem.

### EXERCISES

*For these exercises, assume that $\sin x$, $\cos x$, and $e^x$ are continuous on $\mathbf{R}$.*

1. For each of the following, prove that there is at least one $x \in \mathbf{R}$ which satisfies the given equation.
   a) $e^x = x^2$.
   b) $e^x = \cos x + 1$.
   c) $2^x = 1 - x$.

2. Use limit theorems to show the following functions are continuous on $[0,1]$.

   a) 
$$f(x) = xe^{x^2} + 5.$$

b)
$$f(x) = \frac{1-x}{1+x}.$$

c)
$$f(x) = \begin{cases} \sqrt{x}\sin\dfrac{1}{x} & x \neq 0 \\ 0 & x = 0. \end{cases}$$

d)
$$f(x) = \sqrt{1-x}.$$

e)
$$f(x) = \frac{\sin(e^x)}{x^2 + x - 6}.$$

**3.** If $f : [a,b] \to \mathbf{R}$ is continuous, prove that $\sup_{x \in [a,b]} |f(x)|$ is finite.

**4.** Suppose $f$ is a real-valued function of a real variable. If $f$ is continuous at $a$ with $f(a) < M$ for some $M \in \mathbf{R}$, prove there is an open interval $I$ containing $a$ such that $f(x) < M$ for all $x \in I$.

**5.** Show that there exist nowhere continuous functions $f$ and $g$ whose sum $f + g$ is continuous on $\mathbf{R}$. Show that the same is true for the product of functions.

**6.** Let
$$f(x) = \begin{cases} \cos\dfrac{1}{x} & x \neq 0 \\ 0 & x = 0. \end{cases}$$

a) Prove $f$ is continuous on $(0, \infty)$ and $(-\infty, 0)$ but discontinuous at 0.

b) Suppose $g : [0, 2/\pi] \to \mathbf{R}$ is continuous on $(0, 2/\pi)$ and that there is a positive constant $C > 0$ such that $|g(x)| \leq C\sqrt{x}$ for all $x \in (0, 2/\pi)$. Prove $f(x)g(x)$ is continuous on $[0, 2/\pi]$.

**7.** Suppose $a \in \mathbf{R}$, $I$ is an open interval containing $a$, $f, g : I \to \mathbf{R}$, and $f$ is continuous at $a$.

a) Prove that $g$ is continuous at $a$ if and only if $f + g$ is continuous at $a$.

b) Make and prove an analogous statement for the product $fg$. Show by example that the hypothesis about $f$ you added cannot be dropped.

**8.** Suppose $f : \mathbf{R} \to \mathbf{R}$ satisfies $f(x+y) = f(x) + f(y)$ for each $x, y \in \mathbf{R}$.

a) Show that $f(nx) = nf(x)$ for all $x \in \mathbf{R}$ and $n \in \mathbf{Z}$.

b) Prove $f(qx) = qf(x)$ for all $x \in \mathbf{R}$ and $q \in \mathbf{Q}$.

c) Prove that $f$ is continuous at 0 if and only if $f$ is continuous on $\mathbf{R}$.

d) Prove that if $f$ is continuous at 0, then there is an $m \in \mathbf{R}$ such that $f(x) = mx$ for all $x \in \mathbf{R}$.

**9**. **This exercise is used in Section 7.4**

Suppose that $f : \mathbf{R} \to (0, \infty)$ satisfies $f(x + y) = f(x)f(y)$. Modifying the outline in Exercise 8, show that if $f$ is continuous at 0, then there is an $a \in (0, \infty)$ such that $f(x) = a^x$ for all $x \in \mathbf{R}$. [Note: You may assume that the function $a^x$ is continuous on $\mathbf{R}$.]

**10**. If $f : \mathbf{R} \to \mathbf{R}$ is continuous and

$$\lim_{x \to \infty} f(x) = \lim_{x \to -\infty} f(x) = \infty,$$

prove that $f$ has a minimum on $\mathbf{R}$; i.e., there is an $x_m \in \mathbf{R}$ such that

$$f(x_m) = \inf_{x \in \mathbf{R}} f(x) < \infty.$$

*If f is differentiable on (a,b) with f' bounded on (a, b), then f is uniformly continuous on (a, b). (slop is not ∞ or −∞).*

## 3.4   UNIFORM CONTINUITY

The following concept is very important and will be used many times in the rest of the course.

**3.35 DEFINITION.** Let $E$ be a nonempty subset of $\mathbf{R}$ and $f : E \to \mathbf{R}$. Then $f$ is said to be *uniformly continuous* on $E$ (notation: $f : E \to \mathbf{R}$ is uniformly continuous) if for every $\varepsilon > 0$ there is a $\delta > 0$ such that

(9)             $|x - a| < \delta$   and   $x, a \in E$   imply   $|f(x) - f(a)| < \varepsilon$.

*$|x_1 - x_2| < \delta$ and $x_1, x_2 \in E$ imply $|f(x_1) - f(x_2)| < \varepsilon$.*

Notice that the $\delta$ in Definition 3.35 depends on $\varepsilon$ and $f$, but not on $a$ and $x$. This issue needs to be addressed when one proves a given function is uniformly continuous on a specific set (e.g., by determining $\delta$ before $a$ is mentioned).

**3.36 Example.** Prove $f(x) = x^2$ is uniformly continuous on the interval $(0, 1)$.

PROOF. Given $\varepsilon > 0$, set $\delta = \varepsilon/3$. If $x, a \in (0, 1)$, then $|x + a| \le |x| + |a| \le 2$. Therefore, if $x, a \in (0, 1)$ and $|x - a| < \delta$, then

$$|f(x) - f(a)| = |x^2 - a^2| = |x - a|\,|x + a| \le 2|x - a| \le \frac{2\varepsilon}{3} < \varepsilon. \quad \blacksquare$$

The definitions of continuity and uniform continuity are very similar. In fact, the only difference is that for a continuous function, the parameter $\delta$ may depend on both $\varepsilon$ and $a$, whereas for a uniformly continuous function, $\delta$ must be chosen independently of $a$. In particular, every function uniformly continuous on $E$ is also continuous on $E$. The following example shows that the converse of this statement is false unless some restriction is made on $E$.

**3.37 Example.** Show that $f(x) = x^2$ is not uniformly continuous on $\mathbf{R}$.

PROOF. Suppose to the contrary that $f$ is uniformly continuous on $\mathbf{R}$. Then there is a $\delta > 0$ such that $|x - a| < \delta$ implies $|f(x) - f(a)| < 1$ for all $x, a \in \mathbf{R}$.

*$f(x) = x$ is uniformly continuous on $\mathbf{R}$.*

*$f(x) = \sin x$ is uniformly continuous on $\mathbf{R}$*

By the Archimedean Principle, choose $n \in \mathbf{N}$ so large that $n\delta > 1$. Set $a = n$ and $x = n + \delta/2$. Then $|x - a| < \delta$ and

$$1 > |f(x) - f(a)| = |x^2 - a^2| = n\delta + \frac{\delta^2}{4} > n\delta > 1.$$

This contradiction proves that $f$ is not uniformly continuous on **R**. ∎

Here is a key which unlocks the difference between continuity and uniform continuity.

**3.38 Lemma.** *Suppose $E \subset \mathbf{R}$ and $f : E \to \mathbf{R}$ is uniformly continuous. If $x_n \in E$ is Cauchy, then $f(x_n)$ is Cauchy.*

PROOF. Let $\varepsilon > 0$ and choose $\delta > 0$ such that (9) holds. Since $\{x_n\}$ is Cauchy, choose $N \in \mathbf{N}$ such that $n, m \geq N$ implies $|x_n - x_m| < \delta$. Then $n, m \geq N$ implies $|f(x_n) - f(x_m)| < \varepsilon$. ∎ If $f$ is continuous and $x_n$ is cauchy, but $f(x_n)$ is not Cauchy, Then $f$ is not uniformly continuous

Notice that $f(x) = 1/x$ is continuous on $(0, 1)$ and $x_n = 1/n$ is Cauchy but $f(x_n)$ is not. In particular, $1/x$ is continuous but not uniformly continuous on the open interval $(0, 1)$. Notice how the graph of $y = 1/x$ corroborates this fact. Indeed, as $a$ gets closer to 0, the value of $\delta$ gets smaller (compare $\delta_1$ to $\delta_0$ in Figure 3.5), hence, cannot be chosen independently of $a$.

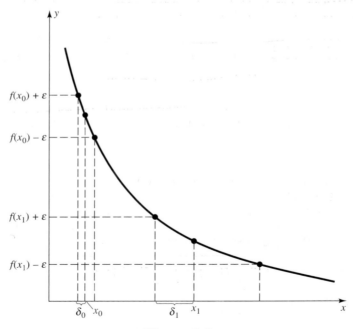

**Figure 3.5**

Thus, for some functions on the open interval $(0, 1)$, continuity and uniform continuity are different. This is not the case for functions on the closed interval

$[0, 1]$. Indeed, the next result shows that on all closed, bounded intervals, continuity and uniform continuity are always equivalent.

**3.39 THEOREM.** *Suppose $I$ is a closed, bounded interval. If $f : I \to \mathbf{R}$ is continuous on $I$, then $f$ is uniformly continuous on $I$.*

PROOF. Suppose to the contrary that $f$ is continuous but not uniformly continuous on $I$. Then there is an $\varepsilon_0 > 0$ and points $x_n, y_n \in I$ such that $|x_n - y_n| < 1/n$ and

$$(10) \qquad\qquad |f(x_n) - f(y_n)| \geq \varepsilon_0, \qquad n \in \mathbf{N}.$$

By the Bolzano–Weierstrass Theorem and the Comparison Theorem, the sequence $\{x_n\}$ has a subsequence, say $x_{n_k}$, which converges, as $k \to \infty$, to some $x \in I$. Similarly, the sequence $\{y_{n_k}\}_{k \in \mathbf{N}}$ has a convergent subsequence, say $y_{n_{k_j}}$, which converges, as $j \to \infty$, to some $y \in I$. Since $x_{n_{k_j}} \to x$ as $j \to \infty$ and $f$ is continuous, it follows from (10) that $|f(x) - f(y)| \geq \varepsilon_0$, i.e., $f(x) \neq f(y)$. But $|x_n - y_n| < 1/n$ for all $n \in \mathbf{N}$ so Theorem 2.9 (the Squeeze Theorem) implies $x = y$. Therefore, $f(x) = f(y)$, a contradiction. ∎

Using this result, we can obtain the following useful and simple characterization of uniform continuity on bounded open intervals.

**3.40 THEOREM.** *Let $(a, b)$ be a bounded, open, nonempty interval and $f : (a, b) \to \mathbf{R}$. Then $f$ is uniformly continuous on $(a, b)$ if and only if $f$ can be continuously extended to $[a, b]$, i.e., if and only if there is a continuous function $g : [a, b] \to \mathbf{R}$ which satisfies*

$$(11) \qquad\qquad f(x) = g(x), \qquad x \in (a, b).$$

PROOF. Suppose $f$ is uniformly continuous on $(a, b)$. Let $x_n \in (a, b)$ converge to $b$ as $n \to \infty$. Then $\{x_n\}$ is Cauchy; hence, by the Lemma 3.38, so is $\{f(x_n)\}$. In particular,

$$g(b) := \lim_{n \to \infty} f(x_n)$$

exists. This value does not change if we use a different sequence to approximate $b$. Indeed, let $y_n \in (a, b)$ be another sequence which converges to $b$ as $n \to \infty$. Given $\varepsilon > 0$, choose $\delta > 0$ such that (9) holds for $E = (a, b)$. Since $x_n - y_n \to 0$, choose $N \in \mathbf{N}$ so that $n \geq N$ implies $|x_n - y_n| < \delta$. By (9), then, $|f(x_n) - f(y_n)| < \varepsilon$ for all $n \geq N$. Taking the limit of this inequality as $n \to \infty$, we obtain

$$\left| \lim_{n \to \infty} f(x_n) - \lim_{n \to \infty} f(y_n) \right| \leq \varepsilon$$

for all $\varepsilon > 0$. It follows from Theorem 1.9 that

$$\lim_{n \to \infty} f(x_n) = \lim_{n \to \infty} f(y_n).$$

Thus, $g(b)$ is well defined. A similar argument defines $g(a)$.

Set $g(x) = f(x)$ for $x \in (a, b)$. Then $g$ is defined on $[a, b]$, satisfies (11), and is continuous on $[a, b]$ by the Sequential Characterization of Limits. Thus, $f$ can be "continuously extended" to $g$ as required.

Conversely, suppose that there is a function $g$ continuous on $[a, b]$ which satisfies (11). By Theorem 3.39, $g$ is uniformly continuous on $[a, b]$; hence, $g$ is uniformly continuous on $(a, b)$. We conclude that $f$ is uniformly continuous on $(a, b)$. ∎

Let $f$ be continuous on a bounded nonempty interval $(a, b)$. Notice that $f$ is continuously extendable to $[a, b]$ if and only if the one-sided limits of $f$ exist at $a$ and $b$. Indeed, when they exist, we can always define $g$ at $a$ and $b$ to be the values of these limits. In particular, we can prove that $f$ is uniformly continuous without using $\varepsilon$'s and $\delta$'s.

**3.41 Example.** Prove that $f(x) = (x - 1)/\log x$ is uniformly continuous on $(0, 1)$.

SOLUTION It is clear that $f(x) \to 0$ as $x \to 0+$. Moreover, by l'Hôpital's Rule,

$$\lim_{x \to 1-} f(x) = \lim_{x \to 1-} \frac{1}{1/x} = 1.$$

Hence $f$ is continuously extendable to $[0, 1]$, so by Theorem 3.40, $f$ is uniformly continuous on $(0, 1)$. ∎

## EXERCISES

**1.** Using Definition 3.35, prove that each of the following functions is uniformly continuous on $(0, 1)$.

   a) $f(x) = x^3$.
   b) $f(x) = x^2 - x$.
   c) $f(x) = x \sin 2x$.

**2.** Prove that each of the following functions is uniformly continuous on $(0, 1)$.

   a)
$$f(x) = \frac{x^3 - 1}{x - 1}.$$

   b)
$$f(x) = x \sin \frac{1}{x}.$$

   c)                 $f(x)$ is any polynomial.

   d)
$$f(x) = \frac{\sin x}{x}.$$

e)
$$f(x) = x^2 \log x.$$

[You may use l'Hôpital's Rule (see Theorem 4.18) on parts d) and e).]

**3.** Find all real $\alpha$ such that $x^\alpha \sin(1/x)$ is uniformly continuous on the open interval $(0, 1)$.

**4.** a) Suppose $f : [0, \infty) \to \mathbf{R}$ is continuous and there is an $L \in \mathbf{R}$ such that $f(x) \to L$ as $x \to \infty$. Prove $f$ is uniformly continuous on $[0, \infty)$.

   b) Prove that $f(x) = 1/(x^2 + 1)$ is uniformly continuous on $\mathbf{R}$.

**5.** a) Let $I$ be a bounded interval. Prove that if $f : I \to \mathbf{R}$ is uniformly continuous on $I$, then $f$ is bounded on $I$.

   b) Prove a) may be false if $I$ is unbounded.

**6.** Suppose $\alpha \in \mathbf{R}$, $E$ is a nonempty subset of $\mathbf{R}$, and $f, g : E \to \mathbf{R}$ are uniformly continuous on $E$.

   a) Prove that $f + g$ and $\alpha f$ are uniformly continuous on $E$.

   b) Suppose $f, g$ are bounded on $E$. Prove $fg$ is uniformly continuous on $E$.

   c) Show there exist functions $f, g$ uniformly continuous on $\mathbf{R}$ such that $fg$ is not uniformly continuous on $\mathbf{R}$.

   d) Suppose $f$ is bounded on $E$ and that there is a positive constant $\varepsilon_0$ such that $g(x) \geq \varepsilon_0$ for all $x \in E$. Prove that $f/g$ is uniformly continuous on $E$.

   e) Show there exist functions $f, g$, uniformly continuous on the interval $(0, 1)$, with $g(x) > 0$ for all $x \in (0, 1)$, such that $f/g$ is not uniformly continuous on $(0, 1)$.

   f) Prove that if $f, g$ are uniformly continuous on a closed interval $[a, b]$ and $g(x) \neq 0$ for $x \in [a, b]$, then $f/g$ is uniformly continuous on $[a, b]$.

**7.** Let $E \subseteq \mathbf{R}$. A function $f : E \to \mathbf{R}$ is said to be *increasing* on $E$ if $x_1, x_2 \in E$ and $x_1 < x_2$ imply $f(x_1) \leq f(x_2)$. Suppose $f$ is increasing and bounded on an open, bounded, nonempty interval $(a, b)$.

   a) Prove that $f(a+)$ and $f(b-)$ both exist and are finite.

   b) Prove that $f$ is continuous on $(a, b)$ if and only if $f$ is uniformly continuous on $(a, b)$.

   c) Show b) is false if $f$ is unbounded. Indeed, find an increasing function $g : (0, 1) \to \mathbf{R}$ which is continuous on $(0, 1)$ but not uniformly continuous on $(0, 1)$.

**8.** Suppose that $f$ is continuous on $[0, 1]$ and set
$$I_k = \left[ \frac{k-1}{2^n}, \frac{k}{2^n} \right]$$

for $k = 1, 2, \ldots, 2^n$. Prove that given $\varepsilon > 0$ there is an $N \in \mathbf{N}$ such that $n \geq N$ implies
$$\sup_{x \in I_k} f(x) - \inf_{x \in I_k} f(x) < \varepsilon, \qquad k = 1, 2, \ldots, 2^n.$$

**9.** Prove that a polynomial of degree $n$ is uniformly continuous on $\mathbf{R}$ if and only if $n = 0$ or $1$.

# Chapter 4

# Differentiability on $\mathbb{R}$

## 4.1 THE DERIVATIVE

For many applications, one needs to compute the slope of a tangent line of some function $f$. The following concept is useful in this regard.

**4.1 DEFINITION.** Let $f$ be a real function.

    i) $f$ is said to be *differentiable* at $a$ if $f$ is defined on some open interval containing $a$ and

(1)
$$f'(a) := \lim_{h \to 0} \frac{f(a+h) - f(a)}{h}$$

       exists, in which case $f'(a)$ is called the *derivative* of $f$ at $a$.

    ii) $f$ is said to be *differentiable* if it is differentiable at each $a$ in its domain.

The assumption that $f$ be defined on an open interval containing $a$ is made so that the quotients in (1) are defined for all $h \neq 0$ sufficiently small.

You may recall that the graph of $y = f(x)$ has a tangent line at the point $(a, f(a))$ if and only if $f$ has a derivative at $a$, in which case the slope of that tangent line is $f'(a)$. To see why this connection makes sense, let us consider a geometric interpretation of (1). Suppose $f$ is differentiable at $a$. A *secant line* of the graph $y = f(x)$ is a line passing through at least two points on the graph, and a *chord* is a line segment which runs from one point on the graph to another. Let $x_0 = a$ and $x = a + h$, and observe that the slope of the chord passing through the points $(x, f(x))$ and $(x_0, f(x_0))$ is given by $(f(x) - f(x_0))/(x - x_0)$. Now, by the choice of $x$ and $x_0$, (1) becomes

$$f'(x_0) = \lim_{x \to x_0} \frac{f(x) - f(x_0)}{x - x_0}.$$

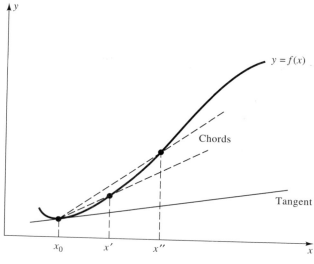

**Figure 4.1**

Hence, as $x \to x_0$ the slopes of the chords through $(x, f(x))$ and $(x_0, f(x_0))$ approximate the slope of the tangent line of $y = f(x)$ at $x = x_0$ (see Figure 4.1), and in the limit, the slope of the tangent line to $y = f(x)$ at $x = x_0$ is precisely $f'(x_0)$. Thus, we shall say that the graph of $y = f(x)$ *has a unique tangent line* at a point $(a, f(a))$ if and only if $f'(a)$ exists.

If $f$ is differentiable at each point in a set $E$, then $f'$ is a function on $E$. This function is denoted several ways:

$$D_x f = \frac{df}{dx} = f^{(1)} = f'.$$

When $y = f(x)$ we shall also use the notation $dy/dx$ or $y'$ for $f'$. Higher-order derivatives are defined recursively; i.e., if $n \in \mathbf{N}$, then $f^{(n+1)}(a) := (f^{(n)}(a))'$, provided these derivatives exist. Higher-order derivatives are also denoted several ways, including $D_x^n f$, $d^n f/dx^n$, $f^{(n)}$, and by $d^n y/dx^n$ and $y^{(n)}$ when $y = f(x)$. The *second derivatives* $f^{(2)}$ (respectively, $y^{(2)}$) are usually written as $f''$ (respectively, $y''$). When $f''$ exists at some point $a$, we shall say that $f$ is *twice differentiable* at $a$.

Here are two characterizations of differentiability which we shall use to study derivatives. The first one, which characterizes the derivative in terms of the "chord function"

(2) $$f^*(x) := \frac{f(x) - f(a)}{x - a} \qquad x \neq a,$$

will be used to prove the Chain Rule below.

**4.2 THEOREM.** *A real function $f$ is differentiable at some point $a \in \mathbf{R}$ if and only if there exist an open interval $I$, such that $f$ is defined on $I$ and $a \in I$, and a* function $f^* : I \to \mathbf{R}$, continuous at $a$, such that

(3) $$f(x) = f^*(x)(x - a) + f(a)$$

holds for all $x \in I$, in which case $f^*(a) = f'(a)$.

PROOF. Notice once and for all that for $x \in I \setminus \{a\}$, (2) and (3) are equivalent. Suppose $f$ is differentiable at $a$. Then $f$ is defined on some open interval $I$ containing $a$, and the limit in (1) exists. Define $f^*$ on $I$ by (2) if $x \neq a$, and by $f^*(a) := f'(a)$. Then (3) holds for all $x \in I$, and $f^*$ is continuous at $a$ since $f'(a)$ exists.

Conversely, if (3) holds, then (2) holds for all $x \in I$, $x \neq a$. Taking the limit of (2) as $x \to a$, bearing in mind that $f^*$ is continuous at $a$, we conclude that $f^*(a) = f'(a)$. ∎

The second characterization of differentiability, in terms of *linear approximations* (i.e., how well $f(a + h) - f(a)$ can be approximated by a straight line through the origin) will be used in Chapter 8 to define the derivative of a function of several variables.

**4.3 THEOREM.** *Let $f : \mathbf{R} \to \mathbf{R}$. Then $f$ is differentiable at $a$ if and only if there is a function $T$ of the form $T(x) := mx$ such that*

(4) $$\lim_{h \to 0} \frac{|f(a + h) - f(a) - T(h)|}{|h|} = 0.$$

PROOF. Suppose $f$ is differentiable, and set $m := f'(a)$. Then by (1),

$$\frac{f(a + h) - f(a) - T(h)}{h} = \frac{f(a + h) - f(a)}{h} - f'(a) \to 0$$

as $h \to 0$.

Conversely, if (4) holds for $T(x) := mx$, then

$$\lim_{h \to 0} \frac{f(a + h) - f(a)}{h} - m = \lim_{h \to 0} \frac{f(a + h) - f(a) - mh}{h}$$
$$= \lim_{h \to 0} \frac{f(a + h) - f(a) - T(h)}{h} = 0.$$

Since the limit of a difference is the difference of its limits, it follows that (1) holds with $m = f'(a)$. ∎

Our first application of Theorem 4.2 answers the question: Are differentiability and continuity related?

**4.4 THEOREM.** *If $f$ is differentiable at $a$, then $f$ is continuous at $a$.*

PROOF. Suppose $f$ is differentiable at $a$. By Theorem 4.2, there is an open interval $I$ and a function $f^*$, continuous at $a$, such that $f(x) - f(a) = f^*(x)(x - a)$ for all $x \in I$. Taking the limit of this equality as $x \to a$, we see that

$$\lim_{x \to a} f(x) - f(a) = f^*(a) \cdot 0 = 0.$$

In particular, $f(x) \to f(a)$ as $x \to a$, i.e., $f$ is continuous at $a$. ∎

Thus any function which fails to be continuous at $a$ cannot be differentiable at $a$. The following example shows that the converse of Theorem 4.4 is false.

Using the point-slope form of equation of a line
if $f'(a)$ exists, then an equation of the tangent line to the curve $y = f(a)$
at the point $(a, f(a))$ is   $y - f(a) = f'(a)(x-a)$

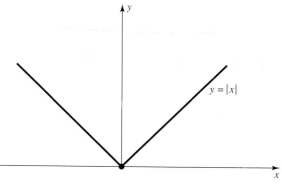

**Figure 4.2**

**4.5 Example.** Show that $f(x) = |x|$ is continuous at 0 but not differentiable there.

PROOF. Since $x \to 0$ implies $|x| \to 0$, $f$ is continuous at 0. On the other hand, since $|h| = h$ when $h > 0$ and $|h| = -h$ when $h < 0$, we have

$$\lim_{h \to 0+} \frac{f(h) - f(0)}{h} = 1 \quad \text{and} \quad \lim_{h \to 0-} \frac{f(h) - f(0)}{h} = -1.$$

Since a limit exists if and only if its one-sided limits exist and are equal (Theorem 3.14), it follows that the limit in (1) does not exist when $a = 0$ and $f(x) = |x|$. Therefore, $f$ is not differentiable at 0. ∎

This example reflects the conventional wisdom about the difference between differentiable and continuous functions. Since a function differentiable at $a$ always has a unique tangent line at $(a, f(a))$, the graph of a differentiable function on an interval is "smooth" with no corners, cusps, or kinks. On the contrary, although the graph of a continuous function on an interval is unbroken (has no holes or jumps), it may well have corners, cusps, or kinks. In particular, the graph of $f(x) = |x|$, continuous but not differentiable at $x = 0$, is unbroken but has a corner at the point $(0, 0)$ (see Figure 4.2).

By Definition 4.1, if $f$ is differentiable at $a$, then $f$ must be defined on an open interval containing $a$, i.e., on both sides of $a$. As with the theory of limits, it is convenient to define "one-sided" derivatives to deal with functions whose domains are closed intervals (see Example 4.7 below). Here is a brief discussion of what it means for a real function to be differentiable *on* an interval (as opposed to being differentiable at every point *in* an interval). This concept will be used in Sections 5.3, 5.6, and 11.1.

**4.6 DEFINITION.** Let $I$ be a nondegenerate interval and $a \in I$.

   i) A function $f : I \to \mathbf{R}$ is said to be *differentiable* on $I$ if

$$f'_I(a) := \lim_{\substack{x \to a \\ x \in I}} \frac{f(x) - f(a)}{x - a}$$

exists and is finite for every $a \in I$.

ii) $f$ is said to be *continuously differentiable* on $I$ if $f_I'$ exists and is continuous on $I$.

Notice that when $a$ is not an endpoint of $I$, $f_I'(a)$ is the same as $f'(a)$. Because of this, we usually drop the subscript on $f_I'$. In particular, if $f$ is differentiable on $[a, b]$ then

$$f'(a) := \lim_{h \to 0+} \frac{f(a + h) - f(a)}{h} \quad \text{and} \quad f'(b) := \lim_{h \to 0-} \frac{f(b + h) - f(b)}{h}.$$

The following example shows that Definition 4.6 enlarges the collection of differentiable functions.

**4.7 Example.** The function $f(x) = x^{3/2}$ is differentiable on $[0, \infty)$ and $f'(x) = 3\sqrt{x}/2$ for all $x \in [0, \infty)$.

PROOF. By the Power Rule (see Exercise 8 in Section 4.2), $f'(x) = 3\sqrt{x}/2$ for all $x \in (0, \infty)$. And by definition,

$$f'(0) = \lim_{h \to 0+} \frac{h^{3/2} - 0}{h} = \lim_{h \to 0+} \sqrt{h} = 0. \quad \blacksquare$$

Here is notation widely used in conjunction with Definition 4.6. Let $I$ be a nondegenerate interval. For each $n \in \mathbf{N}$, $\mathcal{C}^n(I)$ will denote the collection of real functions $f$ whose $n$th derivatives exist and are continuous on $I$. (Thus $\mathcal{C}^1(I)$ is precisely the collection of real functions which are continuously differentiable on $I$.) We shall also denote the collection of $f$ which belong to $\mathcal{C}^n(I)$ for all $n \in \mathbf{N}$ by $\mathcal{C}^\infty(I)$. [When dealing with specific intervals, we shall drop the outer set of parentheses; e.g., we shall write $\mathcal{C}^n[a, b]$ for $\mathcal{C}^n([a, b])$.]

By modifying the proof of Theorem 4.4, we can show that if $f$ is differentiable on $I$, then $f$ is continuous on $I$. Thus, $\mathcal{C}^m(I) \subset \mathcal{C}^n(I)$ when $m > n$.

The following example shows that not every differentiable function belongs to $\mathcal{C}^1(\mathbf{R})$.

**4.8 Example.** The function

$$f(x) = \begin{cases} x^2 \sin(1/x) & x \neq 0 \\ 0 & x = 0 \end{cases}$$

is differentiable on **R** but not continuously differentiable on any interval which contains the origin.

PROOF. By definition,

$$f'(0) = \lim_{h \to 0} h \sin\left(\frac{1}{h}\right) = 0 \quad \text{and} \quad f'(x) = 2x \sin\left(\frac{1}{x}\right) - \cos\left(\frac{1}{x}\right)$$

for $x \neq 0$. Thus $f$ is differentiable on **R** but $\lim_{x \to 0} f'(x)$ does not exist. In particular, $f'$ is not continuous on any interval which contains the origin. $\blacksquare$

It is important to notice that a function which is differentiable on two sets is not necessarily differentiable on their union.

**4.9 Remark.** $f(x) = |x|$ *is differentiable on* $[0,1]$ *and on* $[-1,0]$ *but not on* $[-1,1]$.

PROOF. Since $f(x) = x$ when $x > 0$ and $= -x$ when $x < 0$, it is clear that $f$ is differentiable on $[-1,0) \cup (0,1]$ (with $f'(x) = 1$ for $x > 0$ and $f'(x) = -1$ for $x < 0$). By Example 4.5, $f$ is not differentiable at $x = 0$. However,

$$f'_{[0,1]}(0) = \lim_{h \to 0+} \frac{|h|}{h} = 1 \quad \text{and} \quad f'_{[-1,0]}(0) = \lim_{h \to 0-} \frac{|h|}{h} = -1.$$

Therefore, $f$ is differentiable on $[0,1]$ and on $[-1,0]$. ∎

## EXERCISES

**1.** For each of the following real functions, use Definition 4.1 directly to prove that $f'(a)$ exists.

   a) $f(x) = x^2$, $a \in \mathbf{R}$.
   b) $f(x) = 1/x$, $a \neq 0$.
   c) $f(x) = \sqrt{x}$, $a > 0$.

**2.** Let $I$ be an open interval which contains 0 and $f : I \to \mathbf{R}$. If there exists an $\alpha > 1$ such that $|f(x)| \leq |x|^\alpha$ for all $x \in I$, prove that $f$ is differentiable at 0. What happens when $\alpha = 1$?

**3.** Let $I$ be an open interval, $f : I \to \mathbf{R}$, and $x_0 \in I$. The function $f$ is said to have a *local maximum* at $x_0$ if there is a $\delta > 0$ such that $f(x_0) \geq f(x)$ holds for all $|x - x_0| < \delta$.

   a) If $f$ has a local maximum at $x_0$, prove

$$\frac{f(x_0 + h) - f(x_0)}{h} \leq 0 \quad \text{and} \quad \frac{f(x_0 + H) - f(x_0)}{H} \geq 0$$

   for $h > 0$ and $H < 0$ sufficiently small.
   b) If $f$ is differentiable at $x_0$ and has a local maximum at $x_0$, prove $f'(x_0) = 0$.
   c) Make and prove analogous statements for local minima.
   d) Show by example that the converses of the statements in parts b) and c) are false. Namely, find an $f$ such that $f'(0) = 0$ but $f$ has neither a local maximum nor a local minimum at 0.

**4.** Using elementary geometry and the definition of $\sin x, \cos x$, one can show for every $x, y \in \mathbf{R}$ (see Appendix B) that

   (i) $\qquad |\sin x| \leq 1, \qquad |\cos x| \leq 1, \qquad \sin(0) = 0, \qquad \cos(0) = 1,$

   (ii) $\qquad \sin(-x) = -\sin x, \qquad \cos(-x) = \cos x,$

   (iii) $\qquad \sin^2 x + \cos^2 x = 1, \qquad \cos x = 1 - 2\sin^2\left(\frac{x}{2}\right),$

(iv)          $\sin(x \pm y) = \sin x \cos y \pm \cos x \sin y.$

Moreover, if $x$ is measured in radians then

(v)          $\cos x = \sin\left(\dfrac{\pi}{2} - x\right), \qquad \sin x = \cos\left(\dfrac{\pi}{2} - x\right),$

and

(vi)          $0 < x \cos x < \sin x < x, \qquad 0 < x \le \dfrac{\pi}{2}.$

Using these properties and the Chain Rule when necessary, prove each of the following statements.

a) The functions $\sin x$ and $\cos x$ are continuous at 0.

b) The functions $\sin x$ and $\cos x$ are continuous on **R**.

c) The limits

$$\lim_{x \to 0} \frac{\sin x}{x} = 1 \quad \text{and} \quad \lim_{x \to 0} \frac{1 - \cos x}{x} = 0$$

exist.

d) The function $\sin x$ is differentiable on **R** with $(\sin x)' = \cos x$.

e) The functions $\cos x$ and $\tan x := \sin x / \cos x$ are differentiable on **R** with $(\cos x)' = -\sin x$ and $(\tan x)' = \sec^2 x$.

**5.** Suppose that $f : (0, \infty) \to \mathbf{R}$ satisfies $f(x) - f(y) = f(x/y)$ for all $x, y \in (0, \infty)$ and $f(1) = 0$.

a) Prove that $f$ is continuous on $(0, \infty)$ if and only if $f$ is continuous at 1.

b) Prove that $f$ is differentiable on $(0, \infty)$ if and only if $f$ is differentiable at 1.

c) Prove that if $f$ is differentiable at 1, then $f'(x) = f'(1)/x$ for all $x \in (0, \infty)$. (Note: If $f'(1) = 1$, then $f(x) = \log x$.)

**6**. **This exercise is used in Section 4.2.**

a) Prove $(x^n)' = nx^{n-1}$ for $n \in \mathbf{N}$ and $x \in \mathbf{R}$.

b) Prove that if $f(x) = x^\alpha$, where $\alpha = 1/n$ for some $n \in \mathbf{N}$, then $y = f(x)$ is differentiable and $f'(x) = \alpha \cdot x^{\alpha-1}$ for every $x \in (0, \infty)$.

**7.** Suppose that

$$f_\alpha(x) = \begin{cases} |x|^\alpha \sin \dfrac{1}{x} & x \ne 0 \\ 0 & x = 0. \end{cases}$$

Show that $f_\alpha(x)$ is continuous at $x = 0$ when $\alpha > 0$ and differentiable at $x = 0$ when $\alpha > 1$. Graph these functions for $\alpha = 1$ and $\alpha = 2$ and give a geometric interpretation of your results.

**8.** Suppose $I$ is a an open interval, $a \in I$, and $f, g$ are functions from $I$ to **R**.

a) Prove that if $f$ and $g$ are differentiable at $a$, then $f + g$ is differentiable at $a$ with

$$(f + g)'(a) = f'(a) + g'(a).$$

b) Prove that if $f$ is differentiable at $a$ and $\alpha \in \mathbf{R}$, $\alpha f$ is differentiable at $a$ with

$$(\alpha f)'(a) = \alpha f'(a).$$

## 4.2  DIFFERENTIABILITY THEOREMS

In this section we prove several familiar results about derivatives.

**4.10 THEOREM.** *Let $f$ and $g$ be real functions and $\alpha \in \mathbf{R}$. If $f$ and $g$ are differentiable at $a$, then $f+g$, $\alpha f$, $f \cdot g$, and (when $g(a) \neq 0$) $f/g$ are all differentiable at $a$. In fact,*

(5)
$$(f + g)'(a) = f'(a) + g'(a),$$

(6)
$$(\alpha f)'(a) = \alpha f'(a),$$

(7)
$$(f \cdot g)'(a) = g(a)f'(a) + f(a)g'(a).$$

(8)
$$\left(\frac{f}{g}\right)'(a) = \frac{g(a)f'(a) - f(a)g'(a)}{g^2(a)}.$$

Proof. The proofs of these rules are similar. We provide the details only for (7). By adding and subtracting $f(a)g(x)$ in the numerator of the left side of the following expression, we can write

$$\frac{f(x)g(x) - f(a)g(a)}{x - a} = g(x)\frac{f(x) - f(a)}{x - a} + f(a)\frac{g(x) - g(a)}{x - a}.$$

Since $g$ is continuous (see Theorem 4.4), it follows from Definition 4.1 that

$$\lim_{x \to a} \frac{f(x)g(x) - f(a)g(a)}{x - a} = g(a)f'(a) + f(a)g'(a)$$

as required. ∎

Formula (5) is called the *Sum Rule*, (6) is sometimes called the *Homogeneous Rule*, (7) is called the *Product Rule*, and (8) is called the *Quotient rule*.

Next, we show what the derivative does to a composition of two functions.

**4.11 THEOREM** [Chain Rule]. *Let $f$ and $g$ be real functions. If $f$ is differentiable at $a$ and $g$ is differentiable at $f(a)$, then $g \circ f$ is differentiable at $a$ with*

(9)
$$(g \circ f)'(a) = g'(f(a))f'(a).$$

Proof. By Theorem 4.2, there exist open intervals $I$ and $J$, and functions $f^* : I \to \mathbf{R}$, continuous at $a$, and $g^* : J \to \mathbf{R}$, continuous at $f(a)$, such that $f^*(a) = f'(a)$, $g^*(f(a)) = g'(f(a))$,

(10)
$$f(x) = f^*(x)(x - a) + f(a), \qquad x \in I,$$

and

(11) $$g(y) = g^*(y)(y - f(a)) + g(f(a)), \qquad y \in J.$$

Since $f$ is continuous at $a$ we may assume (by making $I$ smaller if necessary) that $f(x) \in J$ for all $x \in I$.

Fix $x \in I$. Apply (11) to $y = f(x)$ and (10) to $x$ to write

$$(g \circ f)(x) = g(f(x)) = g^*(f(x))(f(x) - f(a)) + g(f(a))$$
$$= g^*(f(x))f^*(x)(x - a) + (g \circ f)(a).$$

Set $h^*(x) = g^*(f(x))f^*(x)$ for $x \in I$. Since $f^*$ is continuous at $a$ and $g^*$ is continuous at $f(a)$, it is clear that $h^*$ is continuous at $a$. Moreover,

$$h^*(a) = g^*(f(a))f^*(a) = g'(f(a))f'(a).$$

It follows from Theorem 4.2, therefore, that $(g \circ f)'(a) = g'(f(a))f'(a)$. ∎

## EXERCISES

**1.** For each of the following functions, find all $x$ for which $f'(x)$ exists and find a formula for $f'$. (You may use Exercise 8 below.)

  a) $f(x) = (x^3 - 2x^2 + 3x)/\sqrt{x}$.

  b) $f(x) = 1/(x^2 + x - 1)$.

  c) $f(x) = x^x$.

  d) $f(x) = |x^3 + 2x^2 - x - 2|$.

*[handwritten: $f(x) = x^x = e^{x \cdot \ln x}$]*

**2.** Suppose $f$ and $g$ are differentiable at 2 and 3 with $f'(2) = a$, $f'(3) = b$, $g'(2) = c$, and $g'(3) = d$. If $f(2) = 1$, $f(3) = 2$, $g(2) = 3$, and $g(3) = 4$, evaluate each of the following derivatives.

  a) $(fg)'(2)$.

  b) $(f/g)'(3)$.

  c) $(g \circ f)'(3)$.

  d) $(f \circ g)'(2)$.

**3.** Assuming $e^x$ is differentiable on **R**, prove that

$$f(x) = \begin{cases} \dfrac{x}{1 + e^{1/x}} & x \neq 0 \\ 0 & x = 0 \end{cases}$$

is differentiable on $(0, \infty)$. Is $f$ differentiable at 0?

**4.** Using Exercise 6 in Section 4.1, prove every polynomial belongs to $\mathcal{C}^\infty(\mathbf{R})$.

**5.** [RECIPROCAL RULE]  Suppose $f$ is differentiable at $a$ and $f(a) \neq 0$.

  a) Show that for $h$ sufficiently small, $f(a + h) \neq 0$.

  b) Using Definition 4.1 directly, prove that $1/f(x)$ is differentiable at $x = a$ and

$$\left(\frac{1}{f}\right)'(a) = -\frac{f'(a)}{f^2(a)}.$$

*[handwritten:]*
1. $\ln(xy) = \ln x + \ln y$.
2. $\ln\left(\frac{x}{y}\right) = \ln x - \ln y$
3. $\ln(x^r) = r \ln x$.

$-\ e^{\ln x} = x$.
$-\ \ln(e^x) = x$
$-\ a^x = e^{x \ln a}$

6. Use Exercise 5 and the Product Rule to prove the Quotient Rule.

7. Suppose $n \in \mathbf{N}$ and $f, g$ are real functions of a real variable whose $n$th derivatives $f^{(n)}$, $g^{(n)}$ exist at a point $a$. Prove Leibniz's generalization of the Product Rule:

$$(fg)^{(n)}(a) = \sum_{k=0}^{n} \binom{n}{k} f^{(k)}(a) g^{(n-k)}(a).$$

$\boxed{8}$. **This exercise is used in Section 5.3**.

    a) Prove that if $f(x) = x^{m/n}$ for some $m, n \in \mathbf{N}$, then $y = f(x)$ is differentiable and satisfies $ny^{n-1}y' = mx^{m-1}$ for every $x \in (0, \infty)$.

    b) [POWER RULE] Prove $x^q$ is differentiable on $(0, \infty)$ for every $q \in \mathbf{Q}$ and $(x^q)' = qx^{q-1}$.

9. Consider the following outline to a proof of the Chain Rule for real functions. Let $y = f(x)$, $y_0 = f(x_0)$, and observe that $y \to y_0$ as $x \to x_0$. Thus

$$\lim_{x \to x_0} \frac{g \circ f(x) - g \circ f(x_0)}{x - x_0} = \lim_{x \to x_0} \frac{g(f(x)) - g(f(x_0))}{f(x) - f(x_0)} \frac{f(x) - f(x_0)}{x - x_0}$$

$$= \left( \lim_{y \to y_0} \frac{g(y) - g(y_0)}{y - y_0} \right) \left( \lim_{x \to x_0} \frac{f(x) - f(x_0)}{x - x_0} \right)$$

$$= g'(y_0) f'(x_0) = g'(f(x_0)) f'(x_0).$$

    a) Find the flaw in this argument.

    b) Write down a statement which this argument does prove.

## 4.3   THE MEAN VALUE THEOREM

The Mean Value Theorem makes a precise statement about the relationship between the derivative of a function and the slope of one of its chords. It was discovered by the following geometric reasoning. Suppose $f$ is differentiable on $(a, b)$. Since the graph of $f$ on $(a, b)$ has a tangent at each of its points, it seems likely that the slope of the chord through the points $(a, f(a))$ and $(b, f(b))$ equals the slope $f'(x_0)$ for some value of $x_0 \in (a, b)$ (see Figure 4.3 below).

    We begin with a special case.

**4.12 Lemma** [ROLLE'S THEOREM]. *Suppose $f$ is continuous on a closed, bounded, nondegenerate interval $[a, b]$ and differentiable on $(a, b)$. If $f(a) = f(b)$, then $f'(x_0) = 0$ for some $x_0 \in (a, b)$.*

PROOF. By the Extreme Value Theorem, $f$ has a finite maximum $M$ and a finite minimum $m$ on $[a, b]$. If $M = m$, then $f$ is constant on $(a, b)$ and $f'(x) = 0$ for all $x \in (a, b)$.

Suppose $M \neq m$. Since $f(a) = f(b)$, $f$ must assume one of the values $M$ or $m$ at some point $x_0 \in (a, b)$. By symmetry, we may suppose that $f(x_0) = M$. (That is, if we can prove the theorem when $f(x_0) = M$, then a similar proof establishes the theorem when $f(x_0) = m$.) Since $M$ is the maximum of $f$ on $[a, b]$, we have

$$f(x_0 + h) - f(x_0) \le 0$$

for all $h$ which satisfy $x_0 + h \in (a, b)$. In the case $h > 0$ this implies

$$f'(x_0) = \lim_{h \to 0+} \frac{f(x_0 + h) - f(x_0)}{h} \le 0,$$

and in the case $h < 0$ this implies

$$f'(x_0) = \lim_{h \to 0-} \frac{f(x_0 + h) - f(x_0)}{h} \ge 0.$$

It follows that $f'(x_0) = 0$. ∎

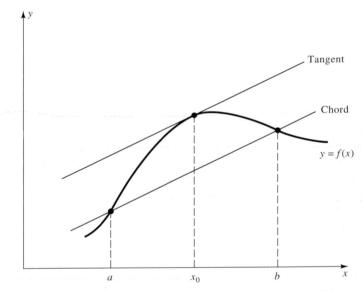

**Figure 4.3**

**4.13 Remark.** *The continuity hypothesis in Rolle's Theorem cannot be relaxed at even one point in $[a, b]$.*

PROOF. The function

$$f(x) = \begin{cases} x & x \in [0, 1) \\ 0 & x = 1 \end{cases}$$

is continuous on $[0, 1)$ and differentiable on $(0, 1)$, $f(0) = f(1) = 0$, but $f'(x)$ is never zero. ∎

*(margin handwriting: f must be differentiable on open interval.)*

**4.14 Remark.** *The differentiability hypothesis in Rolle's Theorem cannot be relaxed at even one point in $(a, b)$.*

PROOF. The function $f(x) = |x|$ is continuous on $[-1, 1]$ and differentiable on $(-1, 1) \setminus \{0\}$ and $f(-1) = f(1)$, but $f'(x)$ is never zero. ∎

We shall use Rolle's Theorem to obtain several useful results. The first is pair of "Mean Value Theorems."

**4.15 THEOREM.** *Suppose $[a, b]$ is a closed, bounded, nondegenerate interval.*

  i)  [GENERALIZED MEAN VALUE THEOREM] *If $f, g$ are continuous on $[a, b]$ and differentiable on $(a, b)$, then there is an $x_0 \in (a, b)$ such that*

$$f'(x_0)(g(b) - g(a)) = g'(x_0)(f(b) - f(a)).$$

*(margin handwriting: $\dfrac{g(b) - g(a)}{f(b) - f(a)} = \dfrac{g'(x_0)}{f'(x_0)}$)*

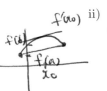

  ii) [MEAN VALUE THEOREM] *If $f$ is continuous on $[a, b]$ and differentiable on $(a, b)$, then there is an $x_0 \in (a, b)$ such that*

$$f(b) - f(a) = f'(x_0)(b - a).$$

*(margin handwriting: $f'(x_0) = \dfrac{f(b) - f(a)}{b - a}$.)*

PROOF. i) Set $h(x) = f(x)(g(b) - g(a)) - g(x)(f(b) - f(a))$. Since $h'(x) = f'(x)(g(b) - g(a)) - g'(x)(f(b) - f(a))$, it is clear that $h$ is continuous on $[a, b]$, differentiable on $(a, b)$, and $h(a) = h(b)$. Thus, by Rolle's Theorem, $h'(x_0) = 0$ for some $x_0 \in (a, b)$.

ii) Set $g(x) = x$ and apply part i). (For a geometric interpretation of this result, see the opening paragraph of this section and Figure 4.3 above.) ∎

The Generalized Mean Value Theorem is also called *Cauchy's Mean Value Theorem*. It is crucial when comparing derivatives of two functions simultaneously (e.g., see Theorem 4.18 below), for studying certain kinds of generalized derivatives (e.g., see Remark 14.33), and for using higher order derivatives to approximate a given function (e.g., see Taylor's Formula, Theorem 7.44).

The Mean Value Theorem is most often used to extract information about $f$ from $f'$ (see, for example, Theorem 4.17 and Exercises 5 through 9 below). It is also useful for comparing the relative strengths of two functions.

**4.16 Example.** *Prove that $1 + x < e^x$ for all $x > 0$.*

PROOF. Let $f(x) = e^x - x - 1$ and fix $x > 0$. By the Mean Value Theorem,

$$e^x - x - 1 = f(x) - f(0) = xf'(x_0)$$

for some $x_0$ between 0 and $x$. But $x_0 > 0$ implies $f'(x_0) = e^{x_0} - 1 > 0$. Hence $e^x - x - 1 = xf'(x_0) > 0$, i.e., $e^x > x + 1$. ∎

Here is another application of the Mean Value Theorem.

**4.17 THEOREM** [BERNOULLI'S INEQUALITY]. *Let $\alpha$ be a positive real number and $\delta \geq -1$. If $0 < \alpha \leq 1$, then*

$$(1 + \delta)^{\alpha} \leq 1 + \alpha\delta,$$

*and if $\alpha \geq 1$, then*

$$(1 + \delta)^{\alpha} \geq 1 + \alpha\delta.$$

PROOF. The proofs of these inequalities are similar. We present the details only for the case $0 < \alpha \leq 1$. Let $f(x) = x^{\alpha}$. By the Mean Value Theorem,

$$f(1 + \delta) = f(1) + \alpha\delta x_0^{\alpha-1}$$

for some $x_0$ between 1 and $1 + \delta$. If $\delta > 0$, then $x_0 > 1$. Since $0 < \alpha \leq 1$, it follows that $x_0^{\alpha-1} < 1$ (see Exercise 5 in Section 5.3); hence, $\delta x_0^{\alpha-1} < \delta$. On the other hand, if $-1 \leq \delta \leq 0$, then $x_0^{\alpha-1} \geq 1$ and again $\delta x_0^{\alpha-1} \leq \delta$. Therefore,

$$(1 + \delta)^{\alpha} = f(1 + \delta) = f(1) + \alpha\delta x_0^{\alpha-1} \leq 1 + \alpha\delta. \quad \blacksquare$$

Another application of the Mean Value Theorem is the following technique for evaluating limits of certain quotients. (Our statement is general enough to include one-sided limits and limits at infinity.)

**4.18 THEOREM** [L'HÔPITAL'S RULE]. *Let $x_0$ be an extended real number and $I$ be an open interval which either contains $x_0$ or has $x_0$ as an endpoint. Suppose that $f$ and $g$ are differentiable on $I \setminus \{x_0\}$, and $g(x) \neq 0 \neq g'(x)$ for all $x \in I \setminus \{x_0\}$. Suppose further that*

$$A := \lim_{\substack{x \to x_0 \\ x \in I}} f(x) = \lim_{\substack{x \to x_0 \\ x \in I}} g(x)$$

*is either 0 or $\infty$. If*

$$B := \lim_{\substack{x \to x_0 \\ x \in I}} \frac{f'(x)}{g'(x)}$$

*exists as an extended real number, then*

$$\lim_{\substack{x \to x_0 \\ x \in I}} \frac{f(x)}{g(x)} = \lim_{\substack{x \to x_0 \\ x \in I}} \frac{f'(x)}{g'(x)}.$$

PROOF. Let $0 < \varepsilon < 1$, $\gamma \in \mathbf{R}$ (to be specified below), and suppose first that $B \in \mathbf{R}$. Then regardless of whether $x_0$ is finite or not, there is a $y_0 \in I$ such that

(12) $$\left| \frac{f'(c)}{g'(c)} - B \right| < \frac{\varepsilon}{\gamma}$$

for every $c$ which lies between $y_0$ and $x_0$.

Fix $x$ between $y_0$ and $x_0$. Let $x_n$ be a sequence between $y_0$ and $x_0$ which converges to $x_0$. By the Generalized Mean Value Theorem, for each $n \in \mathbf{N}$ there is a point $c_n$ between $x_n$ and $x$ such that

$$(13) \qquad g'(c_n)(f(x_n) - f(x)) = f'(c_n)(g(x_n) - g(x)).$$

*Case 1.* $A = 0$. Let $\gamma = 1$. Since $g'$ is never zero on $I$, we have by the Mean Value Theorem that $g(x_n) \neq g(x)$ for all $n \in \mathbf{N}$. Hence it follows from (12) and (13) that

$$B - \varepsilon < \frac{f(x_n) - f(x)}{g(x_n) - g(x)} < B + \varepsilon$$

for all $n \in \mathbf{N}$. Since $f(x_n)$ and $g(x_n)$ converge to zero as $n \to \infty$, it follows that

$$B - \varepsilon \leq \frac{f(x)}{g(x)} \leq B + \varepsilon$$

for all $x$ which lie between $y_0$ and $x_0$. Hence by definition, $f(x)/g(x) \to B$ as $x \to x_0$ through $I$.

*Case 2.* $A = \infty$. This time choose $\gamma$ such that

$$\frac{2}{\gamma} + \frac{B}{\gamma} + \frac{1}{\gamma^2} < 1,$$

and let $x_n \to x_0$ satisfy (13). By the Sequential Characterization of Limits, it suffices to prove that $f(x_n)/g(x_n) \to B$ as $n \to \infty$.

Since neither $g(x_n)$ nor $g'(c_n)$ is zero for any $n \in \mathbf{N}$, we can rewrite (13) as

$$(14) \qquad \frac{f(x_n)}{g(x_n)} = \frac{f(x)}{g(x_n)} + \left(1 - \frac{g(x)}{g(x_n)}\right) \frac{f'(c_n)}{g'(c_n)}.$$

Since $g(x_n) \to \infty$ as $n \to \infty$, both $f(x)/g(x_n)$ and $g(x)/g(x_n)$ converge to zero as $n \to \infty$. Thus choose $N \in \mathbf{N}$ such that

$$\left| \frac{f(x)}{g(x_n)} \right| \quad \text{and} \quad \left| \frac{g(x)}{g(x_n)} \right| < \frac{\varepsilon}{\gamma}.$$

(Recall that $x$ is fixed.) Since the factor $1 - g(x)/g(x_n)$ is positive for large $n$, it follows from (14), (12), the choice of $\gamma$, and the fact that $\varepsilon < 1$ that

$$\frac{f(x_n)}{g(x_n)} < \frac{\varepsilon}{\gamma}\left(1 + \frac{\varepsilon}{\gamma}\right)\left(B + \frac{\varepsilon}{\gamma}\right) < B + \varepsilon$$

for large $n$. Similarly, $f(x_n)/g(x_n) > B - \varepsilon$ for large $n$. Thus $f(x_n)/g(x_n) \to B$ as $n \to \infty$. This proves l'Hôpital's Rule when $B \in \mathbf{R}$. A similar proof handles the case when $B = \pm\infty$ (see Exercise 10). ∎

l'Hôpital's Rule can be used to compare the relative rates of growth of two functions. For example, the next result shows that as $x \to \infty$, $e^x$ converges to $\infty$ much faster that $x^2$ does.

**4.19 Example.** Prove $\lim_{x \to \infty} x^2/e^x = 0$.

PROOF. Since the limits of $x^2/e^x$ and $x/e^x$ are of the form $\infty/\infty$, we apply l'Hôpital's Rule twice to verify

$$\lim_{x \to \infty} \frac{x^2}{e^x} = \lim_{x \to \infty} \frac{2x}{e^x} = \lim_{x \to \infty} \frac{2}{e^x} = 0. \quad \blacksquare$$

For each subsequent application of l'Hôpital's Rule, it is important to check that the hypotheses still hold. For example,

$$\lim_{x \to 0} \frac{x^2}{x^2 + \sin x} = \lim_{x \to 0} \frac{2x}{2x + \cos x} = 0 \neq 1 = \lim_{x \to 0} \frac{2}{2 - \sin x}.$$

Notice that the middle limit is not of the form $0/0$.

l'Hôpital's Rule can be used to evaluate limits of the form $0 \cdot \infty$.

**4.20 Example.** Find $\lim_{x \to 0+} x \log x$.

SOLUTION. By writing $x$ as $1/(1/x)$, we see that the limit in question is of the form $\infty/\infty$. Hence, by l'Hôpital's Rule,

$$\lim_{x \to 0+} x \log x = \lim_{x \to 0+} \frac{\log x}{1/x} = \lim_{x \to 0+} \frac{1/x}{-1/x^2} = 0. \quad \blacksquare$$

The next two examples show that l'Hôpital's Rule can also be used to evaluate limits of the form $1^\infty$ and $0^0$.

**4.21 Example.** Prove that the sequence $(1 + 1/n)^n$ is increasing, as $n \to \infty$, and its limit $e$ satisfies $e \leq 3$ and $\log e = 1$.

PROOF. The sequence $(1 + 1/n)^n$ is increasing, since by Bernoulli's Inequality,

$$\left(1 + \frac{1}{n}\right)^{n/(n+1)} \leq \left(1 + \frac{1}{n+1}\right).$$

To prove that this sequence is bounded above, observe by the Binomial Formula that

$$\left(1 + \frac{1}{n}\right)^n = \sum_{k=0}^{n} \binom{n}{k} \left(\frac{1}{n}\right)^k.$$

Now,

$$\binom{n}{k} \left(\frac{1}{n}\right)^k = \frac{n(n-1)\dots(n-k+1)}{n^k} \cdot \frac{1}{k!} \leq \frac{1}{k!} \leq \frac{1}{2^{k-1}}$$

for all $k \in \mathbf{N}$. It follows from Exercise 1c) in Section 1.2 that

$$\left(1 + \frac{1}{n}\right)^n \leq 1 + 1 + \sum_{k=1}^{n} \frac{1}{2^k} = 3 - \frac{1}{2^n} < 3$$

for all $n \in \mathbf{N}$. Hence, by the Monotone Convergence Theorem, the limit defining $e$ exists.

To verify $\log e = 1$, use l'Hôpital's Rule:

$$\log e = \lim_{n \to \infty} \frac{\log(1 + 1/n)}{1/n} = \lim_{n \to \infty} \frac{(n/(n+1))(-1/n^2)}{-1/n^2} = 1. \quad \blacksquare$$

**4.22 Example.** Find $L = \lim_{x \to 1}(\log x)^{1-x}$.

SOLUTION. Since $\log L = \lim_{x \to 1}(1 - x)\log\log x$ (if this limit exists) is of the form $0 \cdot \infty$, we have by l'Hôpital's Rule that

$$\log L = \lim_{x \to 1} \frac{\log\log x}{1/(1-x)} = \lim_{x \to 1} \frac{1/(x \log x)}{1/(1-x)^2} = \lim_{x \to 1} \frac{-2(1-x)}{1 + \log x} = 0.$$

Hence, $L = e^0 = 1. \quad \blacksquare$

## EXERCISES

**1.** Evaluate the following limits.

a)
$$\lim_{x \to 0} \frac{\sin(3x)}{x}.$$

b)
$$\lim_{x \to 0+} \frac{\cos x - e^x}{\log(1 + x^2)}.$$

c)
$$\lim_{x \to 0} \left(\frac{x}{\sin x}\right)^{1/x^2}.$$

d)
$$\lim_{x \to 0+} x^x.$$

e)
$$\lim_{x \to 1} \frac{\log x}{\sin(\pi x)}.$$

f)
$$\lim_{x \to \infty} x\left(\arctan x - \frac{\pi}{2}\right).$$

(For the derivative of $\arctan x$ see Exercise 7 in Section 4.4.)

**2.** Prove each of the following inequalities.

a) $\sqrt{1 + 2x} < 1 + x$ for all $x > 0$.
b) $\log x \leq x - 1$ for all $x \geq 1$.
c) $7(x - 1) < e^x$ for all $x \geq 2$.
d) $\sin^2 x \leq 2|x|$ for all $x \in \mathbf{R}$.

$\boxed{3}$. **This exercise is used in Sections 7.4 and 12.5.** Assume that $e^x$ is differentiable on **R** with $(e^x)' = e^x$.

a) Show that the derivative of

$$f(x) = \begin{cases} e^{-1/x^2} & x \neq 0 \\ 0 & x = 0 \end{cases}$$

exists and is continuous on **R** with $f'(0) = 0$.

b) Do analogous statements hold for $f^{(n)}(x)$ when $n = 2, 3, \ldots$?

$\boxed{4}$. **This exercise is used in Sections 5.4, 6.3, and elsewhere.**

a) Using $(e^x)' = e^x$, $(\log x)' = 1/x$, and $x^\alpha = e^{\alpha \log x}$, show that $(x^\alpha)' = \alpha x^{\alpha-1}$ for all $x > 0$.

b) Let $\alpha > 0$. Prove that $\log x \leq x^\alpha$ for $x$ large. Prove that there exists a constant $C_\alpha$ such that $\log x \leq C_\alpha x^\alpha$ for all $x \in [1, \infty)$, $C_\alpha \to \infty$ as $\alpha \to 0+$, and $C_\alpha \to 0$ as $\alpha \to \infty$.

c) Obtain an analogue of b) valid for $e^x$ and $x^\alpha$ in place of $\log x$ and $x^\alpha$.

**5.** Suppose $f$ is differentiable on **R**.

a) If $f'(x) = 0$ for all $x \in \mathbf{R}$, prove that $f(x) = f(0)$ for all $x \in \mathbf{R}$.

b) If $f(0) = 1$ and $|f'(x)| \leq 1$ for all $x \in \mathbf{R}$, prove that $|f(x)| \leq |x| + 1$ for all $x \in \mathbf{R}$.

c) If $f'(x) \geq 0$ for all $x \in \mathbf{R}$, prove $a < b$ implies $f(a) \leq f(b)$.

**6.** Suppose $f$ is differentiable on a nonempty, open interval $(a, b)$ with $f'$ bounded on $(a, b)$. Prove $f$ is uniformly continuous on $(a, b)$.

**7.** Suppose $f$ is differentiable on $(a, b)$, continuous on $[a, b]$, and $f(a) = f(b) = 0$. Prove that if $f(c) > 0$ for some $c \in (a, b)$, then there exist $x_1, x_2 \in (a, b)$ such that $f'(x_1) > f'(x_2)$.

**8.** Suppose $f$ is twice differentiable on $(a, b)$ and that there are points $x_1 < x_2 < x_3$ in $(a, b)$ such that $f(x_1) > f(x_2)$ and $f(x_3) > f(x_2)$. Prove that there is a point $c \in (a, b)$ such that $f''(c) > 0$.

**9.** Suppose $f$ is differentiable on $(0, \infty)$. If $L = \lim_{x \to \infty} f'(x)$ and $\lim_{n \to \infty} f(n)$ both exist and are finite, prove that $L = 0$.

**10.** Prove l'Hôpital's Rule for the case $|B| = \infty$ by using (13) and a modified version of (12) to account for the fact that $B$ is no longer finite.

**11.** Suppose $f : [a, b] \to \mathbf{R}$ is continuous and increasing (see Definition 4.23 below). Prove that $\sup f(E) = f(\sup E)$ for every nonempty set $E \subseteq [a, b]$.

**12.** Let $(a, b)$ be an open interval, $f : (a, b) \to \mathbf{R}$, and $x_0 \in (a, b)$. The function $f$ is said to have a *proper local maximum* at $x_0$ if there is a $\delta > 0$ such that $f(x_0) > f(x)$ for all $0 < |x - x_0| < \delta$.

a) If $f$ is differentiable on $(a, b)$ and has a proper local maximum at $x_0 \in (a, b)$, prove that $f'(x_0) = 0$ and that given $\delta > 0$, there exist $x_1 < x_0 < x_2$ such that $f'(x_1) > 0$, $f'(x_2) < 0$ and $|x_j - x_0| < \delta$ for $j = 1, 2$.

b) Make and prove an analogous statement for a proper local minimum.

## 4.4   MONOTONE FUNCTIONS AND THE INVERSE FUNCTION THEOREM

Monotone functions (i.e., those which either increase or decrease on their domain) are important from both a theoretical and a practical point of view (e.g., see Theorem 5.34). In this section we study monotone functions and the role they play in the Inverse Function Theorem.

**4.23 DEFINITION.** Let $E$ be a nonempty subset of $\mathbf{R}$ and $f : E \to \mathbf{R}$.

i) $f$ is said to be *increasing* (respectively, *strictly increasing*) on $E$ if $x_1, x_2 \in E$ and $x_1 < x_2$ imply $f(x_1) \leq f(x_2)$ (respectively, $f(x_1) < f(x_2)$).

ii) $f$ is said to be *decreasing* (respectively, *strictly decreasing*) on $E$ if $x_1, x_2 \in E$ and $x_1 < x_2$ imply $f(x_1) \geq f(x_2)$ (respectively, $f(x_1) > f(x_2)$).

iii) $f$ is said to be *monotone* (respectively, *strictly monotone*) on $E$ if $f$ is either decreasing or increasing (respectively, either strictly decreasing or strictly increasing) on $E$.

Thus, although $f(x) = x^2$ is strictly monotone on $[0, 1]$, and on $[-1, 0]$, it is not monotone on $[-1, 1]$.

The derivative gives a simple method for finding where a differentiable function is monotone.

**4.24 THEOREM.** *Suppose $f$ is continuous on a closed, bounded, nondegenerate interval $[a, b]$ and differentiable on $(a, b)$.*

i) *If $f'(x) > 0$ (respectively, $f'(x) < 0$) for all $x \in (a, b)$, then $f$ is strictly increasing (respectively, strictly decreasing) on $[a, b]$.*

ii) *If $f'(x) = 0$ for all $x \in (a, b)$, then $f$ is constant on $[a, b]$.*

PROOF. Let $a \leq x_1 < x_2 \leq b$. By the Mean Value Theorem, there is an $x_0 \in (a, b)$ such that

$$f(x_2) - f(x_1) = f'(x_0)(x_2 - x_1).$$

Thus, $f(x_2) > f(x_1)$ when $f'(x_0) > 0$ and $f(x_2) < f(x_1)$ when $f'(x_0) < 0$. This proves part i).

To prove part ii), let $a \leq x \leq b$. By the Mean Value Theorem and hypothesis there is an $x_0 \in (a, b)$ such that

$$f(x) - f(a) = f'(x_0)(x - a) = 0.$$

Thus, $f(x) = f(a)$ for all $x \in [a, b]$. ∎

**4.25 Remark.** *If $f$ and $g$ are continuous on a closed, bounded, nondegenerate interval $[a, b]$, differentiable on $(a, b)$, and $f'(x) = g'(x)$ for all $x \in (a, b)$, then $f - g$ is constant on $[a, b]$.*

PROOF. Apply Theorem 4.24ii) to the function $f - g$. ∎

Let $f$ be a real function. Recall (see Figure 1.2) that if $f$ has an inverse function $f^{-1}$, then the graph of $y = f^{-1}(x)$ is a reflection of the graph of $y = f(x)$ about the line $y = x$. Thus, it is not difficult to imagine that $f^{-1}$ is as smooth as $f$. This is the subject of the next two theorems.

*Thm 4.26 is false if "continuous" is dropped from hypotheses.*

$f(2) = \frac{11}{10}$
$f(1) = 1$

*f is 1-1, but not continuous. f is not strictly monotone.*

**4.26 THEOREM.** *If $f$ is 1–1 and continuous on a closed, bounded interval $[a, b]$, then $f$ is strictly monotone on $[a, b]$ and $f^{-1}$ is continuous and strictly monotone on the interval whose endpoints are $f(a)$ and $f(b)$.*

PROOF. We may suppose that $a < b$. There are two cases: either $f(a) < f(b)$ or $f(a) > f(b)$. We supply the details of the proof only for the first case. The proof for the second case is similar.

We first show that if $f(a) < f(b)$, then $f$ is strictly increasing on $[a, b]$. Indeed, suppose $x_1, x_2 \in (a, b)$ and $x_1 < x_2$. If $f(x_1) \geq f(b)$ then $f(b)$ lies between $f(a)$ and $f(x_1)$. Hence, by the Intermediate Value Theorem, there is an $x_0 \in (a, x_1)$ which satisfies $f(x_0) = f(b)$, a contradiction of the fact that $f$ is 1–1. Similarly, $f(x_1) \leq f(a)$ also leads to a contradiction. Therefore, $f(a) < f(x_1) < f(b)$. Applying this same argument with $x_1$ in place of $a$ and $x_2$ in place of $x_1$, we see that $f(x_1) < f(x_2) < f(b)$. This proves that $f$ is strictly increasing on $[a, b]$.

Next, let $c = f(a)$ and $d = f(b)$. By the Intermediate Value Theorem, $f$ takes $[a, b]$ onto $[c, d]$. Hence, $f$ is 1–1 from $[a, b]$ onto $[c, d]$, and it follows from Theorem 1.31 that $f$ has an inverse function $f^{-1}$ on $[c, d]$. Suppose $f^{-1}$ is not strictly increasing on $[c, d]$. Then we can choose $y_1 < y_2$ such that

$$x_1 := f^{-1}(y_1) \geq x_2 := f^{-1}(y_2).$$

But $f$ is increasing, so $y_1 = f(x_1) \geq f(x_2) = y_2$, a contradiction. Thus, $f^{-1}$ is strictly increasing on $[c, d]$.

By symmetry, it remains to show that $f^{-1}(y_0+) = f^{-1}(y_0)$ for each $y_0 \in [c, d)$. Fix such a $y_0$ and let $\varepsilon > 0$. Choose $x_0 \in [a, b)$ such that $f(x_0) = y_0$, and choose $0 < \varepsilon_0 < \varepsilon$ so small that $x_0 + \varepsilon_0 \in [a, b)$. Set $\delta = f(x_0 + \varepsilon_0) - f(x_0)$ and suppose that $0 \leq y - y_0 < \delta$. If $x = f^{-1}(y)$, then the choices of $y$ and $\delta$ imply $y_0 \leq y < y_0 + \delta$, i.e.,

$$f(x_0) \leq f(x) < f(x_0 + \varepsilon_0).$$

Since $f^{-1}$ is strictly increasing, it follows that $x_0 \leq x < x_0 + \varepsilon_0$, i.e., $0 \leq x - x_0 < \varepsilon_0$. Therefore,

$$0 \leq f^{-1}(y) - f^{-1}(y_0) < \varepsilon_0 < \varepsilon$$

for all $0 \leq y - y_0 < \delta$. We conclude that $f^{-1}(y_0+)$ exists and equals $f^{-1}(y_0)$. ∎

**4.27 THEOREM** [INVERSE FUNCTION THEOREM]. *Let $f$ be 1–1 and continuous on an open interval $I$. If $f'(x_0)$ exists and is nonzero for some $x_0 \in I$, then $f^{-1}$ is differentiable at $y_0 = f(x_0)$ and*

$$(f^{-1})'(y_0) = \frac{1}{f'(x_0)}.$$

PROOF. Choose $a, b \in$ **R** such that $x_0 \in (a, b)$ and $[a, b] \subset I$. By Theorem 4.26, $f$ is strictly monotone, say strictly increasing on $[a, b]$, and $f^{-1}$ exists, is continuous, and strictly increasing on $[f(a), f(b)]$. Since $x_0 \in (a, b)$, choose $y_0 \in (f(a), f(b))$

such that $y_0 = f(x_0)$ and let $h \neq 0$ be so small that $f^{-1}(y_0 + h)$ is defined. Set $x = f^{-1}(y_0 + h)$ and observe that

$$f(x) - f(x_0) = y_0 + h - y_0 = h.$$

Since $f^{-1}$ is continuous, $x \to x_0$ if and only if $h \to 0$. Therefore,

$$\lim_{h \to 0} \frac{f^{-1}(y_0 + h) - f^{-1}(y_0)}{h} = \lim_{x \to x_0} \frac{x - x_0}{f(x) - f(x_0)} = \frac{1}{f'(x_0)}. \quad \blacksquare$$

This theorem is usually presented in elementary calculus texts in a form more easily remembered: if $y = f(x)$ and $x = f^{-1}(y)$, then

$$\frac{dx}{dy} = \frac{1}{dy/dx}.$$

Notice that, by using this formula, we do not need to solve explicitly for $f^{-1}$ to be able to compute $(f^{-1})'$ (see Exercises 2, 3, and 7 below).

*We close this section with several optional results which delve a little deeper into differentiability of real functions.*

Recall (see Examples 3.31 and 3.32) that there exist functions which have neither right nor left limits at a given point. The following result shows that monotone functions never behave this badly.

*4.28 Lemma.* *Suppose $f$ is increasing on $[a, b]$.*
   i) *If $x_0 \in [a, b)$ then $f(x_0+)$ exists and $f(x_0) \leq f(x_0+)$.*
   ii) *If $x_0 \in (a, b]$ then $f(x_0-)$ exists and $f(x_0-) \leq f(x_0)$.*

PROOF. Fix $x_0 \in (a, b]$. By symmetry it suffices to show that $f(x_0-)$ exists and satisfies $f(x_0-) \leq f(x_0)$. Set $E = \{f(x) : a < x < x_0\}$ and $s = \sup E$. Since $f$ is increasing, $f(x_0)$ is an upper bound of $E$. Hence, $s$ is a finite real number which satisfies $s \leq f(x_0)$. Given $\varepsilon > 0$, choose by the Approximation Property an $x_1 \in (a, x_0)$ such that $s - \varepsilon < f(x_1) \leq s$. Since $f$ is increasing,

$$s - \varepsilon < f(x_1) \leq f(x) \leq s$$

for all $x_1 < x < x_0$. Therefore, $f(x_0-)$ exists and satisfies $f(x_0-) = s \leq f(x_0)$. $\quad \blacksquare$

We have seen (Example 3.32) that a function can be nowhere continuous, i.e., can have uncountably many points of discontinuity. How many points of discontinuity can a monotone function have?

*4.29 THEOREM.* *If $f$ is monotone on an interval $I$, then $f$ has at most countably many points of discontinuity on $I$.*

PROOF. Without loss of generality, we may suppose that $f$ is increasing. Since the countable union of countable sets is countable (Theorem 1.38), it suffices to show that the set of points of discontinuity of $f$ can be written as a countable union of countable sets. Since **R** is the union of closed intervals $[-n, n]$, $n \in \mathbf{N}$, we may suppose that $I$ is a closed, bounded interval $[a, b]$.

Let $E$ represent the set of points of discontinuity of $f$ on $(a, b)$. By Lemma 4.28, $f(x-) \le f(x) \le f(x+)$ for all $x \in (a, b)$. Thus, $f$ is discontinuous at such an $x$ if and only if $f(x+) - f(x-) > 0$. It follows that

$$E = \bigcup_{j=1}^{\infty} A_j,$$

where for each $j \in \mathbf{N}$, $A_j := \{x \in \mathbf{R} : f(x+) - f(x-) \ge 1/j\}$. We will complete the proof by showing each $A_j$ is finite.

Suppose to the contrary that $A_{j_0}$ is infinite for some $j_0$. Set $y_0 := j_0(f(b) - f(a))$ and observe that since $f$ is finite-valued on $I$, $y_0$ is a finite real number. On the other hand, since $A_{j_0}$ is infinite, there exist $x_1 < x_2 < \ldots$ in $[a, b]$ such that $f(x_k+) - f(x_k-) \ge 1/j_0$ for $k \in \mathbf{N}$. Since $f$ is monotone, it follows that

$$f(b) - f(a) \ge \sum_{k=1}^{n}(f(x_k+) - f(x_k-)) \ge \frac{n}{j_0},$$

i.e., $y_0 = j_0(f(b) - f(a)) \ge n$ for all $n \in \mathbf{N}$. Taking the limit of this last inequality as $n \to \infty$, we see that $y_0 = +\infty$. With this contradiction, the proof of the theorem is complete. ∎

Although a differentiable function might not be continuously differentiable, the following result shows that its derivative does satisfy an intermediate value theorem. (This result is sometimes called Darboux's Theorem.)

**\*4.30 THEOREM** [MEAN VALUE THEOREM FOR DERIVATIVES]. *Suppose $f$ is differentiable on $[a, b]$ with $f'(a) \ne f'(b)$. If $y_0$ is a real number which lies between $f'(a)$ and $f'(b)$, then there is an $x_0 \in (a, b)$ such that $f'(x_0) = y_0$.*

STRATEGY: Let $F(x) := f(x) - y_0 x$. We must find an $x_0 \in (a, b)$ such that $F'(x_0) := f'(x_0) - y_0 = 0$. Since local extrema of a differentiable function $F$ occurs where the derivative of $F$ is zero (e.g., see the proof of Rolle's Theorem), it suffices to show that $F$ has a local extremum at some $x_0 \in (a, b)$.

PROOF. Suppose $y_0$ lies between $f'(a)$ and $f'(b)$. By symmetry, we may suppose that $f'(a) < y_0 < f'(b)$. Set $F(x) = f(x) - y_0 x$ for $x \in [a, b]$, and observe that $F$ is differentiable on $[a, b]$. Hence, by the Extreme Value Theorem, $F$ has an absolute minimum, say $F(x_0)$, on $[a, b]$. Now $F'(a) = f'(a) - y_0 < 0$, so $F(a+h) - F(a) < 0$ for $h > 0$ sufficiently small. Hence $F(a)$ is NOT the absolute minimum of $F$ on $[a, b]$. Similarly, $F(b)$ is not the absolute minimum of $F$ on $[a, b]$. Hence, the absolute minimum $F(x_0)$ must occur on $(a, b)$, i.e., $x_0 \in (a, b)$ and $F'(x_0) = 0$. ∎

Hence, any differentiable function takes intervals in its domain to intervals in its range.

## EXERCISES

**1.** a) Find all $a \in \mathbf{R}$ such that $x^3 + ax^2 + 3x + 15$ is strictly increasing near $x = 1$.
   b) Find all $a \in \mathbf{R}$ such that $ax^2 + 3x + 5$ is strictly increasing on the interval $(1, 2)$.
   c) Find where $f(x) = 2|x - 1| + 5\sqrt{x^2 + 9}$ is strictly increasing and where $f(x)$ is strictly decreasing.

**2.** Suppose $f$ and $g$ are 1–1 and continuous on $\mathbf{R}$. If $f(0) = 2$, $g(1) = 2$, $f'(0) = \pi$, and $g'(1) = e$, compute the following derivatives.
   a) $(f^{-1})'(2)$.
   b) $(g^{-1})'(2)$.
   c) $(f^{-1} \cdot g^{-1})'(2)$.

**3.** Let $f(x) = x^2 e^{x^2}$, $x \in \mathbf{R}$.
   a) Show that $f^{-1}$ exists and is differentiable on $(0, \infty)$.
   b) Compute $(f^{-1})'(e)$.

**4.** Suppose that $f'$ exists and is continuous on a nonempty, open interval $(a, b)$ with $f'(x) \neq 0$ for all $x \in (a, b)$.
   a) Prove $f$ is 1–1 on $(a, b)$ and takes $(a, b)$ onto some open interval $(c, d)$.
   b) Show that $(f^{-1})'$ exists and is continuous on $(c, d)$.
   c) Using the function $f(x) = x^3$, show that b) is false if the assumption $f'(x) \neq 0$ fails to hold for some $x \in (a, b)$.

**5.** Suppose $f$ is continuous on a closed, bounded interval $[a, b]$.
   a) If $f$ is differentiable on $(a, b)$ and $f'(x) \geq \varepsilon_0 > 0$ for all $x \in (a, b)$, prove that $(f^{-1})'$ exists and is bounded on $(f(a), f(b))$.
   b) If $f$ is continuously differentiable on $(a, b)$ and $f'(x_0) \neq 0$ for some $x_0 \in (a, b)$, prove that there are intervals $I$ and $J$ such that $f$ is 1–1 from $I$ onto $J$ and $f^{-1}$ is continuously differentiable on $J$.

**6.** Let $[a, b]$ be a closed, bounded, nondegenerate interval. Find all functions $f$ which satisfy the following conditions for some fixed $\alpha > 0$: $f$ is continuous and 1–1 on $[a, b]$, $f'(x) \neq 0$ and $f'(x) = \alpha(f^{-1})'(f(x))$ for all $x \in (a, b)$.

**7.** Using the Inverse Function Theorem, prove $(\arcsin x)' = 1/\sqrt{1 - x^2}$ for $x \in (-1, 1)$ and $(\arctan x)' = 1/(1 + x^2)$ for $x \in (-\infty, \infty)$.

**\*8.** Let $I$ be an interval and $n \in \mathbf{N}$. Show that if $f_j : I \to \mathbf{R}$ are monotone functions and $f = \sum_{j=1}^{n} \alpha_j f_j$ for some $\alpha_j \in \mathbf{R}$, then $f$ has at most countably many points of discontinuity on $I$.

**\*9.** Suppose $f$ is differentiable at every point in a closed, bounded interval $[a, b]$. Prove that if $f'$ is 1–1 on $[a, b]$, then $f'$ is strictly monotone on $[a, b]$.

**\*10.** Suppose $f$ is differentiable at every point in a closed, bounded interval $[a, b]$. Prove that if $f'$ is increasing on $(a, b)$ then $f'$ is continuous on $(a, b)$.

# Chapter 5

# Integrability on **R**

## 5.1 THE RIEMANN INTEGRAL

In this chapter we shall study integration of real functions. We begin our discussion by introducing the following terminology.

**5.1 DEFINITION.** Let $[a, b]$ be a closed, bounded, nondegenerate interval.

    i) A *partition* of $[a, b]$ is a set of points $P = \{x_0, x_1, \ldots x_n\}$ such that

$$a = x_0 < x_1 < \cdots < x_n = b.$$

    ii) The *norm* of a partition $P = \{x_0, x_1, \ldots x_n\}$ is the number

$$\|P\| = \max_{1 \le j \le n} |x_j - x_{j-1}|.$$

    iii) A *refinement* of a partition $P = \{x_0, x_1, \ldots x_n\}$ is a partition $Q$ of $[a, b]$ which satisfies $Q \supseteq P$. In this case we say that $Q$ is *finer* than $P$.

**5.2 Example.** [The Dyadic Partition]. Prove that for each $n \in \mathbf{N}$, $P_n = \{j/2^n : j = 0, 1, \ldots, 2^n\}$ is a partition of the interval $[0, 1]$, and $P_m$ is finer than $P_n$ when $m > n$.

    PROOF. Fix $n \in \mathbf{N}$. If $x_j = j/2^n$, then $0 = x_0 < x_1 < \cdots < x_{2^n} = 1$. Thus $P_n$ is a partition of $[0, 1]$. Let $m > n$ and set $p = m - n$. If $0 \le j \le 2^n$, then $j/2^n = j2^p/2^m$ and $0 \le j2^p \le 2^m$. Thus $P_m$ is finer than $P_n$. ∎

It is clear that by definition, if $P$ and $Q$ are partitions of $[a, b]$, then $P \cup Q$ is finer than both $P$ and $Q$. (Note that "finer" does not rule out the possibility that $P \cup Q = Q$, which would be the case if $Q$ were a refinement of $P$.) And if $Q$ is a refinement of $P$, then $\|Q\| \le \|P\|$. We shall use these observations often.

106

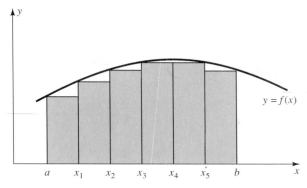

**Figure 5.1**

Let $f$ be nonnegative on an interval $[a, b]$. You may recall that the integral of $f$ over $[a, b]$ (when this integral exists) is the area of the region bounded by the curves $y = f(x)$, $y = 0$, $x = a$, and $x = b$. This area, $A$, can be approximated by rectangles whose bases lie in $[a, b]$ and whose heights approximate $f$ (see Figure 5.1). If the tops of these rectangles lie above the curve $y = f(x)$, the resulting approximation is larger than $A$. If the tops of these rectangles lie below the curve $y = f(x)$, the resulting approximation is smaller than $A$. Hence, we make the following definition.

**5.3 DEFINITION.** Let $[a, b]$ be a closed bounded interval, $P = \{x_0, x_1, \ldots x_n\}$ be a partition of $[a, b]$, and $f : [a, b] \to \mathbf{R}$ be bounded.

i) The *upper Riemann sum* of $f$ over $P$ is the number

$$U(f, P) := \sum_{j=1}^{n} M_j(f)(x_j - x_{j-1}),$$

where

$$M_j(f) := \sup_{x \in [x_{j-1}, x_j]} f(x).$$

ii) The *lower Riemann sum* of $f$ over $P$ is the number

$$L(f, P) := \sum_{j=1}^{n} m_j(f)(x_j - x_{j-1}),$$

where

$$m_j(f) := \inf_{x \in [x_{j-1}, x_j]} f(x).$$

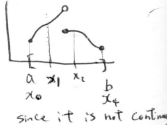

since it is not continuous
it does not achieve max.

[Note: We assumed $f$ is bounded so that the numbers $M_j(f)$ and $m_j(f)$ would exist and be finite.]    not necessary to achieve max and min.

Some upper and lower Riemann sums can be evaluated with the help of the following elementary observation.

*Telescope:*

**5.4 Remark.** *If* $g : \mathbf{N} \to \mathbf{R}$*, then*

$$\sum_{k=m}^{n} (g(k+1) - g(k)) = g(n+1) - g(m)$$

*for all* $n \geq m$ *in* **N**.

$\sum_{j=1}^{n} (x_j - x_{j-1}) = (x_1 - x_0) + (x_2 - x_1) + \cdots + (x_n - x_{n-1})$

$= x_n - x_0 = b - a.$

PROOF. The proof is by induction on $n$. The formula holds for $n = m$. If it holds for some $n - 1 \geq m$, then

$$\sum_{k=m}^{n} (g(k+1) - g(k)) = (g(n) - g(m)) + (g(n+1) - g(n)) = g(n+1) - g(m). \quad \blacksquare$$

We shall refer to this algebraic identity by saying the sum *telescopes* to $g(n+1) - g(m)$.

Before we define what it means for a function to be integrable, we make several elementary observations concerning upper and lower sums.

**5.5 Remark.** *If* $f(x) = \alpha$ *is constant on* $[a,b]$*, then*

$$U(f, P) = L(f, P) = \alpha(b - a)$$

*for all partitions* $P$ *of* $[a, b]$.    $\sum_{j=1}^{n} M_j(f)(x_j - x_{j-1}) = \alpha(b-a)$

PROOF. Since $M_j(f) = m_j(f) = \alpha$ for all $j$, the sums $U(f, P)$ and $L(f, P)$ telescope to $\alpha(b - a)$. $\quad \blacksquare$    $\sum_{j=1}^{n} m_j(f)(x_j - x_{j-1}) = \alpha(b-a)$

**5.6 Remark.** $L(f, P) \leq U(f, P)$ *for all partitions* $P$ *and all bounded functions* $f$.

PROOF. By definition, $m_j(f) \leq M_j(f)$ for all $j$. $\quad \blacksquare$

The next result shows that as the partitions get finer, the upper and lower Riemann sums get nearer each other. *if refine partition, Lower sum increase and Upper sum decrease.*

**5.7 Remark.** *If* $P$ *is any partition of* $[a, b]$ *and* $Q$ *is a refinement of* $P$*, then*

$$L(f, P) \leq L(f, Q) \leq U(f, Q) \leq U(f, P).$$

PROOF. Let $P = \{x_0, x_1, \ldots x_n\}$ be a partition of $[a, b]$. Since $Q$ is finer than $P$, $Q$ can be obtained from $P$ in a finite number of steps by adding one point at a time. Hence it suffices to prove the inequalities above for the special case $Q = \{c\} \bigcup P$ for some $c \in (a, b)$. Moreover, by symmetry and Remark 5.6, we need only show $U(f, Q) \leq U(f, P)$.

We may suppose that $c \notin P$. Hence, choose an index $j_0$ such that $x_{j_0-1} < c < x_{j_0}$. By definition, it is clear that

$$U(f, Q) - U(f, P) = M^{(\ell)}(c - x_{j_0-1}) + M^{(r)}(x_{j_0} - c) - M(x_{j_0} - x_{j_0-1}),$$

*Lemma: Let $f \in B[a,b]$ where $m \leq f(x) \leq M$ for all $x \in [a,b]$.*
*If $P$ is any partition of $[a,b]$, then $m(b-a) \leq L(f,P) \leq U(f,P) \leq M(b-a)$*
*$m = \inf(f) \quad M = \sup(f)$.*

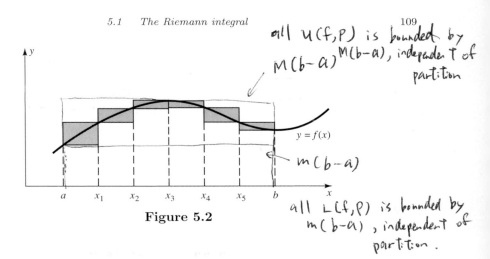

*all $U(f,P)$ is bounded by $M(b-a)$, independent of partition*

$y = f(x)$

$m(b-a)$

**Figure 5.2**

*all $L(f,P)$ is bounded by $m(b-a)$, independent of partition.*

where

$$M^{(\ell)} = \sup_{x \in [x_{j_0-1}, c]} f(x), \quad M^{(r)} = \sup_{x \in [c, x_{j_0}]} f(x), \quad \text{and} \quad M = \sup_{x \in [x_{j_0-1}, x_{j_0}]} f(x).$$

By the Monotone Property of Suprema, $M^{(\ell)}$ and $M^{(r)}$ are both less than or equal to $M$. Therefore,

$$U(f, Q) - U(f, P) \leq M(c - x_{j_0-1}) + M(x_{j_0} - c) - M(x_{j_0} - x_{j_0-1}) = 0. \quad \blacksquare$$

*any lower sum is smaller than any upper sum.*

**5.8 Remark.** *If $P$ and $Q$ are any partitions of $[a, b]$, then*

$$L(f, P) \leq U(f, Q).$$

PROOF. Since $P \cup Q$ is a refinement of $P$ and $Q$, it follows from Remark 5.7 that

$$L(f, P) \leq L(f, P \cup Q) \leq U(f, P \cup Q) \leq U(f, Q)$$

for any pair of partitions $P, Q$, whether $Q$ is a refinement of $P$ or not. $\blacksquare$

We now use the connection between area and integration to motivate the definition of "integrable." Suppose $f(x)$ is nonnegative on $[a, b]$ and the region bounded by the curves $y = f(x)$, $y = 0$, $x = a$, and $x = b$ has a well-defined area $A$. By Definition 5.3, every upper Riemann sum is an overestimate of $A$, and every lower Riemann sum is an underestimate of $A$ (see Figure 5.1). Since the estimates $U(f, P)$ and $L(f, P)$ should get nearer to $A$ as $P$ gets finer, the differences $U(f, P) - L(f, P)$ should get smaller. (The shaded area in Figure 5.2 represents the difference $U(f, P) - L(f, P)$ for a particular $P$.) This leads us to the following definition (see also Exercise 9 below).

**5.9 DEFINITION.** Let $[a, b]$ be a closed, bounded, nondegenerate interval. A function $f : [a, b] \to \mathbf{R}$ is said to be (*Riemann*) *integrable* on $[a, b]$ if $f$ is bounded on $[a, b]$, and for every $\varepsilon > 0$ there is a partition $P$ of $[a, b]$ such that $U(f, P) - L(f, P) < \varepsilon$.

Notice that this definition makes sense whether or not $f$ is nonnegative. The connection between nonnegative functions and area was only a convenient vehicle to motivate Definition 5.9. Also notice that, by Remark 5.6, $U(f, P) - L(f, P) = |U(f, P) - L(f, P)|$ for all partitions $P$. Hence, $U(f, P) - L(f, P) < \varepsilon$ is equivalent to $|U(f, P) - L(f, P)| < \varepsilon$.

This section provides a good illustration of how mathematics works. The connection between area and integration leads directly to Definition 5.9. This definition, however, is not easy to apply in concrete situations. Thus, we search for conditions which imply integrability *and* are easy to apply. In view of Figure 5.2, it seems reasonable that a function is integrable if its graph does not jump around too much (so that it can be covered by thinner and thinner rectangles). Since the graph of a continuous function does not jump at all, we are lead to the following simple criterion that is sufficient (but not necessary) for integrability.

**5.10 THEOREM.** *If $f$ is continuous on a closed, bounded, nondegenerate interval $[a, b]$, then $f$ is integrable on $[a, b]$.*

PROOF. Let $\varepsilon > 0$. Since $f$ is uniformly continuous on $[a, b]$, choose $\delta > 0$ such that

$$(1) \qquad |x - y| < \delta \quad \text{implies} \quad |f(x) - f(y)| < \frac{\varepsilon}{b - a}.$$

Let $P = \{x_0, x_1, \ldots, x_n\}$ be any partition of $[a, b]$ which satisfies $\|P\| < \delta$. Fix an index $j$ and notice, by the Extreme Value Theorem, that there are points $x_m$ and $x_M$ in $[x_{j-1}, x_j]$ such that

$$f(x_m) = m_j(f) \quad \text{and} \quad f(x_M) = M_j(f).$$

Since $\|P\| < \delta$, we also have $|x_M - x_m| < \delta$. Hence by (1), $M_j(f) - m_j(f) < \varepsilon/(b-a)$. In particular,

$$U(f, P) - L(f, P) = \sum_{j=1}^{n} (M_j(f) - m_j(f))(x_j - x_{j-1}) < \frac{\varepsilon}{b - a} \sum_{j=1}^{n} (x_j - x_{j-1}) = \varepsilon.$$

(The last step comes from telescoping.) ∎

Although the converse of Theorem 5.10 is false (see Exercise 7 in Section 5.2 and Exercises 3 and 8 below), there is a close connection between integrability and continuity. Indeed, we shall see (Theorem 9.50) that a function is integrable if and only if it has relatively few discontinuities. This principle is illustrated by the following examples because the nonintegrable function in Example 5.11 is nowhere continuous (hence has many discontinuities) but the integrable function in Example 5.12 has only one discontinuity (hence has few discontinuities).

**5.11 Example.** The Dirichlet function

$$f(x) = \begin{cases} 1 & x \in \mathbf{Q} \\ 0 & x \notin \mathbf{Q} \end{cases}$$

is not Riemann integrable on $[0, 1]$.

PROOF. Clearly, $f$ is bounded on $[0, 1]$. By Theorem 1.24 and Exercise 3 in Section 1.3 (Density of Rationals and Irrationals), the supremum of $f$ over any nondegenerate interval is 1, and the infimum of $f$ over any nondegenerate interval is 0. Therefore, $U(f, P) - L(f, P) = 1 - 0 = 1$ for any partition $P$ of the interval $[0, 1]$; i.e., $f$ is not integrable on $[0, 1]$. ∎

**5.12 Example.** The function

$$f(x) = \begin{cases} 0 & 0 \le x < 1/2 \\ 1 & 1/2 \le x \le 1 \end{cases}$$

is integrable on $[0, 1]$.

PROOF. Let $\varepsilon > 0$ and set

$$P = \left\{ 0, \frac{1 - \varepsilon}{2}, \frac{1 + \varepsilon}{2}, 1 \right\}.$$

We may suppose that $\varepsilon < 1/2$, i.e., that $P$ is a partition of $[0, 1]$. Since $m_1(f) = 0 = M_1(f)$, $m_2(f) = 0 < 1 = M_2(f)$, and $m_3(f) = 1 = M_3(f)$, it is easy to see that $U(f, P) - L(f, P) = \varepsilon$. Therefore, $f$ is integrable on $[0, 1]$. ∎

We have defined integrability, but not the value of the integral. We remedy this situation by using the Riemann sums $U(f, P)$ and $L(f, P)$ to define upper and lower integrals.

**5.13 DEFINITION.** Let $[a, b]$ be a closed, bounded, nondegenerate interval and $f : [a, b] \to \mathbf{R}$ be bounded.

i) The *upper integral* of $f$ on $[a, b]$ is the number

$$(U) \int_a^b f(x)\, dx := \inf\{U(f, P) : P \text{ is a partition of } [a, b]\}.$$

ii) The *lower integral* of $f$ on $[a, b]$ is the number

$$(L) \int_a^b f(x)\, dx := \sup\{L(f, P) : P \text{ is a partition of } [a, b]\}.$$

iii) If the upper and lower integrals of $f$ on $[a, b]$ are equal, we define the *integral* of $f$ on $[a, b]$ to be the common value

$$\int_a^b f(x)\, dx := (U) \int_a^b f(x)\, dx = (L) \int_a^b f(x)\, dx.$$

Although a bounded function might not be integrable (see Example 5.11 above), the following result shows that the upper and lower integrals of a bounded function always exist.

**5.14 Remark.** *If* $f : [a, b] \to \mathbf{R}$ *is bounded, then its upper and lower integrals exist and are finite, and satisfy*

$$(L) \int_a^b f(x)\,dx \leq (U) \int_a^b f(x)\,dx.$$

PROOF. By Remark 5.8, $L(f, P) \leq U(f, Q)$ for all partitions $P$ and $Q$ of $[a, b]$. Taking the supremum of this inequality over all partitions $P$ of $[a, b]$, we have

$$(L) \int_a^b f(x)\,dx \leq U(f, Q);$$

i.e., the lower integral exists and is finite. Taking the infimum of this last inequality over all partitions $Q$ of $[a, b]$, we conclude that the upper integral is also finite and greater than or equal to the lower integral. ∎

Suppose $f$ is bounded and nonnegative on $[a, b]$. Since the upper and lower sums of $f$ approximate the "area" of the region bounded by the curves $y = f(x)$, $y = 0$, $x = a$, and $x = b$, we guess that $f$ is integrable if and only if the upper and lower integrals of $f$ are equal. The following result shows this guess is true whether or not $f$ is nonnegative.

**5.15 THEOREM.** *Let* $[a, b]$ *be a closed, bounded, nondegenerate interval and* $f : [a, b] \to \mathbf{R}$ *be bounded. Then* $f$ *is integrable on* $[a, b]$ *if and only if*

$$(2) \qquad (L) \int_a^b f(x)\,dx = (U) \int_a^b f(x)\,dx.$$

PROOF. Suppose $f$ is integrable. Let $\varepsilon > 0$ and choose a partition $P$ of $[a, b]$ such that

$$(3) \qquad U(f, P) - L(f, P) < \varepsilon.$$

By definition, $(U) \int_a^b f(x)\,dx \leq U(f, P)$ and the opposite inequality holds for the lower integral and the lower sum $L(f, P)$. Therefore, it follows from Remark 5.14 and (3) that

$$\left| (U) \int_a^b f(x)\,dx - (L) \int_a^b f(x)\,dx \right| = (U) \int_a^b f(x)\,dx - (L) \int_a^b f(x)\,dx$$

$$\leq U(f, P) - L(f, P) < \varepsilon.$$

Since this is valid for all $\varepsilon > 0$, (2) holds as promised.

Conversely, suppose (2) holds. Let $\varepsilon > 0$ and choose, by the Approximation Property, partitions $P_1$ and $P_2$ of $[a, b]$ such that

$$U(f, P_1) < (U) \int_a^b f(x)\,dx + \frac{\varepsilon}{2}$$

and
$$\sup L(f, P).$$

$$L(f, P_2) > (L) \int_a^b f(x)\,dx - \frac{\varepsilon}{2}.$$

Set $P = P_1 \cup P_2$. Since $P$ is a refinement of both $P_1$ and $P_2$, it follows from Remark 5.7, the choices of $P_1$ and $P_2$, and (2) that

$$U(f, P) - L(f, P) \le U(f, P_1) - L(f, P_2)$$

$$\le (U) \int_a^b f(x)\,dx + \frac{\varepsilon}{2} - (L) \int_a^b f(x)\,dx + \frac{\varepsilon}{2} = \varepsilon. \ \blacksquare$$

Since the integral has been defined only on nondegenerate intervals $[a, b]$, we have tacitly assumed that $a < b$. We shall use the conventions

$$\int_b^a f(x)\,dx = - \int_a^b f(x)\,dx \quad \text{and} \quad \int_a^a f(x)\,dx = 0$$

to extend the integral to the cases $a > b$ and $a = b$. In particular, if $f(x)$ is integrable and nonpositive on $[a, b]$, then the area of the region bounded by the curves $y = f(x)$, $y = 0$, $x = a$, and $x = b$ is given by $\int_b^a f(x)\,dx$.

In the next section we shall use the machinery of upper and lower sums to prove several familiar theorems about the Riemann integral. We close this section with one more result which reinforces the connection between integration and area.

**5.16 THEOREM.** *If $f(x) = \alpha$ is constant on $[a, b]$, then*

$$\int_a^b f(x)\,dx = \alpha(b - a).$$

PROOF. By Theorem 5.10, $f$ is integrable on $[a, b]$. Hence, it follows from Theorem 5.15 and Remark 5.5 that

$$\int_a^b f(x)\,dx = (U) \int_a^b f(x)\,dx = \inf_P U(f, P) = \alpha(b - a). \ \blacksquare$$


## EXERCISES

**1.** For each of the following, compute $U(f, P)$, $L(f, P)$, and $\int_0^1 f(x)\,dx$, where

$$P = \left\{ 0, \frac{2}{5}, \frac{1}{2}, \frac{3}{5}, 1 \right\}.$$

Find out whether the lower sum or the upper sum is a better approximation to the integral. Graph $f$ and explain why this is so.

 a) $f(x) = 1 - x^2$.
 b) $f(x) = \sin x$.
 c) $f(x) = x^2 - x$.

**2.** a) Prove that for each $n \in \mathbf{N}$,

$$P_n := \left\{ \frac{j}{n} : j = 0, 1, \ldots, n \right\}$$

is a partition of $[0, 1]$.

b) For each of the following functions, use Exercise 1 in Section 1.2 to compute the upper and lower sums $U(f, P_n)$, $L(f, P_n)$ and show directly that

(*)                     $$\lim_{n \to \infty} L(f, P_n) = \lim_{n \to \infty} U(f, P_n).$$

$\alpha)$                                 $f(x) = x.$

$\beta)$                                 $f(x) = x^2.$

$\gamma)$                     $f(x) = \begin{cases} 1 & 0 \le x < 1/2 \\ 2 & 1/2 \le x \le 1. \end{cases}$

c) Prove that any bounded $f$ which satisfies (*) is integrable on $[0, 1]$.

**3.** Let $E = \{1/n : n \in \mathbf{N}\}$. Prove that the function

$$f(x) = \begin{cases} 1 & x \in E \\ 0 & \text{otherwise} \end{cases}$$

is integrable on $[0, 1]$. What is the value of $\int_0^1 f(x)\, dx$?

**4.** **This exercise is used in Section** $^e$**14.2.** Suppose $[a, b]$ is a closed, bounded, nondegenerate interval and $f : [a, b] \to \mathbf{R}$ is bounded.

a) Prove that if $f$ is continuous at $x_0 \in [a, b]$ and $f(x_0) \neq 0$, then

(L) $$\int_a^b |f(x)|\, dx > 0.$$

b) Show that if $f$ is continuous on $[a, b]$, then $\int_a^b |f(x)|\, dx = 0$ if and only if $f(x) = 0$ for all $x \in [a, b]$.

c) Does part b) hold if the absolute values are removed? If it does, prove it. If it does not, provide a counterexample.

**5.** Suppose $f$ is continuous on a nondegenerate interval $[a, b]$. Show that

$$\int_a^c f(x)\, dx = 0$$

for all $c \in [a, b]$ if and only if $f(x) = 0$ for all $x \in [a, b]$. (Compare with Exercise

4, and notice that $f$ need not be nonnegative here.)

**6.** Let $f$ be integrable on $[a, b]$ and $E$ be a finite subset of $[a, b]$. Show that if $g$ is a bounded function which satisfies $g(x) = f(x)$ for all $x \in [a, b] \setminus E$, then $g$ is integrable on $[a, b]$ and

$$\int_a^b g(x)\, dx = \int_a^b f(x)\, dx.$$

**7** . **This exercise is used in Section 12.3.** Let $f, g$ be bounded on $[a, b]$.

a) Prove

$$(U) \int_a^b (f(x) + g(x))\, dx \le (U) \int_a^b f(x)\, dx + (U) \int_a^b g(x)\, dx$$

and

$$(L) \int_a^b (f(x) + g(x))\, dx \ge (L) \int_a^b f(x)\, dx + (L) \int_a^b g(x)\, dx.$$

b) Prove

$$(U) \int_a^b f(x)\, dx = (U) \int_a^c f(x)\, dx + (U) \int_c^b f(x)\, dx$$

and

$$(L) \int_a^b f(x)\, dx = (L) \int_a^c f(x)\, dx + (L) \int_c^b f(x)\, dx$$

for $a < c < b$.

**8** . **This exercise is used in Sections $^e$5.5 and $^e$7.5.** Suppose $f : [a, b] \to \mathbf{R}$.

a) If $f$ is increasing on $[a, b]$ and $P = \{x_0, \ldots, x_n\}$ is any partition of $[a, b]$, prove that

$$\sum_{j=1}^n (M_j(f) - m_j(f))(x_j - x_{j-1}) \le (f(b) - f(a)) \|P\|.$$

b) Prove that if $f$ is monotone on $[a, b]$, then $f$ is integrable on $[a, b]$.
   [*Note: By Theorem 4.29, $f$ has only countably many (i.e., relatively few) discontinuities on $[a, b]$. This has nothing to do with the proof of part b), but points out a general principle which will be discussed in Section 9.5.*]

**9.** Let $f$ be bounded on a nondegenerate interval $[a, b]$. Prove that $f$ is integrable on $[a, b]$ if and only if given $\varepsilon > 0$ there is a partition $P_\varepsilon$ of $[a, b]$ such that

$$P \supseteq P_\varepsilon \quad \text{implies} \quad |U(f, P) - L(f, P)| < \varepsilon.$$

## 5.2  RIEMANN SUMS

There is another definition of the Riemann integral frequently found in elementary calculus texts.

**5.17 DEFINITION.** Let $P = \{x_0, x_1, \ldots, x_n\}$ be a partition of $[a, b]$ and $f : [a, b] \to \mathbf{R}$.

i) A *Riemann sum* of $f$ with respect to $P$ is a sum of the form

$$\sum_{j=1}^{n} f(t_j)(x_j - x_{j-1}),$$

where the choice of $t_j \in [x_{j-1}, x_j]$ is arbitrary.

ii) The Riemann sums of $f$ are said to *converge* to $I(f)$ as $\|P\| \to 0$ if given $\varepsilon > 0$ there is a partition $P_\varepsilon$ of $[a, b]$ such that

$$P \supseteq P_\varepsilon \quad \text{implies} \quad \left| \sum_{j=1}^{n} f(t_j)(x_j - x_{j-1}) - I(f) \right| < \varepsilon$$

for all choices of $t_j \in [x_{j-1}, x_j]$, $j = 1, 2, \ldots, n$. In this case we shall use the notation

$$I(f) = \lim_{\|P\| \to 0} \sum_{j=1}^{n} f(t_j)(x_j - x_{j-1}).$$

The following result shows that for bounded functions this definition of the Riemann integral is the same as the one using upper and lower integrals.

**5.18 THEOREM.** *Let $[a, b]$ be a closed, bounded, nondegenerate interval and $f : [a, b] \to \mathbf{R}$ be bounded. Then $f$ is Riemann integrable on $[a, b]$ if and only if*

$$I(f) = \lim_{\|P\| \to 0} \sum_{j=1}^{n} f(t_j)(x_j - x_{j-1})$$

*exists, in which case*

$$I(f) = \int_a^b f(x)\, dx.$$

**PROOF.** Suppose $f$ is integrable on $[a, b]$ and $\varepsilon > 0$. By the Approximation Property, there is a partition $P_\varepsilon$ of $[a, b]$ such that

$$(4) \qquad L(f, P_\varepsilon) > \int_a^b f(x)\, dx - \varepsilon \quad \text{and} \quad U(f, P_\varepsilon) < \int_a^b f(x)\, dx + \varepsilon.$$

Let $P = \{x_0, x_1, \ldots, x_n\} \supseteq P_\varepsilon$. Then (4) holds with $P$ in place of $P_\varepsilon$. But $m_j(f) \le f(t_j) \le M_j(f)$ for any choice of $t_j \in [x_{j-1}, x_j]$. Hence,

$$\int_a^b f(x)\, dx - \varepsilon < L(f, P) \le \sum_{j=1}^{n} f(t_j)(x_j - x_{j-1}) \le U(f, P) < \int_a^b f(x)\, dx + \varepsilon.$$

Thm: Let f be a function which is bounded on [a,b]. Then f is Riemann integrable iff f is continuous at almost every point in [a,b].; ie the set of points of discontinuity has measure zero.

In particular,

$$\left| \sum_{j=1}^{n} f(t_j)(x_j - x_{j-1}) - \int_a^b f(x)\, dx \right| < \varepsilon$$

for all partitions $P \supseteq P_\varepsilon$ and all choices of $t_j \in [x_{j-1}, x_j]$, $j = 1, 2, \ldots, n$.

Conversely, suppose the Riemann sums of $f$ converge to $I(f)$. Let $\varepsilon > 0$ and choose a partition $P = \{x_0, x_1, \ldots, x_n\}$ of $[a, b]$ such that

$$(5) \qquad \left| \sum_{j=1}^{n} f(t_j)(x_j - x_{j-1}) - I(f) \right| < \frac{\varepsilon}{3}$$

for all choices of $t_j \in [x_{j-1}, x_j]$. By the Approximation Property, choose $t_j, u_j \in [x_{j-1}, x_j]$ such that

$$f(t_j) - f(u_j) > M_j(f) - m_j(f) - \frac{\varepsilon}{3(b-a)}.$$

By (5) and telescoping, we have

$$U(f, P) - L(f, P) = \sum_{j=1}^{n} (M_j(f) - m_j(f))(x_j - x_{j-1})$$

$$< \sum_{j=1}^{n} (f(t_j) - f(u_j))(x_j - x_{j-1}) + \frac{\varepsilon}{3(b-a)} \sum_{j=1}^{n} (x_j - x_{j-1})$$

$$\leq \left| \sum_{j=1}^{n} f(t_j)(x_j - x_{j-1}) - I(f) \right|$$

$$+ \left| I(f) - \sum_{j=1}^{n} f(u_j)(x_j - x_{j-1}) \right| + \frac{\varepsilon}{3(b-a)} \sum_{j=1}^{n} (x_j - x_{j-1})$$

$$< \frac{2\varepsilon}{3} + \frac{\varepsilon}{3} = \varepsilon.$$

Therefore, $f$ is integrable on $[a, b]$. ∎

The next two results show that Riemann integrals of complicated functions can be broken into simpler pieces.

**5.19 THEOREM** [LINEAR PROPERTY]. If $f, g$ are integrable on $[a, b]$ and $\alpha \in \mathbf{R}$, then $f + g$ and $\alpha f$ are integrable on $[a, b]$. In fact,

$$(6) \qquad \int_a^b (f(x) + g(x))\, dx = \int_a^b f(x)\, dx + \int_a^b g(x)\, dx$$

and

$$(7) \qquad \int_a^b (\alpha f(x))\, dx = \alpha \int_a^b f(x)\, dx.$$

ie. - All finite sets have measure zero.
- The set of rationals has measure zero ( any countable set)
- The set of irrationals does not have measure zero

PROOF. Let $\varepsilon > 0$ and choose $P_\varepsilon$ such that for any partition $P = \{x_0, x_1, \ldots, x_n\} \supseteq P_\varepsilon$ of $[a, b]$ and any choice of $t_j \in [x_{j-1}, x_j]$, we have

$$\left| \sum_{j=1}^n f(t_j)(x_j - x_{j-1}) - \int_a^b f(x)\,dx \right| < \frac{\varepsilon}{2}$$

and

$$\left| \sum_{j=1}^n g(t_j)(x_j - x_{j-1}) - \int_a^b g(x)\,dx \right| < \frac{\varepsilon}{2}.$$

By the triangle inequality,

$$\left| \sum_{j=1}^n f(t_j)(x_j - x_{j-1}) + \sum_{j=1}^n g(t_j)(x_j - x_{j-1}) - \int_a^b f(x)\,dx - \int_a^b g(x)\,dx \right| < \varepsilon$$

for any choice of $t_j \in [x_{j-1}, x_j]$. Hence, (6) follows directly from Theorem 5.18.
   Similarly, if $P_\varepsilon$ is chosen so that

$$P \supseteq P_\varepsilon \quad \text{implies} \quad \left| \sum_{j=1}^n f(t_j)(x_j - x_{j-1}) - \int_a^b f(x)\,dx \right| < \frac{\varepsilon}{|\alpha| + 1},$$

then

$$\left| \sum_{j=1}^n \alpha f(t_j)(x_j - x_{j-1}) - \alpha \int_a^b f(x)\,dx \right| < |\alpha| \frac{\varepsilon}{|\alpha| + 1} < \varepsilon$$

for any choice of $t_j \in [x_{j-1}, x_j]$. We conclude by Theorem 5.18 that (7) holds. ∎

**5.20 THEOREM.** *If $f$ is integrable on $[a, b]$, then $f$ is integrable on each subinterval $[c, d]$ of $[a, b]$. Moreover,*

(8)
$$\int_a^b f(x)\,dx = \int_a^c f(x)\,dx + \int_c^b f(x)\,dx$$

*for all $c \in (a, b)$.*

PROOF. Let $\varepsilon > 0$ and choose a partition $P$ of $[a, b]$ such that

(9)
$$U(f, P) - L(f, P) < \varepsilon.$$

Let $P' = P \cup \{c\}$ and $P_1 = P \cap [a, c]$. Since $P_1$ is a partition of $[a, c]$ and $P'$ is a refinement of $P$, we have by (9) that

$$U(f, P_1) - L(f, P_1) \le U(f, P') - L(f, P') \le U(f, P) - L(f, P) < \varepsilon.$$

Therefore, $f$ is integrable on $[a, c]$. A similar argument proves that $f$ is integrable on any subinterval $[c, d]$ of $[a, b]$.

For $a, b, c \in \mathbb{R}$, if $f$ is integrable on $[a, c]$ then $\int_a^b f(x)\,dx = \int_a^c f(x)\,dx + \int_c^b f(x)\,dx$

To verify (8), suppose $P$ is any partition of $[a, b]$. Let $P_0 = P \cup \{c\}$, $P_1 = P_0 \cap [a, c]$, and $P_2 = P_0 \cap [c, b]$. Then $P_0 = P_1 \cup P_2$ and by definition

$$U(f, P) \geq U(f, P_0) = U(f, P_1) + U(f, P_2)$$

$$\geq (U) \int_a^c f(x) \, dx + (U) \int_c^b f(x) \, dx = \int_a^c f(x) \, dx + \int_c^b f(x) \, dx.$$

(This last equality follows from the fact that $f$ is integrable on both $[a, c]$ and $[c, b]$.) Taking the infimum of

$$U(f, P) \geq \int_a^c f(x) \, dx + \int_c^b f(x) \, dx$$

over all partitions $P$ of $[a, b]$, we obtain

$$\int_a^b f(x) \, dx = (U) \int_a^b f(x) \, dx \geq \int_a^c f(x) \, dx + \int_c^b f(x) \, dx.$$

A similar argument using lower integrals shows

$$\int_a^b f(x) \, dx \leq \int_a^c f(x) \, dx + \int_c^b f(x) \, dx. \quad \blacksquare$$

Using the conventions

$$\int_a^b f(x) \, dx = - \int_b^a f(x) \, dx \quad \text{and} \quad \int_a^a f(x) \, dx = 0,$$

it is easy to see that (8) holds whether or not $c$ lies between $a$ and $b$, provided $f$ is integrable on the union of these intervals (see Exercise 4).

**5.21 THEOREM** [COMPARISON THEOREM FOR INTEGRALS]. *If $f, g$ are integrable on $[a, b]$ and $f(x) \leq g(x)$ for all $x \in [a, b]$, then*

$$\int_a^b f(x) \, dx \leq \int_a^b g(x) \, dx.$$

*In particular, if $m \leq f(x) \leq M$ for $x \in [a, b]$, then*

$$m(b - a) \leq \int_a^b f(x) \, dx \leq M(b - a).$$

PROOF. Let $P$ be a partition of $[a, b]$. By hypothesis, $M_j(f) \leq M_j(g)$ whence $U(f, P) \leq U(g, P)$. It follows that

$$\int_a^b f(x) \, dx = (U) \int_a^b f(x) \, dx \leq U(g, P)$$

for all partitions $P$ of $[a, b]$. Taking the infimum of this inequality over all partitions $P$ of $[a, b]$, we obtain

$$\int_a^b f(x) \, dx \leq \int_a^b g(x) \, dx.$$

If $m \leq f(x) \leq M$, then (by what we just proved and by Theorem 5.16)

$$m(b - a) = \int_a^b m \, dx \leq \int_a^b f(x) \, dx \leq \int_a^b M \, dx = M(b - a). \quad \blacksquare$$

We shall use the following result nearly every time we need to estimate an integral.

*( If |f| is improperly integrable, then f is improperly integrable )*

**5.22 THEOREM.** *If f is (Riemann) integrable on $[a,b]$, then $|f|$ is integrable on $[a,b]$ and*

$$\left| \int_a^b f(x)\, dx \right| \le \int_a^b |f(x)|\, dx.$$

PROOF. Let $P = \{x_0, x_1, \ldots, x_n\}$ be a partition of $[a,b]$. We claim that

(10) $$M_j(|f|) - m_j(|f|) \le M_j(f) - m_j(f)$$

holds for $j = 1, 2, \ldots, n$. Indeed, let $x, y \in [x_{j-1}, x_j]$. If $f(x)$, $f(y)$ have the same sign, say both are nonnegative, then

$$|f(x)| - |f(y)| = f(x) - f(y) \le M_j(f) - m_j(f).$$

If $f(x)$, $f(y)$ have opposite signs, say $f(x) \ge 0 \ge f(y)$, then $m_j(f) \le 0$, hence

$$|f(x)| - |f(y)| = f(x) + f(y) \le M_j(f) + 0 \le M_j(f) - m_j(f).$$

Thus (10) holds in any event.

Let $\varepsilon > 0$ and choose a partition $P$ of $[a,b]$ such that $U(f, P) - L(f, P) < \varepsilon$. Since (10) implies $U(|f|, P) - L(|f|, P) \le U(f, P) - L(f, P)$, it follows that

$$U(|f|, P) - L(|f|, P) < \varepsilon.$$

Thus $|f|$ is integrable on $[a,b]$. Since $-|f(x)| \le f(x) \le |f(x)|$ holds for any $x \in [a,b]$, we conclude by Theorem 5.21 that

$$-\int_a^b |f(x)|\, dx \le \int_a^b f(x)\, dx \le \int_a^b |f(x)|\, dx. \quad \blacksquare$$

By Theorem 5.19, the sum of integrable functions is integrable. What about the product?      *( not true for improperly integrable functions f and g.)*

**5.23 COROLLARY.** *If f and g are (Riemann) integrable on $[a,b]$, then so is $fg$.*

PROOF. Suppose for a moment that the square of any integrable function is integrable. Then, by hypothesis, $f^2$, $g^2$, and $(f+g)^2$ are integrable on $[a,b]$. Since

$$fg = \frac{(f+g)^2 - f^2 - g^2}{2},$$

it follows from Theorem 5.19 that $fg$ is integrable on $[a,b]$.

It remains to prove that $f^2$ is integrable on $[a,b]$. Since $M_j(f^2) = (M_j(|f|))^2$ and $m_j(f^2) = (m_j(|f|))^2$, it is clear that

$$\begin{aligned} M_j(f^2) - m_j(f^2) &= (M_j(|f|))^2 - (m_j(|f|))^2 \\ &= (M_j(|f|) + m_j(|f|))(M_j(|f|) - m_j(|f|)) \\ &\le 2M(M_j(|f|) - m_j(|f|)), \end{aligned}$$

where $M = \sup_{x \in [a,b]} |f(x)|$. Multiplying this inequality by $(x_j - x_{j-1})$ and summing over $j = 1, 2, \ldots, n$, we have

$$U(f^2, P) - L(f^2, P) \le 2M(U(|f|, P) - L(|f|, P)).$$

Hence, it follows from Theorem 5.22 that $f^2$ is integrable on $[a,b]$. $\blacksquare$

We close this section with two integral analogues of the Mean Value Theorem.

**5.24 THEOREM** [First Mean Value Theorem for Integrals]. *Suppose that $f$ and $g$ are integrable on $[a, b]$ with $g(x) \geq 0$ for all $x \in [a, b]$. If*

$$m = \inf_{x \in [a,b]} f(x) \quad and \quad M = \sup_{x \in [a,b]} f(x),$$

*then there is a number $c \in [m, M]$ such that*

$$\int_a^b f(x)g(x)\, dx = c \int_a^b g(x)\, dx.$$

*In particular, if $f$ is continuous on $[a, b]$, then there is an $x_0 \in [a, b]$ which satisfies*

$$\int_a^b f(x)g(x)\, dx = f(x_0) \int_a^b g(x)\, dx.$$

PROOF. Since $g \geq 0$ on $[a, b]$, Theorem 5.21 implies

$$m \int_a^b g(x)\, dx \leq \int_a^b f(x)g(x)\, dx \leq M \int_a^b g(x)\, dx.$$

If $\int_a^b g(x)\, dx = 0$, then $\int_a^b f(x)g(x)\, dx = 0$ and there is nothing to prove. Otherwise, set

$$c = \frac{\int_a^b f(x)g(x)\, dx}{\int_a^b g(x)\, dx}$$

and note that $c \in [m, M]$. If $f$ is continuous, then (by the Intermediate Value Theorem) we can choose $x_0 \in [a, b]$ such that $f(x_0) = c$. ∎

Before we state the Second Mean Value Theorem we introduce an idea which will be used in the next section to prove the Fundamental Theorem of Calculus. If $f$ is integrable on $[a, b]$, then $f$ can be used to define a new function

$$F(x) := \int_a^x f(t)\, dt, \qquad x \in [a, b].$$

**5.25 Example.** Find $F(x) = \int_0^x f(t)\, dt$ if

$$f(x) = \begin{cases} 1 & x \geq 0 \\ -1 & x < 0. \end{cases}$$

SOLUTION. By Theorem 5.16,

$$F(x) = \int_0^x f(t)\, dt = \begin{cases} x & x \geq 0 \\ -x & x < 0. \end{cases}$$

Hence, $F(x) = |x|$. ∎

Notice in Example 5.25 that the integral $F$ of $f$ is continuous even though $f$ itself is not. The following result shows this is a general principle.

**5.26 THEOREM.** *If $f$ is (Riemann) integrable on $[a, b]$, then $F(x) = \int_a^x f(t)\,dt$ exists and is continuous on $[a, b]$.*

PROOF. By Theorem 5.20, $F(x)$ exists for all $x \in [a, b]$. To prove that $F$ is continuous on $[a, b]$, it suffices to show that $F(x+) = F(x)$ for all $x \in [a, b)$ and $F(x-) = F(x)$ for all $x \in (a, b]$. Fix $x_0 \in [a, b)$. By definition, $f$ is bounded on $[a, b]$. Thus, choose $M \in \mathbf{R}$ such that $|f(t)| \le M$ for all $t \in [a, b]$. Let $\varepsilon > 0$ and set $\delta = \varepsilon/M$. If $0 \le x - x_0 < \delta$, then by Theorem 5.22,

$$|F(x) - F(x_0)| = \left| \int_{x_0}^x f(t)\,dt \right| \le \int_{x_0}^x |f(t)|\,dt \le M|x - x_0| < \varepsilon.$$

Hence, $F(x_0+) = F(x_0)$. A similar argument shows that $F(x_0-) = F(x_0)$ for all $x_0 \in (a, b]$ ∎

**5.27 THEOREM** [SECOND MEAN VALUE THEOREM FOR INTEGRALS]. *Suppose $f, g$ are integrable on $[a, b]$ with $f$ increasing and $g$ nonnegative on $[a, b]$. If $m \le f(a+)$ and $M \ge f(b-)$, then there is an $x_0 \in [a, b]$ such that*

$$\int_a^b f(x)g(x)\,dx = m \int_a^{x_0} g(x)\,dx + M \int_{x_0}^b g(x)\,dx.$$

*In particular, if $f$ is also nonnegative on $[a, b]$, then there is an $x_0 \in [a, b]$ which satisfies*

$$\int_a^b f(x)g(x)\,dx = M \int_{x_0}^b g(x)\,dx.$$

PROOF. Let

$$F(x) = m \int_a^x g(t)\,dt + M \int_x^b g(t)\,dt$$

for $x \in [a, b]$, and observe by Theorem 5.26 that $F$ is continuous on $[a, b]$. Since $f$ is increasing and $g$ is nonnegative, we have $mg(t) \le f(t)g(t) \le Mg(t)$ for all $t \in (a, b)$. Hence, the Comparison Theorem (Theorem 5.21) implies

$$F(b) = m \int_a^b g(t)\,dt \le \int_a^b f(t)g(t)\,dt \le M \int_a^b g(t)\,dt = F(a).$$

Since $F$ is continuous, it follows from the Intermediate Value Theorem that there is an $x_0 \in (a, b)$ such that

$$F(x_0) = \int_a^b f(t)g(t)\,dt.$$

This proves the first statement. The second statement follows from the first since we may use $m = 0$ when $f \ge 0$. ∎

When $g(x) = 1$ and $f(x) \ge 0$, these mean value theorems have simple geometric interpretations. Indeed, let $A$ represent the area bounded by the curves $y = f(x)$,

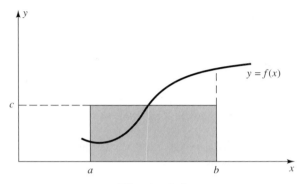

**Figure 5.3**

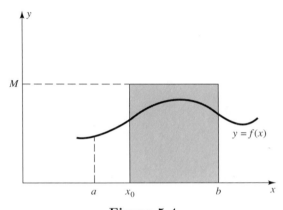

**Figure 5.4**

$y = 0$, $x = a$, and $x = b$. By the First Mean Value Theorem, there is a $c \in (m, M)$ such that the area of the rectangle of height $c$ and base $b - a$ equals $A$ (see Figure 5.3). And by the Second Mean Value Theorem, given $M \geq f(b-)$ there is an $x_0 \in [a, b]$ such that the area of the rectangle of height $M$ and base $b - x_0$ equals $A$ (see Figure 5.4).

## EXERCISES

1. Using the connection between integrals and area, evaluate each of the following integrals.

a)
$$\int_0^1 |x - 0.5| \, dx.$$

b)
$$\int_0^a \sqrt{a^2 - x^2} \, dx, \qquad a > 0.$$

c)
$$\int_{-2}^{2} (|x+1| + |x|)\, dx.$$

d)
$$\int_{a}^{b} (3x+1)\, dx, \qquad a < b.$$

**2.** Prove that if $f$ and $g$ are integrable on $[a,b]$, then so are $f \vee g$ and $f \wedge g$ (see Exercise 9 in Section 3.1).

**3.** Prove that if $f$ is integrable on $[0,1]$ and $\beta > 0$, then

$$\lim_{n \to \infty} n^{\alpha} \int_{0}^{1/n^{\beta}} f(x)\, dx = 0$$

for all $\alpha < \beta$.

**4.** Suppose that $a < b < c$ and $f$ is integrable on $[a,c]$. Prove

$$\int_{a}^{b} f(x)\, dx = \int_{a}^{c} f(x)\, dx + \int_{c}^{b} f(x)\, dx.$$

**5. a)** Suppose that $g_n \geq 0$ is a sequence of integrable functions which satisfies

$$\lim_{n \to \infty} \int_{a}^{b} g_n(x)\, dx = 0.$$

Show that if $f : [a,b] \to \mathbf{R}$ is integrable on $[a,b]$, then

$$\lim_{n \to \infty} \int_{a}^{b} f(x)g_n(x)\, dx = 0.$$

**b)** Prove that if $f$ is integrable on $[0,1]$, then

$$\lim_{n \to \infty} \int_{0}^{1} x^n f(x)\, dx = 0.$$

**6. a)** Prove that if $f$ is integrable on $[0,1]$, then

$$\int_{0}^{1} f(x)\, dx = \lim_{n \to \infty} \sum_{k=0}^{n} \int_{1/2^{k+1}}^{1/2^{k}} f(x)\, dx.$$

**b)** Suppose $f$ is integrable on $[a,b]$, $x_0 = a$, and $x_n$ is a sequence of numbers in $[a,b]$ such that $x_n \uparrow b$ as $n \to \infty$. Prove

$$\int_{a}^{b} f(x)\, dx = \lim_{n \to \infty} \sum_{k=0}^{n} \int_{x_k}^{x_{k+1}} f(x)\, dx.$$

**7.** Let $f : [a, b] \to \mathbf{R}$, $a = x_0 < x_1 < \cdots < x_n = b$, and suppose that $f(x_k+)$ exists and is finite for $k = 0, 1, \ldots, n-1$ and $f(x_k-)$ exists and is finite for $k = 1, \ldots, n$. Show that if $f$ is continuous on each subinterval $(x_{k-1}, x_k)$, then $f$ is integrable on $[a, b]$ and

$$\int_a^b f(x)\, dx = \sum_{k=1}^n \int_{x_{k-1}}^{x_k} f(x)\, dx.$$

**8.** Let $f$ be continuous on a closed, bounded, nondegenerate interval $[a, b]$ and set

$$M = \sup_{x \in [a,b]} |f(x)|.$$

a) Prove that if $M > 0$, then for every $\varepsilon > 0$ there is a nondegenerate interval $I \subset [a, b]$ such that

$$(M - \varepsilon)^n |I| \leq \int_a^b |f(x)|^n\, dx \leq M^n (b - a).$$

b) Prove

$$\lim_{n \to \infty} \left( \int_a^b |f(x)|^n\, dx \right)^{1/n} = M.$$

## 5.3   THE FUNDAMENTAL THEOREM OF CALCULUS

Let $f$ be integrable on $[a, b]$ and $F(x) = \int_0^x f(t)\, dt$. By Theorem 5.26, $F$ is continuous on $[a, b]$. The next result shows that if $f$ is continuous, then $F$ is continuously differentiable.

**5.28 THEOREM** [FUNDAMENTAL THEOREM OF CALCULUS]. *Let $[a, b]$ be a closed, bounded, nondegenerate interval and $f : [a, b] \to \mathbf{R}$.*

i) *If $f$ is continuous on $[a, b]$ and $F(x) = \int_a^x f(t)\, dt$, then $F \in \mathcal{C}^1[a, b]$ and*

$$\frac{d}{dx} \int_a^x f(t)\, dt := F'(x) = f(x)$$

*for each $x \in [a, b]$.*

ii) *If $f$ is differentiable on $[a, b]$ and $f'$ is integrable on $[a, b]$, then*

$$\int_a^x f'(t)\, dt = f(x) - f(a)$$

*for each $x \in [a, b]$.*

PROOF. Let

$$F(x) = \int_a^x f(t)\, dt, \qquad x \in [a, b].$$

By symmetry, it suffices to show that if $f(x_0+) = f(x_0)$ for some $x_0 \in [a, b)$, then

(11) $$F'(x_0) = \lim_{h \to 0+} \frac{F(x_0 + h) - F(x_0)}{h} = f(x_0)$$

*[handwritten: if f is diff. on (a, b), then $f'(a) = \lim_{h \to 0^+} \frac{f(a+h) - f(h)}{h}$, and $f'(b) = \lim_{h \to 0^-} \frac{f(b+h) - f(h)}{h}$]*

(see Definition 4.6). Let $\varepsilon > 0$ and choose a $\delta > 0$ such that $x_0 \le t < x_0 + \delta$ implies $|f(t) - f(x_0)| < \varepsilon$. Fix $0 < h < \delta$. Notice that by Theorem 5.20,

*[handwritten: Since f is cont.]*

*[handwritten: $F'(x_0^+)$]* $$\frac{F(x_0 + h) - F(x_0)}{h} = \frac{\int_{x_0}^{x_0+h} f(t)\, dt}{h}$$

and that by Theorem 5.16,

$$f(x_0) = \frac{1}{h} \int_{x_0}^{x_0+h} f(x_0)\, dt.$$

Therefore,

$$\left| \frac{F(x_0 + h) - F(x_0)}{h} - f(x_0) \right| = \frac{1}{h} \left| \int_{x_0}^{x_0+h} (f(t) - f(x_0))\, dt \right| \le \frac{1}{h} \int_{x_0}^{x_0+h} |f(t) - f(x_0)|\, dt$$

*[handwritten: $< \frac{1}{h} \int_{x_0}^{x_0+h} \varepsilon \, dt = \frac{1}{h} \varepsilon (x_0 + h - x_0) = \varepsilon$]*

Since $0 < h < \delta$, it follows from Theorem 5.22 and the choice of $\delta$ that

$$\left| \frac{F(x_0 + h) - F(x_0)}{h} - f(x_0) \right| \le \frac{1}{h} \int_{x_0}^{x_0+h} |f(t) - f(x_0)|\, dt \le \varepsilon.$$

*[handwritten: f is cont.]*

This verifies (11) and the proof of part i) is complete.

To prove part ii) we may suppose that $x = b$. Let $\varepsilon > 0$. Since $f'$ is integrable, choose a partition $P = \{x_0, x_1, \ldots, x_n\}$ of $[a, b]$ such that

$$\left| \sum_{j=1}^n f'(t_j)(x_j - x_{j-1}) - \int_a^b f'(x)\, dx \right| < \varepsilon$$

*[handwritten: if f is continuous on $[x_{j-1}, x_j]$ and differentiable on $(x_{j-1}, x_j)$]*

for any choice of points $t_j \in [x_{j-1}, x_j]$. Use the Mean Value Theorem to choose points $t_j \in [x_{j-1}, x_j]$ such that $f(x_j) - f(x_{j-1}) = f'(t_j)(x_j - x_{j-1})$. It follows by telescoping that

*[handwritten: f is cont. since f is differentiable.]*

$$\left| f(b) - f(a) - \int_a^b f'(t)\, dt \right| = \left| \sum_{j=1}^n (f(x_j) - f(x_{j-1})) - \int_a^b f'(t)\, dt \right| < \varepsilon. \qquad \blacksquare$$

By the Fundamental Theorem of Calculus, integration is the inverse of differentiation in the following sense. If $f'$ is integrable, then

$$\int_a^b f'(x)\, dx = f(x) \Big|_a^b := f(b) - f(a).$$

In particular,

$$\int_a^b x^\alpha \, dx = \frac{x^{\alpha+1}}{\alpha+1} \, \Big|_a^b$$

for each $\alpha \geq 0$, and for each $\alpha < 0$, provided $\alpha \neq -1$ and $[a, b]$ does not contain 0 (see Exercise 8 in Section 4.2 and Exercise 5e) below). This result is sometimes called the *Power Rule*.

The Fundamental Theorem of Calculus can be used to evaluate many integrals.

**5.29 Example.** Find $\int_0^1 (3x - 2)^2 \, dx$.

SOLUTION. Since $(3x - 2)^2 = 9x^2 - 12x + 4$, we have by the Power Rule that

$$\int_0^1 (3x - 2)^2 \, dx = 3x^3 - 6x^2 + 4x \, \Big|_0^1 = 1. \quad \blacksquare$$

**5.30 Example.** Find $\int_0^{\pi/2} (1 + \sin x) \, dx$.

SOLUTION. Since $(\cos x)' = -\sin x$, we have by the Fundamental Theorem of Calculus that

$$\int_0^{\pi/2} (1 + \sin x) \, dx = x - \cos x \, \Big|_0^{\pi/2} = \frac{\pi}{2} + 1. \quad \blacksquare$$

Combining the Product Rule and the Fundamental Theorem of Calculus, we have another tool for evaluating integrals.

**5.31 THEOREM** [INTEGRATION BY PARTS]. *Suppose that $f, g$ are differentiable on $[a, b]$ with $f', g'$ integrable on $[a, b]$. Then*

$$\int_a^b f'(x)g(x) \, dx = f(b)g(b) - f(a)g(a) - \int_a^b f(x)g'(x) \, dx.$$

PROOF. By the Product Rule, $(f(x)g(x))' = f'(x)g(x) + f(x)g'(x)$ for $x \in [a, b]$. Since $f, g$ are continuous on $[a, b]$ and $f', g'$ are integrable on $[a, b]$, it follows that $(fg)'$ is a sum of products of integrable functions, hence, integrable on $[a, b]$. Thus, by the Fundamental Theorem of Calculus,

$$f(b)g(b) - f(a)g(a) = \int_a^b f'(x)g(x) \, dx + \int_a^b f(x)g'(x) \, dx. \quad \blacksquare$$

This rule is sometimes abbreviated as

$$\int u \, dv = uv - \int v \, du,$$

where it is understood that if $w = h(x)$ for some differentiable function $h$, then the *Leibnizian differential dw* is defined by $dw = h'(x) \, dx$.

Integration by parts can be used to reduce the exponent $n$ on an expression of the form $(ax + b)^n f(x)$ when $f$ is integrable.

**5.32 Example.** Find $\int_0^{\pi/2} x \sin x \, dx$.

SOLUTION. Let $u = x$ and $dv = \sin x \, dx$. Then $du = dx$ and $v = -\cos x$. Hence, by parts,

$$\int_0^{\pi/2} x \sin x = -x \cos x \Big|_0^{\pi/2} - \int_0^{\pi/2} (-\cos x) \, dx = \sin x \Big|_0^{\pi/2} = 1. \quad \blacksquare$$

Integration by parts is also very effective on integrals involving products of polynomials and logarithms.

**5.33 Example.** Find $\int_1^3 \log x \, dx$.

SOLUTION. Let $u = \log x$ and $dv = dx$. Then $du = dx/x$ and $v = x$. Hence, by parts,

$$\int_1^3 \log x \, dx = x \log x \Big|_1^3 - \int_1^3 dx = 3 \log 3 - 2. \quad \blacksquare$$

Complicated problems can frequently be reduced to simpler ones by changing variables. The following result shows how to change variables in a Riemann integral on **R**.

**5.34 THEOREM** [CHANGE OF VARIABLES]. *Let $\phi$ be continuously differentiable on a closed, bounded interval $[a, b]$. If*

(12) $\qquad\qquad\qquad$ *$f$ is continuous on $\phi([a, b])$,*

*or if*

(13) $\qquad$ *$\phi$ is strictly increasing on $[a, b]$ and $f$ is integrable on $[\phi(a), \phi(b)]$,*

*then*

(14) $$\int_{\phi(a)}^{\phi(b)} f(t) \, dt = \int_a^b f(\phi(x)) \phi'(x) \, dx.$$

PROOF. Suppose first that (12) holds. By the Fundamental Theorem of Calculus, the functions

$$G(x) := \int_a^x f(\phi(t)) \phi'(t) \, dt, \quad x \in [a, b], \quad \text{and} \quad F(u) := \int_{\phi(a)}^u f(t) \, dt, \quad u \in \phi([a, b]),$$

satisfy $G'(x) = f(\phi(x)) \phi'(x)$ and $F'(u) = f(u)$. Hence, by the Chain Rule,

$$\frac{d}{dx}(G(x) - F(\phi(x))) = 0$$

for all $x \in [a, b]$. It follows from Theorem 4.24ii) that $G(x) - F(\phi(x))$ is constant on $[a, b]$. Evaluation at $x = a$ shows that this constant is zero. Thus $G(x) = F(\phi(x))$ for all $x \in [a, b]$, in particular, when $x = b$. This proves (14) under hypothesis (12).

The theorem is more difficult to prove under hypotheses (13), but the idea behind the proof is simple.

STRATEGY: Since $\phi$ is increasing, if $P = \{x_0, x_1, \ldots, x_n\}$ is a partition of $[a, b]$, then $\widetilde{P} = \{\phi(x_0), \ldots, \phi(x_n)\}$ is a partition of $I := [\phi(a), \phi(b)]$. By the Intermediate Value Theorem, given $u_j \in [\phi(x_{j-1}), \phi(x_j)]$, there is an $s_j \in [x_{j-1}, x_j]$ such that $u_j = \phi(s_j)$. It follows from the Mean Value Theorem that

$$(15) \qquad \sum_{j=1}^{n} f(u_j)(\phi(x_j) - \phi(x_{j-1})) = \sum_{j=1}^{n} f(\phi(s_j))\phi'(c_j)(x_j - x_{j-1})$$

for some $c_j \in [x_{j-1}, x_j]$. Therefore, a Riemann sum of $f$ on $\widetilde{P}$ is almost a Riemann sum of $(f \circ \phi) \cdot \phi'$ on $P$. Since $\phi'$ is continuous, not much is changed if we replace $c_j$ by $s_j$, i.e., make the right side of (15) exactly a Riemann sum on $P$. Hence, the integral of $f$ on $I$ should equal the integral of $(f \circ \phi) \cdot \phi'$ on $[a, b]$. Here are the details.

Let $\varepsilon > 0$. Since $f$ is bounded, there is an $M \in (0, \infty)$ such that $|f(x)| \leq M$ for all $x \in I$. Since $\phi'$ is uniformly continuous on $[a, b]$, choose $\delta > 0$ such that

$$|\phi'(s_j) - \phi'(c_j)| < \frac{\varepsilon}{2M(b - a)},$$

i.e.,

$$(16) \qquad |f(\phi(s_j))\phi'(s_j) - f(\phi(s_j))\phi'(c_j)| < \frac{\varepsilon}{2(b - a)}$$

for all $s_j, c_j \in [a, b]$ with $|s_j - c_j| < \delta$.

Next, since $f$ is integrable on $I$, choose a partition $\widetilde{P}_\varepsilon = \{w_0, w_1, \ldots, w_m\}$ of $I$ such that if $\widetilde{P} = \{t_0, t_1, \ldots t_n\}$ is finer than $\widetilde{P}_\varepsilon$, then

$$(17) \qquad \left| \sum_{j=1}^{n} f(u_j)(t_j - t_{j-1}) - \int_{\phi(a)}^{\phi(b)} f(t)\, dt \right| < \frac{\varepsilon}{2}$$

for any $u_j \in [t_{j-1}, t_j]$. By Theorem 4.26, $\phi^{-1}$ exists, is continuous, and is strictly increasing on $I$. Hence,

$$P_\varepsilon = \{\phi^{-1}(w_0), \phi^{-1}(w_1), \ldots, \phi^{-1}(w_m)\}$$

is a partition of $[a, b]$ and, by making $\widetilde{P}_\varepsilon$ finer if necessary, we may suppose $\|P_\varepsilon\| < \delta$.

Let $P = \{x_0, x_1, \ldots, x_n\}$ be any partition of $[a, b]$ finer than $P_\varepsilon$ and let $s_j \in [x_{j-1}, x_j]$. For each $j \in \{1, \ldots, n\}$, set $u_j = \phi(s_j)$, $t_j = \phi(x_j)$, and choose $c_j \in [x_{j-1}, x_j]$ by the Mean Value Theorem such that

$$\phi(x_j) - \phi(x_{j-1}) = \phi'(c_j)(x_j - x_{j-1}).$$

Then $\widetilde{P} = \{t_0, t_1, \ldots, t_n\}$ is a partition of $I$ finer than $\widetilde{P}_\varepsilon$ and (17) holds. It follows from (15), (16), (17), and the choice of the $t_j$'s, that

$$\left| \sum_{j=1}^{n} f(\phi(s_j))\phi'(s_j)(x_j - x_{j-1}) - \int_{\phi(a)}^{\phi(b)} f(t)\,dt \right|$$

$$\leq \left| \sum_{j=1}^{n} (f(\phi(s_j))\phi'(s_j) - f(\phi(s_j))\phi'(c_j))(x_j - x_{j-1}) \right|$$

$$+ \left| \sum_{j=1}^{n} f(u_j)(\phi(x_j) - \phi(x_{j-1})) - \int_{\phi(a)}^{\phi(b)} f(t)\,dt \right|$$

$$< \frac{\varepsilon}{2} + \left| \sum_{j=1}^{n} f(u_j)(t_j - t_{j-1}) - \int_{\phi(a)}^{\phi(b)} f(t)\,dt \right| < \varepsilon$$

for any choice of $s_j \in [x_{j-1}, x_j]$. Hence, by Theorem 5.18, $(f \circ \phi) \cdot \phi'$ is integrable on $[a, b]$ and (14) holds. ∎

The Change of Variables Formula can be remembered as a substitution if we use the Leibnizian differentials introduced above: $t = \phi(x)$ implies $dt = \phi'(x)\,dx$.

The following example illustrates a typical application of the Change of Variables Formula.

**5.35 Example.** Find $\int_0^1 e^{\sqrt{x+1}}/\sqrt{x+1}\,dx$.

SOLUTION. Let $t = \sqrt{x+1}$ and observe that

$$dt = \frac{dx}{2\sqrt{x+1}}.$$

Therefore,

$$\int_0^1 \frac{e^{\sqrt{x+1}}}{\sqrt{x+1}}\,dx = 2\int_0^1 e^{\sqrt{x+1}}\frac{dx}{2\sqrt{x+1}} = 2\int_1^{\sqrt{2}} e^t\,dt = 2(e^{\sqrt{2}} - e). \quad ∎$$

Please notice that when changing variables, you must also change the limits of integration, e.g., from $x = 0$ to $t = \sqrt{0+1} = 1$.

It is interesting to note that hypothesis (12) does not require that $\phi$ be 1–1. This observation is used in the following example.

**5.36 Example.** Evaluate

$$\int_{-1}^{1} x f(x^2)\, dx$$

for any $f$ continuous on $[0, 1]$.

SOLUTION. Let $\phi(x) = x^2$ and observe that $f$ is continuous on $\phi([-1, 1]) = [0, 1]$. Hence, by Theorem 5.34,

$$\int_{-1}^{1} x f(x^2)\, dx = \frac{1}{2}\int_{-1}^{1} f(\phi(x))\phi'(x)\, dx = \frac{1}{2}\int_{1}^{1} f(t)\, dt = 0. \quad \blacksquare$$

**EXERCISES**

**1.** Compute each of the following integrals.

a)
$$\int_{-3}^{3} |x^2 + x - 2|\, dx.$$

b)
$$\int_{1}^{4} \frac{\sqrt{x} - 1}{\sqrt{x}}\, dx.$$

c)
$$\int_{0}^{1} (3x + 1)^{99}\, dx.$$

d)
$$\int_{1}^{e} x \log x\, dx.$$

e)
$$\int_{0}^{\pi/2} e^x \sin x\, dx.$$

f)
$$\int_{0}^{1} \sqrt{\frac{4x^2 - 4x + 1}{x^2 - x + 3}}\, dx.$$

**2.** Use the First Mean Value Theorem for Integrals to prove the following version of the Mean Value Theorem for Derivatives. If $f \in \mathcal{C}^1[a, b]$, then there is an $x_0 \in (a, b)$ such that

$$f(b) - f(a) = (b - a)f'(x_0).$$

**3. a)** If $f : [0, \infty) \to \mathbf{R}$ is continuous, find

$$\frac{d}{dx} \int_1^{x^2} f(t)\, dt.$$

**b)** If $h : \mathbf{R} \to \mathbf{R}$ is continuous, find

$$\frac{d}{dt} \int_{\cos t}^{t} h(x)\, dx.$$

**c)** If $g : \mathbf{R} \to \mathbf{R}$ is continuous, find

$$\frac{d}{dt} \int_0^t g(x - t)\, dx.$$

**d)** If $f(x) = \int_0^{x^3} e^{t^2}\, dt$, show

$$6 \int_0^1 x^2 f(x)\, dx - 2 \int_0^1 e^{x^2}\, dx = 1 - e.$$

**4.** **This exercise is used in Sections 5.4 and 6.1.** Define $L : (0, \infty) \to \mathbf{R}$ by

$$L(x) = \int_1^x \frac{dt}{t}.$$

**a)** Prove $L$ is differentiable and strictly increasing on $(0, \infty)$, with $L'(x) = 1/x$ and $L(1) = 0$.

**b)** Prove that $L(x) \to \infty$ as $x \to \infty$ and $L(x) \to -\infty$ as $x \to 0+$. (You may wish to prove

$$L(2^n) = \sum_{k=1}^{n} \int_{2^{k-1}}^{2^k} \frac{dt}{t} > \sum_{k=1}^{n} 2^{-k} \left(2^k - 2^{k-1}\right) = \frac{n}{2}$$

for all $n \in \mathbf{N}$.)

**c)** Using the fact that $(x^q)' = qx^{q-1}$ for $x > 0$ and $q \in \mathbf{Q}$ (see Exercise 8 in Section 4.2), prove $L(x^q) = qL(x)$ for all $q \in \mathbf{Q}$ and $x > 0$.

$L(x) = -L(\frac{1}{x})$ **d)** Prove $L(xy) = L(x) + L(y)$ for all $x, y \in (0, \infty)$.

**e)** Let $e = \lim_{n \to \infty} (1 + 1/n)^n$. Use L'Hôpital's Rule to show that $L(e) = 1$. ($L(x)$ is the *natural logarithm* function $\log x$.)

**5.** **This exercise was used in Section 4.3.** Let $E = L^{-1}$, where $L$ is defined in Exercise 4.

**a)** Use the Inverse Function Theorem to show that $E$ is differentiable and strictly increasing on $\mathbf{R}$ with $E'(x) = E(x)$, $E(0) = 1$, and $E(1) = e$.

**b)** Prove $E(x) \to \infty$ as $x \to \infty$ and $E(x) \to 0$ as $x \to -\infty$.

c) Prove $E(xq) = (E(x))^q$ and $E(q) = e^q$ for all $q \in \mathbf{Q}$ and $x \in \mathbf{R}$.

d) Prove $E(x + y) = E(x)E(y)$ for all $x, y \in \mathbf{R}$.

e) For each $\alpha \in \mathbf{R}$ define $e^\alpha = E(\alpha)$. Let $x > 0$ and define $x^\alpha = e^{\alpha \log x} :=$
$E(\alpha L(x))$. Prove that $0 < x < y$ implies $x^\alpha < y^\alpha$ for $\alpha > 0$ and $x^\alpha > y^\alpha$ for
$\alpha < 0$. Also prove

$$x^{\alpha + \beta} = x^\alpha x^\beta, \qquad x^{-\alpha} = \frac{1}{x^\alpha}, \quad \text{and} \quad (x^\alpha)' = \alpha x^{\alpha - 1}$$

for all $\alpha, \beta \in \mathbf{R}$ and $x > 0$.

**6.** a) Suppose $g$ is integrable and nonnegative on $[1, 3]$ with $\int_1^3 g(x)\, dx = 1$. Prove
that

$$\frac{1}{\pi} \int_1^9 g(\sqrt{x})\, dx < 2.$$

b) Suppose $h$ is integrable and nonnegative on $[1, 11]$ with $\int_1^{11} h(x)\, dx = 3$.
Prove that

$$\int_0^2 h(1 + 3x + 3x^2 - x^3)\, dx \leq 1.$$

**7.** Suppose $f : [a, b] \to \mathbf{R}$ is continuously differentiable and 1–1 on $[a, b]$. Prove
that

$$\int_a^b f(x)\, dx + \int_{f(a)}^{f(b)} f^{-1}(x)\, dx = bf(b) - af(a).$$

**8.** If $f$ is continuous on $[a, b]$ and there exist numbers $\alpha \neq \beta$ such that

$$\alpha \int_a^c f(x)\, dx + \beta \int_c^b f(x)\, dx = 0$$

holds for all $c \in (a, b)$, prove $f(x) = 0$ for all $x \in [a, b]$.

**9.** Let $0 \leq x \leq \pi/2$.

a) Use $0 \leq \cos x \leq 1$ and the Comparison Theorem for integrals to prove
$0 \leq \sin x \leq x$.

b) For each nonnegative integer $m$, set

$$s_m(x) := \sum_{k=0}^m \frac{(-1)^k x^{2k+1}}{(2k+1)!} \quad \text{and} \quad c_m(x) := \sum_{k=0}^m \frac{(-1)^k x^{2k}}{(2k)!}.$$

Prove[1]

$$s_{2n+1}(x) \leq \sin x \leq s_{2n}, \quad s_{2n+1} \leq \sin x \leq s_{2n+2},$$

$$c_{2n+1}(x) \leq \cos x \leq c_{2n}(x), \quad \text{and} \quad c_{2n+1}(x) \leq \cos x \leq c_{2n+2}(x)$$

hold for $n = 0, 1, 2, \cdots$.

---

[1]This exercise is due to Deng Bo ("A Simple Derivation of the Maclaurin Series for Sine and
Cosine," *American Mathematical Monthly*, vol. 97 (1990), 836.

**10.** Suppose $g$ is differentiable on $[a, b]$ and $g'$ is integrable on $[a, b]$.

a) Prove that if $f$ is continuously differentiable and increasing on $[a, b]$ and $g$ is positive on $(a, b)$ with $g(b) = g(a) = 0$, then

$$\int_a^b f(x)g'(x)\,dx = 0$$

if and only if $f$ is constant on $[a, b]$.

b) Show that a) is false if "$g$ is positive on $[a, b]$" is replaced by "$g$ is nonnegative on $[a, b]$ and positive on some subinterval of $(a, b)$."

$\boxed{11}$. **This exercise is used in Section 12.4.** Suppose $\phi$ is continuously differentiable on a closed, bounded interval $[a, b]$ with $\phi'(x) \neq 0$ for all $x \in [a, b]$. Prove that if $[c, d] = \phi([a, b])$ and $f$ is integrable on $[c, d]$, then

$$\int_c^d f(t)\,dt = \int_a^b f(\phi(x))\,|\phi'(x)|\,dx.$$

## 5.4  IMPROPER RIEMANN INTEGRATION

To extend the Riemann integral to unbounded intervals or unbounded functions, we begin with an elementary observation.

**5.37 Remark.** *If $f$ is integrable on $[a, b]$, then*

$$\int_a^b f(x)\,dx = \lim_{c \to a+} \left( \lim_{d \to b-} \int_c^d f(x)\,dx \right).$$

PROOF. By Theorem 5.26,

$$F(x) = \int_a^x f(t)\,dt$$

is continuous on $[a, b]$. Thus

$$\int_a^b f(x)\,dx = F(b) - F(a) = \lim_{c \to a+} (\lim_{d \to b-} (F(d) - F(c)))$$

$$= \lim_{c \to a+} \left( \lim_{d \to b-} \int_c^d f(x)\,dx \right). \ \blacksquare$$

This leads to the following generalization of the Riemann integral.

**5.38 DEFINITION.** Let $(a, b)$ be a nonempty, open (possibly unbounded) interval and $f : (a, b) \to \mathbf{R}$.

   i) $f$ is said to be *locally integrable* on $(a, b)$ if $f$ is integrable on each closed subinterval $[c, d]$ of $(a, b)$.

   ii) $f$ is said to be *improperly integrable* on $(a, b)$ if $f$ is locally integrable on $(a, b)$ and if

(18)
$$\int_a^b f(x)\, dx := \lim_{c \to a+} \left( \lim_{d \to b-} \int_c^d f(x)\, dx \right)$$

exists and is finite. This limit is called the *improper (Riemann) integral* of $f$ over $(a, b)$.

**5.39 Remark.** *The order of the limits in* (18) *does not matter. In particular, if the limit in* (18) *exists, then*

$$\int_a^b f(x)\, dx = \lim_{d \to b-} \left( \lim_{c \to a+} \int_c^d f(x)\, dx \right).$$

PROOF. Let $x_0 \in (a, b)$ be fixed. By Theorems 5.20 and 3.8,

$$\lim_{c \to a+} \left( \lim_{d \to b-} \int_c^d f(x)\, dx \right) = \lim_{c \to a+} \left( \int_c^{x_0} f(x)\, dx + \lim_{d \to b-} \int_{x_0}^d f(x)\, dx \right)$$

$$= \lim_{c \to a+} \int_c^{x_0} f(x)\, dx + \lim_{d \to b-} \int_{x_0}^d f(x)\, dx$$

$$= \lim_{d \to b-} \left( \lim_{c \to a+} \int_c^d f(x)\, dx \right). \quad \blacksquare$$

Thus we shall use the notation

$$\lim_{\substack{c \to a+ \\ d \to b-}} \int_c^d f(x)\, dx$$

to represent the limit in (18). If the integral is not improper at one of the endpoints, e.g., if $f$ is Riemann integrable on closed subintervals of $(a, b]$, we shall say that $f$ is improperly integrable on $(a, b]$ and simplify the notation even further by writing

$$\int_a^b f(x)\, dx = \lim_{c \to a+} \int_c^b f(x)\, dx.$$

The following example shows that an improperly integrable function need not be bounded.

**5.40 Example.** Show that $f(x) = 1/\sqrt{x}$ is improperly integrable on $(0, 1]$.

SOLUTION. By definition,

$$\int_0^1 \frac{1}{\sqrt{x}}\, dx = \lim_{a \to 0+} \int_a^1 \frac{1}{\sqrt{x}}\, dx = \lim_{a \to 0+} (2 - 2\sqrt{a}) = 2. \quad \blacksquare$$

The following example shows that a function can be improperly integrable on an unbounded interval.

**5.41 Example.** Show that $f(x) = 1/x^2$ is improperly integrable on $[1, \infty)$.

SOLUTION. By definition,

$$\int_1^\infty \frac{1}{x^2}\, dx = \lim_{d \to \infty} \int_1^d \frac{1}{x^2}\, dx = \lim_{d \to \infty} \left( 1 - \frac{1}{d} \right) = 1. \quad \blacksquare$$

Because an improper integral is a limit of Riemann integrals, many of the results we proved earlier in this chapter have analogues for the improper integral. The next two results illustrate this principle.

**5.42 THEOREM.** *If $f, g$ are improperly integrable on $(a, b)$ and $\alpha, \beta \in$* **R***, then $\alpha f + \beta g$ is improperly integrable on $(a, b)$ and*

$$\int_a^b (\alpha f(x) + \beta g(x))\, dx = \alpha \int_a^b f(x)\, dx + \beta \int_a^b g(x)\, dx.$$

PROOF. By Theorem 5.19 (the Linear Property for Riemann Integrals),

$$\int_c^d (\alpha f(x) + \beta g(x))\, dx = \alpha \int_c^d f(x)\, dx + \beta \int_c^d g(x)\, dx$$

for all $a < c < d < b$. Taking the limit as $c \to a+$ and $d \to b-$ finishes the proof. $\blacksquare$

**5.43 THEOREM** [COMPARISON THEOREM FOR IMPROPER INTEGRALS]. *Suppose $f, g$ are locally integrable on $(a, b)$. If $0 \le f(x) \le g(x)$ for $x \in (a, b)$, and $g$ is improperly integrable on $(a, b)$, then $f$ is improperly integrable on $(a, b)$ and*

$$\int_a^b f(x)\, dx \le \int_a^b g(x)\, dx.$$

PROOF. Fix $c \in (a, b)$. Let $F(d) = \int_c^d f(x)\, dx$ and $G(d) = \int_c^d g(x)\, dx$ for $d \in [c, b)$. By the Comparison Theorem for Integrals, $F(d) \le G(d)$. Since $f \ge 0$, the function $F$ is increasing on $[c, b]$, hence $F(b-)$ exists. Thus, by definition, $f$ is improperly integrable on $(c, b)$ and

$$\int_c^b f(x)\, dx = F(b-) \le G(b-) = \int_c^b g(x)\, dx.$$

*(handwritten annotations: $\lim_{d \to b-} F(d)$; (Monotone Convergence Thm); monotone and bounded, then limit exists and finite.)*

A similar argument works for the case $c \to a+$. ∎

This test is frequently used in conjunction with the following inequalities: $|\sin x| \le |x|$ for all $x \in \mathbf{R}$ (see Appendix B); for every $\alpha > 0$ there exists a constant $B_\alpha > 1$ such that $|\log x| \le x^\alpha$ for all $x \ge B_\alpha$ (see Exercise 4 in Section 4.3). Here are two typical examples.

**5.44 Example.** Prove that $f(x) = |\sin x / \sqrt{x^3}|$ is improperly integrable on $(0, 1]$.

PROOF. Since $0 \le f(x) = |\sin x / \sqrt{x^3}| \le |x| / x^{3/2} = 1/\sqrt{x}$ on $(0, 1]$, and this last function is improperly integrable on $(0, 1]$ by Example 5.40, it follows from the Comparison Test that $f(x)$ is improperly integrable on $(0, 1]$. ∎

**5.45 Example.** Prove that $f(x) = |\log x / \sqrt{x^5}|$ is improperly integrable on $[1, \infty)$.

PROOF. Since $f$ is continuous on $(0, \infty)$, $f$ is integrable on $[1, C]$ for any $C \in \mathbf{R}$. Since $0 \le f(x) = |\log x / \sqrt{x^5}| \le x^{1/2} / x^{5/2} = 1/x^2$ for $x \ge C := B_{1/2}$, and this last function is improperly integrable on $[1, \infty)$ by Example 5.41, it follows from the Comparison Test that $f(x)$ is improperly integrable on $[1, \infty)$. ∎

Although improperly integrable functions are not closed under multiplication (see Exercise 5), the Comparison Theorem can be used to show some kinds of products are improperly integrable.

**5.46 Remark.** *If $f$ is bounded and locally integrable on $(a, b)$ and $|g|$ is improperly integrable on $(a, b)$, then $|fg|$ is improperly integrable on $(a, b)$.*

PROOF. Let $M = \sup_{x \in (a,b)} |f(x)|$. Then $0 \le |f(x)g(x)| \le M|g(x)|$ for all $x \in (a, b)$. Hence, by Theorem 5.43, $|fg|$ is improperly integrable on $(a, b)$. ∎

For the Riemann integral, we proved that $|f|$ is integrable when $f$ is (see Theorem 5.22). This is not the case for the improper integral (see Example 5.49 below). For this reason we introduce the following concepts.

**5.47 DEFINITION.** Let $(a, b)$ be a nonempty, open interval and $f : (a, b) \to \mathbf{R}$.

    i) $f$ is said to be *absolutely integrable* on $(a, b)$ if $|f|$ is improperly integrable on $(a, b)$.

    ii) $f$ is said to be *conditionally integrable* on $(a, b)$ if $f$ is improperly integrable but not absolutely integrable on $(a, b)$.

The following result, an analogue of Theorem 5.22 for absolutely integrable functions, shows that absolute integrability implies improper integrability.

**5.48 THEOREM.** *If $f$ is absolutely integrable on $(a, b)$, then $f$ is improperly integrable on $(a, b)$ and*

$$\left| \int_a^b f(x)\, dx \right| \le \int_a^b |f(x)|\, dx.$$

PROOF. Since $0 \le |f(x)| + f(x) \le 2|f(x)|$, we have by Theorem 5.43 that $|f| + f$ is improperly integrable on $[a, b]$. Hence, by Theorem 5.42, so is $f = (|f| + f) - |f|$.

Moreover,

$$\left| \int_c^d f(x)\,dx \right| \le \int_c^d |f(x)|\,dx$$

for every $a < c < d < b$. We finish the proof by taking the limit of this last inequality as $c \to a+$ and $d \to b-$. ∎

The converse of Theorem 5.48, however, is false.

**5.49 Example.**  Prove that the function $\sin x/x$ is conditionally integrable on $[1, \infty)$.

PROOF. Integrating by parts, we have

$$\int_1^d \frac{\sin x}{x}\,dx = -\frac{\cos x}{x}\bigg|_1^d - \int_1^d \frac{\cos x}{x^2}\,dx$$

$$= \cos(1) - \frac{\cos d}{d} - \int_1^d \frac{\cos x}{x^2}\,dx.$$

Since $1/x^2$ is absolutely integrable on $[1, \infty)$, it follows from Remark 5.46 that $\cos x/x^2$ is absolutely integrable on $[1, \infty)$. Therefore, $\sin x/x$ is improperly integrable on $[1, \infty)$ and

$$\int_1^\infty \frac{\sin x}{x}\,dx = \cos(1) - \int_1^\infty \frac{\cos x}{x^2}\,dx.$$

To show $\sin x/x$ is not absolutely integrable on $[1, \infty)$, notice that

$$\int_1^{n\pi} \frac{|\sin x|}{x}\,dx \ge \sum_{k=2}^n \int_{(k-1)\pi}^{k\pi} \frac{|\sin x|}{x}\,dx$$

$$\ge \sum_{k=2}^n \frac{1}{k\pi} \int_{(k-1)\pi}^{k\pi} |\sin x|\,dx$$

$$= \sum_{k=2}^n \frac{2}{k\pi} = \frac{2}{\pi} \sum_{k=2}^n \frac{1}{k}$$

for each $n \in \mathbf{N}$. Since

$$\sum_{k=2}^n \frac{1}{k} \ge \sum_{k=2}^n \int_k^{k+1} \frac{1}{x}\,dx = \int_2^{n+1} \frac{1}{x}\,dx = \log(n+1) - \log 2 \to \infty$$

as $n \to \infty$, it follows from the Squeeze Theorem that

$$\lim_{n \to \infty} \int_1^{n\pi} \frac{|\sin x|}{x}\,dx = \infty.$$

Thus, $\sin x/x$ is not absolutely integrable on $[1, \infty)$. ∎

## EXERCISES

**1.** Evaluate the following improper integrals.

a)
$$\int_1^\infty \frac{1+x}{x^3}\,dx.$$

b)
$$\int_{-\infty}^\infty \frac{1}{1+x^2}\,dx.$$

c)
$$\int_0^{\pi/2} \frac{\cos x}{\sqrt[3]{\sin x}}\,dx.$$

d)
$$\int_0^\infty \sqrt{x}e^{-\sqrt{x}}\,dx.$$

**2.** For each of the following, find all values of $p \in \mathbf{R}$ for which $f$ is improperly integrable on $I$.
   a) $f(x) = 1/x^p$, $I = (1, \infty)$.
   b) $f(x) = 1/x^p$, $I = (0, 1)$.
   c) $f(x) = 1/(x \log^p x)$, $I = (e, \infty)$.
   d) $f(x) = 1/(1 + x^p)$, $I = (0, \infty)$.

**3.** Show that for each $p > 0$, $\sin x/x^p$ is improperly integrable on $[1, \infty)$ and $\cos x/\log^p x$ is improperly integrable on $(e, \infty)$.

**4.** Decide which of the following functions are improperly integrable on $I$.
   a) $f(x) = \sin x$, $I = (0, \infty)$.
   b) $f(x) = 1/x^2$, $I = [-1, 1]$.
   c) $f(x) = x^{-1}\sin(x^{-1})$, $I = (1, \infty)$.
   d) $f(x) = \log x$, $I = (0, 1)$.
   e) $f(x) = (1 - \cos x)/x^2$, $I = (0, \infty)$.

**5.** Use the examples provided by Exercise 2b) to show that the product of two improperly integrable functions might not be improperly integrable.

**6.** Suppose that $f, g$ are nonnegative and locally integrable on $[a, b)$ and
$$L := \lim_{x \to b-} \frac{f(x)}{g(x)}$$
exists as an extended real number.

a) Show that if $0 \le L < \infty$ and $g$ is improperly integrable on $[a, b)$, then so is $f$.

b) Show that if $0 < L \le \infty$ and $g$ is not improperly integrable on $[a, b)$, then neither is $f$.

**7.** a) Suppose $f$ is improperly integrable on $[0, \infty)$. Prove that if $L = \lim_{x \to \infty} f(x)$ exists, then $L = 0$.

b) Let
$$f(x) = \begin{cases} 1 & n \le x < n + 2^{-n}, \ n \in \mathbf{N} \\ 0 & \text{otherwise.} \end{cases}$$

Prove that $f$ is improperly integrable on $[0, \infty)$ but $\lim_{x \to \infty} f(x)$ does not exist.

**8.** Prove that if $f$ is absolutely integrable on $[1, \infty)$, then

$$\lim_{n \to \infty} \int_1^\infty f(x^n)\, dx = 0.$$

**9.** Assuming $e = \lim_{n \to \infty} \sum_{k=0}^n 1/k!$ (see Example 7.47), prove

$$\lim_{n \to \infty} \left( \frac{1}{n!} \int_1^\infty x^n e^{-x}\, dx \right) = 1.$$

**10.** a) Prove

$$\int_0^{\pi/2} e^{-a \sin x}\, dx \le \frac{2}{a}$$

for all $a > 0$.

b) What happens if $\cos x$ replaces $\sin x$?

$^e$**5.5  FUNCTIONS OF BOUNDED VARIATION**   *This section uses no material from any other enrichment section.*

In this section we study functions which do not wiggle too much. These functions, which play a prominent role in the theory of Fourier series (see Sections $^e$14.3 and $^e$14.4) and probability theory, are important tools for theoretical as well as applied mathematics.

Let $\phi : [a, b] \to \mathbf{R}$. To measure how much $\phi$ wiggles on an interval $[a, b]$, set

$$V(\phi, P) = \sum_{j=1}^n |\phi(x_j) - \phi(x_{j-1})|$$

for each partition $P = \{x_0, x_1, \ldots, x_n\}$ of $[a, b]$. The *variation* of $\phi$ is defined by

(19)                    $\mathrm{Var}\,(\phi) := \sup\{V(\phi, P) : P \text{ is a partition of } [a, b]\}.$

**5.50 DEFINITION.** Let $[a, b]$ be a closed, bounded, nondegenerate interval and $\phi : [a, b] \to \mathbf{R}$. Then $\phi$ is said to be of *bounded variation* on $[a, b]$ if $\mathrm{Var}\,(\phi) < \infty$.

The following three remarks show how the collection of functions of bounded variation is related to other collections of functions we have studied.

**5.51 Remark.** *If $\phi \in C^1[a,b]$, then $\phi$ is of bounded variation on $[a,b]$. However, there exist functions of bounded variation which are not continuously differentiable.*

PROOF. Let $P = \{x_0, x_1, \ldots, x_n\}$ be a partition of $[a,b]$. By the Extreme Value Theorem, there is an $M > 0$ such that $|\phi'(x)| \le M$ for all $x \in [a,b]$. Therefore, it follows from the Mean Value Theorem that for each $k$ between 1 and $n$ there is a point $c_k$ between $x_{k-1}$ and $x_k$ such that

$$|\phi(x_k) - \phi(x_{k-1})| = |\phi'(c_k)|(x_k - x_{k-1}) \le M(x_k - x_{k-1}).$$

By telescoping, we obtain $V(\phi, P) \le M(b-a)$ for any partition $P$ of $[a,b]$. Therefore,

$$\mathrm{Var}\,(\phi) \le M(b-a).$$

On the other hand, $x^2 \sin(1/x)$ is of bounded variation on $[0,1]$ (see Exercise 2 below) but does not belong to $C^1[0,1]$ (see Example 4.8). ∎

**5.52 Remark.** *If $\phi$ is monotone on $[a,b]$, then $\phi$ is of bounded variation on $[a,b]$. However, there exist functions of bounded variation which are not monotone.*

PROOF. Let $\phi$ be increasing on $[a,b]$ and $P = \{x_0, x_1, \ldots, x_n\}$ be a partition of $[a,b]$. Then by telescoping,

$$\sum_{j=1}^{n} |\phi(x_j) - \phi(x_{j-1})| = \sum_{j=1}^{n} (\phi(x_j) - \phi(x_{j-1}))$$

$$= \phi(x_n) - \phi(x_0) = \phi(b) - \phi(a) =: M < \infty.$$

Thus, $\mathrm{Var}\,(f) \le M$. On the other hand, by Remark 5.51, $\phi(x) = x^2$ is of bounded variation on $[-1,1]$. ∎

**5.53 Remark.** *If $\phi$ is of bounded variation on $[a,b]$, then $\phi$ is bounded on $[a,b]$. However, there exist bounded functions which are not of bounded variation.*

PROOF. Let $x \in [a,b]$ and note by definition that

$$|\phi(x) - \phi(a)| \le |\phi(x) - \phi(a)| + |\phi(b) - \phi(x)| \le \mathrm{Var}\,(\phi).$$

Hence, by the triangle inequality,

$$|\phi(x)| \le |\phi(a)| + \mathrm{Var}\,(\phi).$$

To find a bounded function which is not of bounded variation, consider

$$\phi(x) := \begin{cases} \sin(1/x) & x \ne 0 \\ 0 & x = 0. \end{cases}$$

Clearly, $\phi$ is bounded by 1. On the other hand, if

$$x_j = \begin{cases} 0 & j = 0 \\ \dfrac{2}{(n-j)\pi} & 0 < j < n, \end{cases}$$

then

$$\sum_{j=1}^{n} |\phi(x_j) - \phi(x_{j-1})| = 2n \to \infty$$

as $n \to \infty$. Thus $\phi$ is not of bounded variation on $[0, 2/\pi]$. ∎

The following result and Exercise 3 below are partial answers to the question: Is the class of functions of bounded variation preserved by algebraic operations?

**5.54 THEOREM.** *If $\phi$ and $\psi$ are of bounded variation on a closed, bounded interval $[a, b]$, then so are $\phi + \psi$ and $\phi - \psi$.*

PROOF. Let $a = x_0 < x_1 < \cdots < x_n = b$. Then

$$\sum_{j=1}^{n} |\phi(x_j) \pm \psi(x_j) - (\phi(x_{j-1}) \pm \psi(x_{j-1}))|$$

$$\leq \sum_{j=1}^{n} |\phi(x_j) - \phi(x_{j-1})| + \sum_{j=1}^{n} |\psi(x_j) - \psi(x_{j-1})|$$

$$\leq \mathrm{Var}\,(\phi) + \mathrm{Var}\,(\psi).$$

Therefore, $\mathrm{Var}\,(\phi \pm \psi) \leq \mathrm{Var}\,(\phi) + \mathrm{Var}\,(\psi)$. ∎

It turns out that there is a close connection between functions of bounded variation and monotone functions (see Corollary 5.57 below). To make this connection clear, we introduce the following concept.

**5.55 DEFINITION.** Let $\phi$ be of bounded variation on a closed, bounded interval $[a, b]$. The *total variation* of $\phi$ is the function $\Phi$ defined on $[a, b]$ by

$$\Phi(x) := \sup \left\{ \sum_{j=1}^{k} |\phi(x_j) - \phi(x_{j-1})| : \{x_0, x_1, \ldots, x_k\} \text{ is a partition of } [a, x] \right\}.$$

**5.56 THEOREM.** *Let $\phi$ be of bounded variation on $[a, b]$ and $\Phi$ be its total variation. Then*

  i) *$|\phi(y) - \phi(x)| \leq \Phi(y) - \Phi(x)$ for all $a \leq x < y \leq b$,*
  ii) *$\Phi$ and $\Phi - \phi$ are increasing on $[a, b]$, and*
  iii) *$\mathrm{Var}\,(\phi) \leq \mathrm{Var}\,(\Phi)$.*

PROOF. i) Let $x < y$ belong to $[a, b]$ and $\{x_0, x_1, \ldots, x_k\}$ be a partition of $[a, x]$. Then $\{x_0, x_1, \ldots, x_k, y\}$ is a partition of $[a, y]$, and we have by Definition 5.55 that

$$\sum_{j=1}^{k} |\phi(x_j) - \phi(x_{j-1})| \leq \sum_{j=1}^{k} |\phi(x_j) - \phi(x_{j-1})| + |\phi(y) - \phi(x)| \leq \Phi(y).$$

Taking the supremum of this inequality over all partitions $\{x_0, x_1, \ldots, x_k\}$ of $[a, x]$, we obtain

$$\Phi(x) \leq \Phi(x) + |\phi(y) - \phi(x)| \leq \Phi(y).$$

ii) By the Monotone Property of Suprema, $\Phi$ is increasing on $[a,b]$. To show $\Phi - \phi$ is increasing, suppose $a \leq x < y \leq b$. By part i),

$$\phi(y) - \phi(x) \leq |\phi(y) - \phi(x)| \leq \Phi(y) - \Phi(x).$$

Therefore, $\Phi(x) - \phi(x) \leq \Phi(y) - \phi(y)$.

iii) Let $P = \{x_0, x_1, \ldots, x_n\}$ be a partition of $[a,b]$. By part i) and Definition 5.50,

$$\sum_{j=1}^{n} |\phi(x_j) - \phi(x_{j-1})| \leq \sum_{j=1}^{n} |\Phi(x_j) - \Phi(x_{j-1})| \leq \mathrm{Var}\,(\Phi).$$

Taking the supremum of this inequality over all partitions $P$ of $[a,b]$, we obtain $\mathrm{Var}\,(\phi) \leq \mathrm{Var}\,(\Phi)$. ∎

**5.57 COROLLARY.** *Let $[a,b]$ be a closed, bounded interval. Then $\phi$ is of bounded variation on $[a,b]$ if and only if there exist increasing functions $f, g$ on $[a,b]$ such that*

$$\phi(x) = f(x) - g(x), \qquad x \in [a,b].$$

PROOF. Suppose $\phi$ is of bounded variation, let $\Phi$ represent the total variation of $\phi$, $f = \Phi$, and $g = \Phi - \phi$. By Theorem 5.56, $f$ and $g$ are increasing, and by construction, $\phi = f - g$.

Conversely, suppose $\phi = f - g$ for some increasing $f, g$ on $[a,b]$. Then by Remark 5.52 and Theorem 5.54, $\phi$ is of bounded variation on $[a,b]$. ∎

In particular, if $f$ is of bounded variation on $[a,b]$ then

i) $f(x+)$ exists for each $x \in [a,b)$ and $f(x-)$ exists for each $x \in (a,b]$ (see Lemma 4.28),

ii) $f$ has at most countably many points of discontinuity in $[a,b]$ (see Theorem 4.29), and

iii) $f$ is integrable on $[a,b]$ (see Exercise 8 in Section 5.1).

## EXERCISES

**1.** a) Show $4k/(4k^2 - 1) > 1/k$ for $k \in \mathbf{N}$.

   b) Prove

$$\sum_{k=1}^{2^n - 1} \frac{1}{k} > \sum_{k=0}^{n-1} \left( \sum_{j=2^k}^{2^{k+1}-1} \frac{1}{2^{k+1}} \right) = \frac{n}{2}$$

   for $n \in \mathbf{N}$.

   c) Prove that

$$\phi(x) = \begin{cases} x^2 \sin \dfrac{1}{x^2} & x \neq 0 \\ 0 & x = 0 \end{cases}$$

   is not of bounded variation on $[0,1]$.

**2. a)** Show $(8k^2 + 2)/(4k^2 - 1)^2 < 1/k^2$ for $k = 2, 3, \ldots$

**b)** Prove

$$\sum_{k=1}^{n} \frac{1}{k^2} \leq 1 + \sum_{k=1}^{n-1} \left( \frac{1}{k} - \frac{1}{k+1} \right) = 2 - \frac{1}{n}$$

for $n \in \mathbf{N}$.

**c)** Prove that

$$\phi(x) = \begin{cases} x^2 \sin \dfrac{1}{x} & x \neq 0 \\ 0 & x = 0 \end{cases}$$

is of bounded variation on any bounded interval $[a, b]$.

**3.** **This exercise is used in Section** $^e$**14.3.** Suppose $\phi$ and $\psi$ are of bounded variation on a closed, bounded interval $[a, b]$.

**a)** Prove that $\alpha\phi$ is of bounded variation on $[a, b]$ for every $\alpha \in \mathbf{R}$.

**b)** Prove that $\phi\psi$ is of bounded variation on $[a, b]$.

**c)** If there is an $\varepsilon_0 > 0$ such that

$$\phi(x) \geq \varepsilon_0, \qquad x \in [a, b],$$

prove that $1/\phi$ is of bounded variation on $[a, b]$.

**4.** Suppose $\phi$ is of bounded variation on a closed, bounded interval $[a, b]$. Prove $\phi$ is continuous on $(a, b)$ if and only if $\phi$ is uniformly continuous on $(a, b)$.

**5. a)** If $\phi$ is continuous on a closed, bounded, nondegenerate interval $[a, b]$, differentiable on $(a, b)$, and $\phi'$ is bounded on $(a, b)$, prove that $\phi$ is of bounded variation on $[a, b]$.

**b)** Show that $\phi(x) = \sqrt[3]{x}$ is of bounded variation on $[-1, 1]$ but $\phi'$ is unbounded at some point in $(-1, 1)$.

**6.** Let $P$ be a polynomial of degree $N$.

**a)** Show that $P$ is of bounded variation on any closed, bounded interval $[a, b]$.

**b)** Obtain an estimate for $\text{Var}\,(P)$ on $[a, b]$, using values of the derivative $P'(x)$ at no more than $N$ points.

**7.** Let $\phi$ be a function of bounded variation on $[a, b]$ and $\Phi$ be its total variation function. Prove that if $\Phi$ is continuous at some point $x_0 \in (a, b)$, then $\phi$ is continuous at $x_0$.

**8.** **This exercise is used in Section** $^e$**14.4.** If $f$ is integrable on $[a, b]$, prove

$$F(x) = \int_a^x f(t)\, dt$$

is of bounded variation on $[a, b]$.

**9.** Suppose $f'$ exists and is integrable on $[a, b]$. Prove that $f$ is of bounded variation and

$$\text{Var}\,(f) = \int_a^b |f'(x)|\, dx.$$

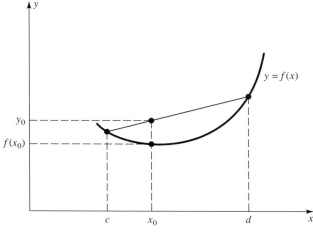

**Figure 5.5**

What happens to this result if $f'$ is bounded but not necessarily integrable?

$^e$**5.6  CONVEX FUNCTIONS**    *The last half page of this section uses Theorems 4.29 and 4.30, optional results from Section 4.4.*

In this section we examine another collection of functions which is important for certain applications, especially for Fourier analysis, functional analysis, numerical analysis, and probability theory.

**5.58 DEFINITION.**  Let $I$ be an interval and $f : I \to \mathbf{R}$.

    i) $f$ is said to be *convex* on $I$ if

$$f(\alpha x + (1 - \alpha)y) \leq \alpha f(x) + (1 - \alpha)f(y)$$

       for all $0 \leq \alpha \leq 1$ and all $x, y \in I$.

    ii) $f$ is said to be *concave* on $I$ if $-f$ is convex on $I$.

Notice that by definition, a function $f$ is convex on an interval $I$ if and only if $f$ is convex on every closed subinterval of $I$.

It is easy to check that $f(x) = mx + b$ is both convex and concave on any interval (see also Exercise 3) but in general it is difficult to apply Definition 5.58 directly. For this reason, we include the following simple geometric characterizations of convexity.

**5.59 Remark.**  *Let $I$ be an interval and $f : I \to \mathbf{R}$. Then $f$ is convex on $I$ if and only if given any $[c, d] \subseteq I$, the chord through the points $(c, f(c))$, $(d, f(d))$ lies on or above the graph of $y = f(x)$ for all $x \in [c, d]$. (See Figure 5.5.)*

PROOF.  Suppose $f$ is convex on $I$ and $x_0 \in [c, d]$.  Choose $0 \leq \alpha \leq 1$ such that $x_0 = \alpha c + (1 - \alpha)d$. The chord from $(c, f(c))$ to $(d, f(d))$ has slope $(f(d) - f(c))/(d - c)$. Hence, the point on this chord which has the form $(x_0, y_0)$ must

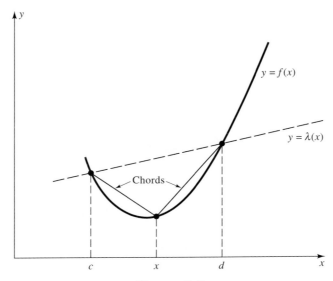

**Figure 5.6**

satisfy $y_0 = \alpha f(c) + (1 - \alpha)f(d)$. Since $f$ is convex, it follows that $f(x_0) \leq y_0$, i.e., the point $(x_0, y_0)$ lies on or above the point $(x_0, f(x_0))$. A similar argument establishes the reverse implication. ∎

Thus both $f(x) = |x|$ and $f(x) = x^2$ are convex on any interval.

**5.60 Remark.** *A function $f$ is convex on a nonempty, open interval $(a, b)$ if and only if the slope of the chord always increases, i.e.,*

$$a < c < x < d < b \quad \text{implies} \quad \frac{f(x) - f(c)}{x - c} \leq \frac{f(d) - f(x)}{d - x}.$$

PROOF. Fix $a < c < x < d < b$ and let $\lambda(x)$ be the equation of the chord to $f$ through the points $(c, f(c))$ and $(d, f(d))$. If $f$ is convex, then $f(x) \leq \lambda(x)$ (see Figure 5.6). Therefore,

$$\frac{f(x) - f(c)}{x - c} \leq \frac{\lambda(x) - \lambda(c)}{x - c} = \frac{\lambda(d) - \lambda(x)}{d - x} \leq \frac{f(d) - f(x)}{d - x}.$$

Conversely, if $f$ is not convex, then $\lambda(x) < f(x)$ for some $x \in (c, d)$. It follows that

$$\frac{f(x) - f(c)}{x - c} > \frac{\lambda(x) - \lambda(c)}{x - c} = \frac{\lambda(d) - \lambda(x)}{d - x} > \frac{f(d) - f(x)}{d - x}.$$

Therefore, the slope of the chord decreases. ∎

This leads us to a characterization of differentiable convex functions.

**5.61 THEOREM.** *Suppose $f$ is differentiable on a nonempty, open interval $(a, b)$. Then $f$ is convex on $(a, b)$ if and only if $f'$ is increasing on $(a, b)$.*

PROOF. Suppose $f$ is convex and that $c, d \in (a, b)$ satisfy $c < d$. Choose $h$ so small that $a < c - h$, $c + h < d$, and $d + h < b$. Then by Remark 5.60,

$$\frac{f(c + h) - f(c)}{h} \le \frac{f(d + h) - f(d)}{h}.$$

In particular, $f'(c) \le f'(d)$.

Conversely, suppose $f'$ is increasing. Let $a < c < x < d < b$ and use the Mean Value Theorem to choose $x_0$ (between $c$ and $x$) and $x_1$ (between $x$ and $d$) such that

$$\frac{f(x) - f(c)}{x - c} = f'(x_0) \quad \text{and} \quad \frac{f(d) - f(x)}{d - x} = f'(x_1).$$

Since $x_0 < x_1$ it follows that $f'(x_0) \le f'(x_1)$. In particular, we conclude by Remark 5.60 that $f$ is convex on $(a, b)$. ∎

Combining Theorems 4.24 and 5.61, we obtain the usual convexity criterion in terms of the second derivative: If $f$ is twice differentiable on $(a, b)$, then $f$ is convex on $(a, b)$ if and only if $f''(x) \ge 0$ for all $x \in (a, b)$. In particular, convexity is what elementary calculus texts call *concave upward* and concavity is what elementary calculus texts call *concave downward*.

On open intervals, convex functions are always continuous. (The statements and proofs of the next two results come from Zygmund [15].)

**5.62 THEOREM.** *If $f$ is convex on some nonempty, open interval $(a, b)$, then $f$ is continuous on $(a, b)$.*

PROOF. Let $x_0 \in (a, b)$. By symmetry, it suffices to show that $f(x) \to f(x_0)$ as $x \to x_0+$. Let $a < c < x_0 < x < d < b$, $y = g(x)$ represent the equation of the chord through $(c, f(c))$, $(x_0, f(x_0))$, and $y = h(x)$ represent the equation of the chord through $(x_0, f(x_0))$, $(d, f(d))$. Since $f$ is convex, we have by Remark 5.59 that $f(x) \le h(x)$. Since $f(x_0)$ lies on or below the chord from $(c, f(c))$ to $(x, f(x))$, we also have that $g(x) \le f(x)$. Consequently,

$$g(x) \le f(x) \le h(x), \qquad x \in (x_0, b).$$

Both chords $y = g(x)$ and $y = h(x)$ pass through the point $(x_0, f(x_0))$, so $g(x) \to f(x_0)$ and $h(x) \to f(x_0)$ as $x \to x_0+$. Hence, it follows from the Squeeze Theorem that $f(x) \to f(x_0)$ as $x \to x_0+$. ∎

Theorem 5.62 does not hold for closed intervals $[a, b]$. Indeed, the function

$$f(x) := \begin{cases} 0 & 0 \le x < 1 \\ 1 & x = 1 \end{cases}$$

is convex on $[0, 1]$ but not continuous there.

A function $f$ is said to have a *proper maximum* (respectively, *proper minimum*) at $x_0$ if there exists a $\delta > 0$ such that $f(x) < f(x_0)$ (respectively, $f(x) > f(x_0)$) for all $0 < |x - x_0| < \delta$. As far as proper extrema are concerned, convex functions behave like strictly increasing functions.

**5.63 THEOREM.** i) *If $f$ is convex on a nonempty, open interval $(a, b)$, then $f$ has no proper maximum on $(a, b)$.*

ii) *If $f$ is convex on $[0, \infty)$ and has a proper minimum, then $f(x) \to \infty$ as $x \to \infty$.*

PROOF. i) Suppose $x_0 \in (a, b)$ and $f(x_0)$ is a proper maximum of $f$. Then there exist $c < x_0 < d$ such that $f(x) < f(x_0)$ for $c < x < d$. In particular, the chord through $(c, f(c))$, $(d, f(d))$ must lie below $f(x_0)$ for $c, d$ near $x_0$, a contradiction.

ii) Suppose $x_0 \in (a, b)$ and $f(x_0)$ is a proper minimum of $f$. Fix $x_1 > x_0$. Let $y = g(x)$ represent the equation of the chord through $(x_0, f(x_0))$ and $(x_1, f(x_1))$. Since $f(x_0)$ is a proper minimum, $f(x_1) > f(x_0)$, hence, $g$ has positive slope. Moreover, by the proof of Theorem 5.62, $g(x) \le f(x)$ for all $x \in (x_1, \infty)$. Since $g(x) \to \infty$ as $x \to \infty$, we conclude that $f(x) \to \infty$ as $x \to \infty$. ∎

Another important result about convex functions addresses the question: what happens when we interchange the order of a convex function and an integral sign?

**5.64 THEOREM** [JENSEN'S INEQUALITY]. *Let $\phi$ be convex on a closed, bounded interval $[a, b]$ and $f : [0, 1] \to [a, b]$. If $f$ and $\phi \circ f$ are integrable on $[0, 1]$, then*

$$(20) \qquad \phi \left( \int_0^1 f(x) \, dx \right) \le \int_0^1 (\phi \circ f)(x) \, dx.$$

PROOF. Set

$$c = \int_0^1 f(x) \, dx$$

and observe that

$$(21) \qquad \phi \left( \int_0^1 f(x) \, dx \right) = \phi(c) + s \left( \int_0^1 f(x) \, dx - c \right)$$

for all $s \in \mathbf{R}$. (Note: Since $a \le f(x) \le b$ for each $x \in [0, 1]$, $c$ must belong to the interval $[a, b]$ by the Comparison Theorem for Integrals. Thus $\phi(c)$ is defined.)

Let

$$s = \sup_{x \in [a, c)} \frac{\phi(c) - \phi(x)}{c - x}.$$

By Remark 5.60, $s \le (\phi(u) - \phi(c))/(u - c)$ for all $u \in (c, b]$, i.e.,

$$(22) \qquad \phi(c) + s(u - c) \le \phi(u)$$

for all $u \in [c, b]$. On the other hand, if $u \in [a, c)$, we have by the definition of $s$ that

$$s \ge \frac{\phi(c) - \phi(u)}{c - u}.$$

Thus (22) holds for all $u \in [a, b]$. Applying (22) to $u = f(x)$, we obtain

$$\phi(c) + s(f(x) - c) \le (\phi \circ f)(x).$$

Integrating this inequality as $x$ runs from 0 to 1, we obtain

$$\phi(c) + s \left( \int_0^1 f(x)\, dx - c \right) \le \int_0^1 (\phi \circ f)(x)\, dx.$$

Combining this inequality with (21), we conclude that (20) holds. ∎

What about differentiability of convex functions? To answer this question we introduce the following concepts (compare with Definition 4.6).

**5.65 DEFINITION.** Let $f : (a, b) \to \mathbf{R}$ and $x \in (a, b)$.

   i) $f$ is said to have a *right-hand derivative* at $x$ if

$$D_R f(x) := \lim_{h \to 0+} \frac{f(x + h) - f(x)}{h},$$

     exists as an extended real number.

   ii) $f$ is said to have a *left-hand derivative* at $x$ if

$$D_L f(x) := \lim_{h \to 0-} \frac{f(x + h) - f(x)}{h},$$

     exists as an extended real number.

The following result is a simple consequence of the definition of differentiability and the characterization of two-sided limits by one-sided limits (see Theorem 3.14).

**5.66 Remark.** *A function $f$ is differentiable at $x$ if and only if both $D_R f(x)$ and $D_L f(x)$ exist, are finite, and equal, in which case $f'(x) = D_R f(x) = D_L f(x)$.*

The next result shows that the left-hand and right-hand derivatives of a convex function are remarkably well-behaved.

**5.67 THEOREM.** *Let $f$ be convex on an open interval $(a, b)$. Then the left-hand and right-hand derivatives of $f$ exist, are increasing on $(a, b)$, and satisfy*

$$-\infty < D_L f(x) \le D_R f(x) < \infty$$

*for all $x \in (a, b)$.*

PROOF. Let $h < 0$ and notice that the slope of the chord through the points $(x, f(x))$ and $(x + h, f(x + h))$ is $(f(x + h) - f(x))/h$. By Remark 5.60, these slopes increase as $h \to 0-$. Since increasing functions have a limit (which may be $+\infty$), it follows that $D_L f(x)$ exists and satisfies $-\infty < D_L f(x) \le \infty$. Similarly, $D_R f(x)$ exists and satisfies $-\infty \le D_R f(x) < \infty$. Remark 5.60 also implies

$$(23) \qquad\qquad\qquad D_L f(x) \le D_R f(x).$$

Hence, both numbers are finite and by symmetry it remains to show that $D_R f(x)$ is increasing on $(a, b)$.

Let $x_1 < u < t < x_2$ be points which belong to $(a, b)$. Then

$$\frac{f(u) - f(x_1)}{u - x_1} \leq \frac{f(x_2) - f(t)}{x_2 - t}.$$

Taking the limit of this inequality as $u \to x_1+$ and $t \to x_2-$, we conclude by (23) that

$$(24) \qquad\qquad D_R f(x_1) \leq D_L f(x_2) \leq D_R f(x_2). \quad \blacksquare$$

The next proof uses Theorem 4.29, an optional result from Section 4.4.

*$^*$**5.68 COROLLARY.** If $f$ is convex on an open interval $(a, b)$, then $f$ is differentiable at all but countably many points of $(a, b)$; i.e., there is a countable set $E \subset (a, b)$ such that $f'(x)$ exists for all $x \in (a, b) \setminus E$.*

Proof. Let $E$ be the set where either $D_L f(x)$ or $D_R f(x)$ is discontinuous. By Theorems 5.67 and 4.29, the set $E$ is countable. Suppose $x_0 \in (a, b) \setminus E$ and $x < x_0$. By (24),

$$D_R f(x) \leq D_L f(x_0) \leq D_R f(x_0).$$

Let $x \to x_0$. Since both $D_L f(x)$ and $D_R f(x)$ are continuous at $x_0$, we obtain $D_R f(x_0) \leq D_L f(x_0) \leq D_R f(x_0)$. In particular, $f'(x_0)$ exists for all $x_0 \in (a, b) \setminus E$. $\blacksquare$

How useful is a statement about $f'(x)$ which holds for all but countably many points $x$? We address this question by proving a generalization of Theorem 4.24. (The proof here uses Theorem 4.30, an optional result from Section 4.4.)

*$^*$**5.69 THEOREM.** Suppose $f$ is continuous on a closed, bounded interval $[a, b]$ and differentiable on $(a, b)$. If $f'(x) \geq 0$ for all but countably many $x \in (a, b)$, then $f$ is increasing on $[a, b]$.*

Proof. Suppose $f'(x_1) < 0$ for some $x_1 \in (a, b)$ and let $y \in (f'(x_1), 0)$. By Theorem 4.30 (the Intermediate Value Theorem for derivatives), there is an $x = x(y) \in (a, b)$ such that $f'(x) = y < 0$. It follows that if $f'(x) < 0$ for one $x \in (a, b)$, then $f'(x) < 0$ for uncountably many $x \in (a, b)$, a contradiction. Therefore, $f'(x) \geq 0$ for all $x \in (a, b)$; hence, by Theorem 4.24, $f$ is increasing on $(a, b)$. $\blacksquare$

*$^*$**5.70 COROLLARY.** If $f$ is continuous on a closed, bounded interval $[a, b]$ and differentiable on $(a, b)$ with $f'(x) = 0$ for all but countably many $x \in (a, b)$, then $f$ is constant on $[a, b]$.*

## EXERCISES

**1.** Suppose $f, g$ are convex on an interval $I$. Prove that $f + g$ and $cf$ are convex on $I$ for any $c \geq 0$.

**2.** Suppose that $f_n$ is a sequence of functions convex on an interval $I$ and

$$f(x) = \lim_{n \to \infty} f_n(x)$$

exists for each $x \in I$. Prove that $f$ is convex on $I$.

**3.** Prove that a function $f$ is both convex and concave on $I$ if and only if there exist $m, b \in \mathbf{R}$ such that $f(x) = mx + b$ for $x \in I$.

**4.** Prove that $f(x) = x^p$ is convex on $[0, \infty)$ for $p \geq 1$, and concave on $[0, \infty)$ for $0 < p \leq 1$.

**5.** Show that if $f$ is increasing on $[a, b]$, then

$$F(x) = \int_a^x f(t)\, dt$$

is convex on $[a, b]$.

(Recall that by Exercise 8 in Section 5.1, $f$ is integrable on $[a, b]$.)

**6.** Suppose $f : [0, 1] \to [a, b]$ is integrable on $[0, 1]$.

a) Prove

$$e^{\int_0^1 f(x)\, dx} \leq \int_0^1 e^{f(x)}\, dx.$$

b) Prove

$$\left( \int_0^1 |f(x)|^r\, dx \right)^{1/r} \leq \int_0^1 |f(x)|\, dx$$

for all $0 < r \leq 1$.

c) Prove

$$\int_0^1 |f(x)|\, dx \leq \left( \int_0^1 |f(x)|^p\, dx \right)^{1/p}$$

for all $p \geq 1$.

**7.** If $f : [a, b] \to \mathbf{R}$ is integrable on $[a, b]$, prove

$$\int_a^b |f(x)|\, dx \leq (b - a)^{1/2} \left( \int_a^b f^2(x)\, dx \right)^{1/2}.$$

**\*8.** Let $f$ be continuous on a closed, bounded interval $[a, b]$ and suppose $D_R f(x)$ exists for all $x \in (a, b)$.

a) Show that if $f(b) < y_0 < f(a)$, then

$$x_0 := \sup\{x \in [a, b] : f(x) > y_0\}$$

satisfies $f(x_0) = y_0$ and $D_R f(x_0) \leq 0$.

b) Prove that if $f(b) < f(a)$, then there are uncountably many points $x$ which satisfy $D_R f(x) \leq 0$.

c) Prove that if $D_R f(x) > 0$ for all but countably many points $x \in (a, b)$, then $f$ is increasing on $[a, b]$.

d) Prove that if $D_R f(x) \geq 0$ and $g(x) = f(x) + x/n$ for some $n \in \mathbf{N}$, then $D_R g(x) > 0$.

e) Prove that if $D_R f(x) \geq 0$ for all but countably many points $x \in (a, b)$, then $f$ is increasing on $[a, b]$.

# Chapter 6

# Infinite Series of Real Numbers

Infinite series are one of the most widely used tools of analysis. They are used to approximate numbers and functions. (Series of Ramanujan type have been used to compute more than a billion digits of the decimal expansion of $\pi$.) They are used to approximate solutions of differential equations. (You may have used power series to solve ordinary differential equations with nonconstant coefficients.) They even form the basis for some very practical applications including pattern recognition (e.g., reading zip codes), image enhancement (e.g., removing raindrop clutter from a radar scan), and data compression (e.g., transmission of hundreds of TV programs through a single, photonic, fiber optic cable). Other applications of infinite series can be found in Section 7.5. In view of the variety of these applications, it should come as no surprise that the subject matter of this chapter (and the next) is of fundamental importance.

## 6.1 INTRODUCTION

Let $\{a_k\}_{k \in \mathbf{N}}$ be a sequence of numbers. We shall call an expression of the form

$$(1) \qquad \sum_{k=1}^{\infty} a_k$$

an *infinite series* with *terms* $a_k$. (No convergence is assumed at this point. This is merely a formal expression.)

**6.1 DEFINITION.** Let $S = \sum_{k=1}^{\infty} a_k$ be an infinite series whose terms $a_k$ belong to $\mathbf{R}$.

    i) The *partial sums of $S$ of order $n$* are the numbers, defined for each $n \in \mathbf{N}$, by

$$s_n := \sum_{k=1}^{n} a_k.$$

ii) $S$ is said to *converge* if its sequence of partial sums $\{s_n\}$ converges to some $s \in \mathbf{R}$ as $n \to \infty$, i.e., if for every $\varepsilon > 0$ there is an $N \in \mathbf{N}$ such that $n \geq N$ implies $|s_n - s| < \varepsilon$. In this case we shall write

(2)
$$\sum_{k=1}^{\infty} a_k = s$$

and call $s$ the *sum*, or *value*, of the series $\sum_{k=1}^{\infty} a_k$.

iii) $S$ is said to *diverge* if its sequence of partial sums $\{s_n\}$ does not converge as $n \to \infty$. When $s_n$ diverges to $+\infty$ as $n \to \infty$, we shall also write

$$\sum_{k=1}^{\infty} a_k = \infty.$$

(We shall deal with series of functions in Chapter 7.)

You are already familiar with one type of infinite series, decimal expansions. Every decimal expansion of a number $x \in (0, 1)$ is a series of the form $\sum_{k=1}^{\infty} x_k/10^k$, where the $x_k$'s are integers in $[0, 9]$. For example, when we write $1/3 = 0.333\ldots$ we mean

$$\frac{1}{3} = \sum_{k=1}^{\infty} \frac{3}{10^k}.$$

In particular, the partial sums $.3, .33, .333, \ldots$ are approximations to $1/3$ which get closer and closer to $1/3$ as more terms of the decimal expansion are taken.

One way to determine if a given series converges is to find a formula for its partial sums simple enough so that we can decide whether or not they converge. Here are two examples.

**6.2 Example.** Prove $\sum_{k=1}^{\infty} 2^{-k} = 1$.

PROOF. By induction, we can show that the partial sums $s_n = \sum_{k=1}^{n} 1/2^k$ satisfy $s_n = 1 - 2^{-n}$ for $n \in \mathbf{N}$. Thus $s_n \to 1$ as $n \to \infty$. ∎

**6.3 Example.** Prove $\sum_{k=1}^{\infty} (-1)^k$ diverges.

PROOF. The partial sums $s_n = \sum_{k=1}^{n} (-1)^k$ satisfy

$$s_n = \begin{cases} -1 & \text{if } n \text{ is odd} \\ 0 & \text{if } n \text{ is even.} \end{cases}$$

Thus $s_n$ does not converge as $n \to \infty$. ∎

Another way to show a series diverges is to estimate its partial sums.

**6.4 Example.** [THE HARMONIC SERIES]. Prove that the sequence $1/k$ converges but the series $\sum_{k=1}^{\infty} 1/k$ diverges to $+\infty$.

PROOF. The sequence $1/k$ converges to zero (by Example 2.2). On the other hand, by the Comparison Theorem for Integrals,

$$\sum_{k=1}^{n} \frac{1}{k} \geq \sum_{k=1}^{n} \int_{k}^{k+1} \frac{1}{x}\, dx = \int_{1}^{n+1} \frac{1}{x}\, dx = \log(n+1).$$

We conclude that $s_n \to \infty$ as $n \to \infty$. ∎

(Note: This example shows that the terms of a divergent series may converge. In particular, a series does not converge just because its terms converge.)

There are many methods for deciding whether or not a given series converges. The following method is simple but useful in a variety of situations.

**6.5 THEOREM** [DIVERGENCE TEST]. *Let $\{a_k\}_{k \in \mathbf{N}}$ be a sequence of real numbers. If $a_k$ does not converge to zero, then the series $\sum_{k=1}^{\infty} a_k$ diverges.*

PROOF. Suppose to the contrary that $\sum_{k=1}^{\infty} a_k$ converges to some $s \in \mathbf{R}$. By definition, the sequence of partial sums $s_n := \sum_{k=1}^{n} a_k$ converges to $s$ as $n \to \infty$. Therefore, $a_k = s_k - s_{k-1} \to s - s = 0$ as $k \to \infty$, a contradiction. ∎

The proof of this result establishes a property interesting in its own right: if $\sum_{k=1}^{\infty} a_k$ converges, then $a_k \to 0$ as $k \to \infty$. It is important to realize from the beginning that the converse of this statement is false, i.e., Theorem 6.5 is a test for divergence, not a test for convergence. Indeed, the harmonic series is a divergent series whose terms converge to zero.

Finding the sum of a convergent series is usually difficult. The following two results show this is not the case for two special kinds of series.

**6.6 THEOREM** [TELESCOPIC SERIES]. *If $\{a_k\}$ is a convergent real sequence, then*

$$\sum_{k=1}^{\infty} (a_k - a_{k+1}) = a_1 - \lim_{k \to \infty} a_k.$$

PROOF. By telescoping, we have

$$s_n := \sum_{k=1}^{n} (a_k - a_{k+1}) = a_1 - a_{n+1}.$$

Hence, $s_n \to a_1 - \lim_{k \to \infty} a_k$ as $n \to \infty$. ∎

**6.7 THEOREM** [GEOMETRIC SERIES]. *The series $\sum_{k=1}^{\infty} x^k$ converges if and only if $|x| < 1$, in which case*

$$\sum_{k=1}^{\infty} x^k = \frac{x}{1-x}.$$

(See also Exercise 1 below.)

PROOF. If $|x| \geq 1$, then $\sum_{k=1}^{\infty} x^k$ diverges by the Divergence Test. If $|x| < 1$, then set $s_n = \sum_{k=1}^{n} x^k$ and observe by telescoping that

$$(1-x)s_n = (1-x)(x + x^2 + \cdots + x^n)$$
$$= x + x^2 + \cdots + x^n - x^2 - x^3 - \cdots - x^{n+1} = x - x^{n+1}.$$

Hence,

$$s_n = \frac{x}{1 - x} - \frac{x^{n+1}}{1 - x}$$

for all $n \in \mathbf{N}$. Since $x^{n+1} \to 0$ as $n \to \infty$ for all $|x| < 1$ (see Example 2.20), we conclude that $s_n \to x/(1 - x)$ as $n \to \infty$. ∎

(Note: In everyday speech, the words *sequence* and *series* are considered synonyms. Example 6.4 shows that in mathematics, this is not the case. In particular, you must not apply a result valid for sequences to series and vice versa. Nevertheless, because convergence of an infinite series is defined in terms of convergence of its sequence of partial sums, any result about sequences contains a result about infinite series. The following three theorems illustrate this principle.)

**6.8 THEOREM** [THE CAUCHY CRITERION]. *Let $\{a_k\}$ be a real sequence. Then the infinite series $\sum_{k=1}^{\infty} a_k$ converges if and only if for every $\varepsilon > 0$ there is an $N \in \mathbf{N}$ such that*

$$m > n \geq N \quad implies \quad \left| \sum_{k=n}^{m} a_k \right| < \varepsilon.$$

PROOF. Let $s_n$ represent the sequence of partial sums of $\sum_{k=1}^{\infty} a_k$ and set $s_0 = 0$. By Cauchy's Theorem (Theorem 2.29), $s_n$ converges if and only if given $\varepsilon > 0$ there is an $N \in \mathbf{N}$ such that $m, n \geq N$ imply $|s_m - s_{n-1}| < \varepsilon$. Since

$$s_m - s_{n-1} = \sum_{k=n}^{m} a_k$$

for all integers $m > n \geq 1$, the proof is complete. ∎

**6.9 COROLLARY.** *Let $\{a_k\}$ be a real sequence. Then the infinite series $\sum_{k=1}^{\infty} a_k$ converges if and only if given $\varepsilon > 0$ there is an $N \in \mathbf{N}$ such that*

$$n \geq N \quad implies \quad \left| \sum_{k=n}^{\infty} a_k \right| < \varepsilon.$$

**6.10 THEOREM.** *Let $\{a_k\}$ and $\{b_k\}$ be real sequences. If $\sum_{k=1}^{\infty} a_k$ and $\sum_{k=1}^{\infty} b_k$ are convergent series, then*

$$\sum_{k=1}^{\infty} (a_k + b_k) = \sum_{k=1}^{\infty} a_k + \sum_{k=1}^{\infty} b_k$$

*and*

$$\sum_{k=1}^{\infty} (\alpha a_k) = \alpha \sum_{k=1}^{\infty} a_k$$

*for any $\alpha \in \mathbf{R}$.*

PROOF. Both identities are corollaries of Theorem 2.12; we provide the details only for the first identity.

Let $s_n$ represent the partial sums of $\sum_{k=1}^{\infty} a_k$ and $t_n$ represent the partial sums of $\sum_{k=1}^{\infty} b_k$. Since real addition is commutative, we have

$$\sum_{k=1}^{n}(a_k + b_k) = s_n + t_n, \qquad n \in \mathbf{N}.$$

Taking the limit of this identity as $n \to \infty$, we conclude by Theorem 2.12 that

$$\sum_{k=1}^{\infty}(a_k + b_k) = \lim_{n \to \infty} s_n + \lim_{n \to \infty} t_n = \sum_{k=1}^{\infty} a_k + \sum_{k=1}^{\infty} b_k. \quad \blacksquare$$

*Thm2.12:*  $\lim\limits_{n \to \infty} (x_n + y_n) = \lim\limits_{n \to \infty} x_n + \lim\limits_{n \to \infty} y_n$.

## EXERCISES

$\lim\limits_{n \to \infty} (\alpha x_n) = \alpha \lim\limits_{n \to \infty} x_n$.

**1.** Show that

$$\sum_{k=n}^{\infty} x^k = \frac{x^n}{1 - x}$$

for $|x| < 1$ and $n = 0, 1, \dots$

**2.** Prove that each of the following series converges and find its value.

a) $\displaystyle\sum_{k=1}^{\infty} \frac{(-1)^{k+1}}{\pi^k}.$   b) $\displaystyle\sum_{k=1}^{\infty} \frac{(-1)^k + 4}{5^k}.$

c) $\displaystyle\sum_{k=1}^{\infty} \frac{3^k}{7^{k-1}}.$   d) $\displaystyle\sum_{k=0}^{\infty} 2^k e^{-k}.$

**3.** Represent each of the following series as a telescopic series and find its value.

a) $\displaystyle\sum_{k=1}^{\infty} \frac{1}{k(k+1)}.$   b) $\displaystyle\sum_{k=2}^{\infty} \log\left(\frac{k(k+2)}{(k+1)^2}\right).$

c) $\displaystyle\sum_{k=1}^{\infty} \sqrt[k]{\frac{\pi}{4}\left(1 - \left(\frac{\pi}{4}\right)^{j_k}\right)}$, where $j_k = -1/(k(k+1))$ for $k \in \mathbf{N}$.

**4.** Find all $x \in \mathbf{R}$ for which

$$\sum_{k=1}^{\infty} 3(x^k - x^{k-1})(x^k + x^{k-1})$$

converges. For each such $x$, find the value of this series.

**5.** Prove that each of the following series diverges.

a) $\displaystyle\sum_{k=1}^{\infty} \cos\left(\frac{1}{k^2}\right).$   b) $\displaystyle\sum_{k=1}^{\infty}\left(1 - \frac{1}{k}\right)^k.$   c) $\displaystyle\sum_{k=1}^{\infty} \frac{k+1}{k^2}.$

$\lim\limits_{n \to \infty} a_n \neq 0$

**6.** a) Prove that if $\sum_{k=1}^{\infty} a_k$ converges, then its partial sums $s_n$ are bounded.

b) Show that the converse of part a) is false. Namely, show that a series $\sum_{k=1}^{\infty} a_k$ may have bounded partial sums and still diverge.

if $\lim\limits_{n \to \infty} a_n \neq 0$, $\sum\limits_{n=1}^{\infty} a_n$ diverges.

**7.** Let $\{b_k\}$ be a real sequence and $b \in \mathbf{R}$.

a) Suppose there is an $N \in \mathbf{N}$ such that $|b - b_k| \leq M$ for all $k \geq N$. Prove that

$$\left| nb - \sum_{k=1}^{n} b_k \right| \leq \sum_{k=1}^{N} |b_k - b| + M(n - N)$$

for all $n > N$.

b) Prove that if $b_k \to b$ as $k \to \infty$, then

$$\frac{b_1 + b_2 + \cdots + b_n}{n} \to b$$

as $n \to \infty$.

c) Show that the converse of b) is false.

**8.** A series $\sum_{k=0}^{\infty} a_k$ is said to be *Cesàro summable* to $L$ if

$$\sigma_n := \sum_{k=0}^{n-1} \left( 1 - \frac{k}{n} \right) a_k$$

converges to $L$ as $n \to \infty$.

a) Let $s_n = \sum_{k=0}^{n-1} a_k$. Prove that

$$\sigma_n = \frac{s_1 + \cdots + s_n}{n}$$

for each $n \in \mathbf{N}$.

b) Prove that if $a_k \in \mathbf{R}$ and $\sum_{k=0}^{\infty} a_k = L$ converges, then $\sum_{k=0}^{\infty} a_k$ is Cesàro summable to $L$.

c) Prove that $\sum_{k=0}^{\infty} (-1)^k$ is Cesàro summable to $1/2$; hence the converse of b) is false.

d) [TAUBER]. Prove that if $a_k \geq 0$ for $k \in \mathbf{N}$ and $\sum_{k=0}^{\infty} a_k$ is Cesàro summable to $L$, then $\sum_{k=0}^{\infty} a_k = L$.

**9.** a) Suppose $\{a_k\}$ is a decreasing sequence of real numbers. Prove that if $\sum_{k=1}^{\infty} a_k$ converges, then $ka_k \to 0$ as $k \to \infty$.

b) Let $s_n = \sum_{k=1}^{n} (-1)^{k+1}/k$ for $n \in \mathbf{N}$. Prove that $s_{2n}$ is strictly increasing, $s_{2n+1}$ is strictly decreasing, and $s_{2n+1} - s_{2n} \to 0$ as $n \to \infty$.

c) Prove that part a) is false if "decreasing" is removed.

**10.** Suppose $a_k \geq 0$ for $k$ large and $\sum_{k=1}^{\infty} a_k/k$ converges. Prove

$$\lim_{j \to \infty} \sum_{k=1}^{\infty} \frac{a_k}{j + k} = 0.$$

## 6.2  SERIES WITH NONNEGATIVE TERMS

Although we obtained exact values in the previous section for telescopic series and geometric series, finding exact values of a given series is frequently difficult, if not impossible. Fortunately, for many applications it is not as important to be able to find the value of a series as it is to know that the series converges. When it does converge, we can use its partial sums to approximate its value as accurately as we wish (up to the limitations of whatever computing device we are using). Therefore, much of this chapter is devoted to establishing tests which can be used to decide whether a given series converges or whether it diverges.

Let $\mathcal{P}_k$ be a statement which depends on $k \in \mathbf{N}$. We shall say that $\mathcal{P}_k$ holds for *large k* if there is an $N \in \mathbf{N}$ such that $\mathcal{P}_k$ is true for $k \geq N$.

The partial sums of a divergent series may be bounded (like $\sum_{k=1}^{\infty}(-1)^k$) or unbounded (like $\sum_{k=1}^{\infty} 1/k$). When the terms of a divergent series are nonnegative, the former cannot happen.

**6.11 THEOREM.** *Suppose $a_k \geq 0$ for $k \geq N$. Then $\sum_{k=1}^{\infty} a_k$ converges if and only if its sequence of partial sums $\{s_n\}$ is bounded, i.e., if and only if there exists a finite number $M > 0$ such that*

$$\left| \sum_{k=1}^{n} a_k \right| \leq M \text{ for all } n \in \mathbf{N}.$$

PROOF. Set $s_n = \sum_{k=1}^{n} a_k$ for $n \in \mathbf{N}$. If $\sum_{k=1}^{\infty} a_k$ converges, then $s_n$ converges as $n \to \infty$. Since every convergent sequence is bounded (Theorem 2.8), $\sum_{k=1}^{\infty} a_k$ has bounded partial sums.

Conversely, suppose $|s_n| \leq M$ for $n \in \mathbf{N}$. Since $a_k \geq 0$ for $k \geq N$, $s_n$ is an increasing sequence when $n \geq N$. Hence by the Monotone Convergence Theorem (Theorem 2.19), $s_n$ converges. ∎

If $a_k \geq 0$ for large $k$, we shall write $\sum_{k=1}^{\infty} a_k < \infty$ when the series is convergent and $\sum_{k=1}^{\infty} a_k = \infty$ when the series is divergent.

In some cases, integration can be used to test convergence of a series. The idea behind this test is that

$$\int_{1}^{\infty} f(x)\, dx = \sum_{k=1}^{\infty} \int_{k}^{k+1} f(x)\, dx \approx \sum_{k=1}^{\infty} f(k)$$

when $f$ is almost constant on each interval $[k, k+1]$. This will surely be the case for large $k$ if $f(k) \downarrow 0$ as $k \to \infty$ (see Figure 6.1). This observation leads us to the following result.

**6.12 THEOREM** [INTEGRAL TEST]. *Suppose $f : [1, \infty) \to \mathbf{R}$ is positive and decreasing on $[1, \infty)$. Then $\sum_{k=1}^{\infty} f(k)$ converges if and only if $f$ is improperly integrable on $[1, \infty)$, i.e., if and only if*

$$\int_{1}^{\infty} f(x)\, dx < \infty.$$

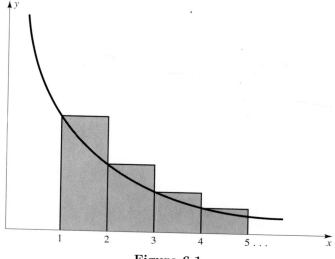

**Figure 6.1**

PROOF.    Let $s_n = \sum_{k=1}^{n} f(k)$ and $t_n = \int_1^n f(x)\,dx$ for $n \in \mathbf{N}$. Since $f$ is decreasing, $f$ is locally integrable on $[1, \infty)$ and $f(k+1) \le f(x) \le f(k)$ for all $x \in [k, k+1]$. Hence, by the Comparison Theorem for Integrals,

$$f(k+1) \le \int_k^{k+1} f(x)\,dx \le f(k)$$

for $k \in \mathbf{N}$. Summing over $k = 1, \ldots, n-1$, we obtain

$$s_n - f(1) = \sum_{k=2}^{n} f(k) \le \int_1^n f(x)\,dx = t_n \le \sum_{k=1}^{n-1} f(k) = s_n - f(n)$$

for all $n \ge N$. In particular,

(3) $$f(n) \le \sum_{k=1}^{n} f(k) - \int_1^n f(x)\,dx \le f(1) \quad \text{for} \ \ n \in \mathbf{N}.$$

By (3) it is clear that $\{s_n\}$ is bounded if and only if $\{t_n\}$ is. Since $f(x) \ge 0$ implies both $s_n$ and $t_n$ are increasing sequences, it follows from the Monotone Convergence Theorem that $s_n$ converges if and only if $t_n$ converges, as $n \to \infty$. ∎

This test works best on series for which the integral of $f$ can be easily computed or estimated. For example, to find out whether $\sum_{k=1}^{\infty} 1/(1 + k^2)$ converges or diverges, let $f(x) = 1/(1 + x^2)$ and observe that $f$ is positive on $[1, \infty)$. Since $f'(x) = -2x/(1 + x^2)^2$ is negative on $[1, \infty)$, it is also clear that $f$ is decreasing. Since

$$\int_1^\infty \frac{dx}{1 + x^2} = \arctan x \ \Big|_1^\infty = \frac{\pi}{2} - \arctan(1) < \infty,$$

it follows from the Integral Test that $\sum_{k=1}^{\infty} 1/(1 + k^2)$ converges.

The Integral Test is most widely used in the following special case.

**6.13 COROLLARY** [*p*-SERIES TEST]. *The series*

(4)
$$\sum_{k=1}^{\infty} \frac{1}{k^p}$$

*converges if and only if $p > 1$.*

PROOF. If $p = 1$ or $p \le 0$, the series diverges. If $p > 0$ and $p \neq 1$, set $f(x) = x^{-p}$ and observe that $f'(x) = -px^{-p-1} < 0$ for all $x \in [1, \infty)$. Hence, $f$ is nonnegative and decreasing on $[1, \infty)$. Since

$$\int_1^{\infty} \frac{dx}{x^p} = \lim_{n \to \infty} x^{1-p} \Big|_1^n = \lim_{n \to \infty} \frac{n^{1-p} - 1}{1 - p}$$

has a finite limit if and only if $1 - p < 0$, it follows from the Integral Test that (4) converges if and only if $p > 1$. ∎

The Integral Test, which requires $f$ to satisfy some very restrictive hypotheses, has limited applications. The following test can be used in a much broader context.

**6.14 THEOREM** [COMPARISON TEST]. *Suppose $0 \le a_k \le b_k$ for large $k$.*

  i) *If $\sum_{k=1}^{\infty} b_k < \infty$, then $\sum_{k=1}^{\infty} a_k < \infty$.*   (use Cauchy Criterion)
  ii) *If $\sum_{k=1}^{\infty} a_k = \infty$, then $\sum_{k=1}^{\infty} b_k = \infty$.*

PROOF. By hypothesis, choose $N \in \mathbf{N}$ so large that $0 \le a_k \le b_k$ for $k > N$. Set $s_n = \sum_{k=1}^n a_k$ and $t_n = \sum_{k=1}^n b_k$, $n \in \mathbf{N}$. Then $0 \le s_n - s_N \le t_n - t_N$ for all $n \ge N$. Since $N$ is fixed, it follows that $s_n$ is bounded when $t_n$ is, and $t_n$ is unbounded when $s_n$ is. Apply Theorem 6.11 and the proof of the theorem is complete. ∎

The Comparison Test is used to compare one series with another whose convergence property is already known, e.g., a $p$-series or a geometric series. Frequently, the inequalities $|\sin x| \le |x|$ for all $x \in \mathbf{R}$ (see Appendix B) and $|\log x| \le x^{\alpha}$ for each $\alpha > 0$ provided $x$ is sufficiently large (see Exercise 4 in Section 4.3) are helpful in this regard. Although there is no simple algorithm for this process, the idea is to examine the terms of the given series, ignoring the superfluous factors, and to replace the more complicated factors by simpler ones. Here is a typical example.

**6.15 Example.** Determine whether the series

(5)
$$\sum_{k=1}^{\infty} \frac{3k}{k^2 + k} \sqrt{\frac{\log k}{k}}$$

converges or diverges.

SOLUTION. The $k$th term of this series can be written by using three factors:

$$\frac{1}{k} \frac{3k}{k+1} \sqrt{\frac{\log k}{k}}.$$

The factor $3k/(k+1)$ is approximately 3 for large $k$ and can be ignored. Since $\log k \leq \sqrt{k}$ for large $k$, the factor $\sqrt{\log k/k}$ satisfies

$$\sqrt{\frac{\log k}{k}} \leq \sqrt{\frac{\sqrt{k}}{k}} = \frac{1}{\sqrt[4]{k}}$$

for large $k$. Therefore, the terms of (5) are dominated by $3/k^{5/4}$. Since $\sum_{k=1}^{\infty} 3/k^{5/4}$ converges by the $p$-Series Test, it follows from the Comparison Test that (5) converges. ∎

The Comparison Test may not be easy to apply to a given series, even when we know which series it should be compared with, because the process of comparison often involves use of delicate inequalities. For situations like this, the following test is usually more efficient.

**6.16 THEOREM** [LIMIT COMPARISON TEST]. *Suppose $a_k \geq 0$ and $b_k > 0$ for large $k$ and $L := \lim_{n \to \infty} a_n/b_n$ exists as an extended real number.*
   i) *If $0 \leq L < \infty$ and $\sum_{k=1}^{\infty} b_k$ converges, then $\sum_{k=1}^{\infty} a_k$ converges.*
   ii) *If $0 < L \leq \infty$ and $\sum_{k=1}^{\infty} b_k$ diverges, then $\sum_{k=1}^{\infty} a_k$ diverges.*
   iii) *If $0 < L < \infty$, then $\sum_{k=1}^{\infty} a_k$ converges if and only if $\sum_{k=1}^{\infty} b_k$ converges.*

PROOF. If $L$ is finite and nonzero, then there is an $N \in \mathbf{N}$ such that

$$\frac{L}{2} b_k < a_k < \frac{3L}{2} b_k$$

for $k \geq N$. Hence, part iii) follows immediately from the Comparison Test and Theorem 6.10. Similar arguments establish parts i) and ii)—see Exercise 8 below. ∎

In general, the Limit Comparison Test is used to replace a series $\sum_{k=1}^{\infty} a_k$ by $\sum_{k=1}^{\infty} b_k$ when $a_k \approx C b_k$ for $k$ large and some absolute fixed constant $C$. For example, to determine whether or not the series

$$S := \sum_{k=1}^{\infty} \frac{k}{\sqrt{4k^4 + k^2 + 5k}}$$

converges, notice that its terms are approximately $1/(2k)$ for $k$ large. This leads us to compare $S$ with the harmonic series $\sum_{k=1}^{\infty} 1/k$. Since the harmonic series diverges and

$$\frac{k/(\sqrt{4k^4 + k^2} + 5k)}{1/k} = \frac{k^2}{\sqrt{4k^4 + k^2} + 5k} \to \frac{1}{2} > 0$$

as $k \to \infty$, it follows from the Limit Comparison Test that $S$ diverges.

Here is another application of the Limit Comparison Test.

**6.17 Example.** Let $a_k \to 0$ as $k \to \infty$. Prove that $\sum_{k=1}^{\infty} \sin |a_k|$ converges if and only if $\sum_{k=1}^{\infty} |a_k|$ converges.

PROOF. By l'Hôpital's Rule,

$$\lim_{k \to \infty} \frac{\sin |a_k|}{|a_k|} = \lim_{x \to 0+} \frac{\sin x}{x} = 1.$$

Hence, by the Limit Comparison Test, $\sum_{k=1}^{\infty} \sin |a_k|$ converges if and only if $\sum_{k=1}^{\infty} |a_k|$ converges. ∎

## EXERCISES

**1.** Prove that each of the following series converges.

a) $\displaystyle\sum_{k=1}^{\infty} \frac{k-3}{k^3+k+1}.$     b) $\displaystyle\sum_{k=1}^{\infty} \frac{k-1}{k2^k}.$     c) $\displaystyle\sum_{k=1}^{\infty} \frac{\log k}{k^p}, \quad p > 1.$

d) $\displaystyle\sum_{k=1}^{\infty} \frac{1}{\sqrt{k}3^{k-1}}.$     e) $\displaystyle\sum_{k=1}^{\infty} \left(10 + \frac{1}{k}\right)k^{-e}.$     f) $\displaystyle\sum_{k=1}^{\infty} \frac{3k^2 - \sqrt{k}}{k^4 - k^2 + 1}.$

**2.** Prove that each of the following series diverges.

a) $\displaystyle\sum_{k=1}^{\infty} \frac{\sqrt[k]{k}}{k}.$     b) $\displaystyle\sum_{k=1}^{\infty} \frac{1}{\log^p(k+1)}, \quad p > 0.$

c) $\displaystyle\sum_{k=1}^{\infty} \frac{k^2+2k+3}{k^3-2k^2+\sqrt{2}}.$     d) $\displaystyle\sum_{k=2}^{\infty} \frac{1}{k\log^p k}, \quad p \le 1.$

**3.** Find all $p \ge 0$ such that the following series converges.

$$\sum_{k=1}^{\infty} \frac{1}{k \log^p(k+1)}.$$

**4.** If $a_k \ge 0$ is a bounded sequence, prove that

$$\sum_{k=1}^{\infty} \frac{a_k}{(k+1)^p}$$

converges for all $p > 1$.

**5.** Suppose $a_k \in [0,1)$ and $a_k \to 0$ as $k \to \infty$. Prove that $\sum_{k=1}^{\infty} \arcsin a_k$ converges if and only if $\sum_{k=1}^{\infty} a_k$ converges.

**6.** If $\sum_{k=1}^{\infty} |a_k|$ converges, prove that

$$\sum_{k=1}^{\infty} \frac{|a_k|}{k^p}$$

converges for all $p \ge 0$. What happens if $p < 0$?

**7.** Suppose $a_k$ and $b_k$ are nonnegative for all $k \in \mathbf{N}$.

a) Prove that if $\sum_{k=1}^{\infty} a_k$ and $\sum_{k=1}^{\infty} b_k$ converge, then $\sum_{k=1}^{\infty} a_k b_k$ also converges.

b) Improve this result by replacing convergence of one of the series by something else.

**8.** Prove Theorem 6.16i) and ii).

**9.** Suppose $a, b \in \mathbf{R}$ satisfy $b/a \in \mathbf{R} \setminus \mathbf{Z}$. Find all $q > 0$ such that

$$\sum_{k=1}^{\infty} \frac{1}{(ak + b)q^k}$$

converges.

**10.** Suppose $a_k \to 0$. Prove that $\sum_{k=1}^{\infty} a_k$ converges if and only if the series $\sum_{k=1}^{\infty} (a_{2k} + a_{2k+1})$ converges.

## 6.3  ABSOLUTE CONVERGENCE

In this section we investigate what happens to a convergent series when its terms are replaced by their absolute values. We begin with some terminology.

**6.18 DEFINITION.** Let $S = \sum_{k=1}^{\infty} a_k$ be an infinite series.

   i) $S$ is said to *converge absolutely* if $\sum_{k=1}^{\infty} |a_k| < \infty$.

   ii) $S$ is said to *converge conditionally* if $S$ converges but not absolutely.

The Cauchy Criterion gives us the following test for absolute convergence.

**6.19 Remark.** *A series $\sum_{k=1}^{\infty} a_k$ converges absolutely if and only if for every $\varepsilon > 0$ there is an $N \in \mathbf{N}$ such that*

$$(6) \qquad\qquad m > n \geq N \quad implies \quad \sum_{k=n}^{m} |a_k| < \varepsilon.$$

As was the case for improper integrals, absolute convergence is stronger than convergence.

**6.20 Remark.** *If $\sum_{k=1}^{\infty} a_k$ converges absolutely, then $\sum_{k=1}^{\infty} a_k$ converges, but not conversely. In particular, there exist conditionally convergent series.*

PROOF. Suppose $\sum_{k=1}^{\infty} a_k$ converges absolutely. Given $\varepsilon > 0$, choose $N \in \mathbf{N}$ so that (6) holds. Then

$$\left| \sum_{k=n}^{m} a_k \right| \leq \sum_{k=n}^{m} |a_k| < \varepsilon$$

for $m > n \geq N$. Hence, by the Cauchy Criterion, $\sum_{k=1}^{\infty} a_k$ converges.

We shall finish the proof by showing that $\sum_{k=1}^{\infty} (-1)^k/k$ converges conditionally. It surely does not converge absolutely since the harmonic series diverges. On the other hand, notice that

$$\sum_{j=k}^{\infty} \frac{(-1)^j}{j} = (-1)^k \left( \frac{1}{k} - \frac{1}{k+1} + \frac{1}{k+2} - \frac{1}{k+3} + \dots \right).$$

By grouping pairs of terms together, it is easy to see that the sum inside the parentheses is greater than 0 but less than $1/k$, i.e.,

$$\left| \sum_{j=k}^{\infty} \frac{(-1)^j}{j} \right| < \frac{1}{k}.$$

Hence $\sum_{k=1}^{\infty} (-1)^k/k$ converges by Corollary 6.9. ∎

We shall presently see that it is important to be able to identify absolutely conver-gent series. Since every result about series with nonnegative terms can be applied to the series $\sum_{k=1}^{\infty} |a_k|$, we already have three tests for absolute convergence (the Integral Test, the Comparison Test, and the Limit Comparison Test). We now develop two additional tests for absolute convergence which are arguably the most practical tests presented in this chapter.

Before we state these tests, we need to introduce another concept. (If you covered Section 2.5, you may skip the next half page and proceed directly to Theorem 6.23.) An extended real number $a$ is called an *adherent point* of a real sequence $\{x_k\}$ if there is a subsequence of $\{x_k\}$ which satisfies $x_{k_j} \to a$ as $j \to \infty$. For example, 1 and $-1$ are adherent points of $\{(-1)^k\}$ and $\infty$ is an adherent point of $\{\log k\}$.

Notice once and for all that if $a$ is an adherent point of a subsequence of $\{x_k\}$, then it is an adherent point of $\{x_k\}$. Also notice that every real sequence has at least one adherent point. Indeed, if the sequence if unbounded, then by definition, either $\infty$ or $-\infty$ is an adherent point. On the other hand, if it is bounded, then by the Bolzano–Weierstrass Theorem, it has a finite adherent point. Hence, the following concept makes sense.

**6.21 Definition.** The *limit supremum* of a sequence $\{x_k\}$, denoted by $\limsup_{k \to \infty} x_k$, is the largest adherent point of $\{x_k\}$.

Thus the limit supremum of $(-1)^k$ is 1, of $3 + (-1)^k$ is 4, and of $-2 - (-1)^k$ is $-1$. (Definition 2.32, a more sophisticated definition of this concept which explains the etymology of the word "limit supremum," is equivalent to Definition 6.21 by Remark 2.37.)

The only thing we need to know about limits supremum (for now) is the following result.

**6.22 Remark.** *Let $x \in \mathbf{R}$ and $\{x_k\}$ be a real sequence.*

  i) *If $\limsup_{k \to \infty} x_k < x$ then $x_k < x$ for large $k$.*
  ii) *If $\limsup_{k \to \infty} x_k > x$ then $x_k > x$ for infinitely many $k$.*
  iii) *If $x_k \to x$ as $k \to \infty$, then $\limsup_{k \to \infty} x_k = x$.*

PROOF. i) Let $s := \limsup_{k \to \infty} x_k < x$ but suppose to the contrary that there exist natural numbers $k_1 < k_2 < \dots$ such that $x_{k_j} \geq x$ for $j \in \mathbf{N}$. If $\{x_{k_j}\}$ is unbounded above, then $\infty$ is an adherent point of $\{x_k\}$ so $s = \infty$, a contradiction. If $\{x_{k_j}\}$ is bounded above (by $C$), then it is bounded (since $x \leq x_{k_j} \leq C$ for all

$j \in \mathbf{N}$). Hence, by the Bolzano–Weierstrass Theorem and the fact that $x_{k_j} \geq x$, $\{x_k\}$ has an adherent point $\geq x$, i.e., $s \geq x$, another contradiction.

ii) If $s > x$, then by definition there is a subsequence $\{x_{k_j}\}$ which converges to $s$, i.e., $x_{k_j} > x$ for large $j$.

iii) If $x_k$ converges to $x$, then any subsequence $x_{k_j}$ also converges to $x$ (see Theorem 2.6). ∎

The limit supremum gives a very useful and efficient test for absolute convergence.

**6.23 THEOREM** [ROOT TEST]. *Let $a_k \in \mathbf{R}$ and $r := \limsup_{k\to\infty} |a_k|^{1/k}$.*

  i) *If $r < 1$, then $\sum_{k=1}^{\infty} a_k$ converges absolutely.*

  ii) *If $r > 1$, then $\sum_{k=1}^{\infty} a_k$ diverges.*

$$r = \lim_{k\to\infty} |a_k|^{\frac{1}{k}}$$

PROOF. i) Suppose $r < 1$. Let $r < x < 1$ and notice that the geometric series $\sum_{k=1}^{\infty} x^k$ converges. By Remark 6.22 or Exercise 3 in Section 2.5,

$$|a_k|^{1/k} < x$$

for large $k$. Hence, $|a_k| < x^k$ for large $k$ and it follows from the Comparison Test that $\sum_{k=1}^{\infty} |a_k|$ converges.

ii) Suppose $r > 1$. By Remark 6.22 or Exercise 3 in Section 2.5,

$$|a_k|^{1/k} > 1$$

for infinitely many $k \in \mathbf{N}$. Hence, $|a_k| > 1$ for infinitely many $k$ and it follows from the Divergence Test that $\sum_{k=1}^{\infty} a_k$ diverges. ∎

Note by Remark 6.22iii) or Theorem 2.36, if $\lim_{k\to\infty} |a_k|^{1/k}$ exists, then (by the Root Test) $\sum_{k=1}^{\infty} a_k$ converges absolutely when $r < 1$ and diverges when $r > 1$.

The following test is weaker than the Root Test (see Exercise 9) but is easier to use when the terms of $\sum_{k=1}^{\infty} a_k$ are made up of products (e.g., factorials).

**6.24 THEOREM** [RATIO TEST]. *Let $a_k \in \mathbf{R}$ with $a_k \neq 0$ for large $k$ and suppose that*

$$r = \lim_{k\to\infty} \frac{|a_{k+1}|}{|a_k|}$$

*exists as an extended real number.*

  i) *If $r < 1$, then $\sum_{k=1}^{\infty} a_k$ converges absolutely.*

  ii) *If $r > 1$, then $\sum_{k=1}^{\infty} a_k$ diverges.*

PROOF. If $r > 1$, then $|a_{k+1}| \geq |a_k|$ for $k$ large and thus $a_k$ cannot converge to zero. Hence, by the Divergence Test, $\sum_{k=1}^{\infty} a_k$ diverges.

If $r < 1$, then observe for any $x \in (r, 1)$ that

$$\frac{|a_{k+1}|}{|a_k|} < x = \frac{x^{k+1}}{x^k}$$

for $k$ large. Hence, the sequence $|a_k|/x^k$ is decreasing for large $k$ and thus bounded. In particular, there is an $M > 0$ such that $|a_k| \leq Mx^k$ for all $k \in \mathbf{N}$. Since $x < 1$, it follows from the Comparison Test that $\sum_{k=1}^{\infty} |a_k|$ converges. ∎

**6.25 Remark.** *The Root and Ratio Tests are inconclusive when $r = 1$.*

For example, under the Ratio Test $\sum_{k=1}^{\infty} 1/k$ and $\sum_{k=1}^{\infty} 1/k^2$ both yield $r = 1$. Nevertheless, the first series diverges whereas the second converges absolutely.

There are two ways to proceed when $r = 1$. There are tests which conclude that a series converges provided its ratios converge to 1 rapidly enough. (Three of these tests are covered in Section 6.6 and its exercises.) And, there is a very useful asymptotic estimate of $k!$ (called Stirling's Formula–see Theorem 12.74) which you may find useful on series which factors of the form $k!/k^k$ (see Exercise 6e below and Exercise 2c in Section 6.6).

It is natural to assume that the usual laws of algebra hold for infinite series, e.g., associativity and commutativity. Is this assumption warranted? We have "inserted parentheses" (i.e., grouped terms together) to aid evaluation of some series (e.g., to evaluate some telescopic series and to prove $\sum_{k=1}^{\infty}(-1)^k/k$ converges conditionally). This is valid for convergent series (absolutely or conditionally) because if the sequence of partial sums $s_n$ converges to $s$, then any subsequence $s_{n_k}$ also converges to $s$. The situation is more complicated when we start changing the order of the terms (compare Theorem 6.27 with Theorem 6.29). To describe what happens, we introduce the following terminology.

**6.26 DEFINITION.** A series $\sum_{j=1}^{\infty} b_j$ is called a *rearrangement* of a series $\sum_{k=1}^{\infty} a_k$ if there is a 1–1 function $f$ from $\mathbf{N}$ onto $\mathbf{N}$ such that

$$b_{f(k)} = a_k, \qquad k \in \mathbf{N}.$$

The following result demonstrates why absolutely convergent series are so important.

**6.27 THEOREM.** *If $\sum_{k=1}^{\infty} a_k$ converges absolutely and $\sum_{j=1}^{\infty} b_j$ is any rearrangement of $\sum_{k=1}^{\infty} a_k$, then $\sum_{j=1}^{\infty} b_j$ converges and*

$$\sum_{k=1}^{\infty} a_k = \sum_{j=1}^{\infty} b_j.$$

PROOF. Let $\varepsilon > 0$. Set $s_n = \sum_{k=1}^{n} a_k$, $s = \sum_{k=1}^{\infty} a_k$, and $t_m = \sum_{j=1}^{m} b_j$, $n, m \in \mathbf{N}$. Since $\sum_{k=1}^{\infty} a_k$ converges absolutely, we can choose $N \in \mathbf{N}$ (see Corollary 6.9) such that

$$(9) \qquad \sum_{k=N+1}^{\infty} |a_k| < \frac{\varepsilon}{2}.$$

Thus

$$(10) \qquad |s_N - s| = \left| \sum_{k=N+1}^{\infty} a_k \right| \le \sum_{k=N+1}^{\infty} |a_k| < \frac{\varepsilon}{2}.$$

Let $f$ be a 1–1 function from $\mathbf{N}$ onto $\mathbf{N}$ which satisfies

$$b_{f(k)} = a_k, \qquad k \in \mathbf{N}$$

and set $M = \max\{f(1), \ldots, f(N)\}$. Notice that

$$\{a_1, \ldots, a_N\} \subseteq \{b_1, \ldots, b_M\}.$$

Let $m \geq M$. Then $t_m - s_N$ contains only $a_k$'s whose indices satisfy $k > N$. Thus, it follows from (9) that

$$|t_m - s_N| \leq \sum_{k=N+1}^{\infty} |a_k| < \frac{\varepsilon}{2}.$$

Hence, by (10),

$$|t_m - s| \leq |t_m - s_N| + |s_N - s| < \frac{\varepsilon}{2} + \frac{\varepsilon}{2} = \varepsilon$$

for $m \geq M$. Therefore,

$$s = \sum_{j=1}^{\infty} b_j. \quad \blacksquare$$

*The rest of this section, which is used nowhere else in this book, is optional.*

We now show that Theorem 6.27 fails in a catastrophic way for conditionally convergent series (see Theorem 6.29 below). To facilitate our discussion, recall (see Exercise 1 in Section 1.1) that the *positive and negative parts* of an $a \in \mathbf{R}$ are defined by

$$a^+ := \frac{|a| + a}{2} = \begin{cases} a & a \geq 0 \\ 0 & a < 0 \end{cases}$$

and

$$a^- := \frac{|a| - a}{2} = \begin{cases} 0 & a \geq 0 \\ -a & a < 0. \end{cases}$$

Notice that

(7) $$a^+ \geq 0, \quad a^- \geq 0,$$

and

(8) $$a = a^+ - a^-, \qquad |a| = a^+ + a^-$$

for all $a \in \mathbf{R}$.

**\*6.28 Lemma.** *Suppose $a_k \in \mathbf{R}$ for $k \in \mathbf{N}$.*

i) *If $\sum_{k=1}^{\infty} a_k$ converges absolutely, then so do $\sum_{k=1}^{\infty} a_k^+$ and $\sum_{k=1}^{\infty} a_k^-$. In fact,*

$$\sum_{k=1}^{\infty} |a_k| = \sum_{k=1}^{\infty} a_k^+ + \sum_{k=1}^{\infty} a_k^- \quad \text{and} \quad \sum_{k=1}^{\infty} a_k = \sum_{k=1}^{\infty} a_k^+ - \sum_{k=1}^{\infty} a_k^-.$$

ii) *If $\sum_{k=1}^{\infty} a_k$ converges conditionally, then*

$$\sum_{k=1}^{\infty} a_k^+ = \sum_{k=1}^{\infty} a_k^- = \infty.$$

PROOF. By definition, $a_k^+ = (|a_k| + a_k)/2$. Since both $\sum_{k=1}^{\infty} |a_k|$ and $\sum_{k=1}^{\infty} a_k$ converge, it follows from Theorem 6.10 that

$$\sum_{k=1}^{\infty} a_k^+ = \frac{1}{2} \sum_{k=1}^{\infty} |a_k| + \frac{1}{2} \sum_{k=1}^{\infty} a_k$$

converges. Similarly,

$$\sum_{k=1}^{\infty} a_k^- = \frac{1}{2} \sum_{k=1}^{\infty} |a_k| - \frac{1}{2} \sum_{k=1}^{\infty} a_k$$

converges. This proves part i).

Suppose part ii) is false. By symmetry we may suppose that $\sum_{k=1}^{\infty} a_k^+$ converges. Since $\sum_{k=1}^{\infty} a_k$ converges, it follows from (8) that

$$\sum_{k=1}^{\infty} a_k^- = \sum_{k=1}^{\infty} a_k^+ - \sum_{k=1}^{\infty} a_k$$

converges. Thus,

$$\sum_{k=1}^{\infty} |a_k| = \sum_{k=1}^{\infty} a_k^+ + \sum_{k=1}^{\infty} a_k^-$$

converges, a contradiction. ∎

We are prepared to show that Theorem 6.27 is false if the hypothesis "absolutely convergent" is dropped. In fact, as the following result shows, rearrangements of conditionally convergent series can converge to anything one wishes (see also Exercise 10 below).

**\*6.29 THEOREM** [RIEMANN]. *Let $x \in \mathbf{R}$. If $\sum_{k=1}^{\infty} a_k$ is conditionally convergent, then there is a rearrangement of $\sum_{k=1}^{\infty} a_k$ which converges to $x$.*

STRATEGY: The idea behind the proof is simple. Since $\sum_{k=1}^{\infty} a_k^+ = \sum_{k=1}^{\infty} a_k^- = \infty$ by Theorem 6.28, begin by adding enough $a_j^+$'s until the resulting partial sum is

$> x$. Then subtract enough $a_k^-$'s until the resulting partial sum is $< x$, and continue adding and subtracting. Since $a_k \to 0$ as $k \to \infty$, the resulting partial sums should be getting closer to $x$. We now make this precise.

PROOF. Since $\sum_{k=1}^{\infty} a_k^+ = \infty$, let $k_1$ be the smallest integer which satisfies

$$s_{k_1} := a_1^+ + a_2^+ + \cdots + a_{k_1}^+ > x.$$

Since $k_1$ is least, $s_{k_1-1} := a_1^+ + a_2^+ + \cdots + a_{k_1-1}^+ \le x$, so $s_{k_1} \le x + a_{k_1}^+$. Therefore,

$$(11) \qquad |s_{k_j} - x| \le a_{k_j}^+$$

for $j = 1$. Similarly, since $\sum_{k=1}^{\infty} a_k^- = \infty$, let $r_1 > k_1$ be the smallest integer which satisfies

$$s_{r_1} := s_{k_1} - a_1^- - \cdots - a_{r_1-k_1}^- < x$$

and observe that

$$(12) \qquad |s_{r_j} - x| \le a_{r_j-k_j}^-$$

for $j = 1$.

Continuing, we generate integers $k_1 < r_1 < k_2 < \ldots$ least so that

$$s_{k_{j+1}} := s_{r_j} + a_{k_j+1}^+ \cdots + a_{k_{j+1}}^+ > x$$

and

$$s_{r_{j+1}} := s_{k_j} - a_{r_j-k_j+1}^- - \cdots - a_{r_{j+1}-k_{j+1}}^- < x,$$

and so that (11) and (12) hold. Since each $a_k^+$ and $-a_k^-$ is either $a_k$ or 0, it is clear (after deleting the zero terms) that the $s_n$'s are the partial sums of a rearrangement of $\sum_{k=1}^{\infty} a_k$. Moreover, since $a_k \to 0$ as $k \to \infty$, (11) and (12) together with the Squeeze Theorem imply that both $s_{k_j}$ and $s_{r_j}$ converge to $x$ as $j \to \infty$.

Suppose $n \in \mathbf{N}$. Then there is a $j \in \mathbf{N}$ such that either $k_j \le n < r_j$ or $r_j \le n < k_{j+1}$. In the former case, since $s_n$ is formed from $s_{k_j}$ by adding negative terms,

$$s_{r_j} < s_n \le s_{k_j}.$$

Similarly, in the latter case we have

$$s_{r_j} \le s_n < s_{k_{j+1}}.$$

We conclude by the Squeeze Theorem that $s_n \to x$ as $n \to \infty$. ∎

## EXERCISES

**1.** Prove that each of the following series converges.

a) $\displaystyle\sum_{k=1}^{\infty} \frac{1}{k!}.$    b) $\displaystyle\sum_{k=1}^{\infty} \frac{1}{k^k}.$    c) $\displaystyle\sum_{k=1}^{\infty} \frac{2^k}{k!}.$    d) $\displaystyle\sum_{k=1}^{\infty} \left(\frac{k}{k+1}\right)^{k^2}.$

2. Decide, using results covered so far in this chapter, which of the following series converge and which diverge.

a) $\displaystyle\sum_{k=1}^{\infty} \frac{k^2}{\pi^k}.$    b) $\displaystyle\sum_{k=1}^{\infty} \frac{k!}{2^k}.$    c) $\displaystyle\sum_{k=1}^{\infty} \left(\frac{k+1}{2k+3}\right)^k.$    d) $\displaystyle\sum_{k=1}^{\infty} \left(\pi - \frac{1}{k}\right) k^{-1}.$

e) $\displaystyle\sum_{k=1}^{\infty} \left(\frac{k!}{(k+2)!}\right)^{k^2}.$    f) $\displaystyle\sum_{k=1}^{\infty} \left(\frac{3+(-1)^k}{3}\right)^k.$    g) $\displaystyle\sum_{k=1}^{\infty} \frac{(1+(-1)^k)^k}{e^k}.$

3. Using Exercise 9 in Section 5.3, prove that

$$\sin x = \sum_{k=0}^{\infty} \frac{(-1)^k x^{2k+1}}{(2k+1)!} \quad \text{and} \quad \cos x = \sum_{k=0}^{\infty} \frac{(-1)^k x^{2k}}{(2k)!}$$

for all $x \in [0, \pi/2]$.

4. Define $a_k$ recursively by $a_1 = 1$ and

$$a_k = (-1)^k \left(1 + k \sin\left(\frac{1}{k}\right)\right)^{-1} a_{k-1}, \qquad k > 1.$$

Prove that $\sum_{k=1}^{\infty} a_k$ converges absolutely.

5. Suppose $a_k \geq 0$ and $a_k^{1/k} \to a$ as $k \to \infty$. Prove that $\sum_{k=1}^{\infty} a_k x^k$ converges absolutely for all $|x| < 1/a$ if $a \neq 0$ and for all $x \in \mathbf{R}$ if $a = 0$.

6. For each of the following, find all values of $p \in \mathbf{R}$ for which the given series converges absolutely.

a) $\displaystyle\sum_{k=2}^{\infty} \frac{1}{k \log^p k}.$    b) $\displaystyle\sum_{k=2}^{\infty} \frac{1}{\log^p k}.$    c) $\displaystyle\sum_{k=1}^{\infty} \frac{k^p}{p^k}.$

d) $\displaystyle\sum_{k=2}^{\infty} \frac{1}{\sqrt{k}(k^p - 1)}.$    e) $\displaystyle\sum_{k=1}^{\infty} \frac{2^{kp} k!}{k^k}.$    f) $\displaystyle\sum_{k=1}^{\infty} (\sqrt{k^{2p}+1} - k^p).$

7. Suppose $a_{kj} \geq 0$ for $k, j \in \mathbf{N}$. Set

$$A_k = \sum_{j=1}^{\infty} a_{kj}$$

for each $k \in \mathbf{N}$, and suppose $\sum_{k=1}^{\infty} A_k$ converges.

a) Prove that

$$\sum_{j=1}^{\infty} \left(\sum_{k=1}^{\infty} a_{kj}\right) \leq \sum_{k=1}^{\infty} \left(\sum_{j=1}^{\infty} a_{kj}\right).$$

b) Show

$$\sum_{j=1}^{\infty} \left( \sum_{k=1}^{\infty} a_{kj} \right) = \sum_{k=1}^{\infty} \left( \sum_{j=1}^{\infty} a_{kj} \right).$$

c) Prove that b) may not hold if $a_{kj}$ has both positive and negative values.
Hint: Consider

$$a_{kj} = \begin{cases} 1 & j = k \\ -1 & j = k+1 \\ 0 & \text{otherwise.} \end{cases}$$

**8.** a) Suppose $\sum_{k=1}^{\infty} a_k$ converges absolutely. Prove that $\sum_{k=1}^{\infty} |a_k|^p$ converges for all $p \geq 1$.

b) Suppose $\sum_{k=1}^{\infty} a_k$ converges conditionally. Prove that $\sum_{k=1}^{\infty} k^p a_k$ diverges for all $p > 1$.

**9.** a) Let $a_n > 0$ for $n \in \mathbf{N}$. Set $b_0 = b_1 = 0$, $b_2 = \log(a_2/a_1)$, and

$$b_k = \log\left(\frac{a_k}{a_{k-1}}\right) - \log\left(\frac{a_{k-1}}{a_{k-2}}\right), \qquad k = 3, 4, \ldots$$

Prove that if

$$r = \lim_{n \to \infty} \frac{a_{n+1}}{a_n}$$

exists and is positive, then

$$\lim_{n \to \infty} \log(a_n^{1/n}) = \lim_{n \to \infty} \sum_{k=0}^{n} \left(1 - \frac{k}{n}\right) b_k = \sum_{k=0}^{\infty} b_k = \log r.$$

b) Prove that if $a_n \in \mathbf{R} \setminus \{0\}$ and $|a_{n+1}/a_n| \to r$ as $n \to \infty$, for some $r > 0$, then $|a_n|^{1/n} \to r$ as $n \to \infty$.

***10.** Let $x \leq y$ be any pair of extended real numbers. Prove that if $\sum_{k=1}^{\infty} a_k$ be conditionally convergent, then there is a rearrangement $\sum_{j=1}^{\infty} b_j$ of $\sum_{k=1}^{\infty} a_k$ whose partial sums $s_n$ satisfy

$$\liminf_{n \to \infty} s_n = x \quad \text{and} \quad \limsup_{n \to \infty} s_n = y.$$

## 6.4  ALTERNATING SERIES

We have identified many tests for absolute convergence, but have said little about conditionally convergent series. In this section we derive two tests to use on series whose terms are of mixed sign.

Both tests rely on the following algebraic observation. (This result will also be used in Chapter 7 to prove that limits of power series are continuous.)

**6.30 THEOREM** [ABEL'S FORMULA]. *Let $\{a_k\}_{k\in\mathbf{N}}$ and $\{b_k\}_{k\in\mathbf{N}}$ be real sequences, and for each pair of integers $n \geq m \geq 1$ set*

$$A_{n,m} = \sum_{k=m}^{n} a_k, \qquad n \geq m.$$

*Then*

$$\sum_{k=m}^{n} a_k b_k = A_{n,m} b_n - \sum_{k=m}^{n-1} A_{k,m}(b_{k+1} - b_k)$$

*for all integers $n > m \geq 1$.*

PROOF. Since $A_{k,m} - A_{(k-1),m} = a_k$ for $k > m$ and $A_{m,m} = a_m$, we have

$$\sum_{k=m}^{n} a_k b_k = a_m b_m + \sum_{k=m+1}^{n} (A_{k,m} - A_{(k-1),m})b_k$$

$$= a_m b_m + \sum_{k=m+1}^{n} A_{k,m} b_k - \sum_{k=m}^{n-1} A_{k,m} b_{k+1}$$

$$= a_m b_m + \sum_{k=m+1}^{n-1} A_{k,m} b_k + A_{n,m} b_n - \sum_{k=m+1}^{n-1} A_{k,m} b_{k+1} - A_{m,m} b_{m+1}$$

$$= A_{n,m} b_n - a_m(b_{m+1} - b_m) - \sum_{k=m+1}^{n-1} A_{k,m}(b_{k+1} - b_k)$$

$$= A_{n,m} b_n - \sum_{k=m}^{n-1} A_{k,m}(b_{k+1} - b_k). \blacksquare$$

This result is somewhat easier to remember using the following analogy. If $f : [1, N] \to \mathbf{R}$ for some $N \in \mathbf{N}$, then the *summation* $\sum_{k=1}^{N-1} f(k)$ is an approximation to $\int_1^N f(x)\,dx$ and the *finite difference* $f(k+1) - f(k)$ is an approximation to $f'(k)$ for $k = 1, 2, \ldots, N-1$. In particular, summation is an analogue of integration and finite difference is an analogue of differentiation. In this context, Abel's formula can be interpreted as a discrete analogue of integration by parts.

Our first application of Abel's Formula is the following test. (Notice that it does not require the $a_k$'s to be nonnegative.)

**6.31 THEOREM** [DIRICHLET'S TEST]. *Let $a_k, b_k \in \mathbf{R}$ for $k \in \mathbf{N}$. If the sequence of partial sums $s_n = \sum_{k=1}^{n} a_k$ is bounded and $b_k \downarrow 0$ as $k \to \infty$, then $\sum_{k=1}^{\infty} a_k b_k$ converges.*

PROOF. Choose $M \in \mathbf{R}$ such that

$$|s_n| = \Big| \sum_{k=1}^{n} a_k \Big| \leq M, \qquad n \in \mathbf{N}.$$

By the triangle inequality,

$$
(13) \qquad |A_{n,m}| = \left| \sum_{k=m}^{n} a_k \right| = |s_n - s_{m-1}| \le 2M
$$

for $n > m > 1$.

Let $\varepsilon > 0$ and choose $N \in \mathbf{N}$ so that $|b_k| < \varepsilon/(4M)$ for $k \ge N$. Since $\{b_k\}$ is decreasing, we find by Abel's Formula and (13) that

$$
\left| \sum_{k=m}^{n} a_k b_k \right| \le |A_{n,m}| |b_n| + \sum_{k=m}^{n-1} |A_{k,m}| (b_k - b_{k+1})
$$
$$
< \frac{\varepsilon}{2} + 2M(b_m - b_n) < \varepsilon
$$

for all $n > m \ge N$. ∎

The following special case of Dirichlet's Test is widely used.

**6.32 COROLLARY** [ALTERNATING SERIES TEST]. *If $a_k \downarrow 0$ as $k \to \infty$, then*

$$
\sum_{k=1}^{\infty} (-1)^k a_k
$$

*converges.*

PROOF. Since the partial sums of $\sum_{k=1}^{\infty} (-1)^k$ are bounded, $\sum_{k=1}^{\infty} (-1)^k a_k$ converges by Dirichlet's Test. ∎

We note that the series $\sum_{k=1}^{\infty} (-1)^k / k$, used in Remark 6.20, is an alternating series. Here is another example.

**6.33 Example.** Prove that $\sum_{k=1}^{\infty} (-1)^k / \log k$ converges.

PROOF. Since $1/\log k \downarrow 0$ as $k \to \infty$, this follows immediately from the Alternating Series Test. ∎

The Dirichlet Test can be used for more than just alternating series.

*6.34 Example.** Prove $S(x) = \sum_{k=1}^{\infty} \sin(kx)/k$ converges for each $x \in \mathbf{R}$.

PROOF. Since $\phi(x) = \sin(kx)$ is periodic of period $2\pi$ (i.e., $\phi(x + 2\pi) = \phi(x)$ for all $x \in \mathbf{R}$) and has value identically zero when $x = 0$ or $2\pi$, we need only show that $S(x)$ converges for each $x \in (0, 2\pi)$. By Dirichlet's Test, it suffices to show

$$
(14) \qquad \widetilde{D}_n(x) := \sum_{k=1}^{n} \sin(kx), \qquad n \in \mathbf{N}
$$

is a bounded sequence for each fixed $x \in (0, 2\pi)$.

This proof, originally discovered by Dirichlet, involves a clever trick which leads to a formula for $\widetilde{D}_n$. Indeed, applying a sum-angle formula (see Appendix B) and telescoping, we have

$$2\sin(\frac{x}{2})\widetilde{D}_n(x) = \sum_{k=1}^{n} 2\sin(\frac{x}{2})\sin(kx)$$

$$= \sum_{k=1}^{n} \left( \cos((k-\frac{1}{2})x) - \cos((k+\frac{1}{2})x) \right)$$

$$= \cos(\frac{x}{2}) - \cos((n+\frac{1}{2})x).$$

Therefore,

$$\mid \widetilde{D}_n(x) \mid = \left| \frac{\cos(\frac{x}{2}) - \cos\left((n+\frac{1}{2})x\right)}{2\sin(\frac{x}{2})} \right| \leq \frac{1}{\mid\sin(\frac{x}{2})\mid}$$

for all $n \in \mathbf{N}$. ∎

## EXERCISES

**1.** Prove that each of the following series converges.

a) $\displaystyle\sum_{k=1}^{\infty}(-1)^k \left( \frac{\pi}{2} - \arctan k \right).$    b) $\displaystyle\sum_{k=1}^{\infty} \frac{(-1)^k k^2}{2^k}.$    c) $\displaystyle\sum_{k=1}^{\infty} \frac{(-1)^k}{k^p}, \quad p > 0.$

d) $\displaystyle\sum_{k=1}^{\infty} \frac{\sin(kx)}{k^p} \quad x \in \mathbf{R}, \quad p > 0.$    e) $\displaystyle\sum_{k=1}^{\infty} \frac{(-1)^k}{k^2} \frac{2 \cdot 4 \cdots (2k)}{1 \cdot 3 \cdots (2k-1)}.$

**2.** For each of the following, find all values $x \in \mathbf{R}$ for which the given series converges.

a) $\displaystyle\sum_{k=1}^{\infty} \frac{x^k}{k}.$    b) $\displaystyle\sum_{k=1}^{\infty} \frac{x^{3k}}{2^k}.$    c) $\displaystyle\sum_{k=1}^{\infty} \frac{(-1)^k x^k}{\sqrt{k^2+1}}.$    d) $\displaystyle\sum_{k=1}^{\infty} \frac{(x+2)^k}{k\sqrt{k+1}}.$

**3.** Using any test covered in this chapter, find out which of the following series converge absolutely, which converge conditionally, and which diverge.

a) $\displaystyle\sum_{k=1}^{\infty} \frac{(-1)^k k^3}{(k+1)!}.$    b) $\displaystyle\sum_{k=1}^{\infty} \frac{(-1)(-3)\ldots(1-2k)}{1 \cdot 4 \ldots (3k-2)}.$

c) $\displaystyle\sum_{k=1}^{\infty} \frac{(k+1)^k}{p^k k!}, \quad p > e.$    d) $\displaystyle\sum_{k=1}^{\infty} \frac{(-1)^{k+1}\sqrt{k}}{k+1}.$    e) $\displaystyle\sum_{k=1}^{\infty} \frac{(-1)^k \sqrt{k+1}}{\sqrt{k}\,k^k}.$

**4.** [ABEL'S TEST] Suppose that $\sum_{k=1}^{\infty} a_k$ converges and $b_k \downarrow b$ as $k \to \infty$. Prove that $\sum_{k=1}^{\infty} a_k b_k$ converges.

***5.** Prove that

$$\sum_{k=1}^{\infty} a_k \cos(kx)$$

converges for every $x \in (0, 2\pi)$ and every $a_k \downarrow 0$. What happens when $x = 0$?

***6.** Suppose $a_k \downarrow 0$ as $k \to \infty$. Prove that

$$\sum_{k=1}^{\infty} a_k \sin((2k+1)x)$$

converges for all $x \in \mathbf{R}$.

**7.** Show that under the hypotheses of Dirichlet's Test,

$$\sum_{k=1}^{\infty} a_k b_k = \sum_{k=1}^{\infty} s_k (b_k - b_{k+1}).$$

**8.** Suppose $\{a_k\}$ and $\{b_k\}$ are real sequences such that $a_k \to 0$ as $k \to \infty$,

$$\sum_{k=1}^{\infty} |a_{k+1} - a_k| < \infty, \quad \text{and} \quad \Big| \sum_{k=1}^{n} b_k \Big| \le M \qquad n \in \mathbf{N}.$$

Prove that $\sum_{k=1}^{\infty} a_k b_k$ converges.

**9.** Suppose $\sum_{k=1}^{\infty} a_k$ converges. Prove that if $b_k \uparrow \infty$ and $\sum_{k=1}^{\infty} a_k b_k$ converges, then

$$b_m \sum_{k=m}^{\infty} a_k \to 0$$

as $m \to \infty$.

## *e*6.5   ESTIMATION OF SERIES

In practice, one estimates a convergent series by truncation, i.e., by adding finitely many terms of the given series. In this section we show how to estimate the error associated with such a truncation.

The proofs of several of our earlier tests actually contain estimates of the truncation error. Here is what we can get from the Integral Test.

**6.35 THEOREM.** *Suppose* $f : [1, \infty) \to \mathbf{R}$ *is positive and decreasing on* $[1, \infty)$. *Then*

$$f(n) \le \sum_{k=1}^{n} f(k) - \int_{1}^{n} f(x)\, dx \le f(1) \quad \text{for } n \in \mathbf{N}.$$

*Moreover, if $\sum_{k=1}^{\infty} f(k)$ converges, then*

$$0 \leq \sum_{k=1}^{n} f(k) + \int_n^{\infty} f(x)\,dx - \sum_{k=1}^{\infty} f(k) \leq f(n)$$

*for all $n \in \mathbf{N}$.*

PROOF. The first set of inequalities have already been verified (see (3) above). To establish the second set, let $u_k = s_k - t_k$ for $k \in \mathbf{N}$, and observe, since $f$ is decreasing, that

$$0 \leq u_k - u_{k+1} = \int_k^{k+1} f(x)\,dx - f(k+1) \leq f(k) - f(k+1).$$

Summing these inequalities over $k \geq n$ and telescoping, we have

$$0 \leq u_n - \lim_{j \to \infty} u_j = \sum_{k=n}^{\infty} (u_k - u_{k+1}) \leq \sum_{k=n}^{\infty} (f(k) - f(k+1)) = f(n).$$

Since $u_j \to \sum_{k=1}^{\infty} f(k) - \int_1^{\infty} f(x)\,dx$ as $j \to \infty$, we conclude that

$$0 \leq \sum_{k=1}^{n} f(k) + \int_n^{\infty} f(x)\,dx - \sum_{k=1}^{\infty} f(k) \leq f(n). \quad \blacksquare$$

The following example shows how to use this result to estimate the accuracy of a truncation of a series to which the Integral Test applies.

**6.36 Example.** Prove $\sum_{k=1}^{\infty} ke^{-k^2}$ converges and estimate its value to an accuracy of $10^{-3}$.

PROOF. Let $f(x) = xe^{-x^2}$. Since $f'(x) = e^{-x^2}(1 - 2x^2) \leq 0$ for $x \geq 1$, $f$ is decreasing on $[1, \infty)$. Since

$$\int_1^{\infty} xe^{-x^2}\,dx = \frac{1}{2} \int_1^{\infty} e^{-u}\,du = \frac{1}{2e} < \infty,$$

it follows from the Integral Test that $\sum_{k=1}^{\infty} ke^{-k^2}$ converges. To estimate the value $s$ of this series, notice that $f(2) = 0.036631$ and $f(3) = 0.000370$. Therefore, by Theorem 6.35, $s$ is approximately equal to

$$\sum_{k=1}^{3} ke^{-k^2} + \int_3^{\infty} xe^{-x^2}\,dx = \frac{1}{e} + \frac{2}{e^4} + \frac{3}{e^9} + \frac{1}{2e^9} \approx 0.4049427$$

with an error no more than 0.000370. $\blacksquare$

The next example shows that Theorem 6.35 can be used to estimate divergent series as well.

**6.37 Example.** Prove that there exist numbers $C_n \in (0, 1]$ such that

$$\sum_{k=1}^{n} \frac{1}{k} = \log n + C_n$$

for all $n \in \mathbf{N}$.

PROOF. Clearly, $f(x) = 1/x$ is positive, decreasing, and locally integrable on $[1, \infty)$. Hence, by Theorem 6.35,

$$\frac{1}{n} \le \sum_{k=1}^{n} \frac{1}{k} - \int_{1}^{n} \frac{1}{x}\, dx = \sum_{k=1}^{n} \frac{1}{k} - \log n \le 1. \quad \blacksquare$$

Next, we see what the Alternating Series Test has to say about truncation error.

**6.38 THEOREM.** *Suppose* $a_k \downarrow 0$ *as* $k \to \infty$. *If* $s = \sum_{k=1}^{\infty} (-1)^k a_k$ *and* $s_n = \sum_{k=1}^{n} (-1)^k a_k$, *then*

$$0 \le |s - s_n| \le a_{n+1}$$

*for all* $n \in \mathbf{N}$.

PROOF. Suppose first that $n$ is even, say $n = 2m$. Then

$$0 \ge (-a_{2m+1} + a_{2m+2}) + (-a_{2m+3} + a_{2m+4}) + \cdots$$

$$= \sum_{k=2m+1}^{\infty} (-1)^k a_k = s - s_n$$

$$= -a_{2m+1} + (a_{2m+2} - a_{2m+3}) + (a_{2m+4} - a_{2m+5}) + \cdots$$

$$\ge -a_{2m+1},$$

i.e., $0 \ge s - s_n \ge -a_{n+1}$. A similar argument proves that $0 \le s - s_n \le a_{n+1}$ when $n$ is odd. $\blacksquare$

This result can be used to estimate the error of a truncation of any alternating series.

**6.39 Example.**  For each $\alpha > 0$, prove that the series $\sum_{k=1}^{\infty} (-1)^k k/(k^2 + \alpha)$ converges. If $s_n$ represents its $n$th partial sum and $s$ its value, find an $n$ so large that $s_n$ approximates $s$ to an accuracy of $10^{-2}$.

PROOF.  Let $f(x) = x/(x^2 + \alpha)$ and note that $f(x) \to 0$ as $x \to \infty$. Since $f'(x) = (\alpha - x^2)/(x^2 + \alpha)^2$ is negative for $x > \sqrt{|\alpha|}$, it follows that $k/(k^2 + \alpha) \downarrow 0$ as $k \to \infty$. Hence, the given series converges by the Alternating Series Test.

By Theorem 6.38, $s_n$ will estimate $s$ to an accuracy of $10^{-2}$ if $f(n) < 10^{-2}$, i.e., if $n^2 - 100n + \alpha > 0$. When $\alpha > 50^2$, this last quadratic has no real roots; hence, the inequality is always satisfied and we may choose $n = 1$. When $\alpha \le 50^2$, the quadratic has roots $50 \pm \sqrt{50^2 - \alpha}$. Hence, choose any $n$ which satisfies $n > 50 + \sqrt{50^2 - \alpha}$. $\blacksquare$

Finally, we examine what information the proofs of the Root and Ratio Tests contain about accuracy of truncations.

**6.40 THEOREM.** *Suppose* $\sum_{k=1}^{\infty} a_k$ *converges absolutely and* $s$ *is the value of* $\sum_{k=1}^{\infty} |a_k|$.

　　i) *If there exist numbers* $x \in (0,1)$ *and* $N \in \mathbf{N}$ *such that*

$$|a_k|^{1/k} \leq x$$

　　*for all* $k > N$, *then*

$$0 \leq s - \sum_{k=1}^{n} |a_k| \leq \frac{x^{n+1}}{1-x}$$

　　*for all* $n \geq N$.

　　ii) *If there exist numbers* $x \in (0,1)$ *and* $N \in \mathbf{N}$ *such that*

$$\frac{|a_{k+1}|}{|a_k|} \leq x$$

　　*for* $k > N$, *then*

$$0 \leq s - \sum_{k=1}^{n} |a_k| \leq \frac{|a_N| x^{n-N+1}}{1-x}$$

　　*for all* $n \geq N$.

PROOF. Let $n \geq N$. Since $|a_k| \leq x^k$ for $k > N$, we have, by summing a geometric series, that

$$0 \leq s - \sum_{k=1}^{n} |a_k| = \sum_{k=n+1}^{\infty} |a_k| \leq \sum_{k=n+1}^{\infty} x^k = \frac{x^{n+1}}{1-x}$$

for all $n \geq N$. This proves part i). The proof of part ii) is left as an exercise. ∎

**6.41 Example.** Prove $\sum_{k=1}^{\infty} k^{2k}/(3k^2+k)^k$ converges absolutely. If $s_n$ represents its $n$th partial sum and $s$ its value, find an $n$ so large that $s_n$ approximates $s$ to an accuracy of $10^{-2}$.

SOLUTION. Since

$$\left( \frac{k^{2k}}{(3k^2+k)^k} \right)^{1/k} = \frac{k^2}{3k^2+k} \leq \frac{1}{3}$$

for all $k \geq N := 1$, the series converges absolutely by the Root Test. Since $(1/3)^{n+1}/(1 - 1/3) \leq 10^{-2}$ for $n \geq 4$, we conclude by Theorem 6.40i) that it takes at most four terms to approximate the value of this series to an accuracy of $10^{-2}$. ∎

## EXERCISES

**1.** For each of the following series, let $s_n$ represent its partial sums and $s$ its value. Prove that $s$ is finite and find an $n$ so large that $s_n$ approximates $s$ to an accuracy of $10^{-2}$.

a) $\displaystyle\sum_{k=1}^{\infty}(-1)^k\left(\frac{\pi}{2}-\arctan k\right)$.
   b) $\displaystyle\sum_{k=1}^{\infty}\frac{(-1)^k k^2}{2^k}$.
   c) $\displaystyle\sum_{k=1}^{\infty}\frac{(-1)^k}{k^2}\frac{2\cdot 4\cdots(2k)}{1\cdot 3\cdots(2k-1)}$.

**2. a)** Find all $p\geq 0$ such that the following series converges.

$$\sum_{k=1}^{\infty}\frac{1}{k\log^p(k+1)}.$$

**b)** For each such $p$, prove that the partial sums of this series $s_n$ and its value $s$ satisfy

$$|s-s_n|\leq\frac{n+p-1}{n(p-1)}\left(\frac{1}{\log^{p-1}(n)}\right)$$

for all $n\geq 2$.

**3.** For each of the following series, let $s_n$ represent its partial sums, $s$ represent its value. Prove that $s$ is finite and find an $n$ so large that $s_n$ approximates $s$ to an accuracy of $10^{-2}$.

a) $\displaystyle\sum_{k=1}^{\infty}\frac{1}{k!}$.
   b) $\displaystyle\sum_{k=1}^{\infty}\frac{1}{k^k}$.
   c) $\displaystyle\sum_{k=1}^{\infty}\frac{2^k}{k!}$.
   d) $\displaystyle\sum_{k=1}^{\infty}\left(\frac{k}{k+1}\right)^{k^2}$.

**4.** Prove Theorem 6.40ii).

## $^e$6.6  ADDITIONAL TESTS

If the Ratio or Root Test yields a value $r=1$, then no conclusion can be made. There are some tests designed to handle just that situation (see Exercise 3 below). We cover two of them in this section (see also Exercise 5 below).

The first test compares the growth of the terms of a series with the growth of the logarithm function.

**6.42 THEOREM** [THE LOGARITHMIC TEST]. *Suppose $a_k\neq 0$ for large $k$ and*

$$p=\lim_{k\to\infty}\frac{\log(1/|a_k|)}{\log k}$$

*exists as an extended real number. If $p>1$, then $\sum_{k=1}^{\infty}a_k$ converges absolutely. If $p<1$, then $\sum_{k=1}^{\infty}|a_k|$ diverges.*

PROOF. Suppose $p>1$. Fix $q\in(1,p)$ and choose $N\in\mathbf{N}$ so that $k\geq N$ implies $\log(1/|a_k|)>q\log k=\log(k^q)$. Since the logarithm function is monotone increasing, it follows that $1/|a_k| > k^q$, i.e., that $|a_k| < k^{-q}$ for $k \geq N$. Hence, by the Comparison Test, $\sum_{k=1}^{\infty} |a_k|$ converges.

Similarly, if $p < 1$, then $|a_k| > 1/k$ for large $k$. Hence, by the Comparison Test, $\sum_{k=1}^{\infty} |a_k|$ diverges. ∎

The final test works by examining how rapidly the ratios of $a_{k+1}/a_k$ converge to $r = 1$ (see also Exercise 5 below).

**6.43 THEOREM** [RAABE'S TEST]. *Suppose there is a constant $C$ and a parameter $p$ such that*

(15)
$$\left| \frac{a_{k+1}}{a_k} \right| \leq 1 - \frac{p}{k + C}$$

*for large $k$. If $p > 1$, then $\sum_{k=1}^{\infty} a_k$ converges absolutely.*

PROOF. Set $x_k = k + C - 1$ for $k \in \mathbf{N}$ and choose $N \in \mathbf{N}$ such that $x_k > 1$ and (15) hold for $k \geq N$. By the $p$-Series Test and the Limit Comparison Test,

(16)
$$\sum_{k=N}^{\infty} x_k^{-p} < \infty.$$

By (15) and Bernoulli's Inequality,

$$\left| \frac{a_{k+1}}{a_k} \right| \leq 1 - \frac{p}{x_{k+1}} \leq \left( 1 - \frac{1}{x_{k+1}} \right)^p = \frac{x_k^p}{x_{k+1}^p}.$$

Hence, the sequence $\{|a_k| \, x_k^p\}_{k=N}^{\infty}$ is decreasing and bounded above. In particular, there is an $M > 0$ such that $|a_k| \leq M x_k^{-p}$ for $k \geq N$. We conclude by (16) that $\sum_{k=1}^{\infty} a_k$ converges. ∎

## EXERCISES

**1.** Using any test covered in this chapter, find out which of the following series converge absolutely, which converge conditionally, and which diverge.

a) $\displaystyle\sum_{k=1}^{\infty} \frac{3 \cdot 5 \cdots (2k + 1)}{2 \cdot 4 \cdots 2k}.$    b) $\displaystyle\sum_{k=1}^{\infty} \frac{1 \cdot 3 \cdots (2k - 1)}{5 \cdot 7 \cdots (2k + 3)}.$

c) $\displaystyle\sum_{k=2}^{\infty} \frac{1}{(\log k)^{\log \log k}}.$    d) $\displaystyle\sum_{k=1}^{\infty} \left( \frac{\sqrt{k} - 1}{\sqrt{k}} \right)^k.$

**2.** For each of the following, find all values of $p \in \mathbf{R}$ for which the given series converges absolutely, for which it converges conditionally, and for which it diverges.

a) $\displaystyle\sum_{k=1}^{\infty} ke^{-kp}$.     b) $\displaystyle\sum_{k=2}^{\infty} \frac{1}{(\log k)^{p\log k}}$.     c) $\displaystyle\sum_{k=1}^{\infty} \frac{(pk)^k}{k!}$.

**3.** a) Prove that the Root Test applied to the series

$$\sum_{k=2}^{\infty} \frac{1}{(\log k)^{\log k}}$$

yields $r = 1$. Use the Logarithmic Test to prove this series converges.

b) Prove that the Ratio Test applied to the series

$$\sum_{k=1}^{\infty} \frac{1\cdot 3\cdots(2k-1)}{4\cdot 6\cdots(2k+2)}$$

yields $r = 1$. Use Raabe's Test to prove this series converges.

**4.** Suppose $f : \mathbf{R} \to (0,\infty)$ is differentiable, $f(x) \to 0$ as $x \to \infty$, and

$$\alpha := \lim_{x\to\infty} \frac{xf'(x)}{f(x)}$$

exists. If $\alpha < -1$, prove that $\sum_{k=1}^{\infty} f(k)$ converges.

**5.** Suppose that $\{a_k\}$ is a sequence of nonzero real numbers and

$$p = \lim_{k\to\infty} k\left(1 - \left|\frac{a_{k+1}}{a_k}\right|\right)$$

exists as an extended real number. Prove that $\sum_{k=1}^{\infty} a_k$ converges absolutely when $p > 1$.

# Chapter 7

# Infinite Series of Functions

## 7.1 UNIFORM CONVERGENCE OF SEQUENCES

You are familiar with what it means for a sequence of numbers to converge. In this section we examine what it means for a sequence of functions to converge. It turns out there are several different ways to define *convergence* of a sequence of functions. We begin with the simplest way.

**7.1 DEFINITION.** Let $E$ be a nonempty subset of $\mathbf{R}$. A sequence of functions $f_n : E \to \mathbf{R}$ is said to *converge pointwise* on $E$ (notation: $f_n \to f$ pointwise on $E$ as $n \to \infty$) if $f(x) = \lim_{n \to \infty} f_n(x)$ exists for each $x \in E$.

Because $\{f_n\}$ converges pointwise on a set $E$ if and only if the sequence of real numbers $\{f_n(x)\}$ converges for each $x \in E$, every result about convergence of real numbers contains a result about pointwise convergence of functions. Here is a typical example.

**7.2 Remark.** *Let $E$ be a nonempty subset of $\mathbf{R}$ and $f_n : E \to \mathbf{R}$ be a sequence of functions. Then $f_n \to f$ pointwise on $E$ if and only if for every $\varepsilon > 0$ and $x \in E$ there is an $N \in \mathbf{N}$ (which may depend on $x$ as well as $\varepsilon$) such that*

$$n \geq N \quad implies \quad |f_n(x) - f(x)| < \varepsilon.$$

PROOF. By Definition 7.1, $f_n \to f$ pointwise on $E$ if and only if $f_n(x) \to f(x)$ for all $x \in E$. This occurs, by Definition 2.1, if and only if for every $\varepsilon > 0$ and $x \in E$ there is an $N \in \mathbf{N}$ such that $n \geq N$ implies $|f_n(x) - f(x)| < \varepsilon$. $\blacksquare$

If $f_n \to f$ pointwise on $[a, b]$, it is natural to ask: What does $f$ inherit from $f_n$? The next four remarks show that, in general, the answer to this question is not much.

**7.3 Remark.** *The pointwise limit of continuous (respectively, differentiable) functions is not necessarily continuous (respectively, differentiable).*

PROOF. Let $f_n(x) = x^n$ and

$$f(x) = \begin{cases} 0 & 0 \leq x < 1 \\ 1 & x = 1. \end{cases}$$

Then $f_n \to f$ pointwise on $[0, 1]$ (see Figure 7.3 below), each $f_n$ is continuous and differentiable on $[0, 1]$, but $f$ is neither differentiable nor continuous at $x = 1$. ∎

**7.4 Remark.** *The pointwise limit of integrable functions is not necessarily integrable.*

PROOF. Set

$$f_n(x) = \begin{cases} 1 & x = p/m \in \mathbf{Q}, \text{ written in reduced form, where } m \leq n \\ 0 & \text{otherwise}, \end{cases}$$

for $n \in \mathbf{N}$ and

$$f(x) = \begin{cases} 1 & x \in \mathbf{Q} \\ 0 & \text{otherwise}. \end{cases}$$

Then $f_n \to f$ pointwise on $[0, 1]$, each $f_n$ is integrable on $[0, 1]$ (with integral zero), but $f$ is not integrable on $[0, 1]$ (see Example 5.11). ∎

**7.5 Remark.** *There exist differentiable functions $f_n$ and $f$ such that $f_n \to f$ pointwise on $[0, 1]$ but*

$$(1) \qquad \lim_{n \to \infty} f_n'(x) \neq \left( \lim_{n \to \infty} f_n(x) \right)'$$

for $x = 1$.

PROOF. Let $f_n(x) = x^n/n$ and $f(x) = 0$. Then $f_n \to f$ pointwise on $[0, 1]$, each $f_n$ is differentiable with $f_n'(x) = x^{n-1}$. Thus the left side of (1) is 1 at $x = 1$ but the right side of (1) is zero. ∎

**7.6 Remark.** *There exist continuous functions $f_n$ and $f$ such that $f_n \to f$ pointwise on $[0, 1]$ but*

$$(2) \qquad \lim_{n \to \infty} \int_0^1 f_n(x)\, dx \neq \int_0^1 \left( \lim_{n \to \infty} f_n(x) \right) dx.$$

PROOF. Let $f_1(x) = 1$ and

$$f_n(x) = \begin{cases} n^2 x & 0 < x < 1/n \\ 2n - n^2 x & 1/n \leq x < 2/n \\ 0 & 2/n \leq x \leq 1 \end{cases}$$

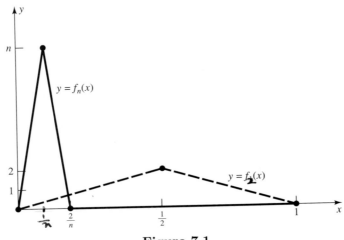

**Figure 7.1**

for $n = 2, 3, \ldots$ (see Figure 7.1). Then $f_n \to 0$ pointwise on $[0, 1]$ and, since the area of a triangle is one-half base times altitude, $\int_0^1 f_n(x)\, dx = 1$ for each $n \in \mathbf{N}$. Thus, the left side of (2) is 1 but the right side is zero. ∎

In view of the preceding examples, it is clear that pointwise convergence is of limited value for the calculus of limits of sequences. It turns out that the following concept, discovered independently by Stokes, Cauchy, and Weierstrass around 1850, is much more useful in this context.

**7.7 DEFINITION.** Let $E$ be a nonempty subset of $\mathbf{R}$. A sequence of functions $f_n : E \to \mathbf{R}$ is said to *converge uniformly* on $E$ to a function $f$ (notation: $f_n \to f$ uniformly on $E$ as $n \to \infty$) if for every $\varepsilon > 0$ there is an $N \in \mathbf{N}$ such that

$$n \geq N \quad \text{implies} \quad |f_n(x) - f(x)| < \varepsilon$$

for all $x \in E$.

Comparing Definition 7.7 with Remark 7.2 above, we see that the only difference between uniform convergence and pointwise convergence is that, for uniform convergence, the integer $N$ must be chosen independently of $x$ (see Figure 7.2). Notice how this is similar to the difference between uniform continuity and continuity (see the discussion following Example 3.37).

By definition, if $f_n$ converges uniformly on $E$, then $f_n$ converges pointwise on $E$. The following example shows that the converse of this statement is false. (This example also shows how to prove $f_n \to f$ uniformly on a set $E$: dominate $|f_n(x) - f(x)|$ by constants $b_n$, independent of $x \in E$, which converge to zero as $n \to \infty$.)

**7.8 Example.** Prove that $x^n \to 0$ uniformly on $[0, b]$ for any $b < 1$, and pointwise, but not uniformly, on $[0, 1)$.

PROOF. By Example 2.20, $x^n \to 0$ pointwise on $[0, 1)$. Let $b < 1$. Given $\varepsilon > 0$

✦ Def: A sequence $\{f_n\}$ converge uniformly on $E$ to a function $f$
iff given $\varepsilon > 0$ there is
$C_n = \sup |f_n(x) - f(x)|$ converges to 0 as $n \to \infty$.

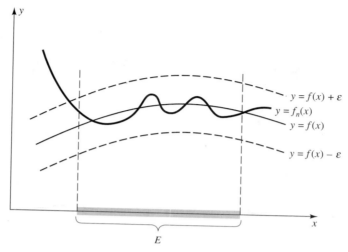

**Figure 7.2**

choose $N \in \mathbf{N}$ such that $n \geq N$ implies $b^n < \varepsilon$. Then $x \in [0, b]$ and $n \geq N$ imply $|x^n| \leq b^n < \varepsilon$, i.e., $x^n \to 0$ uniformly for $x \in [0, b]$.

Finally, suppose that $x^n$ converges to 0 uniformly on $[0, 1)$. Then given $0 < \varepsilon < 1/2$, there is an $N \in \mathbf{N}$ such that $|x^N| < \varepsilon$ for all $x \in [0, 1)$. On the other hand, since $x^N \to 1$ as $x \to 1-$, we can choose an $x_0 \in (0, 1)$ such that $x_0^N > \varepsilon$ (see Figure 7.3). Thus $\varepsilon < x_0^N < \varepsilon$, a contradiction. ∎

The next several results show that if $f_n \to f$ or $f_n' \to f'$ uniformly, then $f$ inherits much from $f_n$.

**7.9 THEOREM.** *Suppose $f_n \to f$ uniformly on an open interval $(a, b)$. If each $f_n$ is continuous at some $x_0 \in (a, b)$, then $f$ is continuous at $x_0 \in (a, b)$.*

PROOF. Let $\varepsilon > 0$ and choose $N \in \mathbf{N}$ such that

$$n \geq N \quad \text{and} \quad x \in (a, b) \quad \text{imply} \quad |f_n(x) - f(x)| < \frac{\varepsilon}{3}.$$

Since $f_N$ is continuous at $x_0 \in (a, b)$, choose $\delta > 0$ such that

$$|x - x_0| < \delta \quad \text{and} \quad x \in (a, b) \quad \text{imply} \quad |f_N(x) - f_N(x_0)| < \frac{\varepsilon}{3}.$$

Suppose $|x - x_0| < \delta$ and $x \in (a, b)$. Then

$$|f(x) - f(x_0)| \leq |f(x) - f_N(x)| + |f_N(x) - f_N(x_0)| + |f_N(x_0) - f(x_0)| < \varepsilon.$$

Thus $f$ is continuous at $x_0 \in (a, b)$. ∎

(For a generalization of this result, see Exercise 6. For a converse of this result when the sequence $f_n$ is pointwise monotone, see Theorem 9.41.)

Here is an important theorem about interchanging a limit sign and an integral sign.

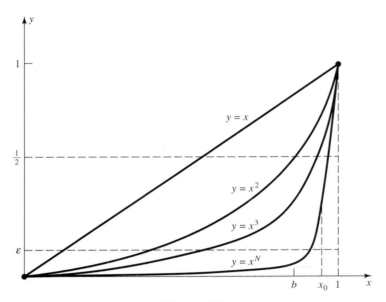

$y = x$

$y = x^2$

$y = x^3$

$y = x^N$

**Figure 7.3**

**7.10 THEOREM.** *Suppose $f_n \to f$ uniformly on a closed interval $[a, b]$. If each $f_n$ is integrable on $[a, b]$, then so is $f$ and*

$$\lim_{n \to \infty} \int_a^b f_n(x)\, dx = \int_a^b \left( \lim_{n \to \infty} f_n(x) \right) dx. \;\; = \int_a^b f(x)\, dx.$$

*In fact, $\lim_{n \to \infty} \int_a^x f_n(t)\, dt = \int_a^x \left( \lim_{n \to \infty} f_n(t) \right) dt$ uniformly for $x \in [a, b]$.*

PROOF. Let $\varepsilon > 0$ and choose $N \in \mathbf{N}$ such that

$$(3) \qquad\qquad n \geq N \quad \text{implies} \quad |f(x) - f_n(x)| < \frac{\varepsilon}{3(b-a)} \qquad \frac{-\varepsilon}{3(b-a)} < f(x) - f_n(x) < \frac{\varepsilon}{3(b-a)}$$

for all $x \in [a, b]$. Using this inequality for $n = N$, we see that by the definition of upper and lower sums,

$$U(f - f_N, P) \leq \frac{\varepsilon}{3} \quad \text{and} \quad L(f - f_N, P) \geq -\frac{\varepsilon}{3}$$

for any partition $P$ of $[a, b]$. Since $f_N$ is integrable, choose a partition $P$ such that

$$U(f_N, P) - L(f_N, P) < \frac{\varepsilon}{3}.$$

It follows that $\qquad f = (f - f_N) + f_N \qquad \sup f \leq \sup (f - f_N) + \sup f_N$

$$U(f, P) - L(f, P) \leq U(f - f_N, P) + U(f_N, P) - L(f_N, P) - L(f - f_N, P)$$
$$< \frac{\varepsilon}{3} + \frac{\varepsilon}{3} + \frac{\varepsilon}{3} = \varepsilon,$$

i.e., $f$ is integrable on $[a,b]$. We conclude by Theorem 5.22 and (3) that

$$\left| \int_a^x f_n(t)\, dt - \int_a^x f(t)\, dt \right| \leq \int_a^x |f_n(t) - f(t)|\, dt \leq \frac{\varepsilon(x-a)}{3(b-a)} < \varepsilon$$

for all $x \in [a,b]$ and $n \geq N$. ∎

Here is a Cauchy Criterion for uniform convergence.

**7.11 Lemma** [UNIFORM CAUCHY CRITERION]. *Let $E$ be a nonempty subset of* **R** *and $f_n : E \to$ **R** be a sequence of functions. Then $f_n$ converges uniformly on $E$ if and only if for every $\varepsilon > 0$ there is an $N \in$ **N** such that*

$$(4) \qquad\qquad n, m \geq N \quad implies \quad |f_n(x) - f_m(x)| < \varepsilon$$

*for all $x \in E$.*

PROOF. Suppose first that $f_n \to f$ uniformly on $E$ as $n \to \infty$. Let $\varepsilon > 0$ and choose $N \in$ **N** such that

$$n \geq N \quad \text{implies} \quad |f_n(x) - f(x)| < \frac{\varepsilon}{2}$$

for $x \in E$. Since $|f_n(x) - f_m(x)| \leq |f_n(x) - f(x)| + |f(x) - f_m(x)|$, it is clear that (4) holds for all $x \in E$.

Conversely, if (4) holds for $x \in E$, then $\{f_n(x)\}_{n \in \mathbf{N}}$ is Cauchy for each $x \in E$. Hence, by Cauchy's Theorem for sequences (Theorem 2.29),

$$f(x) := \lim_{n \to \infty} f_n(x)$$

exists for each $x \in E$. Take the limit of the second inequality in (4) as $m \to \infty$. We obtain $|f_n(x) - f(x)| \leq \varepsilon$ for all $n \geq N$ and $x \in E$. Hence, by definition, $f_n \to f$ uniformly on $E$. ∎

Here is a result about interchanging a limit sign and the derivative sign. (The proof presented here comes from Apostol [1].)

**7.12 THEOREM.** *Let $(a,b)$ be a bounded interval and $f_n$ be a sequence of functions which converges at some $x_0 \in (a,b)$. If each $f_n$ is differentiable on $(a,b)$, and $f_n'$ converges uniformly on $(a,b)$ as $n \to \infty$, then $f_n$ converges uniformly on $(a,b)$ and*

$$\lim_{n \to \infty} f_n'(x) = \left( \lim_{n \to \infty} f_n(x) \right)' \; = \; f(x)'$$

*for each $x \in (a,b)$.*

PROOF. Fix $c \in (a,b)$ and define

$$g_n(x) = \begin{cases} \dfrac{f_n(x) - f_n(c)}{x - c} & x \neq c \\[2mm] f_n'(c) & x = c \end{cases}$$

for $n \in \mathbf{N}$. Clearly,

$$(5) \qquad\qquad f_n(x) = f_n(c) + (x - c)g_n(x)$$

for $n \in \mathbf{N}$ and $x \in (a, b)$.

We claim that for any $c \in (a, b)$, the sequence $g_n$ converges uniformly on $(a, b)$. Let $\varepsilon > 0$, $n, m \in \mathbf{N}$, and $x \in (a, b)$ with $x \neq c$. By the Mean Value Theorem, there is a $\xi$ between $x$ and $c$ such that

$$g_n(x) - g_m(x) = \frac{f_n(x) - f_m(x) - (f_n(c) - f_m(c))}{x - c} = f'_n(\xi) - f'_m(\xi).$$

Since $f'_n$ converges uniformly on $(a, b)$, it follows that there is an $N \in \mathbf{N}$ such that

$$n, m \geq N \quad \text{implies} \quad |g_n(x) - g_m(x)| < \varepsilon$$

for $x \in (a, b)$ with $x \neq c$. This implication also holds for $x = c$ because $g_n(c) = f'_n(c)$ for all $n \in \mathbf{N}$. This proves the claim.

To show $f_n$ converges uniformly on $(a, b)$, notice that by the claim, $g_n$ converges uniformly as $n \to \infty$ and (5) holds for $c = x_0$. Since $f_n(x_0)$ converges as $n \to \infty$ by hypothesis, it follows from (5) and $b - a < \infty$ that $f_n$ converges uniformly on $(a, b)$ as $n \to \infty$.

Fix $c \in (a, b)$. Define $f, g$ on $(a, b)$ by $f(x) := \lim_{n \to \infty} f_n(x)$ and $g(x) := \lim_{n \to \infty} g_n(x)$. We need to show

$$(6) \qquad\qquad f'(c) = \lim_{n \to \infty} f'_n(c).$$

Since each $g_n$ is continuous at $c$, the claim implies $g$ is continuous at $c$. Since $g_n(c) = f'_n(c)$, it follows that the right side of (6) can be written as

$$\lim_{n \to \infty} f'_n(c) = \lim_{n \to \infty} g_n(c) = g(c) = \lim_{x \to c} g(x).$$

On the other hand, if $x \neq c$ we have by definition that

$$\frac{f(x) - f(c)}{x - c} = \lim_{n \to \infty} \frac{f_n(x) - f_n(c)}{x - c} = \lim_{n \to \infty} g_n(x) = g(x).$$

Therefore, the left side of (6) reduces to

$$f'(c) = \lim_{x \to c} \frac{f(x) - f(c)}{x - c} = \lim_{x \to c} g(x).$$

This verifies (6), and the proof of the theorem is complete. ∎

## EXERCISES

**1.** a) Prove $x/n \to 0$ uniformly, as $n \to \infty$, on any closed bounded interval $[a, b]$.

   b) Prove $1/(nx) \to 0$ pointwise but not uniformly on $(0, 1)$ as $n \to \infty$.

**2.** Prove that the following limits exist and evaluate them.

a) $\displaystyle\lim_{n\to\infty} \int_{-1}^{1} e^{x^2/n}\, dx$   b) $\displaystyle\lim_{n\to\infty} \int_{0}^{3} \frac{nx^2+3}{x^3+nx}\, dx$   c) $\displaystyle\lim_{n\to\infty} \int_{0}^{\pi/2} \sqrt{\sin(\frac{x}{n}) + \cos(\frac{x}{n})}\, dx.$

**3.** Suppose that $f_n \to f$ and $g_n \to g$, as $n \to \infty$, uniformly on some set $E \subseteq \mathbf{R}$.

  a) Prove that $f_n + g_n \to f + g$ and $\alpha f_n \to \alpha f$, as $n \to \infty$, uniformly on $E$ for all $\alpha \in \mathbf{R}$.

  b) Prove that $f_n g_n \to fg$ pointwise on $E$.    *— then $f_n, g_n$ are bounded for large $n$*

  c) Prove that if $f$ and $g$ are bounded on $E$, then $f_n g_n \to fg$ uniformly on $E$.

  d) Show c) may be false when $g$ is unbounded.    *stronger assumption than 3 c).*

**4.** Let $f, g$ be continuous on a closed bounded interval $[a,b]$ with $|g(x)| > 0$ for $x \in [a,b]$. Suppose $f_n \to f$ and $g_n \to g$ as $n \to \infty$, uniformly on $[a,b]$.

*f, g bounded by extreme value Thm*

  a) Prove $1/g_n$ is defined for large $n$ and $f_n/g_n \to f/g$ uniformly on $[a,b]$ as $n \to \infty$.

  b) Show that a) is false if $[a,b]$ is replaced by $(a,b)$.

**5.** A sequence of functions $f_n$ is said to be *uniformly bounded* on a set $E$ if there is an $M > 0$ such that

$$|f_n(x)| \le M$$

for all $x \in E$ and all $n \in \mathbf{N}$. Suppose each $f_n$ is a bounded function on a set $E$ and $f_n \to f$ uniformly on $E$. Prove $\{f_n\}$ is uniformly bounded on $E$ and $f$ is a bounded function on $E$.

**6.** **This exercise is used in Section 7.2.** Suppose $f_n \to f$ uniformly on $E$ as $n \to \infty$, where $E$ is any nonempty subset of $\mathbf{R}$.

  a) Prove that if each $f_n$ is continuous on $E$, then $f$ is continuous on $E$.

  b) Prove that if each $f_n$ is uniformly continuous on $E$, then $f$ is uniformly continuous on $E$.

**7.** Suppose $b > a > 0$. Prove that

$$\lim_{n\to\infty} \int_{a}^{b} \left(1 + \frac{x}{n}\right)^n e^{-x}\, dx = b - a.$$

**8.** Let $[a,b]$ be a closed bounded interval, $f : [a,b] \to \mathbf{R}$ be bounded, and $g : [a,b] \to \mathbf{R}$ be continuous with $g(a) = g(b) = 0$. Let $f_n$ be a uniformly bounded sequence of functions on $[a,b]$ (see Exercise 5 above). Prove that if $f_n \to f$ uniformly on all closed intervals $[c,d] \subset (a,b)$, then $f_n g \to fg$ uniformly on $[a,b]$.

**9.** Let $f_n$ be integrable on $[0,1]$ and $f_n \to f$ uniformly on $[0,1]$. Show that if $b_n \uparrow 1$ as $n \to \infty$, then

$$\lim_{n\to\infty} \int_{0}^{b_n} f_n(x)\, dx = \int_{0}^{1} f(x)\, dx.$$

**10.** Let $E$ be a nonempty subset of $\mathbf{R}$ and $f$ be a real-valued function defined on $E$. Suppose $f_n$ is a sequence of bounded functions on $E$ which converges to $f$

uniformly on $E$. Prove that

$$\frac{f_1(x) + \cdots + f_n(x)}{n} \to f(x)$$

uniformly on $E$ as $n \to \infty$ (compare with Exercise 7 in Section 6.1).

## 7.2 UNIFORM CONVERGENCE OF SERIES

In this section we extend the concepts introduced in Section 7.1 from sequences to series.

**7.13 DEFINITION.** Let $f_k$ be a sequence of functions defined on some set $E$ and set

$$s_n(x) := \sum_{k=1}^{n} f_k(x), \qquad x \in E, \ n \in \mathbf{N}.$$

i) The series $\sum_{k=1}^{\infty} f_k$ is said to *converge pointwise* on $E$ if the sequence $s_n(x)$ converges pointwise on $E$ as $n \to \infty$.

ii) The series $\sum_{k=1}^{\infty} f_k$ is said to *converge uniformly* on $E$ if the sequence $s_n(x)$ converges uniformly on $E$ as $n \to \infty$.

iii) The series $\sum_{k=1}^{\infty} f_k$ is said to *converge absolutely pointwise* on $E$ if $\sum_{k=1}^{\infty} |f_k(x)|$ converges for each $x \in E$.

Since convergence of series is defined in terms of convergence of sequences of partial sums, every result about convergence of sequences of functions contains a result about convergence of series of functions. For example, the following result is an immediate consequence of Theorems 7.9, 7.10, and 7.12.

**7.14 THEOREM.** Let $\{f_k\}$ be a sequence of real-valued functions defined on an interval $I$.

*so the sum of continuous functions are continuous*

$\sum_{k=1}^{n} f_k$ is cont., or each $S_n$ is conti.

i) Suppose $I = (a, b)$ is open, $x_0 \in I$, and each $f_k$ is continuous at $x_0 \in (a, b)$. If $f = \sum_{k=1}^{\infty} f_k$ converges uniformly on $(a, b)$, then $f$ is continuous at $x_0 \in (a, b)$.

*sum of integrable functions are integrable*

ii) [TERM-BY-TERM INTEGRATION]. Suppose $I = [a, b]$ is closed and bounded, and each $f_k$ is integrable on $[a, b]$. If $f = \sum_{k=1}^{\infty} f_k$ converges uniformly on $[a, b]$, then $f$ is integrable on $[a, b]$ and

$$\int_a^b \sum_{k=1}^{\infty} f_k(x) \, dx = \sum_{k=1}^{\infty} \int_a^b f_k(x) \, dx.$$

$\int_a^b \lim_{n \to \infty} S_n(x)\, dx = \lim_{n \to \infty} \int_a^b S_n(x)\, dx.$

iii) [TERM-BY-TERM DIFFERENTIATION]. Suppose $I = (a, b)$ is a bounded, open interval, and each $f_k$ is differentiable on $I$. If $f = \sum_{k=1}^{\infty} f_k$ converges at some $x_0 \in (a, b)$, and $\sum_{k=1}^{\infty} f_k'$ converges uniformly on $(a, b)$, then $\sum_{k=1}^{\infty} f_k$ converges uniformly to $f$, $f$ is differentiable on $(a, b)$, and

$$\left( \sum_{k=1}^{\infty} f_k(x) \right)' = \sum_{k=1}^{\infty} f_k'(x)$$

for $x \in (a, b)$.

Here are two much-used tests for uniform convergence of series. (The second test, and its example, are optional because we do not use it elsewhere in this text.)

**7.15 THEOREM** [WEIERSTRASS $M$-TEST]. *Let $f_k$ be defined on a set $E$ and suppose $M_k \geq 0$ satisfies $\sum_{k=1}^{\infty} M_k < \infty$. If $|f_k(x)| \leq M_k$ for $k \in \mathbf{N}$ and $x \in E$, then $\sum_{k=1}^{\infty} f_k$ converges absolutely and uniformly on $E$.*

PROOF. Let $\varepsilon > 0$ and use the Cauchy Criterion to choose $N \in \mathbf{N}$ such that $m > n \geq N$ implies $\sum_{k=n}^{m} M_k < \varepsilon$. Thus, by hypothesis,

$$\left| \sum_{k=n}^{m} f_k(x) \right| \leq \sum_{k=n}^{m} |f_k(x)| \leq \sum_{k=n}^{m} M_k < \varepsilon$$

for $m > n \geq N$ and $x \in E$. Hence, the partial sums of $\sum_{k=1}^{\infty} f_k$ are uniformly Cauchy and the partial sums of $\sum_{k=1}^{\infty} |f_k(x)|$ are Cauchy for each $x \in E$. ∎

*7.16 THEOREM** [DIRICHLET'S TEST FOR UNIFORM CONVERGENCE]. *Let $E$ be a nonempty subset of $\mathbf{R}$ and $f_k, g_k : E \to \mathbf{R}$. If*

$$\left| \sum_{k=1}^{n} f_k(x) \right| \leq M < \infty$$

*for $n \in \mathbf{N}$ and $x \in E$, and if $g_k \downarrow 0$ uniformly on $E$ as $k \to \infty$, then $\sum_{k=1}^{\infty} f_k g_k$ converges uniformly on $E$.*

PROOF. Let

$$F_{n,m}(x) = \sum_{k=m}^{n} f_k(x), \qquad m, n \in \mathbf{N}, \ n \geq m, \ x \in E$$

and fix integers $n > m > 0$. By Abel's Formula and the hypothesis,

$$\left| \sum_{k=m}^{n} f_k(x) g_k(x) \right| = \left| F_{n,m}(x) g_n(x) + \sum_{k=m}^{n-1} F_{k,m}(x)(g_k(x) - g_{k+1}(x)) \right|$$

$$\leq 2M g_n(x) + 2M \sum_{k=m}^{n-1} (g_k(x) - g_{k+1}(x))$$

$$= 2M g_m(x)$$

for all $x \in E$. Since $g_m(x) \to 0$ uniformly on $E$, as $m \to \infty$, it follows from the Uniform Cauchy Criterion that $\sum_{k=1}^{\infty} f_k(x)g_k(x)$ converges uniformly on $E$. ∎

Here is a typical application of Dirichlet's Test.

*7.17 Example. Prove that if $a_k \downarrow 0$ as $k \to \infty$, then $\sum_{k=0}^{\infty} a_k \cos kx$ converges uniformly on any closed subinterval $[a, b]$ of $(0, 2\pi)$.

PROOF. Let $f_k(x) = \cos kx$ and $g_k(x) = a_k$ for $k \in \mathbf{N}$. By the technique used in Example 6.34, we can show that

$$D_n(x) := \sum_{k=0}^{n} \cos kx = \frac{\sin\left(\dfrac{x}{2}\right) + \sin\left(\left(n + \dfrac{1}{2}\right)x\right)}{2\sin\left(\dfrac{x}{2}\right)}$$

for $n \in \mathbf{N}$ and $x \in (0, 2\pi)$. Hence the partial sums of $\sum_{k=0}^{\infty} f_k(x)$ satisfy

$$|D_n(x)| = \left| \frac{\sin\left(\dfrac{x}{2}\right) + \sin\left(\left(n + \dfrac{1}{2}\right)x\right)}{2\sin\left(\dfrac{x}{2}\right)} \right| \leq \frac{1}{\left|\sin\left(\dfrac{x}{2}\right)\right|}$$

for $x \in (0, 2\pi)$. If $\delta = \min\{2\pi - b, a\}$ and $x \in [a, b]$, then $\sin(x/2) \geq \sin(\delta/2)$ (see Figure 7.4). Therefore, $\sum_{k=1}^{\infty} a_k \cos kx$ converges uniformly on $[a, b]$ by Dirichlet's Test. ∎

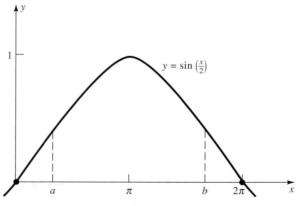

**Figure 7.4**

This example can be used to show that uniform convergence of a series alone is not sufficient for term-by-term differentiation. Indeed, although $\sum_{k=1}^{\infty} \cos kx/k$ converges uniformly on $[\pi/2, 3\pi/2]$, its term-by-term derivative $\sum_{k=1}^{\infty}(-\sin kx)$ converges at no point in $[\pi/2, 3\pi/2]$.

A *double series* is a series of numbers or functions of the form

$$\sum_{k=1}^{\infty}\left(\sum_{j=1}^{\infty}a_{kj}\right).$$

Such a double series is said to converge if $\sum_{j=1}^{\infty}a_{kj}$ converges for each $k \in \mathbf{N}$ and

$$\sum_{k=1}^{\infty}\sum_{j=1}^{\infty}a_{kj} := \lim_{N\to\infty}\sum_{k=1}^{N}\left(\sum_{j=1}^{\infty}a_{kj}\right)$$

exists and is finite.

When working with double series, one frequently wants to be able to change the order of summation. We already know that the order of summation can be changed when $a_{kj} \geq 0$ (see Exercise 7 in Section 6.3). We now prove a more general result. (The elegant proof given here, which comes from Rudin [11][1], uses uniform convergence.)

**7.18 THEOREM.** *Let $a_{kj} \in \mathbf{R}$ for $k, j \in \mathbf{N}$ and suppose*

$$A_j = \sum_{k=1}^{\infty}|a_{kj}| < \infty$$

*for each $j \in \mathbf{N}$. If $\sum_{j=1}^{\infty}A_j$ converges (i.e., the double sum converges absolutely), then*

$$\sum_{k=1}^{\infty}\sum_{j=1}^{\infty}a_{kj} = \sum_{j=1}^{\infty}\sum_{k=1}^{\infty}a_{kj}.$$

PROOF. Let $E = \{0, 1, \frac{1}{2}, \frac{1}{3}, \dots\}$. For each $j \in \mathbf{N}$, define a function $f_j$ on $E$ by

$$f_j(0) = \sum_{k=1}^{\infty}a_{kj}, \quad f_j(\frac{1}{n}) = \sum_{k=1}^{n}a_{kj}, \qquad n \in \mathbf{N}.$$

By hypothesis, $f_j(0)$ exists and by the definition of series convergence,

$$\lim_{n\to\infty}f_j(\frac{1}{n}) = f_j(0);$$

i.e., $f_j$ is continuous at $0 \in E$ for each $j \in \mathbf{N}$. Moreover, since $|f_j(x)| \leq A_j$ for all $x \in E$ and $j \in \mathbf{N}$, the Weierstrass $M$-Test implies that

$$f(x) := \sum_{j=1}^{\infty}f_j(x)$$

[1]Walter Rudin, *Principles of Mathematical Analysis*, 3rd ed. (New York: McGraw-Hill Book Co., 1976). Reprinted with permission of McGraw-Hill Book Co.

converges uniformly on $E$. Hence (see Exercise 6 in Section 7.1), $f$ is continuous at $0 \in E$. It follows from the sequential characterization of continuity (Theorem 3.21) that $f(1/n) \to f(0)$ as $n \to \infty$. Therefore,

$$\sum_{k=1}^{\infty}\sum_{j=1}^{\infty} a_{kj} = \lim_{n\to\infty} \sum_{k=1}^{n}\sum_{j=1}^{\infty} a_{kj} = \lim_{n\to\infty} \sum_{j=1}^{\infty}\sum_{k=1}^{n} a_{kj}$$

$$= \lim_{n\to\infty} \sum_{j=1}^{\infty} f_j(\frac{1}{n}) = \lim_{n\to\infty} f(\frac{1}{n}) = f(0) = \sum_{j=1}^{\infty}\sum_{k=1}^{\infty} a_{kj}. \quad \blacksquare$$

## EXERCISES

**1.** a) Prove $\sum_{k=1}^{\infty} \cos(kx)/k^2$ converges uniformly on $\mathbf{R}$.

b) Prove $\sum_{k=1}^{\infty} \sin(x/k^2)$ converges uniformly on any closed bounded interval $[a, b]$.

**2.** Prove that the geometric series

$$\sum_{k=0}^{\infty} x^k = \frac{1}{1-x}$$

converges uniformly on any closed interval $[a, b] \subset (-1, 1)$.

**3.** Let $E(x) = \sum_{k=0}^{\infty} x^k/k!$.

a) Prove that the series defining $E(x)$ converges uniformly on any closed bounded interval $[a, b]$.

b) Prove that

$$\int_a^b E(x)\, dx = E(b) - E(a)$$

for all $a, b \in \mathbf{R}$.

c) Prove that the function $y = E(x)$ satisfies the initial value problem

$$y' - y = 0, \qquad y(0) = 1.$$

(We shall see in Section 7.4 that $E(x) = e^x$.)

**4.** Suppose

$$f(x) = \sum_{k=1}^{\infty} \frac{\cos(kx)}{k^2}.$$

Prove

$$\int_0^{\pi/2} f(x)\, dx = \sum_{k=0}^{\infty} \frac{(-1)^k}{(2k+1)^3}.$$

**5.** Show that

$$f(x) = \sum_{k=1}^{\infty} \frac{1}{k} \sin\left(\frac{x}{k+1}\right)$$

converges, pointwise on $\mathbf{R}$ and uniformly on each bounded interval in $\mathbf{R}$, to a differentiable function $f$ which satisfies

$$|f(x)| \leq |x| \quad \text{and} \quad |f'(x)| \leq 1$$

for all $x \in \mathbf{R}$.

**6.** Prove that

$$\left| \sum_{k=1}^{\infty} (1 - \cos(1/k)) \right| \leq 2.$$

**7.** Suppose $f = \sum_{k=1}^{\infty} f_k$ converges uniformly on a set $E \subseteq \mathbf{R}$. If $g_1$ is bounded on $E$ and $g_k(x) \geq g_{k+1}(x) \geq 0$ for all $x \in E$ and $k \in \mathbf{N}$, prove that $\sum_{k=1}^{\infty} f_k g_k$ converges uniformly on $E$.

**8.** Let $n \geq 0$ be a fixed nonnegative integer and recall that $0! := 1$. The *Bessel function* of order $n$ is the function defined by

$$B_n(x) := \sum_{k=0}^{\infty} \frac{(-1)^k}{(k!)(n+k)!} \left(\frac{x}{2}\right)^{n+2k}.$$

a) Show $B_n(x)$ converges pointwise on $\mathbf{R}$ and uniformly on any closed bounded interval $[a, b]$.

b) Prove that $y = B_n(x)$ satisfies the differential equation

$$x^2 y'' + xy' + (x^2 - n^2)y = 0$$

for $x \in \mathbf{R}$.

c) Prove that

$$(x^n B_n(x))' = x^n B_{n-1}(x)$$

for $n \in \mathbf{N}$ and $x \in \mathbf{R}$.

**\*9.** Suppose $a_k \downarrow 0$ as $k \to \infty$. Prove $\sum_{k=1}^{\infty} a_k \sin kx$ converges uniformly on any closed interval $[a, b] \subset (0, 2\pi)$.

## 7.3 POWER SERIES

Polynomials are functions of the form $P(x) = \sum_{k=0}^{n} a_k x^k$, where $a_k \in \mathbf{R}$ and $n \geq 0$. In this section we investigate a natural generalization of polynomials, namely, series of the form $\sum_{k=0}^{\infty} a_k x^k$.

Actually, we shall consider a slightly more general class of series. A *power series* (centered at $x_0$) is a series of the form

$$\sum_{k=0}^{\infty} a_k (x - x_0)^k.$$

Since such a series is identically $a_0$ when $x = x_0$, it is clear that a power series always converges at at least one point. The following result shows that this may be the only point.

**7.19 Remark.** *There exist power series which converge only at one point.*

PROOF. For each $x \neq 0$, $(k^k |x|^k)^{1/k} = k|x| \to \infty$ as $k \to \infty$. Therefore, by the Root Test, the series $\sum_{k=1}^{\infty} k^k x^k$ diverges when $x \neq 0$. ∎

Applying the Root Test to a general power series (see the proof of Theorem 7.21 below), we are led naturally to the following idea.

**7.20 DEFINITION.** The *radius of convergence* of a power series $\sum_{k=0}^{\infty} a_k (x - x_0)^k$ is the extended real number

$$R := \frac{1}{\limsup_{k \to \infty} |a_k|^{1/k}}. \quad = \frac{1}{\overline{\lim} |a_k|^{1/k}}$$

(Here, we interpret $1/\infty = 0$ and $1/0 = \infty$.)

In general, a series of functions can converge at several isolated points. (For example, the series $\sum_{k=1}^{\infty} \sin(kx)$ converges only when $x = n\pi$ for some $n \in \mathbf{Z}$.) The next result shows this cannot happen for power series.

**7.21 THEOREM.** *Let $S(x)$ be a power series centered at $x_0$ with radius of convergence $R$.*

   i) *$S(x)$ converges absolutely for each $x \in (x_0 - R, x_0 + R)$.*
   ii) *$S(x)$ converges uniformly on any closed interval $[a, b] \subset (x_0 - R, x_0 + R)$.*
   iii) *If $R$ is finite, then $S(x)$ diverges for each $x \notin [x_0 - R, x_0 + R]$.*

PROOF. Suppose $S(x) = \sum_{k=0}^{\infty} a_k (x - x_0)^k$ has radius of convergence $R$. Then

$$r := \limsup_{k \to \infty} |a_k (x - x_0)^k|^{1/k} = \frac{|x - x_0|}{R}.$$

By the Root Test, $S(x)$ converges absolutely if $r < 1$ and diverges if $r > 1$. If $R = \infty$, then $r = 0$ is always $< 1$. Hence, by the Root Test, $S(x)$ converges absolutely for all $x \in \mathbf{R}$. On the other hand, if $R$ is finite, then $r < 1$ if and only if $|x - x_0| < R$. Hence by the Root Test again, $S(x)$ converges absolutely if $|x - x_0| < R$ and diverges if $|x - x_0| > R$. This proves parts i) and iii).

To prove part ii), let $[a, b] \subset (x_0 - R, x_0 + R)$. Choose an $x_1 \in (x_0 - R, x_0 + R)$ such that $x \in [a, b]$ implies $|x - x_0| \leq |x_1 - x_0|$ (see Figure 7.5). Set $M_k = |a_k| |x_1 - x_0|^k$ and observe by part i) that $\sum_{k=0}^{\infty} M_k$ converges. Since $|a_k (x - x_0)^k| \leq M_k$ for $x \in [a, b]$ and $k \in \mathbf{N}$, it follows from the Weierstrass $M$-Test that $S(x)$ converges uniformly on $[a, b]$. ∎

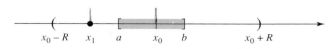

**Figure 7.5**

The following result provides another way to compute the radius of convergence of a power series (see also Exercise 4 below).

**7.22 THEOREM.** *If the limit*

$$R = \lim_{k \to \infty} \frac{|a_k|}{|a_{k+1}|}$$

Ratio Test

$$\frac{1}{R} = \lim_{k \to \infty} \frac{|a_{k+1}|}{|a_k|}$$

*exists as an extended real number, then $R$ is the radius of convergence of the power series $S(x) = \sum_{k=0}^{\infty} a_k(x - x_0)^k$.*

PROOF. Repeat the proof of Theorem 7.21, using the Ratio Test instead of the Root Test, to find that $S(x)$ converges absolutely on $(x_0 - R, x_0 + R)$ and diverges for each $x \notin [x_0 - R, x_0 + R]$. By Theorem 7.21, $R$ must be the radius of convergence of $S(x)$. ∎

**7.23 DEFINITION.** The *interval of convergence* of a power series $S(x)$ is the largest interval on which $S(x)$ converges.

By Theorem 7.21, for a given power series $S = \sum_{k=0}^{\infty} a_k(x - x_0)^k$, there are only three possibilities:

   i) $R = \infty$, in which case the interval of convergence of $S$ is $(-\infty, \infty)$,
   ii) $R = 0$, in which case the interval of convergence of $S$ is $\{x_0\}$, and
   iii) $0 < R < \infty$, in which case the interval of convergence of $S$ is $(x_0 - R, x_0 + R)$, $[x_0 - R, x_0 + R)$, $(x_0 - R, x_0 + R]$, or $[x_0 - R, x_0 + R]$.

To find the interval of convergence of a power series, therefore, one needs to compute the radius of convergence $R$ first. If $0 < R < \infty$, one must also check both endpoints $x_0 - R$ and $x_0 + R$ to see whether the interval of convergence is closed, open, or half open/closed.

**7.24 Example.** Find the interval of convergence of $S(x) = \sum_{k=1}^{\infty} x^k / \sqrt{k}$.

SOLUTION. By Theorem 7.22,

$$R = \lim_{k \to \infty} \frac{\sqrt{k}}{\sqrt{k+1}} = \sqrt{\lim_{k \to \infty} \frac{k}{k+1}} = 1.$$

Thus, the interval of convergence has endpoints 1 and $-1$. $S(x)$ diverges at $x = 1$ by the $p$-Series Test and converges at $x = -1$ by the Alternating Series Test. Thus, the interval of convergence of $S(x)$ is $[-1, 1)$. ∎

**7.25 Remark.** *The interval of convergence may contain none, one, or both its endpoints.*

PROOF. By Theorem 7.22, the radius of convergence of each of the series

$$\sum_{k=1}^{\infty} x^k, \qquad \sum_{k=1}^{\infty} \frac{x^k}{k}, \qquad \sum_{k=1}^{\infty} \frac{x^k}{k^2}$$

is 1, but by the Divergence Test, the Alternating Series Test, and the $p$-Series Test, the intervals of convergence of these series are $(-1,1)$, $[-1,1)$, and $[-1,1]$, respectively. ∎

We now pass from convergence properties of power series to the calculus of power series. The next several results answer the question: What properties (e.g., continuity, differentiability, integrability) does the limit of a power series satisfy?

**7.26 THEOREM.** *If $f(x) = \sum_{k=0}^{\infty} a_k(x - x_0)^k$ is a power series with positive radius of convergence $R$, then $f$ is continuous on $(x_0 - R, x_0 + R)$.*

PROOF. Let $x \in (x_0 - R, x_0 + R)$ and choose $a, b \in \mathbf{R}$ such that $x \in (a, b)$ and $[a, b] \subset (x_0 - R, x_0 + R)$. By Theorems 7.21ii) and 7.14i), $f$ is continuous on $(a, b)$ hence at $x$. ∎     *Since $a_k(x - x_0)^k$ is continuous,* (each)

The following result shows that continuity of the limit extends to the endpoints when they belong to the interval of convergence.

**7.27 THEOREM** [ABEL'S THEOREM]. *Suppose $[a, b]$ is a nondegenerate, closed, bounded interval. If $f(x) := \sum_{k=0}^{\infty} a_k(x - x_0)^k$ converges on $[a, b]$, then $f(x)$ is continuous and converges uniformly on $[a, b]$.*

PROOF. By Theorems 7.21ii) and 7.26, we may suppose that $f$ has a positive, finite radius of convergence $R$, and $b = x_0 + R$ or $a = x_0 - R$. By symmetry, it suffices to prove it for $a = x_0$ and $b = x_0 + R$. Thus, suppose $f(x)$ converges at $x = x_0 + R$ and fix $x_1 \in (x_0, x_0 + R]$. Set $b_k = a_k R^k$ and $c_k = (x_1 - x_0)^k / R^k$ for $k \in \mathbf{N}$. By hypothesis, $\sum_{k=1}^{\infty} b_k$ is convergent. Hence, given $\varepsilon > 0$, there is an integer $N > 1$ such that

$$k > m \geq N \quad \text{implies} \quad \left| \sum_{j=m}^{k} b_j \right| < \frac{\varepsilon}{c_1}.$$

Since $0 < x_1 - x_0 \leq R$, the sequence $\{c_k\}$ is decreasing. Applying Abel's Formula and telescoping, we have

$$\left| \sum_{k=m}^{n} a_k(x_1 - x_0)^k \right| = \left| \sum_{k=m}^{n} b_k c_k \right|$$

$$= \left| c_n \sum_{k=m}^{n} b_k + \sum_{k=m}^{n-1} (c_k - c_{k+1}) \sum_{j=m}^{k} b_j \right|$$

$$< c_n \frac{\varepsilon}{c_1} + (c_m - c_n) \frac{\varepsilon}{c_1} = \frac{c_m}{c_1} \varepsilon.$$

Since $c_m \leq c_1$ it follows that

$$\left| \sum_{k=m}^{n} a_k(x_1 - x_0)^k \right| < \varepsilon$$

for all $x_1 \in (x_0, x_0 + R]$. Since this inequality also holds for $x_1 = x_0$, we conclude that $\sum_{k=0}^{\infty} a_k(x - x_0)^k$ converges uniformly on $[x_0, x_0 + R]$. ∎

**7.28 Remark.** *If a power series $S(x) = \sum_{k=0}^{\infty} a_k(x - x_0)^k$ converges at some $x_1 > x_0$, then $S(x)$ converges uniformly on $[x_0, x_1]$ and absolutely on $[x_0, x_1)$. It need not converge absolutely at $x = x_1$.*

PROOF. By Theorems 7.21 and 7.27, $S(x)$ converges uniformly on $[x_0, x_1]$ and absolutely on $[x_0, x_1)$. The power series $\sum_{k=1}^{\infty} (-x)^k/k$ converges uniformly on $[0, 1]$ but not absolutely at $x = 1$. ∎

To discuss differentiability of the limit of a power series, we first show that the radius of convergence of a power series is not changed by term-by-term differentiation (see also Exercise 6 in Section 2.5).

**7.29 Lemma.** *If $a_n \in \mathbf{R}$ for $n \in \mathbf{N}$, then*

$$x := \limsup_{n \to \infty} (n|a_n|)^{1/n} = \limsup_{n \to \infty} |a_n|^{1/n} =: y.$$

PROOF. Let $\varepsilon > 0$. Since $n^{1/n} \to 1$ as $n \to \infty$, choose $N \in \mathbf{N}$ so that $n \geq N$ implies $1 - \varepsilon < n^{1/n} < 1 + \varepsilon$, i.e.,

$$(1 - \varepsilon)|a_n|^{1/n} < (n|a_n|)^{1/n} < (1 + \varepsilon)|a_n|^{1/n}$$

for large $n$. Since $y$ is the largest adherent point of $|a_n|^{1/n}$, the right-most inequality above implies $x \leq (1 + \varepsilon)y$, i.e., $x \leq y$. Similarly, since $x$ is the largest adherent point of $(n|a_n|)^{1/n}$, the left-most inequality above implies $y \leq x$. ∎

We use this result to prove that each power series with a positive radius of convergence is term-by-term differentiable.

**7.30 THEOREM.** *If $f(x) = \sum_{k=0}^{\infty} a_k(x - x_0)^k$ is a power series with positive radius of convergence $R$, then*

$$f'(x) = \sum_{k=1}^{\infty} ka_k(x - x_0)^{k-1}$$

*for $x \in (x_0 - R, x_0 + R)$.*

PROOF. Let $x \in (x_0 - R, x_0 + R)$. Choose a closed subinterval $[a, b]$ of $(x_0 - R, x_0 + R)$ such that $x \in (a, b)$. By Lemma 7.29, the radius of convergence of the derived series $\sum_{k=1}^{\infty} ka_k(x - x_0)^{k-1}$ is also $R$. Hence by Theorem 7.21 the derived series converges uniformly on $[a, b]$. We conclude by Theorem 7.14iii) that the series $f(x)$ is term-by-term differentiable on $(a, b)$, hence at $x$. ∎

Recall that for each nonempty, open interval $(a, b)$, $\mathcal{C}^{\infty}(a, b)$ represents the set of functions $f$ such that $f^{(k)}$ exists and is continuous on $(a, b)$ for all $k \in \mathbf{N}$. The following result generalizes Theorem 7.30.

**7.31 COROLLARY.** *If $f(x) = \sum_{k=0}^{\infty} a_k(x - x_0)^k$ has a positive radius of convergence $R$, then $f \in C^{\infty}(x_0 - R, x_0 + R)$ and*

$$(7) \qquad f^{(k)}(x) = \sum_{n=k}^{\infty} \frac{n!}{(n-k)!} a_n (x - x_0)^{n-k}$$

*for $x \in (x_0 - R, x_0 + R)$ and $k \in \mathbf{N}$.*

PROOF. The proof is by induction on $k$. By Theorem 7.30 and the fact that $0! := 1$, (7) holds for $k = 1$ and $x \in (x_0 - R, x_0 + R)$. If (7) holds for some $k \in \mathbf{N}$ and all $x \in (x_0 - R, x_0 + R)$, then $f^{(k)}$ is a power series with radius of convergence $R$. It follows from Theorem 7.30 that

$$f^{(k+1)}(x) = (f^{(k)}(x))' = \left( \sum_{n=k}^{\infty} \frac{n!}{(n-k)!} a_n (x - x_0)^{n-k} \right)'$$

$$= \sum_{n=k+1}^{\infty} \frac{n!}{(n-k-1)!} a_n (x - x_0)^{n-k-1}$$

for all $x \in (x_0 - R, x_0 + R)$. Hence, (7) holds for $k + 1$ in place of $k$. ∎

The following result shows that each power series with a positive radius of convergence can also be integrated term-by-term.

**7.32 THEOREM.** *Let $f(x) = \sum_{k=0}^{\infty} a_k(x - x_0)^k$ be a power series and $a < b$ be real numbers.*

i) *If $f(x)$ converges on $[a, b]$, then $f$ is integrable on $[a, b]$ and*

$$\int_a^b f(x)\, dx = \sum_{k=0}^{\infty} a_k \int_a^b (x - x_0)^k\, dx.$$

ii) *If $f(x)$ converges on $[a, b)$ and if $\sum_{k=0}^{\infty} a_k(b - x_0)^{k+1}/(k+1)$ converges, then $f$ is improperly integrable on $[a, b)$ and*

$$\int_a^b f(x)\, dx = \sum_{k=0}^{\infty} a_k \int_a^b (x - x_0)^k\, dx.$$

PROOF. i) By Abel's Theorem, $f(x)$ converges uniformly on $[a, b]$. Hence, by Theorem 7.14ii), $f(x)$ is term-by-term integrable on $[a, b]$.

ii) Let $a \le t < b$ and set $A = \sum_{k=0}^{\infty} a_k(a - x_0)^{k+1}/(k + 1)$. By part i),

$$\int_a^t f(x)\, dx = \sum_{k=0}^{\infty} a_k \int_a^t (x - x_0)^k\, dx = \sum_{k=0}^{\infty} \frac{a_k}{k+1}(t - x_0)^{k+1} - A.$$

The left-most term of this last difference is a power series which by hypothesis converges at $t = b$. Thus, by the definition of improper integration and Abel's Theorem,

$$
\begin{aligned}
\int_a^b f(x)\,dx &= \lim_{t \to b-} \int_a^t f(x)\,dx \\
&= \lim_{t \to b-} \sum_{k=0}^{\infty} \frac{a_k}{k+1}(t - x_0)^{k+1} - A \\
&= \sum_{k=0}^{\infty} \frac{a_k}{k+1}(b - x_0)^{k+1} - A = \sum_{k=0}^{\infty} a_k \int_a^b (x - x_0)^k\,dx. \quad \blacksquare
\end{aligned}
$$

The following result shows that the product of two power series is a power series. (For a result on the division of power series, see Taylor [13], p. 619.)

**7.33 THEOREM.** If $f(x) = \sum_{k=0}^{\infty} a_k x^k$ and $g(x) = \sum_{k=0}^{\infty} b_k x^k$ converge on $(-r, r)$ and

$$
c_k = \sum_{j=0}^{k} a_j b_{k-j}, \qquad k = 0, 1, \ldots,
$$

then $\sum_{k=0}^{\infty} c_k x^k$ converges on $(-r, r)$ and converges to $f(x)g(x)$.

PROOF. Fix $x \in (-r, r)$ and for each $n \in \mathbf{N}$, set

$$
f_n(x) = \sum_{k=0}^{n} a_k x^k, \qquad g_n(x) = \sum_{k=0}^{n} b_k x^k, \quad \text{and} \quad h_n(x) = \sum_{k=0}^{n} c_k x^k.
$$

By changing the order of summation, we see that

$$
\begin{aligned}
h_n(x) &= \sum_{k=0}^{n} \sum_{j=0}^{k} a_j b_{k-j} x^j x^{k-j} = \sum_{j=0}^{n} a_j x^j \sum_{k=j}^{n} b_{k-j} x^{k-j} \\
&= \sum_{j=0}^{n} a_j x^j g_{n-j}(x) = g(x) f_n(x) + \sum_{j=0}^{n} a_j x^j (g_{n-j}(x) - g(x)).
\end{aligned}
$$

Thus, it suffices to show

$$
\lim_{n \to \infty} \sum_{j=0}^{n} a_j x^j (g_{n-j}(x) - g(x)) = 0.
$$

Let $\varepsilon > 0$. Since $g_n(x)$ converges as $n \to \infty$, choose $M > 0$ such that

$$
|g_{n-j}(x) - g(x)| \le M
$$

for all integers $n > j > 0$. Since $C = \sum_{k=0}^{\infty} |a_k x^k| + 1$ is finite, choose $N \in \mathbf{N}$ such that

$$\ell \geq N \quad \text{implies} \quad |g_\ell(x) - g(x)| < \frac{\varepsilon}{2C} \quad \text{and} \quad \sum_{j=N+1}^{\infty} |a_j x^j| < \frac{\varepsilon}{2M}.$$

Let $n > 2N$. Then

$$\Big| \sum_{j=0}^{n} a_j x^j (g_{n-j}(x) - g(x)) \Big|$$

$$= \Big| \sum_{j=0}^{N} a_j x^j (g_{n-j}(x) - g(x)) + \sum_{j=N+1}^{n} a_j x^j (g_{n-j}(x) - g(x)) \Big|$$

$$< \frac{\varepsilon}{2C} \sum_{j=0}^{N} |a_j x^j| + M \sum_{j=N+1}^{n} |a_j x^j| < \frac{\varepsilon}{2} + \frac{\varepsilon}{2} = \varepsilon. \quad \blacksquare$$

**7.34 COROLLARY.** *Suppose $a_k, b_k \in \mathbf{R}$ and $c_k := \sum_{j=0}^{k} a_j b_{k-j}$ for $k = 0, 1, \ldots$. If either*

i) $\sum_{k=0}^{\infty} a_k$ *and* $\sum_{k=0}^{\infty} b_k$ *both converge, and at least one of them converges absolutely,*

ii) *or, if* $\sum_{k=0}^{\infty} a_k$, $\sum_{k=0}^{\infty} b_k$, *and* $\sum_{k=0}^{\infty} c_k$ *all converge,*

*then*

$$(8) \qquad \sum_{k=0}^{\infty} c_k = \left( \sum_{k=0}^{\infty} a_k \right) \left( \sum_{k=0}^{\infty} b_k \right).$$

PROOF. i) Repeat the proof of Theorem 7.33 with $x = 1$.

ii) By hypothesis, the radii of convergence of $\sum_{k=0}^{\infty} a_k x^k$, $\sum_{k=0}^{\infty} b_k x^k$, and $\sum_{k=0}^{\infty} c_k x^k$ are all at least 1; hence, by Theorem 7.33,

$$(9) \qquad \sum_{k=0}^{\infty} c_k x^k = \left( \sum_{k=0}^{\infty} a_k x^k \right) \left( \sum_{k=0}^{\infty} b_k x^k \right)$$

for $x \in (-1, 1)$. But by Abel's Theorem (7.27), the limit of (9) as $x \uparrow 1$ is (8). $\blacksquare$

The hypotheses of Corollary 7.34 cannot be relaxed.

**7.35 Example.** If $a_k = b_k = (-1)^k/\sqrt{k}$ for $k \in \mathbf{N}$ and $a_0 = b_0 = 0$, then $\sum_{k=0}^{\infty} c_k$ diverges.

PROOF. If $\sum_{k=0}^{\infty} c_k$ converges, then $c_k \to 0$ as $k \to \infty$. But for $k > 1$ odd,

$$|c_k| = \sum_{j=1}^{k-1} \frac{1}{\sqrt{j}\sqrt{k-j}} = 2 \sum_{j=1}^{(k-1)/2} \frac{1}{\sqrt{j}\sqrt{k-j}}$$

$$\geq 2 \left( \frac{k-1}{2} \right) \left( \frac{1}{\sqrt{(k-1)/2}} \right) \left( \frac{1}{\sqrt{(k-1)}} \right) = \sqrt{2}$$

is bounded away from zero as $k \to \infty$, a contradiction. ∎

We close this section with some optional material on finding exact values of convergent power series. Namely, we show how term-by-term differentiation and integration can be used in conjunction with geometric series to obtain simple formulas for certain kinds of power series. Such formulas are called *closed forms*.

*7.36 Example.* Find a closed form of the power series

$$f(x) = \sum_{k=1}^{\infty} k x^k.$$

SOLUTION. Since the interval of convergence of this power series is $(-1, 1)$, we have by Theorems 7.32 and 6.7 (the Geometric Series) that

$$\int_0^x \frac{f(t)}{t} \, dt = \sum_{k=1}^{\infty} k \int_0^x t^{k-1} \, dt = \sum_{k=1}^{\infty} x^k = \frac{x}{1-x}$$

for each $x \in (-1, 1)$. (Note that $f(x)/x$ is defined at $x = 0$ and has value 1.) Hence, by the Fundamental Theorem of Calculus,

$$\frac{f(x)}{x} = \left( \frac{x}{1-x} \right)' = \frac{1}{(1-x)^2}$$

and it follows that

$$f(x) = \frac{x}{(1-x)^2}, \qquad x \in (-1, 1). \quad ∎$$

*7.37 Example.* Find a closed form of the power series

$$g(x) = \sum_{k=0}^{\infty} \frac{x^k}{k+1}.$$

SOLUTION. Since the interval of convergence of this power series is $[-1, 1)$, we have by Theorem 7.30 that

$$(xg(x))' = \sum_{k=0}^{\infty} \left( \frac{x^{k+1}}{k+1} \right)' = \sum_{k=0}^{\infty} x^k = \frac{1}{1-x}$$

for $x \in (-1, 1)$. Hence, by the Fundamental Theorem of Calculus,

$$xg(x) = \int_0^x \frac{dt}{1-t} = -\log(1-x)$$

for $x \in (-1, 1)$. Since $g(-1)$ exists and $\log(1 - x)$ is continuous at $x = -1$, we conclude by Abel's Theorem that

$$g(x) = -\frac{\log(1 - x)}{x}, \qquad x \in [-1, 1), \quad x \neq 0. \quad \blacksquare$$

## EXERCISES

**1.** Find the interval of convergence of each of the following power series.

a) $\displaystyle\sum_{k=0}^{\infty} \frac{x^k}{2^k}$.      b) $\displaystyle\sum_{k=0}^{\infty} ((-1)^k + 3)^k (x - 1)^k$.

c) $\displaystyle\sum_{k=1}^{\infty} \log\left(\frac{k+1}{k}\right) x^k$.      *d) $\displaystyle\sum_{k=1}^{\infty} \frac{1 \cdot 3 \dots (2k - 1)}{(k + 1)!} x^{2k}$.

*2. Find a closed form for each of the following series and the largest set on which this formula is valid.      *discontinous at x=0 for function.*

a) $\displaystyle\sum_{k=1}^{\infty} 3x^{3k-1}$.      b) $\displaystyle\sum_{k=2}^{\infty} kx^{k-2}$.      c) $\displaystyle\sum_{k=1}^{\infty} \frac{2k}{k+1} (1 - x)^k$.      d) $\displaystyle\sum_{k=0}^{\infty} \frac{x^{3k}}{k+1}$.

**3.** Use Theorems 7.30 and 7.33 to give two different proofs of the following identity:

$$\frac{1}{(1 - x)^2} = \sum_{k=0}^{\infty} (k + 1)x^k, \qquad x \in (-1, 1).$$

*4. If $\sum_{k=1}^{\infty} a_k x^k$ has radius of convergence $R$ and $a_k \neq 0$ for large $k$, prove

$$\liminf_{k \to \infty} \left| \frac{a_k}{a_{k+1}} \right| \leq R \leq \limsup_{k \to \infty} \left| \frac{a_k}{a_{k+1}} \right|.$$

**5.** Suppose $|a_k| \leq |b_k|$ for large $k$. Prove that if $\sum_{k=1}^{\infty} b_k x^k$ converges on an open interval $I$, then $\sum_{k=1}^{\infty} a_k x^k$ also converges on $I$. Is this result true if "open" is omitted?

**6.** Suppose $\{a_k\}_{k=0}^{\infty}$ is a bounded sequence of real numbers.

a) Prove

$$f_{x_0}(x) = \sum_{k=0}^{\infty} a_k (x - x_0)^k$$

has a positive radius of convergence.

b) Prove that $f_{0.5}$ is integrable on $[0, 1]$ and

$$\int_0^1 f_{0.5}(x) \, dx = \sum_{k=0}^{\infty} \frac{a_{2k}}{(2k + 1)2^{2k}}.$$

**7.** A series $\sum_{k=0}^{\infty} a_k$ is said to be *Abel summable* to $L$ if

$$\lim_{r \to 1-} \sum_{k=0}^{\infty} a_k r^k = L.$$

a) Prove that if $\sum_{k=0}^{\infty} a_k$ converges to $L$, then $\sum_{k=0}^{\infty} a_k$ is Abel summable to $L$.

b) Find the Abel sum of $\sum_{k=0}^{\infty} (-1)^k$.

*8. Prove

$$f(x) = \sum_{k=0}^{\infty} \left( \frac{x}{(-1)^k + 4} \right)^k$$

is differentiable on $(-3, 3)$ and

$$|f'(x)| \leq \frac{3}{(3 - x)^2}$$

for $0 \leq x < 3$.

**9.** Suppose $a_k \downarrow 0$ as $k \to \infty$. Prove that given $\varepsilon > 0$ there is a $\delta > 0$ such that

$$\left| \sum_{k=0}^{\infty} (-1)^k a_k (x^k - y^k) \right| < \varepsilon$$

for all $x, y \in [0, 1]$ which satisfy $|x - y| < \delta$.

*10. a) Prove the following weak form of Stirling's Formula (compare with Theorem 12.74).

$$\frac{n^n}{e^{n-1}} < n! < \frac{n^{n+1}}{e^{n-1}}.$$

b) Find all $x \in \mathbf{R}$ for which the power series

$$\sum_{k=0}^{\infty} \frac{k^k}{k!} x^k$$

converges absolutely.

## 7.4  ANALYTIC FUNCTIONS

In this section we study functions which can be represented by power series. (For a discussion of how to represent functions by trigonometric series instead of power series, see Chapter 14.) We begin with the following definition.

**7.38 DEFINITION.** A real-valued function $f$ is said to be *analytic* on a nonempty, open interval $(a,b)$ if given $x_0 \in (a,b)$ there is a power series centered at $x_0$ which converges to $f$ near $x_0$, i.e., if there exist coefficients $\{a_k\}_{k=0}^\infty$ and points $c, d \in (a,b)$ such that $c < x_0 < d$ and

$$f(x) = \sum_{k=0}^\infty a_k(x-x_0)^k$$

for all $x \in (c,d)$.

We shall develop several techniques for showing that a given function is analytic. To simplify statements of results, we shall use the conventions $f^{(0)} := f$ and $0! := 1$.

First, it is important to realize that if $f$ can be represented by a power series $S$, then $f$ is locally smooth and the coefficients of $S$ can be computed using derivatives of $f$.

**7.39 THEOREM** [UNIQUENESS]. *Let $c < d$ be extended real numbers, $x_0 \in (c,d)$, and $f : (c,d) \to \mathbf{R}$. If $f(x) = \sum_{k=0}^\infty a_k(x-x_0)^k$ for each $x \in (c,d)$, then $f \in \mathcal{C}^\infty(c,d)$ and*

$$a_k = \frac{f^{(k)}(x_0)}{k!}, \qquad k = 0, 1, \cdots.$$

PROOF. Clearly, $f(x_0) = a_0$. Fix $k \in \mathbf{N}$. By hypothesis, the radius of convergence $R$ of the power series $\sum_{k=0}^\infty a_k(x-x_0)^k$ is positive and $(c,d) \subseteq (x_0-R, x_0+R)$. Hence, by Corollary 7.31, $f \in \mathcal{C}^\infty(c,d)$ and

$$(10) \qquad f^{(k)}(x) = \sum_{n=k}^\infty \frac{n!}{(n-k)!} a_n(x-x_0)^{n-k}$$

for $x \in (c,d)$. Apply this to $x = x_0$. The terms on the right side of (10) are zero when $n > k$ and $k!a_k$ when $n = k$. Hence, $f^{(k)}(x_0) = k!a_k$ for each $k \in \mathbf{N}$. ∎

In particular, if $f$ is analytic on $(a,b)$, then for each $x_0 \in (a,b)$ there is only one power series centered at $x_0$ which represents $f$ near $x_0$. This power series has a special name.

**7.40 DEFINITION.** Let $f \in \mathcal{C}^\infty(a,b)$ and $x_0 \in (a,b)$. The *Taylor expansion* (or *Taylor series*) of $f$ centered at $x_0$ is the series

$$\sum_{k=0}^\infty \frac{f^{(k)}(x_0)}{k!}(x-x_0)^k.$$

(No convergence is implied or assumed.) The Taylor expansion of $f$ centered at $x_0 = 0$ is usually called the *Maclaurin expansion* (or *Maclaurin series*) of $f$.

By Theorem 7.39, every analytic function is a $\mathcal{C}^\infty$ function. The next remark shows that the converse of this statement is false.

**7.41 Remark.** [CAUCHY]. *The function*

$$f(x) = \begin{cases} e^{-1/x^2} & x \neq 0 \\ 0 & x = 0 \end{cases}$$

*belongs to* $\mathcal{C}^\infty(-\infty, \infty)$ *but is not analytic on any interval which contains* $x = 0$.

PROOF. It is easy to see (Exercise 3 in Section 4.3) that $f \in \mathcal{C}^\infty(-\infty, \infty)$ and $f^{(k)}(0) = 0$ for all $k \in \mathbf{N}$. Thus the Taylor expansion of $f$ about the point $x_0 = 0$ is identically zero but $f(x) = 0$ only when $x = 0$. ∎

One of the aims of this section is to prove that many of the classical $\mathcal{C}^\infty$ functions used in elementary calculus are analytic on their domain. Since, by Theorem 7.39, a $\mathcal{C}^\infty$ function is analytic on an open interval $I$ if and only if its Taylor expansion at each $x_0 \in I$ converges to $f$ near $x_0$, the following concept is useful in this regard.

**7.42 DEFINITION.** Let $f \in \mathcal{C}^\infty(a, b)$ and $x_0 \in (a, b)$. The *remainder term of order* $n$ of the Taylor expansion of $f$ centered at $x_0$ is the function

$$R_n(x) = R_n^{f,x_0}(x) := f(x) - \sum_{k=0}^{n-1} \frac{f^{(k)}(x_0)}{k!}(x - x_0)^k.$$

In fact, the remainder term completely determines analyticity of a given function in the following way.

**7.43 THEOREM.** *A function $f \in \mathcal{C}^\infty(a, b)$ is analytic on $(a, b)$ if and only if given $x_0 \in (a, b)$ there is an interval $(c, d)$ containing $x_0$ such that the remainder term $R_n^{f,x_0}(x)$ converges to zero for all $x \in (c, d)$.*

PROOF. By Theorem 7.39, $f$ is analytic on $(a, b)$ if and only if given $x_0 \in (a, b)$ there is an interval $(c, d)$ containing $x_0$ such that the Taylor expansion of $f$ centered at $x_0$ converges to $f$ pointwise on $(c, d)$. By Definition 7.42, this happens if and only if $R_n^{f,x_0} \to 0$, as $n \to \infty$, for every $x \in (c, d)$. ∎

Thus, to decide whether a given $f \in \mathcal{C}^\infty(a, b)$ is analytic on $(a, b)$ we need to estimate the corresponding remainder terms. We shall prove two results (see Theorems 7.44 and 7.52 below) which can be used to estimate remainder terms in concrete situations.

To motivate the first result, notice that $R_1^{f,x_0} = f(x) - f(x_0)$, the remainder term of order 1, can always be estimated using the Mean Value Theorem. The proof of the following result shows that the remainder term of order $n$ can be estimated by the Generalized Mean Value Theorem.

**7.44 THEOREM** [TAYLOR'S FORMULA]. *Let $n \in \mathbf{N}$, $f : (a, b) \to \mathbf{R}$, and suppose $f^{(n)}$ exists on $(a, b)$. Then for each pair of distinct points $x, x_0 \in (a, b)$ there is a number $c$ between $x$ and $x_0$ such that*

$$R_n^{f,x_0}(x) = \frac{f^{(n)}(c)}{n!}(x - x_0)^n.$$

*In particular,*

$$f(x) = \sum_{k=0}^{n-1} \frac{f^{(k)}(x_0)}{k!}(x - x_0)^k + \frac{f^{(n)}(c)}{n!}(x - x_0)^n$$

*for some number $c$ between $x$ and $x_0$.*

PROOF. Without loss of generality, suppose $x_0 < x$. Define

$$F(t) = \frac{(x - t)^n}{n!} \quad \text{and} \quad G(t) = R_n^{f,t}(x) = f(x) - \sum_{k=0}^{n-1} \frac{f^{(k)}(t)}{k!}(x - t)^k$$

for each $t \in (a, b)$. In order to apply the Generalized Mean Value Theorem to $F$ and $G$, we need to be sure the hypotheses of that result hold.

Notice by the Chain Rule that

(11)
$$F'(t) = -\frac{(x - t)^{n-1}}{(n - 1)!}$$

for $t \in \mathbf{R}$. Also notice that

$$\frac{d}{dt}\left(\frac{f^{(k)}(t)}{k!}(x - t)^k\right) = \frac{f^{(k+1)}(t)}{k!}(x - t)^k - \frac{f^{(k)}(t)}{(k - 1)!}(x - t)^{k-1}$$

for $t \in (a, b)$ and $k \in \mathbf{N}$. By telescoping, we obtain

(12)
$$G'(t) = -\frac{f^{(n)}(t)}{(n - 1)!}(x - t)^{n-1}$$

for $t \in (a, b)$. Thus, $F$ and $G$ are differentiable on $(x_0, x)$ and continuous on $[x_0, x]$.

By the Generalized Mean Value Theorem and the fact that $F(x) = G(x) = 0$, there is a number $c \in (x_0, x)$ such that

$$-F(x_0)G'(c) = (F(x) - F(x_0))G'(c) = (G(x) - G(x_0))F'(c) = -G(x_0)F'(c).$$

Hence, it follows from (11) and (12) that

$$\frac{(x - x_0)^n}{n!}\left(\frac{f^{(n)}(c)(x - c)^{n-1}}{(n - 1)!}\right) = R_n^{f,x_0}(x)\frac{(x - c)^{n-1}}{(n - 1)!}.$$

Solving this equation for $R_n^{f,x_0}$ completes the proof. ∎

The following theorem, a corollary of Taylor's Formula, is the first of several results which identify conditions on the derivatives $f^{(n)}$ of an $f \in \mathcal{C}^\infty$ sufficient for $f$ to be analytic on an interval $(a, b)$.

**7.45 THEOREM.** *Let $f \in \mathcal{C}^\infty(a,b)$. If there is an $M > 0$ such that*

$$\left| f^{(n)}(x) \right| \le M^n$$

*for all $x \in (a,b)$ and $n \in \mathbf{N}$, then $f$ is analytic on $(a,b)$. In fact, for each $x_0 \in (a,b)$,*

$$f(x) = \sum_{k=0}^{\infty} \frac{f^{(k)}(x_0)}{k!} (x - x_0)^k$$

*holds for all $x \in (a,b)$.*

Proof. Fix $x_0 \in (a,b)$ and set $C = \max\{M|a - x_0|, M|b - x_0|\}$. By Theorem 7.44,

$$|R_n(x)| := |R_n^{f,x_0}(x)| \le \frac{M^n |x - x_0|^n}{n!} \le \frac{C^n}{n!}$$

for all $n \in \mathbf{N}$. But $C^n/n! \to 0$ as $n \to \infty$ for any $C \in \mathbf{R}$ (being terms of a convergent series by the Ratio Test). Thus, by the Squeeze Theorem, the remainder term $R_n(x)$ converges to zero for every $x \in (a,b)$. ∎

**7.46 Example.** Prove that $\sin x$ and $\cos x$ are analytic on $\mathbf{R}$ and have Maclaurin expansions

$$\sin x = \sum_{k=0}^{\infty} \frac{(-1)^k x^{2k+1}}{(2k+1)!}, \qquad \cos x = \sum_{k=0}^{\infty} \frac{(-1)^k x^{2k}}{(2k)!}.$$

Proof. Set $f(x) = \sin x$. It is easy to see that

$$f^{(n)}(x) = \begin{cases} \sin x & n = 4j, \\ \cos x & n = 4j + 1, \\ -\sin x & n = 4j + 2, \\ -\cos x & n = 4j + 3 \end{cases}$$

for $j \in \mathbf{N}$. Hence $|f^{(n)}(x)| \le 1$ for all $x \in \mathbf{R}$, and

$$f^{(n)}(0) = \begin{cases} (-1)^k & n = 2k - 1 \\ 0 & n = 2k \end{cases}$$

for $k \in \mathbf{N}$. It follows from Theorem 7.45 that $f(x) = \sin x$ is analytic on $\mathbf{R}$ and its Maclaurin expansion has the promised form. A similar argument verifies the result for $\cos x$. ∎

**7.47 Example.** Prove that $e^x$ is analytic on $\mathbf{R}$ and has Maclaurin expansion

$$e^x = \sum_{k=0}^{\infty} \frac{x^k}{k!}.$$

PROOF. Fix $C > 0$ and set $M = e^C$. If $f(x) = e^x$, then $f^{(n)}(x) = e^x$ for all $x \in \mathbf{R}$. Hence, $|f^{(n)}(x)| \le M \le M^n$ for $n = 0, 1, \ldots$ and $x \in [-C, C]$. It follows from Theorem 7.45 that the Maclaurin series of $f$ converges to $f$ on $[-C, C]$. Since $f^{(n)}(0) = 1$ for all $n \in \mathbf{N}$, and $C > 0$ was arbitrary, we conclude that $\sum_{k=0}^{\infty} x^k / k!$ converges to $e^x$ for all $x \in \mathbf{R}$. ∎

Sometimes, it is impractical to get the kind of global estimates on the derivatives of $f$ necessary to apply Theorem 7.45. The following result, which shows that the center of a power series can be changed within its interval of convergence, is sometimes used to shortcut this process.

**7.48 THEOREM.** *Suppose $I$ is an open interval centered at $c$ and*

$$f(x) = \sum_{k=0}^{\infty} a_k (x - c)^k, \qquad x \in I.$$

*If $x_0 \in I$ and $r > 0$ satisfy $(x_0 - r, x_0 + r) \subseteq I$, then*

$$f(x) = \sum_{k=0}^{\infty} \frac{f^{(k)}(x_0)}{k!} (x - x_0)^k$$

*for all $x \in (x_0 - r, x_0 + r)$. In particular, if $f$ is a $C^\infty$ function whose Taylor series expansion converges to $f$ on some open interval $J$, then $f$ is analytic on $J$.*

PROOF. It suffices to prove the first statement. By making the change of variables $w = x - c$, we may suppose that $c = 0$ and $I = (-R, R)$, i.e., that $f(x) = \sum_{k=0}^{\infty} a_k x^k$, for all $x \in (-R, R)$. Suppose $(x_0 - r, x_0 + r) \subseteq (-R, R)$ and fix $x \in (x_0 - r, x_0 + r)$. By hypothesis and the Binomial Formula,

$$(13) \quad f(x) = \sum_{k=0}^{\infty} a_k x^k = \sum_{k=0}^{\infty} a_k ((x - x_0) + x_0)^k = \sum_{k=0}^{\infty} a_k \sum_{j=0}^{k} \binom{k}{j} x_0^{k-j} (x - x_0)^j.$$

Since $\sum_{k=0}^{\infty} a_k y^k$ converges absolutely at $y := |x - x_0| + |x_0| < R$, we have

$$\sum_{k=0}^{\infty} \left| a_k \sum_{j=0}^{k} \binom{k}{j} x_0^{k-j} (x - x_0)^j \right| \le \sum_{k=0}^{\infty} |a_k| \sum_{j=0}^{k} \binom{k}{j} |x_0|^{k-j} |x - x_0|^j$$

$$= \sum_{k=0}^{\infty} |a_k| (|x - x_0| + |x_0|)^k < \infty.$$

Hence, by (13), Theorem 7.18, and Corollary 7.31,

$$f(x) = \sum_{k=0}^{\infty} a_k \sum_{j=0}^{k} \binom{k}{j} x_0^{k-j} (x - x_0)^j$$

$$= \sum_{j=0}^{\infty} \left( \sum_{k=j}^{\infty} \binom{k}{j} a_k x_0^{k-j} \right) (x - x_0)^j$$

$$= \sum_{j=0}^{\infty} \left( \sum_{k=j}^{\infty} \frac{k!}{(k-j)!} a_k (x_0 - 0)^{k-j} \right) \frac{(x - x_0)^j}{j!} = \sum_{j=0}^{\infty} \frac{f^{(j)}(x_0)}{j!} (x - x_0)^j. \ \blacksquare$$

**7.49 Example.** Prove that $\arctan x$ is analytic on $(-1, 1)$ and has Maclaurin expansion

$$\arctan x = \sum_{k=0}^{\infty} \frac{(-1)^k x^{2k+1}}{2k+1} \qquad x \in (-1, 1).$$

PROOF. For each $0 < x < 1$, the geometric series $\sum_{k=0}^{\infty}(-1)^k t^{2k}$ converges uniformly on $[-x, x]$ to $1/(1+t^2)$ (see Exercise 1 in Section 6.1). Thus, by Theorem 7.32,

$$\arctan x = \int_0^x \frac{dt}{1+t^2} = \int_0^x \sum_{k=0}^{\infty}(-1)^k t^{2k}\, dt = \sum_{k=0}^{\infty}\frac{(-1)^k x^{2k+1}}{2k+1}.$$

By uniqueness, this is the Maclaurin expansion of $\arctan x$. Since this expansion converges on $(-1, 1)$, it follows from Theorem 7.48 that $\arctan x$ is analytic on $(-1, 1)$. ∎

In Examples 7.46 and 7.47, we found the Taylor expansion of a given $f$ by computing the derivatives of $f$ and estimating the remainder term. In the previous example, we found the Taylor expansion of $\arctan x$ without computing its derivatives. This can be done in general, using term by term differentiation or integration or products of power series, when the function in question can be written as an integral or derivative or product of functions whose Taylor series are known. Here are two more examples of this type.

**7.50 Example.** Find the Maclaurin expansion of $\arctan x/(1 - x)$.

PROOF. By Theorem 7.33 and Example 7.49, for each $|x| < 1$,

$$\left(\frac{\arctan x}{1-x}\right) = \left(\sum_{k=0}^{\infty} x^k\right)\left(\sum_{k=0}^{\infty}\frac{(-1)^k x^{2k+1}}{2k+1}\right)$$

$$= \sum_{k=1}^{\infty}\left(\sum_{j \in A_k}\frac{(-1)^j}{2j+1}\right)x^k,$$

where $A_k := \{j \in \mathbf{N} : 0 \le j \le (k-1)/2\}$. ∎

**7.51 Example.** Show that the Taylor expansion of $\log x$ centered at $x_0 = 1$ is

$$\log x = \sum_{k=1}^{\infty}\frac{(-1)^{k+1}}{k}(x-1)^k \qquad x \in (0, 2).$$

PROOF. By Theorem 7.32, for each $x \in (0, 2)$,

$$\log x = \int_1^x \frac{dt}{t} = \int_1^x \frac{dt}{1-(1-t)}$$

$$= \int_1^x \sum_{k=0}^{\infty}(1-t)^k\, dt = \sum_{k=1}^{\infty}\frac{(-1)^{k+1}}{k}(x-1)^k. \quad ∎$$

In some situations it is useful to have an integral form of the remainder term. This requires a slightly stronger hypothesis than Theorem 7.44 but can yield a sharper estimate.

**7.52 THEOREM** [LAGRANGE]. *Let* $n \in \mathbf{N}$. *If* $f \in \mathcal{C}^n(a,b)$, *then*

$$R_n(x) := R_n^{f, x_0}(x) = \frac{1}{(n-1)!} \int_{x_0}^{x} (x-t)^{n-1} f^{(n)}(t)\, dt$$

*for all* $x, x_0 \in (a,b)$.

PROOF. The proof is by induction on $n$. If $n = 1$, the formula holds by the Fundamental Theorem of Calculus.

Suppose the formula holds for some $n \in \mathbf{N}$. Since

$$R_{n+1}(x) = R_n(x) - \frac{f^{(n)}(x_0)}{n!}(x-x_0)^n \quad \text{and} \quad \frac{(x-x_0)^n}{n!} = \frac{1}{(n-1)!} \int_{x_0}^{x} (x-t)^{n-1}\, dt$$

it follows that

$$R_{n+1}(x) = \frac{1}{(n-1)!} \int_{x_0}^{x} (x-t)^{n-1} \left( f^{(n)}(t) - f^{(n)}(x_0) \right) dt.$$

Let $u(t) = f^{(n)}(t) - f^{(n)}(x_0)$, $v(t) = (x-t)^n/n$ and integrate the right side of the identity above by parts. Since $u(x_0) = 0$ and $v(x) = 0$, we have

$$R_{n+1}(x) = -\frac{1}{(n-1)!} \int_{x_0}^{x} u'(t) v(t)\, dt = \frac{1}{n!} \int_{x_0}^{x} (x-t)^n f^{(n+1)}(t)\, dt.$$

Hence, the formula holds for $n+1$. ∎

In order to generalize the Binomial Formula from integer exponents to real exponents (compare Theorem 1.15 with Theorem 7.54 below), we introduce the following notation. Let $\alpha \in \mathbf{R}$ and $k$ be a nonnegative integer. The *generalized binomial coefficient* $\alpha$ *over* $k$ is defined by

$$\binom{\alpha}{k} := \begin{cases} \dfrac{\alpha(\alpha-1)\ldots(\alpha-k+1)}{k!} & k \neq 0 \\ 1 & k = 0. \end{cases}$$

Notice that when $\alpha \in \mathbf{N}$, these generalized binomial coefficients coincide with the usual binomial coefficients, because in this case $\binom{\alpha}{k} = 0$ for $k > \alpha$.

**7.53 Lemma.** *For all* $\alpha \in \mathbf{R}$ *and all integers* $k \geq 0$,

$$\sum_{j=0}^{k} \binom{\alpha}{k-j} \binom{\beta}{j} = \binom{\alpha+\beta}{k}.$$

PROOF. The formula holds for $k = 0$ and $k = 1$. If it holds for some $k \geq 1$, then by the inductive hypothesis and the definition of the generalized binomial coefficients,

$$\binom{\alpha + \beta}{k + 1} = \binom{\alpha + \beta}{k} \frac{\alpha + \beta - k}{k + 1}$$

$$= \sum_{j=0}^{k} \binom{\alpha}{k - j} \binom{\beta}{j} \left( \frac{\alpha - k + j}{k + 1} + \frac{\beta - j}{k + 1} \right)$$

$$= \sum_{j=0}^{k} \left( \frac{k - j + 1}{k + 1} \right) \binom{\alpha}{k - j + 1} \binom{\beta}{j} + \left( \frac{j + 1}{k + 1} \right) \binom{\alpha}{k - j} \binom{\beta}{j + 1}$$

$$= \binom{\alpha}{k + 1} + \sum_{j=1}^{k} \left( \frac{k - j + 1}{k + 1} + \frac{j}{k + 1} \right) \binom{\alpha}{k - j + 1} \binom{\beta}{j} + \binom{\beta}{k + 1}$$

$$= \sum_{j=0}^{k+1} \binom{\alpha}{k + 1 - j} \binom{\beta}{j}. \quad \blacksquare$$

With this ugly calculation out of the way, we are prepared to generalize the Binomial Formula.

**7.54 THEOREM** [THE BINOMIAL SERIES]. *If $\alpha \in \mathbf{R}$ and $|x| < 1$, then*

$$(1 + x)^\alpha = \sum_{k=0}^{\infty} \binom{\alpha}{k} x^k.$$

*In particular, $(1 + x)^\alpha$ is analytic on $(-1, 1)$ for all $\alpha \in \mathbf{R}$.*

PROOF. Fix $|x| < 1$ and consider the series $F(\alpha) := \sum_{k=0}^{\infty} \binom{\alpha}{k} x^k$. Since

$$\left| \frac{\binom{\alpha}{k + 1} x^{k+1}}{\binom{\alpha}{k} x^k} \right| = \left| \frac{\alpha - k}{k + 1} \right| |x| \to |x| < 1$$

is independent of $\alpha$, it follows from the proof of the Ratio Test that $F$ converges absolutely and uniformly on $\mathbf{R}$. Hence, $F$ is continuous. Moreover, by Theorem 7.33 and Lemma 7.53,

$$F(\alpha)F(\beta) = \sum_{k=0}^{\infty} \binom{\alpha}{k} x^k \sum_{k=0}^{\infty} \binom{\beta}{k} x^k$$

$$= \sum_{k=0}^{\infty} \sum_{j=0}^{k} \binom{\alpha}{k - j} \binom{\beta}{j} x^k$$

$$= \sum_{k=0}^{\infty} \binom{\alpha + \beta}{k} x^k = F(\alpha + \beta).$$

Hence, it follows from Exercise 9 in Section 3.3 that $F(\alpha) = F(1)^{\alpha}$. Since

$$F(1) = \sum_{k=0}^{\infty} \binom{1}{k} x^k = 1 + x,$$

we conclude that $F(x) = (1+x)^{\alpha}$ for all $|x| < 1$. ∎

*The rest of this section contains some additional (but optional) material on analytic functions.*

Lagrange's Theorem gives us another condition on the derivatives of $f$ sufficient to conclude that $f$ is analytic.

**\*7.55 THEOREM** [BERNSTEIN]. *If $f \in C^{\infty}(a, b)$ and $f^{(n)}(x) \geq 0$ for all $x \in (a, b)$ and $n \in \mathbf{N}$, then $f$ is analytic on $(a, b)$. In fact, if $x_0 \in (a, b)$ and $f^{(n)}(x) \geq 0$ for $x \in [x_0, b)$ and $n \in \mathbf{N}$, then*

(14)
$$f(x) = \sum_{k=0}^{\infty} \frac{f^{(k)}(x_0)}{k!} (x - x_0)^k$$

*for all $x \in [x_0, b)$.*

PROOF. Fix $x_0 < x < b$ and $n \in \mathbf{N}$. Use Lagrange's Theorem and a change of variables $t = (x - x_0)u + x_0$ to write

(15)    $$R_n(x) = R_n^{f, x_0}(x) = \frac{(x - x_0)^n}{(n-1)!} \int_0^1 (1 - u)^{n-1} f^{(n)}((x - x_0)u + x_0) \, du.$$

Since $f^{(n)} \geq 0$, (15) implies $R_n(x) \geq 0$. On the other hand, by definition and hypothesis,

$$R_n(x) = f(x) - \sum_{k=0}^{n-1} \frac{f^{(k)}(x_0)}{k!} (x - x_0)^k \leq f(x).$$

Therefore,

(16)                    $$0 \leq R_n(x) \leq f(x)$$

for all $x \in (x_0, b)$.

Let $b_0 \in (x_0, b)$ and notice that it suffices to verify (14) for $x_0 \leq x < b_0$. (We introduce the parameter $b_0$ in order to handle the cases $b \in \mathbf{R}$ and $b = \infty$ simultaneously.) Since $R_n(x_0) = 0$ for all $n \in \mathbf{N}$, we need only show that $R_n(x) \to 0$ as $n \to \infty$ for each $x \in (x_0, b_0)$.

By hypothesis, $f^{(n+1)}(t) \geq 0$ for $t \in [x_0, b)$, so $f^{(n)}$ is increasing on $[x_0, b)$. Since $x < b_0 < b$, we have by (15) and (16) that

$$0 \leq R_n(x) = \frac{(x - x_0)^n}{(n-1)!} \int_0^1 (1 - u)^{n-1} f^{(n)}((x - x_0)u + x_0) \, du$$

$$\leq \frac{(x - x_0)^n}{(n-1)!} \int_0^1 (1 - u)^{n-1} f^{(n)}((b_0 - x_0)u + x_0) \, du$$

$$= \left( \frac{x - x_0}{x - b_0} \right)^n R_n(b_0).$$

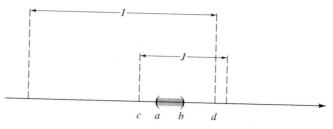

**Figure 7.6**

Since $(x - x_0)/(x - b_0) < 1$ and, by (16), $R_n(b_0) \leq f(b_0)$, we conclude by the Squeeze Theorem that $R_n(x) \to 0$ as $n \to \infty$. ∎

**\*7.56 Example.** Prove that $a^x$ is analytic on **R** for each $a > 0$.

PROOF. First suppose $a \geq 1$. Since $f^{(n)}(x) = (\log a)^n \cdot a^x \geq 0$ for all $x \in \mathbf{R}$ and $n \in \mathbf{N}$, $a^x$ is analytic on **R** by Bernstein's Theorem. If $0 < a < 1$, then by what we just proved and a change of variables,

$$a^x = (a^{-1})^{-x} = \sum_{k=0}^{\infty} \frac{\log^k(a^{-1})(-x)^k}{k!} = \sum_{k=0}^{\infty} \frac{\log^k a \cdot x^k}{k!}.$$

Hence by Theorem 7.48, $a^x$ is analytic on **R**. ∎

Our final theorem shows that an analytic function cannot be extended in an arbitrary way to produce another analytic function. We first prove the following special case.

**\*7.57 Lemma.** *Suppose $f, g$ are analytic on an open interval $(c, d)$ and $x_0 \in (c, d)$. If $f(x) = g(x)$ for $x \in (c, x_0)$, then there is a $\delta > 0$ such that $f(x) = g(x)$ for all $x \in (x_0 - \delta, x_0 + \delta)$.*

PROOF. By Theorem 7.39 and Definition 7.38, there is a $\delta > 0$ such that

$$(17) \qquad f(x) = \sum_{k=0}^{\infty} \frac{f^{(k)}(x_0)}{k!}(x - x_0)^k \quad \text{and} \quad g(x) = \sum_{k=0}^{\infty} \frac{g^{(k)}(x_0)}{k!}(x - x_0)^k$$

for all $x \in (x_0 - \delta, x_0 + \delta)$. By hypothesis, $f, g$ are continuous at $x_0$ and

$$(18) \qquad f(x_0) = \lim_{x \to x_0-} f(x) = \lim_{x \to x_0-} g(x) = g(x_0).$$

Similarly, $f^{(k)}(x_0) = g^{(k)}(x_0)$ for $k \in \mathbf{N}$. We conclude from (17) that $f(x) = g(x)$ for all $x \in (x_0 - \delta, x_0 + \delta)$. ∎

**\*7.58 THEOREM** [ANALYTIC CONTINUATION]. *Suppose $I$ and $J$ are open intervals, $f$ is analytic on $I$, $g$ is analytic on $J$, and $a < b$ are points in $I \cap J$. If $f(x) = g(x)$ for $x \in (a, b)$, then $f(x) = g(x)$ for all $x \in I \cap J$.*

PROOF. We assume for simplicity that $I$ and $J$ are bounded intervals. Since $I \cap J \neq \emptyset$, choose $c, d \in \mathbf{R}$ such that $I \cap J = (c, d)$ (see Figure 7.6).

Consider the set $E = \{t \in (a,d) : f(x) = g(x) \text{ for all } x \in (a,t)\}$. By our assumption, $d < \infty$ and by hypothesis $b \in E$. Thus $E$ is bounded and nonempty. Let $x_0 = \sup E$. If $x_0 < d$, then by Lemma 7.57 there is a $\delta > 0$ such that $f(x) = g(x)$ for all $x \in (x_0 - \delta, x_0 + \delta)$. This contradicts the choice of $x_0$. Therefore, $x_0 = d$, i.e., $f(x) = g(x)$ for all $x \in (a,d)$. A similar argument proves $f(x) = g(x)$ for all $x \in (c,b)$. ∎

## EXERCISES

**1.** Prove that each of the following functions is analytic on **R** and find its Maclaurin expansion.

  a) $\cos(3x)$.     b) $2^x$.     c) $\cos^2 x$.     d) $\sin^2 x + \cos^2 x$.     e) $x^3 e^{x^2}$.

**2.** Prove that each of the following functions is analytic on $(-1,1)$ and find its Maclaurin expansion.

  a) $\log(1-x)$.     b) $\dfrac{x^2}{1-x^3}$.     c) $\dfrac{e^x}{1-x}$.     d) $\dfrac{x^3}{(1-x)^2}$.     *e) $\arcsin x$.

**3.** Prove that each of the following functions is analytic on **R** and find its Maclaurin expansion.

$$\text{a) } (x^2-1)e^x. \qquad \text{b) } e^x \cos x. \qquad \text{c) } \frac{\sin x}{e^x}.$$

  d) $f(x) := \begin{cases} (a^x - 1)/x & x \neq 0 \\ \log a & x = 0 \end{cases}$     (where $a > 0$ fixed).

**4.** For each of the following functions, find its Taylor expansion centered at $x_0 = 1$ and determine the largest interval on which it converges.

$$\text{a) } \log_{10} x. \qquad \text{b) } x^2 + 2x - 1. \qquad \text{c) } e^x.$$

**5.** a) Prove that for all $x \in [0,1]$,

$$1 + x + \frac{x^2}{2} + \frac{x^3}{6} \leq e^x \leq \frac{9}{8} + x + \frac{x^2}{2} + \frac{x^3}{6}.$$

  b) Prove that for all $x \in [0,1]$,

$$x - \frac{x^3}{3} \leq \arctan x \leq \frac{1}{4} + x - \frac{x^3}{3}.$$

  c) Prove that for all $x \in [1,2]$ and $y = x - 1$,

$$y - \frac{y^2}{2} + \frac{y^3}{3} - \frac{y^4}{4} \leq \log x \leq y - \frac{y^2}{2} + \frac{y^3}{3} - \frac{y^4}{64}.$$

**6. a)** Prove

$$|\delta + \sin(\delta + \pi)| \le \frac{\delta^3}{3!}$$

for all $\delta > 0$.

**b)** Prove that if $|x - \pi| \le \delta$, then $|x + \sin x - \pi| \le \delta^3/3!$.

**7.** Suppose $f \in \mathcal{C}^\infty(-\infty, \infty)$ and

$$\lim_{n\to\infty} \frac{1}{n!} \int_0^a x^n f^{(n+1)}(a - x)\, dx = 0$$

for all $a \in \mathbf{R}$. Prove $f$ is analytic on $(-\infty, \infty)$ and

$$f(x) = \sum_{k=0}^\infty \frac{f^{(k)}(0)}{k!} x^k, \qquad x \in \mathbf{R}.$$

**8. a)** Prove that

$$\left| \int_0^1 e^{x^2}\, dx - \sum_{k=0}^{n-1} \frac{1}{(2k+1)k!} \right| \le \frac{3}{n!}$$

for $n \in \mathbf{N}$.

**b)** Show that

$$2.9253 < \int_{-1}^1 e^{x^2}\, dx < 2.9254.$$

**9.** Let $f \in \mathcal{C}^\infty(a, b)$. Prove that $f$ is analytic on $(a, b)$ if and only if $f'$ is analytic on $(a, b)$.

***10.** Suppose $f$ is analytic on $(-\infty, \infty)$ and

$$\int_a^b |f(x)|\, dx = 0$$

for some $a \ne b$ in $\mathbf{R}$. Prove that $f(x) = 0$ for all $x \in \mathbf{R}$.

***11.** Prove that

$$\left( \sum_{k=1}^\infty |a_k|^\beta \right)^{1/\beta} \le \sum_{k=1}^\infty |a_k|$$

for all $a_k \in \mathbf{R}$ and all $\beta > 1$.

## $^e 7.5$  APPLICATIONS  *This section uses no material from any other enrichment section.*

The theory of infinite series is a potent tool for both pure and applied mathematics. In this section we give several examples to back up this claim.

We begin with a nontrivial theorem from number theory. Recall that an integer $n \geq 2$ is called *prime* if the only factors of $n$ in $\mathbf{N}$ are 1 and $n$. Also recall that given $n \in \mathbf{N}$ there are primes $p_1, p_2, \ldots, p_k$ and exponents $\alpha_1, \alpha_2, \ldots, \alpha_k$ such that

$$n = p_1^{\alpha_1} p_2^{\alpha_2} \ldots p_k^{\alpha_k}.$$

**7.59 THEOREM** [EUCLID'S THEOREM; EULER'S PROOF]. *There are infinitely many primes in $\mathbf{N}$.*

PROOF. Suppose to the contrary that $p_1, p_2, \ldots, p_k$ represent all the primes in $\mathbf{N}$. Fix $N \in \mathbf{N}$ and set $\alpha = \sup\{\alpha_1, \ldots, \alpha_k\}$, where this supremum is taken over all $\alpha_j$'s which satisfy $n = p_1^{\alpha_1} p_2^{\alpha_2} \ldots p_k^{\alpha_k}$ for some $n \leq N$. Since every integer $j \in [1, N]$ must have the form $j = p_1^{e_1} \ldots p_k^{e_k}$ for some choice of integers $0 \leq e_i \leq \alpha$, we have

$$\left(1 + \frac{1}{p_1} + \cdots + \frac{1}{p_1^{\alpha}}\right) \left(1 + \frac{1}{p_2} + \cdots + \frac{1}{p_2^{\alpha}}\right) \cdots \left(1 + \frac{1}{p_k} + \cdots + \frac{1}{p_k^{\alpha}}\right)$$

$$= \sum_{0 \leq e_i \leq \alpha} 1 \cdot \frac{1}{p_1^{e_1}} \cdots \frac{1}{p_k^{e_k}} \geq \sum_{j=1}^{N} \frac{1}{j}.$$

On the other hand, for each integer $i \in [1, k]$, we have by Theorem 6.7 that

$$1 + \frac{1}{p_i} + \cdots + \frac{1}{p_i^{\alpha}} \leq \sum_{k=1}^{\infty} \left(\frac{1}{p_i}\right)^k = \frac{p_i}{p_i - 1}.$$

Consequently,

$$\sum_{j=1}^{N} \frac{1}{j} \leq \left(\frac{p_1}{p_1 - 1}\right) \cdots \left(\frac{p_k}{p_k - 1}\right) = M < \infty.$$

Taking the limit of this inequality as $N \to \infty$, we conclude that $\sum_{j=1}^{\infty} 1/j \leq M < \infty$, a contradiction. ∎

Our next application, a result used to approximate roots of twice differentiable functions, shows that if an initial guess $x_0$ is close enough to a root of a suitably well-behaved function $f$, then the sequence $x_n$ generated by (19) converges to a root of $f$.

**7.60 THEOREM** [NEWTON]. *Suppose $f : [a, b] \to \mathbf{R}$ is continuous on $[a, b]$ and $f(c) = 0$ for some $c \in (a, b)$. If $f''$ exists and is bounded on $(a, b)$ and there is an $\varepsilon_0 > 0$ such that $|f'(x)| \geq \varepsilon_0$ for all $x \in (a, b)$, then there is a closed interval $I \subseteq (a, b)$ containing $c$ such that given $x_0 \in I$, the sequence $\{x_n\}_{n \in \mathbf{N}}$ defined by*

$$(19) \qquad\qquad x_n = x_{n-1} - \frac{f(x_{n-1})}{f'(x_{n-1})} \qquad n \in \mathbf{N},$$

*satisfies $x_n \in I$ and $x_n \to c$ as $n \to \infty$.*

PROOF. Choose $M > 0$ such that $|f''(x)| \leq M$ for $x \in (a, b)$. Choose $r_0 \in (0, 1)$ so small that $I = [c - r_0, c + r_0]$ is a subinterval of $(a, b)$ and $r_0 < \varepsilon_0 / M$. Suppose

$x_0 \in I$ and define the sequence $\{x_n\}$ by (19). Set $r := r_0 M/\varepsilon_0$ and observe by the choice of $r_0$ that $r < 1$. Thus it suffices to show that

(20)
$$|x_n - c| \le r^n |x_0 - c|$$

and

(21)
$$|x_n - c| < r_0$$

hold for all $n \in \mathbf{N}$.

The proof is by induction on $n$. Clearly, (20) and (21) hold for $n = 0$. Fix $n \in \mathbf{N}$ and suppose

(22)
$$|x_{n-1} - c| \le r^{n-1} |x_0 - c|$$

and

(23)
$$|x_{n-1} - c| < r_0.$$

Use Taylor's Formula to choose a point $\xi$ between $c$ and $x_{n-1}$ such that

$$-f(x_{n-1}) = f(c) - f(x_{n-1}) = f'(x_{n-1})(c - x_{n-1}) + \frac{1}{2} f''(\xi)(c - x_{n-1})^2.$$

Since (19) implies $-f(x_{n-1}) = f'(x_{n-1})(x_n - x_{n-1})$, it follows that

$$f'(x_{n-1})(x_n - c) = \frac{1}{2} f''(\xi)(c - x_{n-1})^2.$$

Solving this equation for $x_n - c$, we have by the choice of $M$ and $\varepsilon_0$ that

(24)
$$|x_n - c| = \left| \frac{f''(\xi)}{2f'(x_{n-1})} \right| |x_{n-1} - c|^2 < \frac{M}{2\varepsilon_0} |x_{n-1} - c|^2.$$

Since $M/\varepsilon_0 < 1/r_0$, it follows from (24) and (23) that

$$|x_n - c| < \frac{M}{\varepsilon_0} |x_{n-1} - c|^2 < \frac{1}{r_0} |r_0|^2 = r_0.$$

This proves (21). Again, by (24), (22), and the choice of $r$, we have

$$|x_n - c| \le \frac{M}{\varepsilon_0} (r^{n-1} |x_0 - c|)^2 = \frac{r}{r_0} (r^{2n-2} |x_0 - c|^2) < r^{2n-1} |x_0 - c|.$$

Since $r < 1$ and $2n - 1 \ge n$ imply $r^{2n-1} \le r^n$, we conclude that

$$|x_n - c| < r^{2n-1} |x_0 - c| \le r^n |x_0 - c|. \quad \blacksquare$$

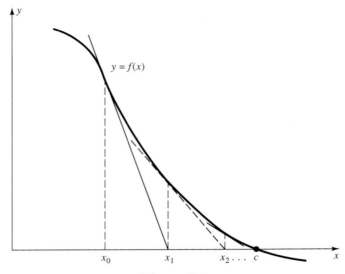

**Figure 7.7**

Notice if $x_{n-1}$ and $x_n$ satisfy (19), then $x_n$ is the $x$-intercept of the tangent line to $y = f(x)$ at the point $(x_{n-1}, f(x_{n-1}))$ (see Exercise 4). Thus, Newton's method is based on a simple geometric principle (see Figure 7.7). Also notice that, by (24), this method converges very rapidly. Indeed, the number of decimal places of accuracy nearly doubles with each successive approximation.

As a general rule, it is extremely difficult to show a given non-algebraic number is irrational. The next result shows how to use infinite series to give an easy proof that certain kinds of numbers are irrational.

**7.61 THEOREM** [HERMITE]. *The number $e$ is irrational.*

PROOF. Suppose to the contrary that $e = p/q$ for some $p, q \in \mathbf{N}$. By Example 7.47,

$$\frac{q}{p} = e^{-1} = \sum_{k=0}^{\infty} \frac{(-1)^k}{k!}.$$

Breaking this sum into two pieces and multiplying by $(-1)^{p+1}(p-1)!$, we have

$$x := (-1)^{p+1} \left( q(p-1)! - \sum_{k=0}^{p} \frac{(-1)^k p!}{k!} \right) = y := \sum_{k=p+1}^{\infty} (-1)^{k+p+1} \frac{p!}{k!}.$$

Since $p!/k! \in \mathbf{N}$ for all integers $k \le p$, the number $x$ must be an integer. On the other hand,

$$y = \frac{1}{p+1} - \frac{1}{(p+1)(p+2)} + \frac{1}{(p+1)(p+2)(p+3)} - \cdots$$

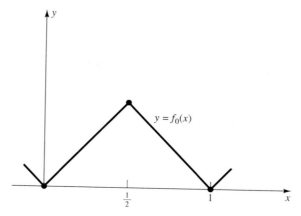

**Figure 7.8**

lies between $1/(p+1)$ and $1/(p+1) - 1/(p+1)(p+2)$. Therefore, $y$ is a number which satisfies $0 < y < 1$. In particular, $x \neq y$, a contradiction. ∎

We know that a continuous function can fail to be differentiable at one point (e.g., $f(x) = |x|$). Hence, it is not difficult to see that, given any finite set of points $E$, there is a continuous function which fails to be differentiable at every point in $E$. We shall now show that there is a continuous function which fails to be differentiable at any point in $\mathbf{R}$. Once again, here is a clear indication that, although we use sketches to motivate proofs and to explain results, we cannot rely on sketches to give a complete picture of the general situation.

**7.62 THEOREM** [WEIERSTRASS]. *There is a function $f$ continuous on $\mathbf{R}$ which is not differentiable at any point in $\mathbf{R}$.*

[Note: Such functions are called *nowhere differentiable*.]

PROOF. Let

$$f_0(x) = \begin{cases} x & 0 \leq x < 1/2 \\ 1 - x & 1/2 \leq x < 1 \end{cases}$$

and extend $f_0$ to $\mathbf{R}$ by periodicity of period 1, i.e., so that $f_0(x) = f_0(x+1)$ for all $x \in \mathbf{R}$ (see Figure 7.8). Set $f_k(x) = f_0(2^k x)/2^k$ for $x \in \mathbf{R}$ and $k \in \mathbf{N}$ and consider the function

$$f(x) = \sum_{k=0}^{\infty} f_k(x), \qquad x \in \mathbf{R}.$$

Normalizing $f_k$ by $2^k$ has two consequences. First, since $f_0'(y) = \pm 1$ for each $y$ which satisfies $2y \notin \mathbf{Z}$, it is easy to see that

(25)             $f_k'(y) = \pm 1$   for each $y$ which satisfies $2^{k+1} y \notin \mathbf{Z}$.

Second, by the Weierstrass $M$-Test, $f$ converges uniformly, hence, is continuous on $\mathbf{R}$.

Since $f$ is periodic of period 1, it suffices to show that $f$ is not differentiable at any $x \in [0,1)$. Suppose to the contrary that $f$ is differentiable at some $x \in [0,1)$. For each $n \in \mathbf{N}$, choose $p \in \mathbf{Z}$ such that $x \in [\alpha_n, \beta_n)$ for $\alpha_n = p/2^n$ and $\beta_n = (p+1)/2^n$. Notice that if $h$ is any function differentiable at $x$, then

$$h'(x) = \lim_{n \to \infty} \frac{h(\beta_n) - h(\alpha_n)}{\beta_n - \alpha_n}.$$

Hence, by (25) and the fact that $f'_k(x) = \pm 1$ when it exists, it follows that

(26)
$$\frac{f_k(\beta_n) - f_k(\alpha_n)}{\beta_n - \alpha_n} = \pm 1$$

for each $n > k$. On the other hand, since $f_0(y) = 0$ if and only if $y \in \mathbf{Z}$, it is clear that $f_k(\beta_n) = f_k(\alpha_n) = 0$ for $k \geq n$. Therefore, $f(\beta_n) = \sum_{k=0}^{n-1} f_k(\beta_n)$ and $f(\alpha_n) = \sum_{k=0}^{n-1} f_k(\alpha_n)$. Since $f'(x)$ exists, it follows from (26) that

$$f'(x) = \lim_{n \to \infty} \frac{f(\beta_n) - f(\alpha_n)}{\beta_n - \alpha_n} = \lim_{n \to \infty} \sum_{k=0}^{n-1} \frac{f_k(\beta_n) - f_k(\alpha_n)}{\beta_n - \alpha_n} = \sum_{k=0}^{\infty} f'_k(x).$$

Hence, by the Divergence Test, $\pm 1 = f'_k(x) \to 0$ as $k \to \infty$, a contradiction. ∎

## EXERCISES

**1.** Using a calculator and Theorem 7.60, approximate all real roots of $f(x) = x^3 + 3x^2 + 4x + 1$ to five decimal places.

**2.** a) Using the proof of Theorem 7.60, prove that (20) holds if $r/2$ replaces $r$.
   b) Use part a) to estimate the difference $|x_4 - \pi|$, where $x_0 = 3$, $f(x) = \sin x$, and $x_n$ is defined by (19). Evaluate $x_4$ directly, and verify that $x_4$ is actually closer than our theory predicts.

**3.** Prove that given any $n \in \mathbf{N}$, there is a function $f \in C^n(\mathbf{R})$ such that $f^{(n+1)}(x)$ does not exist for any $x \in \mathbf{R}$.

**4.** Prove that if $x_{n-1}, x_n$ satisfy (19), then $x_n$ is the $x$–intercept of the tangent line to $y = f(x)$ at the point $(x_{n-1}, f(x_{n-1}))$.

**5.** Prove that $\cos(1)$ is irrational.

**6.** Suppose $f : \mathbf{R} \to \mathbf{R}$. If $f''$ exists and is bounded on $\mathbf{R}$, and there is an $\varepsilon_0 > 0$ such that $|f'(x)| \geq \varepsilon_0$ for all $x \in \mathbf{R}$, prove that there exists a $\delta > 0$ such that if $|f(x_0)| \leq \delta$ for some $x_0 \in \mathbf{R}$, then $f$ has a root, i.e., that $f(c) = 0$ for some $c \in \mathbf{R}$.

# Chapter 8

# Euclidean Spaces

The world we live in is at least four-dimensional: three spatial dimensions together with the time dimension. Moreover, certain problems from engineering, physics, chemistry, and economics force us to consider even higher dimensions. For example, guidance systems for missiles frequently require as many as 100 variables (longitude, latitude, altitude, velocity, time after launch, pitch, yaw, fuel on board, etc.). Another example, the state of a gas in a closed container, can best be described by a function of $6m$ variables, where $m$ is the number of molecules in the system. (Six enters the picture because each molecule of gas is described by three space variables and three momentum variables.) Thus, there are practical reasons for studying functions of more than one variable.

## 8.1 ALGEBRAIC STRUCTURE

For each $n \in \mathbf{N}$, let $\mathbf{R}^n$ denote the $n$–fold cartesian product of $\mathbf{R}$ with itself, i.e.,

$$\mathbf{R}^n := \{(x_1, x_2, \ldots, x_n) : x_j \in \mathbf{R} \text{ for } j = 1, 2, \ldots, n\}.$$

By a *Euclidean space* we shall mean $\mathbf{R}^n$ together with the "Euclidean inner product" defined in Definition 8.1 below. The integer $n$ is called the *dimension* of $\mathbf{R}^n$, elements $\boldsymbol{x} = (x_1, x_2, \ldots, x_n)$ of $\mathbf{R}^n$ are called *points* or *vectors* or *ordered n-tuples*, and the numbers $x_j$ are called *coordinates*, or *components*, of $\boldsymbol{x}$. Two vectors $\boldsymbol{x}, \boldsymbol{y}$ are said to be *equal* if their components are equal, i.e., if $x_j = y_j$ for $j = 1, 2, \ldots, n$. The *zero vector* is the vector whose components are all zero, i.e., $\mathbf{0} := (0, 0, \ldots, 0)$. When $n = 2$ (respectively, $n = 3$), we usually denote the components of $\boldsymbol{x}$ by $x, y$ (respectively, by $x, y, z$).

We have already encountered the sets $\mathbf{R}^n$ for small $n$. $\mathbf{R}^1 = \mathbf{R}$ is the real line; we shall call its elements *scalars*. $\mathbf{R}^2$ is the $xy$ plane used to graph functions of the form $y = f(x)$. And $\mathbf{R}^3$ is the $xyz$ space used to graph functions of the form $z = f(x, y)$.

We began our study of functions of one variable by examining the algebraic structure of $\mathbf{R}$. In this section we begin our study of functions of several variables by examining the algebraic structure of $\mathbf{R}^n$. That structure is described in the following definition.

**8.1 DEFINITION.** Let $\boldsymbol{x} = (x_1, \ldots, x_n)$, $\boldsymbol{y} = (y_1, \ldots, y_n) \in \mathbf{R}^n$ be vectors and $\alpha \in \mathbf{R}$ be a scalar.

  i) The *sum* of $\boldsymbol{x}$ and $\boldsymbol{y}$ is the vector

$$\boldsymbol{x} + \boldsymbol{y} := (x_1 + y_1, x_2 + y_2, \ldots, x_n + y_n).$$

  ii) The *difference* of $\boldsymbol{x}$ and $\boldsymbol{y}$ is the vector

$$\boldsymbol{x} - \boldsymbol{y} := (x_1 - y_1, x_2 - y_2, \ldots, x_n - y_n).$$

  iii) The *product* of a scalar $\alpha$ and a vector $\boldsymbol{x}$ is the vector

$$\alpha\boldsymbol{x} := (\alpha x_1, \alpha x_2, \ldots, \alpha x_n).$$

  iv) The (*Euclidean*) *dot product* (or *scalar product* or *inner product*) of $\boldsymbol{x}$ and $\boldsymbol{y}$ is the scalar

$$\boldsymbol{x} \cdot \boldsymbol{y} := x_1 y_1 + x_2 y_2 + \cdots + x_n y_n.$$

These algebraic operations are analogues of addition, subtraction, and multiplication on $\mathbf{R}$. It is natural to ask: Do the usual laws of algebra hold in $\mathbf{R}^n$? An answer to this question is contained in the following result.

**8.2 THEOREM.** Let $\boldsymbol{x}, \boldsymbol{y}, \boldsymbol{z} \in \mathbf{R}^n$ and $\alpha, \beta \in \mathbf{R}$. Then $\alpha\boldsymbol{0} = \boldsymbol{0}$, $0\boldsymbol{x} = \boldsymbol{0}$, $1\boldsymbol{x} = \boldsymbol{x}$, $\alpha(\beta\boldsymbol{x}) = \beta(\alpha\boldsymbol{x}) = (\alpha\beta)\boldsymbol{x}$, $\alpha(\boldsymbol{x} \cdot \boldsymbol{y}) = (\alpha\boldsymbol{x}) \cdot \boldsymbol{y} = \boldsymbol{x} \cdot (\alpha\boldsymbol{y})$, $\alpha(\boldsymbol{x} + \boldsymbol{y}) = \alpha\boldsymbol{x} + \alpha\boldsymbol{y}$, $\boldsymbol{0} + \boldsymbol{x} = \boldsymbol{x}$, $\boldsymbol{x} - \boldsymbol{x} = \boldsymbol{0}$, $0 \cdot \boldsymbol{x} = 0$, $\boldsymbol{x} + (\boldsymbol{y} + \boldsymbol{z}) = (\boldsymbol{x} + \boldsymbol{y}) + \boldsymbol{z}$, $\boldsymbol{x} + \boldsymbol{y} = \boldsymbol{y} + \boldsymbol{x}$, $\boldsymbol{x} \cdot \boldsymbol{y} = \boldsymbol{y} \cdot \boldsymbol{x}$, and $\boldsymbol{x} \cdot (\boldsymbol{y} + \boldsymbol{z}) = \boldsymbol{x} \cdot \boldsymbol{y} + \boldsymbol{x} \cdot \boldsymbol{z}$.

PROOF. These properties are direct consequences of Definition 8.1 and corresponding properties of real numbers. We will prove that vector addition is associative, and leave the proof of the rest of these properties as an exercise.

By definition and associativity of addition on $\mathbf{R}$ (see Postulate 1 in Section 1.1),

$$\boldsymbol{x} + (\boldsymbol{y} + \boldsymbol{z}) = (x_1, \ldots, x_n) + (y_1 + z_1, \ldots, y_n + z_n)$$
$$= (x_1 + (y_1 + z_1), \ldots, x_n + (y_n + z_n))$$
$$= ((x_1 + y_1) + z_1, \ldots, (x_n + y_n) + z_n) = (\boldsymbol{x} + \boldsymbol{y}) + \boldsymbol{z}. \ \blacksquare$$

Thus (with the exception of the closure of the dot product and the existence of the multiplicative identity and multiplicative inverses), $\mathbf{R}^n$ satisfies the same algebraic laws, listed in Postulate 1, that $\mathbf{R}$ does.

Is there an analogue of the absolute value for $\mathbf{R}^n$? The following definition illustrates the fact that there are many such analogues.

**8.3 DEFINITION.**  Let $\boldsymbol{x} \in \mathbf{R}^n$.

i) The (*Euclidean*) *norm* of $\boldsymbol{x}$ is the scalar

$$\|\boldsymbol{x}\| := \sqrt{\sum_{k=1}^{n} |x_k|^2}.$$

*Euclidean norm:*

$$\|\bar{x}\| = \sqrt{\bar{x} \cdot \bar{x}}$$

ii) The $\ell^1$–*norm* (read L–one–norm) of $\boldsymbol{x}$ is the scalar

$$\|\boldsymbol{x}\|_1 := \sum_{k=1}^{n} |x_k|.$$

*Euclidean distance:*

$$\|\bar{x} - \bar{y}\|$$

iii) The *sup–norm* of $\boldsymbol{x}$ is the scalar

$$\|\boldsymbol{x}\|_\infty := \max\{|x_1|, \dots, |x_n|\}.$$

(Note: The subscript $\infty$ is frequently used for supremum norms because the supremum of a continuous function on a closed bounded interval can be computed by a certain limit as $p \to \infty$—see Exercise 8 in Section 5.2. For relationships between these three norms, see Remark 8.6 below.)

Since $\|x\| = \|x\|_1 = \|x\|_\infty = |x|$, when $n = 1$, each norm defined above is an extension of the absolute value from $\mathbf{R}$ to $\mathbf{R}^n$. The most important, and in some senses the most natural, of these norms is the Euclidean norm. This is true for at least two reasons. First, by definition,

$$\|\boldsymbol{x}\|^2 = \boldsymbol{x} \cdot \boldsymbol{x} \quad \text{for all} \quad \boldsymbol{x} \in \mathbf{R}^n.$$

(This aids in many calculations; see, for example, the proofs of Theorems 8.4 and 8.5 below.) Secondly, if $\Delta$ is a triangle in $\mathbf{R}^2$ with vertices $(0,0)$, $\boldsymbol{x} := (a, b)$, and $(a, 0)$, then by the Pythagorean Theorem, the hypotenuse of $\Delta$ is $\|\boldsymbol{x}\| = \sqrt{a^2 + b^2}$. Hence we define the (*Euclidean*) *distance* between two points $\boldsymbol{a}, \boldsymbol{b} \in \mathbf{R}^n$ by

$$\text{dist}\,(\boldsymbol{a}, \boldsymbol{b}) := \|\boldsymbol{a} - \boldsymbol{b}\|.$$

Although the norm is not multiplicative, the following fundamental inequality can be used as a replacement for the multiplicative property in most proofs. (Some authors call this the Cauchy–Schwarz–Bunyakovsky Inequality.)

**8.4 THEOREM**  [CAUCHY–SCHWARZ INEQUALITY].  *If $\boldsymbol{x}, \boldsymbol{y} \in \mathbf{R}^n$, then*

$$|\boldsymbol{x} \cdot \boldsymbol{y}| \leq \|\boldsymbol{x}\|\,\|\boldsymbol{y}\|.$$

PROOF. The inequality is trivial when $\boldsymbol{y} = \boldsymbol{0}$. Suppose $\boldsymbol{y} \neq \boldsymbol{0}$. By definition,

$$(1) \qquad 0 \leq \|\boldsymbol{x} - t\boldsymbol{y}\|^2 = (\boldsymbol{x} - t\boldsymbol{y}) \cdot (\boldsymbol{x} - t\boldsymbol{y}) = \|\boldsymbol{x}\|^2 - 2t(\boldsymbol{x} \cdot \boldsymbol{y}) + t^2\|\boldsymbol{y}\|^2$$

holds for any scalar $t$. Substituting the value $t = (\boldsymbol{x} \cdot \boldsymbol{y})/\|\boldsymbol{y}\|^2$ into (1), we obtain

$$0 \leq \|\boldsymbol{x}\|^2 - t(\boldsymbol{x} \cdot \boldsymbol{y}) = \|\boldsymbol{x}\|^2 - \frac{(\boldsymbol{x} \cdot \boldsymbol{y})^2}{\|\boldsymbol{y}\|^2},$$

i.e., $0 \leq \|\boldsymbol{x}\|^2 - (\boldsymbol{x} \cdot \boldsymbol{y})^2/\|\boldsymbol{y}\|^2$. Solving this inequality for $(\boldsymbol{x} \cdot \boldsymbol{y})^2$, we conclude that

$$(\boldsymbol{x} \cdot \boldsymbol{y})^2 \leq \|\boldsymbol{x}\|^2\|\boldsymbol{y}\|^2. \quad \blacksquare$$

The analogy between the absolute value and the Euclidean norm is further reinforced by the following result. (See also Exercise 10 below.)

**8.5 THEOREM.** *Let $x, y \in \mathbf{R}^n$. Then*

*Two nonzero vectors are orthogonal if $\bar{x} \cdot \bar{z} = 0$*

i) $\|x\| \geq 0$ *with equality only when* $x = \mathbf{0}$,

ii) $\|\alpha x\| = |\alpha| \|x\|$ *for all scalars $\alpha$,*   $\|\bar{z}\| = \|-\bar{z}\|$

iii) [TRIANGLE INEQUALITIES]. $\|x + y\| \leq \|x\| + \|y\|$ *and* $\|x - y\| \geq \|x\| - \|y\|$.

PROOF. Statements i) and ii) are obvious.   $\|\bar{x}\| - \|\bar{z}\| \leq \|x + \bar{z}\| \leq \|\bar{x}\| + \|\bar{z}\|$

To prove iii), observe that by Definition 8.3, Theorem 8.2, and the Cauchy–Schwarz Inequality,   *For each* $\bar{x} = (x_1, x_2, \cdots, x_n) \in \mathbb{R}$, $|x_j| \leq \|\bar{x}\| \leq \sum_{j=1}^{m} |x_i|$   $j = 1, \cdots, n.$

$$\|x + y\|^2 = (x + y) \cdot (x + y) = x \cdot x + 2x \cdot y + y \cdot y$$
$$= \|x\|^2 + 2x \cdot y + \|y\|^2 \leq \|x\|^2 + 2\|x\| \|y\| + \|y\|^2 = (\|x\| + \|y\|)^2.$$

This establishes the first inequality in iii). By modifying the proof of Theorem 1.7, we can also establish the second inequality in iii). ∎

For some estimates, it is convenient to relate the Euclidean norm to the $\ell^1$–norm and the sup–norm.

**8.6 Remark.** *Let $x \in \mathbf{R}^n$. Then*

i) $|x_j| \leq \|x\| \leq \sqrt{n} \|x\|_\infty$ *for each $j = 1, 2, \ldots, n$, and*

ii) $\|x\| \leq \|x\|_1$.

PROOF. i) By definition,

$$|x_\ell|^2 \leq \|x\|^2 = x_1^2 + \cdots + x_n^2 \leq n(\max_{1 \leq j \leq n} |x_j|)^2.$$

ii) Observe that

$$(|x_1| + \cdots + |x_n|)^2 = |x_1|^2 + \cdots + |x_n|^2 + 2 \sum_{(i,j) \in A} |x_i| |x_j|,$$

where $A = \{(i, j) : 1 \leq i, j \leq n \text{ and } i < j\}$. Since

$$\sum_{(i,j) \in A} |x_i| |x_j| \geq 0,$$

we conclude that

$$\|x\|^2 = x_1^2 + \cdots + x_n^2 \leq (|x_1| + \cdots + |x_n|)^2. \quad \blacksquare$$

We have referred to elements of $\mathbf{R}^n$ as vectors. For engineers, a vector is a directed line segment which begins at a point $a$ and ends at a point $b$. How can we reconcile these two points of view? Two "engineering" vectors are said to be equivalent if they have the same length and point in the same direction. Thus every engineering vector is equivalent to a vector *in standard position*, i.e., whose "tail" sits at the origin and whose "head" is a point in $\mathbf{R}^n$. If we identify each engineering vector

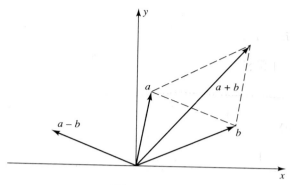

**Figure 8.1**

with the point which sits at the "head" of its equivalent representative in standard position, then the engineering vector from $\boldsymbol{a}$ to $\boldsymbol{b}$, equivalent to the engineering vector which begins at $\boldsymbol{0}$ and ends at $\boldsymbol{b}-\boldsymbol{a}$, will be identified with the point $\boldsymbol{b}-\boldsymbol{a}$ (see Figure 8.1).

We make no distinction between points and vectors, but in each situation we adopt the interpretation which proves most useful. For example, let $\boldsymbol{a},\boldsymbol{b} \in \mathbf{R}^n$ with $\boldsymbol{a} \neq \boldsymbol{b}$. Thinking of $\boldsymbol{a}$ and $\boldsymbol{b}$ as points, we define the *line segment* from $\boldsymbol{a}$ to $\boldsymbol{b}$ to be the set

$$L(\boldsymbol{a};\boldsymbol{b}) = \{\phi(t) := (1-t)\boldsymbol{a} + t\boldsymbol{b} : t \in [0,1]\},$$

and the *parallelogram* associated with $\boldsymbol{a}$ and $\boldsymbol{b}$ to be the set

$$\mathcal{P}(\boldsymbol{a};\boldsymbol{b}) := \{\psi(u,v) := u\boldsymbol{a} + v\boldsymbol{b} : u,v \in [0,1]\}.$$

Notice that $L(\boldsymbol{a};\boldsymbol{b})$ runs from $\phi(0) = \boldsymbol{a}$ to $\phi(1) = \boldsymbol{b}$, and that $\mathcal{P}(\boldsymbol{a};\boldsymbol{b})$ contains the points $\psi(0,0) = \boldsymbol{0}$, $\psi(1,0) = \boldsymbol{a}$, $\psi(0,1) = \boldsymbol{b}$, and $\psi(1,1) = \boldsymbol{a}+\boldsymbol{b}$. Similarly, given $\boldsymbol{a} \in \mathbf{R}^n$ and $r > 0$, we define the *(n–dimensional) open ball* centered at $\boldsymbol{a}$ of *radius* $r$ by

$$B_r(\boldsymbol{a}) := \{\boldsymbol{x} \in \mathbf{R}^n : \|\boldsymbol{x} - \boldsymbol{a}\| < r\}$$

and the *closed ball* by $\{\boldsymbol{x} \in \mathbf{R}^n : \|\boldsymbol{x} - \boldsymbol{a}\| \leq r\}$. (We shall draw pictures for arbitrary Euclidean spaces, using the model $\mathbf{R}^2$. For example, open balls will be drawn as disks with dashed circumferences (see, for example, Figure 9.1), and closed balls will be drawn as disks with solid circumferences.)

On the other hand, thinking of $\boldsymbol{a}$ and $\boldsymbol{b}$ as vectors, we can define the angle $\theta \in [0, \pi]$ between $\boldsymbol{a}$ and $\boldsymbol{b}$ as follows (see Figure 8.1). First, notice in $\mathbf{R}^2$ that $\theta$ belongs to a triangle whose sides have length $\|\boldsymbol{a}\|$, $\|\boldsymbol{b}\|$, and $\|\boldsymbol{a} - \boldsymbol{b}\|$. By the Law of Cosines (see Appendix B),

$$\|\boldsymbol{a} - \boldsymbol{b}\|^2 = \|\boldsymbol{a}\|^2 + \|\boldsymbol{b}\|^2 - 2\|\boldsymbol{a}\| \, \|\boldsymbol{b}\| \cos \theta.$$

Since Theorem 8.2 implies $\|\boldsymbol{a} - \boldsymbol{b}\|^2 = (\boldsymbol{a} - \boldsymbol{b}) \cdot (\boldsymbol{a} - \boldsymbol{b}) = \|\boldsymbol{a}\|^2 - 2\boldsymbol{a} \cdot \boldsymbol{b} + \|\boldsymbol{b}\|^2$, it follows that $-2\boldsymbol{a} \cdot \boldsymbol{b} = -2\|\boldsymbol{a}\| \, \|\boldsymbol{b}\| \cos \theta$, i.e., if $\boldsymbol{a}$ and $\boldsymbol{b}$ are nonzero, then

$$(2) \qquad\qquad\qquad \cos \theta = \frac{\boldsymbol{a} \cdot \boldsymbol{b}}{\|\boldsymbol{a}\| \, \|\boldsymbol{b}\|}.$$

Motivated by this identity, we DEFINE the *angle* between two nonzero vectors $\boldsymbol{a}, \boldsymbol{b} \in \mathbf{R}^n$ (for any $n \in \mathbf{N}$) to be the number $\theta \in [0, \pi]$ determined by (2). Notice that by the Cauchy–Schwarz Inequality, the right side of (2) always belongs to the interval $[-1, 1]$. Hence, for each pair of nonzero vectors $\boldsymbol{a}, \boldsymbol{b} \in \mathbf{R}^n$, there is a unique angle $\theta \in [0, \pi]$ which satisfies (2). Two cases deserve special mention. The vectors $\boldsymbol{a}$ and $\boldsymbol{b}$ are said to be *parallel* when $\theta = 0$, i.e., if there is a scalar $t \in \mathbf{R}$ such that $\boldsymbol{a} = t\boldsymbol{b}$, and are said to be *orthogonal* when $\theta = \pi/2$, i.e., if $\boldsymbol{a} \cdot \boldsymbol{b} = 0$ (see Exercise 8 below). Again, thinking of $\mathbf{R}^n$ as a collection of vectors, we define the *usual basis* of $\mathbf{R}^n$ to be the collection $\{\boldsymbol{e}_1, \ldots, \boldsymbol{e}_n\}$, where $\boldsymbol{e}_j$ is the point in $\mathbf{R}^n$ whose $j$th coordinate is 1, and all other coordinates are 0. By definition, then, each $\boldsymbol{x} = (x_1, \ldots, x_n) \in \mathbf{R}^n$ can be written as a linear combination of the $\boldsymbol{e}_j$'s:

$$\boldsymbol{x} = \sum_{j=1}^{n} x_j \boldsymbol{e}_j.$$

Notice that the usual basis $\{\boldsymbol{e}_j\}$ consists of pairwise orthogonal vectors, i.e., $\boldsymbol{e}_j \cdot \boldsymbol{e}_k = 0$ when $j \neq k$. In particular, the usual basis is an *orthogonal basis*.

In $\mathbf{R}^2$ or $\mathbf{R}^3$, $\boldsymbol{e}_1$ is denoted by $\mathbf{i}$, $\boldsymbol{e}_2$ is denoted by $\mathbf{j}$, and, in $\mathbf{R}^3$, $\boldsymbol{e}_3$ is denoted by $\mathbf{k}$. Thus, in $\mathbf{R}^3$, $\mathbf{i} := (1, 0, 0)$, $\mathbf{j} := (0, 1, 0)$, and $\mathbf{k} := (0, 0, 1)$. In particular, vectors in $\mathbf{R}^2$ have the form $x\mathbf{i} + y\mathbf{j}$, and vectors in $\mathbf{R}^3$ have the form $x\mathbf{i} + y\mathbf{j} + z\mathbf{k}$.

We shall not discuss other bases of $\mathbf{R}^n$ or the more general concept of "vector spaces," which can be introduced using postulates similar in spirit to Postulate 1 in Chapter 1. Instead, we have introduced just enough algebraic machinery in $\mathbf{R}^n$ to develop the calculus of multivariable functions. For more information about $\mathbf{R}^n$ and abstract vector spaces, see Noble and Daniel [9].

Interpreting elements of $\mathbf{R}^n$ as both vectors and points may sound confusing and sloppy, but it is no different from letting $1/2$ represent $2/4$, $3/6$, $4/8$, etc. (In both cases, there is an underlying equivalence relation, and we are using one member of an equivalence class to represent all of its members. For vectors, we are using the representative which lies in standard position; for rationals, we are using the representative which is in reduced form.)

**8.7 Remark.** *The identification of vectors with points is a powerful tool for at least three reasons.*

1) It provides us with a geometric interpretation of the algebraic structure of $\mathbf{R}^n$. For example, in $\mathbf{R}^2$ the sum $\boldsymbol{a} + \boldsymbol{b}$ is the vector which begins at the origin and ends at the opposite vertex of $\mathcal{P}(\boldsymbol{a}; \boldsymbol{b})$. The difference $\boldsymbol{a} - \boldsymbol{b}$ is equivalent to the engineering vector which begins at $\boldsymbol{b}$ and ends at $\boldsymbol{a}$ (see Figure 8.1). The norm of $\boldsymbol{a}$ is the length or magnitude of the vector which begins at $\boldsymbol{0}$ and ends at $\boldsymbol{a}$. In this context, the triangle inequality $\|\boldsymbol{a} + \boldsymbol{b}\| \leq \|\boldsymbol{a}\| + \|\boldsymbol{b}\|$ states that the length of one side of a triangle (namely, the triangle whose vertices are $\boldsymbol{0}$, $\boldsymbol{a}$, and $\boldsymbol{a} + \boldsymbol{b}$) is less than or equal to the sum of the lengths of its other two sides.

2) It frequently suggests and clarifies methods of proof. For example, the mysterious choice of $t$ in the proof of the Cauchy–Schwarz Inequality above is based on a simple geometric observation. Inequality (1) is trivial for large $t$ but is sharpest

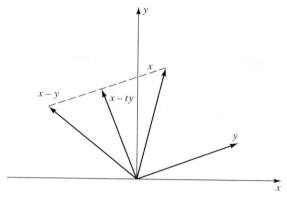

**Figure 8.2**

when $\|\boldsymbol{x} - t\boldsymbol{y}\|$ is minimal. To identify a value of $t$ which minimizes $\|\boldsymbol{x} - t\boldsymbol{y}\|$, look at the case $n = 2$, and use the vector interpretation. The set of points $\boldsymbol{x} - t\boldsymbol{y}$, $t \in [0, 1]$, is a line segment from $\boldsymbol{x}$ to $\boldsymbol{x} - \boldsymbol{y}$, i.e., a line segment which is parallel to the vector $\boldsymbol{y}$ (see Figure 8.2). The length of the vector $\boldsymbol{x} - t\boldsymbol{y}$ is minimized when $\boldsymbol{x} - t\boldsymbol{y}$ is orthogonal to $\boldsymbol{y}$, i.e.,

$$0 = (\boldsymbol{x} - t\boldsymbol{y}) \cdot \boldsymbol{y} = \boldsymbol{x} \cdot \boldsymbol{y} - t\boldsymbol{y} \cdot \boldsymbol{y} = \boldsymbol{x} \cdot \boldsymbol{y} - t\|\boldsymbol{y}\|^2.$$

It follows that $t = \boldsymbol{x} \cdot \boldsymbol{y}/\|\boldsymbol{y}\|^2$. In particular, the choice of $t$ in the proof of Theorem 8.4 is a natural one under the interpretation of $\mathbf{R}^n$ as a collection of vectors.

3) It can be used to extend concepts from $\mathbf{R}^2$ or $\mathbf{R}^3$ to $\mathbf{R}^n$. We already demonstrated this above, while defining the angle between two vectors. For another example, we define the *hyperplane* (or simply *plane* when $n = 3$) passing through a point $\boldsymbol{a} \in \mathbf{R}^n$ with normal $\boldsymbol{b} \neq \boldsymbol{0}$ to be the set

$$\Pi_{\boldsymbol{b}}(\boldsymbol{a}) := \{\boldsymbol{x} \in \mathbf{R}^n : (\boldsymbol{x} - \boldsymbol{a}) \cdot \boldsymbol{b} = 0\}.$$

Notice that by definition, $\Pi_{\boldsymbol{b}}(\boldsymbol{a})$ is the set of all points $\boldsymbol{x}$ such that $\boldsymbol{x} - \boldsymbol{a}$ and $\boldsymbol{b}$ are orthogonal. (Several such points $\boldsymbol{x}$ are shown in Figure 8.3.) Hence we have built "flatness" into the definition of hyperplanes.

By an *equation* of a hyperplane $\Pi$ we mean an expression of the form $F(\boldsymbol{x}) = 0$, where $F : \mathbf{R}^n \to \mathbf{R}$ and a point $\boldsymbol{x}$ belongs to $\Pi$ if and only if $F(\boldsymbol{x}) = 0$. By definition, then, an equation of the hyperplane $\Pi_{\boldsymbol{b}}(\boldsymbol{a})$ is given by

$$b_1 x_1 + b_2 x_2 + \cdots + b_n x_n = d,$$

where $\boldsymbol{b} = (b_1, \ldots, b_n)$ and $d = b_1 a_1 + b_2 a_2 + \cdots + b_n a_n$. In particular, planes in $\mathbf{R}^3$ have equations of the form

$$ax + by + cz = d.$$

Since $\boldsymbol{x} \cdot \boldsymbol{y}$ is a scalar, the dot product in $\mathbf{R}^n$ does not satisfy the closure property for any $n > 1$. Here is another product, defined only on $\mathbf{R}^3$, which does satisfy the closure property. (As we shall see below, this product allows us to exploit the geometry of $\mathbf{R}^3$ in many unique ways.)

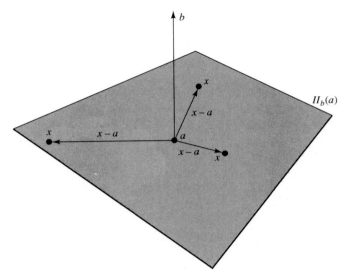

**Figure 8.3**

**8.8 DEFINITION.** The *cross product* of two vectors $\boldsymbol{x} = (x_1, x_2, x_3)$ and $\boldsymbol{y} = (y_1, y_2, y_3)$ in $\mathbf{R}^3$ is the vector defined by

$$\boldsymbol{x} \times \boldsymbol{y} := (x_2 y_3 - x_3 y_2, x_3 y_1 - x_1 y_3, x_1 y_2 - x_2 y_1).$$

Using the usual basis $\mathbf{i} = \boldsymbol{e}_1$, $\mathbf{j} = \boldsymbol{e}_2$, $\mathbf{k} = \boldsymbol{e}_3$, and the determinant operator (see Appendix C), we can give the cross product a more easily remembered form:

$$\boldsymbol{x} \times \boldsymbol{y} = \det \begin{bmatrix} \mathbf{i} & \mathbf{j} & \mathbf{k} \\ x_1 & x_2 & x_3 \\ y_1 & y_2 & y_3 \end{bmatrix}.$$

The following result shows that the cross product satisfies some, but not all, of the usual laws of algebra. (Specifically, notice that the cross product satisfies neither the commutative property nor the associative property.)

**8.9 THEOREM.** *Let* $\boldsymbol{x}, \boldsymbol{y}, \boldsymbol{z} \in \mathbf{R}^3$ *be vectors and* $\alpha$ *be a scalar. Then*

i) $$\boldsymbol{x} \times \boldsymbol{x} = \boldsymbol{0}, \qquad \boldsymbol{x} \times \boldsymbol{y} = -\boldsymbol{y} \times \boldsymbol{x},$$

ii) $$(\alpha \boldsymbol{x}) \times \boldsymbol{y} = \alpha(\boldsymbol{x} \times \boldsymbol{y}) = \boldsymbol{x} \times (\alpha \boldsymbol{y}),$$

iii) $$\boldsymbol{x} \times (\boldsymbol{y} + \boldsymbol{z}) = (\boldsymbol{x} \times \boldsymbol{y}) + (\boldsymbol{x} \times \boldsymbol{z}),$$

iv)
$$(\boldsymbol{x} \times \boldsymbol{y}) \cdot \boldsymbol{z} = \boldsymbol{x} \cdot (\boldsymbol{y} \times \boldsymbol{z}) = \det \begin{bmatrix} x_1 & x_2 & x_3 \\ y_1 & y_2 & y_3 \\ z_1 & z_2 & z_3 \end{bmatrix},$$

v)
$$\boldsymbol{x} \times (\boldsymbol{y} \times \boldsymbol{z}) = (\boldsymbol{x} \cdot \boldsymbol{z})\boldsymbol{y} - (\boldsymbol{x} \cdot \boldsymbol{y})\boldsymbol{z},$$

and

vi)
$$\|\boldsymbol{x} \times \boldsymbol{y}\|^2 = (\boldsymbol{x} \cdot \boldsymbol{x})(\boldsymbol{y} \cdot \boldsymbol{y}) - (\boldsymbol{x} \cdot \boldsymbol{y})^2,$$

PROOF. These properties follow immediately from the definitions. We will prove properties iv) and v) and leave the rest as an exercise.

To prove iv), notice that by definition,

$$(\boldsymbol{x} \times \boldsymbol{y}) \cdot \boldsymbol{z} = (x_2 y_3 - x_3 y_2) z_1 + (x_3 y_1 - x_1 y_3) z_2 + (x_1 y_2 - x_2 y_1) z_3$$
$$= x_1 (y_2 z_3 - y_3 z_2) + x_2 (y_3 z_1 - y_1 z_3) + x_3 (y_1 z_2 - y_2 z_1).$$

Since this last expression is both the scalar $\boldsymbol{x} \cdot (\boldsymbol{y} \times \boldsymbol{z})$ and the value of the determinant on the right side of iv) (expanded along the first row), this verifies iv).

To prove v), notice that since

$$\boldsymbol{x} \times (\boldsymbol{y} \times \boldsymbol{z}) = (x_1, x_2, x_3) \times (y_2 z_3 - y_3 z_2, y_3 z_1 - y_1 z_3, y_1 z_2 - y_2 z_1),$$

the first component of $\boldsymbol{x} \times (\boldsymbol{y} \times \boldsymbol{z})$ is

$$x_2 y_1 z_2 - x_2 y_2 z_1 - x_3 y_3 z_1 + x_3 y_1 z_3 = (x_1 z_1 + x_2 z_2 + x_3 z_3) y_1 - (x_1 y_1 + x_2 y_2 + x_3 y_3) z_1.$$

This proves that the first components of $\boldsymbol{x} \times (\boldsymbol{y} \times \boldsymbol{z})$ and $(\boldsymbol{x} \cdot \boldsymbol{z})\boldsymbol{y} - (\boldsymbol{x} \cdot \boldsymbol{y})\boldsymbol{z}$ are equal. A similar argument shows that the second and third components are also equal. ∎

By (2), there is a close connection between dot products and cosines. The following result shows there is a similar connection between cross products and sines. (For a connection between cross products and area or volume, see Exercise 7 below.)

**8.10 Remark.** *Let $\boldsymbol{x}, \boldsymbol{y}$ be nonzero vectors in $\mathbf{R}^3$ and $\theta$ be the angle between $\boldsymbol{x}$ and $\boldsymbol{y}$. Then*

$$\|\boldsymbol{x} \times \boldsymbol{y}\| = \|\boldsymbol{x}\| \, \|\boldsymbol{y}\| \, \sin \theta.$$

PROOF. By Theorem 8.9vi) and (2),

$$\|\boldsymbol{x} \times \boldsymbol{y}\|^2 = (\|\boldsymbol{x}\| \, \|\boldsymbol{y}\|)^2 - (\|\boldsymbol{x}\| \, \|\boldsymbol{y}\| \cos \theta)^2$$
$$= (\|\boldsymbol{x}\| \, \|\boldsymbol{y}\|)^2 (1 - \cos^2 \theta) = (\|\boldsymbol{x}\| \, \|\boldsymbol{y}\|)^2 \sin^2 \theta. \quad \blacksquare$$

In particular, the cross product of two nonzero, nonparallel vectors is always nonzero. This observation can be used to construct a vector orthogonal to any two nonzero, nonparallel vectors $\boldsymbol{x}, \boldsymbol{y}$ in $\mathbf{R}^3$. Indeed, such a vector is given by $\boldsymbol{x} \times \boldsymbol{y}$

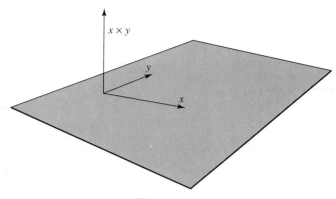

**Figure 8.4**

(see Figure 8.4) since by Theorem 8.9, $\boldsymbol{x} \cdot (\boldsymbol{x} \times \boldsymbol{y}) = (\boldsymbol{x} \times \boldsymbol{x}) \cdot \boldsymbol{y} = \boldsymbol{0}$ and, similarly, $\boldsymbol{y} \cdot (\boldsymbol{x} \times \boldsymbol{y}) = \boldsymbol{0}$.

## EXERCISES

**1.** a) Find all nonzero vectors orthogonal to $(1, -1, 0)$ which lie in the plane $z = x$.
 b) Find all nonzero vectors orthogonal to the vector $(3, 2, -5)$ whose components sum to 4.
 c) Find an equation of the plane containing the point $(1, 0, 1)$ with normal $(-1, 2, 1)$.

**2.** Using Postulate 1 in Section 1.1 and Definition 8.1, prove Theorem 8.2.

**3.** Use the proof of Theorem 8.4 to show that equality in the Cauchy–Schwarz Inequality holds if and only if $\boldsymbol{x} = \boldsymbol{0}$, $\boldsymbol{y} = \boldsymbol{0}$ or $\boldsymbol{x}$ is parallel to $\boldsymbol{y}$.

**4.** Prove Theorem 8.9, parts i) through iii) and vi).

**5.** Suppose that $\boldsymbol{a}, \boldsymbol{b}, \boldsymbol{c} \in \mathbf{R}^3$ are three points which do not lie on the same straight line and $\Pi$ is the plane which contains the points $\boldsymbol{a}, \boldsymbol{b}, \boldsymbol{c}$. Prove that an equation of $\Pi$ is given by

$$\det \begin{bmatrix} x - a_1 & y - a_2 & z - a_3 \\ b_1 - a_1 & b_2 - a_2 & b_3 - a_3 \\ c_1 - a_1 & c_2 - a_2 & c_3 - a_3 \end{bmatrix} = 0.$$

**6.** Suppose $\{a_k\}$ and $\{b_k\}$ are sequences of real numbers which satisfy

$$\sum_{k=1}^{\infty} a_k^2 < \infty \quad \text{and} \quad \sum_{k=1}^{\infty} b_k^2 < \infty.$$

Prove that the infinite series $\sum_{k=1}^{\infty} a_k b_k$ converges absolutely.

**7.** **This exercise is used in Appendix E.** Recall that the area of a parallelogram with base $b$ and altitude $h$ is given by $bh$, and the volume of a parallelepiped is given by the area of its base times its altitude.

a) Let $\boldsymbol{a}, \boldsymbol{b} \in \mathbf{R}^3$ be nonzero vectors and $\mathcal{P}$ represent the parallelogram

$$\{(x, y, z) = u\boldsymbol{a} + v\boldsymbol{b} : u, v \in [0, 1]\}.$$

Prove that the area of $\mathcal{P}$ is $\|\boldsymbol{a} \times \boldsymbol{b}\|$.

b) Let $\boldsymbol{a}, \boldsymbol{b}, \boldsymbol{c} \in \mathbf{R}^3$ be nonzero vectors and $\mathcal{P}$ represent the parallelepiped

$$\{(x, y, z) = t\boldsymbol{a} + u\boldsymbol{b} + v\boldsymbol{c} : t, u, v \in [0, 1]\}.$$

Prove that the volume of $\mathcal{P}$ is $|(\boldsymbol{a} \times \boldsymbol{b}) \cdot \boldsymbol{c}|$.

**8.** Let $\boldsymbol{a}$ and $\boldsymbol{b}$ be nonzero vectors in $\mathbf{R}^n$.

a) Prove that $\boldsymbol{a}$ and $\boldsymbol{b}$ are parallel if and only if there is a scalar $t \in \mathbf{R}$ such that $\boldsymbol{a} = t\boldsymbol{b}$.

b) Prove that $\boldsymbol{a}$ and $\boldsymbol{b}$ are orthogonal if and only if $\boldsymbol{a} \cdot \boldsymbol{b} = 0$.

**9.** The distance from a point $\boldsymbol{x}_0 = (x_0, y_0, z_0)$ to a plane $\Pi$ in $\mathbf{R}^3$ is defined to be

$$\operatorname{dist}(\boldsymbol{x}_0, \Pi) := \begin{cases} 0 & \boldsymbol{x}_0 \in \Pi \\ \|\boldsymbol{v}\| & \boldsymbol{x}_0 \notin \Pi, \end{cases}$$

where $\boldsymbol{v} := (x_0 - x_1, y_0 - y_1, z_0 - z_1)$ for some $(x_1, y_1, z_1) \in \Pi$, and $\boldsymbol{v}$ is orthogonal to $\Pi$, i.e., parallel to its normal. Sketch $\Pi$ and $\boldsymbol{x}_0$ for a typical plane $\Pi$, and convince yourself that this is the correct definition. Prove that this definition does not depend on the choice of $\boldsymbol{v}$, by showing the distance from $\boldsymbol{x}_0 = (x_0, y_0, z_0)$ to the plane $\Pi$ described by $ax + by + cz = d$ is

$$\operatorname{dist}(\boldsymbol{x}_0, \Pi) = \frac{|ax_0 + by_0 + cz_0 - d|}{\sqrt{a^2 + b^2 + c^2}}.$$

**10.** a) Prove that the $\ell^1$–norm and the sup–norm also satisfy Theorem 8.5.

b) Describe what the open balls in $\mathbf{R}^2$ look like with respect to each of these "nonEuclidean" norms.

## 8.2 LIMITS OF SEQUENCES

Using the analogy between norms and the absolute value, we can introduce convergent, bounded, and Cauchy sequences to arbitrary Euclidean spaces as follows.

**8.11 DEFINITION.** Let $\{\boldsymbol{x}_k\}$ be a sequence points in $\mathbf{R}^n$.

i) $\{\boldsymbol{x}_k\}$ is said to *converge* to some point $\boldsymbol{a} \in \mathbf{R}^n$ (called the *limit* of $\boldsymbol{x}_k$) if for every $\varepsilon > 0$ there is an $N \in \mathbf{N}$ such that

$$k \geq N \quad \text{implies} \quad \|\boldsymbol{x}_k - \boldsymbol{a}\| < \varepsilon.$$

ii) $\{\boldsymbol{x}_k\}$ is said to be *bounded* if there is an $M > 0$ such that $\|\boldsymbol{x}_k\| \leq M$ for all $k \in \mathbf{N}$.

iii) $\boldsymbol{x}_k \in \mathbf{R}^n$ is said to be *Cauchy* if for every $\varepsilon > 0$ there is an $N \in \mathbf{N}$ such that

$$k, m \geq N \quad \text{imply} \quad \|\boldsymbol{x}_k - \boldsymbol{x}_m\| < \varepsilon.$$

For each $k \in \mathbf{N}$, let $\bar{P}_k = (\bar{P}_{k1}, \bar{P}_{k2}, \cdots, \bar{P}_{kn})$ and $P = (P_1, P_2, \cdots, P_n)$.
Then $\bar{P}_k \to \bar{P}$ iff $\bar{P}_{kj} \to \bar{P}_j$ for each $j = 1, \cdots, n$

$$\bar{P}_1 = (P_{11}, P_{12}, \cdots, P_{1n})$$
$$\bar{P}_2 = (P_{21}, P_{22}, \cdots, P_{2n})$$
$$\bar{P}_k = (P_{k1}, P_{k2}, \cdots, P_{kn}) \downarrow$$
$$\bar{P} = (P_1, P_2, \cdots, P_n)$$

The following result shows that to evaluate the limit of a specific sequence in $\mathbf{R}^n$ we need only take the limits of the component sequences.

**8.12 THEOREM.** Let $\boldsymbol{a}, \boldsymbol{x}_k \in \mathbf{R}^n$ for $k \in \mathbf{N}$. Denote the $j$th component of $\boldsymbol{x}_k$ by $x_k(j)$ and the $j$th component of $\boldsymbol{a}$ by $a(j)$. Then the sequence $\{\boldsymbol{x}_k\}_{k \in \mathbf{N}}$ converges to $\boldsymbol{a}$ in $\mathbf{R}^n$, as $k \to \infty$, if and only if for each $j = 1, 2, \ldots, n$, the component sequence $\{x_k(j)\}_{k \in \mathbf{N}}$ converges to $a(j)$ in $\mathbf{R}$, as $k \to \infty$.

PROOF. By Theorem 8.5,

$$|x_k(\ell) - x(\ell)| \leq \|\boldsymbol{x}_k - \boldsymbol{a}\| \leq \sqrt{n} \max_{1 \leq j \leq n} |x_k(j) - a(j)|$$

for $\ell = 1, 2, \ldots, n$. Hence, by the Squeeze Theorem, $x_k(j) \to a(j)$ as $k \to \infty$ for all $1 \leq j \leq n$ if and only if the real sequence $\|\boldsymbol{x}_k - \boldsymbol{a}\| \to 0$ as $k \to \infty$. Since $\|\boldsymbol{x}_k - \boldsymbol{a}\| \to 0$ if and only if $\boldsymbol{x}_k \to \boldsymbol{a}$, as $k \to \infty$, the proof of the theorem is complete. ∎

This result can be used to obtain the following analogue of the Density of Rationals (Theorem 1.24).

**8.13 THEOREM.** For each $\boldsymbol{a} \in \mathbf{R}^n$ there is a sequence $\boldsymbol{x}_k \in \mathbf{Q}^n$ such that $\boldsymbol{x}_k \to \boldsymbol{a}$ as $k \to \infty$.

PROOF. Let $\boldsymbol{a} := (a_1, \ldots, a_n) \in \mathbf{R}^n$. For each $1 \leq j \leq n$, choose by Theorem 1.24 sequences $r_k^{(j)} \in \mathbf{Q}$ such that $r_k^{(j)} \to a_j$ (in $\mathbf{R}$) as $k \to \infty$. By Theorem 8.12, $\boldsymbol{x}_k := (r_k^{(1)}, \ldots, r_k^{(n)})$ converges to $\boldsymbol{a}$ (in $\mathbf{R}^n$) as $k \to \infty$. Moreover, $\boldsymbol{x}_k \in \mathbf{Q}^n$ for each $k \in \mathbf{N}$. ∎

A set $E$ is said to be *separable* if there is a countable subset $Z$ of $E$ such that to each $\boldsymbol{a} \in E$ there corresponds a sequence $\boldsymbol{x}_k \in Z$ such that $\boldsymbol{x}_k \to \boldsymbol{a}$ as $k \to \infty$. Since the argument illustrated in Figure 1.3 can be extended to show that $\mathbf{Q}^n$ is countable, it follows from Theorem 8.13 that $\mathbf{R}^n$ is separable.

Theorem 8.13 illustrates a general principle. As long as we stay away from results about monotone sequences (which have no analogue in $\mathbf{R}^n$ when $n > 1$), we can extend most of the results found in Chapter 2 from $\mathbf{R}$ to $\mathbf{R}^n$. Since the proofs of these results require little more than replacing $|x - y|$ in the real case by $\|\boldsymbol{x} - \boldsymbol{y}\|$ in the Euclidean space case, we will summarize what is true and leave most of the details to the reader.

**8.14 THEOREM.** Let $X$ be a Euclidean space.

   i) A sequence in $X$ can have at most one limit.

  ii) If $\{\boldsymbol{x}_n\}_{n \in \mathbf{N}}$ is a sequence in $X$ which converges to $\boldsymbol{a}$ and $\{\boldsymbol{x}_{n_k}\}_{k \in \mathbf{N}}$ is any subsequence of $\{\boldsymbol{x}_n\}_{n \in \mathbf{N}}$, then $\boldsymbol{x}_{n_k}$ converges to $\boldsymbol{a}$ as $k \to \infty$.

 iii) Every convergent sequence in $X$ is bounded, but not conversely.

 iv) Every convergent sequence in $X$ is Cauchy.

  v) If $\{\boldsymbol{x}_n\}$ and $\{\boldsymbol{y}_n\}$ are convergent sequences in $X$ and $\alpha \in \mathbf{R}$, then

$$\lim_{n \to \infty} (\boldsymbol{x}_n + \boldsymbol{y}_n) = \lim_{n \to \infty} \boldsymbol{x}_n + \lim_{n \to \infty} \boldsymbol{y}_n,$$

$$\lim_{n \to \infty} (\alpha \boldsymbol{x}_n) = \alpha \lim_{n \to \infty} \boldsymbol{x}_n,$$

*and*

$$\lim_{n \to \infty} (\boldsymbol{x}_n \cdot \boldsymbol{y}_n) = (\lim_{n \to \infty} \boldsymbol{x}_n) \cdot (\lim_{n \to \infty} \boldsymbol{y}_n).$$

Notice once and for all that (since $\|\boldsymbol{x}_n\|^2 = \boldsymbol{x}_n \cdot \boldsymbol{x}_n$), the last equation above contains the following corollary. If $\boldsymbol{x}_n$ converges, then

$$\lim_{n \to \infty} \|\boldsymbol{x}_n\| = \| \lim_{n \to \infty} \boldsymbol{x}_n \|.$$

As in the real case, the converse of part iv) is also true. In order to prove that, we need an $n$-dimensional version of the Bolzano–Weierstrass Theorem.

**8.15 THEOREM** [Bolzano–Weierstrass Theorem for $\mathbf{R}^n$]. *Every bounded sequence in $\mathbf{R}^n$ has a convergent subsequence.*

PROOF. Suppose $\{\boldsymbol{x}_k\}$ is bounded in $\mathbf{R}^n$. For each $j \in \{1, \ldots, n\}$, let $x_k(j)$ represent the $j$th component of the vector $\boldsymbol{x}_k$. By hypothesis, the sequence $\{x_k(j)\}_{k \in \mathbf{N}}$ is bounded in $\mathbf{R}$ for each $j = 1, 2, \ldots, n$.

Let $j = 1$. By the one-dimensional Bolzano–Weierstrass Theorem, there is a sequence of integers $1 \le k(1,1) < k(1,2) < \cdots$ and a number $x(1)$ such that $x_{k(1,\nu)}(1) \to x(1)$ as $\nu \to \infty$.

Let $j = 2$. Again, since the sequence $\{x_{k(1,\nu)}(2)\}_{\nu \in \mathbf{N}}$ is bounded in $\mathbf{R}$, there is a subsequence $\{k(2,\nu)\}_{\nu \in \mathbf{N}}$ of $\{k(1,\nu)\}_{\nu \in \mathbf{N}}$ and a number $x(2)$ such that $x_{k(2,\nu)}(2) \to x(2)$ as $\nu \to \infty$. Since $\{k(2,\nu)\}_{\nu \in \mathbf{N}}$ is a subsequence of $\{k(1,\nu)\}_{\nu \in \mathbf{N}}$, we also have $x_{k(2,\nu)}(1) \to x(1)$ as $\nu \to \infty$. Thus, $x_{k(2,\nu)}(\ell) \to x(\ell)$ as $\nu \to \infty$ for all $1 \le \ell \le j = 2$.

Continuing this process until $j = n$, we choose a subsequence $k_\nu = k(n,\nu)$ and points $x(\ell)$ such that

$$\lim_{\nu \to \infty} x_{k_\nu}(\ell) = x(\ell)$$

for $1 \le \ell \le j = n$. Set $\boldsymbol{x} = (x(1), x(2), \ldots, x(n))$. Then by Theorem 8.12, $\boldsymbol{x}_{k_\nu}$ converges to $\boldsymbol{x}$ as $\nu \to \infty$. ∎

Since the Bolzano–Weierstrass Theorem holds for $\mathbf{R}^n$, we can modify proof of Theorem 2.29 to establish the following result.

**8.16 THEOREM.** *A sequence $\{\boldsymbol{x}_k\}$ in $\mathbf{R}^n$ is Cauchy if and only if it converges.*

## EXERCISES

**1.** Using Definition 8.11i), prove that the following limits exist.

a)
$$\boldsymbol{x}_k = \left( \frac{1}{k}, 1 - \frac{1}{k^2} \right).$$

b)
$$\boldsymbol{x}_k = \left(\frac{k}{k+1}, \sin\frac{1}{k}\right).$$

c)
$$\boldsymbol{x}_k = \left(\log(k+1) - \log k, 2^{-k}\right).$$

**2.** Using limit theorems, find the limit of each of the following vector sequences.

a)
$$\boldsymbol{x}_k = \left(\frac{1}{k}, \frac{k - 3k^2}{k + k^2}\right).$$

b)
$$\boldsymbol{x}_k = \left(1, \sin\pi k, \cos\frac{1}{k}\right).$$

c)
$$\boldsymbol{x}_k = \left(k - \sqrt{k^2 + k}, k^{1/k}, \frac{1}{k}\right).$$

$$\|\bar{x}_k\| = \sqrt{\sum_{j=1}^{n} x_{kj}^2}$$

**3.** Suppose $\boldsymbol{x}_k \to \boldsymbol{0}$ in $\mathbf{R}^n$ as $k \to \infty$ and $\boldsymbol{y}_k$ is bounded in $\mathbf{R}^n$. Prove $\boldsymbol{x}_k \cdot \boldsymbol{y}_k \to 0$ as $k \to \infty$.   $r_k = \bar{x}_k \cdot \bar{y}_k = \sum_{j=1}^{n} x_{kj} \cdot y_{kj}$   $\{r_k\} \to 0$ as $k \to \infty$.

**4.** Find convergent subsequences of
$$\boldsymbol{x}_k = \left((-1)^k, \frac{1}{k}, (-1)^{3k}\right)$$

which converge to different limits. Prove your limits exist.

**5.** a) Prove Theorem 8.14i) and ii).
  b) Prove Theorem 8.14iii) and iv).

**6.** Prove Theorem 8.14v).

**7.** Prove Theorem 8.16.

**8.** a) A subset $E$ of $\mathbf{R}^n$ is said to be *sequentially compact* if every sequence $\boldsymbol{x}_k \in E$ has a convergent subsequence whose limit belongs to $E$. Prove that every closed ball in $\mathbf{R}^n$ is sequentially compact.
  b) Prove that $\mathbf{R}^n$ is not sequentially compact.

**9.** a) Let $E$ be a subset of $\mathbf{R}^n$. A point $\boldsymbol{a} \in \mathbf{R}^n$ is called a *cluster point* of $E$ if $E \cap B_r(\boldsymbol{a})$ contains infinitely many points for every $r > 0$. Prove that $\boldsymbol{a}$ is a cluster point of $E$ if and only if for each $r > 0$, $E \cap B_r(\boldsymbol{a}) \setminus \{\boldsymbol{a}\}$ is nonempty.
  b) Prove that every bounded infinite subset of $\mathbf{R}^n$ has at least one cluster point.

## 8.3  LIMITS OF FUNCTIONS

We now turn our attention to vector-valued functions. Suppose $f$ takes some subset of $\mathbf{R}^n$ into $\mathbf{R}^m$. The largest subset of $\mathbf{R}^n$ on which $f$ is defined, called the *domain*

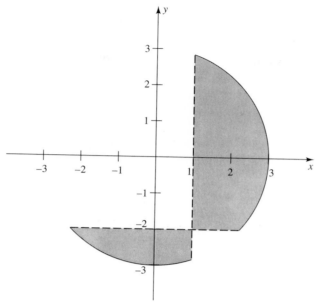

**Figure 8.5**

of $f$, will be denoted by $\mathrm{Dom}\,(f)$. Since $f(\boldsymbol{x}) \in \mathbf{R}^m$ for each $\boldsymbol{x} \in \mathrm{Dom}\,(f)$, there are functions $f_j : \mathrm{Dom}\,(f) \to \mathbf{R}$ (called the *component functions* of $f$) such that $f(\boldsymbol{x}) = (f_1(\boldsymbol{x}), \ldots, f_m(\boldsymbol{x}))$ for each $\boldsymbol{x} \in \mathrm{Dom}\,(f)$. When $m = 1$, $f$ has only one component and we shall call $f$ *real-valued*.

For the most part, domains of functions of one variable were fairly simple, e.g., unions of intervals. For functions of several variables, domains exhibit much more variety.

**8.17 Example.** Find the domain of $f(x, y) = (\log(xy - y + 2x - 2), \sqrt{9 - x^2 - y^2})$.

SOLUTION. This function has two components: $f_1(x, y) = \log(xy - y + 2x - 2)$ and $f_2(x, y) = \sqrt{9 - x^2 - y^2}$. Since the logarithm is real-valued only when its argument is positive, the domain of $f_1$ is the set of points $(x, y)$ which satisfy

$$0 < xy - y + 2x - 2 = (x - 1)(y + 2).$$

Since the square root function is real-valued if and only if its argument is non-negative, the domain of $f_2$ is the set of points $(x, y)$ which satisfy $x^2 + y^2 \le 9$. Thus

$$\mathrm{Dom}\,(f) = \{(x, y) : x^2 + y^2 \le 9 \text{ and } (x - 1)(y + 2) > 0\}.$$

(The $\mathrm{Dom}\,(f)$ is graphed in Figure 8.5.) ∎

**8.18 Example.** Find the domain of

$$f(x, y) = (\sqrt{1 - x^2}, \log(x^2 - y^2), \sin x \cos y).$$

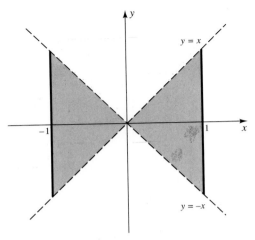

**Figure 8.6**

SOLUTION. This function has three component functions: $f_1(x,y) = \sqrt{1-x^2}$, $f_2(x,y) = \log(x^2 - y^2)$, and $f_3(x,y) = \sin x \cos y$. $f_1$ is real-valued when $1 - x^2 \geq 0$, i.e., $-1 \leq x \leq 1$. $f_2$ is real-valued when $x^2 - y^2 > 0$, i.e., when $-|x| < y < |x|$. The domain of $f_3$ is all of $\mathbf{R}^2$. Thus

$$\text{Dom}\,(f) = \{(x,y) : -1 \leq x \leq 1 \text{ and } -|x| < y < |x|\}.$$

(The Dom$\,(f)$ is graphed in Figure 8.6.) ∎

Again, using the analogy between the norm on $\mathbf{R}^n$ and the absolute value on $\mathbf{R}$, we can extend much of the theory of limits of functions developed in Chapter 3 to the Euclidean space setting. Here is a brief summary of the details.

**8.19 DEFINITION.** Let $X$, $Y$ be Euclidean spaces, $\boldsymbol{a} \in X$, and $\boldsymbol{L} \in Y$. Suppose $V$ is an open ball containing $\boldsymbol{a}$ and $f : V \setminus \{\boldsymbol{a}\} \to Y$. Then $f(\boldsymbol{x})$ is said to *converge to $\boldsymbol{L}$, as $\boldsymbol{x}$ approaches $\boldsymbol{a}$*, if for every $\varepsilon > 0$ there is a $\delta > 0$ (which in general depends on $\varepsilon$, $f$, and $\boldsymbol{a}$) such that

$$0 < \|\boldsymbol{x} - \boldsymbol{a}\| < \delta \quad \text{implies} \quad \|f(\boldsymbol{x}) - \boldsymbol{L}\| < \varepsilon.$$

In this case we write

$$\boldsymbol{L} = \lim_{\boldsymbol{x} \to \boldsymbol{a}} f(\boldsymbol{x})$$

and call $\boldsymbol{L}$ the *limit* of $f(\boldsymbol{x})$ as $\boldsymbol{x}$ approaches $\boldsymbol{a}$.

Let $E \subseteq \mathbf{R}^n$ and $f, g : E \to \mathbf{R}^m$. For each $\boldsymbol{x} \in E$, the *scalar product* of an $\alpha \in \mathbf{R}$ with $f$ is defined by

$$(\alpha f)(\boldsymbol{x}) := \alpha f(\boldsymbol{x}),$$

the *sum* of $f$ and $g$ is defined by

$$(f + g)(\boldsymbol{x}) := f(\boldsymbol{x}) + g(\boldsymbol{x}),$$

the (*Euclidean*) *dot product* of $f$ and $g$ is defined by

$$(f \cdot g)(\boldsymbol{x}) := f(\boldsymbol{x}) \cdot g(\boldsymbol{x}),$$

and (when $m = 3$) the *cross product* of $f$ and $g$ is defined by

$$(f \times g)(\boldsymbol{x}) := f(\boldsymbol{x}) \times g(\boldsymbol{x}).$$

(Notice that when $m = 1$, the dot product of two functions is the pointwise product defined in Section 3.1.)

**8.20 THEOREM.** *Let* $X, Y$ *be Euclidean spaces,* $\boldsymbol{a} \in X$, $V$ *be an open ball which contains* $\boldsymbol{a}$*, and* $f, g : V \setminus \{\boldsymbol{a}\} \to Y$.

  i) *If* $f(\boldsymbol{x}) = g(\boldsymbol{x})$ *for all* $\boldsymbol{x} \in V \setminus \{\boldsymbol{a}\}$ *and* $f(\boldsymbol{x})$ *has a limit as* $\boldsymbol{x} \to \boldsymbol{a}$*, then* $g(\boldsymbol{x})$ *also has a limit as* $\boldsymbol{x} \to \boldsymbol{a}$*, and*

$$\lim_{\boldsymbol{x} \to \boldsymbol{a}} g(\boldsymbol{x}) = \lim_{\boldsymbol{x} \to \boldsymbol{a}} f(\boldsymbol{x}).$$

  ii) [SEQUENTIAL CHARACTERIZATION OF LIMITS]. $L = \lim_{\boldsymbol{x} \to \boldsymbol{a}} f(\boldsymbol{x})$ *exists if and only if* $f(\boldsymbol{x}_n) \to L$ *as* $n \to \infty$ *for every sequence* $\boldsymbol{x}_n \in V \setminus \{\boldsymbol{a}\}$ *which converges to* $\boldsymbol{a}$ *as* $n \to \infty$.

  iii) *Suppose* $\alpha \in \mathbf{R}$*. If* $f(\boldsymbol{x})$ *and* $g(\boldsymbol{x})$ *have limits, as* $\boldsymbol{x}$ *approaches* $\boldsymbol{a}$*, then so do* $(f + g)(\boldsymbol{x})$*,* $(\alpha f)(\boldsymbol{x})$*,* $(f \cdot g)(\boldsymbol{x})$*, and* $\|f(\boldsymbol{x})\|$*. In fact,*

$$\lim_{\boldsymbol{x} \to \boldsymbol{a}} (f + g)(\boldsymbol{x}) = \lim_{\boldsymbol{x} \to \boldsymbol{a}} f(\boldsymbol{x}) + \lim_{\boldsymbol{x} \to \boldsymbol{a}} g(\boldsymbol{x}),$$

$$\lim_{\boldsymbol{x} \to \boldsymbol{a}} (\alpha f)(\boldsymbol{x}) = \alpha \lim_{\boldsymbol{x} \to \boldsymbol{a}} f(\boldsymbol{x}),$$

$$\lim_{\boldsymbol{x} \to \boldsymbol{a}} (f \cdot g)(\boldsymbol{x}) = \left( \lim_{\boldsymbol{x} \to \boldsymbol{a}} f(x) \right) \cdot \left( \lim_{\boldsymbol{x} \to \boldsymbol{a}} g(\boldsymbol{x}) \right),$$

*and*

$$\left\| \lim_{\boldsymbol{x} \to \boldsymbol{a}} f(\boldsymbol{x}) \right\| = \lim_{\boldsymbol{x} \to \boldsymbol{a}} \|f(\boldsymbol{x})\|.$$

*Moreover, when* $m = 3$,

$$\lim_{\boldsymbol{x} \to \boldsymbol{a}} (f \times g)(\boldsymbol{x}) = \left( \lim_{\boldsymbol{x} \to \boldsymbol{a}} f(\boldsymbol{x}) \right) \times \left( \lim_{\boldsymbol{x} \to \boldsymbol{a}} g(\boldsymbol{x}) \right),$$

*and when* $m = 1$ *and the limit of* $g$ *is nonzero,*

$$\lim_{\boldsymbol{x} \to \boldsymbol{a}} f(\boldsymbol{x})/g(\boldsymbol{x}) = \left( \lim_{\boldsymbol{x} \to \boldsymbol{a}} f(\boldsymbol{x}) \right) / \left( \lim_{\boldsymbol{x} \to \boldsymbol{a}} g(\boldsymbol{x}) \right).$$

  iv) [SQUEEZE THEOREM FOR FUNCTIONS]. *Suppose* $f, g, h : V \setminus \{\boldsymbol{a}\} \to \mathbf{R}$ *and* $g(\boldsymbol{x}) \leq h(\boldsymbol{x}) \leq f(\boldsymbol{x})$ *for all* $\boldsymbol{x} \in V \setminus \{\boldsymbol{a}\}$*. If*

$$\lim_{\boldsymbol{x} \to \boldsymbol{a}} f(\boldsymbol{x}) = \lim_{\boldsymbol{x} \to \boldsymbol{a}} g(\boldsymbol{x}) = L,$$

*then the limit of* $h$ *also exists, as* $\boldsymbol{x} \to \boldsymbol{a}$*, and*

$$\lim_{\boldsymbol{x} \to \boldsymbol{a}} h(\boldsymbol{x}) = L.$$

How do we actually compute the limit of a given vector-valued function? The following result shows that evaluation of such limits reduces to the real-valued case, i.e., the case where the range is one-dimensional. Consequently, our examples will be almost exclusively real-valued.

**8.21 THEOREM.** *Let $\boldsymbol{a} \in \mathbf{R}^n$, $V$ be an open ball which contains $\boldsymbol{a}$, $f : V \setminus \{\boldsymbol{a}\} \to \mathbf{R}^m$, and $\boldsymbol{L} = (L_1, L_2, \ldots, L_m) \in \mathbf{R}^m$. Then*

(3)
$$\boldsymbol{L} = \lim_{\boldsymbol{x} \to \boldsymbol{a}} f(\boldsymbol{x})$$

*exists if and only if*

(4)
$$L_j = \lim_{\boldsymbol{x} \to \boldsymbol{a}} f_j(\boldsymbol{x})$$

*exists for each $j = 1, 2, \ldots, m$.*

PROOF. By the Sequential Characterization of Limits, we must show that for all sequences $\boldsymbol{x}_k \in V \setminus \{\boldsymbol{a}\}$ which converge to $\boldsymbol{a}$, $f(\boldsymbol{x}_k) \to \boldsymbol{L}$ as $k \to \infty$ if and only if $f_j(\boldsymbol{x}_k) \to L_j$, as $k \to \infty$, for each $1 \le j \le n$. But this last statement is obviously true by Theorem 8.12. Therefore, (3) holds if and only if (4) holds. ∎

Using Theorem 8.20, it is easy to prove (see Exercise 4 below) that the limit of any polynomial $P$ on $\mathbf{R}^n$ exists at any point $\boldsymbol{a} \in \mathbf{R}^n$ and satisfies

$$\lim_{\boldsymbol{x} \to \boldsymbol{a}} P(\boldsymbol{x}) = P(\boldsymbol{a}).$$

This observation is often used in conjunction with Theorem 8.21 to evaluate simple limits like the following.

**8.22 Example.** Find
$$\lim_{(x,y) \to (0,0)} (3x + 1, e^y + 2).$$

SOLUTION. By Theorem 8.21, this limit is $(0 + 1, e^0 + 2) = (1, 3)$. ∎

**8.23 Example.** Prove that the function

$$f(x, y) = \frac{2 + x - y}{1 + 2x^2 + 3y^2}$$

has a limit as $(x, y) \to (0, 0)$.

PROOF. The polynomial $2 + x - y$ (respectively, $1 + 2x^2 + 3y^2$) converges to 2 (respectively, to 1) as $(x, y) \to (0, 0)$. Hence,

$$\lim_{(x,y) \to (0,0)} \frac{2 + x - y}{1 + 2x^2 + 3y^2} = \frac{2}{1} = 2$$

by Theorem 8.20. ∎

We could use Theorem 8.20 in this example because the limit quotient was not of the form $0/0$. Proving that a limit of the form $0/0$ exists often involves showing that $\|f(\boldsymbol{x}) - \boldsymbol{L}\|$ is *dominated by* (i.e., less than or equal to) some nonnegative function $g$ which satisfies $g(\boldsymbol{x}) \to 0$ as $\boldsymbol{x} \to \boldsymbol{a}$. Here is a typical example.

**8.24 Example.** Prove that

$$f(x,y) = \frac{3x^2 y}{x^2 + y^2}$$

converges as $(x,y) \to (0,0)$.

PROOF. Since the numerator is a polynomial of degree 3 and the denominator is a polynomial of degree 2, we expect the numerator to overpower the denominator, i.e., the limit to be 0 as $(x,y) \to (0,0)$. To prove this, we must estimate $f(x,y)$ near $(0,0)$. Since $2|xy| \le x^2 + y^2$ for all $(x,y) \in \mathbf{R}^2$, it is easy to check that

$$|f(x,y)| \le \frac{3}{2}|x|$$

for all $(x,y) \ne (0,0)$. Let $\varepsilon > 0$ and set $\delta = \varepsilon/2$. If $0 < \|(x,y)\| < \delta$, then $|f(x,y)| \le 2|x| \le 2\|(x,y)\| < 2\delta = \varepsilon$. Thus, by definition,

$$\lim_{(x,y)\to(0,0)} f(x,y) = 0. \quad \blacksquare$$

It is important to realize that by Definition 8.19, if $f$ converges to $\boldsymbol{L}$ as $\boldsymbol{x} \to \boldsymbol{a}$, then $\|f(\boldsymbol{x}) - \boldsymbol{L}\|$ is small for all $\boldsymbol{x}$ near $\boldsymbol{a}$. In particular, $f(\boldsymbol{x}) \to \boldsymbol{L}$ as $\boldsymbol{x} \to \boldsymbol{a}$, no matter what path $\boldsymbol{x}$ takes. The next two examples show how to use this observation to prove that a limit does not exist.

**8.25 Example.** Prove that the function

$$f(x,y) = \frac{2xy}{x^2 + y^2}$$

has no limit as $(x,y) \to (0,0)$.

PROOF. Suppose $f$ has a limit $L$, as $(x,y) \to (0,0)$. If $(x,y)$ approaches $(0,0)$ along a vertical path, e.g., if $x = 0$ and $y \downarrow 0$, then $L = 0$ (because $f(0,y) = 0$ for all $y \ne 0$). If $(x,y)$ approaches $(0,0)$ along a "diagonal" path, e.g., if $y = x$ and $x \downarrow 0$, then $L = 1$ (because $f(x,x) = 1$ for all $x \ne 0$). Since $0 \ne 1$, $f$ has no limit at $(0,0)$. $\blacksquare$

In the solution to Example 8.25, the diagonal path was chosen so that the denominator of $f(x,y)$ would collapse to a single term. This same strategy is used in the next example.

**8.26 Example.** Determine whether

$$f(x,y) = \frac{xy^2}{x^2 + y^4}$$

has a limit as $(x,y) \to (0,0)$.

SOLUTION. The vertical path $x = 0$ gives $f(0, y) = 0$ even before we take the limit as $y \to 0$. On the other hand, the parabolic path $x = y^2$ gives

$$f(y^2, y) = \frac{y^4}{2y^4} = \frac{1}{2} \neq 0.$$

Therefore, $f$ cannot have a limit as $(x, y) \to (0, 0)$. ∎

(Notice that if $y = mx$, then

$$f(x, y) = \frac{m^2 x^3}{x^2 + m^4 x^4} \to 0$$

as $x \to 0$. Thus, Example 8.26 shows that the two-dimensional limit of a function might not exist even when its limit along every linear path exists and gives the same value.)

When asked whether the limit of a function $f(\boldsymbol{x})$ exists, it is natural to begin by taking the limit as each variable moves independently. Comparing Examples 8.23 and 8.25, we see that this strategy works for some functions but not all. To look at this problem more closely, we introduce the following terminology. Let $V$ be an open ball in $\mathbf{R}^2$, $(a, b) \in V$, and $f : V \setminus \{(a, b)\} \to \mathbf{R}^m$. The *iterated limits* of $f$ at $(a, b)$ are defined to be

$$\lim_{x \to a} \lim_{y \to b} f(x, y) := \lim_{x \to a} \left( \lim_{y \to b} f(x, y) \right), \quad \lim_{y \to b} \lim_{x \to a} f(x, y) := \lim_{y \to b} \left( \lim_{x \to a} f(x, y) \right),$$

when they exist.

The iterated limits of a given function might not exist. Even when they do, we cannot be sure that the corresponding two-dimensional limit exists. Indeed, although the iterated limits of the function $f$ in Example 8.25 above exist and are both zero at $(0, 0)$, $f$ has no limit as $(x, y) \to (0, 0)$.

It is even possible for both iterated limits to exist but give different values.

**8.27 Example.** Evaluate the iterated limits of

$$f(x, y) = \frac{x^2}{x^2 + y^2}$$

at $(0, 0)$.

SOLUTION. For each $x \neq 0$, $x^2/(x^2 + y^2) \to 1$ as $y \to 0$. Therefore,

$$\lim_{x \to 0} \lim_{y \to 0} \frac{x^2}{x^2 + y^2} = \lim_{x \to 0} \frac{x^2}{x^2} = 1.$$

On the other hand,

$$\lim_{y \to 0} \lim_{x \to 0} \frac{x^2}{x^2 + y^2} = \lim_{y \to 0} \frac{0}{y^2} = 0. \quad ∎$$

This leads us to ask: When are the iterated limits equal? The following result shows that if $f$ has a limit as $(x, y) \to (a, b)$ and both iterated limits exist, then these limits must be equal.

**8.28 Remark.** *Suppose I and J are open intervals, $a \in I$, $b \in J$, and $f : I \times J \to$*
**R**. *If $\lim_{x \to a} f(x, y_0)$ exists for each $y_0 \in J$, $\lim_{y \to b} f(x_0, y)$ exists for each $x_0 \in I$,*
*and $f(x, y) \to L$ as $(x, y) \to (a, b)$ (in $\mathbf{R}^2$), then*

$$L = \lim_{x \to a} \lim_{y \to b} f(x, y) = \lim_{y \to b} \lim_{x \to a} f(x, y).$$

PROOF. Let

$$g(x) := \lim_{y \to b} f(x, y)$$

for $x \in I$. Given $\varepsilon > 0$, choose $\delta > 0$ such that

$$0 < \|(x, y) - (a, b)\| < \delta \quad \text{implies} \quad |f(x, y) - L| < \varepsilon.$$

Suppose $x \in I$ and $0 < |x - a| < \delta/\sqrt{2}$. Then for any $y$ which satisfies $0 < |y - b| < \delta/\sqrt{2}$, we have $0 < \|(x, y) - (a, b)\| < \delta$, hence

$$|g(x) - L| \leq |g(x) - f(x, y)| + |f(x, y) - L| < |g(x) - f(x, y)| + \varepsilon.$$

Taking the limit of this inequality as $y \to b$, we find that $|g(x) - L| \leq \varepsilon$ for all $x \in I$ which satisfy $|x - a| < \delta/\sqrt{2}$. It follows that $g(x) \to L$ as $x \to a$, i.e.,

$$L = \lim_{x \to a} \lim_{y \to b} f(x, y).$$

A similar argument proves that the other iterated limit also exists and equals $L$. ∎

Notice by Example 8.27 that the conclusion of Remark 8.28 might not hold if the hypothesis "$f(x, y) \to L$ as $(x, y) \to (a, b)$" is omitted. In particular, if the limit of a function does not exist, we must be careful about the order in which an iterated limit is taken.

## EXERCISES

**1.** For each of the following functions, find the domain of $f$, prove that the limit of $f$ exists as $(x, y) \to (a, b)$, and find the value of that limit. (Note: You can prove that the limit exists without using $\varepsilon$'s and $\delta$'s—see Example 8.23.)

a) $\qquad f(x, y) = \left( \dfrac{x - 1}{y - 1}, \dfrac{x^2 + x - 2}{x - 1} \right), \qquad (a, b) = (1, -1).$

b) $\qquad f(x, y) = \left( \dfrac{y \sin x}{x}, \tan \dfrac{x}{y}, x^2 + y^2 - xy \right), \qquad (a, b) = (0, 1).$

c)       $f(x, y) = \left( \dfrac{x^4 + y^4}{x^2 + y^2}, \dfrac{\sqrt{|xy|}}{\sqrt[3]{x^2 + y^2}} \right)$,       $(a, b) = (0, 0)$.

d)      $f(x, y) = \left( \dfrac{x^2 - 1}{y^2 + 1}, \dfrac{x^2 y - 2xy + y - (x - 1)^2}{x^2 + y^2 - 2x - 2y + 2} \right)$,       $(a, b) = (1, 1)$.

**2.** Compute the iterated limits at $(0, 0)$ of each of the following functions. Determine which of these functions has a limit as $(x, y) \to (0, 0)$ in $\mathbf{R}^2$, and prove that the limit exists.

a)                              $f(x, y) = \dfrac{\sin x \sin y}{x^2 + y^2}$.

b)                              $f(x, y) = \dfrac{x^2 + y^4}{x^2 + 2y^4}$.

c)                              $f(x, y) = \dfrac{x - y}{(x^2 + y^2)^\alpha}$,       $\alpha < \dfrac{1}{2}$.

**3.** Prove that each of the following functions has a limit as $(x, y) \to (0, 0)$.

a)               $f(x, y) = \dfrac{x^3 - y^3}{x^2 + y^2}$,       $(x, y) \neq (0, 0)$.

b)               $f(x, y) = \dfrac{x^\alpha y^4}{x^2 + y^4}$,       $(x, y) \neq (0, 0)$,

   where $\alpha$ is ANY positive number.

**4.** A *polynomial* on $\mathbf{R}^n$ is a function of the form

$$P(x_1, x_2, \dots, x_n) = \sum_{j_1=0}^{N_1} \cdots \sum_{j_n=0}^{N_n} a_{j_1,\dots,j_n} x_1^{j_1} \dots x_n^{j_n},$$

where $a_{j_1,\dots,j_n}$ are scalars and $N_1, \dots, N_n$ are nonnegative integers. Prove that if $P$ is a polynomial on $\mathbf{R}^n$ and $\boldsymbol{a} \in \mathbf{R}^n$, then $\lim_{\boldsymbol{x} \to \boldsymbol{a}} P(\boldsymbol{x}) = P(\boldsymbol{a})$.

**5.** Prove Theorem 8.20i).

**6.** Prove Theorem 8.20ii).

**7.** Prove Theorem 8.20iii).

**8.** Prove Theorem 8.20iv).

## 8.4 THE TOTAL DERIVATIVE

In this section we define what it means for a vector-valued function to be differentiable. First, notice that the one variable definition,

$$f'(a) = \lim_{h \to 0} \frac{f(a+h) - f(a)}{h},$$

cannot simply be adapted to $\mathbf{R}^n$ by replacing $a$ by $\boldsymbol{a}$ and $h$ by $\boldsymbol{h}$. Division by $\boldsymbol{h}$ is not defined on $\mathbf{R}^n$ when $n > 1$. Even if we rewrite this limit in the form

(5)
$$\lim_{h \to 0} \frac{|f(a+h) - f(a) - f'(a)h|}{|h|} = 0$$

(to avoid division by $h$), it is not clear how to proceed. What kind of "product" should take the place of $f'(a)h$; equivalently, what kind of object (scalar?, vector?) should $f'(\boldsymbol{a})$ be?

To answer these questions, recall that in the one variable case (see Theorem 4.3), $f$ is differentiable at $a$ if and only if there is a function of the form $T(x) = mx$ (i.e., a straight line through the origin) such that (5) holds with $T(h)$ in place of $f'(a)h$. Such functions $T$ have a simple characterization, which also makes sense in the multidimensional setting.

**8.29 Remark.** *Let $T : \mathbf{R} \to \mathbf{R}$. Then $T(x) = mx$ for some $m \in \mathbf{R}$ if and only if $T$ satisfies*

(6)
$$T(x+y) = T(x) + T(y) \quad and \quad T(\alpha x) = \alpha T(x)$$

*for all $x, y, \alpha \in \mathbf{R}$.*

PROOF. If $T(x) = mx$, then $T$ satisfies (6) since the distributive and commutative laws hold on $\mathbf{R}$. Conversely, if $T$ satisfies (6), set $m := T(1)$. Then (let $\alpha = x$),

$$T(x) = T(x \cdot 1) = xT(1) = mx$$

for all $x \in \mathbf{R}$. ∎

Accordingly, we make the following definition.

**8.30 DEFINITION.** A function $T : \mathbf{R}^n \to \mathbf{R}^m$ is said to be *linear* (notation: $T \in \mathcal{L}(\mathbf{R}^n; \mathbf{R}^m)$) if it satisfies

$$T(\boldsymbol{x} + \boldsymbol{y}) = T(\boldsymbol{x}) + T(\boldsymbol{y}) \quad and \quad T(\alpha \boldsymbol{x}) = \alpha T(\boldsymbol{x})$$

for all $\boldsymbol{x}, \boldsymbol{y} \in \mathbf{R}^n$ and all scalars $\alpha$.

Notice once and for all that if $T$ is a linear function, then

(7)
$$T(\mathbf{0}) = \mathbf{0}.$$

Indeed, by definition, $T(\mathbf{0}) = T(\mathbf{0}+\mathbf{0}) = T(\mathbf{0}) + T(\mathbf{0})$. Hence (7) can be obtained by subtracting $T(\mathbf{0})$ from both sides of this last equation.

Combining Remark 8.29 with Theorem 4.3, we are lead to the following definition.

**8.31 DEFINITION.** Let $X$ and $Y$ be Euclidean spaces and $f$ be a function such that $\text{Dom}\,(f) \subseteq X$ and $f : \text{Dom}\,(f) \to Y$.

 i) $f$ is said to be *differentiable* at a point $\boldsymbol{a}$ if there exist an open ball $V$ which contains $\boldsymbol{a}$ such that $V \subset \text{Dom}\,(f)$, and a linear function $T \in \mathcal{L}(X;Y)$ (called the *total derivative* of $f$ at $\boldsymbol{a}$) such that

(8) $$\lim_{\boldsymbol{h}\to\boldsymbol{0}} \frac{\|f(\boldsymbol{a}+\boldsymbol{h}) - f(\boldsymbol{a}) - T(\boldsymbol{h})\|}{\|\boldsymbol{h}\|} = 0.$$

 ii) $f$ is said to be *differentiable* if it is differentiable at each $\boldsymbol{a}$ in its domain.

The assumption that $f$ be defined on an open ball containing $\boldsymbol{a}$ is made so that the quotients in (8) are defined for $\|\boldsymbol{h}\|$ sufficiently small. We shall see in Section 11.2 that this definition generalizes both the analytic and geometric character of the one-dimensional derivative. In the meantime, we make a few elementary observations about linear functions themselves which will be used in future chapters.

The following result justifies our calling $T$ *the* total derivative of $f$. (It is called the *total derivative* to distinguish it from "partial" derivatives which will be defined in Section 11.1.)

**8.32 Remark.** [UNIQUENESS OF THE TOTAL DERIVATIVE] *If $f$ is differentiable at $\boldsymbol{a}$, then there is only one linear function $T$ which satisfies (8).*

PROOF. Let $S, T \in \mathcal{L}(X;Y)$, and suppose (8) holds for $T$, and when $T$ is replaced by $S$. We must show that $S(\boldsymbol{x}) = T(\boldsymbol{x})$ for all $\boldsymbol{x} \in X$.

Fix $\boldsymbol{x} \in X$. If $\boldsymbol{x} = \boldsymbol{0}$, then by (7), $S(\boldsymbol{0}) = \boldsymbol{0} = T(\boldsymbol{0})$. If $\boldsymbol{x} \neq \boldsymbol{0}$, then set $\boldsymbol{h} = \alpha\boldsymbol{x}$ for some $\alpha > 0$. Since $S$ and $T$ are linear and $\alpha > 0$, we have

$$\frac{S(\boldsymbol{x}) - T(\boldsymbol{x})}{\|\boldsymbol{x}\|} = \frac{\alpha S(\boldsymbol{x}) - \alpha T(\boldsymbol{x})}{\alpha\|\boldsymbol{x}\|} = \frac{S(\boldsymbol{h}) - T(\boldsymbol{h})}{\|\boldsymbol{h}\|}$$

$$= \frac{S(\boldsymbol{h}) - f(\boldsymbol{a}+\boldsymbol{h}) + f(\boldsymbol{a})}{\|\boldsymbol{h}\|} + \frac{f(\boldsymbol{a}+\boldsymbol{h}) - f(\boldsymbol{a}) - T(\boldsymbol{h})}{\|\boldsymbol{h}\|}$$

$$=: I_1(\boldsymbol{h}) + I_2(\boldsymbol{h}).$$

Therefore,

$$\|S(\boldsymbol{x}) - T(\boldsymbol{x})\| \le \|I_1(\boldsymbol{h}) + I_2(\boldsymbol{h})\|\,\|\boldsymbol{x}\|.$$

We shall take the limit of this inequality as $\alpha \to 0+$. First, notice (since $\boldsymbol{x}$ is fixed) that the left side is independent of $\alpha$. Next, notice that $\boldsymbol{h} \to \boldsymbol{0}$ in $X$ if and only if $\alpha \to 0+$ in $\mathbf{R}$. Since by our assumption, $\|I_j(\boldsymbol{h})\| \to 0$ as $\boldsymbol{h} \to \boldsymbol{0}$ for each $j = 1, 2$, it follows that from the Squeeze Theorem that $\|S(\boldsymbol{x}) - T(\boldsymbol{x})\| \le 0 \cdot \|\boldsymbol{x}\| = 0$. Since the norm is positive definite, we conclude that $S(\boldsymbol{x}) = T(\boldsymbol{x})$ for all $\boldsymbol{x} \in X$. ∎

We shall usually represent the total derivative of an $f$ at $\boldsymbol{a}$ by $Df(\boldsymbol{a})$, and frequently denote $T(\boldsymbol{h})$ and $(Df(\boldsymbol{a}))(\boldsymbol{h})$ by $Df(\boldsymbol{a})(\boldsymbol{h})$.

We have still evaded the questions above: Given a differentiable function $f :$ $\mathbf{R}^n \to \mathbf{R}^m$, what kind of object IS the derivative of $f$? In the case when $m = n = 1$, the derivative is a number (see Remark 8.29). Can Remark 8.29 be extended to $\mathbf{R}^n$? To answer this question, we use the following half page to review some elementary linear algebra.

Recall that an $m \times n$ *matrix* $B$ is a rectangular array which has $m$ rows and $n$ columns:

$$B = [b_{ij}]_{m \times n} := \begin{bmatrix} b_{11} & b_{12} & \cdots & b_{1n} \\ b_{21} & b_{22} & \cdots & b_{2n} \\ \vdots & \vdots & \ddots & \vdots \\ b_{m1} & b_{m2} & \cdots & b_{mn} \end{bmatrix}.$$

For us, the *entries* $b_{ij}$ of a matrix $B$ will usually be numbers or real-valued functions. Let $B = [b_{ij}]_{m \times n}$ and $C = [c_{\nu k}]_{p \times q}$ be such matrices. Recall that the *product* of $B$ and a scalar $\alpha$ is defined by

$$\alpha B = [\alpha b_{ij}]_{m \times n},$$

the *sum* of $B$ and $C$ is defined (when $m = p$ and $n = q$) by

$$B + C = [b_{ij} + c_{ij}]_{m \times n},$$

and the *product* of $B$ and $C$ is defined (when $n = p$) by

$$BC = \left[ \sum_{\nu=1}^{n} b_{i\nu} c_{\nu j} \right]_{m \times q}.$$

Also recall that most of the usual laws of algebra hold for addition and multiplication of matrices (see Theorem C.1 in Appendix C). One glaring exception is that matrix multiplication is not commutative.

We shall identify points $\boldsymbol{x} = (x_1, x_2, \ldots, x_n) \in \mathbf{R}^n$ with $1 \times n$ row matrices or $n \times 1$ column matrices by setting

$$[\boldsymbol{x}] = [\, x_1 \quad x_2 \quad \cdots \quad x_n \,] \quad \text{or} \quad [\boldsymbol{x}] = [\, x_1 \quad x_2 \quad \cdots \quad x_n \,]^T := \begin{bmatrix} x_1 \\ x_2 \\ \vdots \\ x_n \end{bmatrix},$$

where $B^T$ represents the transpose of a matrix $B$ (see Appendix C). Abusing the notation slightly, we shall usually represent the product of an $m \times n$ matrix $B$ and an $n \times 1$ column matrix $[\boldsymbol{x}]$ by $B\boldsymbol{x}$. This notation is justified, as the following result shows, since the function $\boldsymbol{x} \longmapsto [\boldsymbol{x}]$ takes vector addition to matrix addition, the dot product to matrix multiplication, and scalar multiplication to scalar multiplication.

**8.33 Remark.** *If $\boldsymbol{x}, \boldsymbol{y} \in \mathbf{R}^n$ and $\alpha$ is a scalar, then*

$$[\boldsymbol{x} + \boldsymbol{y}] = [\boldsymbol{x}] + [\boldsymbol{y}], \qquad [\boldsymbol{x} \cdot \boldsymbol{y}] = [\boldsymbol{x}][\boldsymbol{y}]^T, \qquad and \quad [\alpha \boldsymbol{x}] = \alpha[\boldsymbol{x}].$$

PROOF. These laws follow immediately from the definitions of addition and multiplication of matrices and vectors. For example,

$$[\boldsymbol{x}+\boldsymbol{y}] = [\, x_1 + y_1 \quad x_2 + y_2 \quad \ldots \quad x_n + y_n \,]$$
$$= [\, x_1 \quad x_2 \quad \ldots \quad x_n \,] + [\, y_1 \quad y_2 \quad \ldots \quad y_n \,] = [\boldsymbol{x}] + [\boldsymbol{y}]. \ \blacksquare$$

The following result shows that each $m \times n$ matrix gives rise to a linear function from $\mathbf{R}^n$ to $\mathbf{R}^m$.

**8.34 Remark.** *Let $B = [b_{ij}]$ be an $m \times n$ matrix whose entries are real numbers and let $\boldsymbol{e}_1, \ldots, \boldsymbol{e}_n$ represent the usual basis of $\mathbf{R}^n$. If*

$$(9) \qquad\qquad\qquad T(\boldsymbol{x}) = B\boldsymbol{x}, \qquad \boldsymbol{x} \in \mathbf{R}^n,$$

*then $T$ is a linear function from $\mathbf{R}^n$ to $\mathbf{R}^m$ and*

$$(10) \qquad\qquad T(\boldsymbol{e}_j) = (b_{1j}, b_{2j}, \ldots, b_{mj}), \qquad j = 1, 2, \ldots, n.$$

PROOF. Notice, first, that (10) holds by (9) and the definition of matrix multiplication. Next, observe by Remark 8.33 and the distributive law of matrix multiplication (see Theorem C.1) that

$$T(\boldsymbol{x}+\boldsymbol{y}) = B[\boldsymbol{x}+\boldsymbol{y}] = B([\boldsymbol{x}] + [\boldsymbol{y}]) = B[\boldsymbol{x}] + B[\boldsymbol{y}] = T(\boldsymbol{x}) + T(\boldsymbol{y})$$

for all $\boldsymbol{x}, \boldsymbol{y} \in \mathbf{R}^n$. Similarly, $T(\alpha\boldsymbol{x}) = B[\alpha\boldsymbol{x}] = B(\alpha[\boldsymbol{x}]) = \alpha B[\boldsymbol{x}] = \alpha T(\boldsymbol{x})$ for all $\boldsymbol{x} \in \mathbf{R}^n$ and $\alpha \in \mathbf{R}$. Thus $T \in \mathcal{L}(\mathbf{R}^n; \mathbf{R}^m)$. $\blacksquare$

Remark 8.34 would barely be worth mentioning were it not the case that ALL linear functions from $\mathbf{R}^n$ to $\mathbf{R}^m$ have this form.

**8.35 THEOREM.** *For each $T \in \mathcal{L}(\mathbf{R}^n; \mathbf{R}^m)$ there is a matrix $B = [b_{ij}]_{m \times n}$ such that (9) holds. Moreover, the matrix $B$ is unique. Specifically, for each fixed $T$ there is only one $B$ which satisfies (9), and the entries of that $B$ are defined by (10).*

PROOF. Uniqueness has been established in Remark 8.34. To prove existence, suppose $T \in \mathcal{L}(\mathbf{R}^n; \mathbf{R}^m)$. Define $B$ by (10). Then

$$T(\boldsymbol{x}) = T\left( \sum_{j=1}^{n} x_j \boldsymbol{e}_j \right)$$

$$= \sum_{j=1}^{n} x_j T(\boldsymbol{e}_j) = \sum_{j=1}^{n} x_j (b_{1j}, b_{2j}, \ldots, b_{mj})$$

$$= \left( \sum_{j=1}^{n} x_j b_{1j}, \sum_{j=1}^{n} x_j b_{2j}, \ldots, \sum_{j=1}^{n} x_j b_{mj} \right) = B\boldsymbol{x}. \ \blacksquare$$

The unique matrix $B$ which satisfies (9) is called the *matrix which represents $T$*.

Combining Theorem 8.35 with Definition 8.31, we see that the total derivative of a differentiable function $f : \mathbf{R}^n \to \mathbf{R}^m$ is an $m \times n$ matrix. In particular, $f$ is differentiable at $\boldsymbol{a}$ if and only if $f$ is defined in some open ball containing $\boldsymbol{a}$ and there is an $m \times n$ matrix $B$ such that

$$(11) \qquad \lim_{\boldsymbol{h} \to 0} \frac{f(\boldsymbol{a} + \boldsymbol{h}) - f(\boldsymbol{a}) - B\boldsymbol{h}}{\|\boldsymbol{h}\|} = \boldsymbol{0}.$$

(This matrix is easy to compute, using partial derivatives of $f$—see Theorem 11.12).

Much of Chapter 11 is devoted to the following question. How many results about derivatives of real functions can be generalized to the vector-valued case? In this regard, the following concept will be used several times.

**8.36 DEFINITION.** Let $X$ and $Y$ be Euclidean spaces, and $T : X \to Y$. The *operator norm* of $T$ is the extended real number

$$\|T\| := \inf\{C > 0 : \|T(\boldsymbol{x})\| \le C\|\boldsymbol{x}\| \quad \text{for all} \quad \boldsymbol{x} \in X\}.$$

One interesting corollary of Theorem 8.35 is that the operator norm of a linear function is always finite.

**8.37 THEOREM.** Let $T \in \mathcal{L}(\mathbf{R}^n; \mathbf{R}^m)$. Then the operator norm of $T$ is finite, and satisfies

$$(12) \qquad \|T(\boldsymbol{x})\| \le \|T\|\, \|\boldsymbol{x}\|$$

for all $\boldsymbol{x} \in \mathbf{R}^n$.

PROOF. Let $B$ be the $m \times n$ matrix which represents $T$, and suppose the rows of $B$ are given by $\boldsymbol{b}_1, \ldots, \boldsymbol{b}_m$. By the definition of matrix multiplication and our identification of $\mathbf{R}^m$ with $m \times 1$ matrices,

$$T(\boldsymbol{x}) = (\boldsymbol{b}_1 \cdot \boldsymbol{x}, \ldots, \boldsymbol{b}_m \cdot \boldsymbol{x}).$$

Hence, by the Cauchy–Schwarz inequality, the square of the Euclidean norm of $T(\boldsymbol{x})$ satisfies

$$\begin{aligned}
\|T(\boldsymbol{x})\|^2 &= (\boldsymbol{b}_1 \cdot \boldsymbol{x})^2 + \cdots + (\boldsymbol{b}_m \cdot \boldsymbol{x})^2 \\
&\le (\|\boldsymbol{b}_1\|\, \|\boldsymbol{x}\|)^2 + \cdots + (\|\boldsymbol{b}_m\|\, \|\boldsymbol{x}\|)^2 \\
&\le m \cdot \max\{\|\boldsymbol{b}_j\|^2 : 1 \le j \le m\}\, \|\boldsymbol{x}\|^2 =: C\, \|\boldsymbol{x}\|^2.
\end{aligned}$$

Thus the set defining $\|T\|$ is nonempty. Since it is bounded below (by 0), it follows from the Completeness Axiom that $\|T\|$ exists and is finite. In particular, there are $C_k > 0$ such that $C_k \downarrow \|T\|$ and $\|T(\boldsymbol{x})\| \le C_k\|\boldsymbol{x}\|$ for all $\boldsymbol{x} \in \mathbf{R}^n$. Taking the limit of this last inequality as $k \to \infty$, we obtain (12). ∎

The operator norm, an analogue of the Cauchy–Schwarz inequality, will be used to estimate differentiable functions. The following proof of the multidimensional analogue of Theorem 4.12 is a typical example.

**8.38 THEOREM.** *If $f$ is differentiable at $\boldsymbol{a} \in \mathbf{R}^n$, then $f$ is continuous at $\boldsymbol{a}$.*

PROOF. Suppose $f$ is differentiable at $\boldsymbol{a}$. By Definition 8.31, choose $\delta > 0$ such that

$$\|\boldsymbol{h}\| < \delta \quad \text{implies} \quad \|f(\boldsymbol{a} + \boldsymbol{h}) - f(\boldsymbol{a}) - Df(\boldsymbol{a})(\boldsymbol{h})\| < \|\boldsymbol{h}\|.$$

By the triangle inequality (see Theorem 8.5) and the definition of the operator norm,

$$\|f(\boldsymbol{a} + \boldsymbol{h}) - f(\boldsymbol{a})\| \leq \|Df(\boldsymbol{a})(\boldsymbol{h})\| + \|\boldsymbol{h}\| \leq (\|Df(\boldsymbol{a})\| + 1)\|\boldsymbol{h}\|.$$

It follows from the Squeeze Theorem that $f(\boldsymbol{a} + \boldsymbol{h}) \to f(\boldsymbol{a})$ as $\boldsymbol{h} \to \boldsymbol{0}$, i.e., $f$ is continuous at $\boldsymbol{a}$. ∎

We close this section with an optional result which shows that under the identification of linear functions with matrices, function composition is taken to matrix multiplication. This, in fact, is why matrix multiplication is defined the way it is.

*\*8.39 Remark. If $T : \mathbf{R}^n \to \mathbf{R}^m$ and $U : \mathbf{R}^m \to \mathbf{R}^p$ are linear, then so is $U \circ T$. In fact, if $B$ is the $m \times n$ matrix which represents $T$, and $C$ is the $p \times m$ matrix which represents $U$, then $CB$ is the matrix which represents $U \circ T$.*

PROOF. Let $\boldsymbol{e}_1, \ldots, \boldsymbol{e}_n$ be the usual basis of $\mathbf{R}^n$, $\boldsymbol{u}_1, \ldots, \boldsymbol{u}_m$ be the usual basis of $\mathbf{R}^m$, and $\boldsymbol{w}_1, \ldots, \boldsymbol{w}_p$ be the usual basis of $\mathbf{R}^p$. If $B = [b_{ij}]_{m \times n}$ represents $T$ and $C = [c_{\nu k}]_{p \times m}$ represents $U$, then by Theorem 8.35,

$$\sum_{k=1}^{m} b_{kj}\boldsymbol{u}_k = (b_{1j}, \ldots, b_{mj}) = T(\boldsymbol{e}_j), \qquad j = 1, 2, \ldots, n,$$

and

$$\sum_{\nu=1}^{p} c_{\nu k}\boldsymbol{w}_\nu = (c_{1k}, \ldots, c_{pk}) = U(\boldsymbol{u}_k), \qquad k = 1, 2, \ldots, m.$$

Hence

$$(U \circ T)(\boldsymbol{e}_j) = U(T(\boldsymbol{e}_j)) = U(\sum_{k=1}^{m} b_{kj}\boldsymbol{u}_k) = \sum_{k=1}^{m} b_{kj}U(\boldsymbol{u}_k)$$

$$= \sum_{k=1}^{m}\sum_{\nu=1}^{p} b_{kj}c_{\nu k}\boldsymbol{w}_\nu = (\sum_{k=1}^{m} b_{kj}c_{1k}, \ldots, \sum_{k=1}^{m} b_{kj}c_{pk})$$

for each $1 \leq j \leq n$. Since this last vector is the $j$th column of the matrix $CB$, it follows that $CB$ is the matrix which represents $U \circ T$. ∎

### EXERCISES

**1.** Let $f(x, y) = (x^2, y^2)$. Using Definition 8.31, prove that $f$ is differentiable on

$\mathbf{R}^2$ and its total derivative is given by

$$Df(x,y) = \begin{bmatrix} 2x & 0 \\ 0 & 2y \end{bmatrix}.$$

**2.** Let $f(x,y) = (xy, x+y, x^2 - y^2)$. Using Definition 8.31, prove that $f$ is differentiable on $\mathbf{R}^2$ and its total derivative is given by

$$Df(x,y) = \begin{bmatrix} y & x \\ 1 & 1 \\ 2x & -2y \end{bmatrix}.$$

**3.** Let $\phi(r,\theta) = (r\cos\theta, r\sin\theta)$. Using Definition 8.31, prove that $\phi$ is differentiable at $(0,\theta)$ for all $\theta \in \mathbf{R}$ and its total derivative is given by

$$D\phi(0,\theta) = \begin{bmatrix} \cos\theta & 0 \\ \sin\theta & 0 \end{bmatrix}.$$

**4.** Let $f(t) = (f_1(t), \ldots, f_m(t))$, where each $f_j$ is a real function differentiable on $\mathbf{R}$. Prove that $f$ is differentiable on $\mathbf{R}$ and its total derivative is given by

$$Df(t) = \begin{bmatrix} f_1'(t) \\ \vdots \\ f_n'(t) \end{bmatrix}.$$

**5.** Prove that every linear function from $\mathbf{R}^n$ to $\mathbf{R}^m$ is continuous.

**6.** Suppose $X$ and $Y$ are Euclidean spaces, $\boldsymbol{a} \in X$, and $f, g$ are functions from some subset of $X$ to $Y$.

a) Prove that if $f$ and $g$ are differentiable at $\boldsymbol{a}$, then $f + g$ is differentiable at $\boldsymbol{a}$ with

$$D(f+g)(\boldsymbol{a}) = Df(\boldsymbol{a}) + Dg(\boldsymbol{a}).$$

b) Prove that if $f$ is differentiable at $\boldsymbol{a}$ and $\alpha \in \mathbf{R}$, $\alpha f$ is differentiable at $\boldsymbol{a}$ with

$$D(\alpha f)(\boldsymbol{a}) = \alpha Df(\boldsymbol{a}).$$

**7.** [RECIPROCAL RULE] Suppose $f : \mathbf{R}^n \to \mathbf{R}$ is differentiable at $\boldsymbol{a}$ and $f(\boldsymbol{a}) \neq 0$.

a) Show that for $\|\boldsymbol{h}\|$ sufficiently small, $f(\boldsymbol{a}+\boldsymbol{h}) \neq 0$.

b) Prove that $Df(\boldsymbol{a})(\boldsymbol{h})/\|\boldsymbol{h}\|$ is bounded for all $\boldsymbol{h} \in \mathbf{R}^n \setminus \{\boldsymbol{0}\}$.

c) If $T := -Df(\boldsymbol{a})/f^2(\boldsymbol{a})$, show that

$$\frac{1}{f(\boldsymbol{a}+\boldsymbol{h})} - \frac{1}{f(\boldsymbol{a})} - T(\boldsymbol{h}) = \frac{f(\boldsymbol{a}) - f(\boldsymbol{a}+\boldsymbol{h}) + Df(\boldsymbol{a})(\boldsymbol{h})}{f(\boldsymbol{a})f(\boldsymbol{a}+\boldsymbol{h})}$$

$$+ \frac{(f(\boldsymbol{a}+\boldsymbol{h}) - f(\boldsymbol{a}))Df(\boldsymbol{a})(\boldsymbol{h})}{f^2(\boldsymbol{a})f(\boldsymbol{a}+\boldsymbol{h})}$$

for $\|\boldsymbol{h}\|$ sufficiently small.

d) Prove that $1/f(\boldsymbol{x})$ is differentiable at $\boldsymbol{x} = \boldsymbol{a}$ and

$$D\left(\frac{1}{f}\right)(\boldsymbol{a}) = -\frac{Df(\boldsymbol{a})}{f^2(\boldsymbol{a})}.$$

**8**. **This exercise is used several times in Chapter 11.**
Suppose $X$ and $Y$ are Euclidean spaces. Prove that if $T \in \mathcal{L}(X;Y)$, then $T$ is differentiable everywhere on $X$ with

$$DT(\boldsymbol{a}) = T \quad \text{for all} \quad \boldsymbol{a} \in X.$$

**9**. [ROTATIONS IN $\mathbf{R}^2$]. **This exercise is used in Section $^e$15.1.** Let

$$B = \begin{bmatrix} \cos\theta & -\sin\theta \\ \sin\theta & \cos\theta \end{bmatrix}$$

for some $\theta \in \mathbf{R}$.

a) Prove that $\|B(x,y)\| = \|(x,y)\|$ for all $(x,y) \in \mathbf{R}^2$.

b) Let $(x,y) \in \mathbf{R}^2$ be a nonzero vector and $\varphi$ represent the angle between $B(x,y)$ and $(x,y)$. Prove $\cos\varphi = \cos\theta$. Thus, show that $B$ rotates $\mathbf{R}^2$ through an angle $\theta$. (When $\theta > 0$, we shall call $B$ *counterclockwise rotation* about the origin through the angle $\theta$.)

# Chapter 9

## Topology of Euclidean Spaces

Topology, a study of geometric objects which emphasizes how they are put together over their exact shape and proportion, is based on the fundamental concepts of open and closed sets. In this chapter, we introduce these concepts in $\mathbf{R}^n$, and explore how they can be used to characterize limits and continuity without using distance explicitly. This additional step in abstraction will yield powerful benefits, as we shall see in Section 9.5 and in Chapter 11 when we begin to study the calculus of functions of several variables.

*If you want a more abstract introduction to the topology of Euclidean spaces, skip this chapter and begin Chapter 10 now.*

### 9.1 INTERIOR, CLOSURE, AND BOUNDARY

Observe that each element of an open interval $I$ lies "inside" $I$, i.e., is surrounded by other points in $I$. On the other hand, although closed intervals do NOT satisfy this property, their complements do. Accordingly, we make the following definition.

**9.1 DEFINITION.** Let $X$ be a Euclidean space.
   i) A set $V$ in $X$ is said to be *open* if for every $\boldsymbol{a} \in V$ there is an $\varepsilon > 0$ such that $B_\varepsilon(\boldsymbol{a}) := \{\boldsymbol{x} \in X : \|\boldsymbol{x} - \boldsymbol{a}\| < \varepsilon\} \subseteq V$.
   ii) A set $E$ in $X$ is said to be *closed* if $E^c := X \setminus E$ is open.

The following result shows that every "open" ball is open. (Closed balls are also closed—see Exercise 3 below.)

253

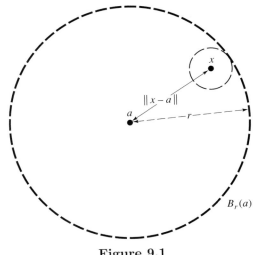

**Figure 9.1**

**9.2 Remark.** *For every $\boldsymbol{x} \in B_r(\boldsymbol{a})$ there is an $\varepsilon > 0$ such that $B_\varepsilon(\boldsymbol{x}) \subseteq B_r(\boldsymbol{a})$.*

PROOF. Let $\boldsymbol{x} \in B_r(\boldsymbol{a})$. Using Figure 9.1 for guidance, we set $\varepsilon = r - \|\boldsymbol{x} - \boldsymbol{a}\|$. If $\boldsymbol{y} \in B_\varepsilon(\boldsymbol{x})$, then by the triangle inequality, assumption, and the choice of $\varepsilon$,

$$\|\boldsymbol{y} - \boldsymbol{a}\| \le \|\boldsymbol{y} - \boldsymbol{x}\| + \|\boldsymbol{x} - \boldsymbol{a}\| < \varepsilon + \|\boldsymbol{x} - \boldsymbol{a}\| = r.$$

Thus by definition, $\boldsymbol{y} \in B_r(\boldsymbol{a})$. In particular, $B_\varepsilon(\boldsymbol{x}) \subseteq B_r(\boldsymbol{a})$. ∎

(This proof illustrates a valuable technique. Drawing diagrams in $\mathbf{R}^2$ sometimes leads to a proof valid for all Euclidean spaces.)

Here are more examples of open sets and closed sets.

**9.3 Remark.** *If $\boldsymbol{a} \in \mathbf{R}^n$, then $\mathbf{R}^n \setminus \{\boldsymbol{a}\}$ is open and $\{\boldsymbol{a}\}$ is closed.*

PROOF. By Definition 9.1, it suffices to prove that the complement of every *singleton* $E := \{\boldsymbol{a}\}$ is open. Let $\boldsymbol{x} \in E^c$ and set $\varepsilon = \|\boldsymbol{x} - \boldsymbol{a}\|$. Then by definition, $\boldsymbol{a} \notin B_\varepsilon(\boldsymbol{x})$, so $B_\varepsilon(\boldsymbol{x}) \subseteq E^c$. Therefore, $E^c$ is open by Definition 9.1. ∎

Students sometimes mistakenly believe that every set is either open or closed. Some sets are neither open nor closed (like the interval $[0, 1)$). And, as the following result shows, every Euclidean space contains two special sets which are both open and closed. (We shall see below that these are the only subsets of $\mathbf{R}^n$ which are simultaneously open and closed in $\mathbf{R}^n$.)

**9.4 Remark.** *For each $n \in \mathbf{N}$, the empty set $\emptyset$ and the whole space $\mathbf{R}^n$ are both open and closed.*

PROOF. Since $\mathbf{R}^n = \emptyset^c$ and $\emptyset = (\mathbf{R}^n)^c$, it suffices by Definition 9.1 to prove that $\emptyset$ and $\mathbf{R}^n$ are both open. Because the empty set contains no points, "every" point $\boldsymbol{x} \in \emptyset$ satisfies $B_\varepsilon(\boldsymbol{x}) \subseteq \emptyset$. (This is called the *vacuous implication*.) Therefore, $\emptyset$ is open. On the other hand, since $B_\varepsilon(\boldsymbol{x}) \subseteq \mathbf{R}^n$ for all $x \in \mathbf{R}^n$ and all $\varepsilon > 0$, it is clear that $\mathbf{R}^n$ is open. ∎

The following result is a first step toward developing a "distanceless" theory of convergence.

**9.5 Remark.** *Let $\boldsymbol{x}_k \in \mathbf{R}^n$. Then $\boldsymbol{x}_k \to \boldsymbol{a}$ as $k \to \infty$ if and only if for every open set $V$ which contains $\boldsymbol{a}$ there is an $N \in \mathbf{N}$ such that $k \geq N$ implies $\boldsymbol{x}_k \in V$.*

PROOF. Suppose $\boldsymbol{x}_k \to \boldsymbol{a}$ and let $V$ be an open set which contains $\boldsymbol{a}$. By Definition 9.1, there is an $\varepsilon > 0$ such that $B_\varepsilon(\boldsymbol{a}) \subseteq V$. Given this $\varepsilon$, use Definition 8.11 to choose an $N \in \mathbf{N}$ such that $k \geq N$ implies $\boldsymbol{x}_k \in B_\varepsilon(\boldsymbol{a})$. By the choice of $\varepsilon$, $\boldsymbol{x}_k \in V$ for all $k \geq N$.

Conversely, let $\varepsilon > 0$ and set $V = B_\varepsilon(\boldsymbol{a})$. Then $V$ is an open set which contains $\boldsymbol{a}$, hence by hypothesis, there is an $N \in \mathbf{N}$ such that $k \geq N$ implies $\boldsymbol{x}_k \in V$. In particular, $\|\boldsymbol{x}_k - \boldsymbol{a}\| < \varepsilon$ for all $k \geq N$. ∎

The next result, which we shall use many times, shows how to use convergent sequences to characterize closed sets.

**9.6 THEOREM.** *Let $E \subseteq \mathbf{R}^n$. Then $E$ is closed if and only if the limit of every convergent sequence $\boldsymbol{x}_k \in E$ satisfies*

$$\lim_{k \to \infty} \boldsymbol{x}_k \in E.$$

PROOF. The theorem is vacuously satisfied if $E$ is the empty set.

Suppose that $E \neq \emptyset$ is closed but some sequence $\boldsymbol{x}_k \in E$ converges to a point $\boldsymbol{x} \in E^c$. Since $E$ is closed, $E^c$ is open. Thus, by Remark 9.5, there is an $N \in \mathbf{N}$ such that $k \geq N$ implies $\boldsymbol{x}_k \in E^c$, a contradiction.

Conversely, suppose $E$ is a nonempty set such that every convergent sequence in $E$ has its limit in $E$. If $E$ is not closed, then by Remark 9.4, $E \neq \mathbf{R}^n$ and by definition, $E^c$ is nonempty and not open. Thus, there is at least one point $\boldsymbol{x} \in E^c$ such that no ball $B_r(\boldsymbol{x})$ is contained in $E^c$. Let $\boldsymbol{x}_k \in B_{1/k}(\boldsymbol{x}) \cap E$ for $k = 1, 2, \ldots$ Then $\boldsymbol{x}_k \in E$ and $\|\boldsymbol{x}_k - \boldsymbol{x}\| < 1/k$ for all $k \in \mathbf{N}$. Now by the Squeeze Theorem, $\|\boldsymbol{x}_k - \boldsymbol{x}\| \to 0$, i.e., $\boldsymbol{x}_k \to \boldsymbol{x}$ as $k \to \infty$. Thus, by hypothesis, $\boldsymbol{x} \in E$, a contradiction. ∎

It is important to recognize that open sets and closed sets behave very differently with respect to unions and intersections.

**9.7 THEOREM.** *Let $X$ be a Euclidean space.*

i) *If $\{V_\alpha\}_{\alpha \in A}$ is any collection of open subsets of $X$, then*

$$\bigcup_{\alpha \in A} V_\alpha$$

*is open.*

ii) *If $\{V_k : k = 1, 2, \ldots, n\}$ is a finite collection of open subsets of $X$, then*

$$\bigcap_{k=1}^n V_k := \bigcap_{k \in \{1,2,\ldots,n\}} V_k$$

*is open.*

iii) *If $\{E_\alpha\}_{\alpha \in A}$ is any collection of closed subsets of $X$, then*

$$\bigcap_{\alpha \in A} E_\alpha$$

*is closed.*

iv) *If $\{E_k : k = 1, 2, \ldots, n\}$ is a finite collection of closed subsets of $X$, then*

$$\bigcup_{k=1}^{n} E_k := \bigcup_{k \in \{1,2,\ldots,n\}} E_k$$

*is closed.*

v) *If $V$ is open and $E$ is closed, then $V \setminus E$ is open and $E \setminus V$ is closed.*

PROOF. i) Let $\boldsymbol{x} \in \bigcup_{\alpha \in A} V_\alpha$. Then $\boldsymbol{x} \in V_\alpha$ for some $\alpha \in A$. Since $V_\alpha$ is open, it follows that there is an $r > 0$ such that $B_r(\boldsymbol{x}) \subseteq V_\alpha$. Thus $B_r(\boldsymbol{x}) \subseteq \bigcup_{\alpha \in A} V_\alpha$; i.e., this union is open.

ii) Let $\boldsymbol{x} \in \bigcap_{k=1}^{n} V_k$. Then $\boldsymbol{x} \in V_k$ for $k = 1, 2, \ldots, n$. Since each $V_k$ is open, it follows that there are numbers $r_k > 0$ such that $B_{r_k}(\boldsymbol{x}) \subseteq V_k$. Let $r = \min\{r_1, \ldots, r_n\}$. Then $r > 0$ and $B_r(\boldsymbol{x}) \subseteq V_k$ for all $k = 1, 2, \ldots, n$; i.e., $B_r(\boldsymbol{x}) \subseteq \bigcap_{k=1}^{n} V_k$. Hence, this intersection is open.

iii) By DeMorgan's Law (Theorem 1.41) and part i),

$$\left(\bigcap_{\alpha \in A} E_\alpha\right)^c = \bigcup_{\alpha \in A} E_\alpha^c$$

is open, so $\bigcap_{\alpha \in A} E_\alpha$ is closed.

iv) By DeMorgan's Law and part ii),

$$\left(\bigcup_{k=1}^{n} E_k\right)^c = \bigcap_{k=1}^{n} E_k^c$$

is open, so $\bigcup_{k=1}^{n} E_k$ is closed.

v) Since $V \setminus E = V \cap E^c$ and $E \setminus V = E \cap V^c$, the former is open by part ii), and the latter is closed by part iii). ∎

The finiteness hypothesis in Theorem 9.7 is crucial, even for the case $X = \mathbf{R}$.

**9.8 Remark.** *Statements ii) and iv) of Theorem 9.7 are false if arbitrary collections are used in place of finite collections.*

PROOF. In the Euclidean space $X = \mathbf{R}$,

$$\bigcap_{k \in \mathbf{N}} \left(-\frac{1}{k}, \frac{1}{k}\right) = \{0\}$$

is closed and

$$\bigcup_{k \in \mathbf{N}} \left[ \frac{1}{k+1}, \frac{k}{k+1} \right] = (0, 1)$$

is open. ∎

Theorem 9.7 has many applications. Our first application is that every set contains a largest open set, and is contained in a smallest closed set. To facilitate our discussion, we introduce the following topological operations.

**9.9 DEFINITION.** Let $E$ be a subset of a Euclidean space $X$.

i) The *interior* of $E$ is the set

$$E^o := \bigcup \{ V : V \subseteq E \text{ and } V \text{ is open in } X \}.$$

ii) The *closure* of $E$ is the set

$$\overline{E} := \bigcap \{ B : B \supseteq E \text{ and } B \text{ is closed in } X \}.$$

Notice that every set $E$ contains the open set $\emptyset$ and is contained in the closed set $X$. Hence, the sets $E^o$ and $\overline{E}$ are well–defined. Also notice that by Theorem 9.7, the interior of a set is always open and the closure of a set is always closed.

The following result shows that $E^o$ is the largest open set contained in $E$, and $\overline{E}$ is the smallest closed set which contains $E$.

**9.10 THEOREM.** *Let $E \subseteq \mathbf{R}^n$. Then*

i) $E^o \subseteq E \subseteq \overline{E}$,

ii) *if $V$ is open and $V \subseteq E$ then $V \subseteq E^o$, and*

iii) *if $C$ is closed and $C \supseteq E$ then $C \supseteq \overline{E}$.*

PROOF. Since every open set $V$ in the union defining $E^o$ is a subset of $E$, it is clear that the union of these $V$'s is a subset of $E$. Thus $E^o \subseteq E$. A similar argument establishes $E \subseteq \overline{E}$. This proves i).

By Definition 9.9, if $V$ is an open subset of $E$, then $V \subseteq E^o$ and if $C$ is a closed set containing $E$, then $\overline{E} \subseteq C$. This proves ii) and iii). ∎

In particular, the interior of a bounded interval with endpoints $a$ and $b$ is $(a, b)$, and its closure is $[a, b]$. In fact, it is evident by parts ii) and iii) that $E = E^o$ if and only if $E$ is open, and $E = \overline{E}$ if and only if $E$ is closed. We shall use this observation many times below.

Let us examine these concepts in the concrete setting $\mathbf{R}^2$.

**9.11 Example.** Find the interior and closure of the set $E = \{(x, y) : -1 \leq x \leq 1 \text{ and } -|x| < y < |x|\}$.

SOLUTION. Graph $y = |x|$ and $x = \pm 1$, and observe that $E$ is a bow-tie shaped region with "solid" vertical edges (see Figure 8.6). Now, by Definition 9.1, any open set in $\mathbf{R}^2$ must contain a disk around each of its points. Since $E^o$ is the largest open set inside $E$, it is clear that

$$E^o = \{(x, y) : -1 < x < 1 \text{ and } -|x| < y < |x|\}.$$

Similarly,

$$\overline{E} = \{(x, y) : -1 \le x \le 1 \text{ and } -|x| \le y \le |x|\}. \quad \blacksquare$$

This illustrates the fact that the interior of a nice enough set $E$ in $\mathbf{R}^2$ can be obtained by removing all its "edges," and the closure of $E$ by adding all its "edges." Here is another example.

**9.12 Example.** Find the interior and closure of the set $E = B_1(-2, 0) \cup B_1(2, 0) \cup \{(x, 0) : -1 \le x \le 1\}$.

SOLUTION. Draw a graph of this region. It turns out to be "dumbbell shaped": two open disks joined by a straight line. Thus $E^o = B_1(-2, 0) \cup B_1(2, 0)$, and

$$\overline{E} = \overline{B_1(-2, 0)} \cup \overline{B_1(2, 0)} \cup \{(x, 0) : -1 \le x \le 1\}. \quad \blacksquare$$

One of the most important results from one-dimensional calculus is the Fundamental Theorem of Calculus. It states that the behavior of a derivative $f'$ on an interval $[a, b]$ is completely determined by the values of $f$ at the endpoints of $[a, b]$. What shall we use for "endpoints" of an arbitrary set in $\mathbf{R}^n$? Notice that the endpoints $a, b$ are the only points which lie near both $[a, b]$ and the complement of $[a, b]$. Using this as a cue, we introduce the following concept.

**9.13 DEFINITION.** Let $E \subseteq \mathbf{R}^n$. The *boundary* of $E$ is the set

$$\partial E := \{\boldsymbol{x} \in \mathbf{R}^n : \text{for all } r > 0, \quad B_r(\boldsymbol{x}) \cap E \neq \emptyset \text{ and } B_r(\boldsymbol{x}) \cap E^c \neq \emptyset\}.$$

[We will refer to the last two conditions in the definition of $\partial E$ by saying $B_r(\boldsymbol{x})$ *intersects $E$ and $E^c$.*]

**9.14 Example.** Describe the boundary of the set

$$E = \{(x, y) : x^2 + y^2 \le 9 \text{ and } (x - 1)(y + 2) > 0\}.$$

SOLUTION. Graph the relations $x^2 + y^2 = 9$ and $(x - 1)(y + 2) = 0$ to see that $E$ is a region with a solid curved edges and dotted straight edges (see Figure 8.5). By definition, then, the boundary of $E$ is the union of these curved and straight edges (all made solid). Rather than describing $\partial E$ analytically (which would involve solving for the intersection points of the straight lines $x = 1$, $y = -2$, and the circle $x^2 + y^2 = 9$), it is easier to describe $\partial E$ by using set algebra.

$$\partial E = \{(x, y) : x^2 + y^2 \le 9 \text{ and } (x - 1)(y + 2) \ge 0\}$$
$$\setminus \{(x, y) : x^2 + y^2 < 9 \text{ and } (x - 1)(y + 2) > 0\}. \quad \blacksquare$$

It turns out that set algebra can be used to describe the boundary of any set.

**9.15 THEOREM.** *Let $E \subseteq \mathbf{R}^n$. Then $\partial E = \overline{E} \setminus E^o$.*

PROOF. By Definition 9.13, it suffices to show that

(1) $\qquad \boldsymbol{x} \in \overline{E}$ if and only if $B_r(\boldsymbol{x}) \cap E \neq \emptyset$ for all $r > 0$, and

(2) $\qquad \boldsymbol{x} \notin E^o$ if and only if $B_r(\boldsymbol{x}) \cap E^c \neq \emptyset$ for all $r > 0$.

We will provide the details for (1) and leave the proof of (2) as an exercise. Suppose $\boldsymbol{x} \in \overline{E}$ but $B_{r_0}(\boldsymbol{x}) \cap E = \emptyset$ for some $r_0 > 0$. Then $(B_{r_0}(\boldsymbol{x}))^c$ is a closed set which contains $E$; hence, by Theorem 9.10iii), $\overline{E} \subseteq (B_{r_0}(\boldsymbol{x}))^c$. It follows that $\overline{E} \cap B_{r_0}(\boldsymbol{x}) = \emptyset$, e.g., $\boldsymbol{x} \notin \overline{E}$, a contradiction.

Conversely, suppose $\boldsymbol{x} \notin \overline{E}$. Since $(\overline{E})^c$ is open, there is an $r_0 > 0$ such that $B_{r_0}(\boldsymbol{x}) \subseteq (\overline{E})^c$. In particular, $\emptyset = B_{r_0}(\boldsymbol{x}) \cap \overline{E} \supseteq B_{r_0}(\boldsymbol{x}) \cap E$ for some $r_0 > 0$. ∎

We have introduced topological operations (interior, closure, and boundary). The following result answers the question: How do these operations interact with the set operations (union and intersection)?

**9.16 THEOREM.** *Let $A, B \subseteq \mathbf{R}^n$. Then*

i) $\qquad (A \cup B)^o \supseteq A^o \cup B^o, \quad (A \cap B)^o = A^o \cap B^o,$

ii) $\qquad \overline{A \cup B} = \overline{A} \cup \overline{B}, \qquad \overline{A \cap B} \subseteq \overline{A} \cap \overline{B},$

iii) $\qquad \partial(A \cup B) \subseteq \partial A \cup \partial B, \quad \text{and} \quad \partial(A \cap B) \subseteq \partial A \cup \partial B.$

PROOF. i) Since the union of two open sets is open, $A^o \cup B^o$ is an open subset of $A \cup B$. Hence, by Theorem 9.10ii), $A^o \cup B^o \subseteq (A \cup B)^o$.

Similarly, $(A \cap B)^o \supseteq A^o \cap B^o$. On the other hand, if $V \subset A \cap B$, then $V \subset A$ and $V \subset B$. Thus $(A \cap B)^o \subseteq A^o \cap B^o$.

ii) Since $\overline{A} \cup \overline{B}$ is closed and contains $A \cup B$, it is clear that, by Theorem 9.10iii), $\overline{A \cup B} \subseteq \overline{A} \cup \overline{B}$. Similarly, $\overline{A \cap B} \subseteq \overline{A} \cap \overline{B}$. To prove the reverse inequality for union, suppose $\boldsymbol{x} \notin \overline{A \cup B}$. Then, by Definition 9.9, there is a closed set $E$ which contains $A \cup B$ such that $\boldsymbol{x} \notin E$. Since $E$ contains both $A$ and $B$, it follows that $\boldsymbol{x} \notin \overline{A}$ and $\boldsymbol{x} \notin \overline{B}$. This proves part ii).

iii) Let $\boldsymbol{x} \in \partial(A \cup B)$; i.e., suppose $B_r(\boldsymbol{x})$ intersects $A \cup B$ and $(A \cup B)^c$ for all $r > 0$. Since $(A \cup B)^c = A^c \cap B^c$, it follows that $B_r(\boldsymbol{x})$ intersects both $A^c$ and $B^c$ for all $r > 0$. Thus $B_r(\boldsymbol{x})$ intersects $A$ and $A^c$ for all $r > 0$, or $B_r(\boldsymbol{x})$ intersects $B$ and $B^c$ for all $r > 0$, i.e., $\boldsymbol{x} \in \partial A \cup \partial B$. This proves the first set inequality in part iii). A similar argument establishes the second inequality in part iii). ∎

The second inequality in part iii) can be improved (see Exercise 10 below).

## EXERCISES

**1.** Find the interior, closure, and boundary of each of the following subsets of $\mathbf{R}$.

    a) $[a, b)$ where $a < b$.

b) $E = \{1/n : n \in \mathbf{N}\}$.

c) $E = \bigcup_{n=1}^{\infty} \left( \dfrac{1}{n+1}, \dfrac{1}{n} \right)$.

d) $E = \bigcup_{n=1}^{\infty}(-n, n)$.

2. Identify which of the following sets are open, which are closed, and which are neither. Find $E^o$, $\overline{E}$, and $\partial E$, and sketch $E$ in each case.

a) $E = \{(x, y) : x^2 + 4y^2 \le 1\}$.

b) $E = \{(x, y) : x^2 - 2x + y^2 = 0\} \cup \{(x, 0) : x \in [2, 3]\}$.

c) $E = \{(x, y) : y \ge x^2,\ 0 \le y < 1\}$.

d) $E = \{(x, y) : x^2 - y^2 < 1,\ -1 < y < 1\}$.

3. Let $X$ be a Euclidean space, $\boldsymbol{a} \in X$, $s < r$,

$$V = \{\boldsymbol{x} \in X : s < \|\boldsymbol{x} - \boldsymbol{a}\| < r\}, \quad \text{and} \quad E = \{x \in X : s \le \|\boldsymbol{x} - \boldsymbol{a}\| \le r\}.$$

Prove that $V$ is open and $E$ is closed.

4. Suppose $A \subseteq B \subseteq \mathbf{R}^n$. Prove that $\overline{A} \subseteq \overline{B}$ and $A^o \subseteq B^o$.

5. **This exercise is used in Section 9.3.** Show that if $E$ is closed in $\mathbf{R}^n$ and $\boldsymbol{a} \notin E$, then

$$\inf_{\boldsymbol{x} \in E} \|\boldsymbol{x} - \boldsymbol{a}\| > 0.$$

6. Prove (2).

7. Show that Theorem 9.16 is best possible in the following sense.

a) There exist sets $A, B$ in $\mathbf{R}$ such that $(A \cup B)^o \ne A^o \cup B^o$.

b) There exist sets $A, B$ in $\mathbf{R}$ such that $\overline{A \cap B} \ne \overline{A} \cap \overline{B}$.

c) There exist sets $A, B$ in $\mathbf{R}$ such that $\partial(A \cup B) \ne \partial A \cup \partial B$ and $\partial(A \cap B) \ne \partial A \cup \partial B$.

8. Let $f : \mathbf{R} \to \mathbf{R}$. Prove that $f$ is continuous on $\mathbf{R}$ if and only if $f^{-1}(I)$ is open in $\mathbf{R}$ for every open interval $I$.

9. Let $V$ be a subset of a Euclidean space $X$.

a) Prove that $V$ is open if and only if there is a collection of open balls $\{B_\alpha : \alpha \in A\}$ such that

$$V = \bigcup_{\alpha \in A} B_\alpha.$$

b) What happens to this result when "open" is replaced by "closed"?

10. Let $A$ and $B$ be subsets of $\mathbf{R}^n$.

a) Show that $A^c \cup (\partial B)^c \subseteq \partial A$.

b) Show that if $\boldsymbol{x} \in \partial(A \cap B)$ and $\boldsymbol{x} \notin (A \cap \partial B) \cup (B \cap \partial A)$, then $\boldsymbol{x} \in \partial A \cap \partial B$.

c) Prove that $\partial(A \cap B) \subseteq (A \cap \partial B) \cup (B \cap \partial A) \cup (\partial A \cap \partial B)$.

d) Show that even in $\mathbf{R}$, there exist sets $A$ and $B$ such that $\partial(A \cap B) \ne (A \cap \partial B) \cup (B \cap \partial A) \cup (\partial A \cap \partial B)$.

## 9.2 COMPACT SETS

In Chapter 3, we proved the Extreme Value Theorem for real functions on closed bounded intervals. In this section we introduce a concept (compact sets) which will allow us to extend the Extreme Value Theorem to functions of several variables. Compact sets, which play the same role that "closed, bounded intervals" played for **R**, will also provide a powerful tool for extending local results to global ones (see especially the proofs of Remark 9.20 and Theorems 9.32 below, and Theorem 12.46).

Since compact sets are defined by how they can be "covered" by a collection of open sets, we begin by introducing the following terminology.

**9.17 DEFINITION.** Let $\mathcal{V} = \{V_\alpha\}_{\alpha \in A}$ be a collection of subsets of a Euclidean space $X$, and suppose $E \subseteq X$.

i) $\mathcal{V}$ is said to *cover* $E$ (or be a *covering* of $E$) if

$$E \subseteq \bigcup_{\alpha \in A} V_\alpha.$$

ii) $\mathcal{V}$ is said to be an *open covering* of $E$ if $\mathcal{V}$ covers $E$ and each $V_\alpha$ is open.

iii) Let $\mathcal{V}$ be a covering of $E$. $\mathcal{V}$ is said to have a *finite* (respectively, *countable*) *subcovering* if there is a finite (respectively, countable) subset $A_0$ of $A$ such that $\{V_\alpha\}_{\alpha \in A_0}$ covers $E$.

Notice that the collections of open intervals

$$\left\{ \left( \frac{1}{k+1}, \frac{k}{k+1} \right) \right\}_{k \in \mathbf{N}} \quad \text{and} \quad \left\{ \left( -\frac{1}{k}, \frac{k+1}{k} \right) \right\}_{k \in \mathbf{N}}$$

are open coverings of the interval $(0, 1)$. The first covering of $(0, 1)$ has no finite subcovering, but any member of the second covering covers $(0, 1)$. Thus, an open covering of an arbitrary set might not have a finite subcovering. Sets which satisfy this special property are important enough to be given a name.

**9.18 DEFINITION.** A subset $H$ of a Euclidean space $X$ is said to be *compact* (in $X$) if every open covering of $H$ has a finite subcovering.

To get a feeling for what this definition means, we make some elementary observations concerning compact sets in general.

**9.19 Remark.** *The empty set and all finite subsets of a Euclidean space are compact.*

PROOF. These statements follow immediately from Definition 9.18. The empty set needs no set to cover it, and any finite set $H$ can be covered by finitely many sets, one set for each element in $H$. ∎

Since the empty set and finite sets are also closed, it is natural to ask whether there is a relationship between compact sets and closed sets in general. The following three results address this question.

**9.20 Remark.** *A compact set is always closed.*

PROOF. Suppose $H$ is compact but not closed. Then $H$ is nonempty and (by Theorem 9.6) there is a convergent sequence $\boldsymbol{x}_k \in H$ whose limit $\boldsymbol{x}$ does not belong to $H$. For each $\boldsymbol{y} \in H$, set $r(\boldsymbol{y}) := \|\boldsymbol{x} - \boldsymbol{y}\|/2$. Since $\boldsymbol{x}$ does not belong to $H$, $r(\boldsymbol{y}) > 0$. Thus each $B_{r(\boldsymbol{y})}(\boldsymbol{y})$ is open and contains $\boldsymbol{y}$; i.e., $\{B_{r(\boldsymbol{y})}(\boldsymbol{y}) : \boldsymbol{y} \in H\}$ is an open covering of $H$. Since $H$ is compact, we can choose points $\boldsymbol{y}_j$ and radii $r_j := r(\boldsymbol{y}_j)$ such that $\{B_{r_j}(\boldsymbol{y}_j) : j = 1, 2, \ldots, N\}$ covers $H$.

Set $r := \min\{r_1, \ldots, r_N\}$. (This is a finite set of positive numbers, so $r$ is also positive.) Since $\boldsymbol{x}_k \to \boldsymbol{x}$ as $k \to \infty$, $\boldsymbol{x}_k \in B_r(\boldsymbol{x})$ for large $k$. But $\boldsymbol{x}_k \in B_r(\boldsymbol{x}) \cap H$ implies $\boldsymbol{x}_k \in B_{r_j}(\boldsymbol{y}_j)$ for some $j \in \mathbf{N}$. Therefore, it follows from the choices of $r_j$ and $r$, and from the triangle inequality, that

$$r_j \geq \|\boldsymbol{x}_k - \boldsymbol{y}_j\| \geq \|\boldsymbol{x} - \boldsymbol{y}_j\| - \|\boldsymbol{x}_k - \boldsymbol{x}\|$$
$$= 2r_j - \|\boldsymbol{x}_k - \boldsymbol{x}\| > 2r_j - r \geq 2r_j - r_j = r_j,$$

a contradiction. ∎

Since $\{(n-1, n+1) : n \in \mathbf{N}\}$ is an open covering of the closed set $E := [1, \infty)$, the converse of Theorem 9.20 is false. The following result shows this is not the case if $E$ is a subset of some compact set (see also Theorem 9.24 below).

**9.21 Remark.** *A closed subset of a compact set is compact.*

PROOF. Let $E$ be a closed subset of $H$, where $H$ is compact in some Euclidean space $X$, and suppose $\mathcal{V} = \{V_\alpha\}_{\alpha \in A}$ is an open covering of $E$. Now $E^c = X \setminus E$ is open. Thus $\mathcal{V} \cup \{E^c\}$ is an open covering of $H$. Since $H$ is compact, there is a finite set $A_0 \subseteq A$ such that

$$H \subseteq E^c \cup \left( \bigcup_{\alpha \in A_0} V_\alpha \right).$$

But $E \cap E^c = \emptyset$. Therefore, $E$ is covered by $\{V_\alpha\}_{\alpha \in A_0}$. ∎

Remark 9.20 can be improved.

**9.22 THEOREM.** *Let $H$ be a subset of a Euclidean space $X$. If $H$ is compact, then $H$ is closed and bounded.*

PROOF. Suppose $H$ is compact. By Remark 9.20, $H$ is closed. It is also bounded. Indeed, since $\{B_n(\mathbf{0}) : n \in \mathbf{N}\}$ covers $H$ and $H$ is compact, we can choose an $N \in \mathbf{N}$ such that

$$H \subset \bigcup_{n=1}^{N} B_n(\mathbf{0}).$$

Evidently, $H$ is a subset of $B_N(\mathbf{0})$; i.e., $H$ is bounded. ∎

The nice thing about Euclidean spaces is that the converse of this result is also true. Before we prove it, we must show that every open covering of a set in a Euclidean space has a countable subcovering.

**9.23 THEOREM** [LINDELÖF]. *Let $E$ be a subset of a Euclidean space $X$. If $\{V_\alpha\}_{\alpha \in A}$ is a collection of open sets and $E \subseteq \cup_{\alpha \in A} V_\alpha$, then there is a countable subset $A_0$ of $A$ such that*

$$E \subseteq \bigcup_{\alpha \in A_0} V_\alpha.$$

PROOF. Let $\mathcal{T}$ be the collection of open balls with rational radii and rational centers, i.e., centers which belong to $\mathbf{Q}^n$. This collection is countable. Moreover, it "approximates" all other open sets in the following sense:

CLAIM. Given any open ball $B_r(\boldsymbol{x}) \subseteq X$, there is a ball $B_q(\boldsymbol{a}) \in \mathcal{T}$ such that $\boldsymbol{x} \in B_q(\boldsymbol{a})$ and $B_q(\boldsymbol{a}) \subseteq B_r(\boldsymbol{x})$.

PROOF OF CLAIM. Let $B_r(\boldsymbol{x}) \subseteq X$. By Theorem 8.13, choose $\boldsymbol{a} \in \mathbf{Q}^n$ such that $\|\boldsymbol{x} - \boldsymbol{a}\| < r/4$, and choose by Theorem 1.24 a rational $q \in \mathbf{Q}$ such that $r/4 < q < r/2$. Since $r/4 < q$, we have $\boldsymbol{x} \in B_q(\boldsymbol{a})$. Moreover, if $\boldsymbol{y} \in B_q(\boldsymbol{a})$, then

$$\|\boldsymbol{x} - \boldsymbol{y}\| \leq \|\boldsymbol{x} - \boldsymbol{a}\| + \|\boldsymbol{a} - \boldsymbol{y}\| < q + \frac{r}{4} < \frac{r}{2} + \frac{r}{4} < r.$$

Therefore, $B_q(\boldsymbol{a}) \subseteq B_r(\boldsymbol{x})$. This establishes the claim.

To prove the theorem, let $\boldsymbol{x} \in E$. By hypothesis, $\boldsymbol{x} \in V_\alpha$ for some $\alpha \in A$. Hence, by the claim, there is a ball $B_{\boldsymbol{x}} \in \mathcal{T}$ such that

$$\boldsymbol{x} \in B_{\boldsymbol{x}} \subseteq V_\alpha.$$

The collection $\mathcal{T}$ is countable, hence, so is the subcollection

$$\{U_1, U_2, \dots\} := \{B_{\boldsymbol{x}} : \boldsymbol{x} \in E\}.$$

By the choice of the balls $B_{\boldsymbol{x}}$, for each $k \in \mathbf{N}$ there is at least one $\alpha_k \in A$ such that $U_k \subseteq V_{\alpha_k}$. Hence, by construction,

$$E \subseteq \bigcup_{\boldsymbol{x} \in E} B_{\boldsymbol{x}} = \bigcup_{k \in \mathbf{N}} U_k \subseteq \bigcup_{k \in \mathbf{N}} V_{\alpha_k}.$$

Thus, set $A_0 := \{\alpha_k : k \in \mathbf{N}\}$. ∎

We are prepared to obtain a converse of Theorem 9.22.

**9.24 THEOREM** [HEINE–BOREL]. *A subset $H$ of $\mathbf{R}^n$ is compact if and only if it is closed and bounded.*

PROOF. By Theorem 9.22, every compact set is closed and bounded.

Conversely, suppose to the contrary that $H$ is closed and bounded but not compact. Let $\mathcal{V}$ be an open covering of $H$ which has no finite subcovering of $H$. By Lindelöf's Theorem, we may suppose that $\mathcal{V} = \{V_k\}_{k \in \mathbf{N}}$, i.e.,

(3)
$$H \subseteq \bigcup_{k \in \mathbf{N}} V_k.$$

By the choice of $\mathcal{V}$, $\cup_{j=1}^{k} V_j$ cannot contain $H$ for any $k \in \mathbf{N}$. Thus, we can choose a point

$$(4) \qquad\qquad \boldsymbol{x}_k \in H \setminus \bigcup_{j=1}^{k} V_j$$

for each $k \in \mathbf{N}$. Since $H$ is bounded, the sequence $\boldsymbol{x}_k$ is bounded. Hence, by the Bolzano–Weierstrass Theorem, there is a subsequence $\boldsymbol{x}_{k_\nu}$ which converges to some $\boldsymbol{x}$ as $\nu \to \infty$. Since $H$ is closed, $\boldsymbol{x} \in H$. Hence, by (3), $\boldsymbol{x} \in V_N$ for some $N \in \mathbf{N}$. But $V_N$ is open, hence there is an $M \in \mathbf{N}$ such that $\nu \geq M$ implies $k_\nu > N$ and $\boldsymbol{x}_{k_\nu} \in V_N$. This contradicts (4). We conclude that $H$ is compact. ∎

### EXERCISES

1. Identify which of the following sets are compact and which are not. If $E$ is not compact, find the smallest compact set $H$ (if there is one) such that $E \subset H$.

   a) $\{1/k : k \in \mathbf{N}\} \cup \{0\}$.
   b) $\{(x,y) \in \mathbf{R}^2 : a \leq x^2 + y^2 \leq b\}$ for real numbers $0 < a < b$.
   c) $\{(x,y) \in \mathbf{R}^2 : y = \sin(1/x) \text{ for some } x \in (0,1]\}$.
   d) $\{(x,y) \in \mathbf{R}^2 : |xy| \leq 1\}$.

2. Let $A, B$ be compact subsets of $\mathbf{R}^n$. Prove that $A \cup B$ and $A \cap B$ are compact.

3. Suppose $E \subseteq \mathbf{R}$ is compact and nonempty. Prove $\sup E, \inf E \in E$.

4. Suppose $\{V_\alpha\}_{\alpha \in A}$ is a collection of nonempty open sets in $\mathbf{R}^n$ which satisfies $V_\alpha \cap V_\beta = \emptyset$ for all $\alpha \neq \beta$ in $A$. Prove that $A$ is countable. What happens to this result when "open" is omitted?

5. Prove that if $V$ is open in $\mathbf{R}^n$, then there are open balls $B_1, B_2, \ldots$ such that

$$V = \bigcup_{j \in \mathbf{N}} B_j.$$

   Prove that every open set in $\mathbf{R}$ is a countable union of open intervals.

6. Let $E \subseteq \mathbf{R}^n$ be closed.

   a) Prove $\partial E \subseteq E$.
   b) Prove that $\partial E = E$ if and only if $E^o = \emptyset$.
   c) Show b) is false if $E$ is not closed.

7. Let $X$ be a Euclidean space.

   a) A subset $E$ of $X$ is said to be *sequentially compact* if every sequence $\boldsymbol{x}_k$ in $E$ has a convergent subsequence $\boldsymbol{x}_{k_j}$ whose limit belongs to $E$. Prove that every compact set is sequentially compact.
   b) Prove that every sequentially compact set is closed and bounded.
   c) Prove that a set $E \subset \mathbf{R}^n$ is sequentially compact if and only if it is compact.

## 9.3 CONNECTED SETS

We have introduced open sets (analogues of open intervals), closed sets (analogues of closed intervals), and compact sets (analogues of closed bounded intervals) in order to develop a calculus of functions of several variables in Chapters 11 through 13 which parallels that developed for functions of a single variable in Chapters 2 through 5. Some of the earlier theory, however, depended on properties of intervals not yet discussed. For example, the proof of the Intermediate Value Theorem tacitly used the fact that an interval is connected, i.e., is unbroken and all of one piece. We shall also use connected sets in Chapter 13 to provide a sufficiently broad definition of surfaces for computational ease. Before describing "connectedness," we introduce the following idea.

**9.25 DEFINITION.** Let $E \subseteq \mathbf{R}^n$.

   i) A set $U$ is said to be *relatively open* in $E$ if there is an open set $A$ such that $U = E \cap A$.

   ii) A set $C$ is said to be *relatively closed* in $E$ if there is a closed set $B$ such that $C = E \cap B$.

For example, the set $E$ of Example 9.11 is relatively open in $X := \{(x,y) : -1 \leq x \leq 1\}$ and relatively closed in $Y := \{(x,y) : -|x| < y < |x|\}$ because $Y$ is open, $X$ is closed, and $E = X \cap Y$.

Using this concept, we define "connected" as follows.

**9.26 DEFINITION.** Let $E$ be a subset of $\mathbf{R}^n$.

   i) A pair of sets $U, V$ is said to *separate* $E$ if $U$ and $V$ are nonempty, relatively open in $E$, $E = U \cup V$, and $U \cap V = \emptyset$.

   ii) $E$ is said to be *connected* if $E$ cannot be separated by any pair of relatively open sets $U, V$.

Loosely speaking, a connected space is all in one piece, i.e., cannot be broken into smaller, nonempty, relatively open pieces which do not share any common points.

The empty set is connected, since it can never be written as the union of nonempty sets. Every singleton $E = \{\boldsymbol{a}\}$ is also connected, since if $E = U \cup V$ where both $U$ and $V$ are nonempty, then $E$ has at least two points. More complicated connected sets in $\mathbf{R}^2$ can be found in Examples 9.12 and 9.14 above.

Notice that by Definitions 9.25 and 9.26, a set $E$ is connected if there are no open sets $A, B$ such that $E \cap A$, $E \cap B$ are nonempty, $E = (E \cap A) \cup (E \cap B)$ and $A \cap B = \emptyset$. Is this statement valid if we replace $E = (E \cap A) \cup (E \cap B)$ by $E \subseteq A \cup B$?

**9.27 Remark.** *Let $E \subseteq \mathbf{R}^n$. If there exists a pair of open sets $A, B$ such that $E \cap A \neq \emptyset$, $E \cap B \neq \emptyset$, $E \subseteq A \cup B$, and $A \cap B = \emptyset$, then $E$ is not connected.*

Proof. Set $U = E \cap A$ and $V = E \cap B$. By hypothesis and Definition 9.25, $U$ and $V$ are relatively open in $E$ and nonempty. Since $U \cap V \subseteq A \cap B = \emptyset$, it suffices by Definition 9.26 to prove that $E = U \cup V$. But $E$ is a subset of $A \cup B$, so $E \subseteq U \cup V$. On the other hand, both $U$ and $V$ are subsets of $E$, so $E \supseteq U \cup V$. We conclude that $E = U \cup V$. ∎

(The converse of this result is also true, but harder to prove—see Theorem 9.29 below.)

In practice, Remark 9.27 is often easier to apply than Definition 9.26. Here are several examples. The set $\mathbf{Q}$ is not connected: set $A = (-\infty, \sqrt{2})$ and $B = (\sqrt{2}, \infty)$. The "bowtie set" in Example 9.11 is not connected: set $A = \{(x, y) : x < 0\}$ and $B = \{(x, y) : x > 0\}$.

Is there a simple description of all connected subsets of $\mathbf{R}$?

**9.28 THEOREM.** *A subset $E$ of $\mathbf{R}$ is connected if and only if $E$ is an interval.*

PROOF. Suppose that $E$ is a connected subset of $\mathbf{R}$. If $E$ is empty or if $E$ contains only one point $c$, then $E$ is one of the intervals $(c, c)$ or $[c, c]$.

Otherwise, $E$ contains at least two points. Set $a = \inf E$ and $b = \sup E$, and observe that $-\infty \le a < b \le \infty$. Suppose for simplicity that $a, b \notin E$, i.e., $E \subseteq (a, b)$. If $E \neq (a, b)$, then there is an $x \in (a, b)$ such that $x \notin E$. By the Approximation Property, $E \cap (a, x) \neq \emptyset$ and $E \cap (x, b) \neq \emptyset$, and by assumption, $E \subseteq (a, x) \cup (x, b)$. Hence, by Remark 9.27, $E$ is not connected, a contradiction.

Conversely, suppose $I$ is an interval which is not connected. Then there are sets $U, V$, relatively open in $I$, which separate $I$, i.e., $I = U \cup V$, $U \cap V = \emptyset$, and there exist points $x_1 \in I \cap U$ and $x_2 \in I \cap V$. We may suppose that $x_1 < x_2$. Consider the set

$$W = \{t \in I : \text{ the interval } (x_1, t) \text{ satisfies } (x_1, t) \subseteq U\}.$$

Notice once and for all that, since the intersection of two open intervals is an open interval, any set which is relatively open in $I$ contains an open interval about each if its points. Since $U$ is relatively open, it follows that $W \neq \emptyset$. Since $V$ is relatively open, it also follows that $x_2 \notin W$ and $W$ is bounded above by some $c < x_2$. Thus $x_3 = \sup W$ is a finite number which belongs to $(x_1, c] \subset I$. In particular, either $x_3 \in U$ or $x_3 \in V$.

Suppose $x_3 \in U$. Since $x_3 > x_1$, we can choose $\delta > 0$ so small that $x_3 - \delta > x_1$ and $(x_3 - \delta, x_3 + \delta) \subset U$. Since $x_3 = \sup W$, we can choose by the Approximation Property a $t \in W$ such that $t > x_3 - \delta$ and $(x_1, t) \subset U$. It follows that $(x_1, x_3 + \delta) = (x_1, t) \cup (x_3 - \delta, x_3 + \delta) \subset U$; i.e., $x_3$ is not the supremum of $W$, a contradiction. On the other hand, if $x_3 \in V$, the same reasoning shows us that there is a $\delta > 0$ such that $(x_3 - \delta, x_3 + \delta) \subset V$ and a $t \in W$ such that $t > x_3 - \delta$ and $(x_3 - \delta, t) \subset U$. It follows that $(x_3 - \delta, t) \subset U \cap V$, i.e., $U \cap V \neq \emptyset$, a contradiction. Thus, the pair $U, V$ does not separate $I$, and $I$ must be connected. ∎

We shall use this result later to prove that a real function is continuous on a closed, bounded interval if and only if its graph is closed and connected (see Theorem 9.52 below).

*We close this section by showing that the converse of Remark 9.27 is also true. This result is optional because we do not use it anywhere else.*

**\*9.29 THEOREM.** *Let $E \subseteq \mathbf{R}^n$. If there exist nonempty, relatively open sets $U, V$ which separate $E$, then there is a pair of open sets $A, B$ such that $A \cap E \neq \emptyset$, $B \cap E \neq \emptyset$, $A \cap B = \emptyset$, and $E \subseteq A \cup B$.*

PROOF. We first show that

$$(5) \qquad\qquad \overline{U} \cap V = \emptyset.$$

Indeed, since $V$ is relatively open in $E$, there is a set $\Omega$, open in $\mathbf{R}^n$, such that $V = E \cap \Omega$. Since $U \cap V = \emptyset$, it follows that $U \subset \Omega^c$. This last set is closed in $\mathbf{R}^n$. Therefore,

$$\overline{U} \subseteq \overline{\Omega^c} = \Omega^c,$$

i.e., (5) holds.

Next, we use (5) to construct the open set $B$. Set

$$\delta_{\boldsymbol{x}} := \inf\{\|\boldsymbol{x} - \boldsymbol{u}\| : \boldsymbol{u} \in \overline{U}\}, \quad \boldsymbol{x} \in V, \quad \text{and} \quad B = \bigcup_{\boldsymbol{x} \in V} B_{\delta_{\boldsymbol{x}}/2}(\boldsymbol{x}).$$

Clearly, $B$ is open in $\mathbf{R}^n$. Since $\delta_{\boldsymbol{x}} > 0$ for each $\boldsymbol{x} \notin \overline{U}$ (see Exercise 5 in Section 9.1), $B$ contains $V$, hence $B \cap E \supseteq V$. The reverse inequality also holds, since by construction $B \cap U = \emptyset$ and by hypothesis $E = U \cup V$. Therefore, $B \cap E = V$. Similarly, we can construct an open set $A$ such that $A \cap E = U$ by setting

$$\varepsilon_{\boldsymbol{y}} := \inf\{\|v - \boldsymbol{y}\| : \boldsymbol{v} \in \overline{V}\}, \quad \boldsymbol{y} \in U \quad \text{and} \quad A = \bigcup_{\boldsymbol{y} \in U} B_{\varepsilon_{\boldsymbol{y}}/2}(\boldsymbol{y}).$$

In particular, $A$ and $B$ are nonempty open sets which satisfy $E \subseteq A \cup B$.

It remains to prove $A \cap B = \emptyset$. Suppose, to the contrary, that there is a point $\boldsymbol{a} \in A \cap B$. Then $\boldsymbol{a} \in B_{\delta_{\boldsymbol{x}}/2}(\boldsymbol{x})$ for some $\boldsymbol{x} \in V$ and $\boldsymbol{a} \in B_{\varepsilon_{\boldsymbol{y}}/2}(\boldsymbol{y})$ for some $\boldsymbol{y} \in U$. We may suppose $\delta_{\boldsymbol{x}} \leq \varepsilon_{\boldsymbol{y}}$. Then

$$\|\boldsymbol{x} - \boldsymbol{y}\| \leq \|\boldsymbol{x} - \boldsymbol{a}\| + \|\boldsymbol{a} - \boldsymbol{y}\| < \frac{\delta_{\boldsymbol{x}}}{2} + \frac{\varepsilon_{\boldsymbol{y}}}{2} \leq \varepsilon_{\boldsymbol{y}}.$$

Therefore, $\|\boldsymbol{x} - \boldsymbol{y}\| < \inf\{\|\boldsymbol{v} - \boldsymbol{y}\| : \boldsymbol{v} \in \overline{V}\}$. Since $\boldsymbol{x} \in V$, this is impossible. We conclude that $A \cap B = \emptyset$. ∎

### EXERCISES

**1.** a) Let $a \leq b$ and $c \leq d$ be real numbers. Sketch a graph of the rectangle

$$[a, b] \times [c, d] := \{(x, y) : x \in [a, b], y \in [c, d]\},$$

and decide whether this set is compact or connected. Explain your answers.

b) Sketch a graph of set

$$B_1(-2, 0) \cup B_1(2, 0) \cup \{(x, 0) : -1 < x < 1\},$$

and decide whether this set is compact or connected. Explain your answers.

**2.** a) Sketch a graph of the set
$$\{(x, y) : x^2 + 2y^2 < 6, \ y \geq 0\},$$

and decide whether this set is relatively open or relatively closed in the subset $\{(x, y) : y \geq 0\}$. Do the same for the subset $\{(x, y) : x^2 + 2y^2 < 6\}$. Explain your answers.

b) Sketch a graph of set
$$\{(x, y) : x^2 + y^2 \leq 1, \ (x - 2)^2 + y^2 < 2\},$$

and decide whether this set is relatively open or relatively closed in the subset $\overline{B_1(0, 0)}$. Do the same for the subset $B_{\sqrt{2}}(2, 0)$. Explain your answers.

**3.** Prove that the intersection of connected sets in $\mathbf{R}$ is connected. Show that this is false if "$\mathbf{R}$" is replaced by "$\mathbf{R}^2$."

**4.** Prove that if $E \subseteq \mathbf{R}$ is connected, then $E^o$ is also connected. Show that this is false if "$\mathbf{R}$" is replaced by "$\mathbf{R}^2$."

**5.** Suppose that $E \subset \mathbf{R}^n$ is connected and $E \subseteq A \subseteq \overline{E}$. Prove that $A$ is connected.

**6.** Suppose $\{E_\alpha\}_{\alpha \in A}$ is a collection of connected sets in a Euclidean space $X$ such that $\cap_{\alpha \in A} E_\alpha \neq \emptyset$. Prove that
$$E = \bigcup_{\alpha \in A} E_\alpha$$

is connected.

**7**. **This exercise is used in Section 9.4.** Let $H \subseteq X$. Prove that $H$ is compact if and only if every cover $\{E_\alpha\}_{\alpha \in A}$ of $H$, where the $E_\alpha$'s are relatively open in $H$, has a finite subcovering.

**8.** A set $A$ is called *clopen* if it is both open and closed.

a) Prove that every Euclidean space has at least two clopen sets.

b) Prove that a proper subset $E$ of $\mathbf{R}^n$ is connected if and only if it contains exactly one relatively clopen set.

**9.** Prove that every nonempty proper subset of $\mathbf{R}^n$ has a nonempty boundary.

**\*10**. **This exercise is used to prove \*Corollary 11.29**.

a) A set $E \subseteq \mathbf{R}^n$ is said to be *polygonally connected* if any two points $\boldsymbol{a}, \boldsymbol{b} \in E$ can be connected by a polygonal path in $E$; i.e., there exist points $\boldsymbol{x}_k \in E$, $k = 1, \dots, N$, such that $\boldsymbol{x}_0 = \boldsymbol{a}$, $\boldsymbol{x}_N = \boldsymbol{b}$ and $L(\boldsymbol{x}_{k-1}; \boldsymbol{x}_k) \subseteq E$ for $k = 1, \dots, N$. Prove every polygonally connected set in $\mathbf{R}^n$ is connected.

b) Let $E \subseteq \mathbf{R}^n$ be open and $\boldsymbol{x}_0 \in E$. Let $U$ be the set of points $\boldsymbol{x} \in E$ which can be polygonally connected in $E$ to $\boldsymbol{x}_0$. Prove that $U$ is open.

c) Prove every open connected set in $\mathbf{R}^n$ is polygonally connected.

## 9.4  CONTINUOUS FUNCTIONS

In this section we examine how open sets, closed sets, compact sets, and connected sets behave under images and inverse images by continuous functions. We shall use these results many times in the sequel.

**9.30 DEFINITION.** Suppose $E$ is a nonempty subset of $\mathbf{R}^n$ and $f : E \to \mathbf{R}^m$.

i) $f$ is said to be *continuous at* $\boldsymbol{a} \in E$ if for every $\varepsilon > 0$ there is a $\delta > 0$ (which in general depends on $\varepsilon$ and $\boldsymbol{a}$) such that

(6)
$$\|\boldsymbol{x} - \boldsymbol{a}\| < \delta \quad \text{and} \quad \boldsymbol{x} \in E \quad \text{imply} \quad \|f(\boldsymbol{x}) - f(\boldsymbol{a})\| < \varepsilon.$$

ii) $f$ is said to be *continuous on* $E$ (notation: $f : E \to \mathbf{R}^m$ is continuous) if $f$ is continuous at every $\boldsymbol{x} \in E$.

iii) $f$ is said to be *continuous* if it is continuous on its domain Dom $(f)$.

Suppose $E$ is a nonempty subset of $\mathbf{R}^n$. It is easy to check that $f$ is continuous at $\boldsymbol{a} \in E$ if and only if $f(\boldsymbol{x}_k) \to f(\boldsymbol{a})$ for all $\boldsymbol{x}_k \in E$ which converge to $\boldsymbol{a}$. Hence, by Theorem 8.14, if $f$ and $g$ are continuous at a point $\boldsymbol{a} \in E$ (respectively, continuous on $E$), then so are $f + g$, $\alpha f$ (for $\alpha \in \mathbf{R}$), $f \cdot g$, $\|f\|$, and (when $m = 3$) $f \times g$. Moreover, if $f : E \to \mathbf{R}^m$ is continuous at $\boldsymbol{a} \in E$ and $g : f(E) \to \mathbf{R}^\ell$ is continuous at $f(\boldsymbol{a}) \in f(E)$, then $g \circ f$ is continuous at $\boldsymbol{a} \in E$.

We shall occasionally need the following stronger version of continuity.

**9.31 DEFINITION.** Let $E$ be a nonempty subset of $\mathbf{R}^n$ and $f : E \to \mathbf{R}^m$. Then $f$ is said to be *uniformly continuous* on $E$ (notation: $f : E \to \mathbf{R}^m$ is uniformly continuous) if for every $\varepsilon > 0$ there is a $\delta > 0$ such that

(7)
$$\|\boldsymbol{x} - \boldsymbol{a}\| < \delta \quad \text{and} \quad \boldsymbol{x}, \boldsymbol{a} \in E \quad \text{imply} \quad \|f(\boldsymbol{x}) - f(\boldsymbol{a})\| < \varepsilon.$$

As in the real case, continuity and uniform continuity are equivalent on closed, bounded subsets of a Euclidean space.

**9.32 THEOREM.** *Suppose $E$ is a compact subset of $\mathbf{R}^n$ and $f : E \to \mathbf{R}^m$. Then $f$ is uniformly continuous on $E$ if and only if $f$ is continuous on $E$.*

PROOF. If $f$ is uniformly continuous on a set, then it is continuous, whether the set is compact or not.

Conversely, suppose $f$ is continuous on $E$. Given $\varepsilon > 0$ and $\boldsymbol{a} \in E$, choose $\delta(\boldsymbol{a}) > 0$ such that

$$\boldsymbol{x} \in B_{\delta(\boldsymbol{a})}(\boldsymbol{a}) \quad \text{and} \quad \boldsymbol{x} \in E \quad \text{imply} \quad \|f(\boldsymbol{x}) - f(\boldsymbol{a})\| < \frac{\varepsilon}{2}.$$

Since $\boldsymbol{a} \in B_\delta(\boldsymbol{a})$ for all $\delta > 0$, it is clear that $\{B_{\delta(\boldsymbol{a})/2}(\boldsymbol{a}) : \boldsymbol{a} \in E\}$ is an open covering of $E$. Since $E$ is compact, choose finitely many points $\boldsymbol{a}_j \in E$ and numbers $\delta_j := \delta(\boldsymbol{a}_j)/2$ such that

(8)
$$E \subseteq \bigcup_{j=1}^{N} B_{\delta_j}(\boldsymbol{a}_j).$$

Set $\delta := \min\{\delta_1, \ldots, \delta_N\}$.

Suppose $\boldsymbol{x}, \boldsymbol{a} \in E$ and $\|\boldsymbol{x} - \boldsymbol{a}\| < \delta$. By (8), $\boldsymbol{x}$ belongs to $B_{\delta_j}(\boldsymbol{a}_j)$ for some $1 \leq j \leq N$. Hence,

$$\|\boldsymbol{a} - \boldsymbol{a}_j\| \leq \|\boldsymbol{a} - \boldsymbol{x}\| + \|\boldsymbol{x} - \boldsymbol{a}_j\| < \delta_j + \delta_j = 2\delta_j = \delta(\boldsymbol{a}_j),$$

i.e., $\boldsymbol{a}$ also belongs to $B_{\delta(\boldsymbol{a}_j)}(\boldsymbol{a}_j)$. It follows, therefore, from the choice of $\delta(\boldsymbol{a}_j)$ that

$$\|f(\boldsymbol{x}) - f(\boldsymbol{a})\| \le \|f(\boldsymbol{x}) - f(\boldsymbol{a}_j)\| + \|f(\boldsymbol{a}_j) - f(\boldsymbol{a})\| < \frac{\varepsilon}{2} + \frac{\varepsilon}{2} = \varepsilon.$$

This proves that $f$ is uniformly continuous on $E$. $\blacksquare$

To describe continuity without using norms, notice by Definition 9.30 that if $f$ is continuous at $\boldsymbol{a} \in E$, then for all $\varepsilon > 0$ there is a $\delta > 0$ such that

$$(9) \qquad\qquad B_\delta(\boldsymbol{a}) \cap E \subseteq f^{-1}(B_\varepsilon(f(\boldsymbol{a}))).$$

This observation can be extended to obtain a powerful characterization of continuous functions (see also Exercise 3).

**9.33 THEOREM.** *Let $E$ be a nonempty subset of $\mathbf{R}^n$, and $f : E \to \mathbf{R}^m$. Then $f$ is continuous on $E$ if and only if $f^{-1}(V) \cap E$ is relatively open in $E$ for every open set $V$ in $\mathbf{R}^m$.*

Proof. Suppose $f$ is continuous on $E$ and $V$ is open in $\mathbf{R}^m$. We may suppose $f^{-1}(V) \cap E$ is nonempty. Let $\boldsymbol{a} \in f^{-1}(V) \cap E$, i.e., $\boldsymbol{a} \in E$ and $f(\boldsymbol{a}) \in V$. Since $V$ is open, choose $\varepsilon > 0$ such that $B_\varepsilon(f(\boldsymbol{a})) \subseteq V$. Since $f$ is continuous at $\boldsymbol{a} \in E$, choose $\delta > 0$ such that (6) holds. Evidently,

$$(10) \qquad\qquad B_\delta(\boldsymbol{a}) \cap E \subseteq f^{-1}(B_\varepsilon(f(\boldsymbol{a}))) \subseteq f^{-1}(V).$$

Set

$$U := \bigcup_{\boldsymbol{a} \in f^{-1}(V)} B_\delta(\boldsymbol{a}).$$

Since $U$ is a union of open sets, $U$ is open. Hence, $A := U \cap E$ is relatively open in $E$, and it remains to prove that $A = f^{-1}(V) \cap E$. On the one hand, $A \subseteq f^{-1}(V) \cap E$ follows directly from (10) and the construction of $U$. On the other hand, since $U$ is a union of $B_\delta(\boldsymbol{a})$ over all $\boldsymbol{a} \in f^{-1}(V)$, it is also clear that $A \supseteq f^{-1}(V) \cap E$. Therefore, $f^{-1}(V) \cap E = A$ is relatively open.

Conversely, let $\varepsilon > 0$ and $\boldsymbol{a} \in E$. The ball $V = B_\varepsilon(f(\boldsymbol{a}))$ is open in $\mathbf{R}^m$; hence, by hypothesis, $f^{-1}(V) \cap E$ is relatively open in $E$, i.e., there is an open set $\Omega$ such that $f^{-1}(V) \cap E = \Omega \cap E$. Since $\boldsymbol{a} \in f^{-1}(V) \cap E$, it follows that there is a $\delta > 0$ such that $B_\delta(\boldsymbol{a}) \cap E \subseteq f^{-1}(V) \cap E$. This means that if $\|\boldsymbol{x} - \boldsymbol{a}\| < \delta$ and $\boldsymbol{x} \in E$, then $\|f(\boldsymbol{x}) - f(\boldsymbol{a})\| < \varepsilon$. Therefore, $f$ is continuous at $\boldsymbol{a} \in E$. $\blacksquare$

We shall refer to Theorem 9.33 by saying that open sets are invariant under inverse images by continuous functions. It is interesting to notice that closed sets are also invariant under inverse images by continuous functions (see Exercises 3 and 4). It is natural to ask whether compact sets and connected sets are invariant under inverse images by continuous functions. The following examples show the answer to this question is no.

**9.34 Examples.** i) If $f(x) = 1/x$ and $H = [0, 1]$, then $f$ is continuous on $(0, \infty)$ and $H$ is compact, but $f^{-1}(H) = [1, \infty)$ is not compact.

ii) If $f(x) = x^2$ and $E = (1, 4)$, then $f$ is continuous on $\mathbf{R}$ and $E$ is connected, but $f^{-1}(E) = (-2, -1) \cup (1, 2)$ is not connected.

The next two results show that compact sets and connected sets are invariant under **images**, rather than inverse images, by continuous functions.

**9.35 THEOREM.** *Let $X$ and $Y$ be Euclidean spaces. If $H$ is compact in $X$ and $f : H \to Y$ is continuous on $H$, then $f(H)$ is compact in $Y$.*

PROOF. Suppose $\{V_\alpha\}_{\alpha \in A}$ is an open covering of $f(H)$. By Theorem 1.43,

$$H \subseteq f^{-1}(f(H)) \subseteq f^{-1}\left(\bigcup_{\alpha \in A} V_\alpha\right) = \bigcup_{\alpha \in A} f^{-1}(V_\alpha).$$

Hence, by Theorem 9.33, $\{f^{-1}(V_\alpha) \cap H\}_{\alpha \in A}$ is a covering of $H$ whose sets are all relatively open in $H$. Since $H$ is compact, there are indices $\alpha_1, \alpha_2, \ldots, \alpha_N$ such that

$$H \subseteq \bigcup_{j=1}^{N} f^{-1}(V_{\alpha_j})$$

(see Exercise 7 in Section 9.3). It follows from Theorem 1.43 that

$$f(H) \subseteq f\left(\bigcup_{j=1}^{N} f^{-1}(V_{\alpha_j})\right) = \bigcup_{j=1}^{N}(f \circ f^{-1})(V_{\alpha_j}) = \bigcup_{j=1}^{N} V_{\alpha_j}.$$

Therefore, $f(H)$ is compact. ∎

**9.36 THEOREM.** *Let $X$ and $Y$ be Euclidean spaces. If $E$ is connected in $X$ and $f : E \to Y$ is continuous on $E$, then $f(E)$ is connected in $Y$.*

PROOF. Suppose $f(E)$ is not connected. By Definition 9.26, there exist a pair of relatively open sets $U, V$ in $f(E)$ which separates $f(E)$; i.e., $U \cap f(E) \neq \emptyset$, $V \cap f(E) \neq \emptyset$, $f(E) = U \cup V$, and $U \cap V = \emptyset$. Set $A := f^{-1}(U) \cap E$ and $B := f^{-1}(V) \cap E$. By Exercise 4 below, $A$ and $B$ are relatively open in $E$. Since $f(E) = U \cup V$, we have

$$E = (f^{-1}(U) \cap E) \cup (f^{-1}(V) \cap E) = A \cup B.$$

Moreover, $U \cap V = \emptyset$ implies $f^{-1}(U) \cap f^{-1}(V) = \emptyset$, so surely $A \cap B = \emptyset$. Thus $A, B$ is a pair of relatively open sets which separates $E$; i.e., $E$ is not connected, a contradiction. ∎

(Note: Theorems 9.35 and 9.36 do not hold if "compact" or "connected" are replaced by "open" or "closed." For example, if $f(x) = x^2$ and $V = (-1, 1)$, then $f$ is continuous on $\mathbf{R}$ and $V$ is open, but $f(V) = [0, 1)$ is neither open nor closed.)

The topological point of view presented above is a powerful tool. To illustrate this fact, we offer the following four results.

**9.37 Remark.** *The graph $y = f(x)$ of a continuous real function $f$ on a closed, bounded interval $[a, b]$ is connected and compact.*

PROOF. The function $F(x) = (x, f(x))$ is continuous from $\mathbf{R}$ into $\mathbf{R}^2$, and the graph of $y = f(x)$ for $x \in [a, b]$ is the image of $[a, b]$ under $F$. Hence the graph of $f$ is compact and connected by Theorems 9.35 and 9.36. ∎

It is interesting to note that this property actually characterizes continuity of real functions (see Theorem 9.52).

**9.38 THEOREM** [EXTREME VALUE THEOREM]. *If $H$ is a nonempty, compact subset of $\mathbf{R}^n$ and $f : H \to \mathbf{R}$ is continuous, then*

$$M := \sup\{f(\boldsymbol{x}) : \boldsymbol{x} \in H\} \quad \text{and} \quad m := \inf\{f(\boldsymbol{x}) : \boldsymbol{x} \in H\}$$

*are finite real numbers and there exist points $\boldsymbol{x}_M, \boldsymbol{x}_m \in H$ such that $M = f(\boldsymbol{x}_M)$ and $m = f(\boldsymbol{x}_m)$.*

PROOF. By symmetry, it suffices to prove the result for $M$. Since $H$ is compact, $f(H)$ is compact. Hence, by the Heine–Borel Theorem, $f(H)$ is closed and bounded. Since $f(H)$ is bounded, $M$ is finite. By the Approximation Property, choose $\boldsymbol{x}_k \in H$ such that $f(\boldsymbol{x}_k) \to M$ as $k \to \infty$. Since $f(H)$ is closed, $M \in f(H)$. Therefore, there is an $\boldsymbol{x}_M \in H$ such that $M = f(\boldsymbol{x}_M)$. ∎

(Note how much simpler this proof is than the non-topological proof of Theorem 3.26. For a multidimensional analogue of Theorem 3.29, see Exercise 6 below.)

The following analogue of Theorem 4.26 will be used in Chapter 13 to examine change of parametrizations of curves and surfaces.

**9.39 THEOREM.** *Let $X$ and $Y$ be Euclidean spaces. If $H$ is a compact subset of $X$ and $f : H \to Y$ is 1–1 and continuous, then $f^{-1}$ is continuous on $f(H)$.*

PROOF. Let $E$ be closed. Then $E \cap H$ is compact (Remark 9.21), $f(E \cap H)$ is compact (Theorem 9.35), hence $f(E \cap H)$ is closed (Heine–Borel Theorem). Since $f$ is 1–1, $f(E \cap H) = f(E) \cap f(H)$ (see Exercise 6 in Section 1.4), so $f(E) \cap f(H)$ is relatively closed in $f(H)$. Since $f(E) = (f^{-1})^{-1}(E)$, it follows that $(f^{-1})^{-1}(E) \cap f(H)$ is relatively closed in $f(H)$. We have proved that the inverse image of a closed set $E$ under $f^{-1}$ is relatively closed in $f(H)$. By Exercise 4a) below, this proves that $f^{-1}$ is continuous on $f(H)$. ∎

The final result of this section shows "rectangles" are connected in $\mathbf{R}^n$.

**9.40 Remark.** *If $a_j \leq b_j$ for $j = 1, 2, \ldots, n$, then*

$$R := \{(x_1, \ldots, x_n) : a_j \leq x_j \leq b_j\}$$

*is connected.*

PROOF. Suppose not. Choose nonempty sets $U$ and $V$, relatively open in $R$ such that $R = U \cup V$ and $U \cap V = \emptyset$. Let $\boldsymbol{a} \in U$ and $\boldsymbol{b} \in V$, and consider the line segment

$$E := \{t\boldsymbol{a} + (1 - t)\boldsymbol{b} : t \in [0, 1]\}.$$

Notice that $E \subset R$ by the definition of $R$. Hence, $U_0 := U \cap E$ and $V_0 := V \cap E$ are nonempty sets, relatively open in $E$, which satisfy $E = U_0 \cup V_0$ and $U_0 \cap V_0 = \emptyset$. It follows that $E$ is not connected. This contradicts the fact that $E$ is connected by Theorems 9.28 and 9.36, since $E$ is a continuous image of the interval $[0, 1]$. ∎

## EXERCISES

**1.** Let $f(x) = \sin x$ and $g(x) = x/|x|$ if $x \neq 0$ and $g(0) = 0$.

   a) Find $f(E)$ and $g(E)$ for $E = (0, \pi)$, $E = [0, \pi]$, $E = (-1, 1)$, and $E = [-1, 1]$, and explain some of your answers by appealing to results in this section.

   b) Find $f^{-1}(E)$ and $g^{-1}(E)$ for $E = (0, 1)$, $E = [0, 1]$, $E = (-1, 1)$, and $E = [-1, 1]$, and explain some of your answers by appealing to results in this section.

**2.** Let $f(x) = \sqrt{x}$ and $g(x) = 1/x$ if $x \neq 0$ and $g(0) = 0$.

   a) Find $f(E)$ and $g(E)$ for $E = (0, 1)$, $E = [0, 1)$, and $E = [0, 1]$, and explain some of your answers by appealing to results in this section.

   b) Find $f^{-1}(E)$ and $g^{-1}(E)$ for $E = (-1, 1)$ and $E = [-1, 1]$, and explain some of your answers by appealing to results in this section.

**3.** Let $f : \mathbf{R}^n \to \mathbf{R}^m$. Prove that the following are equivalent:

   a) $f$ is continuous.

   b) $f^{-1}(V)$ is open in $\mathbf{R}^n$ for every open subset $V$ of $\mathbf{R}^m$.

   c) $f^{-1}(C)$ is closed in $\mathbf{R}^n$ for every closed subset $C$ of $\mathbf{R}^m$.

**4.** Suppose $E \subseteq \mathbf{R}^n$ and $f : E \to \mathbf{R}^m$.

   a) Prove that $f$ is continuous on $E$ if and only if $f^{-1}(A) \cap E$ is relatively closed in $E$ for every closed set $A$ in $\mathbf{R}^m$.

   b) Suppose that $f$ is continuous on $E$. Prove that if $V$ is relatively open in $f(E)$, then $f^{-1}(V) \cap E$ is relatively open in $E$, and if $A$ is relatively closed in $f(E)$, then $f^{-1}(A) \cap E$ is relatively closed in $E$.

**5.** Prove that
$$f(x, y) = \begin{cases} e^{-1/|x-y|} & x \neq y \\ 0 & x = y \end{cases}$$

is continuous on $\mathbf{R}^2$.

**6.** [INTERMEDIATE VALUE THEOREM]. Let $E$ be a connected subset of a Euclidean space $X$. If $f : E \to \mathbf{R}$ is continuous, $f(a) \neq f(b)$ for some $a, b \in E$, and $y$ is a number which lies between $f(a)$ and $f(b)$, then prove there is an $\boldsymbol{x} \in E$ such that $f(\boldsymbol{x}) = y$. (You may use Theorem 9.28.)

**\*7**. **This exercise is used in Section** $^e$**9.5.** Let $X$ and $Y$ be Euclidean spaces and $H$ be a nonempty compact subset of $X$.

   a) Suppose $f : H \to Y$ is continuous. Prove that

$$\|f\|_H := \sup_{\boldsymbol{x} \in H} \|f(x)\|$$

is finite and there exists an $x_0 \in H$ such that $\|f(x_0)\| = \|f\|_H$.

b) A sequence of functions $f_k : H \to Y$ is said to converge uniformly on $H$ to a function $f : H \to Y$ if for every $\varepsilon > 0$ there is an $N \in \mathbf{N}$ such that

$$k \geq N \quad \text{and} \quad x \in H \quad \text{imply} \quad \|f_k(x) - f(x)\| < \varepsilon.$$

Show that $\|f_k - f\|_H \to 0$ as $k \to \infty$ if and only if $f_k \to f$ uniformly on $H$ as $k \to \infty$.

c) Prove that a sequence of functions $f_k$ converges uniformly on $H$ if and only if for every $\varepsilon > 0$ there is an $N \in \mathbf{N}$ such that

$$k, j \geq N \quad \text{implies} \quad \|f_k - f_j\|_H < \varepsilon.$$

**8.** Let $X$ and $Y$ be Euclidean spaces and $D$ be a *dense* subspace of $X$; i.e., $D \subset X$ and $\overline{D} = X$. If $f : D \to Y$ is uniformly continuous on $D$, prove that $f$ has a continuous extension to $X$; i.e., prove that there is a continuous function $g : X \to Y$ such that $g(x) = f(x)$ for all $x \in D$.

## $^e$9.5  APPLICATIONS   *This section requires no material from any other enrichment section.*

We have seen that the topological ideas introduced in this chapter (e.g., closed sets, open sets, compact sets, and connected sets) are powerful theoretical tools. In this section we continue this theme by obtaining three independent theorems (i.e., you may cover them in any order) which further elucidate results we obtained in earlier chapters.

Our first application of topological ideas is a partial converse of Theorem 7.10. A sequence of real-valued functions $\{f_k\}$ is said to be *pointwise increasing* (respectively, *pointwise decreasing*) on a nonempty subset $E$ of $\mathbf{R}^n$ if $f_k(x) \leq f_{k+1}(x)$ (respectively, $f_k(x) \geq f_{k+1}(x)$) for all $x \in E$ and $k \in \mathbf{N}$. A sequence is said to be *pointwise monotone* on $E$ if it is pointwise increasing on $E$ or pointwise decreasing on $E$.

**THEOREM 9.41** [DINI]. *Suppose $H$ is a nonempty, compact subset of $\mathbf{R}^n$ and $f_k : H \to \mathbf{R}$ is a pointwise monotone sequence of continuous functions. If $f_k \to f$ pointwise on $H$ as $k \to \infty$ and $f$ is continuous on $H$, then $f_k \to f$ uniformly on $H$. In particular, if $\phi_k$ is a pointwise monotone sequence of functions continuous on an interval $[a, b]$ which converges pointwise to a continuous function, then*

$$\lim_{k \to \infty} \int_a^b \phi_k(t)\, dt = \int_a^b \left( \lim_{k \to \infty} \phi_k(t) \right) dt.$$

PROOF. By Theorem 7.10, we need only show that $f_k \to f$ uniformly on $H$. We may suppose that $f_k$ is pointwise increasing.

Let $\varepsilon > 0$. For each $x \in H$, choose $N(x) \in \mathbf{N}$ such that

$$k \geq N(x) \quad \text{implies} \quad |f_k(x) - f(x)| < \frac{\varepsilon}{3}.$$

Since $f$ and $f_{N(\boldsymbol{x})}$ are continuous on $H$, choose an $r = r(\boldsymbol{x}) > 0$ such that

$$\boldsymbol{y} \in H \cap B_r(\boldsymbol{x}) \quad \text{implies} \quad |f(\boldsymbol{x}) - f(\boldsymbol{y})| < \frac{\varepsilon}{3} \quad \text{and} \quad |f_{N(\boldsymbol{x})}(\boldsymbol{x}) - f_{N(\boldsymbol{x})}(\boldsymbol{y})| < \frac{\varepsilon}{3}.$$

Since $H$ is compact, choose $\boldsymbol{x}_j \in H$ and $r_j = r(\boldsymbol{x}_j)$ such that

$$H \subset \bigcup_{j=1}^{M} B_{r_j}(\boldsymbol{x}_j).$$

Set $N = \max\{N(\boldsymbol{x}_1), \ldots, N(\boldsymbol{x}_M)\}$, let $\boldsymbol{x} \in H$, and suppose $k \geq N$. Since $\boldsymbol{x} \in B_{r_j}(\boldsymbol{x}_j)$ for some $j \in \{1, \ldots, M\}$ and $k \geq N(\boldsymbol{x}_j)$, it follows that

$$\begin{aligned} |f(\boldsymbol{x}) - f_k(\boldsymbol{x})| &= f(\boldsymbol{x}) - f_k(\boldsymbol{x}) \leq f(\boldsymbol{x}) - f_{N(\boldsymbol{x}_j)}(\boldsymbol{x}) \\ &\leq |f(\boldsymbol{x}) - f(\boldsymbol{x}_j)| + |f(\boldsymbol{x}_j) - f_{N(\boldsymbol{x}_j)}(\boldsymbol{x}_j)| \\ &\quad + |f_{N(\boldsymbol{x}_j)}(\boldsymbol{x}_j) - f_{N(\boldsymbol{x}_j)}(\boldsymbol{x})| \\ &< \frac{\varepsilon}{3} + \frac{\varepsilon}{3} + \frac{\varepsilon}{3} = \varepsilon. \end{aligned}$$

Since this inequality holds for all $\boldsymbol{x} \in H$, we conclude that $f_k \to f$ uniformly on $H$ as $k \to \infty$. ∎

Our next application of topological ideas is a characterization of Riemann integrability of a function $f$ by the size of the set of points of discontinuity of $f$. To measure the size of this set, we make the following definition. (Recall that $|I|$ denotes the length of an interval $I$.)

**9.42 DEFINITION.** i) A set $E \subset \mathbf{R}$ is said to be of *measure zero* if for every $\varepsilon > 0$ there is a countable collection of intervals $\{I_j\}_{j \in \mathbf{N}}$ which covers $E$ such that

$$\sum_{j=1}^{\infty} |I_j| \leq \varepsilon.$$

ii) A function $f : [a, b] \to \mathbf{R}$ is said to be *almost everywhere continuous* on $[a, b]$ if the set of points $x \in [a, b]$ where $f$ is discontinuous is a set of measure zero.

Notice that by definition, if $E$ is of measure zero, then every subset of $E$ is also of measure zero. Loosely speaking, a set is of measure zero if it is so small that it can be covered by a sequence of intervals whose total length is as small as we wish.

It is easy to see that a single point $E = \{x\}$ is a set of measure zero. Indeed, $I_1 := (x - \varepsilon/2, x + \varepsilon/2)$, $I_k := \emptyset$ for $k \geq 2$, cover $E$ and have total length $\varepsilon$. Modifying this technique, we can show that any finite set is a set of measure zero (see also Remark 9.43 below). On the other hand, since $[a, b]$ is compact, any open covering of $[a, b]$ has a finite subcovering; hence, any covering of $[a, b]$ by open intervals must have total length greater than or equal to $b - a$. In particular, a closed nondegenerate interval cannot be of measure zero.

The following result shows that if a set is small in the set theoretical sense, then it is small in the measure theoretical sense.

**9.43 Remark.** *Every countable set of real numbers is a set of measure zero.*

PROOF. We may suppose that $E$ is countably infinite, say $E = \{x_1, x_2, \dots\}$. Given $\varepsilon > 0$ and $j \in \mathbf{N}$, set

$$I_j = (x_j - \varepsilon 2^{-j-1}, x_j + \varepsilon 2^{-j-1}).$$

Then $x_j \in I_j$ and $|I_j| = \varepsilon 2^{-j}$ for $j \in \mathbf{N}$. Therefore, $E \subseteq \cup_{j=1}^{\infty} I_j$ and

$$\sum_{j=1}^{\infty} |I_j| = \varepsilon \sum_{j=1}^{\infty} \frac{1}{2^j} = \varepsilon. \quad \blacksquare$$

The converse of Remark 9.43 is false; i.e., there exist uncountable sets of measure zero (see Exercise 9 below).

The following result shows that the countable union of sets of measure zero is a set of measure zero.

**9.44 Remark.** *If $E_1, E_2, \dots$ is a sequence of sets of measure zero, then*

$$E = \bigcup_{k=1}^{\infty} E_k$$

*is also a set of measure zero.*

PROOF. Let $\varepsilon > 0$. By hypothesis, given $k \in \mathbf{N}$ we can choose a collection of intervals $\{I_j^{(k)}\}_{j \in \mathbf{N}}$ which covers $E_k$ such that

$$\sum_{j=1}^{\infty} |I_j^{(k)}| < \frac{\varepsilon}{2^k}.$$

Then the collection $\{I_j^{(k)}\}_{k,j \in \mathbf{N}}$ is countable, covers $E$, and

$$\sum_{k=1}^{\infty} \sum_{j=1}^{\infty} |I_j^{(k)}| \leq \sum_{k=1}^{\infty} \frac{\varepsilon}{2^k} = \varepsilon.$$

Consequently, $E$ is of measure zero. $\quad \blacksquare$

To facilitate our discussion of points of discontinuity, we introduce the following concepts.

**9.45 DEFINITION.** Let $[a, b]$ be a closed interval and $f : [a, b] \to \mathbf{R}$ be bounded.

i) The *oscillation* of $f$ on an interval $J$ which intersects $[a, b]$ is defined to be

$$\Omega_f(J) := \sup_{x,y \in J \cap [a,b]} (f(x) - f(y)).$$

ii) The *oscillation* of $f$ at a point $t \in [a, b]$ is defined to be

$$\omega_f(t) := \lim_{h \to 0+} \Omega_f((t - h, t + h)),$$

when this limit exists.

**9.46 Remark.** *If $f : [a, b] \to \mathbf{R}$ is bounded, then $\omega_f(t)$ exists for all $t \in [a, b]$ and satisfies $0 \leq \omega_f(t) < \infty$.*

PROOF. Fix $t \in [a, b]$ and for each interval $J$, set

$$M_J = \sup_{x \in J \cap [a,b]} f(x), \qquad m_J = \inf_{x \in J \cap [a,b]} f(x).$$

Since $\sup(-f(x)) = -\inf f(x)$, it is obvious that

(11)
$$\Omega_f(J) = M_J - m_J \geq 0.$$

Suppose for simplicity that $t \in (a, b)$, and choose $h_0$ so small that $(t - h_0, t + h_0) \subset (a, b)$. For each $0 < h < h_0$, set

$$F(h) = \Omega_f((t - h, t + h)).$$

By the Monotone Property of Suprema, $F(h)$ is increasing on $(0, h_0)$, hence, has a finite limit as $h \to 0+$. By (11), $F(h) \geq 0$. Therefore, $\omega_f(t)$ exists, and is both finite and nonnegative. ∎

The next result shows that, by using the oscillation function $\omega_f$, one can represent the set of points of discontinuity of any bounded $f$ as a countable union.

**9.47 Remark.** *Let $f : [a, b] \to \mathbf{R}$ be bounded. If $E$ represents the set of points of discontinuity of $f$ in $[a, b]$, then*

$$E = \bigcup_{j=1}^{\infty} \left\{ t \in [a, b] : \omega_f(t) \geq \frac{1}{j} \right\}.$$

PROOF. By (11), $f$ is continuous at $t \in [a, b]$ if and only if $\omega_f(t) = 0$. Hence, $t$ belongs to $E$ if and only if $\omega_f(t) > 0$. Since, by the Archimedean Principle, $\omega_f(t) > 0$ if and only if $\omega_f(t) \geq 1/j$ for some $j \in \mathbf{N}$, the result follows at once. ∎

We need two technical results about the oscillation of $f$ at a point $t$.

**9.48 Lemma.** *Let $f : [a, b] \to \mathbf{R}$ be bounded. For each $\varepsilon > 0$, the set*

$$H = \{t \in [a, b] : \omega_f(t) \geq \varepsilon\}$$

*is compact.*

PROOF. Suppose to the contrary that $H$ is not compact. Then $H$ is nonempty. Since $H$ is bounded, it cannot be closed. Hence, there are points $t_k \in H$ such that $t_k \to t$ as $k \to \infty$ but $t \notin H$. Since $\omega_f(t) < \varepsilon$, it follows that there is an $h_0 > 0$ such that

(12)
$$\Omega_f((t - h_0, t + h_0)) < \varepsilon.$$

Since $t_k \to t$, choose $N \in \mathbf{N}$ so that

$$\left( t_N - \frac{h_0}{2}, t_N + \frac{h_0}{2} \right) \subset (t - h_0, t + h_0).$$

Then, by (12), $\Omega_f((t_N - h_0/2, t_N + h_0/2)) < \varepsilon$. Therefore, $\omega_f(t_N) < \varepsilon$ which contradicts the fact that $t_N \in H$. ∎

**9.49 Lemma.** *Let $I$ be a closed bounded interval and $f : I \to \mathbf{R}$ be bounded. If $\varepsilon > 0$ and $\omega_f(t) < \varepsilon$ for all $t \in I$, then there is a $\delta > 0$ such that $\Omega_f(J) < \varepsilon$ for all closed intervals $J \subseteq I$ which satisfy $|J| < \delta$.*

PROOF. For each $t \in I$, choose $\delta_t$ such that

$$(13) \qquad\qquad \Omega_f((t - \delta_t, t + \delta_t)) < \varepsilon.$$

Since $\{(t - \delta_t/2, t + \delta_t/2)\}_{t \in I}$ covers $I$, choose $t_1, \ldots, t_N$ such that

$$I \subseteq \bigcup_{j=1}^{N} \left( t_j - \frac{\delta_{t_j}}{2}, t_j + \frac{\delta_{t_j}}{2} \right)$$

and set

$$\delta = \min_{1 \le j \le N} \frac{\delta_{t_j}}{2}.$$

If $J \subseteq I$, then

$$J \cap \left( t_j - \frac{\delta_{t_j}}{2}, t_j + \frac{\delta_{t_j}}{2} \right) \ne \emptyset$$

for some $j \in \{1, \ldots, N\}$. If $J$ also satisfies $|J| < \delta$, then it follows from $2\delta \le \delta_{t_j}$ that $J \subseteq (t_j - \delta_{t_j}, t_j + \delta_{t_j})$. In particular, (13) implies

$$\Omega_f(J) \le \Omega_f((t_j - \delta_{t_j}, t_j + \delta_{t_j})) < \varepsilon. \quad \blacksquare$$

**9.50 THEOREM** [LEBESGUE]. *Let $f : [a, b] \to \mathbf{R}$ be bounded. Then $f$ is Riemann integrable on $[a, b]$ if and only if $f$ is almost everywhere continuous on $[a, b]$. In particular, if $f$ is bounded and has countably many points of discontinuity on $[a, b]$, then $f$ is integrable on $[a, b]$.*

PROOF. Let $E$ be the set of points of discontinuity of $f$ in $[a, b]$. Suppose $f$ is integrable but $E$ is not of measure zero. By Remarks 9.44 and 9.47, there is a $j_0 \in \mathbf{N}$ such that

$$H := \{ t \in [a, b] : \omega_f(t) \ge \frac{1}{j_0} \}$$

is not of measure zero. In particular, there is an $\varepsilon_0 > 0$ such that if $\{I_k\}_{k \in \mathbf{N}}$ is any collection of intervals which covers $H$, then

$$(14) \qquad\qquad \sum_{k=1}^{\infty} |I_k| \ge \varepsilon_0.$$

Let $P = \{x_0, \ldots, x_n\}$ be a partition of $[a, b]$. If $(x_{k-1}, x_k) \cap H \ne \emptyset$, then by definition, $M_k(f) - m_k(f) \ge 1/j_0$. Hence,

$$U(f, P) - L(f, P) = \sum_{k=1}^{n} (M_k(f) - m_k(f))(x_k - x_{k-1})$$

$$\ge \sum_{(x_{k-1}, x_k) \cap H \ne \emptyset} (M_k(f) - m_k(f))(x_k - x_{k-1})$$

$$\ge \frac{1}{j_0} \sum_{(x_{k-1}, x_k) \cap H \ne \emptyset} (x_k - x_{k-1}).$$

But $\{[x_{k-1}, x_k] : (x_{k-1}, x_k) \cap H \neq \emptyset\}$ is a collection of intervals which covers $H$. Hence, it follows from (14) that

$$U(f, P) - L(f, P) \geq \frac{\varepsilon_0}{j_0} > 0.$$

Therefore, $f$ cannot be integrable on $[a, b]$.

Conversely, suppose $E$ is of measure zero. Let $M = \sup_{x \in [a,b]} f(x)$ and $m = \inf_{x \in [a,b]} f(x)$. Given $\varepsilon > 0$, choose $j_0 \in \mathbf{N}$ such that

$$\frac{M - m + b - a}{j_0} < \varepsilon.$$

Since $E$ is of measure zero, so is

$$H = \{t \in [a, b] : \omega_f(t) \geq \frac{1}{j_0}\}.$$

Hence, by Definition 9.42, there exists a collection of intervals which covers $H$, whose lengths sum to a real number less than $1/(2j_0)$. By expanding these intervals slightly, we may suppose there exist open intervals $I_1, I_2, \ldots$ which cover $H$ such that

$$\sum_{\nu=1}^{\infty} |I_\nu| < \frac{1}{j_0}.$$

Hence, by Lemma 9.48, we can choose $N \in \mathbf{N}$ such that $\{I_1, I_2, \ldots, I_N\}$ covers $H$ and

(15)
$$\sum_{\nu=1}^{N} |I_\nu| < \frac{1}{j_0}.$$

We must find a partition $P$ such that $U(f, P) - L(f, P) < \varepsilon$. The endpoints of the $I_\nu$'s form part of this partition. Other points will come from further division of that part of $[a, b]$ not covered by the $I_\nu$'s. Indeed, let $I' \subseteq [a, b] \setminus (\cup_{\nu=1}^{N} I_\nu)$. Since the $I_\nu$'s cover $H$, $\omega_f(t) < 1/j_0$ for all $t \in I'$. Hence, by Lemma 9.49, there is a $\delta > 0$ such that if $J \subseteq I'$ satisfies $|J| < \delta$, then $\Omega_f(J) < 1/j_0$. Subdivide $[a, b] \setminus (\cup_{\nu=1}^{N} I_\nu)$ into intervals $J_\ell$, $\ell = 1, \ldots, s$, such that $|J_\ell| < \delta$. Then

(16)
$$\Omega_f(J_\ell) < \frac{1}{j_0}$$

for $\ell = 1, \ldots, s$.

Let $P = \{x_0, x_1, \ldots, x_n\}$ represent the collection of points $x$ such that $x$ is an endpoint of some $I_\nu$ or of some $J_\ell$. Notice that if $(x_{k-1}, x_k) \cap H \neq \emptyset$, then $x_{k-1}$ and $x_k$ are endpoints of some $I_\nu$, whence by (15),

$$\sum_{(x_{k-1}, x_k) \cap H \neq \emptyset} (M_k(f) - m_k(f))(x_k - x_{k-1}) \leq \frac{M - m}{j_0}.$$

On the other hand, if $(x_{k-1}, x_k) \cap H = \emptyset$, then $x_{k-1}$ and $x_k$ are endpoints of some $J_\ell$, whence by (16),

$$\sum_{(x_{k-1}, x_k) \cap H = \emptyset} (M_k(f) - m_k(f))(x_k - x_{k-1}) \leq \frac{1}{j_0} \sum_{k=1}^{n} (x_k - x_{k-1}) = \frac{b-a}{j_0}.$$

Consequently,

$$U(f, P) - L(f, P) = \sum_{k=1}^{n} (M_k(f) - m_k(f))(x_k - x_{k-1}) \leq \frac{M - m + b - a}{j_0} < \varepsilon.$$

We conclude that $f$ is integrable on $[a, b]$. ∎

Recall that if $\alpha > 0$ and $f(x)$ is positive, then

$$f^\alpha(x) := e^{\alpha \log(f(x))}.$$

Suppose $f$ is Riemann integrable. Although Corollary 5.23 implies that $f^n$ is integrable for each $n \in \mathbf{N}$, we have not yet investigated the integrability of noninteger powers of $f$, e.g., $\sqrt{f}$ and $\sqrt[3]{f}$. The following result shows that Lebesgue's Theorem answers the question of integrability for all positive powers of $f$, rational or irrational.

**9.51 COROLLARY.** If $f : [a, b] \to [0, \infty)$ is Riemann integrable, then so is $f^\alpha$ for every $\alpha > 0$.

In our final application, we use connectivity to characterize the graph of a continuous function.

**9.52 THEOREM** [CLOSED GRAPH THEOREM]. *Let $I$ be a closed interval and $f : I \to \mathbf{R}$. Then $f$ is continuous on $I$ if and only if the graph of $f$ is closed and connected in $\mathbf{R}^2$.*

PROOF. For any interval $J \subseteq I$, let $\mathcal{G}(J)$ represent the graph of $y = f(x)$ for $x \in J$. Suppose $f$ is continuous on $I$. The function $x \longmapsto (x, f(x))$ is continuous from $I$ into $\mathbf{R}^2$, and $I$ is connected in $\mathbf{R}$. Thus $\mathcal{G}(I)$ is connected in $\mathbf{R}^2$ by Theorem 9.36. To prove $\mathcal{G}(I)$ is closed, we shall use Theorem 9.6. Let $x_k \in I$ and $(x_k, f(x_k)) \to (x, y)$ as $k \to \infty$. Then $x_k \to x$ and $f(x_k) \to y$, as $k \to \infty$. Hence, $x \in I$ and since $f$ is continuous, $f(x_k) \to f(x)$. In particular, the graph of $f$ is closed.

Conversely, suppose the graph of $f$ is closed and connected in $\mathbf{R}^2$. We first show that $f$ satisfies the Intermediate Value Theorem on $I$. Indeed, suppose to the contrary that there exist $x_1 < x_2$ in $I$ with $f(x_1) \neq f(x_2)$ and a value $y_0$ between $f(x_1)$ and $f(x_2)$ such that $f(t) \neq y_0$ for all $t \in [x_1, x_2]$. Suppose for simplicity that $f(x_1) < f(x_2)$. Since $f(t) \neq y_0$ for any $t \in [x_1, x_2]$, the open sets

$$U = \{(x, y) : x < x_1\} \cup \{(x, y) : x < x_2, \ y < y_0\},$$

$$V = \{(x, y) : x > x_2\} \cup \{(x, y) : x > x_1, \ y > y_0\}$$

separate $\mathcal{G}(I)$, a contradiction. Therefore, $f$ satisfies the Intermediate Value Theorem on $I$.

If $f$ is not continuous on $I$, then there exist numbers $x_0 \in I$, $\varepsilon_0 > 0$, and $x_k \in I$ such that $x_k \to x_0$ and $|f(x_k) - f(x_0)| > \varepsilon_0$. By symmetry, we may suppose that $f(x_k) > f(x_0) + \varepsilon_0$ for infinitely many $k$'s, say

$$f(x_{k_j}) > f(x_0) + \varepsilon_0 > f(x_0), \qquad j \in \mathbf{N}.$$

By the Intermediate Value Theorem, choose $c_j$ between $x_{k_j}$ and $x_0$ such that $f(c_j) = f(x_0) + \varepsilon_0$. By construction, $(c_j, f(c_j)) \to (x_0, f(x_0) + \varepsilon_0)$ and $c_j \to x_0$ as $j \to \infty$. Hence, the graph of $f$ on $I$ is not closed. ∎

## EXERCISES

**1.** Suppose $f_k : [a, b] \to [0, \infty)$ for $k \in \mathbf{N}$ and

$$f(x) := \sum_{k=1}^{\infty} f_k(x)$$

converges pointwise on $[a, b]$. If $f$ and $f_k$ are continuous on $[a, b]$ for each $k \in \mathbf{N}$, prove that

$$\int_a^b \sum_{k=1}^{\infty} f_k(x) \, dx = \sum_{k=1}^{\infty} \int_a^b f_k(x) \, dx.$$

**2.** Let $E$ be a compact subset of $X$. Suppose $g, f_k, g_k : E \to \mathbf{R}$ are continuous on $E$ with $g_k \geq 0$ and $f_1 \geq f_2 \cdots \geq f_k \geq 0$ for all $k \in \mathbf{N}$. If $g = \sum_{k=1}^{\infty} g_k$ converges pointwise on $E$, prove $\sum_{k=1}^{\infty} f_k g_k$ converges uniformly on $E$.

**3.** Suppose $f, f_k : \mathbf{R} \to \mathbf{R}$ are continuous and nonnegative. Prove that if $f(x) \to 0$ as $x \to \pm\infty$ and $f_k \uparrow f$ everywhere on $\mathbf{R}$, then $f_k \to f$ uniformly on $\mathbf{R}$.

**4.** For each of the following functions, find a formula for $\omega_f(t)$.

a)
$$f(x) = \begin{cases} 1 & x \in \mathbf{Q} \\ 0 & x \notin \mathbf{Q}. \end{cases}$$

b)
$$f(x) = \begin{cases} 1 & x \geq 0 \\ 0 & x < 0. \end{cases}$$

c)
$$f(x) = \begin{cases} \sin(1/x) & x \neq 0 \\ 0 & x = 0. \end{cases}$$

**5.** Prove $(1 - x/k)^k \to e^{-x}$ uniformly on any compact subset of $\mathbf{R}$.

**6.** Show that if $f : [a, b] \to \mathbf{R}$ is integrable and $g : f([a, b]) \to \mathbf{R}$ is continuous, then $g \circ f$ is integrable on $[a, b]$. (Notice by Remark 3.34 that this result is false if $g$ is allowed even one point of discontinuity.)

**7.** Using Theorem 7.10 or Theorem 9.36, prove that each of the following limits exists. Find a value for the limit in each case.

a)
$$\lim_{k \to \infty} \int_0^{\pi/2} \sin x \sqrt{\frac{2k}{4k - 3x}} \, dx.$$

b)
$$\lim_{k \to \infty} \int_0^1 x^2 f\left(\frac{k}{k^2 + x}\right) dx,$$

where $f$ is continuously differentiable on $[0, 1]$ and $f'(0) > 0$.

c)
$$\lim_{k \to \infty} \int_0^1 x^3 \cos\left(\frac{\log k + x}{k + x}\right) dx.$$

d)
$$\lim_{k \to \infty} \int_{-1}^1 \left(1 + \frac{x}{k}\right)^k e^x \, dx.$$

**8. a)** Prove that for every $\varepsilon > 0$ there is a sequence of open intervals $\{I_k\}_{k \in \mathbf{N}}$ which covers $[0, 1] \cap \mathbf{Q}$ such that

$$\sum_{k=1}^\infty |I_k| < \varepsilon.$$

**b)** Prove that if $\{I_k\}_{k \in \mathbf{N}}$ is a sequence of open intervals which covers $[0, 1]$, then there is an $N \in \mathbf{N}$ such that

$$\sum_{k=1}^N |I_k| \geq 1.$$

**9.** Let $E_1$ be the unit interval $[0, 1]$ with its middle third $(1/3, 2/3)$ removed, i.e., $E_1 = [0, 1/3] \cup [2/3, 1]$. Let $E_2$ be $E_1$ with its middle thirds removed, i.e.,

$$E_2 = [0, 1/9] \cup [2/9, 1/3] \cup [2/3, 7/9] \cup [8/9, 1].$$

Continuing in this manner, generate nested sets $E_k$ such that each $E_k$ is the union of $2^k$ closed intervals of length $1/3^k$. The *Cantor set* is the set

$$E := \bigcap_{k=1}^\infty E_k.$$

Assume that every point $x \in [0,1]$ has a binary expansion and a ternary expansion; i.e., there exist $a_k \in \{0,1\}$ and $b_k \in \{0,1,2\}$ such that

$$x = \sum_{k=1}^{\infty} \frac{a_k}{2^k} = \sum_{k=1}^{\infty} \frac{b_k}{3^k}.$$

(For example, if $x = 1/3$, then $a_{2k-1} = 0$, $a_{2k} = 1$ for all $k$ and either $b_1 = 1$, $b_k = 0$ for $k > 1$ or $b_1 = 0$ and $b_k = 1$ for all $k > 1$.)

a) Prove that $E$ is a nonempty compact set of measure zero.

b) Show that a point $x \in [0,1]$ belongs to $E$ if and only if $x$ has a ternary expansion whose digits satisfy $b_k \neq 1$ for all $k \in \mathbf{N}$.

c) Define $f : E \to [0,1]$ by

$$f\left(\sum_{k=1}^{\infty} \frac{b_k}{3^k}\right) = \sum_{k=1}^{\infty} \frac{b_k/2}{2^k}.$$

Prove $f$ is 1–1 from $E$ onto $[0,1]$; i.e., prove $E$ is uncountable.

d) Extend $f$ from $E$ to $[0,1]$ by making $f$ constant on the middle thirds $E_{k-1} \setminus E_k$. Prove that $f : [0,1] \to [0,1]$ is continuous and increasing.

(Note: The function $f$ is almost everywhere constant on $[0,1]$, i.e., constant off a set of measure zero. Yet, it begins at $f(0) = 0$ and ends at $f(1) = 1$.)

# Chapter 10

# Metric Spaces

This chapter, an alternative to Chapter 9, covers topological ideas in a metric space setting. *If you have already covered Chapter 9, skip this one and proceed directly to Chapter 11.*

## 10.1 INTRODUCTION

The following concept shows up in many parts of analysis.

**10.1 DEFINITION.** A *metric space* is a set $X$ together with a function $\rho : X \times X \to \mathbf{R}$ (called the *metric* of $X$), which satisfies the following properties for all $x, y, z \in X$:

$$\text{POSITIVE DEFINITE} \quad \rho(x, y) \geq 0 \text{ with } \rho(x, y) = 0 \text{ if and only if } x = y,$$
$$\text{SYMMETRIC} \quad \rho(x, y) = \rho(y, x),$$
$$\text{TRIANGLE INEQUALITY} \quad \rho(x, y) \leq \rho(x, z) + \rho(z, y).$$

(Notice that by definition, $\rho(x, y)$ is finite–valued for all $x, y \in X$.)

We are already very familiar with a whole class of metric spaces.

**10.2 Example.** For each $n \in \mathbf{N}$, $\mathbf{R}^n$ is a metric space with metric $\rho(\boldsymbol{x}, \boldsymbol{y}) = \|\boldsymbol{x} - \boldsymbol{y}\|$. (We shall call this the *usual metric* on $\mathbf{R}^n$. Unless specified otherwise, we shall always use the usual metric on $\mathbf{R}^n$.)

PROOF. By Theorems 1.7 and 8.5, $\rho$ is a metric on $\mathbf{R}^n$. ∎

We shall develop a theory of convergence (for both sequences and functions) for arbitrary metric spaces. According to Example 10.2, this theory is valid (and will be used by us almost exclusively) on $\mathbf{R}^n$. Why, then, subject ourselves to such stark generality? Why not stick with the concrete Euclidean space case? There are at least three answers to these questions: 1) Economy. You will soon discover that there are many other metric spaces which crop up in analysis, e.g., all Hilbert spaces, all normed linear spaces, and many function spaces, including the space of continuous functions on a closed bounded interval. Our general theory of convergence in metric spaces will be valid for each of these examples too. 2) Visualization. As we mentioned in Section 1.1, analysis has a strong geometric flavor. Working in an abstract metric space only makes that aspect more apparent. 3) Simplicity. Emphasizing the fact that $\mathbf{R}^n$ is a metric space strips $\mathbf{R}$ of all extraneous details (the field operations, the order relation, decimal expansions) so that we can focus our attention on the underlying concept (distance) which governs convergence. Mathematics frequently benefits from such abstraction. Instead of becoming more difficult, generality actually makes the proofs easier to construct.

On the other hand, $\mathbf{R}^2$ provides a good and sufficiently general model for most of the theory of abstract metric spaces (especially, convergence of sequences and continuity of functions). For this reason, we often draw two–dimensional pictures to illustrate ideas and motivate proofs in an arbitrary metric space. (For example, see the proof of Remark 10.9 below.) We must not, however, allow ourselves to be mislead into believing that $\mathbf{R}^2$ provides the complete picture. Metric spaces have such simple structure that they can take on many bizarre forms. With that in mind, we introduce several more examples.

**10.3 Example.** $\mathbf{R}$ is a metric space with metric

$$\sigma(x, y) = \begin{cases} 0 & x = y \\ 1 & x \neq y. \end{cases}$$

(This metric is called the *discrete metric.*)

PROOF. The function $\sigma$ is obviously positive definite and symmetric. To prove $\sigma$ satisfies the triangle inequality, we consider three cases. If $x = z$, then $\sigma(x, y) = 0 + \sigma(z, y) = \sigma(x, z) + \sigma(z, y)$. A similar equality holds if $y = z$. Finally, if $x \neq z$ and $y \neq z$, then $\sigma(x, y) \leq 1 < 2 = \sigma(x, z) + \sigma(z, y)$. ∎

Comparing Examples 10.2 and 10.3, we see that a given set can have more than one metric. Hence, to describe a particular metric space, we must specify both the set $X$ and the metric $\rho$. For the rest of this chapter (unless otherwise stated), $X$ and $Y$ will represent arbitrary metric spaces (with respective metrics $\rho$ and $\tau$).

**10.4 Example.** If $E \subseteq X$, then $E$ is a metric space with metric $\rho$. (We shall call such metric spaces $E$ *subspaces* of $X$.)

PROOF. If the positive definite property, the symmetric property, and the triangle inequality hold for all $x, y \in X$, then they hold for all $x, y \in E$. ∎

A particular example of a subspace is provided by the set of rationals in $\mathbf{R}$.

**10.5 Example. Q** is a metric space with metric $\rho(x,y) = |x - y|$.

Metric spaces are by no means confined to numbers and vectors. Here is an important metric space whose "points" are functions.

**10.6 Example.** Let $\mathcal{C}[a,b]$ represent the collection of continuous $f : [a,b] \to \mathbf{R}$ and

$$\|f\| := \sup_{x \in [a,b]} |f(x)|.$$

Then $\rho(f,g) := \|f - g\|$ is a metric on $\mathcal{C}[a,b]$.

PROOF. By the Extreme Value Theorem, $\|f\|$ is finite for each $f \in \mathcal{C}[a,b]$. By definition, $\|f\| \geq 0$ for all $f$, and $\|f\| = 0$ if and only if $f(x) = 0$ for every $x \in [a,b]$. Thus $\rho$ is positive definite. Since $\rho$ is obviously symmetric, it remains to verify the triangle inequality. But

$$\|f + g\| = \sup_{x \in [a,b]} |f(x) + g(x)| \leq \sup_{x \in [a,b]} |f(x)| + \sup_{x \in [a,b]} |g(x)| = \|f\| + \|g\|. \quad \blacksquare$$

It is interesting to note that convergence in this metric space means uniform convergence (see Exercise 8 in Section 10.2 below).

There are two ways to generalize open and closed intervals to arbitrary metric spaces. One way is to use the metric directly as follows.

**10.7 DEFINITION.** Let $a \in X$ and $r > 0$. The *open ball* (in $X$) with *center a* and *radius r* is the set

$$B_r(a) := \{x \in X : \rho(x,a) < r\},$$

and the *closed ball* (in $X$) with *center a* and *radius r* is the set

$$\overline{B_r(a)} := \{x \in X : \rho(x,a) \leq r\}.$$

Notice by Theorem 1.6 that in $\mathbf{R}$ (with the usual metric), $B_r(a) = (r - a, r + a)$ and $\overline{B_r(a)} = [r - a, r + a]$; i.e., open balls are open intervals and closed balls are closed intervals. With respect to the discrete metric, however, balls look quite different. For example, $B_r(a) = \overline{B_r(a)} = \{a\}$ for all $0 < r < 1$.

The other way to generalize open and closed intervals to $X$ is to specify what "open" and "closed" mean. Notice that every point $x$ in an open interval $I$ is surrounded by points in $I$. The same property holds for complements of closed intervals. This leads us to the following definition.

**10.8 DEFINITION.** i) A set $V \subseteq X$ is said to be *open* if for every $x \in V$ there is an $\varepsilon > 0$ such that the open ball $B_\varepsilon(x)$ is contained in $V$.

ii) A set $E \subseteq X$ is said to be *closed* if $E^c := X \setminus E$ is open.

Our first result about these concepts shows that they are consistent as applied to balls.

**10.9 Remark.** *Every open ball is open, and every closed ball is closed.*

PROOF. Let $B_r(a)$ be an open ball. By definition, we must prove that given $x \in B_r(a)$ there is an $\varepsilon > 0$ such that $B_\varepsilon(x) \subseteq B_r(a)$. Let $x \in B_r(a)$ and set $\varepsilon = r - \rho(x, a)$. (Look at Figure 9.1 in the previous chapter to see why this choice of $\varepsilon$ should work.) If $y \in B_\varepsilon(x)$, then by the triangle inequality, assumption, and the choice of $\varepsilon$,

$$\rho(y, a) \leq \rho(y, x) + \rho(x, a) < \varepsilon + \rho(x, a) = r.$$

Thus by Definition 10.7, $y \in B_r(a)$. In particular, $B_\varepsilon(x) \subseteq B_r(a)$. Similarly, we can show that $\{x \in X : \rho(x, a) > r\}$ is also open. Hence, every closed ball is closed. ∎

Here are more examples of open sets and closed sets.

**10.10 Remark.** *If $a \in X$, then $X \setminus \{a\}$ is open and $\{a\}$ is closed.*

PROOF. By Definition 10.8, it suffices to prove that the complement of every *singleton* $E := \{a\}$ is open. Let $x \in E^c$ and set $\varepsilon = \rho(x, a)$. Then by Definition 10.7, $a \notin B_\varepsilon(x)$, so $B_\varepsilon(x) \subseteq E^c$. Therefore, $E^c$ is open by Definition 10.8. ∎

Students sometimes mistakenly believe that every set is either open or closed. Some sets are neither open nor closed (like the interval $[0, 1)$). And, as the following result shows, every metric space contains two special sets which are both open and closed.

**10.11 Remark.** *In an arbitrary metric space, the empty set $\emptyset$ and the whole space $X$ are both open and closed.*

PROOF. Since $X = \emptyset^c$ and $\emptyset = X^c$, it suffices by Definition 10.8 to prove that $\emptyset$ and $X$ are both open. Because the empty set contains no points, "every" point $x \in \emptyset$ satisfies $B_\varepsilon(x) \subseteq \emptyset$. (This is called the *vacuous implication*.) Therefore, $\emptyset$ is open. On the other hand, since $B_\varepsilon(x) \subseteq X$ for all $x \in X$ and all $\varepsilon > 0$, it is clear that $X$ is open. ∎

For some metric spaces (like $\mathbf{R}^n$), these are the only two sets which are simultaneously open and closed. For other metric spaces, there are many such sets.

**10.12 Example.** In the discrete space $\mathbf{R}$, every set is both open and closed.

PROOF. It suffices to prove that every subset of $\mathbf{R}$ is open (with respect to the discrete metric). Let $E \subseteq \mathbf{R}$. By Remark 10.11, we may assume that $E$ is nonempty. Let $a \in E$. Since $B_1(a) = \{a\}$, some open ball containing $a$ is a subset of $E$. By Definition 10.8, $E$ is open. ∎

To see how these concepts are connected with limits, we examine convergence of sequences in an arbitrary metric space. Using the analogy between the metric $\rho$ and the absolute value, we can transfer much of the theory of limits of sequences from $\mathbf{R}$ to any metric space. Here are the basic definitions.

**10.13 DEFINITION.** Let $\{x_n\}$ be a sequence in a metric space $X$.

  i) $x_n$ *converges* (in $X$) if there is a point $a \in X$ (called the *limit* of $a$) such that for every $\varepsilon > 0$ there is an $N \in \mathbf{N}$ such that

$$n \geq N \quad \text{implies} \quad \rho(x_n, a) < \varepsilon.$$

ii) $x_n$ is *Cauchy* if for every $\varepsilon > 0$ there is an $N \in \mathbf{N}$ such that

$$n, m \geq N \quad \text{implies} \quad \rho(x_n, x_m) < \varepsilon.$$

iii) $x_n$ is *bounded* if there is an $M > 0$ and a $b \in X$ such that $\rho(x_n, b) \leq M$ for all $n \in \mathbf{N}$.

Modifying the proofs in Chapter 2, by doing little more than replacing $|x - y|$ by $\rho(x, y)$, we can establish the following result.

**10.14 THEOREM.** *Let $X$ be a metric space.*

i) *A sequence in $X$ can have at most one limit.*

ii) *If $x_n \in X$ converges to $a$ and $\{x_{n_k}\}$ is any subsequence of $\{x_n\}$, then $x_{n_k}$ converges to $a$ as $k \to \infty$.*

iii) *Every convergent sequence in $X$ is bounded.*

iv) *Every convergent sequence in $X$ is Cauchy.*

The following result shows that, by using open sets, we can describe convergence of sequences in an arbitrary metric space without reference to the distance function. Later in this chapter, we shall use this point of view to great advantage.

**10.15 Remark.** *Let $x_n \in X$. Then $x_n \to a$ as $n \to \infty$ if and only if for every open set $V$ which contains $a$ there is an $N \in \mathbf{N}$ such that $n \geq N$ implies $x_n \in V$.*

PROOF. Suppose $x_n \to a$, and let $V$ be an open set which contains $a$. By Definition 10.8, there is an $\varepsilon > 0$ such that $B_\varepsilon(a) \subseteq V$. Given this $\varepsilon$, use Definition 10.13 to choose an $N \in \mathbf{N}$ such that $n \geq N$ implies $x_n \in B_\varepsilon(a)$. By the choice of $\varepsilon$, $x_n \in V$ for all $n \geq N$.

Conversely, let $\varepsilon > 0$ and set $V = B_\varepsilon(a)$. Then $V$ is an open set which contains $a$; hence, by hypothesis, there is an $N \in \mathbf{N}$ such that $n \geq N$ implies $x_n \in V$. In particular, $\rho(x_n, a) < \varepsilon$ for all $n \geq N$. ∎

The following result, which we shall use many times, shows that convergent sequences can also be used to characterize closed sets.

**10.16 THEOREM.** *Let $E \subseteq X$. Then $E$ is closed if and only if the limit of every convergent sequence $x_k \in E$ satisfies*

$$\lim_{k \to \infty} x_k \in E.$$

PROOF. The theorem is vacuously satisfied if $E$ is the empty set.

Suppose that $E \neq \emptyset$ is closed but some sequence $x_n \in E$ converges to a point $x \in E^c$. Since $E$ is closed, $E^c$ is open. Thus, by Remark 10.15, there is an $N \in \mathbf{N}$ such that $n \geq N$ implies $x_n \in E^c$, a contradiction.

Conversely, suppose $E$ is a nonempty set such that every convergent sequence in $E$ has its limit in $E$. If $E$ is not closed, then by Remark 10.11, $E \neq X$, and by definition, $E^c$ is nonempty and not open. Thus, there is at least one point $x \in E^c$ such that no ball $B_r(x)$ is contained in $E^c$. Let $x_k \in B_{1/k}(x) \cap E$ for $k = 1, 2, \cdots$. Then $x_k \in E$ and $\rho(x_k, x) < 1/k$ for all $k \in \mathbf{N}$. Now $1/k \to 0$ as $k \to \infty$, so it follows from the Squeeze Theorem (these are real sequences) that $\rho(x_k, x) \to 0$ as $k \to \infty$, i.e., $x_k \to x$ as $k \to \infty$. Thus, by hypothesis, $x \in E$, a contradiction. ∎

Notice that the Bolzano–Weierstrass Theorem and Cauchy's Theorem are missing from Theorem 10.14. There is a simple reason for this. As the next two remarks show, neither of these results holds in an arbitrary metric space.

**10.17 Remark.** *The discrete space contains bounded sequences which have no convergent subsequences.*

Proof. Let $X = \mathbf{R}$ be the discrete metric space introduced in Example 10.3. Since $\sigma(0, k) = 1$ for all $k \in \mathbf{N}$, $\{k\}$ is a bounded sequence in $X$. Suppose there exist integers $k_1 < k_2 < \ldots$ and an $x \in X$ such that $k_j \to x$ as $j \to \infty$. Then there is an $N \in \mathbf{N}$ such that $\sigma(k_j, x) < 1$ for $j \geq N$, i.e., $k_j = x$ for all $j \geq N$. This contradiction proves that $\{k\}$ has no convergent subsequences. ∎

**10.18 Remark.** *The metric space $X = \mathbf{Q}$, introduced in Example 10.5, contains Cauchy sequences which do not converge.*

Proof. Choose (by the Density of Rationals) points $q_k \in \mathbf{Q}$ such that $q_k \to \sqrt{2}$. Then $\{q_k\}$ is Cauchy (by Theorem 10.14iv) but does not converge in $X$ since $\sqrt{2} \notin X$. ∎

This leads us to the following concept.

**10.19 DEFINITION.** A metric space $X$ is said to be *complete* if every Cauchy sequence $x_n \in X$ converges to some point in $X$.

By Theorem 8.16, $\mathbf{R}^n$ is complete for all $n \in \mathbf{N}$. What can be said about complete metric spaces in general?

**10.20 Remark.** *By Definition 10.19, a complete metric space $X$ satisfies two properties: 1) every Cauchy sequence in $X$ converges; 2) the limit of every Cauchy sequence in $X$ stays in $X$.*

Property 2), by Theorem 10.16, means that $X$ is closed. Hence, it is natural to ask: Is there a simple relationship between complete subspaces and closed subsets?

**10.21 THEOREM.** *Let $X$ be a complete metric space and $E$ be a subset of $X$. Then $E$ (as a subspace) is complete if and only if $E$ (as a subset) is closed.*

Proof. Suppose $E$ is complete and $x_n \in E$ converges. By Theorem 10.14iv) $\{x_n\}$ is Cauchy. Since $E$ is complete, it follows from Definition 10.19 that the limit of $\{x_n\}$ belongs to $E$. Thus, by Theorem 10.16, $E$ is closed.

Conversely, suppose $E$ is closed and $x_n \in E$ is Cauchy in $E$. Since the metrics on $X$ and $E$ are identical, $\{x_n\}$ is Cauchy in $X$. Since $X$ is complete, it follows that

$x_n \to x$, as $n \to \infty$, for some $x \in X$. But $E$ is closed, so $x$ must belong to $E$. Thus $E$ is complete by definition. ∎

## EXERCISES

**1.** If $a, b \in X$ and $\rho(a, b) < \varepsilon$ for all $\varepsilon > 0$, prove that $a = b$.

**2.** a) Prove that $\{x_k\}$ is bounded in $X$ if and only if $\sup_{k \in \mathbf{N}} \rho(x_k, a) < \infty$ for all $a \in X$.

     b) Prove that $\{\mathbf{x}_k\}$ is bounded in $\mathbf{R}^n$ if and only if there is a $C > 0$ such that $\|\mathbf{x}_k\| \leq C$ for all $k \in \mathbf{N}$.

**3.** Prove Theorem 10.14.

**4.** a) Let $a \in X$. Prove that if $x_n = a$ for every $n \in \mathbf{N}$, then $x_n$ converges. What does it converge to?

     b) Let $X = \mathbf{R}$ with the discrete metric. Prove that $x_n \to a$ as $n \to \infty$ if and only if $x_n = a$ for large $n$.

**5.** a) Let $\{x_n\}$ and $\{y_n\}$ be sequences in $X$ which converge to the same point. Prove that $\rho(x_n, y_n) \to 0$ as $n \to \infty$.

     b) Show that the converse of part a) is false.

**6.** Let $\{x_n\}$ be Cauchy in $X$. Prove that $\{x_n\}$ converges if and only if at least one of its subsequences converges.

**7.** Prove that the discrete space $\mathbf{R}$ is complete.

**8.** a) Prove that every finite subset of a metric space $X$ is closed.

     b) Prove that $\mathbf{Q}$ is not closed in $\mathbf{R}$.

**9.** a) Show that if $x \in B_r(a)$ then there is an $\varepsilon > 0$ such that $\overline{B_\varepsilon(x)} \subset B_r(a)$.

     b) If $a \neq b$ are distinct points in $X$, prove that there is an $r > 0$ such that $B_r(a) \cap B_r(b) = \emptyset$.

     c) Show that given two balls $B_r(a)$ and $B_s(b)$, and a point $x \in B_r(a) \cap B_s(b)$, there are radii $c$ and $d$ such that

$$B_c(x) \subseteq B_r(a) \cap B_s(b) \quad \text{and} \quad B_d(x) \supseteq B_r(a) \cup B_s(b).$$

**10.** a) A subset $E$ of $X$ is said to be *sequentially compact* if every sequence $x_n \in E$ has a convergent subsequence whose limit belongs to $E$. Prove that every sequentially compact set is closed and bounded.

     b) Prove that $\mathbf{R}$ is closed but not sequentially compact.

     c) Prove that every closed bounded subset of $\mathbf{R}$ is sequentially compact.

## 10.2 LIMITS OF FUNCTIONS

In the previous section we used results in Chapter 2 as a model for the theory of limits of sequences in an arbitrary metric space $X$. In this section we use results in Chapter 3 as a model to develop a theory of limits of functions which take one metric space $X$ to another $Y$.

A straightforward adaptation of Definition 3.1 leads us to guess that, in an arbitrary metric space, $f(x) \to L$ as $x \to a$ if for every $\varepsilon > 0$ there is a $\delta > 0$ such that

$$0 < \rho(x, a) < \delta \quad \text{implies} \quad \tau(f(x), L) < \varepsilon.$$

The only problem with this definition is that there may be no $x$ which satisfies $0 < \rho(x, a) < \delta$; e.g., if $X$ is the set $\mathbf{N}$ together with the metric $\rho(x, y) = |x - y|$ and $\delta = 1$. To prevent our theory from collapsing into the vacuous case, we introduce the following idea.

**DEFINITION 10.22.** A point $a$ is said to be a *cluster point* (of $X$) if $B_\delta(a)$ contains infinitely many points for each $\delta > 0$.

For example, every point in any Euclidean space $\mathbf{R}^n$ is a cluster point (of $\mathbf{R}^n$).

Notice that any concept defined on a metric space $X$ is automatically defined on all subsets of $X$. Indeed, since any subset $E$ of $X$ is itself a metric space (see Example 10.4 above), the definition can be applied to $E$ as well as to $X$.

To be more specific, let $E$ is a subspace of $X$, i.e., a nonempty subset of $X$. By Definition 10.7 an open ball in $E$ has the form

$$B_r^E(a) := \{x \in E : \rho(x, a) < r\}.$$

Since the metrics on $X$ and $E$ are the same, it follows that

$$B_r^E(a) = B_r(a) \cap E,$$

where $B_r(a)$ is an open ball in $X$. A similar statement holds for closed balls. We shall call these balls *relative balls* (in $E$). In particular, in the subspace $\mathbf{Q}$ of Example 10.5 above, the relative balls take on the form $B_r(a) = (r - a, r + a) \cap \mathbf{Q}$ and $\overline{B_r(a)} = [r - a, r + a] \cap \mathbf{Q}$.

What, then, does it mean for a set $E$ to have a cluster point? By Definition 10.22, a point $a \in X$ is a cluster point of a nonempty set $E \subseteq X$ if the relative ball $E \cap B_\delta(a)$ contains infinitely many points for each $\delta > 0$.

The etymology of the term *cluster point* is obvious. A cluster point of $E$ is a point near which $E$ "clusters." Cluster points are also called *points of accumulation*.

Notice that by definition, no finite set has cluster points. On the other hand, a set may have infinitely many cluster points. Indeed, by the Density of Rationals (Theorem 1.24), every point of $\mathbf{R}$ is a cluster point of $\mathbf{Q}$.

Here are two more examples of sets and their cluster points.

**10.23 Example.** Show that 0 is the only cluster point of the set

$$E = \left\{ \frac{1}{n} : n \in \mathbf{N} \right\}.$$

SOLUTION. By Theorem 1.22 (the Archimedean Principle), given $\delta > 0$ there is an $N \in \mathbf{N}$ such that $1/N < \delta$. Since $n \geq N$ implies $1/n \leq 1/N$, it follows that $(-\delta, \delta) \cap E$ contains infinitely many points. Thus 0 is a cluster point of $E$.

On the other hand, if $x_0 \neq 0$, then choose $\delta < |x_0|$, and notice that either $x_0 - \delta > 0$ or $x_0 + \delta < 0$. Thus $(x_0 - \delta, x_0 + \delta) \cap E$ contains at most finitely many points; i.e., $x_0$ is not a cluster point of $E$. ∎

**10.24 Example.** Show every point in the interval $[0, 1]$ is a cluster point of the open interval $(0, 1)$.

SOLUTION. Let $x_0 \in [0, 1]$ and $\delta > 0$. Then $x_0 + \delta > 0$ and $x_0 - \delta < 1$. In particular, $(x_0 - \delta, x_0 + \delta) \cap (0, 1)$ is itself a nondegenerate interval, say $(a, b)$. But $(a, b)$ contains infinitely many points, e.g., $(a + b)/2$, $(2a + b)/3$, $(3a + b)/4$, $\cdots$. Therefore, $x_0$ is a cluster point of $(0, 1)$. ∎

We are now prepared to define limits of functions on metric spaces.

**10.25 DEFINITION.** Let $a$ be a cluster point of $X$ and $f : X \setminus \{a\} \to Y$. Then $f(x)$ is said to *converge to $L$, as $x$ approaches $a$,* if for every $\varepsilon > 0$ there is a $\delta > 0$ such that

$$(1) \qquad 0 < \rho(x, a) < \delta \quad \text{implies} \quad \tau(f(x), L) < \varepsilon.$$

In this case we write

$$L = \lim_{x \to a} f(x)$$

and call $L$ the *limit* of $f(x)$ as $x$ approaches $a$.

By modifying the proofs presented in Chapter 3, we can prove the following results about limits in metric spaces.

**10.26 THEOREM.** *Let $a$ be a cluster point of $X$ and $f, g : X \setminus \{a\} \to Y$.*

i) *If $f(x) = g(x)$ for all $x \in X \setminus \{a\}$ and $f(x)$ has a limit as $x \to a$, then $g(x)$ also has a limit as $x \to a$, and*

$$\lim_{x \to a} g(x) = \lim_{x \to a} f(x).$$

ii) [SEQUENTIAL CHARACTERIZATION OF LIMITS]. *The limit*

$$L := \lim_{x \to a} f(x)$$

*exists if and only if $f(x_n) \to L$ as $n \to \infty$ for every sequence $x_n \in X \setminus \{a\}$ which converges to $a$ as $n \to \infty$.*

iii) *Suppose $Y = \mathbf{R}^n$. If $f(x)$ and $g(x)$ have a limit as $x$ approaches $a$, then so do $(f + g)(x)$, $(f \cdot g)(x)$, $(\alpha f)(x)$, and $(f/g)(x)$ (when $Y = \mathbf{R}$ and the limit of $g(x)$ is nonzero). In fact,*

$$\lim_{x \to a} (f + g)(x) = \lim_{x \to a} f(x) + \lim_{x \to a} g(x),$$

$$\lim_{x \to a} (\alpha f)(x) = \alpha \lim_{x \to a} f(x),$$

$$\lim_{x \to a} (f \cdot g)(x) = \lim_{x \to a} f(x) \cdot \lim_{x \to a} g(x),$$

and (when $Y = \mathbf{R}$ and the limit of $g(x)$ is nonzero)

$$\lim_{x \to a} \left( \frac{f}{g} \right)(x) = \frac{\lim_{x \to a} f(x)}{\lim_{x \to a} g(x)}.$$

iv) [SQUEEZE THEOREM FOR FUNCTIONS]. *Suppose* $Y = \mathbf{R}$. *If* $h : X \backslash \{a\} \to \mathbf{R}$ *satisfies* $g(x) \le h(x) \le f(x)$ *for all* $x \in X \backslash \{a\}$, *and*

$$\lim_{x \to a} f(x) = \lim_{x \to a} g(x) = L,$$

*then the limit of* $h$ *exists, as* $x \to a$, *and*

$$\lim_{x \to a} h(x) = L.$$

v) [COMPARISON THEOREM FOR FUNCTIONS]. *Suppose* $Y = \mathbf{R}$. *If* $f(x) \le g(x)$ *for all* $x \in X \backslash \{a\}$, *and* $f$ *and* $g$ *have a limit as* $x$ *approaches* $a$, *then*

$$\lim_{x \to a} f(x) \le \lim_{x \to a} g(x).$$

Here is the metric space version Definition 3.20.

**10.27 DEFINITION.** Let $E$ be a nonempty subset of $X$ and $f : E \to Y$.

i) $f$ is said to be *continuous at a point* $a \in E$ if given $\varepsilon > 0$ there is a $\delta > 0$ such that

$$\rho(x, a) < \delta \quad \text{and} \quad x \in E \quad \text{imply} \quad \tau(f(x), f(a)) < \varepsilon.$$

ii) $f$ is said to be *continuous on* $E$ (notation: $f : E \to Y$ is continuous) if $f$ is continuous at every $x \in E$.

iii) $f$ is said to be *continuous* if it is continuous on its domain $\mathrm{Dom}\,(f)$.

Notice that this definition is valid whether $a$ is a cluster point or not. Modifying corresponding proofs in Chapter 3, we can prove the following results.

**10.28 THEOREM.** Let $E$ be a nonempty subset of $X$ and $f, g : E \to Y$.

i) $f$ *is continuous at* $a \in E$ *if and only if* $f(x_n) \to f(a)$, *as* $n \to \infty$, *for all sequences* $x_n \in E$ *which converge to* $a$.

ii) *Suppose* $Y = \mathbf{R}^n$. *If* $f, g$ *are continuous at a point* $a \in E$ *(respectively, continuous on a set* $E$*), then so are* $f + g$, $f \cdot g$, *and* $\alpha f$ *(for any* $\alpha \in \mathbf{R}$*). Moreover, in the case* $Y = \mathbf{R}$, $f/g$ *is continuous at* $a \in E$ *when* $g(a) \ne 0$ *(respectively, on* $E$ *when* $g(x) \ne 0$ *for all* $x \in E$*).*

The following result shows that the composition of two continuous functions is continuous regardless which metric spaces are involved.

**10.29 THEOREM.** *Suppose $X$, $Y$, and $Z$ are metric spaces, $a$ is a cluster point of $X$, $f : X \to Y$, and $g : f(X) \to Z$. If $f(x) \to L$ as $x \to a$ and $g$ is continuous at $L$, then*

$$\lim_{x \to a} (g \circ f)(x) = g\left(\lim_{x \to a} f(x)\right).$$

We shall examine the metric space analogues of the Extreme Value Theorem, the Intermediate Value Theorem, and uniform continuity in Section 10.4 below.

## EXERCISES

**1.** Find all cluster points of each of the following sets.
   a) $E = \mathbf{R} \setminus \mathbf{Q}$.
   b) $E = [a, b)$, $a, b \in \mathbf{R}$, $a < b$.
   c) $E = \{(-1)^n n : n \in \mathbf{N}\}$.
   d) $E = \{x_n : n \in \mathbf{N}\}$, where $x_n \to x$ as $n \to \infty$.
   e) $E = \{1, 1, 2, 1, 2, 3, 1, 2, 3, 4, \ldots\}$.

**2.** a) A point $a$ in a metric space $X$ is said to be *isolated* if there is an $r > 0$ so small that $B_r(a) = \{a\}$. Show that a point $a \in X$ is not a cluster point of $X$ if and only if $a$ is isolated.
   b) Prove that the discrete space has no cluster points.

**3.** Prove that $a$ is a cluster point for some $E \subseteq X$ if and only if there is a sequence $x_n \in E \setminus \{a\}$ such that $x_n \to a$ as $n \to \infty$.

**4.** a) Let $E$ be a nonempty subset of $X$. Prove that $a$ is a cluster point of $E$ if and only if for each $r > 0$, $E \cap B_r(a) \setminus \{a\}$ is nonempty.
   b) Prove that every bounded infinite subset of $\mathbf{R}$ has at least one cluster point.

**5.** Prove Theorem 10.26.
**6.** Prove Theorem 10.28.
**7.** Prove Theorem 10.29.
**8.** Prove that if $f_n \in \mathcal{C}[a, b]$, then $f_n \to f$ uniformly on $[a, b]$ if and only if $f_n \to f$ in the metric of $\mathcal{C}[a, b]$ (see Example 10.6).
**9.** Suppose $X$ is a metric space which satisfies the following condition.

**10.30 DEFINITION.** $X$ is said to satisfy the *Bolzano–Weierstrass Property* if every bounded sequence $x_n \in X$ has a convergent subsequence.

   a) Prove that if $E$ is a closed, bounded subset of $X$ and $x_n \in E$, then there is an $a \in E$ and a subsequence $x_{n_k}$ of $x_n$ such that $x_{n_k} \to a$ as $k \to \infty$.
   b) If $E$ is closed and bounded in $X$ and $f : E \to \mathbf{R}$ is continuous on $E$, prove $f$ is bounded on $E$.
   c) Prove that under the hypotheses of part b) that there exist points $x_m, x_M \in E$ such that

$$f(x_M) = \sup_{x \in E} f(x) \quad \text{and} \quad f(x_m) = \inf_{x \in E} f(x).$$

## 10.3  INTERIOR, CLOSURE, AND BOUNDARY

Thus far, we have used "open" and "closed" mostly for identification. At this point, we begin to examine these concepts in more depth. Our first result shows that open sets and closed sets behave very differently with respect to unions and intersections.

**10.31 THEOREM.** *Let $X$ be a metric space.*

i) *If $\{V_\alpha\}_{\alpha \in A}$ is any collection of open sets in $X$, then*

$$\bigcup_{\alpha \in A} V_\alpha$$

*is open.*

ii) *If $\{V_k : k = 1, 2, \ldots, n\}$ is a finite collection of open sets in $X$, then*

$$\bigcap_{k=1}^{n} V_k := \bigcap_{k \in \{1,2,\ldots,n\}} V_k$$

*is open.*

iii) *If $\{E_\alpha\}_{\alpha \in A}$ is any collection of closed sets in $X$, then*

$$\bigcap_{\alpha \in A} E_\alpha$$

*is closed.*

iv) *If $\{E_k : k = 1, 2, \ldots, n\}$ is a finite collection of closed sets in $X$, then*

$$\bigcup_{k=1}^{n} E_k := \bigcup_{k \in \{1,2,\ldots,n\}} E_k$$

*is closed.*

v) *If $V$ is open in $X$ and $E$ is closed in $X$, then $V \setminus E$ is open and $E \setminus V$ is closed.*

**PROOF.** i) Let $x \in \bigcup_{\alpha \in A} V_\alpha$. Then $x \in V_\alpha$ for some $\alpha \in A$. Since $V_\alpha$ is open, it follows that there is an $r > 0$ such that $B_r(x) \subseteq V_\alpha$. Thus $B_r(x) \subseteq \bigcup_{\alpha \in A} V_\alpha$; i.e., this union is open.

ii) Let $x \in \bigcap_{k=1}^{n} V_k$. Then $x \in V_k$ for $k = 1, 2, \ldots, n$. Since each $V_k$ is open, it follows that there are numbers $r_k > 0$ such that $B_{r_k}(x) \subseteq V_k$. Let $r = \min\{r_1, \ldots, r_n\}$. Then $r > 0$ and $B_r(x) \subseteq V_k$ for all $k = 1, 2, \ldots, n$, i.e., $B_r(x) \subseteq \bigcap_{k=1}^{n} V_k$. Hence, this intersection is open.

iii) By DeMorgan's Law (Theorem 1.41) and part i),

$$\left( \bigcap_{\alpha \in A} E_\alpha \right)^c = \bigcup_{\alpha \in A} E_\alpha^c$$

is open, so $\bigcap_{\alpha \in A} E_\alpha$ is closed.

iv) By DeMorgan's Law and part ii),

$$\left( \bigcup_{k=1}^{n} E_k \right)^c = \bigcap_{k=1}^{n} E_k^c$$

is open, so $\bigcup_{k=1}^{n} E_k$ is closed.

v) Since $V \setminus E = V \cap E^c$ and $E \setminus V = E \cap V^c$, the former is open by part ii), and the latter is closed by part iii). ∎

The finiteness hypothesis in Theorem 10.31 is critical, even for the case $X = \mathbf{R}$.

**10.32 Remark.** *Statements ii) and iv) of Theorem 10.31 are false if arbitrary collections are used in place of finite collections.*

PROOF. In the metric space $X = \mathbf{R}$,

$$\bigcap_{k \in \mathbf{N}} \left( -\frac{1}{k}, \frac{1}{k} \right) = \{0\}$$

is closed and

$$\bigcup_{k \in \mathbf{N}} \left[ \frac{1}{k+1}, \frac{k}{k+1} \right] = (0, 1)$$

is open. ∎

Theorem 10.31 has many applications. Our first application is that every set contains a largest open set, and is contained in a smallest closed set. To facilitate our discussion, we introduce the following topological operations.

**10.33 DEFINITION.** Let $E$ be a subset of a metric space $X$.

i) The *interior* of $E$ is the set

$$E^o := \bigcup \{V : V \subseteq E \text{ and } V \text{ is open in } X\}.$$

ii) The *closure* of $E$ is the set

$$\overline{E} := \bigcap \{B : B \supseteq E \text{ and } B \text{ is closed in } X\}.$$

Notice that every set $E$ contains the open set $\emptyset$ and is contained in the closed set $X$. Hence, the sets $E^o$ and $\overline{E}$ are well–defined. Also notice that by Theorem 10.31, the interior of a set is always open and the closure of a set is always closed.

The following result shows that $E^o$ is the largest open set contained in $E$, and $\overline{E}$ is the smallest closed set which contains $E$.

**10.34 THEOREM.** *Let $E \subseteq X$. Then*

   i) *$E^o \subseteq E \subseteq \overline{E}$,*

   ii) *if $V$ is open and $V \subseteq E$ then $V \subseteq E^o$, and*

   iii) *if $C$ is closed and $C \supseteq E$ then $C \supseteq \overline{E}$.*

PROOF. Since every open set $V$ in the union defining $E^o$ is a subset of $E$, it is clear that the union of these $V$'s is a subset of $E$. Thus $E^o \subseteq E$. A similar argument establishes $E \subseteq \overline{E}$. This proves i).

By Definition 10.33, if $V$ is an open subset of $E$, then $V \subseteq E^o$ and if $C$ is a closed set containing $E$, then $\overline{E} \subseteq C$. This proves ii) and iii). ∎

In particular, the interior of a bounded interval with endpoints $a$ and $b$ is $(a, b)$, and its closure is $[a, b]$. In fact, it is evident by parts ii) and iii) that $E = E^o$ if and only if $E$ is open and $E = \overline{E}$ if and only if $E$ is closed. We shall use this observation many times below.

The following examples illustrate the fact that the interior of a nice enough set $E$ in $\mathbf{R}^2$ can be obtained by removing all its "edges", and the closure of $E$ by adding all its "edges".

**10.35 Example.** Find the interior and closure of the set $E = \{(x, y) : -1 \le x \le 1$ and $-|x| < y < |x|\}$.

SOLUTION. Graph $y = |x|$ and $x = \pm 1$, and observe that $E$ is a bowtie-shaped region with "solid" vertical edges (see Figure 8.6). Now by Definition 10.8, any open set in $\mathbf{R}^2$ must contain a disk around each of its points. Since $E^o$ is the largest open set inside $E$, it is clear that

$$E^o = \{(x, y) : -1 < x < 1 \text{ and } -|x| < y < |x|\}.$$

Similarly,

$$\overline{E} = \{(x, y) : -1 \le x \le 1 \text{ and } -|x| \le y \le |x|\}. \quad ∎$$

**10.36 Example.** Find the interior and closure of the set $E = B_1(-2, 0) \cup B_1(2, 0) \cup \{(x, 0) : -1 \le x \le 1\}$.

SOLUTION. Draw a graph of this region. It turns out to be "dumbbell shaped": two open disks joined by a straight line. Thus $E^o = B_1(-2, 0) \cup B_1(2, 0)$ and

$$\overline{E} = \overline{B_1(-2, 0)} \cup \overline{B_1(2, 0)} \cup \{(x, 0) : -1 \le x \le 1\}. \quad ∎$$

One of the most important results from one-dimensional calculus is the Fundamental Theorem of Calculus. It states that the behavior of a derivative $f'$ on an interval $[a, b]$ is completely determined by the values of $f$ at the endpoints of $[a, b]$. What shall we use for "endpoints" of an arbitrary set in $X$? Notice that the endpoints $a, b$ are the only points which lie near both $[a, b]$ and the complement of $[a, b]$. Using this as a cue, we introduce the following concept.

**10.37 DEFINITION.** Let $E \subseteq X$. The *boundary* of $E$ is the set

$$\partial E := \{x \in X : \text{for all } r > 0, \quad B_r(x) \cap E \neq \emptyset \text{ and } B_r(x) \cap E^c \neq \emptyset\}.$$

(We will refer to the last two conditions in the definition of $\partial E$ by saying $B_r(x)$ *intersects* $E$ and $E^c$.)

**10.38 Example.** Describe the boundary of the set

$$E = \{(x, y) : x^2 + y^2 \leq 9 \text{ and } (x - 1)(y + 2) > 0\}.$$

SOLUTION. Graph the relations $x^2 + y^2 = 9$ and $(x - 1)(y + 2) = 0$ to obtain a region with solid curved edges and dotted straight edges (see Figure 8.5). By definition, then, the boundary of $E$ is the union of these curved and straight edges (all made solid). Rather than describing $\partial E$ analytically (which would involve solving for the intersection points of the straight lines $x = 1$, $y = -2$, and the circle $x^2 + y^2 = 9$), it is easier to describe $\partial E$ by using set algebra.

$$\partial E = \{(x, y) : x^2 + y^2 \leq 9 \text{ and } (x - 1)(y + 2) \geq 0\}$$
$$\setminus \{(x, y) : x^2 + y^2 < 9 \text{ and } (x - 1)(y + 2) > 0\}. \quad \blacksquare$$

It turns out that set algebra can be used to describe the boundary of any set.

**10.39 THEOREM.** *Let $E \subseteq X$. Then $\partial E = \overline{E} \setminus E^o$.*

PROOF. By Definition 10.37, it suffices to show

(2)                $x \in \overline{E}$ if and only if $B_r(x) \cap E \neq \emptyset$ for all $r > 0$, and

(3)                $x \notin E^o$ if and only if $B_r(x) \cap E^c \neq \emptyset$ for all $r > 0$.

We will provide the details for (2) and leave the proof of (3) as an exercise. Suppose $x \in \overline{E}$ but $B_{r_0}(x) \cap E = \emptyset$ for some $r_0 > 0$. Then $(B_{r_0}(x))^c$ is a closed set which contains $E$; hence, by Theorem 10.34iii), $\overline{E} \subseteq (B_{r_0}(x))^c$. It follows that $\overline{E} \cap B_{r_0}(x) = \emptyset$, e.g., $x \notin \overline{E}$, a contradiction.

Conversely, suppose $x \notin \overline{E}$. Since $(\overline{E})^c$ is open, there is an $r_0 > 0$ such that $B_{r_0}(x) \subseteq (\overline{E})^c$. In particular, $\emptyset = B_{r_0}(x) \cap \overline{E} \supseteq B_{r_0}(x) \cap E$ for some $r_0 > 0$. $\blacksquare$

We have introduced topological operations (interior, closure, and boundary). The following result answers the question: How do these operations interact with the set operations (union and intersection)?

**10.40 THEOREM.** *Let $A, B \subseteq X$. Then*

i)                $(A \cup B)^o \supseteq A^o \cup B^o, \quad (A \cap B)^o = A^o \cap B^o,$

ii)                $\overline{A \cup B} = \overline{A} \cup \overline{B}, \quad \overline{A \cap B} \subseteq \overline{A} \cap \overline{B},$

iii) $\partial(A \cup B) \subseteq \partial A \cup \partial B, \quad$ and $\quad \partial(A \cap B) \subseteq (A \cap \partial B) \cup (B \cap \partial A) \cup (\partial A \cap \partial B).$

PROOF. i) Since the union of two open sets is open, $A^o \cup B^o$ is an open subset of $A \cup B$. Hence, by Theorem 10.34ii), $A^o \cup B^o \subseteq (A \cup B)^o$.

Similarly, $(A \cap B)^o \supseteq A^o \cap B^o$. On the other hand, if $V \subset A \cap B$, then $V \subset A$ and $V \subset B$. Thus, $(A \cap B)^o \subseteq A^o \cap B^o$.

ii) Since $\overline{A} \cup \overline{B}$ is closed and contains $A \cup B$, it is clear that by Theorem 10.34iii), $\overline{A \cup B} \subseteq \overline{A} \cup \overline{B}$. Similarly, $\overline{A \cap B} \subseteq \overline{A} \cap \overline{B}$. To prove the reverse inequality for union, suppose $x \notin \overline{A \cup B}$. Then there is a closed set $E$ which contains $A \cup B$ such that $x \notin E$. Since $E$ contains both $A$ and $B$, it follows that $x \notin \overline{A}$ and $x \notin \overline{B}$. This proves part ii).

iii) Let $x \in \partial(A \cup B)$; i.e., suppose $B_r(x)$ intersects $A \cup B$ and $(A \cup B)^c$ for all $r > 0$. Since $(A \cup B)^c = A^c \cap B^c$, it follows that $B_r(x)$ intersects both $A^c$ and $B^c$ for all $r > 0$. Thus, $B_r(x)$ intersects $A$ and $A^c$ for all $r > 0$, or $B_r(x)$ intersects $B$ and $B^c$ for all $r > 0$, i.e., $x \in \partial A \cup \partial B$. This proves the first set inequality in part iii).

To prove the second set inequality, suppose $x \in \partial(A \cap B)$; i.e., suppose $B_r(x)$ intersects $A \cap B$ and $(A \cap B)^c$ for all $r > 0$. If $x \in (A \cap \partial B) \cup (B \cap \partial A)$, then there is nothing to prove. If $x \notin (A \cap \partial B) \cup (B \cap \partial A)$, then $x \in (A^c \cup (\partial B)^c) \cap (B^c \cup (\partial A)^c)$. Hence, it remains to prove that $A^c \cup (\partial B)^c \subseteq \partial A$ and $B^c \cup (\partial A)^c \subseteq \partial B$. By symmetry, we need only prove the first one. To this end, let $x \in A^c \cup (\partial B)^c$.

*Case 1.* $x \in A^c$. Since $B_r(x)$ intersects $A$, it follows that $x \in \partial A$.

*Case 2.* $x \in (\partial B)^c$. Since $B_r(x)$ intersects $B$, it follows that $B_r(x) \subseteq B$ for small $r > 0$. Since $B_r(x)$ also intersects $A^c \cup B^c$, it must be the case that $B_r(x)$ intersects $A^c$. In particular, $x \in \partial A$. ∎

### EXERCISES

**1.** Find the interior, closure, and boundary of each of the following subsets of $\mathbf{R}$.

a) $[a, b)$ where $a < b$.

b) $E = \{1/n : n \in \mathbf{N}\}$.

c) $E = \bigcup_{n=1}^{\infty} \left( \dfrac{1}{n+1}, \dfrac{1}{n} \right)$.

d) $E = \bigcup(-n, n)$.

**2.** Identify which of the following sets are open, which are closed, and which are neither. Find $E^o$, $\overline{E}$, and $\partial E$ and sketch $E$ in each case.

a) $E = \{(x, y) : x^2 + 4y^2 \leq 1\}$.

b) $E = \{(x, y) : x^2 - 2x + y^2 = 0\} \cup \{(x, 0) : x \in [2, 3]\}$.

c) $E = \{(x, y) : y \geq x^2, \ 0 \leq y < 1\}$.

d) $E = \{(x, y) : x^2 - y^2 < 1, \ -1 < y < 1\}$.

**3.** Let $X$ be a metric space, $a \in X$, $s < r$,

$$V = \{x \in X : s < \rho(x, a) < r\}, \quad \text{and} \quad E = \{x \in X : s \leq \rho(x, a) \leq r\}.$$

Prove that $V$ is open and $E$ is closed.

**4.** Suppose $A \subseteq B \subseteq X$. Prove that $\overline{A} \subseteq \overline{B}$ and $A^o \subseteq B^o$.

**5**. **This exercise is used in Section 10.5.** Show that if $E$ is closed in $X$ and $a \notin E$, then

$$\inf_{x \in E} \rho(x, a) > 0.$$

**6.** Prove (3).

**7.** Show that Theorem 10.40 is best possible in the following sense.

   a) There exist sets $A, B$ in $\mathbf{R}$ such that $(A \cup B)^o \neq A^o \cup B^o$.

   b) There exist sets $A, B$ in $\mathbf{R}$ such that $\overline{A \cap B} \neq \overline{A} \cap \overline{B}$.

   c) There exist sets $A, B$ in $\mathbf{R}$ such that $\partial(A \cup B) \neq \partial A \cup \partial B$ and $\partial(A \cap B) \neq (A \cap \partial B) \cup (B \cap \partial A) \cup (\partial A \cap \partial B)$.

**8**. **This exercise is used many times from Section 10.4 onwards.**

   Let $Y$ be a subspace of $X$.

   a) Show that a set $V$ is open in $Y$ if and only if there is an open set $U$ in $X$ such that $V = U \cap Y$.

   b) Show that a set $E$ is closed in $Y$ if and only if there is a closed set $A$ in $X$ such that $E = A \cap Y$.

**9.** Let $f : \mathbf{R} \to \mathbf{R}$. Prove that $f$ is continuous on $\mathbf{R}$ if and only if $f^{-1}(I)$ is open in $\mathbf{R}$ for every open interval $I$.

**10.** Let $V$ be a subset of a metric space $X$.

   a) Prove that $V$ is open in $X$ if and only if there is a collection of open balls $\{B_\alpha : \alpha \in A\}$ such that

$$V = \bigcup_{\alpha \in A} B_\alpha.$$

   b) What happens to this result if "open" is replaced by "closed"?

## 10.4  COMPACT SETS

In Chapter 3, we proved the Extreme Value Theorem for functions defined on $\mathbf{R}$. In this section we shall extend that result to functions defined on an arbitrary metric space. To replace the hypothesis "closed, bounded interval" used in the real case, we introduce "compactness," a concept which gives us a powerful tool for extending local results to global ones (see especially Remark 10.44 below and Theorem 10.52 below, and Theorem 12.46).

Since compactness of $E$ depends on how $E$ can be "covered" by a collection of open sets, we begin by introducing the following terminology.

**10.41 DEFINITION.** Let $\mathcal{V} = \{V_\alpha\}_{\alpha \in A}$ be a collection of subsets of a metric space $X$ and suppose $E$ is a subset of $X$.

   i) $\mathcal{V}$ is said to *cover* $E$ (or be a *covering* of $E$) if

$$E \subseteq \bigcup_{\alpha \in A} V_\alpha.$$

ii) $\mathcal{V}$ is said to be an *open covering* of $E$ if $\mathcal{V}$ covers $E$ and each $V_\alpha$ is open.

iii) Let $\mathcal{V}$ be a covering of $E$. $\mathcal{V}$ is said to have a *finite* (respectively, *countable*) *subcovering* if there is a finite (respectively, countable) subset $A_0$ of $A$ such that $\{V_\alpha\}_{\alpha \in A_0}$ covers $E$.

Notice that the collections of open intervals

$$\left\{\left(\frac{1}{k+1}, \frac{k}{k+1}\right)\right\}_{k \in \mathbf{N}} \quad \text{and} \quad \left\{\left(-\frac{1}{k}, \frac{k+1}{k}\right)\right\}_{k \in \mathbf{N}}$$

are open coverings of the interval $(0, 1)$. The first covering of $(0, 1)$ has no finite subcover but any member of the second covering covers $(0, 1)$. Thus an open covering of an arbitrary set may or may not have a finite subcovering. Sets which satisfy this special property are important enough to be given a name.

**10.42 DEFINITION.** A subset $H$ of a metric space $X$ is said to be *compact* if every open covering of $H$ has a finite subcover.

To get a feeling for what this definition means, we make some elementary observations concerning compact sets in general.

**10.43 Remark.** *The empty set and all finite subsets of a metric space are compact.*

PROOF. These statements follow immediately from Definition 10.42. The empty set needs no set to cover it, and any finite set $H$ can be covered by finitely many sets, one set for each element in $H$. ∎

Since the empty set and finite sets are also closed, it is natural to ask whether there is a relationship between compact sets and closed sets in general. The following three results address this question in an arbitrary metric space.

**10.44 Remark.** *A compact set is always closed.*

PROOF. Suppose $H$ is compact but not closed. Then $H$ is nonempty and (by Theorem 10.16) there is a convergent sequence $x_k \in H$ whose limit $x$ does not belong to $H$. For each $y \in H$, set $r(y) := \rho(x, y)/2$. Since $x$ does not belong to $H$, $r(y) > 0$; hence, each $B_{r(y)}(y)$ is open and contains $y$; i.e., $\{B_{r(y)}(y) : y \in H\}$ is an open covering of $H$. Since $H$ is compact, we can choose points $y_j$ and radii $r_j := r(y_j)$ such that $\{B_{r_j}(y_j) : j = 1, 2, \ldots, N\}$ covers $H$.

Set $r := \min\{r_1, \ldots, r_N\}$. (This is a finite set of positive numbers, so $r$ is also positive.) Since $x_k \to x$ as $k \to \infty$, $x_k \in B_r(x)$ for large $k$. But $x_k \in B_r(x) \cap H$ implies $x_k \in B_{r_j}(y_j)$ for some $j \in \mathbf{N}$. Therefore, it follows from the choices of $r_j$ and $r$, and from the triangle inequality, that

$$r_j \geq \rho(x_k, y_j) \geq \rho(x, y_j) - \rho(x_k, x)$$
$$= 2r_j - \rho(x_k, x) > 2r_j - r \geq 2r_j - r_j = r_j,$$

a contradiction. ∎

The following result is a partial converse of Remark 10.44 (see also Theorem 10.50 below).

**10.45 Remark.** *A closed subset of a compact set is compact.*

PROOF. Let $E$ be a closed subset of $H$, where $H$ is compact in $X$ and suppose $\mathcal{V} = \{V_\alpha\}_{\alpha \in A}$ is an open covering of $E$. Now $E^c = X \setminus E$ is open; hence, $\mathcal{V} \cup \{E^c\}$ is an open covering of $H$. Since $H$ is compact, there is a finite set $A_0 \subseteq A$ such that

$$H \subseteq E^c \cup \left( \bigcup_{\alpha \in A_0} V_\alpha \right).$$

But $E \cap E^c = \emptyset$. Therefore, $E$ is covered by $\{V_\alpha\}_{\alpha \in A_0}$. $\blacksquare$

Here is the connection between closed bounded sets and compact sets.

**10.46 THEOREM.** *Let $H$ be a subset of a metric space $X$. If $H$ is compact, then $H$ is closed and bounded.*

PROOF. Suppose $H$ is compact. By Remark 10.44, $H$ is closed. It is also bounded. Indeed, fix $b \in X$ and observe that $\{B_n(b) : n \in \mathbf{N}\}$ covers $X$. Since $H$ is compact, it follows that

$$H \subset \bigcup_{n=1}^{N} B_n(b)$$

for some $N \in \mathbf{N}$. Since these balls are nested, we conclude that $H \subset B_N(b)$; i.e., $H$ is bounded. $\blacksquare$

**10.47 Remark.** *The converse of Theorem 10.46 is false for arbitrary metric spaces.*

PROOF. Let $X = \mathbf{R}$ be the discrete metric space introduced in Example 10.3. Since $\sigma(0, x) \leq 1$ for all $x \in \mathbf{R}$, every subset of $X$ is bounded. Since $x_k \to x$ in $X$ implies $x_k = x$ for large $k$, every subset of $X$ is closed. Thus $[0, 1]$ is a closed bounded subset of $X$. Since $\{x\}_{x \in [0,1]}$ is an uncountable open covering of $[0, 1]$ which has no proper subcover, we conclude that $[0, 1]$ is closed and bounded, but not compact. $\blacksquare$

The problem here is that the discrete space has too many open sets. To identify a large class of metric spaces for which the converse of Theorem 10.46 DOES hold, we need a property which cuts down the "number of essential" open sets to a reasonable size.

**10.48 DEFINITION.** A metric space $X$ is said to be *separable* if it contains a countable dense subset, i.e., if there is a countable set $Z$ of $X$ such that for every point $a \in X$ there is a sequence $x_k \in Z$ such that $x_k \to a$ as $k \to \infty$.

We have seen (Theorem 8.13) that all Euclidean spaces are separable. The space $\mathcal{C}[a, b]$ is also separable (see Exercise 7 in Section 14.2). Hence, the hypothesis of separability is not an unusual requirement.

The following result makes clear what we meant above by "number of essential" open sets. It shows that every open covering of a set in a separable metric space has a countable subcovering.

**10.49 THEOREM** [LINDELÖF]. *Let $E$ be a subset of a separable metric space $X$. If $\{V_\alpha\}_{\alpha \in A}$ is a collection of open sets and $E \subseteq \cup_{\alpha \in A} V_\alpha$, then there is a countable subset $A_0$ of $A$ such that*

$$E \subseteq \bigcup_{\alpha \in A_0} V_\alpha.$$

PROOF. Let $Z$ be a countable dense subset of $X$, and consider the collection $\mathcal{T}$ of open balls with centers in $Z$ and rational radii. This collection is countable. Moreover, it "approximates" all other open sets in the following sense:

CLAIM. Given any open ball $B_r(x) \subset X$, there is a ball $B_q(a) \in \mathcal{T}$ such that $x \in B_q(a)$ and $B_q(a) \subseteq B_r(x)$.

PROOF OF CLAIM. Let $B_r(x) \subset X$ be given. By Definition 10.48, choose $a \in Z$ such that $\rho(x, a) < r/4$, and choose by Theorem 1.24 a rational $q \in \mathbf{Q}$ such that $r/4 < q < r/2$. Since $r/4 < q$, we have $x \in B_q(a)$. Moreover, if $y \in B_q(a)$, then

$$\rho(x, y) \le \rho(x, a) + \rho(a, y) < q + \frac{r}{4} < \frac{r}{2} + \frac{r}{4} < r.$$

Therefore, $B_q(a) \subseteq B_r(x)$. This establishes the claim.

To prove the theorem, let $x \in E$. By hypothesis, $x \in V_\alpha$ for some $\alpha \in A$. Hence, by the claim, there is a ball $B_x \in \mathcal{T}$ such that

$$(4) \qquad\qquad x \in B_x \subseteq V_\alpha.$$

The collection $\mathcal{T}$ is countable, hence, so is the subcollection

$$(5) \qquad\qquad \{U_1, U_2, \dots\} := \{B_x : x \in E\}.$$

By (4), for each $k \in \mathbf{N}$ there is at least one $\alpha_k \in A$ such that $U_k \subseteq V_{\alpha_k}$. Hence, by (5),

$$E \subseteq \bigcup_{x \in E} B_x = \bigcup_{k \in \mathbf{N}} U_k \subseteq \bigcup_{k \in \mathbf{N}} V_{\alpha_k}.$$

Thus set $A_0 := \{\alpha_k : k \in \mathbf{N}\}$. ∎

We are prepared to obtain a converse of Theorem 10.46. (For the definition of the Bolzano–Weierstrass Property, see Exercise 9 in Section 10.2.)

**10.50 THEOREM** [HEINE–BOREL]. *Let $X$ be a separable metric space which satisfies the Bolzano–Weierstrass Property and $H$ be a subset of $X$. Then $H$ is compact if and only if it is closed and bounded.*

PROOF. By Theorem 10.46, every compact set is closed and bounded.

Conversely, suppose to the contrary that $H$ is closed and bounded but not compact. Let $\mathcal{V}$ be an open covering of $H$ which has no finite subcover of $H$. By Lindelöf's Theorem, we may suppose that $\mathcal{V} = \{V_k\}_{k \in \mathbf{N}}$, i.e.,

$$(6) \qquad\qquad H \subseteq \bigcup_{k \in \mathbf{N}} V_k.$$

By the choice of $\mathcal{V}$, $\cup_{j=1}^{k} V_j$ cannot contain $H$ for any $k \in \mathbf{N}$. Thus we can choose a point

$$(7) \qquad x_k \in H \setminus \bigcup_{j=1}^{k} V_j$$

for each $k \in \mathbf{N}$. Since $H$ is bounded, the sequence $x_k$ is bounded. Hence, by the Bolzano–Weierstrass Property, there is a subsequence $x_{k_\nu}$ which converges to some $x$ as $\nu \to \infty$. Since $H$ is closed, $x \in H$. Hence, by (6), $x \in V_N$ for some $N \in \mathbf{N}$. But $V_N$ is open; hence, there is an $M \in \mathbf{N}$ such that $\nu \geq M$ implies $k_\nu > N$ and $x_{k_\nu} \in V_N$. This contradicts (7). We conclude that $H$ is compact. ∎

Since $\mathbf{R}^n$ satisfies the hypotheses of Theorem 10.50, it follows that a subset of a Euclidean space is compact if and only if it is closed and bounded.

We now turn our attention to uniform continuity on an arbitrary metric space.

**10.51 DEFINITION.** Let $E$ be a nonempty subset of $X$ and $f : E \to Y$. Then $f$ is said to be *uniformly continuous* on $E$ (notation: $f : E \to Y$ is uniformly continuous) if given $\varepsilon > 0$ there is a $\delta > 0$ such that

$$\rho(x, a) < \delta \quad \text{and} \quad x, a \in E \quad \text{imply} \quad \tau(f(x), f(a)) < \varepsilon.$$

In the real case, we proved that uniform continuity and continuity were equivalent on closed bounded intervals. That result, whose proof relied on the Bolzano–Weierstrass Theorem, is not true in an arbitrary metric space. If we strengthen the hypothesis from closed and bounded to compact, however, the result is valid for any metric space.

**10.51 THEOREM.** *Suppose $E$ is a compact subset of $X$ and $f : X \to Y$. Then $f$ is uniformly continuous on $E$ if and only if $f$ is continuous on $E$.*

PROOF. If $f$ is uniformly continuous on a set, then it is continuous, whether the set is compact or not.

Conversely, suppose $f$ is continuous on $E$. Given $\varepsilon > 0$ and $a \in E$, choose $\delta(a) > 0$ such that

$$x \in B_{\delta(a)}(a) \quad \text{and} \quad x \in E \quad \text{imply} \quad \tau(f(x), f(a)) < \frac{\varepsilon}{2}.$$

Since $a \in B_\delta(a)$ for all $\delta > 0$, it is clear that $\{B_{\delta(a)/2}(a) : a \in E\}$ is an open covering of $E$. Since $E$ is compact, choose finitely many points $a_j \in E$ and numbers $\delta_j := \delta(a_j)$ such that

$$(9) \qquad E \subseteq \bigcup_{j=1}^{N} B_{\delta_j/2}(a_j).$$

Set $\delta := \min\{\delta_1/2, \ldots, \delta_N/2\}$.

Suppose $x, a \in E$ and $\rho(x, a) < \delta$. By (9), $x$ belongs to $B_{\delta_j/2}(a_j)$ for some $1 \leq j \leq N$. Hence,

$$\rho(a, a_j) \leq \rho(a, x) + \rho(x, a_j) < \frac{\delta_j}{2} + \frac{\delta_j}{2} = \delta_j;$$

i.e., $a$ also belongs to $B_{\delta_j}(a_j)$. It follows, therefore, from the choice of $\delta_j$ that

$$\tau(f(x), f(a)) \leq \tau(f(x), f(a_j)) + \tau(f(a_j), f(a)) < \frac{\varepsilon}{2} + \frac{\varepsilon}{2} = \varepsilon.$$

This proves that $f$ is uniformly continuous on $E$.  ∎

### EXERCISES

**1.** Identify which of the following sets are compact and which are not. If $E$ is not compact, find the smallest compact set $H$ (if there is one) such that $E \subset H$.

    a) $\{1/k : k \in \mathbf{N}\} \cup \{0\}$.

    b) $\{(x, y) \in \mathbf{R}^2 : a \leq x^2 + y^2 \leq b\}$ for real numbers $0 < a < b$.

    c) $\{(x, y) \in \mathbf{R}^2 : y = \sin(1/x) \text{ for some } x \in (0, 1]\}$.

    d) $\{(x, y) \in \mathbf{R}^2 : |xy| \leq 1\}$.

**2.** Let $A, B$ be compact subsets of $X$. Prove that $A \cup B$ and $A \cap B$ are compact.

**3.** Suppose $E \subseteq \mathbf{R}$ is compact and nonempty. Prove $\sup E, \inf E \in E$.

**4.** Suppose $\{V_\alpha\}_{\alpha \in A}$ is a collection of nonempty open sets in $X$ which satisfies $V_\alpha \cap V_\beta = \emptyset$ for all $\alpha \neq \beta$ in $A$. Prove that if $X$ is separable then $A$ is countable. What happens to this result when "open" is omitted?

**5.** Prove that if $V$ is open in a separable metric space $X$, then there are open balls $B_1, B_2, \ldots$ such that

$$V = \bigcup_{j \in \mathbf{N}} B_j.$$

    Prove that every open set in $\mathbf{R}$ is a countable union of open intervals.

**6.** Let $E \subseteq X$ be closed.

    a) Prove $\partial E \subseteq E$.

    b) Prove that $\partial E = E$ if and only if $E^o = \emptyset$.

    c) Show b) is false if $E$ is not closed.

**7.** Prove directly that the discrete space $\mathbf{R}$ is not separable.

**8.** a) Prove that Cantor's Intersection Theorem holds for nested compact sets in an arbitrary metric space; i.e., if $H_1, H_2, \ldots$ is a nested sequence of nonempty compact sets in $X$, then

$$\bigcap_{k=1}^{\infty} H_k \neq \emptyset.$$

    b) Prove $(\sqrt{2}, \sqrt{3}) \cap \mathbf{Q}$ is closed and bounded but not compact in the metric space $\mathbf{Q}$ introduced in Example 10.5.

    c) Show that Cantor's Intersection Theorem does not hold in an arbitrary metric space if "compact" is replaced by "closed and bounded."

**9.** Prove that the Bolzano–Weierstrass Property does not hold for $\mathcal{C}[a, b]$ and $\|f\|$ (see Example 10.6). Namely, prove that if $f_n(x) = x^n$, then $\|f_n\|$ is bounded but $\|f_{n_k} - f\|$ does not converge for any $f \in \mathcal{C}[0, 1]$ and any subsequence $\{n_k\}$.

**10.** Let $X$ be a metric space.

    a) Prove that if $E \subseteq X$ is compact, then $E$ is sequentially compact (see Exercise 11 in Section 10.1).

    b) Prove that if $X$ is separable and satisfies the Bolzano–Weierstrass Property, then a set $E \subseteq X$ is sequentially compact if and only if it is compact.

## 10.5  CONNECTED SETS

We have introduced open sets (analogues of open intervals), closed sets (analogues of closed intervals), and compact sets (analogues of closed bounded intervals) in order to develop a calculus of functions of several variables in Chapters 11 through 13, which parallels that developed for functions of a single variable in Chapters 2 through 5. Some of the earlier theory, however, depended on properties of intervals not yet discussed. For example, the proof of the Intermediate Value Theorem tacitly used the fact that an interval is connected, i.e., is unbroken and all of one piece. We shall also use connected sets in Chapter 13 to provide a sufficiently broad definition of surfaces for computational ease. Thus we introduce the following idea.

**10.53 DEFINITION.** Let $X$ be a metric space.

    i) A pair of nonempty open sets $U, V$ in $X$ is said to *separate* $X$ if $X = U \cup V$ and $U \cap V = \emptyset$.

    ii) $X$ is said to be *connected* if $X$ cannot be separated by any pair of open sets $U, V$.

Loosely speaking, a connected space is all in one piece, i.e., cannot be broken into smaller, nonempty, open pieces which do not share any common points. Indeed, we shall see that $\mathbf{R}$, under the usual metric, is connected. On the other hand, under the discrete metric, $\mathbf{R}$ is not connected (since $(-\infty, 0]$ and $(0, \infty)$ are both "open" in the discrete space).

Recall (Example 10.4) that every subset of $X$ is a metric space. Hence Definition 10.53 also defines what it means for a subset $E$ of $X$ to be connected. One can always find two subsets of an arbitrary metric space which are connected: 1) The empty set is connected, since it can never be written as the union of nonempty sets. 2) Every singleton $E = \{a\}$ is also connected since, if $E = U \cup V$ where both $U$ and $V$ are nonempty, then $E$ has at least two points.

To obtain deeper results about connectivity, it is convenient to introduce the following concepts. (These concepts will also be used to study continuous functions in the next section.)

**10.54 DEFINITION.** Let $X$ be a metric space and $E \subseteq X$.

    i) A set $U \subseteq E$ is said to be *relatively open* in $E$ if there is a set $V$ open in $X$ such that $U = E \cap V$.

ii) A set $A \subseteq E$ is said to be *relatively closed* in $E$ if there is a set $C$ closed in $X$ such that $A = E \cap C$.

For example, the set $E$ of Example 10.35 is relatively open in the subspace $Y := \{(x, y) : -1 \le x \le 1\}$ and relatively closed in the subspace $Z := \{(x, y) : -|x| < y < |x|\}$. Indeed, $V = Z$ is open in $\mathbf{R}^2$ (it contains none of its boundary), $A = Y$ is closed in $\mathbf{R}^2$ (it contains all its boundary), and $E = V \cap Y$, $E = A \cap Z$.

Recall (Exercise 8 in Section 10.3) that a subset $A$ of $E$ is open (respectively, closed) in the *subspace* $E$ if and only if it is relatively open (respectively, relatively closed) in the *set* $E$. Thus all Definition 10.54 does is codify the "subspace topology."

By Definition 10.53, then, a set $E$ is connected if there are no nonempty sets $U, V$, relatively open in $E$, such that $E = U \cup V$ and $U \cap V = \emptyset$. The following result, which is usually easier to use than Definition 10.53, shows that when "separating" a nonconnected set, we can use open sets instead of relatively open sets. (The converse of this result is also true, but harder to prove–see Theorem 10.57 below.)

**10.55 Remark.** *Let $E \subseteq X$. If there exists a pair of open sets $A, B$ in $X$ which separate $E$, i.e., if $E \subseteq A \cup B$, $A \cap B = \emptyset$, $A \cap E \ne \emptyset$, and $B \cap E \ne \emptyset$, then $E$ is not connected.*

PROOF. Set $U = A \cap E$ and $V = B \cap E$. It suffices to prove that $U$ and $V$ are relatively open in $E$ and separate $E$. It is clear by hypothesis and the remarks above that $U$ and $V$ are nonempty, they are both relatively open in $E$, and $U \cap V = \emptyset$. It remains to prove that $E = U \cup V$. But $E$ is a subset of $A \cup B$, so $E \subseteq U \cup V$. On the other hand, both $U$ and $V$ are subsets of $E$, so $E \supseteq U \cup V$. We conclude that $E = U \cup V$. ∎

Thus when looking for "separations" of a given set $E \subset X$, we can confine our attention to open sets in $X$. Here are several examples. The set $\mathbf{Q}$ is not connected since the pair $A = (-\infty, \sqrt{2})$, $B = (\sqrt{2}, \infty)$ separate $\mathbf{Q}$. Example 10.35 is not connected since $\{(x, y) : x < 0\}$ and $\{(x, y) : x > 0\}$ are open in $\mathbf{R}^2$ (neither of them contains any of their boundary points) and separate the bowtie set $E$. Notice that Examples 10.36 and 10.38 are both connected in $\mathbf{R}^2$.

There is a simple description of all connected subsets of $\mathbf{R}$.

**10.56 THEOREM.** *A subset $E$ of $\mathbf{R}$ is connected if and only if $E$ is an interval.*

PROOF. Let $E$ be a connected subset of $\mathbf{R}$. If $E$ is empty or contains only one point, then $E$ is a degenerate interval. Hence we may suppose that $E$ contains at least two points.

Set $a = \inf E$ and $b = \sup E$. Notice that $-\infty \le a < b \le \infty$. Suppose for simplicity that $a, b \notin E$, i.e., $E \subseteq (a, b)$. If $E \ne (a, b)$, then there is an $x \in (a, b)$ such that $x \notin E$. By the Approximation Property, $E \cap (a, x) \ne \emptyset$ and $E \cap (x, b) \ne \emptyset$, and by assumption, $E \subseteq (a, x) \cup (x, b)$. Hence, $E$ is separated by the open sets $(a, x)$, $(x, b)$, a contradiction.

Conversely, suppose $I$ is an interval which is not connected. Then there are sets $U, V$, relatively open in $I$, which separate $I$, i.e., $I = U \cup V$, and there are points

$x_1 \in I \cap U$ and $x_2 \in I \cap V$. We may suppose that $x_1 < x_2$. Consider the set

$$W = \{t \in I : \text{ the interval } (x_1, t) \text{ satisfies } (x_1, t) \subseteq U\}.$$

Notice once and for all that since the intersection of two open intervals is an open interval, any set which is relatively open in $I$ contains an interval about each if its points. Since $U$ is relatively open, it follows that $W \neq \emptyset$. Since $V$ is relatively open, it also follows that $x_2 \notin W$ and $W$ is bounded above by some $c < x_2$. Thus $x_3 = \sup W$ is a finite number which belongs to $(x_1, c] \subset I$. In particular, either $x_3 \in U$ or $x_3 \in V$.

Suppose $x_3 \in U$. Since $x_3 > x_1$, we can choose $\delta > 0$ so small that $x_3 - \delta > x_1$ and $(x_3 - \delta, x_3 + \delta) \subset U$. Since $x_3 = \sup W$, we can choose by the Approximation Property a $t \in W$ such that $t > x_3 - \delta$ and $(x_1, t) \subset U$. It follows that $(x_1, x_3 + \delta) = (x_1, t) \cup (x_3 - \delta, x_3 + \delta) \subset U$; i.e., $x_3$ is not the supremum of $W$, a contradiction. On the other hand, if $x_3 \in V$, the same reasoning shows us that there is a $\delta > 0$ such that $(x_3 - \delta, x_3 + \delta) \subset V$ and a $t \in W$ such that $t > x_3 - \delta$ and $(x_3 - \delta, t) \subset U$. It follows that $(x_3 - \delta, t) \subset U \cap V$; i.e., $U \cap V \neq \emptyset$, a contradiction. Thus the pair $U$, $V$ does not separate $I$, and $I$ must be connected. ∎

We can use this result to prove that a real function is continuous on a closed, bounded interval if and only if its graph is closed and connected (see Theorem 9.52 in the previous chapter).

*We close this section by showing that the converse of Remark 10.55 is also true. This result is optional because we do not use it elsewhere.*

**\*10.57 THEOREM.** *Let $E \subseteq X$. If there exist sets $U, V$, relatively open in $E$, such that $U \cap V = \emptyset$, $E = U \cup V$, $U \neq \emptyset$, and $V \neq \emptyset$, then there is a pair of open sets $A, B$ which separates $E$.*

PROOF. We first show that

$$(9) \qquad\qquad \overline{U} \cap V = \emptyset.$$

Indeed, since $V$ is relatively open in $E$, there is a set $\Omega$, open in $X$, such that $V = E \cap \Omega$. Since $U \cap V = \emptyset$, it follows that $U \subset \Omega^c$. This last set is closed in $X$. Therefore,

$$\overline{U} \subseteq \overline{\Omega^c} = \Omega^c,$$

i.e., (9) holds.

Next, we use (9) to construct the set $B$. Set

$$\delta_x = \inf\{\rho(x, u) : u \in \overline{U}\}, \quad x \in V, \quad \text{and} \quad B = \bigcup_{x \in V} B_{\delta_x/2}(x).$$

Clearly, $B$ is open in $X$. Since $\delta_x > 0$ for each $x \notin \overline{U}$ (see Exercise 5 in Section 10.3), $B$ contains $V$; hence, $B \cap E \supseteq V$. The reverse inequality also holds since by construction $B \cap U = \emptyset$ and by hypothesis $E = U \cup V$. Therefore, $B \cap E = V$.

Similarly, we can construct an open set $A$ such that $A \cap E = U$ by setting

$$\varepsilon_y = \inf\{\rho(v, y) : v \in \overline{V}\}, \quad y \in U \quad \text{and} \quad A = \bigcup_{y \in U} B_{\varepsilon_y/2}(y).$$

To prove the pair $A, B$ separates $E$, it remains to prove $A \cap B = \emptyset$. Suppose to the contrary that there is a point $a \in A \cap B$. Then $a \in B_{\delta_x/2}(x)$ for some $x \in V$ and $a \in B_{\varepsilon_y/2}(y)$ for some $y \in U$. We may suppose $\delta_x \leq \varepsilon_y$. Then

$$\rho(x, y) \leq \rho(x, a) + \rho(a, y) < \frac{\delta_x}{2} + \frac{\varepsilon_y}{2} \leq \varepsilon_y.$$

Therefore, $\rho(x, y) < \inf\{\rho(v, y) : v \in \overline{V}\}$. Since $x \in V$, this is impossible. We conclude that $A \cap B = \emptyset$. ∎

### EXERCISES

**1.** a) Let $a \leq b$ and $c \leq d$ be real numbers. Sketch a graph of the rectangle

$$[a, b] \times [c, d] := \{(x, y) : x \in [a, b], y \in [c, d]\},$$

and decide whether this set is compact or connected. Explain your answers.
b) Sketch a graph of set

$$B_1(-2, 0) \cup B_1(2, 0) \cup \{(x, 0) : -1 < x < 1\},$$

and decide whether this set is compact or connected. Explain your answers.

**2.** a) Sketch a graph of the set

$$\{(x, y) : x^2 + 2y^2 < 6, \ y \geq 0\},$$

and decide whether this set is relatively open or relatively closed in the subspace $\{(x, y) : y \geq 0\}$. Do the same for the subspace $\{(x, y) : x^2 + 2y^2 < 6\}$. Explain your answers.
b) Sketch a graph of set

$$\{(x, y) : x^2 + y^2 \leq 1, \ (x - 2)^2 + y^2 < 2\},$$

and decide whether this set is relatively open or relatively closed in the subspace $\overline{B_1(0, 0)}$. Do the same for the subspace $B_{\sqrt{2}}(2, 0)$. Explain your answers.

**3.** Prove that the intersection of connected sets in $\mathbf{R}$ is connected. Show that this is false if "$\mathbf{R}$" is replaced by "$\mathbf{R}^2$."

**4.** Prove that if $E \subseteq \mathbf{R}$ is connected, then $E^o$ is also connected. Show that this is false if "$\mathbf{R}$" is replaced by "$\mathbf{R}^2$."

**5.** Suppose that $E \subset X$ is connected and $E \subseteq A \subseteq \overline{E}$. Prove that $A$ is connected.

**6.** Suppose $\{E_\alpha\}_{\alpha \in A}$ is a collection of connected sets in a metric space $X$ such that $\cap_{\alpha \in A} E_\alpha \neq \emptyset$. Prove that

$$E = \bigcup_{\alpha \in A} E_\alpha$$

is connected.

**7**. **This exercise is used in Section 10.6.**   Let $H \subseteq X$.   Prove that $H$ is compact if and only if every cover $\{E_\alpha\}_{\alpha \in A}$ of $H$, where the $E_\alpha$'s are relatively open in $H$, has a finite subcover.

**8.** A set $E$ in a metric space is called *clopen* if it is both open and closed.

   a) Prove that every metric space has at least two clopen sets.

   b) Prove that a metric space is connected if and only if it contains exactly two clopen sets.

**9.** Let $X$ be a metric space. Prove that $X$ is connected if and only if every nonempty proper subset of $X$ has a nonempty boundary.

**\*10**. **This exercise is used to prove \*Corollary 11.29.**

   a) A set $E \subseteq \mathbf{R}^n$ is said to be *polygonally connected* if any two points $\boldsymbol{a}, \boldsymbol{b} \in E$ can be connected by a polygonal path in $E$; i.e., there exist points $\boldsymbol{x}_k \in E$, $k = 1, \ldots, N$, such that $\boldsymbol{x}_0 = \boldsymbol{a}$, $\boldsymbol{x}_N = \boldsymbol{b}$ and $L(\boldsymbol{x}_{k-1}; \boldsymbol{x}_k) \subseteq E$ for $k = 1, \ldots, N$. Prove every polygonally connected set in $\mathbf{R}^n$ is connected.

   b) Let $E \subseteq \mathbf{R}^n$ be open and $\boldsymbol{x}_0 \in E$. Let $U$ be the set of points $\boldsymbol{x} \in E$ which can be polygonally connected in $E$ to $\boldsymbol{x}_0$. Prove that $U$ is open.

   c) Prove every open connected set in $\mathbf{R}^n$ is polygonally connected.

## 10.6  CONTINUOUS FUNCTIONS

In this section we discuss the behavior of images and inverse images of open sets, closed sets, compact sets, and connected sets under continuous functions. We shall use these results many times in the sequel.

Recall that if $X$ and $Y$ are metric spaces (with respective metrics $\rho$ and $\tau$), then a function $f : X \to Y$ is continuous on $X$ if given $a \in X$ and $\varepsilon > 0$ there is a $\delta > 0$ such that $\rho(x, a) < \delta$ implies $\tau(f(x), f(a)) < \varepsilon$, i.e., such that

$$(10) \qquad\qquad B_\delta(a) \subseteq f^{-1}(B_\varepsilon(f(a))).$$

This observation can be used to give the following simple but powerful characterization of continuous functions which can be stated without using the metric of $X$ (see also Exercise 3).

**10.58 THEOREM.** *Let $X$ be a metric space and $f : X \to Y$. Then $f$ is continuous if and only if $f^{-1}(V)$ is open in $X$ for every open $V$ in $Y$.*

PROOF. Suppose $f$ is continuous on $X$ and $V$ is open in $Y$. We may suppose $f^{-1}(V)$ is nonempty. Let $a \in f^{-1}(V)$, i.e., $f(a) \in V$. Since $V$ is open, choose $\varepsilon > 0$ such that $B_\varepsilon(f(a)) \subseteq V$. Since $f$ is continuous at $a \in E$, choose $\delta > 0$ such that (10) holds. Evidently,

(11) $$B_\delta(a) \subseteq f^{-1}(B_\varepsilon(f(a))) \subseteq f^{-1}(V).$$

Set

$$U = \bigcup_{a \in f^{-1}(V)} B_\delta(a).$$

Since $U$ is a union of open sets, $U$ is open. Since $U$ is a union of $B_\delta(a)$ over all $a \in f^{-1}(V)$, $U$ contains $f^{-1}(V)$. On the other hand, by (11) $U \subseteq f^{-1}(V)$. It follows that $f^{-1}(V) = U$, i.e., $f^{-1}(V)$ is open.

Conversely, let $\varepsilon > 0$ and $a \in X$. The ball $V = B_\varepsilon(f(a))$ is open in $Y$. By hypothesis, $f^{-1}(V)$ is open in $E$. Since $a \in f^{-1}(V)$, it follows that there is a $\delta > 0$ such that $B_\delta(a) \subseteq f^{-1}(V)$. This means that if $\rho(x, a) < \delta$, then $\tau(f(x), f(a)) < \varepsilon$. Therefore, $f$ is continuous at $a \in X$. ∎

By using the subspace (i.e., relative) topology, we see that Theorem 10.58 contains the following criterion for $f$ to be continuous on a subset of $X$.

**10.59 COROLLARY.** *Let $E \subseteq X$ and $f : E \to Y$. Then $f$ is continuous on $E$ if and only if $f^{-1}(V) \cap E$ is relatively open in $E$ for all open sets $V$ in $Y$.*

We shall refer to Theorem 10.58 and its corollary by saying that open sets are invariant under inverse images by continuous functions. It is interesting to notice that closed sets are also invariant under inverse images by continuous functions (see Exercises 3 and 4 below).

It is natural to ask whether compact sets and connected sets are invariant under inverse images by continuous functions. The following examples show the answer to this question is no.

**10.60 Examples.** i) If $f(x) = 1/x$ and $H = [0, 1]$, then $f$ is continuous on $(0, \infty)$ and $H$ is compact, but $f^{-1}(H) = [1, \infty)$ is not compact.

ii) If $f(x) = x^2$ and $E = (1, 4)$, then $f$ is continuous on $\mathbf{R}$ and $E$ is connected, but $f^{-1}(E) = (-2, -1) \cup (1, 2)$ is not connected.

The next two results show that compact sets and connected sets are invariant under *images*, rather than inverse images, by continuous functions.

**10.61 THEOREM.** *If $H$ is compact in $X$ and $f : H \to Y$ is continuous on $H$, then $f(H)$ is compact in $Y$.*

PROOF. Suppose $\{V_\alpha\}_{\alpha \in A}$ is an open covering of $f(H)$. By Theorem 1.43,

$$H \subseteq f^{-1}(f(H)) \subseteq f^{-1}\left(\bigcup_{\alpha \in A} V_\alpha\right) = \bigcup_{\alpha \in A} f^{-1}(V_\alpha).$$

Hence, by Corollary 10.59, $\{f^{-1}(V_\alpha)\}_{\alpha \in A}$ is a covering of $H$ whose sets are all relatively open in $H$. Since $H$ is compact, there are indices $\alpha_1, \alpha_2, \ldots, \alpha_N$ such that

$$H \subseteq \bigcup_{j=1}^{N} f^{-1}(V_{\alpha_j})$$

(see Exercise 7 in Section 10.5). It follows from Theorem 1.43 that

$$f(H) \subseteq f\left(\bigcup_{j=1}^{N} f^{-1}(V_{\alpha_j})\right) = \bigcup_{j=1}^{N} (f \circ f^{-1})(V_{\alpha_j}) = \bigcup_{j=1}^{N} V_{\alpha_j}.$$

Therefore, $f(H)$ is compact. ∎

**10.62 THEOREM.** *If $E$ is connected in $X$ and $f : E \to Y$ is continuous on $E$, then $f(E)$ is connected in $Y$.*

PROOF. Suppose $f(E)$ is not connected. By Definition 10.53, there exist a pair $U, V \subset Y$ of relatively open sets in $f(E)$ which separates $f(E)$. By Exercise 4 below, $f^{-1}(U) \cap E$ and $f^{-1}(V) \cap E$ are relatively open in $E$. Since $f(E) = U \cup V$, we have

$$E = (f^{-1}(U) \cap E) \cup (f^{-1}(V) \cap E).$$

Since $U \cap V = \emptyset$, we also have $f^{-1}(U) \cap f^{-1}(V) = \emptyset$. Thus $f^{-1}(U) \cap E$, $f^{-1}(V) \cap E$ is a pair of relatively open sets which separates $E$. Hence, by Definition 10.53, $E$ is not connected, a contradiction. ∎

(Note: Theorems 10.61 and 10.62 do not hold if "compact" or "connected" are replaced by "open" or "closed." For example, if $f(x) = x^2$ and $V = (-1, 1)$, then $f$ is continuous on $\mathbf{R}$ and $V$ is open, but $f(V) = [0, 1)$ is neither open nor closed.)

Suppose $f$ is a real function continuous on a closed bounded interval $[a, b]$. Then the function $F(x) = (x, f(x))$ is continuous from $\mathbf{R}$ into $\mathbf{R}^2$. Since the graph of $y = f(x)$ for $x \in [a, b]$ is the image of $[a, b]$ under $F$, it follows from Theorems 10.61 and 10.62 that the graph of $f$ is compact and connected. It is interesting to note that this property actually characterizes continuity of real functions (see Theorem 9.52 in the previous chapter).

To illustrate the power of the topological point of view presented above, compare the proofs of the following theorem and Exercise 6 with those of Theorems 3.28 and 3.31.

**10.63 THEOREM** [EXTREME VALUE THEOREM]. *Let $H$ be a nonempty, compact set in a metric space $X$ and suppose $f : H \to \mathbf{R}$ is continuous. Then*

$$M := \sup\{f(x) : x \in H\} \quad \text{and} \quad m := \inf\{f(x) : x \in H\}$$

*are finite real numbers and there exist points $x_M, x_m \in H$ such that $M = f(x_M)$ and $m = f(x_m)$.*

PROOF. By symmetry, it suffices to prove the result for $M$. Since $H$ is compact, $f(H)$ is compact. Hence, by the Theorem 10.46, $f(H)$ is closed and bounded. Since

$f(H)$ is bounded, $M$ is finite. By the Approximation Property, choose $x_k \in H$ such that $f(x_k) \to M$ as $k \to \infty$. Since $f(H)$ is closed, $M \in f(H)$. Therefore, there is an $x_M \in H$ such that $M = f(x_M)$. A similar argument shows that $m$ is finite and attained on $H$. ∎

The following analogue of Theorem 4.26 will be used in Chapter 13 to examine change of parametrizations of curves and surfaces.

**10.64 THEOREM.** *Let $X$ and $Y$ be metric spaces. If $H$ is a compact subset of $X$ and $f : H \to Y$ is 1–1 and continuous, then $f^{-1}$ is continuous on $f(H)$.*

PROOF. By Exercise 4a below, it suffices to show $(f^{-1})^{-1}$ takes closed sets in $X$ to relatively closed sets in $f(H)$. Let $A$ be closed in $X$. Then $E \cap H$ is a closed subset of $H$, so by Remark 10.45, $E \cap H$ is compact. Hence, by Theorem 10.61, $f(E \cap H)$ is compact, in particular closed. Since $f$ is 1–1, $f(E \cap H) = f(E) \cap f(H)$ (see Exercise 6 in Section 1.4). Since $f(E \cap H)$ and $f(H)$ are closed, it follows that $f(E) \cap f(H)$ is relatively closed in $f(H)$. Since $(f^{-1})^{-1} = f$, we conclude that $(f^{-1})^{-1}(E) \cap f(H)$ is relatively closed in $f(H)$. ∎

*If you are interested in how to use these topological ideas to study real functions further, you may read Section 9.5 in the previous chapter now.*

## EXERCISES

**1.** Let $f(x) = \sin x$ and $g(x) = x/|x|$ if $x \neq 0$ and $g(0) = 0$.

   a) Find $f(E)$ and $g(E)$ for $E = (0, \pi)$, $E = [0, \pi]$, $E = (-1, 1)$, and $E = [-1, 1]$, and explain some of your answers by appealing to results in this section.

   b) Find $f^{-1}(E)$ and $g^{-1}(E)$ for $E = (0, 1)$, $E = [0, 1]$, $E = (-1, 1)$, and $E = [-1, 1]$, and explain some of your answers by appealing to results in this section.

**2.** Let $f(x) = \sqrt{x}$ and $g(x) = 1/x$ if $x \neq 0$ and $g(0) = 0$.

   a) Find $f(E)$ and $g(E)$ for $E = (0, 1)$, $E = [0, 1)$, and $E = [0, 1]$, and explain some of your answers by appealing to results in this section.

   b) Find $f^{-1}(E)$ and $g^{-1}(E)$ for $E = (-1, 1)$ and $E = [-1, 1]$, and explain some of your answers by appealing to results in this section.

**3.** Let $X$ be a metric space and $f : X \to Y$. Prove that $f$ is continuous if and only if $f^{-1}(C)$ is closed in $X$ for every set $C$ closed in $Y$.

**4.** Suppose $E \subseteq X$ and $f : E \to Y$.

   a) Let $E \subseteq X$ and $f : E \to Y$. Prove that $f$ is continuous on $E$ if and only if $f^{-1}(A) \cap E$ is relatively closed in $E$ for all closed sets $A$ in $Y$.

   b) Suppose $f$ is continuous on $E$. Prove that if $V$ is relatively open in $f(E)$, then $f^{-1}(V)$ is relatively open in $E$, and if $A$ is relatively closed in $f(E)$, then $f^{-1}(A)$ is relatively closed in $E$.

**5.** [INTERMEDIATE VALUE THEOREM]. Let $E$ be a connected subset of a metric space $X$. If $f : E \to \mathbf{R}$ is continuous, $f(a) \neq f(b)$ for some $a, b \in E$, and $y$ is a number which lies between $f(a)$ and $f(b)$, then prove there is an $x \in E$ such that $f(x) = y$. (You may use Theorem 10.56.)

**6.** Let $X$ be metric space, $Y$ be a Euclidean space, and $H$ be a nonempty compact subset of $X$.

    a) Suppose $f : H \to Y$ is continuous. Prove that

$$\|f\|_H := \sup_{x \in H} \|f(x)\|_Y$$

    is finite and there exists an $x_0 \in H$ such that $\|f(x_0)\|_Y = \|f\|_H$.

    b) A sequence of functions $f_k : H \to Y$ is said to converge uniformly on $H$ to a function $f : H \to Y$ if, given $\varepsilon > 0$, there is an $N \in \mathbf{N}$ such that

$$k \geq N \quad \text{and} \quad x \in H \quad \text{imply} \quad \|f_k(x) - f(x)\|_Y < \varepsilon.$$

    Show that $\|f_k - f\|_H \to 0$ as $k \to \infty$ if and only if $f_k \to f$ uniformly on $H$ as $k \to \infty$.

    c) Prove that a sequence of functions $f_k$ converges uniformly on $H$ if and only if, given $\varepsilon > 0$, there is an $N \in \mathbf{N}$ such that

$$k, j \geq N \quad \text{implies} \quad \|f_k - f_j\|_H < \varepsilon.$$

**7.** Suppose $E$ is a compact subset of a metric space $X$.

    a) If $f, g : E \to \mathbf{R}^n$ are uniformly continuous, prove that $f + g$ and $f \cdot g$ are uniformly continuous. Did you need compactness for both results?

    b) If $g : E \to \mathbf{R}$ is continuous on $E$ and $g(x) \neq 0$ for $x \in E$, prove that $1/g$ is a bounded function.

    c) If $f, g : E \to \mathbf{R}$ are uniformly continuous on $E$ and $g(x) \neq 0$ for $x \in E$, prove that $f/g$ is uniformly continuous on $E$.

**8.** Let $X$ and $Y$ be metric spaces, $E \subseteq X$, and $f : E \to Y$.

    a) If $f$ is uniformly continuous on $E$ and $x_n \in E$ is Cauchy in $X$, prove that $f(x_n)$ is Cauchy in $Y$.

    b) Suppose $D$ is a *dense* subspace of $X$, i.e., $D \subset X$ and $\overline{D} = X$. If $Y$ is complete and $f : D \to Y$ is uniformly continuous on $D$, prove that $f$ has a continuous extension to $X$, i.e., prove that there is a continuous function $g : X \to Y$ such that $g(x) = f(x)$ for all $x \in D$.

# Chapter 11

# Differentiability on $\mathbf{R}^n$

## 11.1 PARTIAL DERIVATIVES AND PARTIAL INTEGRALS

The most natural way to define derivatives and integrals of functions of several variables is to allow one variable to move at a time. The corresponding objects, partial derivatives and partial integrals, are the subject of this section. Our main goal is to identify conditions under which partial derivatives, partial integrals, and evaluation of limits commute with each other, e.g., under which the limit of a partial integral is the partial integral of a limit.

We begin with some notation. The *Cartesian product* of a finite collection of sets $E_1, E_2, \ldots, E_n$ is the set of ordered $n$-tuples defined by

$$E_1 \times E_2 \times \cdots \times E_n := \{(x_1, x_2, \ldots, x_n) : x_j \in E_j \text{ for } j = 1, 2, \ldots, n\}.$$

Thus the Cartesian product of $n$ subsets of $\mathbf{R}$ is a subset of $\mathbf{R}^n$. By a *rectangle* in $\mathbf{R}^n$ (or an *n-dimensional rectangle*) we mean a Cartesian product of $n$ closed bounded intervals. An $n$-dimensional rectangle $R = [a_1, b_1] \times \cdots \times [a_n, b_n]$ is called an *n-dimensional cube with side $s$* if $|b_j - a_j| = s$ for $j = 1, \ldots, n$.

Let $f : \{x_1\} \times \cdots \times \{x_{j-1}\} \times [a, b] \times \{x_{j+1}\} \times \cdots \times \{x_n\} \to \mathbf{R}$. We shall denote the function

$$g(t) := f(x_1, \ldots, x_{j-1}, t, x_{j+1}, \ldots, x_n), \qquad t \in [a, b],$$

by $f(x_1, \ldots, x_{j-1}, \cdot, x_{j+1}, \ldots, x_n)$. If $g$ is integrable on $[a, b]$, then the *partial integral* of $f$ on $[a, b]$ with respect to $x_j$ is defined by

$$\int_a^b f(x_1, \ldots, x_n) \, dx_j := \int_a^b g(t) \, dt.$$

315

If $g$ is differentiable at some $t_0 \in (a, b)$, then the *partial derivative* (or *first-order partial derivative*) of $f$ at $(x_1, \ldots, x_{j-1}, t_0, x_{j+1}, \ldots x_n)$ with respect to $x_j$ is defined by

$$f_{x_j}(x_1, \ldots, x_{j-1}, t_0, x_{j+1}, \ldots, x_n)$$
$$:= \frac{\partial f}{\partial x_j}(x_1, \ldots, x_{j-1}, t_0, x_{j+1}, \ldots, x_n) := g'(t_0).$$

Thus the partial derivative $f_{x_j}$ exists at a point $\boldsymbol{a}$ if and only if the limit

$$\frac{\partial f}{\partial x_j}(\boldsymbol{a}) := \lim_{h \to 0} \frac{f(\boldsymbol{a} + h\boldsymbol{e}_j) - f(\boldsymbol{a})}{h}$$

exists. (Some authors use $f_j$ to denote the partial derivative $f_{x_j}$. To avoid confusing first-order partial derivatives with sequences and components of functions, we will not use this notation.)

We extend partial derivatives to vector-valued functions in the following way. Suppose $\boldsymbol{a} = (a_1, \ldots, a_n) \in \mathbf{R}^n$ and $f = (f_1, f_2, \ldots, f_m) : \{a_1\} \times \cdots \times \{a_{j-1}\} \times I \times \{a_{j+1}\} \times \cdots \times \{a_n\} \to \mathbf{R}^m$, where $j \in \{1, 2, \ldots, n\}$ is fixed and $I$ is an open interval containing $a_j$. If for each $k = 1, 2, \ldots, m$ the first-order partial derivative $\partial f_k / \partial x_j$ exists at $\boldsymbol{a}$, then we define the *first-order partial derivative* of $f$ with respect to $x_j$ to be the vector-valued function

$$f_{x_j}(\boldsymbol{a}) := \frac{\partial f}{\partial x_j}(\boldsymbol{a}) := \left( \frac{\partial f_1}{\partial x_j}(\boldsymbol{a}), \ldots, \frac{\partial f_m}{\partial x_j}(\boldsymbol{a}) \right).$$

Higher-order partial derivatives are defined by iteration. For example, the *second-order partial derivative* of $f$ with respect to $x_j$ and $x_k$ is defined by

$$f_{x_j x_k} := \frac{\partial^2 f}{\partial x_k \partial x_j} := \frac{\partial}{\partial x_k}\left( \frac{\partial f}{\partial x_j} \right)$$

when it exists. Second-order partial derivatives are called *mixed* when $j \neq k$.

This brings us to the following important collection of functions.

**11.1 DEFINITION.** Let $V$ be a nonempty, open subset of $\mathbf{R}^n$, $f : V \to \mathbf{R}^m$, and $p \in \mathbf{N}$.

i) $f$ is said to be $\mathcal{C}^p$ on $V$ if each partial derivative of $f$ of order $k \leq p$ exists and is continuous on $V$.

ii) $f$ is said to be $\mathcal{C}^\infty$ on $V$ if $f$ is $\mathcal{C}^p$ on $V$ for all $p \in \mathbf{N}$.

Clearly, if $f$ is $\mathcal{C}^p$ on $V$ and $q < p$, then $f$ is $\mathcal{C}^q$ on $V$.

For simplicity, we shall state all results in this section for the case $n = 2$ and $m = 1$, using $x$ for $x_1$ and $y$ for $x_2$. It is clear that, with appropriate changes in notation, these results also hold for any $n, m \in \mathbf{N}$.

Since partial derivatives and partial integrals are essentially one-dimensional ideas, each one-dimensional result about derivatives and integrals contains information about partial derivatives and partial integrals. Here are three examples.

By the Product Rule (Theorem 4.10), if $f_x$ and $g_x$ exist, then

$$\frac{\partial}{\partial x}(fg) = f\frac{\partial g}{\partial x} + g\frac{\partial f}{\partial x}.$$

By the Mean Value Theorem (Theorem 4.15), if $f(\cdot, y)$ is continuous on $[a, b]$ and the partial derivative $f_x(\cdot, y)$ exists on $(a, b)$, then there is a point $c \in (a, b)$ (which may depend on $y$) such that

$$f(b, y) - f(a, y) = (b - a)\frac{\partial f}{\partial x}(c, y).$$

And by the Fundamental Theorem of Calculus (Theorem 5.28), if $f(\cdot, y)$ is continuous on $[a, b]$, then

$$\frac{\partial}{\partial x}\int_a^x f(t, y)\, dt = f(x, y),$$

and if the partial derivative $f_x(\cdot, y)$ exists and is integrable on $[a, b]$, then

$$\int_a^b \frac{\partial f}{\partial x}(x, y)\, dx = f(b, y) - f(a, y).$$

Our first result about the commutation of partial derivatives, partial integrals, and evaluation of limits deals with interchanging two first-order partial derivatives (see also Exercise 11 in Section 11.2). For most applications, it is enough to remember that these hypotheses are met if $f \in \mathcal{C}^2(V)$.

**11.2 THEOREM.** *Suppose $V$ is open in $\mathbf{R}^2$, $(a, b) \in V$, and $f : V \to \mathbf{R}$. If $f$ is $\mathcal{C}^1$ on $V$, and if one of the mixed second partial derivatives of $f$ exists on $V$ and is continuous at the point $(a, b)$, then the other mixed second partial derivative exists at $(a, b)$ and*

$$\frac{\partial^2 f}{\partial y\partial x}(a, b) = \frac{\partial^2 f}{\partial x\partial y}(a, b).$$

PROOF. Suppose $f_{yx}$ exists on $V$ and is continuous at the point $(a, b)$. Consider $\Delta(h, k) := f(a+h, b+k) - f(a+h, b) - f(a, b+k) + f(a, b)$, defined for $|h|, |k| < r/\sqrt{2}$, where $r > 0$ is so small that $B_r(a, b) \subset V$. Apply the Mean Value Theorem twice to choose scalars $s, t \in (0, 1)$ such that

$$\Delta(h, k) = k\frac{\partial f}{\partial y}(a+h, b+tk) - k\frac{\partial f}{\partial y}(a, b+tk) = hk\frac{\partial^2 f}{\partial x\partial y}(a+sh, b+tk).$$

Since this last mixed partial derivative is continuous at the point $(a, b)$, we have

(1) $$\lim_{k \to 0}\lim_{h \to 0}\frac{\Delta(h, k)}{hk} = \frac{\partial^2 f}{\partial x\partial y}(a, b).$$

On the other hand, the Mean Value Theorem also implies that there is a scalar $u \in (0, 1)$ such that

$$\Delta(h, k) = f(a + h, b + k) - f(a, b + k) - f(a + h, b) + f(a, b)$$
$$= h\frac{\partial f}{\partial x}(a + uh, b + k) - h\frac{\partial f}{\partial x}(a + uh, b).$$

Hence, it follows from (1) that

$$\lim_{k \to 0} \lim_{h \to 0} \frac{1}{k}\left(\frac{\partial f}{\partial x}(a + uh, b + k) - \frac{\partial f}{\partial x}(a + uh, b)\right)$$
$$= \lim_{k \to 0} \lim_{h \to 0} \frac{\Delta(h, k)}{hk} = \frac{\partial^2 f}{\partial x \partial y}(a, b).$$

Since $f_x$ is continuous on $B_r(a, b)$, we can let $h = 0$ in the first expression. We conclude by definition that

$$\frac{\partial^2 f}{\partial y \partial x}(a, b) = \lim_{k \to 0} \frac{1}{k}\left(\frac{\partial f}{\partial x}(a, b + k) - \frac{\partial f}{\partial x}(a, b)\right) = \frac{\partial^2 f}{\partial x \partial y}(a, b). \quad \blacksquare$$

We shall refer to the conclusion of Theorem 11.2 by saying the first partial derivatives of $f$ *commute*.

The following example shows that Theorem 11.2 is false if the assumption about continuity of the second-order partial derivative is dropped.

**11.3 Example.** Prove that

$$f(x, y) = \begin{cases} xy\left(\dfrac{x^2 - y^2}{x^2 + y^2}\right) & (x, y) \neq \mathbf{0} \\ 0 & (x, y) = \mathbf{0} \end{cases}$$

is $\mathcal{C}^1$ on **R**$^2$, both mixed second partial derivatives of $f$ exist on **R**$^2$, but the first partial derivatives of $f$ do not commute at $(0, 0)$, i.e., $f_{xy}(0, 0) \neq f_{yx}(0, 0)$.

PROOF. By the one-dimensional Product and Quotient Rules,

$$\frac{\partial f}{\partial x}(x, y) = xy\frac{\partial}{\partial x}\left(\frac{x^2 - y^2}{x^2 + y^2}\right) + \frac{\partial}{\partial x}(xy)\left(\frac{x^2 - y^2}{x^2 + y^2}\right)$$
$$= xy\left(\frac{4xy^2}{(x^2 + y^2)^2}\right) + y\left(\frac{x^2 - y^2}{x^2 + y^2}\right)$$

for $(x, y) \neq (0, 0)$. Since $2|xy| \leq x^2 + y^2$, we have $|f_x(x, y)| \leq 2|y|$. Therefore, $f_x(x, y) \to 0$ as $(x, y) \to (0, 0)$. On the other hand, by definition

$$\frac{\partial f}{\partial x}(0, y) = \lim_{h \to 0} y\left(\frac{h^2 - y^2}{h^2 + y^2}\right) = -y$$

for all $y \in \mathbf{R}$; hence, $f_x(0,0) = 0$. This proves that $f_x$ exists and is continuous on $\mathbf{R}^2$ with value $-y$ at $(0, y)$. A similar argument shows that $f_y$ exists and is continuous on $\mathbf{R}^2$ with value $x$ at $(x, 0)$. It follows that the mixed second partial derivatives of $f$ exist on $\mathbf{R}^2$, and

$$\frac{\partial^2 f}{\partial y \partial x}(0,0) = -1 \neq 1 = \frac{\partial^2 f}{\partial x \partial y}(0,0). \quad \blacksquare$$

The following result shows that we can interchange a limit sign and a partial integral sign when the integrand is continuous on a rectangle.

**11.4 THEOREM.** *Let* $H = [a, b] \times [c, d]$ *be a rectangle and* $f : H \to \mathbf{R}$ *be continuous. If*

$$F(y) = \int_a^b f(x, y)\, dx,$$

*then* $F$ *is continuous on* $[c, d]$, *i.e.,*

$$\lim_{\substack{y \to y_0 \\ y \in [c,d]}} \int_a^b f(x, y)\, dx = \int_a^b \lim_{\substack{y \to y_0 \\ y \in [c,d]}} f(x, y)\, dx$$

*for all* $y_0 \in [c, d]$.

Proof. For each $y \in [c, d]$, $f(\cdot, y)$ is continuous on $[a, b]$. Hence, by Theorem 5.10, $F(y)$ exists for $y \in [c, d]$.

Fix $y_0 \in [c, d]$ and let $\varepsilon > 0$. Since $H$ is compact, $f$ is uniformly continuous on $H$. Hence, choose $\delta > 0$ such that $\|(x, y) - (z, w)\| < \delta$ and $(x, y), (z, w) \in H$ imply

$$|f(x, y) - f(z, w)| < \frac{\varepsilon}{b - a}.$$

Since $|y - y_0| = \|(x, y) - (x, y_0)\|$, it follows that

$$|F(y) - F(y_0)| \leq \int_a^b |f(x, y) - f(x, y_0)|\, dx < \varepsilon$$

for all $y \in [c, d]$ which satisfy $|y - y_0| < \delta$. We conclude that $F$ is continuous on $[c, d]$. $\blacksquare$

The following result shows that we can interchange a derivative and an integral sign when the first partial derivative of the integrand is sufficiently smooth. We will refer to this process as *differentiating under the integral sign*. For most applications, it is enough to remember that these hypotheses are met if $f \in \mathcal{C}^1(H)$.

**11.5 THEOREM.** *Let* $H = [a, b] \times [c, d]$ *be a rectangle in* $\mathbf{R}^2$ *and* $f : H \to \mathbf{R}$. *Suppose that* $f(\cdot, y)$ *is integrable on* $[a, b]$ *for each* $y \in [c, d]$, *and that the partial* derivative $f_y(x, \cdot)$ exists on $[c, d]$ for each $x \in [a, b]$. If the two-variable function $f_y(x, y)$ is continuous on $H$, then

$$(2) \qquad \frac{d}{dy} \int_a^b f(x, y)\, dx = \int_a^b \frac{\partial f}{\partial y}(x, y)\, dx$$

for all $y \in [c, d]$.

PROOF. Recall that "$f_y(x, \cdot)$ exists on $[c, d]$" means $f_y(x, \cdot)$ exists on $(c, d)$, and

$$f_y(x, c) := \lim_{h \to 0+} \frac{f(x, c + h) - f(x, c)}{h}, \quad f_y(x, d) := \lim_{h \to 0-} \frac{f(x, d + h) - f(x, d)}{h}$$

both exist (see Definition 4.6). Hence, it suffices to show

$$\lim_{h \to 0+} \int_a^b \frac{f(x, y + h) - f(x, y)}{h}\, dx = \int_a^b \frac{\partial f}{\partial y}(x, y)\, dx$$

for $y \in [c, d)$, and

$$\lim_{h \to 0-} \int_a^b \frac{f(x, y + h) - f(x, y)}{h}\, dx = \int_a^b \frac{\partial f}{\partial y}(x, y)\, dx$$

for $y \in (c, d]$. The arguments are similar; we provide the details only for the first identity.

Fix $x \in [a, b]$ and $y \in [c, d)$, and let $h > 0$ be so small that $y + h \in [c, d)$. By the Mean Value Theorem, choose a point $z(x; h)$ between $y$ and $y + h$ such that

$$\frac{f(x, y + h) - f(x, y)}{h} = \frac{\partial f}{\partial y}(x, z(x; h)).$$

Since $z(x; h) \to y$ as $h \to 0+$, it follows from Theorem 11.4 that

$$\frac{d}{dy} \int_a^b f(x, y)\, dx = \lim_{h \to 0+} \int_a^b \frac{\partial f}{\partial y}(x, z(x; h))\, dx = \int_a^b \frac{\partial f}{\partial y}(x, y)\, dx. \quad \blacksquare$$

*The rest of this section contains optional material which shows what happens to the results above when the improper integral is used.*

We begin by borrowing a concept from the theory of infinite series.

**\*11.7 DEFINITION.** Let $a < b$ be extended real numbers, $I$ be an interval in $\mathbf{R}$, and $f : (a, b) \times I \to \mathbf{R}$. The improper integral

$$\int_a^b f(x, y)\, dx$$

is said to *converge uniformly* on $I$ if $f(\cdot, y)$ is improperly integrable on $(a, b)$ for each $y \in I$, and if given $\varepsilon > 0$ there exist real numbers $A, B \in (a, b)$ such that

$$\left| \int_a^b f(x, y)\, dx - \int_\alpha^\beta f(x, y)\, dx \right| < \varepsilon$$

for all $a < \alpha < A$, $B < \beta < b$, and all $y \in I$.

For most applications, the following simple test for uniform convergence of an improper integral will be used instead of Definition 11.7 (compare with Theorem 7.15).

*11.8 THEOREM** [WEIERSTRASS $M$–TEST]. *Suppose $a < b$ are extended real numbers, $I$ is an interval in $\mathbf{R}$, $f : (a, b) \times I \to \mathbf{R}$, and $f(\cdot, y)$ is locally integrable on the interval $(a, b)$ for each $y \in I$. If there is a function $M : (a, b) \to \mathbf{R}$, absolutely integrable on $(a, b)$, such that*

$$|f(x, y)| \le M(x)$$

*for all $x \in (a, b)$ and $y \in I$, then*

$$\int_a^b f(x, y) \, dx$$

*converges uniformly on $I$.*

PROOF. Let $\varepsilon > 0$. By hypothesis and the Comparison Test for improper integrals, $\int_a^b f(x, y) \, dx$ exists and is finite for each $y \in I$. Moreover, since $M(x)$ is improperly integrable on $(a, b)$, there exist real numbers $A, B$ such that $a < A < B < b$ and

$$\int_a^A M(x) \, dx + \int_B^b M(x) \, dx < \varepsilon.$$

Thus for each $a < \alpha < A < B < \beta < b$ and each $y \in I$, we have

$$\left| \int_a^b f(x, y) \, dx - \int_\alpha^\beta f(x, y) \, dx \right| \le \int_a^\alpha |f(x, y)| \, dx + \int_\beta^b |f(x, y)| \, dx$$

$$\le \int_a^A M(x) \, dx + \int_B^b M(x) \, dx < \varepsilon. \quad \blacksquare$$

The following is an improper integral analogue of Theorem 11.4.

*11.9 THEOREM.** *Suppose $a < b$ are extended real numbers, $c < d$ are finite real numbers, and $f : (a, b) \times [c, d] \to \mathbf{R}$ is continuous. If*

$$F(y) = \int_a^b f(x, y) \, dx$$

*converges uniformly on $[c, d]$, then $F$ is continuous on $[c, d]$; i.e.,*

$$\lim_{\substack{y \to y_0 \\ y \in [c, d]}} \int_a^b f(x, y) \, dx = \int_a^b \lim_{\substack{y \to y_0 \\ y \in [c, d]}} f(x, y) \, dx$$

*for all $y_0 \in [c, d]$.*

PROOF. Let $\varepsilon > 0$ and $y_0 \in [c, d]$. Choose real numbers $A, B$ such that $a < A < B < b$ and

$$\left| F(y) - \int_A^B f(x, y) \, dx \right| < \frac{\varepsilon}{3}$$

for all $y \in [c, d]$. By Theorem 11.4, choose $\delta > 0$ such that

$$\left| \int_A^B (f(x, y) - f(x, y_0)) \, dx \right| < \frac{\varepsilon}{3}$$

for all $y \in [c, d]$ which satisfy $|y - y_0| < \delta$. Then

$$|F(y) - F(y_0)| \le \left| F(y) - \int_A^B f(x, y) \, dx \right| + \left| \int_A^B (f(x, y) - f(x, y_0)) \, dx \right|$$

$$+ \left| F(y_0) - \int_A^B f(x, y_0) \, dx \right|$$

$$< \frac{\varepsilon}{3} + \frac{\varepsilon}{3} + \frac{\varepsilon}{3} = \varepsilon$$

for all $y \in [c, d]$ which satisfy $|y - y_0| < \delta$. $\blacksquare$

The proof of Theorem 11.5 can be modified to prove the following result.

**\*11.10 THEOREM.** *Suppose $a < b$ are extended real numbers, $c < d$ are finite real numbers, $f : (a, b) \times [c, d] \to \mathbf{R}$ is continuous, and the improper integral*

$$F(y) = \int_a^b f(x, y) \, dx$$

*exists for all $y \in [c, d]$. If $f_y(x, y)$ exists and is continuous on $(a, b) \times [c, d]$ and if*

$$\phi(y) = \int_a^b \frac{\partial f}{\partial y}(x, y) \, dx$$

*converges uniformly on $[c, d]$, then $F$ is differentiable on $[c, d]$ and $F'(y) = \phi(y)$; i.e.,*

$$\frac{d}{dy} \int_a^b f(x, y) \, dx = \int_a^b \frac{\partial f}{\partial y}(x, y) \, dx$$

*for all $y \in [c, d]$.*

For a result about interchanging two partial integrals, see Theorem 12.31 and Exercise 10 in Section 12.3.

## EXERCISES

**1.** Compute all mixed second-order partial derivatives of each of the following functions and verify that the mixed partial derivatives are equal.

   a) $f(x, y) = xe^y$.     b) $f(x, y) = \cos(xy)$.     c) $f(x, y) = \dfrac{x + y}{x^2 + 1}$.

**2.** Compute all first-order partial derivatives of each of the following functions and find where they are continuous.

a) $f(x, y) = x^2 + \sin(xy)$.   b) $f(x, y, z) = \dfrac{xy}{1+z}$.   c) $f(x, y) = \sqrt{x^2 + y^2}$.

**3.** For each of the following functions, compute $f_x$, and determine where it is continuous.

a)
$$f(x, y) = \begin{cases} \dfrac{x^4 + y^4}{x^2 + y^2} & (x, y) \neq (0, 0) \\ 0 & (x, y) = (0, 0). \end{cases}$$

b)
$$f(x, y) = \begin{cases} \dfrac{x^2 - y^2}{\sqrt[3]{x^2 + y^2}} & (x, y) \neq (0, 0) \\ 0 & (x, y) = (0, 0). \end{cases}$$

**4.** Suppose $H = [a, b] \times [c, d]$ is a rectangle, $f : H \to \mathbf{R}$ is continuous, and $g : [a, b] \to \mathbf{R}$ is integrable. Prove

$$F(y) = \int_a^b g(x) f(x, y) \, dx$$

is uniformly continuous on $[c, d]$.

**5.** Evaluate each of the following expressions.

a)
$$\lim_{y \to 0} \int_0^1 \cos(x^2 y + xy^2) \, dx.$$

b)
$$\frac{d}{dy} \int_{-1}^1 \sqrt{x^2 y^2 + xy + y + 2} \, dx \qquad \text{at } y = 0.$$

**\*6.** Evaluate each of the following expressions.

a)
$$\lim_{y \to 0+} \int_0^1 \frac{x \cos y}{\sqrt[3]{1 - x + y}} \, dx.$$

b)
$$\frac{d}{dy} \int_\pi^\infty \frac{e^{-xy} \sin x}{x} \, dx \qquad \text{at } y = 1.$$

**\*7. a)** Prove

$$\int_0^1 \frac{\cos(x^2 + y^2)}{\sqrt{x}} \, dx$$

converges uniformly on $(-\infty, \infty)$.

b) Prove $\int_0^\infty e^{-xy} \, dx$ converges uniformly on $[1, \infty)$.

c) Prove $\int_0^\infty y e^{-xy} \, dx$ exists for each $y \in [0, \infty)$, converges uniformly on any $[a, b] \subset (0, \infty)$, but does not converge uniformly on $[0, 1]$.

*11.11 DEFINITION.** The *Laplace transform* of a function $f : (0, \infty) \to \mathbf{R}$ is said to exist at a point $s \in (0, \infty)$ if the integral

$$\mathcal{L}\{f\}(s) := \int_0^\infty e^{-st} f(t)\, dt$$

converges. (Note: This integral is improper at $\infty$ and may be improper at 0.)

*8.** Prove

a) $$\mathcal{L}\{1\}(s) = \frac{1}{s}, \qquad s > 0.$$

b) $$\mathcal{L}\{t^n\}(s) = \frac{n!}{s^{n+1}}, \qquad s > 0, \ n \in \mathbf{N}.$$

c) $$\mathcal{L}\{e^{at}\}(s) = \frac{1}{s-a}, \qquad s > a, \ a \in \mathbf{R}.$$

d) $$\mathcal{L}\{\cos(bt)\}(s) = \frac{s}{s^2 + b^2}, \qquad s > 0, \ b \in \mathbf{R}.$$

e) $$\mathcal{L}\{\sin(bt)\}(s) = \frac{b}{s^2 + b^2}, \qquad s > 0, \ b \in \mathbf{R}.$$

*9.** Suppose $f : (0, \infty) \to \mathbf{R}$ is continuous and bounded and $\mathcal{L}\{f\}$ exists at some $a \in (0, \infty)$. Let

$$\phi(t) = \int_0^t e^{-au} f(u)\, du, \qquad t \in (0, \infty).$$

a) Prove that

$$\int_0^N e^{-st} f(t)\, dt = \phi(N) e^{-(s-a)N} + (s-a) \int_0^N e^{-(s-a)t} \phi(t)\, dt$$

for all $N \in \mathbf{N}$.

b) Prove that the integral $\int_0^\infty e^{-(s-a)t} \phi(t)\, dt$ converges uniformly on $[b, \infty)$ for any $b > a$ and

$$\int_0^\infty e^{-st} f(t)\, dt = (s-a) \int_0^\infty e^{-(s-a)t} \phi(t)\, dt, \qquad s > a.$$

c) Prove that $\mathcal{L}\{f\}$ exists, is continuous on $(a, \infty)$, and satisfies

$$\lim_{s \to \infty} \mathcal{L}\{f\}(s) = 0.$$

d) Let $g(t) = tf(t)$ for $t \in (0, \infty)$. Prove that $\mathcal{L}\{f\}$ is differentiable on $(a, \infty)$ and

$$\frac{d}{ds}\mathcal{L}\{f\}(s) = -\mathcal{L}\{g\}(s)$$

for all $s \in (a, \infty)$.

e) If in addition, $f'$ is continuous and bounded on $(0, \infty)$, prove that

$$\mathcal{L}(f')(s) = s\mathcal{L}(f)(s) - f(0)$$

for all $s \in (a, \infty)$.

*10. Using Exercises 8 and 9, find the Laplace transforms of the each of the functions $te^t$, $t \sin \pi t$, and $t^2 \cos t$.

## 11.2  THE DEFINITION OF DIFFERENTIABILITY

Let $V$ be a nonempty, open subset of $\mathbf{R}^n$ and $f : V \to \mathbf{R}^m$. By Definition 8.31 and Theorem 8.35, $f$ is differentiable at $\boldsymbol{a} \in V$ if and only if there is an $m \times n$ matrix $B$ such that

$$(3) \qquad \qquad \lim_{\boldsymbol{h} \to \boldsymbol{0}} \frac{f(\boldsymbol{a} + \boldsymbol{h}) - f(\boldsymbol{a}) - B\boldsymbol{h}}{\|\boldsymbol{h}\|} = \boldsymbol{0}.$$

Moreover, by the uniqueness of the total derivative (Remark 8.32), if $B$ is an $m \times n$ matrix which satisfies (3), then $B$ is (the matrix representative of) the total derivative $Df(\boldsymbol{a})$. Is there an easy way to compute this matrix?

To answer this question, we introduce the following notation. Let $V$ be open in $\mathbf{R}^n$ and $\boldsymbol{a} \in V$. For each function $f : V \to \mathbf{R}^m$ whose first-order partial derivatives exist at $\boldsymbol{a}$, define the *Jacobian matrix* of $f$ at $\boldsymbol{a}$ by

$$\left[\frac{\partial f_i}{\partial x_j}(\boldsymbol{a})\right]_{m \times n} := \begin{bmatrix} \dfrac{\partial f_1}{\partial x_1}(\boldsymbol{a}) & \cdots & \dfrac{\partial f_1}{\partial x_n}(\boldsymbol{a}) \\ \vdots & \ddots & \vdots \\ \dfrac{\partial f_m}{\partial x_1}(\boldsymbol{a}) & \cdots & \dfrac{\partial f_m}{\partial x_n}(\boldsymbol{a}) \end{bmatrix}.$$

The following result shows that if $f$ is differentiable, then the Jacobian matrix of $f$ represents the total derivative of $f$.

**11.12 THEOREM.** *Let $V$ be nonempty and open in $\mathbf{R}^n$ and $f : V \to \mathbf{R}^m$. If $f$ is differentiable at some $\boldsymbol{a} \in V$, then all first-order partial derivatives of $f$ exist at*

*$\mathbf{a}$, Moreover, the matrix which represents total derivative of $f$ at $\mathbf{a}$ is the Jacobian matrix; i.e.,*

$$Df(\mathbf{a})(\mathbf{h}) = \left[\frac{\partial f_i}{\partial x_j}(\mathbf{a})\right]_{m \times n} \cdot \mathbf{h}$$

*for all $\mathbf{h} \in \mathbf{R}^n$.*

PROOF. Let $B = [b_{ij}]_{m \times n}$ be the matrix which represents $Df(\mathbf{a})$. Fix $1 \leq j \leq n$ and set $\mathbf{h} = u\mathbf{e}_j$ for some $u > 0$. Since $\|\mathbf{h}\| = u$, we have

$$\frac{f(\mathbf{a} + \mathbf{h}) - f(\mathbf{a}) - B\mathbf{h}}{\|\mathbf{h}\|} := \frac{f(\mathbf{a} + u\mathbf{e}_j) - f(\mathbf{a})}{u} - B\mathbf{e}_j.$$

Take the limit of this identity as $u \to 0+$, using (3) and the definition of matrix multiplication. We obtain

$$\lim_{u \to 0+} \frac{f(\mathbf{a} + u\mathbf{e}_j) - f(\mathbf{a})}{u} = B\mathbf{e}_j = (b_{1j}, \ldots, b_{mj}).$$

A similar argument shows that the limit of this quotient as $u \to 0-$ also exists and equals $(b_{1j}, \ldots, b_{mj})$. Since a vector-valued function converges if and only if each of its components converges (Theorem 8.21), it follows that the first-order partial derivative of each component $f_i$ with respect to $x_j$ exists at $\mathbf{a}$ and satisfies

$$\frac{\partial f_i}{\partial x_j}(\mathbf{a}) = b_{ij}$$

for $i = 1, 2, \ldots, m$. In particular, $B$ is the Jacobian matrix of $f$ at $\mathbf{a}$. ∎

Abusing the notation slightly, we shall use $Df(\mathbf{a})$ to represent both the Jacobian matrix and the total derivative of $f$ at $\mathbf{a}$. This is done to make the multidimensional theory look more like the one-dimensional case.

If $n = 1$ or $m = 1$, the Jacobian matrix $Df$ is an $m \times 1$ or $1 \times n$ matrix, hence, can be identified with a vector. Most applied mathematicians represent $Df$ in these cases by a different notation. For the case $n = 1$,

$$Df(a) = \begin{bmatrix} f_1'(a) \\ \vdots \\ f_m'(a) \end{bmatrix}$$

is sometimes denoted in vector notation by

$$f'(a) := (f_1'(a), \ldots, f_m'(a)).$$

For the case $m = 1$,

$$Df(\mathbf{a}) = \begin{bmatrix} \dfrac{\partial f}{\partial x_1}(\mathbf{a}) & \cdots & \dfrac{\partial f}{\partial x_n}(\mathbf{a}) \end{bmatrix}$$

is sometimes denoted in vector notation by

$$\nabla f(\boldsymbol{a}) := \left( \frac{\partial f}{\partial x_1}(\boldsymbol{a}), \dots, \frac{\partial f}{\partial x_n}(\boldsymbol{a}) \right).$$

($\nabla f$ is called the *gradient* of $f$ because it identifies the direction of steepest ascent. For this connection and a relationship between gradients and directional derivatives, see Exercise 11c in Section 11.3.)

Let $f$ be a function from a subset of $\mathbf{R}^n$ into $\mathbf{R}^m$ and $\boldsymbol{a} \in \text{Dom } f$. The following notation will be used in the multidimensional versions of the Inverse Function Theorem and the Change of Variables Formula. For each subset $\{k_1, k_2, \dots, k_n\}$ of $\{1, 2, \dots, m\}$, we define the corresponding *partial Jacobian* by

$$\frac{\partial(f_{k_1}, \dots, f_{k_n})}{\partial(x_1, \dots, x_n)} := \det \left[ \frac{\partial f_{k_i}}{\partial x_j}(\boldsymbol{a}) \right]_{n \times n} = \det \begin{bmatrix} \dfrac{\partial f_{k_1}}{\partial x_1}(\boldsymbol{a}) & \cdots & \dfrac{\partial f_{k_1}}{\partial x_n}(\boldsymbol{a}) \\ \vdots & \ddots & \vdots \\ \dfrac{\partial f_{k_n}}{\partial x_1}(\boldsymbol{a}) & \cdots & \dfrac{\partial f_{k_n}}{\partial x_n}(\boldsymbol{a}) \end{bmatrix}$$

when all these partial derivatives exist. The partial Jacobian for the case $n = m$ comes up so frequently that we give it a separate name and notation: we call it the *Jacobian* of $f$ at $\boldsymbol{a}$, and denote it by $\Delta_f$, i.e.,

$$\Delta_f(\boldsymbol{a}) := \frac{\partial(f_1, \dots, f_n)}{\partial(x_1, \dots, x_n)}.$$

(Note: The Jacobian $\Delta_f$ can be interpreted as a change of volumes (see Exercise 6 in Section 12.4) and will play a prominent role in Chapters 12 and 13.) We shall see that for multivariable functions, $Df$ and $\Delta_f$ play the same role that $f'$ played for one-variable functions.

At this point you may be wondering why the definition of differentiability on $\mathbf{R}^n$ is not simpler. For example, in the one-dimensional case, a function is differentiable if and only if its derivative exists. Working by analogy, we might guess that a function of several variables is differentiable if and only if all its first-order partial derivatives exist. The following example shows that this guess is wrong (see also Exercise 10 in Section 11.3).

**11.13 Example.** Prove that the first-order partial derivatives of

$$f(x, y) = \begin{cases} x + y & x = 0 \quad \text{or} \quad y = 0 \\ 1 & \text{otherwise} \end{cases}$$

exist at $(0, 0)$, but $f$ is not differentiable at $(0, 0)$.

PROOF. Since

$$\lim_{x \to 0} f(x, x) = 1 \neq 0 = f(0, 0),$$

it is clear that $f$ is not continuous at $(0,0)$ (hence, not differentiable by Theorem 8.38). On the other hand, the first-order partial derivatives of $f$ exist since

$$f_x(0,0) = \lim_{h \to 0} \frac{f(h,0) - f(0,0)}{h} = 1$$

and, similarly, $f_y(0,0) = 1$. ∎

The function in Example 11.13 failed to be differentiable because it was discontinuous at $(0,0)$. Still looking for a simpler definition of differentiability, we might modify our guess by insisting that if a function is *continuous* and has first-order partial derivatives, then it is differentiable. The following (optional) example shows that this guess is also wrong.

*11.14 Example.** Prove that

$$f(x,y) = \begin{cases} \dfrac{x^3 - xy^2}{x^2 + y^2} & (x,y) \neq (0,0) \\ 0 & (x,y) = (0,0) \end{cases}$$

is continuous, has first-order partial derivatives everywhere on $\mathbf{R}^2$, but $f$ is not differentiable at $(0,0)$.

PROOF. Clearly, $f$ is continuous and has first-order partial derivatives at every point $(x,y) \neq (0,0)$. What happens at $(0,0)$? Since

$$|f(x,y)| = \frac{|x||x^2 - y^2|}{x^2 + y^2} \leq |x|,$$

it follows from the Squeeze Theorem that $f$ is continuous at $(0,0)$ with $f(0,0) = 0$. Moreover, the function $f$ has first-order partial derivatives at $(0,0)$, since

$$\frac{\partial f}{\partial x}(0,0) = \lim_{h \to 0} \frac{f(h,0) - f(0,0)}{h} = \lim_{h \to 0} \frac{h^3}{h^3} = 1,$$

and

$$\frac{\partial f}{\partial y}(0,0) = \lim_{h \to 0} \frac{f(0,h) - f(0,0)}{h} = \lim_{h \to 0} \frac{0}{h^3} = 0.$$

Finally, if $f$ were differentiable at $(0,0)$, then

$$0 = \lim_{(h,k) \to (0,0)} \frac{f(h,k) - f(0,0) - \nabla f(0,0) \cdot (h,k)}{\sqrt{h^2 + k^2}} = \lim_{(h,k) \to (0,0)} \frac{-2hk^2}{(h^2 + k^2)^{3/2}}.$$

But the path $h = k$ gives a limit of $-1/\sqrt{2} \neq 0$ as $h \to 0+$. Thus $f$ is not differentiable at $(0,0)$. ∎

Although there is no simpler definition of differentiability, there is an uncomplicated condition with wide applicability which *implies* differentiability. Indeed, the following result shows that if $f$ is $\mathcal{C}^1$ on some open set $V$, then $f$ is differentiable on $V$.

**11.15 THEOREM.** Let $V$ be open in $\mathbf{R}^n$, $\mathbf{a} \in V$, and $f : V \to \mathbf{R}^m$. If all first-order partial derivatives of $f$ exist in $V$ and are continuous at $\mathbf{a}$, then $f$ is differentiable at $\mathbf{a}$.

PROOF. Since a function converges if and only if each of its components converge (Theorem 8.21), we may suppose $m = 1$. By the uniqueness of the total derivative, it suffices to show

$$\lim_{\mathbf{h} \to \mathbf{0}} \frac{f(\mathbf{a} + \mathbf{h}) - f(\mathbf{a}) - \nabla f(\mathbf{a}) \cdot \mathbf{h}}{\|\mathbf{h}\|} = 0.$$

Let $\mathbf{a} = (a_1, \dots, a_n)$. Suppose $r > 0$ is so small that $B_r(\mathbf{a}) \subset V$. Fix $\mathbf{h} = (h_1, \dots, h_n) \neq \mathbf{0}$ in $B_r(\mathbf{0})$. By telescoping and using the one-dimensional Mean Value Theorem, we can choose numbers $c_j$ between $a_j$ and $a_j + h_j$ such that

$$
\begin{aligned}
f(\mathbf{a} + \mathbf{h}) - f(\mathbf{a}) &= f(a_1 + h_1, \dots, a_n + h_n) - f(a_1, a_2 + h_2, \dots, a_n + h_n) \\
&\quad + \cdots + f(a_1, \dots, a_{n-1}, a_n + h_n) - f(a_1, \dots, a_n) \\
&= \sum_{j=1}^{n} h_j \frac{\partial f}{\partial x_j}(a_1, \dots, a_{j-1}, c_j, a_{j+1} + h_{j+1}, \dots, a_n + h_n).
\end{aligned}
$$

Therefore,

$$(4) \qquad f(\mathbf{a} + \mathbf{h}) - f(\mathbf{a}) - \nabla f(\mathbf{a}) \cdot \mathbf{h} = \mathbf{h} \cdot \boldsymbol{\delta},$$

where $\boldsymbol{\delta} \in \mathbf{R}^n$ is the vector with components

$$\delta_j = \frac{\partial f}{\partial x_j}(a_1, \dots, a_{j-1}, c_j, a_{j+1} + h_{j+1}, \dots, a_n + h_n) - \frac{\partial f}{\partial x_j}(a_1, \dots, a_n).$$

Since the first-order partial derivatives of $f$ are continuous at $\mathbf{a}$, $\delta_j \to 0$ for each $1 \le j \le n$, i.e., $\|\boldsymbol{\delta}\| \to 0$ as $\mathbf{h} \to \mathbf{0}$. Moreover, by the Cauchy–Schwarz Inequality and (4),

$$(5) \qquad 0 \le \frac{|f(\mathbf{a} + \mathbf{h}) - f(\mathbf{a}) - \nabla f(\mathbf{a}) \cdot \mathbf{h}|}{\|\mathbf{h}\|} = \frac{|\mathbf{h} \cdot \boldsymbol{\delta}|}{\|\mathbf{h}\|} \le \|\boldsymbol{\delta}\|.$$

It follows from the Squeeze Theorem that the first quotient in (5) converges to 0 as $\mathbf{h} \to \mathbf{0}$. Thus $f$ is differentiable at $\mathbf{a}$ by definition. ∎

If all first-order partial derivatives of a function $f$ exist and are continuous at a point $\mathbf{a}$ (respectively, on an open set $V$), we shall call $f$ *continuously differentiable* at $\mathbf{a}$ (respectively, on $V$). By Theorem 11.15, every continuously differentiable function is differentiable. In particular, every function which is $\mathcal{C}^p$ on an open set $V$, for some $1 \le p \le \infty$, is continuously differentiable on $V$.

Although Theorem 11.15 is much simpler to use than (3), the following example shows that there are times when (3) must be used directly.

**11.16 Example.** Prove that

$$f(x,y) = \begin{cases} (x^2 + y^2) \sin \dfrac{1}{\sqrt{x^2 + y^2}} & (x,y) \neq (0,0) \\ 0 & (x,y) = (0,0) \end{cases}$$

is differentiable but not continuously differentiable at $(0,0)$.

PROOF. Clearly, $f$ is $\mathcal{C}^1$, hence differentiable, on $\mathbf{R} \setminus \{(0,0)\}$. To prove that $f$ is differentiable at $(0,0)$, we must verify (3) for $\boldsymbol{a} = (0,0)$ and $B = \nabla f(\boldsymbol{a})$. By definition,

$$f_x(0,0) = \lim_{t \to 0} \frac{f(t,0) - f(0,0)}{t} = \lim_{t \to 0} t \sin \frac{1}{|t|} = 0,$$

and similarly, $f_y(0,0) = 0$. Thus,

$$\frac{f(h,k) - f(0,0) - \nabla f(0,0) \cdot (h,k)}{\|(h,k)\|} = \sqrt{h^2 + k^2} \sin \frac{1}{\sqrt{h^2 + k^2}} \to 0$$

as $(h,k) \to (0,0)$; i.e., $f$ is differentiable at $(0,0)$. On the other hand, if $(x,y) \neq (0,0)$ it follows from the one-dimensional Product Rule that

$$f_x(x,y) = \frac{-x}{\sqrt{x^2 + y^2}} \cos \frac{1}{\sqrt{x^2 + y^2}} + 2x \sin \frac{1}{\sqrt{x^2 + y^2}}.$$

Thus $f_x(x,0)$ has no limit as $x \to 0$, and the partial derivative $f_x$ is not continuous at $(0,0)$. ∎

Combining Theorem 11.15 with Example 11.16, we see that every continuously differentiable function is differentiable, but not conversely.

These results suggest the following procedure to determine whether a function $f$ is differentiable at a point $\boldsymbol{a}$. Compute all first-order partial derivatives of $f$ at $\boldsymbol{a}$. If one of these does not exist, then $f$ is not differentiable at $\boldsymbol{a}$ (Theorem 11.12). If all first-order partial derivatives exist and are continuous at $\boldsymbol{a}$, then $f$ is differentiable at $\boldsymbol{a}$ (Theorem 11.15). If one of the first-order partial derivatives exists but fails to be continuous at $\boldsymbol{a}$, then use (3) directly. This will involve evaluation of a limit of vectors using the methods outlined in Section 8.3.

Let $g : \mathbf{R} \to \mathbf{R}$ represent a generic function of one variable and $f : \mathbf{R}^2 \to \mathbf{R}$ represent a generic function of two variables. We know that $g$ is differentiable at $a$ if and only if the curve $y = g(x)$ has a unique tangent line at $(a, g(a))$, in which case $g'(a)$ is the slope of that tangent line. What happens in the two-dimensional case? Working by analogy, $f$ should be differentiable at a point $(a,b)$ if and only if the surface $z = f(x,y)$ has a unique tangent plane at $(a, b, f(a,b))$. Moreover, it would be nice if the normal vector of that tangent plane is somehow related to the total derivative $\nabla f(a,b)$. We shall show that both of these results are correct, and that the relationship between $\boldsymbol{n}$ and $\nabla f$ is a simple one (see (8) below, and Exercise 8 in Section 11.5). Thus, for the case $n = 2$, $m = 1$, Definition 8.31 captures both the analytic and geometric spirit of the one-dimensional derivative.

First, we show that if $z = f(x,y)$ has a tangent plane $\varPi$, then $\nabla f$ can be used to produce a normal to $\varPi$. (We shall show below that the hypotheses of this result are satisfied if $f$ is differentiable at $(a,b)$.)

**11.17 Remark.** *Suppose $f_x$ and $f_y$ exist at $(a, b)$, and $\Pi$ is a plane passing through the point $(a, b, f(a, b))$. If $\Pi$ is tangent to $z = f(x, y)$ at the point $(a, b, f(a, b))$, then a normal to $\Pi$ is given by*

$$\boldsymbol{n} = (-f_x(a, b), -f_y(a, b), 1),$$

*and an equation of $\Pi$ is given by*

$$z = f_x(a, b)(x - a) + f_y(a, b)(y - b) + f(a, b).$$

Proof. Let $F(x, y, z) = d$ be an equation of $\Pi$. Slice the surface $z = f(x, y)$ at $y = b$. We have not yet defined what it means for a plane to be tangent to a surface (see Definition 11.18 below), but surely whatever definition we use, if $\Pi$ is tangent to $z = f(x, y)$ at $(a, b, f(a, b))$, then the line $F(x, b, z) = d$ will be tangent to the curve $z = f(x, b)$ at the point $(a, b, f(a, b))$. Hence, by one-variable calculus, the slope of the line tangent to the curve $z = f(x, b)$ at $x = a$ is $f_x(a, b)$, so $\boldsymbol{u} = (1, 0, f_x(a, b))$ is a vector parallel to $\Pi$. Similarly, by slicing at $x = a$, we see that $\boldsymbol{v} = (0, 1, f_y(a, b))$ is also a vector parallel to $\Pi$. Therefore (see Remark 8.10),

$$\boldsymbol{n} := \boldsymbol{u} \times \boldsymbol{v} = (-f_x(a, b), -f_y(a, b), 1)$$

is normal to $\Pi$ and

$$\boldsymbol{n} \cdot (x - a, y - b, z - f(a, b)) = 0$$

is an equation of $\Pi$. ∎

Next, we define what it means for a plane to be tangent to a surface, and prove that its existence is equivalent to differentiability. Since we do not use this material in an explicit way elsewhere in the book, the rest of this section is optional.

*11.18 DEFINITION. Let $V$ be open in $\mathbf{R}^2$, $(a, b) \in V$, $f : V \to \mathbf{R}$, and $\Pi$ be a plane which contains $\boldsymbol{c} := (a, b, f(a, b))$. For small $(h, k)$, let $\Delta z = f(a + h, b + k) - f(a, b)$, let $\theta_{h,k}$ represent the angle between any normal of $\Pi$ and the vector*

$$(6) \qquad (h, k, \Delta z) := (a + h, b + k, f(a + h, b + k)) - (a, b, f(a, b)).$$

Then $\Pi$ is said to be tangent to $z = f(x, y)$, or $z = f(x, y)$ is said to have a *tangent plane* at the point $\boldsymbol{c}$, if $\theta_{h,k}$ converges to $\pi/2$ as $(h, k) \to (0, 0)$.

Notice that, by (6), $\Pi$ is tangent to $z = f(x, y)$ if and only if the angle between the nearest vector which lies in $\Pi$, and the vector $(h, k, \Delta z)$ converges to $0$ as $(h, k) \to (0, 0)$ (see Figure 11.1). Thus Definition 11.18 agrees with our intuitive idea of what a tangent plane is.

If the plane $\Pi$ is nonvertical, any normal $(\alpha, \beta, \gamma)$ of $\Pi$ has a nonzero third component $\gamma$. Dividing by $\gamma$, we may suppose that $\Pi$ has a normal of the form $\boldsymbol{n} = (n_1, n_2, 1)$. The following result shows that, in this case, the condition $\theta_{h,k} \to \pi/2$ looks like a differentiability condition.

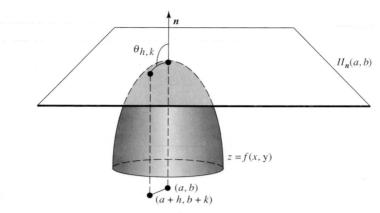

**Figure 11.1**

*11.19 Lemma.* Let $V$ be open in $\mathbf{R}^2$, $(a,b) \in V$, $f : V \to \mathbf{R}$, $\boldsymbol{c} = (a, b, f(a,b))$, and $\boldsymbol{n} = (n_1, n_2, 1)$ for some $n_1, n_2 \in \mathbf{R}$. Then the plane $\Pi_{\boldsymbol{n}}(\boldsymbol{c})$ is tangent to $z = f(x,y)$ at $\boldsymbol{c}$ if and only if

$$(7) \qquad \lim_{(h,k) \to (0,0)} \frac{\Delta z + n_1 h + n_2 k}{\|(h, k, \Delta z)\|} = 0.$$

PROOF. Since $\theta_{h,k}$ is the angle between $\boldsymbol{n}$ and $(h, k, \Delta z)$, we have by (2) in Section 8.1 that

$$\cos \theta_{h,k} = \frac{\boldsymbol{n} \cdot (h, k, \Delta z)}{\|\boldsymbol{n}\| \, \|(h, k, \Delta z)\|}$$

$$= \frac{1}{\|\boldsymbol{n}\|} \cdot \frac{\Delta z + n_1 h + n_2 k}{\|(h, k, \Delta z)\|}.$$

Therefore, (7) holds if and only if $\cos \theta_{h,k} \to 0$ as $(h,k) \to (0,0)$. But $\cos \theta_{h,k} \to 0$ as $(h,k) \to (0,0)$ if and only if $\theta_{h,k} \to \pi/2$. We conclude by Definition 11.18 that $\Pi_{\boldsymbol{n}}(\boldsymbol{c})$ is tangent to $z = f(x,y)$ at $\boldsymbol{c}$ if and only if (7) holds. ∎

Notice that (7) looks very similar to the definition of differentiability. By the uniqueness of the total derivative, we guess that $\nabla f(a,b) = (-n_1, -n_2)$, i.e., that $n_1 = -f_x(a,b)$ and $n_2 = -f_y(a,b)$. (Of course, this agrees with Remark 11.17 above.) The following result shows that this guess is correct. (The proof presented here is based on Taylor [13][1].)

*11.20 THEOREM.* Let $V$ be open in $\mathbf{R}^2$, $(a,b) \in V$, and $f : V \to \mathbf{R}$. Then $z = f(x,y)$ has a nonvertical tangent plane $\Pi$ at $\boldsymbol{c} := (a, b, f(a,b))$ if and only if $f$ is differentiable at $(a,b)$, in which case $\Pi = \Pi_{\boldsymbol{n}}(\boldsymbol{c})$, where

$$(8) \qquad \boldsymbol{n} = (-f_x(a,b), -f_y(a,b), 1).$$

---

[1]Angus E. Taylor, *Advanced Calculus* (Boston: Ginn and Company, 1955). Reprinted with permission of John Wiley & Sons, Inc.

PROOF. Suppose $z = f(x, y)$ has a nonvertical tangent plane $\Pi$ at $\boldsymbol{c}$ with normal $\boldsymbol{n} = (n_1, n_2, 1)$. Fix $r > 0$ such that $B_r(a, b) \subset V$ and for each $(h, k) \in B_r(\boldsymbol{0}) \backslash \{(0, 0)\}$ set

$$\varepsilon = \frac{\Delta z + \delta}{\sqrt{h^2 + k^2 + (\Delta z)^2}},$$

where $\Delta z := f(a + h, b + k) - f(a, b)$ and $\delta := n_1 h + n_2 k$. The expression defining $\varepsilon$ is a quadratic in $\Delta z$. Solving for $\Delta z$, we have

(9)
$$\Delta z = \frac{\delta \pm \varepsilon \sqrt{\delta^2 + (1 - \varepsilon^2)(h^2 + k^2)}}{\varepsilon^2 - 1}.$$

Notice that $|\delta| = |(n_1, n_2) \cdot (h, k)| \leq \|\boldsymbol{n}\| \, \|(h, k)\|$. Hence, it follows from (9) that

(10)
$$|\Delta z + n_1 h + n_2 k| := |\Delta z + \delta| \leq |G(\varepsilon)| \, \|(h, k)\|,$$

where $G(\varepsilon) := (\varepsilon^2 \|\boldsymbol{n}\| + \varepsilon \sqrt{\|\boldsymbol{n}\|^2 + (1 - \varepsilon^2)})/|\varepsilon^2 - 1|$. By Lemma 11.19, $\varepsilon \to 0$ as $(h, k) \to (0, 0)$. Consequently, (10) implies

$$0 \leq \frac{|f(a + h, b + k) - f(a, b) + (n_1, n_2) \cdot (h, k)|}{\|(h, k)\|} \leq |G(\varepsilon)| \to 0$$

as $(h, k) \to (0, 0)$. Therefore, $f$ is differentiable at $\boldsymbol{a}$ and $\nabla f(a, b) = (-n_1, -n_2)$. Conversely, suppose $f$ is differentiable at $(a, b)$ and define $\boldsymbol{n}$ by (8). Since

$$\|(h, k, \Delta z)\| := \|(h, k, f(a + h, b + k) - f(a, b))\| \geq \|(h, k)\|,$$

we have

$$0 \leq \frac{|\Delta z + n_1 h + n_2 k|}{\|(h, k, \Delta z)\|} \leq \frac{|\Delta z - \nabla f(a, b) \cdot (h, k)|}{\|(h, k)\|}.$$

Since $f$ is differentiable, this last quotient converges to zero as $(h, k) \to (0, 0)$. Hence, by the Squeeze Theorem, the middle quotient must also converge to zero as $(h, k) \to (0, 0)$. We conclude by Lemma 11.19 that $\Pi_{\boldsymbol{n}}(\boldsymbol{c})$ is the unique tangent of $z = f(x, y)$ at $\boldsymbol{c}$. $\blacksquare$

### EXERCISES

**1.** For each of the following functions, prove that $f$ is differentiable on its domain and compute $Df$.

  a) $f(x, y) = (\sin x, xy, \cos y)$.    b) $f(s, t, u, v) = (st + u^2, uv - s^2)$.

  c) $f(t) = (\log t, 1/(1 + t))$.    d) $f(r, \theta) = (r \cos \theta, r \sin \theta)$.

**2.** Prove that $f = (f_1, \ldots, f_m)$ is differentiable at a point $\boldsymbol{a} \in \mathbf{R}^n$ if and only if its components $f_j$ are differentiable at $\boldsymbol{a}$ for $j = 1, 2, \ldots, m$.

**3.** Prove that $f(x, y) = \sqrt{|xy|}$ is not differentiable at $(0, 0)$.

**4.** Prove that

$$f(x, y) = \begin{cases} \dfrac{x^2 + y^2}{\sin \sqrt{x^2 + y^2}} & 0 < \|(x, y)\| < \pi \\ 0 & (x, y) = (0, 0) \end{cases}$$

is not differentiable at $(0, 0)$.

**5.** Let $r > 0$, $f : B_r(\mathbf{0}) \to \mathbf{R}$, and suppose there exists an $\alpha > 1$ such that $|f(\mathbf{x})| \leq \|\mathbf{x}\|^\alpha$ for all $\mathbf{x} \in B_r(\mathbf{0})$. Prove $f$ is differentiable at $\mathbf{0}$. What happens to this result when $\alpha = 1$?

**6.** Prove that if $\alpha > 1/2$, then

$$f(x, y) = \begin{cases} (xy)^\alpha \log(x^2 + y^2) & (x, y) \neq (0, 0) \\ 0 & (x, y) = (0, 0) \end{cases}$$

is differentiable on $\mathbf{R}^2$.

**7.** Prove that

$$f(x, y) = \begin{cases} \dfrac{x^4 + y^4}{(x^2 + y^2)^\alpha} & (x, y) \neq (0, 0) \\ 0 & (x, y) = (0, 0) \end{cases}$$

is differentiable on $\mathbf{R}^2$ for all $\alpha < 3/2$.

**8.** For each of the following functions, find an equation of the tangent plane to $z = f(x, y)$ at $\mathbf{c}$.

a) $f(x, y) = x^3 \sin y$, $\mathbf{c} = (0, 0, 0)$.

b) $f(x, y) = x^3 y - xy^3$, $\mathbf{c} = (1, 1, 0)$.

**9.** Find all points on the paraboloid $z = x^2 + y^2$ (see Appendix D) where the tangent plane is parallel to the plane $x + y + z = 1$. Find equations of the corresponding tangent planes. Sketch the graphs of these functions to see that your answer agrees with your intuition.

**10.** Let $\mathcal{H}$ be the hyperboloid of one sheet, given by $x^2 + y^2 - z^2 = 1$.

a) Prove that at every point $(a, b, c) \in \mathcal{H}$, $\mathcal{H}$ has a tangent plane whose normal is given by $(-a, -b, c)$.

b) Find an equation of each plane tangent to $\mathcal{H}$ which is perpendicular to the $xy$ plane.

c) Find an equation of each plane tangent to $\mathcal{H}$ which is parallel to the plane $x + y - z = 1$.

**11.** Let $r > 0$, $(a, b) \in \mathbf{R}^2$, $f : B_r(a, b) \to \mathbf{R}$, and suppose that the first-order partial derivatives $f_x$ and $f_y$ exist in $B_r(a, b)$ and are differentiable at $(a, b)$.

a) Set $\Delta(h) = f(a + h, b + h) - f(a + h, b) - f(a, b + h) + f(a, b))$ and prove for $h$ sufficiently small that

$$\frac{\Delta(h)}{h} = f_y(a + h, b + th) - f_y(a, b) - \nabla f_y(a, b) \cdot (h, th)$$

$$- (f_y(a, b + th) - f_y(a, b) - \nabla f_y(a, b) \cdot (0, th)) + h f_{yx}(a, b)$$

for some $t \in (0, 1)$.

b) Prove that

$$\lim_{h \to 0} \frac{\Delta(h)}{h^2} = f_{yx}(a, b).$$

c) Prove

$$\frac{\partial^2 f}{\partial x \partial y}(a, b) = \frac{\partial^2 f}{\partial y \partial x}(a, b).$$

## 11.3 DIFFERENTIABILITY THEOREMS

In this section we begin to explore the analogy between $Df$ and $f'$. Namely, for real functions we know how the derivative interacts with the algebra of functions (e.g., the derivative of a sum is the sum of its derivatives). How many of these classical rules still hold in the Euclidean space setting?

Before we address this question, we obtain a simple, but highly useful, characterization of differentiability (compare with Theorem 4.2).

**11.21 Lemma.** *Let $V$ be an open set in $\mathbf{R}^n$, $\boldsymbol{a} \in V$, and $f : V \to \mathbf{R}^m$. Then $f$ is differentiable at $\boldsymbol{a}$ if and only if there is a linear function $T \in \mathcal{L}(\mathbf{R}^n; \mathbf{R}^m)$ and a function $\varepsilon : \mathbf{R}^n \to \mathbf{R}^m$ such that $\varepsilon(\boldsymbol{h}) \to \mathbf{0}$ in $\mathbf{R}^m$ as $\boldsymbol{h} \to \mathbf{0}$ in $\mathbf{R}^n$ and*

(11)
$$f(\boldsymbol{a} + \boldsymbol{h}) - f(\boldsymbol{a}) = T(\boldsymbol{h}) + \|\boldsymbol{h}\| \varepsilon(\boldsymbol{h})$$

*for $\boldsymbol{h}$ sufficiently small; i.e., there is an $r > 0$ such that (11) holds for $\|\boldsymbol{h}\| < r$.*

PROOF. Suppose $f$ is differentiable at $\boldsymbol{a}$. Choose, by Remark 9.2 or 10.9, an $r > 0$ such that $B_r(\boldsymbol{a}) \subset V$. Set $T = Df(\boldsymbol{a})$ and

(12)
$$\varepsilon(\boldsymbol{h}) = \frac{f(\boldsymbol{a} + \boldsymbol{h}) - f(\boldsymbol{a}) - T(\boldsymbol{h})}{\|\boldsymbol{h}\|}$$

for $0 < \|\boldsymbol{h}\| < r$, and $\varepsilon(\boldsymbol{h}) = 0$ otherwise. Clearly, (11) holds for $\boldsymbol{h} = \mathbf{0}$. By (12), (11) also holds for $0 < \|\boldsymbol{h}\| < r$. And by Definition 8.31, $\varepsilon(\boldsymbol{h}) \to \mathbf{0}$ as $\boldsymbol{h} \to \mathbf{0}$.

Conversely, if (11) holds for $\|\boldsymbol{h}\| < r$, then (12) holds for $0 < \|\boldsymbol{h}\| < r$. Since $\varepsilon(\boldsymbol{h}) \to \mathbf{0}$ as $\boldsymbol{h} \to \mathbf{0}$, it follows that $f$ is differentiable at $\boldsymbol{a}$. ∎

Using this characterization, we begin to explore the analogy mentioned in the first sentence of this section.

**11.22 THEOREM.** *Let $V$ be an open set in $\mathbf{R}^n$, $\boldsymbol{a} \in V$, $\alpha \in \mathbf{R}$, and $f, g : V \to \mathbf{R}^m$. If $f$ and $g$ are differentiable at $\boldsymbol{a}$, then $f + g$, $\alpha f$, and $f \cdot g$ are all differentiable at $\boldsymbol{a}$. In fact,*

(13)
$$D(f + g)(\boldsymbol{a}) = Df(\boldsymbol{a}) + Dg(\boldsymbol{a}),$$

(14)
$$D(\alpha f)(\boldsymbol{a}) = \alpha Df(\boldsymbol{a}),$$

*and*

(15) $$D(f \cdot g)(\boldsymbol{a}) = g(\boldsymbol{a})Df(\boldsymbol{a}) + f(\boldsymbol{a})Dg(\boldsymbol{a}).$$

(The sum which appears on the right side of (13) represents matrix addition, and the products which appear on the right side of (15) represent matrix multiplication.)

PROOF. The proofs of these rules are similar. We provide the details only for (15). Let

(16) $$T = g(\boldsymbol{a})Df(\boldsymbol{a}) + f(\boldsymbol{a})Dg(\boldsymbol{a}).$$

Since $g(\boldsymbol{a})$ and $f(\boldsymbol{a})$ are $1 \times m$ matrices, and $Df(\boldsymbol{a})$ and $Dg(\boldsymbol{a})$ are $m \times n$ matrices, $T$ is a $1 \times n$ matrix, the right size for the total derivative of $f \cdot g$. By the uniqueness of the total derivative, we need only show

$$\lim_{\boldsymbol{h} \to \boldsymbol{0}} \frac{(f \cdot g)(\boldsymbol{a} + \boldsymbol{h}) - (f \cdot g)(\boldsymbol{a}) - T(\boldsymbol{h})}{\|\boldsymbol{h}\|} = 0.$$

Choose by Lemma 11.21 functions $\varepsilon, \delta : \mathbf{R}^n \to \mathbf{R}^m$ with $\varepsilon(\boldsymbol{h}), \delta(\boldsymbol{h}) \to \boldsymbol{0}$ in $\mathbf{R}^m$ as $\boldsymbol{h} \to \boldsymbol{0}$ in $\mathbf{R}^n$ such that

$$\varepsilon(\boldsymbol{h}) = \frac{f(\boldsymbol{a} + \boldsymbol{h}) - f(\boldsymbol{a}) - Df(\boldsymbol{a})(\boldsymbol{h})}{\|\boldsymbol{h}\|} \quad \text{and} \quad \delta(\boldsymbol{h}) = \frac{g(\boldsymbol{a} + \boldsymbol{h}) - g(\boldsymbol{a}) - Dg(\boldsymbol{a})(\boldsymbol{h})}{\|\boldsymbol{h}\|}$$

both hold for $\boldsymbol{h} \neq \boldsymbol{0}$ sufficiently small. By (16),

$$\begin{aligned}
(f \cdot g)&(\boldsymbol{a} + \boldsymbol{h}) - (f \cdot g)(\boldsymbol{a}) - T(\boldsymbol{h}) \\
&= (f \cdot g)(\boldsymbol{a} + \boldsymbol{h}) - (f \cdot g)(\boldsymbol{a}) - g(\boldsymbol{a})Df(\boldsymbol{a})(\boldsymbol{h}) - f(\boldsymbol{a})Dg(\boldsymbol{a})(\boldsymbol{h}) \\
&= (f(\boldsymbol{a} + \boldsymbol{h}) - f(\boldsymbol{a}) - Df(\boldsymbol{a})(\boldsymbol{h})) \cdot g(\boldsymbol{a} + \boldsymbol{h}) \\
&\quad + (Df(\boldsymbol{a})(\boldsymbol{h})) \cdot (g(\boldsymbol{a} + \boldsymbol{h}) - g(\boldsymbol{a})) \\
&\quad\quad + f(\boldsymbol{a}) \cdot (g(\boldsymbol{a} + \boldsymbol{h}) - g(\boldsymbol{a}) - Dg(\boldsymbol{a})(\boldsymbol{h})) \\
&=: T_1(\boldsymbol{h}) + T_2(\boldsymbol{h}) + T_3(\boldsymbol{h}).
\end{aligned}$$

It remains to verify that $T_j(\boldsymbol{h})/\|\boldsymbol{h}\| \to 0$ as $\boldsymbol{h} \to \boldsymbol{0}$ for $j = 1, 2, 3$. To estimate the first term, observe by the triangle inequality and the choice of $\varepsilon$ that

$$\begin{aligned}
|T_1(\boldsymbol{h})| &\leq |f(\boldsymbol{a} + \boldsymbol{h}) - f(\boldsymbol{a}) - Df(\boldsymbol{a})(\boldsymbol{h})| \, |g(\boldsymbol{a} + \boldsymbol{h})| \\
&= \|\boldsymbol{h}\| \, |\varepsilon(\boldsymbol{h})| \, |g(\boldsymbol{a} + \boldsymbol{h})|.
\end{aligned}$$

Since $g$ is continuous at $\boldsymbol{a}$ (see Theorem 8.38) and $\varepsilon(\boldsymbol{h}) \to \boldsymbol{0}$ as $\boldsymbol{h} \to \boldsymbol{0}$, it follows that $|T_1(\boldsymbol{h})|/\|\boldsymbol{h}\| \to \boldsymbol{0}$ as $\boldsymbol{h} \to \boldsymbol{0}$. A similar argument shows that $|T_3(\boldsymbol{h})|/\|\boldsymbol{h}\| \to \boldsymbol{0}$ as $\boldsymbol{h} \to \boldsymbol{0}$. To estimate the second term, observe by definition of the operator norm (see Theorem 8.37) that

$$|T_2(\boldsymbol{h})| = |Df(\boldsymbol{a})(\boldsymbol{h})| \, |g(\boldsymbol{a} + \boldsymbol{h}) - g(\boldsymbol{a})| \leq \|Df(\boldsymbol{a})\| \, \|\boldsymbol{h}\| \, |g(\boldsymbol{a} + \boldsymbol{h}) - g(\boldsymbol{a})|.$$

Thus $|T_2(\boldsymbol{h})|/\|\boldsymbol{h}\| \leq \|Df(\boldsymbol{a})\| \, |g(\boldsymbol{a} + \boldsymbol{h}) - g(\boldsymbol{a})| \to 0$ as $\boldsymbol{h} \to \boldsymbol{0}$. We conclude that $f \cdot g$ is differentiable at $\boldsymbol{a}$ and its total derivative is $T$. ∎

Formula (13) is called the *Sum Rule*; (14) is sometimes called the *Homogeneous Rule*; and (15) is called the *Dot Product Rule*. (We note that a quotient rule also holds for real-valued functions: see Exercise 7 in Section 8.4.)

Next, we obtain a Euclidean space version of the Chain Rule.

**11.23 THEOREM** [CHAIN RULE]. *Let $V$ be an open set in $\mathbf{R}^n$ and $U$ be an open set in $\mathbf{R}^m$. Suppose further that $g : V \to \mathbf{R}^m$, $f : U \to \mathbf{R}^p$, $\boldsymbol{a} \in V$, and $g(\boldsymbol{a}) \in U$. If $g$ is differentiable at $\boldsymbol{a}$ and $f$ is differentiable at $g(\boldsymbol{a})$, then $f \circ g$ is differentiable at $\boldsymbol{a}$ and*

$$(17) \qquad\qquad D(f \circ g)(\boldsymbol{a}) = Df(g(\boldsymbol{a}))Dg(\boldsymbol{a}).$$

(The product $Df(g(\boldsymbol{a}))Dg(\boldsymbol{a})$ is matrix multiplication.)

PROOF. Set $T = Df(g(\boldsymbol{a}))Dg(\boldsymbol{a})$ and observe that $T$, the product of a $p \times m$ matrix with an $m \times n$ matrix, is a $p \times n$ matrix, the right size for the total derivative of $f \circ g$. By the uniqueness of the total derivative, we must show

$$(18) \qquad\qquad \lim_{\boldsymbol{h} \to \boldsymbol{0}} \frac{f(g(\boldsymbol{a}+\boldsymbol{h})) - f(g(\boldsymbol{a})) - T(\boldsymbol{h})}{\|\boldsymbol{h}\|} = 0.$$

Let $\boldsymbol{b} = g(\boldsymbol{a})$. Choose by Lemma 11.21 functions $\varepsilon : \mathbf{R}^n \to \mathbf{R}^m$ and $\delta : \mathbf{R}^m \to \mathbf{R}^p$, with $\varepsilon(\boldsymbol{h}) \to \boldsymbol{0}$ in $\mathbf{R}^m$ as $\boldsymbol{h} \to \boldsymbol{0}$ in $\mathbf{R}^n$ and $\delta(\boldsymbol{k}) \to \boldsymbol{0}$ in $\mathbf{R}^p$ as $\boldsymbol{k} \to \boldsymbol{0}$ in $\mathbf{R}^m$ such that

$$(19) \qquad\qquad g(\boldsymbol{a}+\boldsymbol{h}) - g(\boldsymbol{a}) = Dg(\boldsymbol{a})(\boldsymbol{h}) + \|\boldsymbol{h}\|\varepsilon(\boldsymbol{h})$$

and

$$(20) \qquad\qquad f(\boldsymbol{b}+\boldsymbol{k}) - f(\boldsymbol{b}) = Df(\boldsymbol{b})(\boldsymbol{k}) + \|\boldsymbol{k}\|\delta(\boldsymbol{k})$$

both hold for $\boldsymbol{h}$ and $\boldsymbol{k}$ sufficiently small. Fix $\boldsymbol{h} \neq \boldsymbol{0}$ and set $\boldsymbol{k} = g(\boldsymbol{a}+\boldsymbol{h}) - g(\boldsymbol{a})$. Since (20) and (19) imply

$$\begin{aligned} f(g(\boldsymbol{a}+\boldsymbol{h})) - f(g(\boldsymbol{a})) &= f(\boldsymbol{b}+\boldsymbol{k}) - f(\boldsymbol{b}) = Df(\boldsymbol{b})(\boldsymbol{k}) + \|\boldsymbol{k}\|\delta(\boldsymbol{k}) \\ &= Df(\boldsymbol{b})(Dg(\boldsymbol{a})(\boldsymbol{h}) + \|\boldsymbol{h}\|\varepsilon(\boldsymbol{h})) + \|\boldsymbol{k}\|\delta(\boldsymbol{k}), \end{aligned}$$

we have

$$\begin{aligned} f(g(\boldsymbol{a}+\boldsymbol{h})) - f(g(\boldsymbol{a})) - T(\boldsymbol{h}) &= \|\boldsymbol{h}\|Df(\boldsymbol{b})(\varepsilon(\boldsymbol{h})) + \|\boldsymbol{k}\|\delta(\boldsymbol{k}) \\ &=: T_1(\boldsymbol{h}) + T_2(\boldsymbol{h}). \end{aligned}$$

It remains to verify that $T_j(\boldsymbol{h})/\|\boldsymbol{h}\| \to \boldsymbol{0}$ as $\boldsymbol{h} \to \boldsymbol{0}$ for $j = 1, 2$.

Since $\varepsilon(\boldsymbol{h}) \to \boldsymbol{0}$ as $\boldsymbol{h} \to \boldsymbol{0}$ and $Df(\boldsymbol{b})(\boldsymbol{h})$ is matrix multiplication, it is clear that $T_1(\boldsymbol{h})/\|\boldsymbol{h}\| \to Df(\boldsymbol{b})(\boldsymbol{0}) = \boldsymbol{0}$ as $\boldsymbol{h} \to \boldsymbol{0}$. On the other hand, by (19), the triangle inequality, and the definition of the operator norm, we have

$$\begin{aligned} \|\boldsymbol{k}\| := \|g(\boldsymbol{a}+\boldsymbol{h}) - g(\boldsymbol{a})\| &= \|Dg(\boldsymbol{a})(\boldsymbol{h}) + \|\boldsymbol{h}\|\varepsilon(\boldsymbol{h})\| \\ &\leq \|\boldsymbol{h}\| \left(\|Dg(\boldsymbol{a})\| + \|\varepsilon(\boldsymbol{h})\|\right). \end{aligned}$$

Thus $\|\boldsymbol{k}\|/\|\boldsymbol{h}\|$ is bounded for $\boldsymbol{h}$ sufficiently small. Since $\boldsymbol{k} \to \boldsymbol{0}$ in $\mathbf{R}^m$ (hence $\delta(\boldsymbol{k}) \to \boldsymbol{0}$ in $\mathbf{R}^p$) as $\boldsymbol{h} \to \boldsymbol{0}$ in $\mathbf{R}^n$, it follows that $\|T_2(\boldsymbol{h})\|/\|\boldsymbol{h}\| = (\|\boldsymbol{k}\|/\|\boldsymbol{h}\|) \|\delta(\boldsymbol{k})\| \to \boldsymbol{0}$ as $\boldsymbol{h} \to \boldsymbol{0}$. We conclude that $f \circ g$ is differentiable at $\boldsymbol{a}$ and the derivative is $Df(g(\boldsymbol{a}))Dg(\boldsymbol{a})$. ∎

The Chain Rule can be used to compute individual partial derivatives without writing out the entire Jacobian matrices $Df$ and $Dg$. For example, suppose $f(u_1, \dots, u_m)$ is differentiable from $\mathbf{R}^m$ to $\mathbf{R}$, $g(x_1, \dots, x_n)$ is differentiable from $\mathbf{R}^n$ to $\mathbf{R}^m$, and $z = f(g(x_1, \dots, x_n))$. Since $Df = \nabla f$ and the $j$th column of $Dg$ consists of first partial derivatives, with respect to $x_j$, of the components $u_j := g_j(x_1, \dots, x_n)$, it follows from the Chain Rule and the definition of matrix multiplication that

$$\frac{\partial z}{\partial x_j} = \frac{\partial f}{\partial u_1}\frac{\partial u_1}{\partial x_j} + \cdots + \frac{\partial f}{\partial u_m}\frac{\partial u_m}{\partial x_j}$$

for $j = 1, 2, \dots, n$. Here are two concrete examples which illustrate this principle.

**11.24 Examples.** a) If $F, G, H : \mathbf{R}^2 \to \mathbf{R}$ are differentiable and $z = F(x, y)$, where $x = G(r, \theta)$, and $y = H(r, \theta)$, then

$$\frac{\partial z}{\partial r} = \frac{\partial z}{\partial x}\frac{\partial x}{\partial r} + \frac{\partial z}{\partial y}\frac{\partial y}{\partial r} \quad \text{and} \quad \frac{\partial z}{\partial \theta} = \frac{\partial z}{\partial x}\frac{\partial x}{\partial \theta} + \frac{\partial z}{\partial y}\frac{\partial y}{\partial \theta}.$$

b) If $f : \mathbf{R}^3 \to \mathbf{R}$ and $\phi, \psi, \sigma : \mathbf{R} \to \mathbf{R}$ are differentiable and $w = f(x, y, z)$, where $x = \phi(t)$, $y = \psi(t)$, and $z = \sigma(t)$, then

$$\frac{dw}{dt} = \frac{\partial w}{\partial x}\frac{dx}{dt} + \frac{\partial w}{\partial y}\frac{dy}{dt} + \frac{\partial w}{\partial z}\frac{dz}{dt}.$$

**EXERCISES**

**1.** For each of the following, find $D(f + g)(\boldsymbol{a})$ and $D(3f - 2g)(\boldsymbol{a})$.

a) $\qquad f(t) = t^2 + 1, \qquad g(t) = \log t - \dfrac{1}{t}, \qquad \boldsymbol{a} = 1.$

b) $\qquad f(x, y) = x - y, \qquad g(x, y) = x^2 - y^2, \qquad \boldsymbol{a} = (1, 1).$

c) $\qquad f(x, y) = xy, \qquad g(x, y) = x\sin x - \cos y, \qquad \boldsymbol{a} = (\pi, \pi).$

d) $f(x, y, z) = (x - z, x + z), \qquad g(x, y, z) = (xyz, x^2 - y^2), \qquad \boldsymbol{a} = (1, 1, 1).$

e) $\qquad f(x, y) = (x, y, \pi^2), \qquad g(x, y) = (y, x, xy), \qquad \boldsymbol{a} = (1, -1).$

**2.** Let $\alpha$ be a scalar, $V$ be open in $\mathbf{R}^n$, $\boldsymbol{a} \in V$, $f, g : V \to \mathbf{R}^3$, and suppose $f$ and $g$

are differentiable at $\boldsymbol{a}$.

   a) [CROSS PRODUCT RULE] For the case $n = 1$, prove that $f \times g$ is differentiable at $\boldsymbol{a}$ and

$$(f \times g)'(\boldsymbol{a}) = f(\boldsymbol{a}) \times g'(\boldsymbol{a}) + f'(\boldsymbol{a}) \times g(\boldsymbol{a}).$$

   b) What happens to part a) when $n > 1$?

   c) Suppose $f(\boldsymbol{a}) = (2, 1, 2)$, $g(\boldsymbol{a}) = (1, 2, 1)$,

$$Df(\boldsymbol{a}) = \begin{bmatrix} 0 & 1 & 0 \\ 1 & 0 & 1 \\ 1 & 1 & 1 \end{bmatrix} \quad \text{and} \quad Dg(\boldsymbol{a}) = \begin{bmatrix} -1 & 0 & 0 \\ 0 & -1 & 1 \\ 1 & 0 & -1 \end{bmatrix}.$$

     Find $D(f \cdot g)(\boldsymbol{a})(1, 1, 1)$ and $D(f \times g)(\boldsymbol{a})(1, 1, 1)$.

**3.** Let $F : \mathbf{R}^3 \to \mathbf{R}$ and $f, g, h : \mathbf{R}^2 \to \mathbf{R}$ be $\mathcal{C}^2$ functions. If $w = F(x, y, z)$, where $x = f(p, q)$, $y = g(p, q)$, and $z = h(p, q)$, find formulas for $w_p$, $w_q$, and $w_{pp}$.

**4.** Let $r > 0$, $\boldsymbol{a} \in \mathbf{R}^n$, and $g : B_r(\boldsymbol{a}) \to \mathbf{R}^m$ be differentiable at $\boldsymbol{a}$.

   a) If $f : B_r(g(\boldsymbol{a})) \to \mathbf{R}$ is differentiable at $g(\boldsymbol{a})$, prove that the partial derivatives of $h = f \circ g$ are given by

$$\frac{\partial h}{\partial x_j}(\boldsymbol{a}) = \nabla f(g(\boldsymbol{a})) \cdot \frac{\partial g}{\partial x_j}(\boldsymbol{a})$$

     for $j = 1, 2, \ldots, n$.

   b) If $n = m$ and $f : B_r(g(\boldsymbol{a})) \to \mathbf{R}^n$ is differentiable at $g(\boldsymbol{a})$, prove $\Delta_{f \circ g}(\boldsymbol{a}) = \Delta_f(g(\boldsymbol{a})) \Delta_g(\boldsymbol{a})$.

**5.** Let $f, g : \mathbf{R} \to \mathbf{R}$ be twice differentiable. Prove that $u(x, y) := f(xy)$ satisfies

$$x \frac{\partial u}{\partial x} - y \frac{\partial u}{\partial y} = 0,$$

and $v(x, y) := f(x - y) + g(x + y)$ satisfies the *wave equation*; i.e.,

$$\frac{\partial^2 v}{\partial x^2} - \frac{\partial^2 v}{\partial y^2} = 0.$$

**6.** Let $u : \mathbf{R} \to [0, \infty)$ be differentiable. Prove that for each $(x, y, z) \neq (0, 0, 0)$,

$$F(x, y, z) := u(\sqrt{x^2 + y^2 + z^2})$$

satisfies

$$\left( \left( \frac{\partial F}{\partial x} \right)^2 + \left( \frac{\partial F}{\partial y} \right)^2 + \left( \frac{\partial F}{\partial z} \right)^2 \right)^{1/2} = |u'(\sqrt{x^2 + y^2 + z^2})|.$$

**7.** Let

$$u(x, t) = \frac{e^{-x^2/4t}}{\sqrt{4\pi t}}, \qquad t > 0, \ x \in \mathbf{R}.$$

   a) Prove that $u$ satisfies the *heat equation*; i.e., $u_{xx} - u_t = 0$ for all $t > 0$ and $x \in \mathbf{R}$.

   b) If $a > 0$, prove that $u(x, t) \to 0$, as $t \to 0+$, uniformly for $x \in [a, \infty)$.

**8.** Suppose $I$ is a nonempty, open interval and $f : I \to \mathbf{R}^m$ is differentiable on $I$. If $f(I) \subseteq \partial B_r(\mathbf{0})$ for some fixed $r > 0$, prove that $f(t)$ is orthogonal to $f'(t)$ for all $t \in I$.

**9.** Suppose $z = F(x, y)$ is differentiable at $(a, b)$, $F_y(a, b) \neq 0$, and $I$ is an open interval containing $a$. Prove that if $f : I \to \mathbf{R}$ is differentiable at $a$, $f(a) = b$, and $F(x, f(x)) = 0$ for all $x \in I$, then

$$\frac{df}{dx}(a) = \frac{-\dfrac{\partial F}{\partial x}(a, b)}{\dfrac{\partial F}{\partial y}(a, b)}.$$

**10.** Let $V$ be open in $\mathbf{R}^n$, $\boldsymbol{a} \in V$, and $f : V \to \mathbf{R}^m$.

**\*11.25 DEFINITION.** If $\boldsymbol{u}$ is a *unit* vector in $\mathbf{R}^n$, i.e., $\|\boldsymbol{u}\| = 1$, then the *directional derivative* of $f$ at $\boldsymbol{a}$ in the direction $\boldsymbol{u}$ is defined by

$$D_{\boldsymbol{u}}f(\boldsymbol{a}) := \lim_{t \to 0} \frac{f(\boldsymbol{a} + t\boldsymbol{u}) - f(\boldsymbol{a})}{t}$$

when this limit exists.

a) Prove that $D_{\boldsymbol{u}}f(\boldsymbol{a})$ exists for $\boldsymbol{u} = \boldsymbol{e}_k$ if and only if $f_{x_k}(\boldsymbol{a})$ exists, in which case

$$D_{\boldsymbol{e}_k}f(\boldsymbol{a}) = \frac{\partial f}{\partial x_k}(\boldsymbol{a}).$$

b) Show if $f$ has directional derivatives at $\boldsymbol{a}$ in all directions $\boldsymbol{u}$, then the first-order partial derivatives of $f$ exist at $\boldsymbol{a}$. Use Example 11.13 to show that the converse of this statement is false.

c) Prove that the directional derivatives of

$$f(x, y) = \begin{cases} \dfrac{x^2 y}{x^4 + y^2} & (x, y) \neq (0, 0) \\ 0 & (x, y) = (0, 0) \end{cases}$$

exist at $(0, 0)$ in all directions $\boldsymbol{u}$, but $f$ is neither continuous nor differentiable at $(0, 0)$.

**11.** Let $V$ be open in $\mathbf{R}^n$, $\boldsymbol{a} \in V$, $f : V \to \mathbf{R}$, and suppose $f$ is differentiable at $\boldsymbol{a}$.

a) Prove the directional derivative $D_{\boldsymbol{u}}f(\boldsymbol{a})$ exists (see Exercise 10 above), for each $\boldsymbol{u} \in \mathbf{R}^n$ such that $\|\boldsymbol{u}\| = 1$, and $D_{\boldsymbol{u}}f(\boldsymbol{a}) = \nabla f(\boldsymbol{a}) \cdot \boldsymbol{u}$.

b) If $\nabla f(\boldsymbol{a}) \neq \boldsymbol{0}$ and $\theta$ represents the angle between $\boldsymbol{u}$ and $\nabla f(\boldsymbol{a})$, prove $D_{\boldsymbol{u}}f(\boldsymbol{a}) = \|\nabla f(\boldsymbol{a})\| \cos \theta$.

c) Show that as $\boldsymbol{u}$ ranges over all unit vectors in $\mathbf{R}^n$, the maximum of $D_{\boldsymbol{u}}f(\boldsymbol{a})$ is $\|\nabla f(\boldsymbol{a})\|$, and it occurs when $\boldsymbol{u}$ is parallel to $\nabla f(\boldsymbol{a})$.

**12.** Let $f, g : \mathbf{R}^2 \to \mathbf{R}$ be differentiable and satisfy the *Cauchy–Riemann equations*, i.e., that

$$\frac{\partial f}{\partial x} = \frac{\partial g}{\partial y} \quad \text{and} \quad \frac{\partial f}{\partial y} = -\frac{\partial g}{\partial x}$$

hold on $\mathbf{R}^2$. If $u(r, \theta) = f(r \cos \theta, r \sin \theta)$, and $v(r, \theta) = g(r \cos \theta, r \sin \theta)$, prove

$$\frac{\partial u}{\partial r} = \frac{1}{r} \frac{\partial v}{\partial \theta}, \qquad \frac{\partial v}{\partial r} = -\frac{1}{r} \frac{\partial u}{\partial \theta}, \qquad r \neq 0.$$

**13.** Let $f : \mathbf{R}^2 \to \mathbf{R}$ be $\mathcal{C}^2$ on $\mathbf{R}^2$ and set $u(r, \theta) = f(r \cos \theta, r \sin \theta)$. If $f$ satisfies the *Laplace equation*, i.e., if

$$\frac{\partial^2 f}{\partial x^2} + \frac{\partial^2 f}{\partial y^2} = 0,$$

prove for each $r \neq 0$ that

$$\frac{1}{r^2} \frac{\partial^2 u}{\partial \theta^2} + \frac{1}{r} \frac{\partial u}{\partial r} + \frac{\partial^2 u}{\partial r^2} = 0.$$

## 11.4   THE MEAN VALUE THEOREM AND TAYLOR'S FORMULA

Using $Df$ as a replacement for $f'$, we guess that the multidimensional analogue of the Mean Value Theorem is

$$f(\boldsymbol{x}) - f(\boldsymbol{a}) = Df(\boldsymbol{c})(\boldsymbol{x} - \boldsymbol{a})$$

for some $\boldsymbol{c}$ "between" $\boldsymbol{x}$ and $\boldsymbol{a}$, i.e., some $\boldsymbol{c} \in L(\boldsymbol{x};\boldsymbol{a})$, the line segment from $\boldsymbol{a}$ to $\boldsymbol{x}$. The following result shows that our guess is wrong for functions $f : \mathbf{R}^n \to \mathbf{R}^m$ when $m > 1$.

**11.26 Remark.** *The function $f(t) = (\cos t, \sin t)$ is differentiable on $\mathbf{R}$ and satisfies $f(2\pi) = f(0)$, but there is no $c \in \mathbf{R}$ such that $Df(c) = (0, 0)$.*

PROOF. $Df(t) = (-\sin t, \cos t)$ exists and is continuous for $t \in \mathbf{R}$ but $(0, 0) \neq (-\sin t, \cos t)$ for $t \in \mathbf{R}$. ∎

The following is a correct version of the Mean Value Theorem for multivariable functions.

**11.27 THEOREM** [THE MEAN VALUE THEOREM ON $\mathbf{R}^n$]. *Let $V$ be nonempty and open in $\mathbf{R}^n$ and $f : V \to \mathbf{R}^m$ be differentiable on $V$. If $\boldsymbol{x}, \boldsymbol{a} \in V$ and $L(\boldsymbol{x};\boldsymbol{a}) \subset V$, then given any $\boldsymbol{u} \in \mathbf{R}^m$ there is a $\boldsymbol{c} \in L(\boldsymbol{x};\boldsymbol{a})$ such that*

$$\boldsymbol{u} \cdot (f(\boldsymbol{x}) - f(\boldsymbol{a})) = \boldsymbol{u} \cdot (Df(\boldsymbol{c})(\boldsymbol{x} - \boldsymbol{a})).$$

PROOF. Let

$$g(t) = \boldsymbol{a} + t(\boldsymbol{x} - \boldsymbol{a}), \qquad t \in \mathbf{R},$$

and notice by Exercise 8 in Section 8.4 that $g : \mathbf{R} \to \mathbf{R}^n$ is differentiable with $Dg(t) = \boldsymbol{x} - \boldsymbol{a}$ for all $t \in \mathbf{R}$. Since $L(\boldsymbol{x};\boldsymbol{a}) \subset V$ and $V$ is open, choose $\delta > 0$ such that $g(t) \in V$ for all $t \in I_\delta := (-\delta, 1 + \delta)$. By the Chain Rule,

$$(21) \qquad\qquad D(f \circ g)(t) = Df(g(t))(\boldsymbol{x} - \boldsymbol{a}), \qquad t \in I_\delta.$$

Fix $\boldsymbol{u} \in \mathbf{R}^m$, and consider the function

$$F(t) = \boldsymbol{u} \cdot (f \circ g)(t), \qquad t \in I_\delta.$$

The function $F$ is a real-valued function on $I_\delta$. By (17) (the Dot Product Rule) and (21), $F$ is differentiable on $I_\delta$ with

$$F'(t) = \boldsymbol{u} \cdot D(f \circ g)(t) = \boldsymbol{u} \cdot (Df(g(t))(\boldsymbol{x} - \boldsymbol{a})).$$

Hence, by the one-dimensional Mean Value Theorem, there is a $t_0 \in (0, 1)$ such that

$$\boldsymbol{u} \cdot (f(\boldsymbol{x}) - f(\boldsymbol{a})) = F(1) - F(0) = F'(t_0) = \boldsymbol{u} \cdot (Df(g(t_0))(\boldsymbol{x} - \boldsymbol{a})).$$

Thus set $\boldsymbol{c} = g(t_0)$. ∎

Sets which satisfy the hypothesis "$L(\boldsymbol{x};\boldsymbol{a}) \subset V$" come up often enough to warrant a name.

**11.28 DEFINITION.** A subset $E$ of $\mathbf{R}^n$ is said to be *convex* if $L(\boldsymbol{x};\boldsymbol{a}) \subseteq E$ for all $\boldsymbol{x}, \boldsymbol{a} \in E$.

Using this terminology, we see that the Mean Value Theorem holds for any convex, open set $V$.

It is easy to see that any ball is convex. For example, if $\boldsymbol{x}, \boldsymbol{a} \in B_r(\boldsymbol{b})$, then

$$\|((1 - t)\boldsymbol{a} + t\boldsymbol{x}) - \boldsymbol{b}\| = \|(1 - t)(\boldsymbol{a} - \boldsymbol{b}) + t(\boldsymbol{x} - \boldsymbol{b})\| < (1 - t)r + tr = r.$$

On the other hand, Figure 11.2 is an example of a nonconvex set in $\mathbf{R}^2$ (because the line segment which joins $\boldsymbol{a}$ to $\boldsymbol{b}$ contains some points outside $V$.)

Our next result shows that the Mean Value Theorem for scalar-valued functions recaptures the simplicity of the one-dimensional version (see also Exercises 1 and 5 below).

**11.29 COROLLARY.** *Let $V$ be nonempty, convex, and open in $\mathbf{R}^n$ and $f : V \to \mathbf{R}$. If $f$ is differentiable on $V$ and $\boldsymbol{a} + \boldsymbol{h}, \boldsymbol{a}$ both belong to $V$, then there is a $0 < t < 1$ such that*

$$(22) \qquad f(\boldsymbol{a} + \boldsymbol{h}) - f(\boldsymbol{a}) = \sum_{j=1}^{n} \frac{\partial f}{\partial x_j}(\boldsymbol{a} + t\boldsymbol{h})h_j = \nabla f(\boldsymbol{a} + t\boldsymbol{h}) \cdot \boldsymbol{h}.$$

PROOF. Let $u$ be a nonzero scalar, and suppose $\boldsymbol{a} + \boldsymbol{h}, \boldsymbol{a}$ both belong to $V$. Since $V$ is convex, $L(\boldsymbol{a} + \boldsymbol{h};\boldsymbol{a}) \subseteq V$. Hence, by Theorem 11.27,

$$u(f(\boldsymbol{a} + \boldsymbol{h}) - f(\boldsymbol{a})) = u(\nabla f(\boldsymbol{c}) \cdot \boldsymbol{h})$$

for some $\boldsymbol{c} \in L(\boldsymbol{a} + \boldsymbol{h};\boldsymbol{a})$. Dividing this inequality by $u$ and choosing $t \in (0, 1)$ such that $\boldsymbol{c} = \boldsymbol{a} + t\boldsymbol{h}$, we conclude that (22) holds. ∎

As in the one-dimensional case, the Mean Value Theorem is used most often to obtain information about a function from properties of its derivative. Here is a typical example.

**11.30 COROLLARY.** *Let $V$ be a nonempty, open set in $\mathbf{R}^n$, $H$ be a compact subset of $V$, and $f : V \to \mathbf{R}^m$ be $\mathcal{C}^1$ on $V$. If $E$ is a convex subset of $H$, then there is a constant $M$ (which depends on $H$ and $f$ but not on $E$) such that*

$$\|f(\boldsymbol{x}) - f(\boldsymbol{a})\| \le M\|\boldsymbol{x} - \boldsymbol{a}\|$$

*for all $\boldsymbol{x}, \boldsymbol{a} \in E$.*

PROOF. Since $H$ is compact and the entries of $Df$ are continuous on $H$, we have by the Extreme Value Theorem (Theorem 9.38 or 10.63) and the proof of Theorem 8.37 that the operator norm of $Df$ is bounded on $H$, i.e., that

$$M := \sup_{\boldsymbol{c} \in H} \|Df(\boldsymbol{c})\|$$

is finite. Notice that $M$ depends only on $H$ and $f$.

Let $\boldsymbol{x}, \boldsymbol{a} \in E$ and $\boldsymbol{u} = f(\boldsymbol{x}) - f(\boldsymbol{a})$. Since $E$ is convex, $L(\boldsymbol{x}; \boldsymbol{a}) \subseteq E$. Hence, by Theorem 11.27, there is a $\boldsymbol{c} \in L(\boldsymbol{x}; \boldsymbol{a})$ such that

$$\|f(\boldsymbol{x}) - f(\boldsymbol{a})\|^2 = \boldsymbol{u} \cdot (f(\boldsymbol{x}) - f(\boldsymbol{a})) = \boldsymbol{u} \cdot (Df(\boldsymbol{c})(\boldsymbol{x} - \boldsymbol{a})) = (f(\boldsymbol{x}) - f(\boldsymbol{a})) \cdot (Df(\boldsymbol{c})(\boldsymbol{x} - \boldsymbol{a})).$$

It follows from the Cauchy–Schwarz Inequality and the definition of the operator norm that

$$\|f(\boldsymbol{x}) - f(\boldsymbol{a})\|^2 \le \|f(\boldsymbol{x}) - f(\boldsymbol{a})\| \, \|Df(\boldsymbol{c})\| \, \|\boldsymbol{x} - \boldsymbol{a}\|.$$

If $\|f(\boldsymbol{x}) - f(\boldsymbol{a})\| = 0$, there is nothing to prove. Otherwise, we can divide the inequality above by $\|f(\boldsymbol{x}) - f(\boldsymbol{a})\|$ to obtain

$$\|f(\boldsymbol{x}) - f(\boldsymbol{a})\| \le \|Df(\boldsymbol{c})\| \, \|\boldsymbol{x} - \boldsymbol{a}\| \le M\|\boldsymbol{x} - \boldsymbol{a}\|. \quad \blacksquare$$

As the following optional result shows, for some applications of the Mean Value Theorem the convexity hypothesis can sometimes be replaced by connectivity. (This is an analogue of the one-dimensional result: if $f' = 0$ on $[a, b]$, then $f$ is constant on $[a, b]$.)

\*11.31 COROLLARY. *Suppose $V$ is a nonempty, open, connected set in $\mathbf{R}^n$ and $f : V \to \mathbf{R}^m$ is differentiable on $V$. If $Df(\boldsymbol{c}) = O$ for all $\boldsymbol{c} \in V$, then $f$ is constant on $V$.*

PROOF. Fix $\boldsymbol{a} \in V$, and let $\boldsymbol{x} \in V$. Since $V$ is open and connected, $V$ is polygonally connected (see Exercise 10 in Section 9.3). Thus, there exist points $\boldsymbol{x}_0 = \boldsymbol{a}, \boldsymbol{x}_1, \dots, \boldsymbol{x}_k = \boldsymbol{x}$ such that $L(\boldsymbol{x}_{j-1}; \boldsymbol{x}_j) \subset V$ for $j = 1, 2, \dots, k$ (see Figure 11.2). Let $\boldsymbol{u} = f(\boldsymbol{x}) - f(\boldsymbol{a})$ and choose by Theorem 11.27 points $\boldsymbol{c}_j \in L(\boldsymbol{x}_{j-1}; \boldsymbol{x}_j)$ such that

$$\boldsymbol{u} \cdot (f(\boldsymbol{x}_j) - f(\boldsymbol{x}_{j-1})) = \boldsymbol{u} \cdot (Df(\boldsymbol{c}_j)(\boldsymbol{x}_j - \boldsymbol{x}_{j-1})) = 0$$

for $j = 1, 2, \dots, k$. Summing over $j$ and telescoping, we see by the choice of $\boldsymbol{u}$ that

$$0 = \sum_{j=1}^{k} \boldsymbol{u} \cdot (f(\boldsymbol{x}_j) - f(\boldsymbol{x}_{j-1})) = \boldsymbol{u} \cdot (f(\boldsymbol{x}) - f(\boldsymbol{a})) = \|f(\boldsymbol{x}) - f(\boldsymbol{a})\|^2.$$

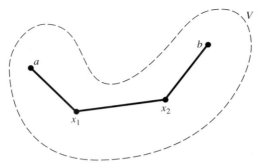

**Figure 11.2**

Therefore, $f(\boldsymbol{x}) = f(\boldsymbol{a})$. ∎

To obtain a multidimensional version of Taylor's Formula, we need to define higher-order differentials. Let $p \geq 1$, $V$ be open in $\mathbf{R}^n$, and $f : V \to \mathbf{R}$. If the $(p-1)$-st order partial derivatives of $f$ exist on $V$ and are differentiable at some $\boldsymbol{a} \in V$, then the $p$th *total differential* of $f$ at $\boldsymbol{a}$ and $\boldsymbol{h} = (h_1, \ldots, h_n)$ is the expression

$$D^{(p)} f(\boldsymbol{a}; \boldsymbol{h}) = \sum_{i_1=1}^{n} \cdots \sum_{i_p=1}^{n} \frac{\partial^p f}{\partial x_{i_1} \ldots \partial x_{i_p}}(\boldsymbol{a}) h_{i_1} \ldots h_{i_p}.$$

Notice that

$$D^{(p)} f(\boldsymbol{a}; \boldsymbol{h}) = D^{(1)}(D^{(p-1)} f)(\boldsymbol{a}; \boldsymbol{h})$$

$$= \sum_{j=1}^{n} \frac{\partial}{\partial x_j} \left( \sum_{i_1=1}^{n} \cdots \sum_{i_{p-1}=1}^{n} \frac{\partial^{p-1} f}{\partial x_{i_1} \ldots \partial x_{i_{p-1}}}(\boldsymbol{a}) h_{i_1} \ldots h_{i_{p-1}} \right) h_j$$

for $p > 1$.

The first total differential of $f$ is essentially the total derivative of $f$; indeed,

$$D^{(1)} f(\boldsymbol{a}; \boldsymbol{h}) := \sum_{j=1}^{n} \frac{\partial f}{\partial x_j}(\boldsymbol{a}) h_j = \nabla f(\boldsymbol{a}) \cdot \boldsymbol{h} = Df(\boldsymbol{a})(\boldsymbol{h}).$$

For the case $n = 2$, this differential has a simple geometric interpretation (see Figure 11.4 below).

Although total differentials look messy to evaluate, when $f$ is a sufficiently smooth function of two variables they are relatively easy to calculate using binomial coefficients (see the next example and Exercise 4 below).

**11.32 Example.** Suppose $f : V \to \mathbf{R}$ is $\mathcal{C}^2$ on $V$. Find a formula for the second total differential of $f$ at $(a, b)$.

SOLUTION. By definition,

$$D^{(2)} f((a, b); (h, k)) = h^2 \frac{\partial^2 f}{\partial x^2}(a, b) + hk \frac{\partial^2 f}{\partial x \partial y}(a, b) + hk \frac{\partial^2 f}{\partial y \partial x}(a, b) + k^2 \frac{\partial^2 f}{\partial y^2}(a, b).$$

But by Theorem 11.2, $f_{xy}(a,b) = f_{yx}(a,b)$. Therefore,

$$D^{(2)}f((a,b);(h,k)) = h^2 \frac{\partial^2 f}{\partial x^2}(a,b) + 2hk \frac{\partial^2 f}{\partial x \partial y}(a,b) + k^2 \frac{\partial^2 f}{\partial y^2}(a,b). \quad \blacksquare$$

Thus the second total differential of $f(x,y) = (xy)^2$ is

$$D^{(2)}f((x,y);(h,k)) = 2y^2 h^2 + 8xyhk + 2x^2 k^2.$$

Here is a multidimensional version of Taylor's Formula. (For most applications, it is enough to remember that the hypotheses are met if $V$ is convex and $f$ is $\mathcal{C}^p$ on $V$.)

**11.33 THEOREM** [Taylor's Formula on $\mathbf{R}^n$]. *Let $p \in \mathbf{N}$, $V$ be an open set in $\mathbf{R}^n$, $\boldsymbol{x}, \boldsymbol{a} \in V$, $f : V \to \mathbf{R}$, and suppose that the partial derivatives of $f$ order $p-1$ exist on $V$. If each $(p-1)$-st order partial derivative of $f$ is differentiable on $V$ and $L(\boldsymbol{x};\boldsymbol{a}) \subset V$, then there is a point $\boldsymbol{c} \in L(\boldsymbol{x};\boldsymbol{a})$ such that*

$$f(\boldsymbol{x}) = f(\boldsymbol{a}) + \sum_{k=1}^{p-1} \frac{1}{k!} D^{(k)} f(\boldsymbol{a};\boldsymbol{h}) + \frac{1}{p!} D^{(p)} f(\boldsymbol{c};\boldsymbol{h})$$

*for $\boldsymbol{h} := \boldsymbol{x} - \boldsymbol{a}$.*

PROOF. Let $\boldsymbol{h} = \boldsymbol{x} - \boldsymbol{a}$. As in the proof of Theorem 11.27, choose $\delta > 0$ so small that $\boldsymbol{a} + t\boldsymbol{h} \subset V$ for $t \in I_\delta := (-\delta, 1 + \delta)$. The function $F(t) = f(\boldsymbol{a} + t\boldsymbol{h})$ is differentiable on $I_\delta$. In fact, by the Chain Rule,

$$F'(t) = Df(\boldsymbol{a} + t\boldsymbol{h})(\boldsymbol{h}) = \sum_{k=1}^{n} \frac{\partial f}{\partial x_k}(\boldsymbol{a} + t\boldsymbol{h}) h_k.$$

Extending this observation by induction, we verify

$$F^{(j)}(t) = \sum_{i_1=1}^{n} \cdots \sum_{i_j=1}^{n} \frac{\partial^j f}{\partial x_{i_1} \ldots \partial x_{i_j}}(\boldsymbol{a} + t\boldsymbol{h}) h_{i_1} \ldots h_{i_j}$$

for $j = 1, 2, \ldots, p$. Thus

(23) $\qquad\qquad F^{(j)}(0) = D^{(j)}f(\boldsymbol{a};\boldsymbol{h}) \quad \text{and} \quad F^{(p)}(t) = D^{(p)}f(\boldsymbol{a} + t\boldsymbol{h};\boldsymbol{h})$

for $j = 1, \ldots, p-1$, and $t \in I_\delta$.

We have proved that $F : I_\delta \to \mathbf{R}$ has a derivative of order $p$ everywhere on $I_\delta \supset [0,1]$. Therefore, by the one-dimensional Taylor Formula and (23),

$$f(\boldsymbol{x}) - f(\boldsymbol{a}) = F(1) - F(0) = \sum_{j=1}^{p-1} \frac{1}{j!} F^{(j)}(0) + \frac{1}{p!} F^{(p)}(t)$$

$$= \sum_{j=1}^{p-1} \frac{1}{j!} D^{(j)}f(\boldsymbol{a};\boldsymbol{h}) + \frac{1}{p!} D^{(p)}f(\boldsymbol{a} + t\boldsymbol{h};\boldsymbol{h})$$

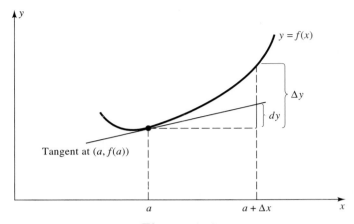

**Figure 11.3**

for some $t \in (0,1)$. Thus set $\boldsymbol{c} = \boldsymbol{a} + t\boldsymbol{h}$. ∎

Applied mathematicians and physicists use a different notation for the first total differential. Suppose $V$ is open in $\mathbf{R}^n$, $\boldsymbol{a} \in V$, $f : V \to \mathbf{R}$ is continuously differentiable at $\boldsymbol{a}$, and $w = f(\boldsymbol{x})$. If $\boldsymbol{dx} := (dx_1, \ldots, dx_n)$, where the $dx_j$'s are small real numbers, then the first total differential $D^{(1)}f(\boldsymbol{a}, \boldsymbol{dx})$ is usually denoted by $dw$. In particular, if $y = f(x)$ then $dy = f'(x)\,dx$, and if $z = f(x,y)$, then $dz = f_x\,dx + f_y\,dy$.

Let $f : \mathbf{R} \to \mathbf{R}$ be differentiable. Recall from single-variable calculus that the change in $y = f(x)$ as $x$ moves from $a$ to $a + \Delta x$ is defined by $\Delta y = f(a + \Delta x) - f(a)$, and if $dx = \Delta x$ is a small real number, then the total differential $dy$ approximates $\Delta y$ (see Figure 11.3). If $z = f(x,y)$, does the total differential $dz$ play an analogous geometric role in $\mathbf{R}^3$?

To answer this question, suppose $f$ is differentiable at $(a,b)$, and set $\Delta z = f(a + \Delta x, b + \Delta y) - f(a,b)$. Then the picture corresponding to Figure 11.3 involves a tangent plane and a wedge-shaped region (see Figure 11.4). Namely, let $z_0 = f(a,b)$ and consider the wedge-shaped region $\mathcal{W}$ with vertical sides parallel to the $xz$ and $yz$ planes whose base has vertices $\boldsymbol{c}_0 := (a,b,z_0)$, $\boldsymbol{c}_1 := (a + \Delta x, b, z_0)$, $\boldsymbol{c}_2 := (a, b + \Delta y, z_0)$, $\boldsymbol{c}_3 := (a + \Delta x, b + \Delta y, z_0)$, and whose top is tangent to $z = f(x,y)$ at $\boldsymbol{c}_0$. Let $A$ represent the length of the vertical edge of $\mathcal{W}$ based at $\boldsymbol{c}_1$, $B$ the length of the edge based at $\boldsymbol{c}_2$, and $C$ the length of the edge based at $\boldsymbol{c}_3$. If $dz$ is to play the same role in Figure 11.4 that $dy$ plays in Figure 11.3, then it must be the case that $C = dz$. This is actually easy to verify. Since the diagonals of rectangles bisect one another, the line segment from the intersection of the diagonals in the base of $\mathcal{W}$ to the intersection of the diagonals in the top of $\mathcal{W}$ must be parallel to the $z$ axis. Thus the length $D$ of this line segment can be computed two ways. On the one hand, $D = C/2$. On the other hand, $D = (A + B)/2$. Therefore, $C = A + B$. But from one-dimensional calculus, $A = f_x(a,b)\,dx$ and $B = f_y(a,b)\,dy$. Consequently,

$$C = A + B = \frac{\partial f}{\partial x}(a,b)\,dx + \frac{\partial f}{\partial y}(a,b)\,dy = dz.$$

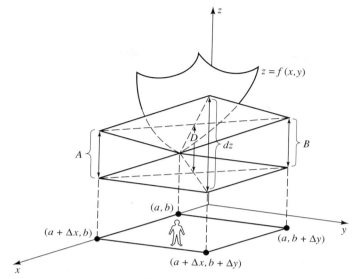

**Figure 11.4**

We close this section with some optional material about applications of the first total differential,

$$(24) \qquad dz := \sum_{j=1}^{n} \frac{\partial f}{\partial x_j}(\boldsymbol{a})\, dx_j.$$

*11.34 Remark.* *Let $f : \mathbf{R}^n \to \mathbf{R}$ be differentiable at $\boldsymbol{a}$ and $\Delta\boldsymbol{x} = (\Delta x_1, \dots, \Delta x_n)$. If the change, $\Delta z$, in $z = f(\boldsymbol{x})$ as $\boldsymbol{x}$ moves from $\boldsymbol{a}$ to $\boldsymbol{a} + \Delta\boldsymbol{x}$, is defined by $\Delta z := f(\boldsymbol{a} + \Delta\boldsymbol{x}) - f(\boldsymbol{a})$, then*

$$\frac{\Delta z - dz}{\|\Delta\boldsymbol{x}\|} \to 0 \qquad as \quad \Delta\boldsymbol{x} \to \mathbf{0}.$$

*In particular, the differential $dz$ approximates $\Delta z$.*

PROOF. By definition, if $f$ is differentiable at $\boldsymbol{a}$, then given $\varepsilon > 0$ there is a $\delta > 0$ such that

$$|f(\boldsymbol{a} + \boldsymbol{h}) - f(\boldsymbol{a}) - \nabla f(\boldsymbol{a}) \cdot \boldsymbol{h}| \le \varepsilon \|\boldsymbol{h}\|$$

for all $\|\boldsymbol{h}\| < \delta$. If $\boldsymbol{h} = \Delta\boldsymbol{x} = (\Delta x_1, \dots, \Delta x_n)$, then $\Delta z = f(\boldsymbol{a} + \boldsymbol{h}) - f(\boldsymbol{a})$ and $dz = \nabla f(\boldsymbol{a}) \cdot \boldsymbol{h}$. Thus $|\Delta z - dz| < \varepsilon \|\Delta\boldsymbol{x}\|$ for $\|\Delta\boldsymbol{x}\|$ sufficiently small. ∎

Hence, if $f$ is differentiable at $\boldsymbol{a}$, then the differential of $f$ can be used to approximate the change of $f$ as $\boldsymbol{x}$ moves from $\boldsymbol{a}$ to $\boldsymbol{a} + \boldsymbol{h}$ for $\boldsymbol{h}$ sufficiently small.

*11.35 Example.* Use differentials to approximate the change of $f(x, y) = x^2 y - y^3$ as $(x, y)$ moves from $(0, 1)$ to $(0.02, 1.01)$.

SOLUTION. Let $z = x^2y - y^3$, $a = 0$, and $b = 1$. Then $dx = 0.02$ and $dy = 0.01$. Since $dz = 2xy\,dx + (x^2 - 3y^2)\,dy$, we have

$$\Delta z \approx 0(0.02) - 3(0.01) = -0.03.$$

Note that $\Delta z = f(0.02, 1.01) - f(0, 1) = -0.029897\ldots$ is very close to $-0.03$. ∎

*11.36 Example.** Use differentials to approximate $(5.97)\sqrt[4]{16.03}$.

SOLUTION. Let $z = y\sqrt[4]{x}$, $a = 16$, and $b = 6$. Then $dx = 0.03$ and $dy = -0.03$. Since

$$dz = \frac{y}{4\sqrt[4]{x^3}}dx + \sqrt[4]{x}\,dy,$$

we have

$$\Delta z \approx \frac{6(0.03)}{4\sqrt[4]{(16)^3}}dx + \sqrt[4]{16}(-0.03) \approx -0.054375.$$

Thus,

$$z \approx 6\sqrt[4]{16} - 0.054375 = 11.945625.$$

Note that the actual value of $5.97\sqrt[4]{16.03}$ is $11.945593\ldots$. Thus our approximation is good to three decimal places. ∎

*11.37 Example.** Find the maximum percentage error for the calculated value of the volume of a right circular cylinder if the radius can be measured with a maximum error of 3% and the altitude can be measured with a maximum error of 2%.

SOLUTION. The volume of a right circular cylinder is $V = \pi r^2 h$, where $r$ is the radius and $h$ is the altitude. Hence, the differential of $V$ is $dV = 2\pi rh\,dr + \pi r^2\,dh$. Thus

$$\frac{dV}{V} = 2\frac{dr}{r} + \frac{dh}{h}.$$

Since the percentage error of a variable $x$ is $\Delta x/x \approx dx/x$, it follows that the maximum percentage error in calculating the volume $V$ is approximately 8%:

$$\frac{dV}{V} = 2(\pm 0.03) + (\pm 0.02) = \pm 0.08. \quad ∎$$

## EXERCISES

1. Let $f : \mathbf{R}^n \to \mathbf{R}$. Suppose that for each unit vector $\boldsymbol{u} \in \mathbf{R}^n$, the directional derivative $D_{\boldsymbol{u}}f(\boldsymbol{a} + t\boldsymbol{u})$ exists for $t \in [0, 1]$ (see Definition 11.25). Prove

$$f(\boldsymbol{a} + \boldsymbol{u}) - f(\boldsymbol{a}) = D_{\boldsymbol{u}}f(\boldsymbol{a} + t\boldsymbol{u})$$

for some $t \in (0, 1)$.

2. Suppose $r$ and $\alpha$ are positive numbers, $E$ is a convex subset of $\mathbf{R}^n$ such that $\overline{E} \subset B_r(\mathbf{0})$, and there exists a sequence $\boldsymbol{x}_k \in E$ such that $\boldsymbol{x}_k \to \mathbf{0}$ as $k \to \infty$. If $f : B_r(\mathbf{0}) \to \mathbf{R}$ continuously differentiable and $|f(\boldsymbol{x})| \le \|\boldsymbol{x}\|^\alpha$ for all $\boldsymbol{x} \in E$, prove that there is an $M > 0$ such that $|f(\boldsymbol{x})| \le M\|\boldsymbol{x}\|$ for $\boldsymbol{x} \in E$.

3. a) Write out an expression in powers of $(x + 1)$ and $(y - 1)$ for $f(x, y) = x^2 + xy + y^2$.
   b) Write Taylor's Formula for $f(x, y) = \sqrt{x} + \sqrt{y}$, $\boldsymbol{a} = (1, 4)$, and $p = 3$.
   c) Write Taylor's Formula for $f(x, y) = e^{xy}$, $\boldsymbol{a} = (0, 0)$, and $p = 4$.

4. Suppose $f : \mathbf{R}^2 \to \mathbf{R}$ is $\mathcal{C}^p$ on $B_r(x_0, y_0)$ for some $r > 0$. Prove that, given $(x, y) \in B_r(x_0, y_0)$, there is a point $(c, d)$ on the line segment between $(x_0, y_0)$ and $(x, y)$ such that

$$
f(x, y) = f(x_0, y_0)
$$
$$
+ \sum_{k=1}^{p-1} \frac{1}{k!} \left( \sum_{j=0}^{k} \binom{k}{j} (x - x_0)^j (y - y_0)^{k-j} \frac{\partial^k f}{\partial x^j \partial y^{k-j}} (x_0, y_0) \right)
$$
$$
+ \frac{1}{p!} \sum_{j=0}^{p} \binom{p}{j} (x - x_0)^j (y - y_0)^{p-j} \frac{\partial^p f}{\partial x^j \partial y^{p-j}} (c, d).
$$

5. Let $r > 0$, $a, b \in \mathbf{R}$, $f : B_r(a, b) \to \mathbf{R}$ be differentiable, and $(x, y) \in B_r(a, b)$.
   a) Let $g(t) = f(tx + (1-t)a, y) + f(a, ty + (1-t)b)$ and compute the derivative of $g$.
   b) Prove there are numbers $c$ between $a$ and $x$, and $d$ between $b$ and $y$ such that
   $$
   f(x, y) - f(a, b) = (x - a) f_x(c, y) + (y - b) f_y(a, d).
   $$

   (This is Exercise 12.20 in Apostol [1].)

6. [INTEGRAL FORM OF TAYLOR'S FORMULA]. Let $p \in \mathbf{N}$, $V$ be an open set in $\mathbf{R}^n$, $\boldsymbol{x}, \boldsymbol{a} \in V$, and $f : V \to \mathbf{R}$ be $\mathcal{C}^p$ on $V$. If $L(\boldsymbol{x}; \boldsymbol{a}) \subset V$ and $\boldsymbol{h} = \boldsymbol{x} - \boldsymbol{a}$, prove

$$
f(\boldsymbol{x}) - f(\boldsymbol{a}) = \sum_{k=1}^{p-1} \frac{1}{k!} D^{(k)} f(\boldsymbol{a}; \boldsymbol{h}) + \frac{1}{(p-1)!} \int_0^1 (1-t)^{p-1} D^{(p)} f(\boldsymbol{a} + t\boldsymbol{h}; \boldsymbol{h}) \, dt.
$$

7. Suppose $V$ is open in $\mathbf{R}^n$, $f : V \to \mathbf{R}$ is $\mathcal{C}^2$ on $V$, and $f_{x_j}(\boldsymbol{a}) = 0$ for some $\boldsymbol{a} \in V$ and all $j = 1, \ldots, n$. Prove that if $H$ is a compact convex subset of $V$, then there is a constant $M$ such that

$$
|f(\boldsymbol{x}) - f(\boldsymbol{a})| \le M\|\boldsymbol{x} - \boldsymbol{a}\|^2
$$

   for all $\boldsymbol{x} \in H$.

8. Suppose $V$ is an open subset of $\mathbf{R}^2$, $(a, b) \in V$, and $f : V \to \mathbf{R}$ is $\mathcal{C}^3$ on $V$. Prove

$$
\lim_{r \to 0} \frac{4}{\pi r^2} \int_0^{2\pi} f(a + r\cos\theta, b + r\sin\theta) \cos(2\theta) \, d\theta = f_{xx}(a, b) - f_{yy}(a, b).
$$

**9.** Suppose $V$ is an open subset of $\mathbf{R}^2$, $H = [a, b] \times [0, c] \subset V$, $u : V \to \mathbf{R}$ is $\mathcal{C}^2$ on $V$, and $u(x_0, t_0) \geq 0$ for all $(x_0, t_0) \in \partial H$.

    a) Show that, given $\varepsilon > 0$, there is a compact set $K \subset H^o$ such that $u(x, t) \geq -\varepsilon$ for all $(x, t) \in H \setminus K$.

    b) Suppose $u(x_1, t_1) = -\ell < 0$ for some $(x_1, t_1) \in H^o$, and choose $r > 0$ so small that $2rt_1 < \ell$. Apply part a) to $\varepsilon := \ell/2 - rt_1$ to choose the compact set $K$, and prove that the minimum of

$$w(x, t) := u(x, t) + r(t - t_1)$$

    on $H$ occurs at some $(x_2, t_2) \in K$.

    c) Prove that if $u$ satisfies the *heat equation*, i.e., $u_{xx} - u_t = 0$ on $V$, and if $u(x_0, t_0) \geq 0$ for all $(x_0, t_0) \in \partial H$, then $u(x, t) \geq 0$ for all $(x, t) \in H$.

**\*10.** Compute the differential of the each of the following functions.

    a) $z = x^2 + y^2$.      b) $z = \sin(xy)$.      c) $z = \dfrac{xy}{1 + x^2 + y^2}$.

**\*11.** Let $w = x^2 y + z$. Use differentials to approximate $\Delta w$ as $(x, y, z)$ moves from $(1, 2, 1)$ to $(1.01, 1.98, 1.03)$. Compare your approximation with the actual value of $\Delta w$.

**\*12.** The time $T$ it takes for a pendulum to complete one full swing is given by

$$T = 2\pi \sqrt{\frac{L}{g}},$$

where $g$ is the acceleration due to gravity and $L$ is the length of the pendulum. If $g$ can be measured with a maximum error of $1\%$, how accurately must $L$ be measured (in terms of percentage error) so that the calculated value of $T$ has a maximum error of $2\%$?

**\*13.** Suppose

$$\frac{1}{w} = \frac{1}{x} + \frac{1}{y} + \frac{1}{z},$$

where each variable $x, y, z$ can be measured with a maximum error of $p\%$. Prove that the calculated value of $w$ also has a maximum error of $p\%$.

**14.** a) Prove that every convex set in $\mathbf{R}^n$ is connected.

    b) Show that the converse of part a) is false.

    \*c) Suppose $f : \mathbf{R} \to \mathbf{R}$. Prove that $f$ is convex (as a function) if and only if $E := \{(x, y) : y \geq f(x)\}$ is convex (as a set in $\mathbf{R}^2$).

## 11.5   THE INVERSE FUNCTION THEOREM

By the one-dimensional Inverse Function Theorem (Theorem 4.27), if $g : \mathbf{R} \to \mathbf{R}$ is 1–1 and differentiable with $g'(x_0) \neq 0$, then $g^{-1}$ is differentiable at $y_0 = g(x_0)$ and

$$(g^{-1})'(y_0) = \frac{1}{g'(x_0)}.$$

In this section we obtain a multivariable analogue of this result, i.e., an Inverse Function Theorem for functions $f : \mathbf{R}^n \to \mathbf{R}^n$ which are differentiable at a point $\boldsymbol{a}$. What shall we use for hypotheses? We needed $g$ to be 1–1 so that the inverse function $g^{-1}$ existed. For the same reason, we shall assume that $f$ is 1–1. We needed $g'(x_0)$ to be nonzero so that we could divide by it. In the multidimensional case, $Df(\boldsymbol{a})$ is a matrix, hence "divisibility" corresponds to invertibility. Since a matrix is invertible if and only if it has a nonzero determinant (see Appendix C), we shall assume that Jacobian $\Delta_f(\boldsymbol{a})$ is nonzero.

The proof of the Inverse Function Theorem on $\mathbf{R}^n$ is not simple. It lies somewhat deeper than the previous results of this chapter, and we precede it by three preliminary results which explore the consequences of the hypothesis $\Delta_f \neq 0$.

If $f^{-1}$ is differentiable, then $f^{-1}$ is continuous; hence, $f = (f^{-1})^{-1}$ must take open sets to open sets (see Theorem 9.33 or 10.58). Our first preliminary result, a step in the right direction, shows that if $f$ is 1–1 and its Jacobian is nonzero at $\boldsymbol{a}$, then $f(\boldsymbol{a})$ is interior to $f(B_r(\boldsymbol{a}))$.

**11.38 Lemma.** *Let $V$ be open in $\mathbf{R}^n$, $f : V \to \mathbf{R}^n$, $\boldsymbol{a} \in V$, and $r > 0$ be so small that $\overline{B_r(\boldsymbol{a})} \subset V$. Suppose $f$ is continuous and 1–1 on $\overline{B_r(\boldsymbol{a})}$, and its first-order partial derivatives exist at every point in $B_r(\boldsymbol{a})$. If $\Delta_f \neq 0$ on $B_r(\boldsymbol{a})$, then there is a $\rho > 0$ such that $B_\rho(f(\boldsymbol{a})) \subset f(B_r(\boldsymbol{a}))$.*

STRATEGY: The idea behind this proof is simple. Let $\boldsymbol{y} \in B_\rho(f(\boldsymbol{a}))$ where $\rho$ is to be determined later. To show $B_\rho(f(\boldsymbol{a})) \subset f(B_r(\boldsymbol{a}))$ we must show that $\boldsymbol{y} = f(\boldsymbol{b})$ for some $\boldsymbol{b} \in B_r(\boldsymbol{a})$, i.e., $f(\boldsymbol{b}) - \boldsymbol{y} = \boldsymbol{0}$. If such a $\boldsymbol{b}$ exists, we should be able to find it by choosing a $\boldsymbol{b} \in B_r(\boldsymbol{a})$, which minimizes $\|f(\boldsymbol{b}) - \boldsymbol{y}\|$. This strategy has a problem: by the Extreme Value Theorem, the continuous function $\|f(\boldsymbol{b}) - \boldsymbol{y}\|$ assumes its minimum on the compact set $\overline{B_r(\boldsymbol{a})}$, not on the open set $B_r(\boldsymbol{a})$; hence, although such a $\boldsymbol{b}$ exists and belongs to the closure of $B_r(\boldsymbol{a})$, it might not belong to $B_r(\boldsymbol{a})$ itself. By controlling $\rho$, we can eliminate this problem. If $\rho < m$, where $m$ is the minimal distance from $f(\partial B_r(\boldsymbol{a}))$ to $f(\boldsymbol{a})$, then $\boldsymbol{b}$ cannot belong to $\partial B_r(\boldsymbol{a})$. Thus $\boldsymbol{b} \in B_r(\boldsymbol{a})$, as required. Here are the details.

PROOF. Let

$$g(\boldsymbol{x}) = \|f(\boldsymbol{x}) - f(\boldsymbol{a})\|, \qquad \boldsymbol{x} \in \overline{B_r(\boldsymbol{a})}.$$

By hypothesis, $g : \overline{B_r(\boldsymbol{a})} \to \mathbf{R}$ is continuous. Since $f$ is 1–1, $g(\boldsymbol{x}) > 0$ for all $\boldsymbol{x} \neq \boldsymbol{a}$. Since $\partial B_r(\boldsymbol{a})$ is compact, it follows that

$$m = \inf_{\boldsymbol{x} \in \partial B_r(\boldsymbol{a})} g(\boldsymbol{x}) > 0.$$

Set $\rho = m/2$ and fix $\boldsymbol{y} \in B_\rho(f(\boldsymbol{a}))$. Since the function $h(\boldsymbol{x}) := \|f(\boldsymbol{x}) - \boldsymbol{y}\|$ is continuous on the compact set $\overline{B_r(\boldsymbol{a})}$, it attains its minimum there. Thus there is a $\boldsymbol{b} \in \overline{B_r(\boldsymbol{a})}$ such that $h(\boldsymbol{b}) \leq h(\boldsymbol{x})$ for all $\boldsymbol{x} \in \overline{B_r(\boldsymbol{a})}$.

To show $\boldsymbol{b} \in B_r(\boldsymbol{a})$, suppose to the contrary that $\boldsymbol{b} \notin B_r(\boldsymbol{a})$. Then $\boldsymbol{b} \in \partial B_r(\boldsymbol{a})$. Since $h(\boldsymbol{a}) = \|f(\boldsymbol{a}) - \boldsymbol{y}\| < \rho$, the minimum $h(\boldsymbol{b})$ must also satisfy $h(\boldsymbol{b}) < \rho$. Since $\boldsymbol{b} \in \partial B_r(\boldsymbol{a})$, it follows from the triangle inequality and the choice of $\rho$ that

$$\rho > h(\boldsymbol{b}) = \|f(\boldsymbol{b}) - \boldsymbol{y}\| \geq \|f(\boldsymbol{b}) - f(\boldsymbol{a})\| - \|f(\boldsymbol{a}) - \boldsymbol{y}\| = g(\boldsymbol{b}) - h(\boldsymbol{a}) > 2\rho - \rho = \rho,$$

a contradiction.

It remains to prove $\boldsymbol{y} = f(\boldsymbol{b})$. Notice that, since $h(\boldsymbol{b}) \geq 0$, $h^2(\boldsymbol{b})$ is the minimum of $h^2$ on $\overline{B_r(\boldsymbol{a})}$. Thus by one-dimensional calculus,

$$\frac{\partial h^2}{\partial x_k}(\boldsymbol{b}) = 0$$

for $k = 1, \ldots, n$. Since $h^2(\boldsymbol{x}) = \sum_{j=1}^n (f_j(\boldsymbol{x}) - y_j)^2$, it follows that

$$0 = \frac{\partial h^2}{\partial x_k}(\boldsymbol{b}) = \sum_{j=1}^n 2(f_j(\boldsymbol{b}) - y_j)\frac{\partial f_j}{\partial x_k}(\boldsymbol{b}).$$

This is a system of $n$ linear equations in $n$ unknowns, $f_j(\boldsymbol{b}) - y_j$. Since the matrix of coefficients of this system has determinant $2^n \Delta_f(\boldsymbol{b}) \neq 0$, it follows from Cramer's Rule (see Appendix C) that this system has only the trivial solution; i.e., $f_j(\boldsymbol{b}) - y_j = 0$ for all $j = 1, \ldots, n$. In particular, $\boldsymbol{y} = f(\boldsymbol{b})$. ∎

Next, we show that $f^{-1}$ is continuous when $f$ is 1–1 and $\Delta_f$ is nonzero.

**11.39 THEOREM.** *Let $V$ be a nonempty, open set in $\mathbf{R}^n$ and $f : V \to \mathbf{R}^n$ be continuous. If $f$ is 1–1 and has first-order partial derivatives on $V$, and if $\Delta_f \neq 0$ on $V$, then $f^{-1}$ is continuous on $f(V)$.*

PROOF. By Theorem 9.33 or 10.58 (applied to $f^{-1}$), it suffices to show that $f(W)$ is open in $\mathbf{R}^n$ for every open $W \subseteq V$ in $\mathbf{R}^n$. Let $\boldsymbol{b} \in f(W)$, i.e., $\boldsymbol{b} = f(\boldsymbol{a})$ for some $\boldsymbol{a} \in W$. Since $W$ is open, choose $q > 0$ such that $B_q(\boldsymbol{a}) \subset W$. Fix $0 < r < q$, and notice that $\overline{B_r(\boldsymbol{a})} \subset W$. Since $f$ is 1–1 on $V \supseteq W$, apply Lemma 11.38 to choose $\rho > 0$ such that

$$B_\rho(\boldsymbol{b}) = B_\rho(f(\boldsymbol{a})) \subset f(B_r(\boldsymbol{a})).$$

Since $f(B_r(\boldsymbol{a})) \subset f(W)$, this proves $f(W)$ is open. ∎

Our final preliminary result shows that if the Jacobian of a continuously differentiable function $f$ is nonzero at a point, then $f$ must be 1–1 near that point. (This will provide a key step in the proof of the Inverse Function Theorem below.)

**11.40 Lemma.** *Let $V$ be nonempty and open in $\mathbf{R}^n$ and $f : V \to \mathbf{R}^n$ be $\mathcal{C}^1$ on $V$. If $\Delta_f(\boldsymbol{a}) \neq 0$ for some $\boldsymbol{a} \in V$, then there is an $r > 0$ such that $B_r(\boldsymbol{a}) \subset V$, $f$ is 1–1 on $B_r(\boldsymbol{a})$, $\Delta_f(\boldsymbol{x}) \neq 0$ for all $\boldsymbol{x} \in B_r(\boldsymbol{a})$, and*

$$\det\left[\frac{\partial f_i}{\partial x_j}(\boldsymbol{c}_i)\right]_{n \times n} \neq 0$$

*for all $\boldsymbol{c}_1, \ldots \boldsymbol{c}_n \in B_r(\boldsymbol{a})$.*

STRATEGY: The idea behind the proof is simple. If $f$ is not 1–1 on some $B_r(\boldsymbol{a})$, then there exist $\boldsymbol{x}, \boldsymbol{y} \in B_r(\boldsymbol{a})$ such that $\boldsymbol{x} \neq \boldsymbol{y}$ and $f(\boldsymbol{x}) = f(\boldsymbol{y})$. Since $L(\boldsymbol{x}; \boldsymbol{y}) \subset B_r(\boldsymbol{a})$, we have by Corollary 11.29 (the Mean Value Theorem) that

$$(25) \qquad 0 = f_i(\boldsymbol{y}) - f_i(\boldsymbol{x}) = \sum_{k=1}^n \frac{\partial f_i}{\partial x_k}(\boldsymbol{c}_i)(y_k - x_k)$$

for $\boldsymbol{x} = (x_1, \ldots, x_n)$, $\boldsymbol{y} = (y_1, \ldots, y_n)$, $\boldsymbol{c}_i \in L(\boldsymbol{x}; \boldsymbol{y})$, and $i = 1, \ldots, n$. Notice that (25) is a system of $n$ linear equations in $n$ unknowns, $(y_k - x_k)$. If we can show, for sufficiently small $r$, that the matrix of coefficients of (25) has nonzero determinant for any choice of $\boldsymbol{c}_i \in B_r(\boldsymbol{a})$, then by Cramer's Rule the linear system (25) has only one solution: $y_k - x_k = 0$ for $k = 1, \ldots, n$. This would imply $\boldsymbol{x} = \boldsymbol{y}$, a contradiction. Here are the details.

PROOF. To show there is an $r > 0$ such that the matrix of coefficients of the linear system (25) is nonzero for all $\boldsymbol{c}_i \in B_r(\boldsymbol{a})$, let $V^{(n)} = V \times \cdots \times V$ represent the $n$-fold Cartesian product of $V$ with itself, and define $h : V^{(n)} \to \mathbf{R}$ by

$$h(\boldsymbol{x}_1, \boldsymbol{x}_2, \ldots, \boldsymbol{x}_n) = \det \left[ \frac{\partial f_i}{\partial x_j}(\boldsymbol{x}_i) \right]_{n \times n}.$$

Since the determinant of a matrix is defined using products and differences of its entries (see Appendix C), we have by hypothesis that $h$ is continuous on $V^{(n)}$. Since $h(\boldsymbol{a}, \ldots, \boldsymbol{a}) = \Delta_f(\boldsymbol{a}) \neq 0$, it follows that there is an $r > 0$ such that $B_r(\boldsymbol{a}) \subset V$ and $h(\boldsymbol{c}_1, \ldots, \boldsymbol{c}_n) \neq 0$ for $\boldsymbol{c}_i \in B_r(\boldsymbol{a})$. In particular, the matrix of coefficients of the linear system (25) is nonzero for all $\boldsymbol{c}_i \in B_r(\boldsymbol{a})$, and $\Delta_f(\boldsymbol{x}) = h(\boldsymbol{x}, \ldots, \boldsymbol{x}) \neq 0$ for all $\boldsymbol{x} \in B_r(\boldsymbol{a})$. ∎

We now prove a multidimensional version of the Inverse Function Theorem.

**11.41 THEOREM** [THE INVERSE FUNCTION THEOREM]. *Let $V$ be open in $\mathbf{R}^n$ and $f : V \to \mathbf{R}^n$ be $\mathcal{C}^1$ on $V$. If $\Delta_f(\boldsymbol{a}) \neq 0$ for some $\boldsymbol{a} \in V$, then there exists an open set $W$ containing $\boldsymbol{a}$ such that*

  i) *$f$ is 1–1 on $W$,*
  ii) *$f^{-1}$ is $\mathcal{C}^1$ on $f(W)$, and*
  iii) *for each $\boldsymbol{y} := f(\boldsymbol{x}) \in f(W)$,*

$$D(f^{-1})(\boldsymbol{y}) = [Df(\boldsymbol{x})]^{-1},$$

*where $[\;]^{-1}$ represents matrix inversion (see Theorem C.5).*

PROOF. By Lemma 11.40, there is an open ball $B$ centered at $\boldsymbol{a}$ such that $f$ is 1–1 and $\Delta_f \neq 0$ on $B$, and

$$\Delta := \det \left[ \frac{\partial f_i}{\partial x_j}(\boldsymbol{c}_i) \right]_{n \times n} \neq 0$$

for all $\boldsymbol{c}_i \in B$. Let $B_0$ be an open ball centered at $\boldsymbol{a}$ which is smaller than $B$; i.e., the radius of $B_0$ is strictly less than the radius of $B$. Then $\overline{B}_0 \subset B$, $f$ is 1–1 on $B_0$ and, by Theorem 11.39, $f^{-1}$ is continuous on $f(B_0)$.

Let $W$ be any open ball centered at $\boldsymbol{a}$ which is smaller than $B_0$. Then $f$ is 1–1 on $W$ and $f(W)$ is open. To show that the first partial derivatives of $f^{-1}$ exist and are continuous on $f(W)$, fix $\boldsymbol{y}_0 \in f(W)$ and $1 \leq i, k \leq n$. Choose $t \in \mathbf{R}$, $t \neq 0$, so small that $\boldsymbol{y}_0 + t\boldsymbol{e}_k \in f(W)$, and choose $\boldsymbol{x}_0, \boldsymbol{x}_1 \in W$ such that $\boldsymbol{x}_0 = f^{-1}(\boldsymbol{y}_0)$ and $\boldsymbol{x}_1 = f^{-1}(\boldsymbol{y}_0 + t\boldsymbol{e}_k)$. Observe that for each $i = 1, 2, \ldots, n$,

$$f_i(\boldsymbol{x}_1) - f_i(\boldsymbol{x}_0) = \begin{cases} t & k = i \\ 0 & k \neq i. \end{cases}$$

Hence by Corollary 11.29 (the Mean Value Theorem), there exist points $\boldsymbol{c}_i \in L(\boldsymbol{x}_0; \boldsymbol{x}_1)$ such that

$$(26) \qquad \nabla f_i(\boldsymbol{c}_i) \cdot \frac{\boldsymbol{x}_1 - \boldsymbol{x}_0}{t} = \frac{f_i(\boldsymbol{x}_1) - f_i(\boldsymbol{x}_0)}{t} = \begin{cases} 1 & k = i \\ 0 & k \neq i \end{cases} \qquad i = 1, 2, \ldots, n.$$

Let $x_0(j)$ (respectively, $x_1(j)$) denote the $j$th component of $\boldsymbol{x}_0$ (respectively, $\boldsymbol{x}_1$). Since (26) is a system of $n$ linear equations in $n$ variables $(x_1(j) - x_0(j))/t$ whose coefficient matrix has determinant $\Delta$ (which is nonzero by the choice of $B$), we see by Cramer's Rule that the solutions of (26) satisfy

$$(27) \qquad \frac{(f^{-1})_j(\boldsymbol{y}_0 + t\boldsymbol{e}_k) - (f^{-1})_j(\boldsymbol{y}_0)}{t} := \frac{x_1(j) - x_0(j)}{t} = \mathcal{Q}_j(t),$$

where $\mathcal{Q}_j(t)$ is a quotient of determinants whose entries are 0's or 1's, or first-order partial derivatives of components of $f$ evaluated at the $\boldsymbol{c}_i$'s. Since $t \to 0$ implies $\boldsymbol{x}_1 \to \boldsymbol{x}_0$, $\boldsymbol{c}_i \to \boldsymbol{x}_0$, and $\boldsymbol{y}_0 + t\boldsymbol{e}_k \to \boldsymbol{y}_0$, it follows that $\mathcal{Q}_j(t)$ converges to $\mathcal{Q}_j$, a quotient of determinants whose entries are 0's or 1's, or first-order partial derivatives of components of $f$ evaluated at $\boldsymbol{x}_0 = f^{-1}(\boldsymbol{y}_0)$. Since $f^{-1}$ is continuous on $f(W)$, it follows that $\mathcal{Q}_j$ is continuous at each $\boldsymbol{y}_0 \in f(W)$. Taking the limit of (27) as $t \to 0$, we see that the first-order partial derivatives of $(f^{-1})_j$ exist at $\boldsymbol{y}_0$ and equal $\mathcal{Q}_j$; i.e., $f^{-1}$ is continuously differentiable on $f(W)$.

It remains to verify iii). Fix $\boldsymbol{y} \in f(W)$, and observe, by the Chain Rule and Exercise 8 in Section 8.4, that

$$I = DI(\boldsymbol{y}) = D(f \circ f^{-1})(\boldsymbol{y}) = Df(f^{-1}(\boldsymbol{y}))Df^{-1}(\boldsymbol{y}).$$

By the uniqueness of matrix inverses, we conclude that

$$D(f^{-1})(\boldsymbol{y}) = [Df(f^{-1}(\boldsymbol{y}))]^{-1}. \quad \blacksquare$$

Of course, the value $Df^{-1}(\boldsymbol{y})$ is not unique because $f^{-1}$ may have several branches. For example, if $f(x) = x^2$, then $f^{-1}(1) = \pm 1$, depending on whether we take the inverse of $f(x)$ near $x = 1$ or $x = -1$ (compare with Example 1.32).

**11.42 Remark.** *The hypothesis "$\Delta_f \neq 0$" in Theorem 11.39 can be relaxed.*

PROOF. If $f(x) = x^3$, then $f : \mathbf{R} \to \mathbf{R}$ and its inverse $f^{-1}(x) = \sqrt[3]{x}$ are continuous on $\mathbf{R}$, but $\Delta_f(0) = f'(0) = 0$. $\blacksquare$

**11.43 Remark.** *The hypothesis "$\Delta_f \neq 0$" in Theorem 11.41 cannot be relaxed. In fact, if $f : B_r(\boldsymbol{a}) \to \mathbf{R}^n$ is differentiable at $\boldsymbol{a}$ and its inverse $f^{-1}$ exists and is differentiable at $f(\boldsymbol{a})$, then $\Delta_f(\boldsymbol{a}) \neq 0$.*

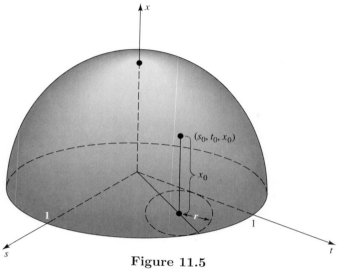

**Figure 11.5**

PROOF. If $f$ were such a function and $\Delta_f(\mathbf{a}) = 0$, then by Exercise 8 in Section 8.4 and the Chain Rule,

$$I = D(f^{-1} \circ f)(\mathbf{a}) = D(f^{-1})(f(\mathbf{a}))Df(\mathbf{a}).$$

Taking the determinant of this identity, we have

$$1 = \Delta_{f^{-1}}(f(\mathbf{a}))\Delta_f(\mathbf{a}) = 0,$$

a contradiction. ∎

**11.44 Remark.** *The hypothesis "$f$ is $\mathcal{C}^1$ on $V$" in Theorem 11.41 cannot be relaxed.*

PROOF. If $f(x) = x + 2x^2\sin(1/x)$, $x \neq 0$, and $f(0) = 0$, then $f : \mathbf{R} \to \mathbf{R}$ is differentiable on $V := (-1, 1)$ and $f'(0) = 1 \neq 0$. However, since

$$f\left(\frac{2}{(4k-1)\pi}\right) < f\left(\frac{2}{(4k+1)\pi}\right) < f\left(\frac{2}{(4k-3)\pi}\right)$$

for $k \in \mathbf{N}$, $f$ is not 1–1 on any open set which contains 0. Therefore, no open subset of $f(V)$ can be chosen on which $f^{-1}$ exists. ∎

Although Theorem 11.41 says $f$ must be 1–1 on some subset $W$ of $V$, it does not say that $f$ is 1–1 on $V$.

**11.45 Remark.** *The set $W$ chosen in Theorem 11.41 is in general a proper subset of $V$, even when $V$ is connected.*

PROOF. Set $f(x, y) = (x^2 - y^2, xy)$ and $V = \mathbf{R}^2 \backslash \{(0, 0)\}$. Then $\Delta_f = 2(x^2 + y^2) \neq 0$ for $(x, y) \in V$, but $f(x, -y) = f(-x, y)$ for all $(x, y) \in \mathbf{R}^2$. Thus $f$ is not 1–1 on $V$. ∎

Sometimes functions from $\mathbf{R}^p$ to $\mathbf{R}^n$ are defined implicitly by relations on $\mathbf{R}^{n+p}$.

**11.46 Example.**   *If* $x_0^2 + s_0^2 + t_0^2 = 1$ *and* $x_0 \neq 0$, *prove there exist an* $r > 0$ *and a function* $g(s, t)$, *continuously differentiable on* $B_r(s_0, t_0)$, *such that* $x_0 = g(s_0, t_0)$ *and*

$$x^2 + s^2 + t^2 = 1$$

*for* $x = g(s, t)$ *and* $(s, t) \in B_r(s_0, t_0)$.

PROOF. Solve $x^2 + s^2 + t^2 = 1$ for $x$ to obtain

$$x = \pm\sqrt{1 - s^2 - t^2}.$$

Which sign shall we take? If $x_0 > 0$, set $g(s, t) = \sqrt{1 - s^2 - t^2}$. By the Chain Rule,

$$\frac{\partial g}{\partial s} = \frac{-s}{\sqrt{1 - s^2 - t^2}} \quad \text{and} \quad \frac{\partial g}{\partial t} = \frac{-t}{\sqrt{1 - s^2 - t^2}}.$$

Thus $g$ is differentiable at any point $(s, t)$ which lies inside the two-dimensional unit ball, i.e., which satisfies $s^2 + t^2 < 1$. Since $x_0^2 + s_0^2 + t_0^2 = 1$ and $x_0 > 0$, $(s_0, t_0, x_0)$ lies on the boundary of the three-dimensional unit ball in $stx$ space a distance $x_0$ units above the $st$ plane (see Figure 11.5). In particular, if $r := 1 - \sqrt{1 - x_0^2}$ and $(s, t) \in B_r(s_0, t_0)$, then $s^2 + t^2 < 1$. Therefore, $g$ is continuously differentiable on $B_r(s_0, t_0)$. If $x_0 < 0$, a similar argument works for $g(s, t) = -\sqrt{1 - s^2 - t^2}$. ∎

Not all relations can be solved explicitly (see Example 11.48 below). For situations like this, the following result can prove useful. (In this theorem we use the notation $(\boldsymbol{x}, \boldsymbol{t})$ to represent the vector $(x_1, \dots, x_n, t_1, \dots, t_p)$.)

**11.47 THEOREM** [THE IMPLICIT FUNCTION THEOREM]. *Suppose* $V$ *is open in* $\mathbf{R}^{n+p}$, *and* $F = (F_1, \dots, F_n) : V \to \mathbf{R}^n$ *is* $\mathcal{C}^1$ *on* $V$. *Suppose further that* $F(\boldsymbol{x}_0, \boldsymbol{t}_0) = \mathbf{0}$ *for some* $(\boldsymbol{x}_0, \boldsymbol{t}_0) \in V$, *where* $\boldsymbol{x}_0 \in \mathbf{R}^n$ *and* $\boldsymbol{t}_0 \in \mathbf{R}^p$. *If*

$$\frac{\partial(F_1, \dots, F_n)}{\partial(x_1, \dots, x_n)}(\boldsymbol{x}_0, \boldsymbol{t}_0) \neq 0,$$

*then there is an open set* $W \subset \mathbf{R}^p$ *containing* $\boldsymbol{t}_0$ *and a unique continuously differentiable function* $g : W \to \mathbf{R}^n$ *such that* $g(\boldsymbol{t}_0) = \boldsymbol{x}_0$, *and* $F(g(\boldsymbol{t}), \boldsymbol{t}) = \mathbf{0}$ *for all* $\boldsymbol{t} \in W$.

STRATEGY: The idea behind the proof is simple. If $F$ took its range in $\mathbf{R}^{n+p}$ instead of $\mathbf{R}^n$ and had nonzero Jacobian, then by the Inverse Function Theorem, $F^{-1}$ would exist and be differentiable on some open set. Presumably, the first $n$ components of $F^{-1}$ would solve $F$ for the variables $x_1, \dots, x_n$. Thus we should extend $F$ (in the simplest possible way) to a function $\widetilde{F}$ which takes its range in $\mathbf{R}^{n+p}$ and has nonzero Jacobian, and apply the Inverse Function Theorem to $\widetilde{F}$. Here are the details.

PROOF. For each $(\boldsymbol{x}, \boldsymbol{t}) \in V$, set

(28)          $$\widetilde{F}(\boldsymbol{x}, \boldsymbol{t}) = (F_1(\boldsymbol{x}, \boldsymbol{t}), \dots, F_n(\boldsymbol{x}, \boldsymbol{t}), t_1, \dots, t_p).$$

Clearly, $\widetilde{F} : V \to \mathbf{R}^{n+p}$ and

$$
D\widetilde{F} = \left[ \begin{array}{cc} \left[ \dfrac{\partial F_i}{\partial x_j} \right]_{n \times n} & B \\ O_{p \times n} & I_{p \times p} \end{array} \right],
$$

where $O_{p \times n}$ represents a zero matrix, $I_{p \times p}$ represents an identity matrix, and $B$ represents a certain $n \times p$ matrix whose entries are first-order partial derivatives of $F_j$'s with respect to $t_k$'s. Expanding the determinant of $D\widetilde{F}$ along the bottom rows first, we see by hypothesis that

$$
\Delta_{\widetilde{F}}(\boldsymbol{x}_0, \boldsymbol{t}_0) = 1 \cdot \frac{\partial(F_1, \dots, F_n)}{\partial(x_1, \dots, x_n)}(\boldsymbol{x}_0, \boldsymbol{t}_0) \neq 0.
$$

Since $\widetilde{F}(\boldsymbol{x}_0, \boldsymbol{t}_0) = (\boldsymbol{0}, \boldsymbol{t}_0)$, it follows from the Inverse Function Theorem that there exist open sets $\Omega_1$ containing $(\boldsymbol{x}_0, \boldsymbol{t}_0)$ and $\Omega_2 := \widetilde{F}(\Omega_1)$ containing $(\boldsymbol{0}, \boldsymbol{t}_0)$ such that $\widetilde{F}$ is 1–1 on $\Omega_1$, and $G := \widetilde{F}^{-1}$ is 1–1 and continuously differentiable on $\Omega_2$.

Let $\phi = (G_1, \dots, G_n)$. Since $G = \widetilde{F}^{-1}$ is 1–1 from $\Omega_2$ onto $\Omega_1$, it is evident by (28) that

(29) $$\phi(\widetilde{F}(\boldsymbol{x}, \boldsymbol{t})) = \boldsymbol{x}$$

for all $(\boldsymbol{x}, \boldsymbol{t}) \in \Omega_1$ and

(30) $$\widetilde{F}(\phi(\boldsymbol{x}, \boldsymbol{t}), \boldsymbol{t}) = (\boldsymbol{x}, \boldsymbol{t})$$

for all $(\boldsymbol{x}, \boldsymbol{t}) \in \Omega_2$. Define $g$ on $W := \{ \boldsymbol{t} \in \mathbf{R}^p : (\boldsymbol{0}, \boldsymbol{t}) \in \Omega_2 \}$ by $g(\boldsymbol{t}) = \phi(\boldsymbol{0}, \boldsymbol{t})$. Since $\Omega_2$ is open in $\mathbf{R}^{n+p}$, $W$ is open in $\mathbf{R}^p$. Since $G$ is continuously differentiable on $\Omega_2$ and $\phi$ represents the first $n$ components of $G$, $g$ is continuously differentiable on $W$. By (28) and (29) we have

$$
g(\boldsymbol{t}_0) = \phi(\boldsymbol{0}, \boldsymbol{t}_0) = \phi(\widetilde{F}(\boldsymbol{x}_0, \boldsymbol{t}_0)) = \boldsymbol{x}_0.
$$

Moreover, by (28) and (30) we have $F(\phi(\boldsymbol{x}, \boldsymbol{t}), \boldsymbol{t}) = \boldsymbol{x}$ for all $(\boldsymbol{x}, \boldsymbol{t}) \in \Omega_2$. Specializing to the case $\boldsymbol{x} = \boldsymbol{0}$, we obtain $F(g(\boldsymbol{t}), \boldsymbol{t}) = \boldsymbol{0}$ for $\boldsymbol{t} \in W$.

It remains to show uniqueness. But if $h : W \to \mathbf{R}^n$ satisfies $F(h(\boldsymbol{t}), \boldsymbol{t}) = \boldsymbol{0} = F(g(\boldsymbol{t}), \boldsymbol{t})$, i.e., $\widetilde{F}(h(\boldsymbol{t}), \boldsymbol{t}) = (\boldsymbol{0}, \boldsymbol{t}) = \widetilde{F}(g(\boldsymbol{t}), \boldsymbol{t})$, then $g(\boldsymbol{t}) = h(\boldsymbol{t})$ for all $\boldsymbol{t} \in W$, since $\widetilde{F}$ is 1–1 om $\Omega_2$. ∎

Theorem 11.47 is an existence theorem. It states that a solution $g$ exists without giving us any idea how to find it. Fortunately, for many applications it is not as important to be able to write an explicit formula for $g$ as it is to know that $g$ exists.

Here is an example for which an explicit solution is unobtainable.

**11.48 Example.** Prove that there is a function $g(s, t)$, continuously differentiable on some $B_r(1, 0)$, such that $1 = g(1, 0)$, and

$$
sx^2 + tx^3 + 2\sqrt{t+s} + t^2 x^4 - x^5 \cos t - x^6 = 1
$$

for $x = g(s,t)$ and $(s,t) \in B_r(1,0)$.

PROOF. If $F(x,s,t) = sx^2 + tx^3 + 2\sqrt{t+s} + t^2 x^4 - x^5 \cos t - x^6 - 1$, then $F(1,1,0) = 0$, and $F_x = 2sx + 3tx^2 + 4t^2 x^3 - 5x^4 \cos t - 6x^5$ is nonzero at the point $(1,1,0)$. Applying the Implicit Function Theorem to $F$, with $n = 1$, $p = 2$, $x_0 = 1$, and $(s_0, t_0) = (1,0)$, we conclude that such a $g$ exists. ∎

Even when an explicit solution is obtainable, it is frequently easier to apply the Implicit Function Theorem than it is to solve a relation explicitly for one or more of its variables. Indeed, consider Example 11.46 again. Let $F(x,s,t) = 1 - x^2 - s^2 - t^2$ and notice that $F_x = -2x$. Thus, by the Implicit Function Theorem, a continuously differentiable solution $x = g(s,t)$ exists for each $x_0 \neq 0$.

The following example shows that the Implicit Function Theorem can be used to show several differentiable solutions exist simultaneously.

**11.49 Example.** Prove there exist functions $u, v : \mathbf{R}^4 \to \mathbf{R}$, continuously differentiable on some ball $B$ centered at the point $(x,y,z,w) = (2,1,-1,-2)$, such that $u(2,1,-1,-2) = 4$, $v(2,1,-1,-2) = 3$, and the equations

$$u^2 + v^2 + w^2 = 29, \qquad \frac{u^2}{x^2} + \frac{v^2}{y^2} + \frac{w^2}{z^2} = 17$$

both hold for all $(x,y,z,w)$ in $B$.

PROOF. Set $n = 2$, $p = 4$, and

$$F(u,v,x,y,z,w) = (u^2 + v^2 + w^2 - 29, u^2/x^2 + v^2/y^2 + w^2/z^2 - 17).$$

Then $F(4,3,2,1,-1,-2) = (0,0)$, and

$$\frac{\partial(F_1, F_2)}{\partial(u,v)} = \det \begin{bmatrix} 2u & 2v \\ 2u/x^2 & 2v/y^2 \end{bmatrix} = 4uv \left( \frac{1}{y^2} - \frac{1}{x^2} \right).$$

This determinant is nonzero when $u = 4$, $v = 3$, $x = 2$, and $y = 1$. Therefore, such functions $u, v$ exist by the Implicit Function Theorem. ∎

## EXERCISES

1. For each of the following functions, prove that $f^{-1}$ exists and is differentiable in some nonempty, open set containing $(a,b)$, and compute $D(f^{-1})(a,b)$.

   a) $f(u,v) = (3u - v, 2u + 5v)$ at $(a,b)$.
   b) $f(u,v) = (u + v, \sin u + \cos v)$ at $(a,b) = (0,1)$.
   c) $f(u,v) = (uv, u^2 + v^2)$ at $(a,b) = (2,5)$.
   d) $f(u,v) = (u^3 - v^2, \sin u - \log v)$ at $(a,b) = (-1,0)$.

2. For each of the following functions, find out whether the given expression can be solved for $z$ in a nonempty, open set $V$ containing $(0,0,0)$. Is the solution

differentiable near $(0,0)$?

a) $xyz + \sin(x + y + z) = 0$.

b) $x^2 + y^2 + z^2 + \sqrt[3]{2xy + 3z + 8} = 2$.

c) $xyz(2\cos y - \cos z) + (z\cos x - x\cos y) = 0$.

d) $x + y + z + g(x, y, z) = 0$, where $g$ is any continuously differentiable function which satisfies $g(0,0,0) = 0$ and $g_z(0,0,0) > 0$.

**3.** Prove there exist functions $u(x,y)$, $v(x,y)$, and $w(x,y)$, and an $r > 0$ such that $u, v, w$ are continuously differentiable and satisfy the equations

$$u^5 + xv^2 - y + w = 0$$
$$v^5 + yu^2 - x + w = 0$$
$$w^4 + y^5 - x^4 = 1$$

on $B_r(1,1)$, and $u(1,1) = 1$, $v(1,1) = 1$, $w(1,1) = -1$.

**4.** Find conditions on a point $(x_0, y_0, u_0, v_0)$ such that there exist real-valued functions $u(x,y)$ and $v(x,y)$ which are continuously differentiable near $(x_0, y_0)$ and satisfy the simultaneous equations

$$xu^2 + yv^2 + xy = 9$$
$$xv^2 + yu^2 - xy = 7.$$

Prove that the solutions satisfy $u^2 + v^2 = 16/(x+y)$.

**5.** Given nonzero numbers $x_0, y_0, u_0, v_0, s_0, t_0$ which satisfy the simultaneous equations

(*)
$$u^2 + sx + ty = 0$$
$$v^2 + tx + sy = 0$$
$$2s^2x + 2t^2y - 1 = 0$$
$$s^2x - t^2y = 0,$$

prove that there exist functions $u(x,y)$, $v(x,y)$, $s(x,y)$, $t(x,y)$, and an open ball $B$ containing $(x_0, y_0)$ such that $u, v, s, t$ are continuously differentiable and satisfy (*) on $B$, and such that $u(x_0, y_0) = u_0$, $v(x_0, y_0) = v_0$, $s(x_0, y_0) = s_0$, and $t(x_0, y_0) = t_0$.

**6.** Let $E = \{(x,y) : 0 < y < x\}$ and set $f(x,y) = (x + y, xy)$ for $(x,y) \in E$.

a) Prove $f$ is 1–1 from $E$ onto $\{(s,t) : s > 2\sqrt{t}, t > 0\}$ and find a formula for $f^{-1}(s,t)$.

b) Use the Inverse Function Theorem to compute $D(f^{-1})(f(x,y))$ for $(x,y) \in E$.

c) Use the formula you obtained in part a) to compute $D(f^{-1})(s,t)$ directly. (Check to see that this answer agrees with the one you found in part b).)

**7.** Suppose $f : \mathbf{R}^2 \to \mathbf{R}^2$ has continuous first-order partial derivatives in some ball $B_r(x_0, y_0)$, $r > 0$. Prove that if $\Delta_f(x_0, y_0) \neq 0$, then

$$\frac{\partial f_1^{-1}}{\partial x}(f(x_0, y_0)) = \frac{\partial f_2 / \partial y(x_0, y_0)}{\Delta_f(x_0, y_0)}, \qquad \frac{\partial f_1^{-1}}{\partial y}(f(x_0, y_0)) = \frac{-\partial f_1 / \partial y(x_0, y_0)}{\Delta_f(x_0, y_0)},$$

and

$$\frac{\partial f_2^{-1}}{\partial x}(f(x_0, y_0)) = \frac{-\partial f_2 / \partial x(x_0, y_0)}{\Delta_f(x_0, y_0)}, \qquad \frac{\partial f_2^{-1}}{\partial y}(f(x_0, y_0)) = \frac{\partial f_1 / \partial x(x_0, y_0)}{\Delta_f(x_0, y_0)}.$$

**8**. **This exercise is used in Section** $^e$**11.6**. Let $F(x, y, z)$ be a relation in $\mathbf{R}^3$ and consider its graph defined by

$$\mathcal{G} := \{(x, y, z) \in \mathbf{R}^3 : F(x, y, z) = 0\}.$$

$\mathcal{G}$ is said to have a tangent plane at a point $(a, b, c)$ if $F(x, y, z) = 0$ can be "solved" for one of the variables in a differentiable way; e.g., there is a function $x = f(y, z)$, differentiable at the point $(b, c)$, such that $F(f(y, z), y, z) = 0$ for all $(y, z)$ in an open ball centered at $(b, c)$. Let $F : \mathbf{R}^3 \to \mathbf{R}$ be continuously differentiable at $(a, b, c)$ with $\nabla F(a, b, c) \neq 0$.

a) Prove $\mathcal{G}$ has a tangent plane at $(a, b, c)$.

b) Show that if $\Pi$ is a plane passing through the point $(a, b, c)$ with equation $\lambda(x, y, z) = d$ such that the line $\lambda(x, b, z) = d$ is tangent to the curve $F(x, b, z) = 0$ at $x = a$, and the line $\lambda(a, y, z) = d$ is tangent to the curve $F(a, y, z) = 0$ at $y = b$, then a normal vector to $\Pi$ is given by $\nabla F(a, b, c)$.

**9**. **This exercise is used in Section 12.4.** If $V$ is open in $\mathbf{R}^n$, $\phi : V \to \mathbf{R}^n$ is $\mathcal{C}^1$ and $\Delta_\phi \neq 0$ on $V$, prove that $\phi(V)$ is open.

### $^e$11.6  OPTIMIZATION  *This section uses no material from any other enrichment section.*

In this section we discuss how to find extreme values of differentiable functions of several variables.

**11.50 DEFINITION.** Let $V$ be an open subset of $\mathbf{R}^n$, $\boldsymbol{a} \in V$, and $f : V \to \mathbf{R}$.

i) $f(\boldsymbol{a})$ is called a *local minimum* of $f$ if there is an $r > 0$ such that $f(\boldsymbol{a}) \leq f(\boldsymbol{x})$ for all $\boldsymbol{x} \in B_r(\boldsymbol{a})$.

ii) $f(\boldsymbol{a})$ is called a *local maximum* of $f$ if there is an $r > 0$ such that $f(\boldsymbol{a}) \geq f(\boldsymbol{x})$ for all $\boldsymbol{x} \in B_r(\boldsymbol{a})$.

iii) $f(\boldsymbol{a})$ is called a *local extremum* of $f$ if $f(\boldsymbol{a})$ is a local maximum or a local minimum of $f$.

The following result shows that, as in the one-dimensional case, extrema of real-valued differentiable functions occur at points where the "derivative" is zero.

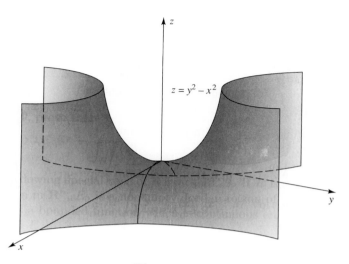

**Figure 11.6**

**11.51 Remark.** *If the first-order partial derivatives of $f$ exist at $\boldsymbol{a}$, and $f(\boldsymbol{a})$ is a local extremum of $f$, then*

$$\nabla f(\boldsymbol{a}) = \boldsymbol{0}.$$

PROOF. The one-dimensional function $g(t) = f(a_1, \ldots, a_{j-1}, t, a_{j+1}, \ldots, a_n)$ has a local extremum at $t = a_j$ for each $j = 1, \ldots, n$. Hence, by the one-dimensional theory,

$$\frac{\partial f}{\partial x_j}(\boldsymbol{a}) = g'(a_j) = 0. \quad \blacksquare$$

As in the one-dimensional case, $\nabla f(\boldsymbol{a}) = \boldsymbol{0}$ is necessary but not sufficient for $f(\boldsymbol{a})$ to be a local extremum.

**11.52 Remark.** *There exist continuously differentiable functions which satisfy $\nabla f(\boldsymbol{a}) = \boldsymbol{0}$ such that $f(\boldsymbol{a})$ is neither a local maximum nor a local minimum.*

PROOF. Consider

$$f(x, y) = y^2 - x^2.$$

Since the first-order partial derivatives of $f$ exist and are continuous everywhere on $\mathbf{R}^2$, $f$ is continuously differentiable on $\mathbf{R}^2$. Moreover, it is evident that $\nabla f(\boldsymbol{0}) = \boldsymbol{0}$, but $f(\boldsymbol{0})$ is not a local extremum (see Figure 11.6). $\blacksquare$

The fact that the graph of this function resembles a saddle motivates the following terminology.

**11.53 DEFINITION.** Let $V$ be open in $\mathbf{R}^n$, $\boldsymbol{a} \in V$, and $f : V \to \mathbf{R}$ be differentiable at $\boldsymbol{a}$. Then $\boldsymbol{a}$ is called a *saddle point* of $f$ if $\nabla f(\boldsymbol{a}) = \boldsymbol{0}$ and there is a $r_0 > 0$ such that given any $0 < \rho < r_0$ there are points $\boldsymbol{x}, \boldsymbol{y} \in B_\rho(\boldsymbol{a})$ which satisfy

$$f(\boldsymbol{x}) < f(\boldsymbol{a}) < f(\boldsymbol{y}).$$

By the Extreme Value Theorem, if $f$ is continuous on a compact set $H$, then it *attains* its maximum and minimum on $H$; i.e., there exist points $\boldsymbol{a}, \boldsymbol{b} \in H$ such that

$$f(\boldsymbol{a}) = \sup_{\boldsymbol{x} \in H} f(\boldsymbol{x}) \quad \text{and} \quad f(\boldsymbol{b}) = \inf_{\boldsymbol{x} \in H} f(\boldsymbol{x}).$$

When $f$ is a function of two variables, these points can be found by combining Remark 11.51 with one-dimensional techniques.

**11.54 Example.** Find the maximum and minimum of $f(x, y) = x^2 - x + y^2 - 2y$ on $H = \overline{B_1(0, 0)}$.

SOLUTION. Since $\nabla f(x, y) = (0, 0)$ implies $(x, y) = (1/2, 1)$, $f$ has no local extrema inside $H$. Thus the extrema of $f$ on $H$ must occur on $\partial H$. Using polar coordinates, we can describe $\partial H$ by $(x, y) = (\cos \theta, \sin \theta)$, where $0 \le \theta < 2\pi$. Set

$$h(\theta) := f(\cos \theta, \sin \theta) = 1 - \cos \theta - 2 \sin \theta.$$

Notice that the derivative of $h$ is zero when $\tan \theta = 2$; i.e., $\theta = \arctan 2 \approx 1.10715$ or $\theta = \arctan 2 + \pi \approx 4.24874$. Therefore, candidates for the extrema of $f$ on $\partial H$ are $(x, y) \approx (0.4472, 0.8944)$ and $(x, y) \approx (-0.4472, -0.8944)$. Checking the sign of $h''(\theta)$, we see that the first point corresponds to a minimum, and the second point corresponds to a maximum. Therefore, the maximum of $f$ on $H$ is $f(-0.4472, -0.8944) \approx 3.236$, and the minimum of $f$ on $H$ is $f(0.4472, 0.8944) \approx -1.236$. ∎

Using the second-order total differential $D^{(2)} f$ introduced in Section 11.4, we can obtain a multidimensional analogue of the Second Derivative Test. First, we prove a technical result.

**11.55 Lemma.** *Let $V$ be open in $\mathbf{R}^n$, $\boldsymbol{a} \in V$, and $f : V \to \mathbf{R}$. If all second-order partial derivatives of $f$ exist at $\boldsymbol{a}$ and $D^{(2)} f(\boldsymbol{a}; \boldsymbol{h}) > 0$ for all $\boldsymbol{h} \neq \boldsymbol{0}$, then there is an $m > 0$ such that*

$$(31) \qquad D^{(2)} f(\boldsymbol{a}; \boldsymbol{x}) \ge m \|\boldsymbol{x}\|^2$$

*for all $\boldsymbol{x} \in \mathbf{R}^n$.*

PROOF. Set $H = \{\boldsymbol{x} \in \mathbf{R}^n : \|\boldsymbol{x}\| = 1\}$ and consider the function

$$g(\boldsymbol{x}) := D^{(2)} f(\boldsymbol{a}; \boldsymbol{x}) := \sum_{j=1}^{n} \sum_{k=1}^{n} \frac{\partial^2 f}{\partial x_k \partial x_j}(\boldsymbol{a}) \, x_j x_k, \qquad \boldsymbol{x} \in \mathbf{R}^n.$$

By hypothesis, $g$ is continuous and positive on $\mathbf{R}^n \setminus \{\boldsymbol{0}\}$, hence, on $H$. Since $H$ is compact, it follows from the Extreme Value Theorem that $g$ has a positive minimum $m$ on $H$.

Clearly, (31) holds for $\boldsymbol{x} = \boldsymbol{0}$. If $\boldsymbol{x} \neq \boldsymbol{0}$, then $\boldsymbol{x}/\|\boldsymbol{x}\| \in H$, and it follows from the choice of $g$ and $m$ that

$$D^{(2)} f(\boldsymbol{a}; \boldsymbol{x}) = \frac{g(\boldsymbol{x})}{\|\boldsymbol{x}\|^2} \|\boldsymbol{x}\|^2 = g\left(\frac{\boldsymbol{x}}{\|\boldsymbol{x}\|}\right) \|\boldsymbol{x}\|^2 \ge m \|\boldsymbol{x}\|^2.$$

We conclude that (31) holds for all $\boldsymbol{x} \in \mathbf{R}^n$. ∎

**11.56 THEOREM** [THE SECOND DERIVATIVE TEST]. *Let $V$ be open in $\mathbf{R}^n$, $\boldsymbol{a} \in V$, and suppose $f : V \to \mathbf{R}$ satisfies $\nabla f(\boldsymbol{a}) = \boldsymbol{0}$. Suppose further that all second-order partial derivatives of $f$ exist on $V$ and are continuous at $\boldsymbol{a}$.*

i) *If $D^{(2)} f(\boldsymbol{a}; \boldsymbol{h}) > 0$ for all $\boldsymbol{h} \neq \boldsymbol{0}$, then $f(\boldsymbol{a})$ is a local minimum of $f$.*
ii) *If $D^{(2)} f(\boldsymbol{a}; \boldsymbol{h}) < 0$ for all $\boldsymbol{h} \neq \boldsymbol{0}$, then $f(\boldsymbol{a})$ is a local maximum of $f$.*
iii) *If $D^{(2)} f(\boldsymbol{a}; \boldsymbol{h})$ takes on both positive and negative values for $\boldsymbol{h} \in \mathbf{R}^n$, then $\boldsymbol{a}$ is a saddle point of $f$.*

PROOF. Choose $r > 0$ such that $B_r(\boldsymbol{a}) \subset V$, and suppose for a moment that there is a function $\varepsilon : B_r(\boldsymbol{0}) \to \mathbf{R}$ such that $\varepsilon(\boldsymbol{h}) \to 0$ as $\boldsymbol{h} \to \boldsymbol{0}$ and

$$(32) \qquad f(\boldsymbol{a} + \boldsymbol{h}) - f(\boldsymbol{a}) = \frac{1}{2} D^{(2)} f(\boldsymbol{a}; \boldsymbol{h}) + \|\boldsymbol{h}\|^2 \varepsilon(\boldsymbol{h})$$

for $\boldsymbol{h}$ sufficiently small. If $D^{(2)} f(\boldsymbol{a}; \boldsymbol{h}) > 0$ for $\boldsymbol{h} \neq \boldsymbol{0}$, then (31) and (32) imply

$$f(\boldsymbol{a} + \boldsymbol{h}) - f(\boldsymbol{a}) \geq \left( \frac{m}{2} + \varepsilon(\boldsymbol{h}) \right) \|\boldsymbol{h}\|^2$$

for $\boldsymbol{h}$ sufficiently small. Since $m > 0$ and $\varepsilon(\boldsymbol{h}) \to 0$ as $\boldsymbol{h} \to \boldsymbol{0}$, it follows that $f(\boldsymbol{a} + \boldsymbol{h}) - f(\boldsymbol{a}) > 0$ for $\boldsymbol{h}$ sufficiently small, i.e., $f(\boldsymbol{a})$ is a local minimum. Similarly, if $D^{(2)} f(\boldsymbol{a}; \boldsymbol{h}) < 0$ for $\boldsymbol{h} \neq \boldsymbol{0}$, then $f(\boldsymbol{a})$ is a local maximum. This proves parts i) and ii).

To prove part iii), fix $\boldsymbol{h} \in \mathbf{R}^n$ and notice that (32) implies

$$f(\boldsymbol{a} + t\boldsymbol{h}) - f(\boldsymbol{a}) = t^2 \left( \frac{1}{2} D^{(2)} f(\boldsymbol{a}; \boldsymbol{h}) + \|\boldsymbol{h}\|^2 \varepsilon(t\boldsymbol{h}) \right)$$

for $t \in \mathbf{R}$. Since $\varepsilon(t\boldsymbol{h}) \to 0$ as $t \to 0$, it follows that $f(\boldsymbol{a} + t\boldsymbol{h}) - f(\boldsymbol{a})$ takes on the same sign as $D^{(2)} f(\boldsymbol{a}; \boldsymbol{h})$ for $t$ small. In particular, if $D^{(2)} f(\boldsymbol{a}; \boldsymbol{h})$ takes on both positive and negative values as $\boldsymbol{h}$ varies, then $\boldsymbol{a}$ is a saddle point.

It remains to find a function $\varepsilon : B_r(\boldsymbol{0}) \to \mathbf{R}$ such that $\varepsilon(\boldsymbol{h}) \to 0$ as $\boldsymbol{h} \to \boldsymbol{0}$, and (32) holds for all $\boldsymbol{h}$ sufficiently small. Set $\varepsilon(\boldsymbol{0}) = 0$ and

$$\varepsilon(\boldsymbol{h}) = \frac{f(\boldsymbol{a} + \boldsymbol{h}) - f(\boldsymbol{a}) - \frac{1}{2} D^{(2)} f(\boldsymbol{a}; \boldsymbol{h})}{\|\boldsymbol{h}\|^2}, \qquad \boldsymbol{h} \in B_r(\boldsymbol{0}), \ \boldsymbol{h} \neq \boldsymbol{0}.$$

By the definition of $\varepsilon(\boldsymbol{h})$, (32) holds for $\boldsymbol{h} \in B_r(\boldsymbol{0})$. Does $\varepsilon(\boldsymbol{h}) \to 0$ as $\boldsymbol{h} \to \boldsymbol{0}$? Fix $\boldsymbol{h} = (h_1, h_2, \dots, h_n) \in B_r(\boldsymbol{0})$. Since $\nabla f(\boldsymbol{a}) = \boldsymbol{0}$, Taylor's Formula implies

$$f(\boldsymbol{a} + \boldsymbol{h}) - f(\boldsymbol{a}) = \frac{1}{2} D^{(2)} f(\boldsymbol{c}; \boldsymbol{h})$$

for some $\boldsymbol{c} \in L(\boldsymbol{a}; \boldsymbol{a} + \boldsymbol{h})$; i.e.,

$$f(\boldsymbol{a} + \boldsymbol{h}) - f(\boldsymbol{a}) - \frac{1}{2} D^{(2)} f(\boldsymbol{a}; \boldsymbol{h}) = \frac{1}{2} \left( D^{(2)} f(\boldsymbol{c}; \boldsymbol{h}) - D^{(2)} f(\boldsymbol{a}; \boldsymbol{h}) \right)$$

$$= \frac{1}{2} \sum_{j=1}^{n} \sum_{k=1}^{n} \left( \frac{\partial^2 f}{\partial x_j \partial x_k} (\boldsymbol{c}) - \frac{\partial^2 f}{\partial x_j \partial x_k} (\boldsymbol{a}) \right) h_j h_k.$$

Since $|h_j h_k| \leq \|\boldsymbol{h}\|^2$ and the second-order partial derivatives of $f$ are continuous at $\boldsymbol{a}$, it follows that

$$0 \leq |\varepsilon(\boldsymbol{h})| \leq \frac{1}{2} \left( \sum_{j=1}^{n} \sum_{k=1}^{n} \left| \frac{\partial^2 f}{\partial x_j \partial x_k}(\boldsymbol{c}) - \frac{\partial^2 f}{\partial x_j \partial x_k}(\boldsymbol{a}) \right| \right) \to 0$$

as $\boldsymbol{h} \to \boldsymbol{0}$. We conclude by the Squeeze Theorem that $\varepsilon(\boldsymbol{h}) \to 0$ as $\boldsymbol{h} \to \boldsymbol{0}$. ∎

The following result shows that the strict inequalities in Theorem 11.56 cannot be relaxed.

**11.57 Remark.** *If $D^{(2)}f(\boldsymbol{a};\boldsymbol{h}) \geq 0$, then $f(\boldsymbol{a})$ can be a local minimum or $\boldsymbol{a}$ can be a saddle point.*

PROOF. $f(0,0)$ is a local minimum of $f(x,y) = x^4 + y^2$, and $(0,0)$ is a saddle point of $f(x,y) = x^3 + y^2$. ∎

In practice, it is not easy to determine the sign of $D^{(2)}f(\boldsymbol{a};\boldsymbol{h})$. For the case $n = 2$, the second total differential $D^{(2)}f(\boldsymbol{a};\boldsymbol{h})$ is a *quadratic form*, i.e., has the form $Ah^2 + 2Bhk + Ck^2$. The following result shows that the sign of a quadratic form is completely determined by the *discriminant $D = B^2 - AC$*.

**11.58 Lemma.** *Let $A, B, C \in \mathbf{R}$, $D = B^2 - AC$, and $\phi(h,k) = Ah^2 + 2Bhk + Ck^2$.*

   i) *If $D < 0$, then $A$ and $\phi(h,k)$ have the same sign for all $(h,k) \neq (0,0)$.*

   ii) *If $D > 0$, then $\phi(h,k)$ takes on both positive and negative values as $(h,k)$ varies over $\mathbf{R}^2$.*

PROOF. i) Suppose $D < 0$. Then $A \neq 0$ and $A\phi(h,k)$ is a sum of two squares:

$$A\phi(h,k) = A^2 h^2 + 2ABhk + ACk^2 = (Ah + Bk)^2 + |D|k^2.$$

Since $A \neq 0 \neq D$, at least one of these squares is positive for each $(h,k) \neq (0,0)$. It follows that $A$ and $\phi(h,k)$ have the same sign for all $(h,k) \neq (0,0)$.

   ii) Suppose $D > 0$. Then either $A \neq 0$ or $B \neq 0$.

If $A \neq 0$, then $A\phi(h,k)$ is a difference of two squares:

$$A\phi(h,k) = (Ah + Bk - \sqrt{D}k)(Ah + Bk + \sqrt{D}k).$$

The lines $Ah + Bk - \sqrt{D}k = 0$ and $Ah + Bk + \sqrt{D}k = 0$ divide the $hk$ plane into four open regions (see Figure 11.7). Since $A\phi(h,k)$ is positive on two of these regions and negative on the other two, it follows that $\phi(h,k)$ takes on both positive and negative values as $(h,k)$ varies over $\mathbf{R}^2$.

If $A = 0$ and $B \neq 0$, then

$$\phi(h,k) = 2Bhk + Ck^2 = (2Bh + Ck)k.$$

Since $B \neq 0$, the lines $2Bh + Ck = 0$ and $k = 0$ divide the $hk$ plane into four open regions. As before, $\phi(h,k)$ takes on both positive and negative values as $(h,k)$ varies over $\mathbf{R}^2$. ∎

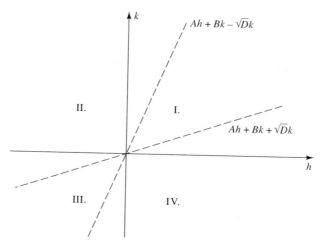

**Figure 11.7**

This result leads us to the following simple test for extrema and saddle points.

**11.59 THEOREM.** *Let $V$ be open in $\mathbf{R}^2$, $(a, b) \in V$, and suppose $f : V \to \mathbf{R}$ satisfies $\nabla f(a, b) = \mathbf{0}$. Suppose further that all second-order partial derivatives of $f$ exist on $V$, are continuous at $(a, b)$, and*

$$D = f_{xy}^2(a, b) - f_{xx}(a, b) f_{yy}(a, b).$$

i) *If $D < 0$ and $f_{xx}(a, b) > 0$, then $f(a, b)$ is a local minimum.*
ii) *If $D < 0$ and $f_{xx}(a, b) < 0$, then $f(a, b)$ is a local maximum.*
iii) *If $D > 0$, then $(a, b)$ is a saddle point.*

PROOF. Set $A = f_{xx}(a, b)$, $B = f_{xy}(a, b)$, and $C = f_{yy}(a, b)$. Apply Theorem 11.56 and Lemma 11.58. ∎

(For a discriminant which works for functions of three variables, see Widder [14], p. 134.)

**11.60 Remark.** *If the discriminant $D = 0$, $f(a, b)$ may be a local maximum, a local minimum, or $(a, b)$ may be a saddle point.*

PROOF. The function $f(x, y) = x^2$ has zero discriminant at $(a, b) = (0, 0)$, and $0 = f(0, 0)$ is a local minimum for $f$. On the other hand, $f(x, y) = x^3$ has zero discriminant at $(a, b) = (0, 0)$, and $(0, 0)$ is a saddle point for $f$. ∎

In practice, one often wishes to optimize a function subject to certain constraints. (For example, we do not simply want to build the cheapest shipping container, but the cheapest shipping container which will fit in a standard railway car and will not fall apart after several trips.)

**11.61 DEFINITION.** Let $V$ be open in $\mathbf{R}^n$, $\boldsymbol{a} \in V$, and $f, g_j : V \to \mathbf{R}$ for $j = 1, 2, \ldots, m$.

    i) $f(\boldsymbol{a})$ is called a *local minimum of $f$ subject to the constraints* $g_j(\boldsymbol{a}) = 0$, $j = 1, \ldots, m$, if there is a $\rho > 0$ such that $\boldsymbol{x} \in B_\rho(\boldsymbol{a})$ and $g_j(\boldsymbol{x}) = 0$ for all $j = 1, \ldots, m$ imply $f(\boldsymbol{x}) \geq f(\boldsymbol{a})$.

    ii) $f(\boldsymbol{a})$ is called a *local maximum of $f$ subject to the constraints* $g_j(\boldsymbol{a}) = 0$, $j = 1, \ldots, m$, if there is a $\rho > 0$ such that $\boldsymbol{x} \in B_\rho(\boldsymbol{a})$ and $g_j(\boldsymbol{x}) = 0$ for all $j = 1, \ldots, m$ imply $f(\boldsymbol{x}) \leq f(\boldsymbol{a})$.

**11.62 Example.** Find all points on the ellipsoid $x^2 + 2y^2 + 3z^2 = 1$ (see Appendix D) which lie closest to or furthest from the origin.

    SOLUTION. We must optimize the distance formula $\sqrt{x^2 + y^2 + z^2}$, equivalently, we must optimize the function $f(x, y, z) = x^2 + y^2 + z^2$, subject to the constraint $g(x, y, z) = x^2 + 2y^2 + 3z^2 - 1 = 0$. Using $g$ to eliminate the variable $x$ in $f$, we see that $f$ takes on the form

$$\phi(y, z) = 1 - y^2 - 2z^2.$$

Solving $\nabla \phi(y, z) = (0, 0)$, we obtain $(y, z) = (0, 0)$, i.e., $x^2 = 1$. Thus, elimination of $x$ leads to the points $(\pm 1, 0, 0)$. Similarly, elimination of $y$ leads to $(0, \pm 1/\sqrt{2}, 0)$, and elimination of $z$ leads to $(0, 0, \pm 1/\sqrt{3})$. Checking the distance formula, we see that the maximum distance is 1, which occurs at the points $(\pm 1, 0, 0)$, and the minimum distance is $1/\sqrt{3}$, which occurs at the points $(0, 0, \pm 1/\sqrt{3})$. (The points $(0, \pm 1/\sqrt{2}, 0)$ are saddle points, i.e., correspond neither to a maximum nor a minimum.) ∎

    Optimizing a function subject to constraints, as above, by eliminating one or more of the variables is called the *direct method*. There is another, more geometric, method for solving Example 11.62. Notice that for the ellipsoid $g(x, y, z) = x^2 + 2y^2 + 3z^2 - 1 = 0$, the points closest to and furthest from the origin occur when the tangent planes of $g(x, y, z)$ and the sphere $f(x, y, z) = 1$ are parallel (see Figure 11.8). Recall that two nonzero vectors $\boldsymbol{a}$ and $\boldsymbol{b}$ are parallel if and only if $\boldsymbol{a} + \lambda \boldsymbol{b} = \boldsymbol{0}$ for some scalar $\lambda \neq 0$. Since normal vectors of the tangent planes of $f(x, y, z) = 1$ and $g(x, y, z) = 0$ are $\nabla f$ and $\nabla g$ (see Exercise 8 in Section 11.5), it follows that points $(x, y, z)$ closest to and furthest from the origin must satisfy

(33) $$\nabla f(x, y, z) + \lambda \nabla g(x, y, z) = \boldsymbol{0}$$

for some $\lambda \neq 0$. For the case at hand, (33) implies $(2x, 2y, 2z) + \lambda(2x, 4y, 6z) = (0, 0, 0)$. Combining this equation with the constraint $g(x, y, z) = 0$, we have four equations in four unknowns:

$$x(\lambda + 1), \qquad y(2\lambda + 1) = 0, \qquad z(3\lambda + 1) = 0, \quad \text{and} \quad x^2 + 2y^2 + 3z^2 = 1.$$

Solving these equations, we obtain three pairs of solutions: $(\pm 1, 0, 0)$ (when $\lambda = -1$), $(0, \pm 1/\sqrt{2}, 0)$ (when $\lambda = -1/2$), and $(0, 0, \pm 1/\sqrt{3})$ (when $\lambda = -1/3$). Hence,

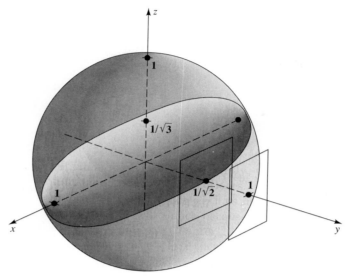

**Figure 11.8**

we obtain the same solutions with the geometric method as we did with the direct method.

The following result shows that the geometric method is valid, even in the case when the functions have nothing to do with spheres and ellipsoids, and even when several constraints are used. This is fortunate since the direct method cannot be used unless the constraints are relatively simple.

**11.63 THEOREM** [LAGRANGE MULTIPLIERS]. *Let $m < n$, $V$ be a nonempty, open set in $\mathbf{R}^n$, and $f, g_j : V \to \mathbf{R}$ be $\mathcal{C}^1$ on $V$ for $j = 1, 2, \ldots, m$. Suppose there is an $\boldsymbol{a} \in V$ such that*

$$\frac{\partial(g_1, \ldots, g_m)}{\partial(x_1, \ldots, x_m)}(\boldsymbol{a}) \neq 0.$$

*If $f(\boldsymbol{a})$ is a local extremum of $f$ subject to the constraints $g_k(\boldsymbol{a}) = 0$, $k = 1, \ldots, m$, then there exist scalars $\lambda_1, \lambda_2, \ldots, \lambda_m$ such that*

(34)
$$\nabla f(\boldsymbol{a}) + \sum_{k=1}^{m} \lambda_k \nabla g_k(\boldsymbol{a}) = \boldsymbol{0}.$$

PROOF. Equation (34) is a system of $n$ equations in $m$ unknowns, $\lambda_1, \lambda_2, \ldots, \lambda_m$:

(35)
$$\sum_{k=1}^{m} \lambda_k \frac{\partial g_k}{\partial x_j}(\boldsymbol{a}) = -\frac{\partial f}{\partial x_j}(\boldsymbol{a}), \qquad j = 1, 2, \ldots, n.$$

If we restrict our attention to the indices $j = 1, 2, \ldots, m$, then we see by hypothesis that the determinant of the matrix of coefficients of the system (35) is nonzero.

Thus, the $\lambda_k$'s are uniquely determined by the first $m$ equations of (35). What remains to be seen is that these $\lambda_k$'s also satisfy (35) for $j = m + 1, \ldots, n$. This is a question about implicit functions.

Let $p = n - m$. As in the proof of the Implicit Function Theorem, write vectors in $\mathbf{R}^{m+p}$ in the form

$$\boldsymbol{x} = (\boldsymbol{y}, \boldsymbol{t}) = (y_1, \ldots, y_m, t_1, \ldots, t_p).$$

We must show

$$(36) \qquad 0 = \frac{\partial f}{\partial t_\ell}(\boldsymbol{a}) + \sum_{k=1}^{m} \lambda_k \frac{\partial g_k}{\partial t_\ell}(\boldsymbol{t}_0)$$

for $\ell = 1, \ldots, p$.

Let $g = (g_1, \ldots, g_m)$, and choose $\boldsymbol{y}_0 \in \mathbf{R}^m$, $\boldsymbol{t}_0 \in \mathbf{R}^p$ such that $\boldsymbol{a} = (\boldsymbol{y}_0, \boldsymbol{t}_0)$. By hypothesis, $g(\boldsymbol{y}_0, \boldsymbol{t}_0) = \boldsymbol{0}$ and the Jacobian of $g$ (with respect to the variables $y_j$) is nonzero at $(\boldsymbol{y}_0, \boldsymbol{t}_0)$. Hence, by the Implicit Function Theorem, there is an open set $W \subset \mathbf{R}^p$ which contains $\boldsymbol{t}_0$, and a function $h : W \to \mathbf{R}^m$ such that $h$ is continuously differentiable on $W$, $h(\boldsymbol{t}_0) = \boldsymbol{y}_0$, and

$$(37) \qquad g(h(\boldsymbol{t}), \boldsymbol{t}) = \boldsymbol{0}, \qquad \boldsymbol{t} \in W.$$

For each $\boldsymbol{t} \in W$ and $k = 1, \ldots, m$, set

$$G_k(\boldsymbol{t}) = g_k(h(\boldsymbol{t}), \boldsymbol{t}) \quad \text{and} \quad F(\boldsymbol{t}) = f(h(\boldsymbol{t}), \boldsymbol{t}).$$

We shall use the functions $G_1, \ldots, G_m$ and $F$ to verify (36) for $\ell = 1, \ldots, p$. Fix such an $\ell$. By (37), each $G_k$ is identically zero on $W$, hence, has derivative zero there. Since $\boldsymbol{t}_0 \in W$ and $(h(\boldsymbol{t}_0), \boldsymbol{t}_0) = (\boldsymbol{y}_0, \boldsymbol{t}_0) = \boldsymbol{a}$, it follows from the Chain Rule that

$$O = DG_k(\boldsymbol{t}_0) = \begin{bmatrix} \dfrac{\partial g_k}{\partial x_1}(\boldsymbol{a}) & \cdots & \dfrac{\partial g_k}{\partial x_n}(\boldsymbol{a}) \end{bmatrix} \begin{bmatrix} \dfrac{\partial h_1}{\partial t_1}(\boldsymbol{t}_0) & \cdots & \dfrac{\partial h_1}{\partial t_p}(\boldsymbol{t}_0) \\ \vdots & \ddots & \vdots \\ \dfrac{\partial h_m}{\partial t_1}(\boldsymbol{t}_0) & \cdots & \dfrac{\partial h_m}{\partial t_p}(\boldsymbol{t}_0) \\ 1 & \cdots & 0 \\ \vdots & \ddots & \vdots \\ 0 & \cdots & 1 \end{bmatrix}.$$

Hence, the $\ell$th component of $DG_k(\boldsymbol{t}_0)$ is given by

$$(38) \qquad 0 = \sum_{j=1}^{m} \frac{\partial g_k}{\partial x_j}(\boldsymbol{a}) \frac{\partial h_j}{\partial t_\ell}(\boldsymbol{t}_0) + \frac{\partial g_k}{\partial t_\ell}(\boldsymbol{a})$$

for $k = 1, 2, \ldots, m$. Multiplying (38) by $\lambda_k$ and adding, we obtain

$$
0 = \sum_{k=1}^{m} \sum_{j=1}^{m} \lambda_k \frac{\partial g_k}{\partial x_j}(\boldsymbol{a}) \frac{\partial h_j}{\partial t_\ell}(\boldsymbol{t}_0) + \sum_{k=1}^{m} \lambda_k \frac{\partial g_k}{\partial t_\ell}(\boldsymbol{a})
$$

$$
= \sum_{j=1}^{m} \left( \sum_{k=1}^{m} \lambda_k \frac{\partial g_k}{\partial x_j}(\boldsymbol{a}) \right) \frac{\partial h_j}{\partial t_\ell}(\boldsymbol{t}_0) + \sum_{k=1}^{m} \lambda_k \frac{\partial g_k}{\partial t_\ell}(\boldsymbol{a}).
$$

Hence, it follows from (35) that

$$
(39) \qquad 0 = -\sum_{j=1}^{m} \frac{\partial f}{\partial x_j}(\boldsymbol{a}) \frac{\partial h_j}{\partial t_\ell}(\boldsymbol{t}_0) + \sum_{k=1}^{m} \lambda_k \frac{\partial g_k}{\partial t_\ell}(\boldsymbol{a}).
$$

Suppose $f(\boldsymbol{a})$ is a local maximum subject to the constraints $g(\boldsymbol{a}) = \boldsymbol{0}$. Set $E_0 = \{\boldsymbol{x} \in V : g(\boldsymbol{x}) = \boldsymbol{0}\}$, and choose an $n$-dimensional open ball $B(\boldsymbol{a})$ such that

$$
(40) \qquad \boldsymbol{x} \in B(\boldsymbol{a}) \cap E_0 \quad \text{implies} \quad f(\boldsymbol{x}) \le f(\boldsymbol{a}).
$$

Since $h$ is continuous, choose a $p$-dimensional open ball $B(\boldsymbol{t}_0)$ such that $\boldsymbol{t} \in B(\boldsymbol{t}_0)$ implies $(h(\boldsymbol{t}), \boldsymbol{t}) \in B(\boldsymbol{a})$. By (40), $F(\boldsymbol{t}_0)$ is a local maximum of $F$ on $B(\boldsymbol{t}_0)$. Hence, $\nabla F(\boldsymbol{t}_0) = \boldsymbol{0}$. Applying the Chain Rule as above, we obtain

$$
(41) \qquad 0 = \sum_{j=1}^{m} \frac{\partial f}{\partial x_j}(\boldsymbol{a}) \frac{\partial h_j}{\partial t_\ell}(\boldsymbol{t}_0) + \frac{\partial f}{\partial t_\ell}(\boldsymbol{a})
$$

(compare with (38)). Adding (41) and (39), we conclude that

$$
0 = \frac{\partial f}{\partial t_\ell}(\boldsymbol{a}) + \sum_{k=1}^{m} \lambda_k \frac{\partial g_k}{\partial t_\ell}(\boldsymbol{a}). \quad \blacksquare
$$

**11.64 Example.** Find all extrema of $x^2 + y^2 + z^2$ subject to the constraints $x - y = 1$ and $y^2 - z^2 = 1$.

SOLUTION. Let $f(x, y, z) = x^2 + y^2 + z^2$, $g(x, y, z) = x - y - 1$, and $h(x, y, z) = y^2 - z^2 - 1$. Then (34) takes on the form $\nabla f + \lambda \nabla g + \mu \nabla h = \boldsymbol{0}$; i.e.,

$$
(2x, 2y, 2z) + \lambda(1, -1, 0) + \mu(0, 2y, -2z) = (0, 0, 0).
$$

In particular, $2x + \lambda = 0$, $2y + 2\mu y - \lambda = 0$, and $2z - 2\mu z = 0$. From this last equation, either $\mu = 1$ or $z = 0$.

If $\mu = 1$, then $\lambda = 4y$. Since $2x + \lambda = 0$, we find that $x = -2y$. From $g = 0$ we obtain $-3y = 1$, i.e., $y = -1/3$. Substituting this into $h = 0$, we obtain $z^2 = -8/9$, a contradiction.

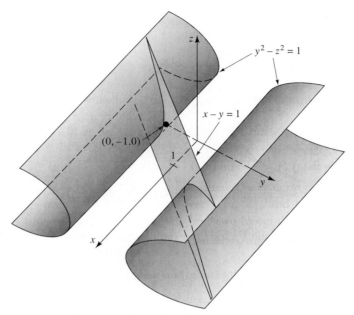

$y^2 - z^2 = 1$

$x - y = 1$

$(0, -1, 0)$

**Figure 11.9**

If $z = 0$, then from $h = 0$ we obtain $y = \pm 1$. Since $g = 0$, we obtain $x = 2$ when $y = 1$, and $x = 0$ when $y = -1$. Thus, the only candidates for extrema of $f$ subject to the constraints $g = 0 = h$ are $f(2, 1, 0) = 5$ and $f(0, -1, 0) = 1$. To decide whether these are maxima, minima, or neither, look at the problem from a geometric point of view. The problem requires us to find points on the intersection of the plane $x - y = 1$ and the hyperbolic cylinder $y^2 - z^2 = 1$ which lie closest to the origin. Evidently, both of these points correspond to local minima, and there is no maximum (see Figure 11.9). In particular, the minimum of $x^2 + y^2 + z^2$ subject to the given constraints is 1, attained at the point $(0, -1, 0)$. ∎

## EXERCISES

**1.** Find all local extrema of each of the following functions.

    a) $f(x, y) = x^2 - xy + y^3 - y$.

    b) $f(x, y) = \sin x + \cos y$.

    c) $f(x, y, z) = e^{x+y} \cos z$.

    d) $f(x, y) = ax^2 + bxy + cy^2$, where $a \neq 0$ and $b^2 - 4ac \neq 0$.

**2.** For each of the following, find the maximum and minimum of $f$ on $H$.

    a) $f(x, y) = x^2 + 2x - y^2$ and $H = \{(x, y) : x^2 + 4y^2 \leq 4\}$.

    b) $f(x, y) = x^2 + 2xy + 3y^2$, and $H$ is the region bounded by the triangle with vertices $(1, 0)$, $(1, 2)$, $(3, 0)$.

    c) $f(x, y) = x^3 + 3xy - y^3$, and $H = [-1, 1] \times [-1, 1]$.

**3.** For each of the following, use Lagrange Multipliers to find all extrema of $f$ subject to the given constraints

  a) $f(x,y) = x + y^2$ and $x^2 + y^2 = 4$.
  b) $f(x,y) = x^2 - 4xy + 4y^2$ and $x^2 + y^2 = 1$.
  c) $f(x,y,z) = xy$, $x^2 + y^2 + z^2 = 1$ and $x + y + z = 0$.
  d) $f(x,y,z,w) = 3x + y + w$, $3x^2 + y + 4z^3 = 1$ and $-x^3 + 3z^4 + w = 0$.

**4.** Suppose $f : \mathbf{R}^n \to \mathbf{R}^m$ is differentiable at $\boldsymbol{a}$, and $g : \mathbf{R}^m \to \mathbf{R}$ is differentiable at $\boldsymbol{b} = f(\boldsymbol{a})$. Prove that if $g(\boldsymbol{b})$ is a local extremum of $g$, then $\nabla(g \circ f)(\boldsymbol{a}) = \boldsymbol{0}$.

**5.** Suppose $V$ is open in $\mathbf{R}^2$, $(a,b) \in V$, and $f : V \to \mathbf{R}$ has second-order partial derivatives on $V$ with $f_x(a,b) = f_y(a,b) = 0$. If the second-order partial derivatives of $f$ are continuous at $(a,b)$ and exactly two of the three numbers $f_{xx}(a,b)$, $f_{xy}(a,b)$, and $f_{yy}(a,b)$ are zero, prove that $(a,b)$ is a saddle point if and only if $f_{xy}(a,b) \neq 0$.

**6.** Suppose $V$ is an open set in $\mathbf{R}^n$, $\boldsymbol{a} \in V$, and $f : V \to \mathbf{R}$ is $\mathcal{C}^2$ on $V$. If $f(\boldsymbol{a})$ is a local minimum of $f$, prove $D^{(2)} f(\boldsymbol{a})(\boldsymbol{h}) \geq 0$ for all $\boldsymbol{h} \in \mathbf{R}^n$.

**7.** Let $a, b, c, D, E$ be real numbers with $c \neq 0$.

  a) If $DE > 0$, find all extrema of $ax + by + cz$ subject to the constraint $z = Dx^2 + Ey^2$. Prove that a maximum occurs when $cD < 0$ and a minimum when $cD > 0$.
  b) What can you say when $DE < 0$?

**8.** [IMPLICIT METHOD].

  a) Suppose $f, g : \mathbf{R}^3 \to \mathbf{R}$ are differentiable at a point $(a,b,c)$, and $f(a,b,c)$ is an extremum of $f$ subject to the constraint $g(x,y,z) = k$, where $k$ is a constant. Prove that

$$\frac{\partial f}{\partial x}(a,b,c)\frac{\partial g}{\partial z}(a,b,c) - \frac{\partial f}{\partial z}(a,b,c)\frac{\partial g}{\partial x}(a,b,c) = 0$$

  and

$$\frac{\partial f}{\partial y}(a,b,c)\frac{\partial g}{\partial z}(a,b,c) - \frac{\partial f}{\partial z}(a,b,c)\frac{\partial g}{\partial y}(a,b,c) = 0.$$

  b) Use part a) to find all extrema of $f(x,y,z) = 4xy + 2xz + 2yz$ subject to the constraint $xyz = 16$.

**9.** **This exercise is used in Section $^e$14.4.**

  a) Let $p > 1$. Find all extrema of $f(\boldsymbol{x}) = \sum_{k=1}^{n} x_k^2$ subject to the constraint $\sum_{k=1}^{n} |x_k|^p = 1$.
  b) Prove that

$$\frac{1}{n^{(2-p)/(2p)}} \left( \sum_{k=1}^{n} |x_k|^p \right)^{1/p} \leq \left( \sum_{k=1}^{n} x_k^2 \right)^{1/2} \leq \left( \sum_{k=1}^{n} |x_k|^p \right)^{1/p}$$

  for all $x_1, \ldots, x_n \in \mathbf{R}$, $n \in \mathbf{N}$, and $1 \leq p \leq 2$.

# Chapter 12

# Integration on $\mathbf{R}^n$

## 12.1 JORDAN REGIONS

In this section we define grids (a multidimensional analogue of partitions) and use them to identify special subsets of $\mathbf{R}^n$, called *Jordan regions*, which have a well–defined volume. In the next section, when we define integrals of multivariable functions on Jordan regions, grids will play the role that partitions did in the one–variable case.

Throughout this chapter, $R$ will represent a nondegenerate $n$-dimensional rectangle; i.e.,

$$(1) \quad R = [a_1, b_1] \times \cdots \times [a_n, b_n] := \{\boldsymbol{x} \in \mathbf{R}^n : \ x_j \in [a_j, b_j] \text{ for } j = 1, \ldots, n\},$$

where $a_j < b_j$ for $j = 1, 2, \ldots, n$. A *grid* on $R$ is a collection of $n$-dimensional rectangles $\mathcal{G} = \{R_1, \ldots, R_p\}$ obtained by subdividing the sides of $R$; i.e., for each $j = 1, \ldots, n$ there are integers $\nu_j \in \mathbf{N}$ and partitions $\mathcal{P}_j = \mathcal{P}_j(\mathcal{G}) = \{x_k^{(j)} : k = 1, \ldots, \nu_j\}$ of $[a_j, b_j]$ such that $\mathcal{G}$ is the collection of rectangles of the form $I_1 \times \cdots \times I_n$, where each $I_j = [x_{k-1}^{(j)}, x_k^{(j)}]$ for some $k = 1, \ldots, \nu_j$ (see Figure 12.1). A grid $\mathcal{G}$ is said to be *finer* than a grid $\mathcal{H}$ if each partition $\mathcal{P}_j(\mathcal{G})$ is finer than the corresponding partition $\mathcal{P}_j(\mathcal{H})$, $j = 1, \ldots, n$.

If $R$ is an $n$-dimensional rectangle of the form (1), then the *volume* of $R$ is defined to be

$$|R| = (b_1 - a_1) \ldots (b_n - a_n).$$

(When $n = 1$, we shall call $|R|$ the *length* of $R$, and when $n = 2$, we shall call $|R|$ the *area* of $R$.) Notice that given $\varepsilon > 0$ there exists a rectangle $R^*$ such that $R \subset (R^*)^o$ and $|R^*| = |R| + \varepsilon$. Indeed, since $b_j - a_j + 2\delta \to b_j - a_j$ as $\delta \to 0$, we can choose $\delta > 0$ so small that $R^* := [a_1 - \delta, b_1 + \delta] \times \cdots \times [a_n - \delta, b_n + \delta]$ satisfies $|R^*| = |R| + \varepsilon$.

We want to define the integral of a multivariable function on a variety of sets, for example, the integral of a function of two variables on rectangles, disks, triangles, ellipses, and the integral of a function of three variables on balls, cones, ellipsoids,

372

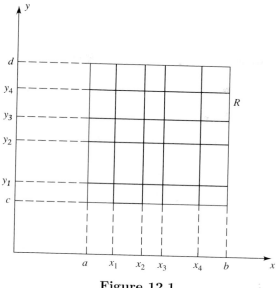

**Figure 12.1**

pyramids, etc. One property these regions all have in common is that they have a well-defined "area" or "volume."

How shall we define the volume of a general set $E$? Not every set will have a well-defined volume, but for those which do, the situation is analogous to the definition of the integral, using upper and lower sums. Given $E \subset \mathbf{R}^n$, a subset of some $n$-dimensional rectangle $R$, and $\mathcal{G} = \{R_j : j = 1, \ldots, p\}$, a grid on $R$, the *outer* and *inner sums* of $E$ with respect to $\mathcal{G}$ are defined by

$$V(E; \mathcal{G}) := \sum_{R_j \cap \overline{E} \neq \emptyset} |R_j| \quad \text{and} \quad v(E; \mathcal{G}) := \sum_{R_j \subset E^o} |R_j|,$$

where the empty sum is interpreted to be zero. Thus $V(\emptyset; \mathcal{G}) = v(\emptyset; \mathcal{G}) = 0$ and $v(E; \mathcal{G}) = 0$ for all grids $\mathcal{G}$ and all sets $E$ satisfying $E^o = \emptyset$.

Inner and outer sums will be used to define inner and outer volume in the same way that upper and lower sums were used to define upper and lower integrals (see Definition 12.4 below). In this regard, notice that if $\mathcal{G}$ is fine enough, each inner and outer sum of a set $E$ is a sum of volumes of rectangles which approximates the "volume" of $E$ if it exists (see Figure 12.2); $V(E; \mathcal{G})$ overestimates the "volume" of $E$ and $v(E; \mathcal{G})$ underestimates the "volume" of $E$. (In Figure 12.2, the underestimate $v(E; \mathcal{G})$ is represented by the lightly shaded rectangles; the overestimate $V(E; \mathcal{G})$ is represented by the union of the lightly and darkly shaded rectangles. You might refine the grid there and revisualize the inner and outer sums to illustrate that these estimates get better as the grid gets finer.)

Before we define inner and outer volume, we prove three elementary results about inner and outer sums. Our first result reveals why the sum $V(E; \mathcal{G})$ is taken over

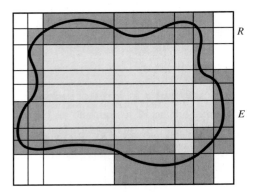

**Figure 12.2**

rectangles which intersect the closure of $E$, and the sum $v(E;\mathcal{G})$ is taken over rectangles which are contained in the interior of $E$.

**12.1 Remark.** *Let $R$ be an $n$-dimensional rectangle and $E$ be a subset of $R$. For all grids $\mathcal{G}$ on $R$,*

$$(2) \qquad\qquad V(E;\mathcal{G}) - v(E;\mathcal{G}) = V(\partial E;\mathcal{G}).$$

PROOF. If $E^o = \emptyset$, then $\partial E = E$ and (2) is obvious. Otherwise, suppose $R_j \in \mathcal{G}$ is a rectangle which appears in the sum represented by the left side of (2); i.e., $R_j$ intersects $\overline{E}$ but $R_j$ is not a subset of $E^o$. If $R_j$ does not appear in the sum represented by the right side of (2), then $R_j \cap \overline{\partial E} = \emptyset$. It follows that the pair $E^o$, $(\mathbf{R}^n \setminus E)^o$ separates $R_j$, a contradiction since all rectangles are connected (see Remark 9.40). Therefore, every rectangle which appears in the sum represented by the left side of (2) also appears in the sum represented by the right side; i.e., $V(E;\mathcal{G}) - v(E;\mathcal{G}) \leq V(\partial E;\mathcal{G})$.

On the other hand, suppose $R_j \in \mathcal{G}$ is a rectangle which appears in the sum represented by the right side of (2); i.e., $R_j \cap \overline{\partial E} \neq \emptyset$. Recall from Theorems 9.15 and 9.7 (or 10.39 and 10.31) that $\partial E = \overline{E} \setminus E^o$ is closed, so $R_j \cap \partial E \neq \emptyset$. It follows that $R_j$ contains points in $E$ and $E^c$, in particular, $R_j$ intersects $\overline{E}$ but $R_j$ is not a subset of $E^o$. Thus every rectangle which appears in the sum represented by the right side of (2) also appears in the sum represented by the left side i.e., $V(E;\mathcal{G}) - v(E;\mathcal{G}) \geq V(\partial E;\mathcal{G})$. $\blacksquare$

The next result shows that as the grid gets finer, the estimates $V(E;\mathcal{G})$ and $v(E;\mathcal{G})$ of the "volume" of $E$ get better.

**12.2 Remark.** *Let $R$ be an $n$-dimensional rectangle, $E$ be a subset of $R$, and $\mathcal{G}$, $\mathcal{H}$ be grids on $R$. If $\mathcal{G}$ is finer than $\mathcal{H}$, then*

$$0 \leq v(E;\mathcal{H}) \leq v(E;\mathcal{G}) \leq V(E;\mathcal{G}) \leq V(E;\mathcal{H}).$$

PROOF. Since $v(E;\mathcal{H})$ is either zero or a sum of nonnegative terms, it is clear that $v(E;\mathcal{H}) \geq 0$ for all grids $\mathcal{H}$.

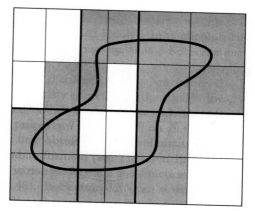

**Figure 12.3**

The third inequality is trivial if $E^o$ is empty because in this case $v(E; \mathcal{G}) = 0$ for all grids $\mathcal{G}$. If $E^o$ is nonempty, then the sum represented by $V(E; \mathcal{G})$ contains every term from the sum represented by $v(E; \mathcal{G})$ because $R_j \subset E^o$ implies $R_j \cap \overline{E} \neq \emptyset$. Thus $v(E; \mathcal{G}) \leq V(E; \mathcal{G})$ in either case.

It remains to verify the second and fourth inequalities. Since their proofs are similar, we supply the details only for the fourth inequality. Since $\mathcal{G}$ is finer than $\mathcal{H}$, each $Q \in \mathcal{H}$ is a finite union of $R_j$'s in $\mathcal{G}$. If $Q \cap \overline{E} \neq \emptyset$, then some of the $R_j$'s in $Q$ intersect $\overline{E}$ and others might not (see Figure 12.3, where the darker lines represent the grid $\mathcal{H}$, the lighter lines represent $\mathcal{G} \setminus \mathcal{H}$, and the $R_j$'s which intersect $\overline{E}$ are shaded). Let $\mathcal{I}_1 = \{R \in \mathcal{G} : R \cap \overline{E} \neq \emptyset\}$ and $\mathcal{I}_2 = \{R \in \mathcal{G} \setminus \mathcal{I}_1 : R \subseteq Q \text{ for some } Q \in \mathcal{H} \text{ with } Q \cap \overline{E} \neq \emptyset\}$. Then

$$V(E; \mathcal{H}) = \sum_{R \in \mathcal{I}_1} |R| + \sum_{R \in \mathcal{I}_2} |R| \geq \sum_{R \in \mathcal{I}_1} |R| = V(E; \mathcal{G}). \quad \blacksquare$$

**12.3 Remark.** *Let $R$ be an n-dimensional rectangle and $E$ be a subset of $R$. If $\mathcal{G}$ and $\mathcal{H}$ are grids on $R$, then*

(3) $$0 \leq v(E; \mathcal{G}) \leq V(E; \mathcal{H}).$$

PROOF. Let $\mathcal{I}$ be a grid finer than both $\mathcal{G}$ and $\mathcal{H}$. (Such a grid can be constructed by taking $\mathcal{P}_j(\mathcal{I}) = \mathcal{P}_j(\mathcal{G}) \cup \mathcal{P}_j(\mathcal{H})$ for $j = 1, \ldots, n$.) By Remark 12.2,

$$0 \leq v(E; \mathcal{G}) \leq v(E; \mathcal{I}) \leq V(E; \mathcal{I}) \leq V(E; \mathcal{H}). \quad \blacksquare$$

Using the sums $v(E; \mathcal{G})$ and $V(E; \mathcal{G})$, we define inner and outer volume. (We assume that $E$ is bounded so that there exist rectangles $R$ which contain $E$. This also guarantees that the inner and outer sums of $E$ are finite.)

**12.4 DEFINITION.** Let $E$ be a bounded subset of $\mathbf{R}^n$ and $R$ be an $n$-dimensional rectangle which satisfies $E \subseteq R$. The *inner volume* of $E$ is defined by

$$\underline{\mathrm{Vol}}\,(E) := \sup\{v(E; \mathcal{G}) : \ \mathcal{G} \text{ ranges over all grids on } R\},$$

and the *outer volume* of $E$ is defined by

$$\overline{\mathrm{Vol}}\,(E) := \inf\{V(E; \mathcal{G}) : \ \mathcal{G} \text{ ranges over all grids on } R\}.$$

There is a logical difficulty with Definition 12.4 which we must address before continuing. Does the definition of inner and outer volume of a set $E$ depend on the rectangle $R$ used to generate the grids $\mathcal{G}$? To show that the answer to this question is no, let $R$ and $Q$ be rectangles which contain $E$. Since the intersection of two rectangles is a rectangle, we may suppose $E \subset Q \subset R$. Choose a rectangle $Q^*$ such that $Q \subset (Q^*)^o \subset R^o$ and $|Q^*| = |Q| + \varepsilon$. Let $\mathcal{G}$ be any grid on $R$, and let $\mathcal{G}_0$ be the grid on $R$ formed by "adding the endpoints of $Q^*$ to $\mathcal{G}$"; i.e., if $Q^* = [c_1, d_1] \times \cdots \times [c_n, d_n]$, then $\mathcal{P}_j(\mathcal{G}_0) = \mathcal{P}_j(\mathcal{G}) \cup \{c_j, d_j\}$. Clearly, $\mathcal{G}_0$ is finer than $\mathcal{G}$, and $\mathcal{H}_0 := \mathcal{G}_0 \cap Q^*$ is a grid on $Q^*$. Since $\overline{E} \subset (Q^*)^o \subset R^o$ implies $V(E; \mathcal{H}_0) = V(E; \mathcal{G}_0)$, it follows from Remark 12.2 that

$$\inf_{\mathcal{H} \text{ on } Q^*} V(E; \mathcal{H}) \leq V(E; \mathcal{H}_0) = V(E; \mathcal{G}_0) \leq V(E; \mathcal{G}).$$

Since this inequality holds for any grid $\mathcal{G}$ on $R$, it follows that

$$\inf_{\mathcal{H} \text{ on } Q} V(E; \mathcal{H}) \leq \inf_{\mathcal{H} \text{ on } Q^*} V(E; \mathcal{H}) \leq \inf_{\mathcal{G} \text{ on } R} V(E; \mathcal{G}).$$

On the other hand, if $\mathcal{H}$ is any grid on $Q^*$, and $\mathcal{H}_0$ is the grid on $R$ formed by "adding the endpoints of $R$ to $\mathcal{H}$," then

$$V(E; \mathcal{H}) = V(E; \mathcal{H}_0) \geq \inf_{\mathcal{G} \text{ on } R} V(E; \mathcal{G}).$$

Since this holds for any grid $\mathcal{H}$ on $Q^*$, and $|Q^*| = |Q| + \varepsilon$, it follows that

$$\inf_{\mathcal{G} \text{ on } R} V(E; \mathcal{G}) \leq \inf_{\mathcal{H} \text{ on } Q^*} V(E; \mathcal{H}) \leq \inf_{\mathcal{H} \text{ on } Q} V(E; \mathcal{H}) + \varepsilon.$$

We have shown that

$$\inf_{\mathcal{H} \text{ on } Q} V(E; \mathcal{H}) \leq \inf_{\mathcal{G} \text{ on } R} V(E; \mathcal{G}) \leq \inf_{\mathcal{H} \text{ on } Q} V(E; \mathcal{H}) + \varepsilon,$$

for every $\varepsilon > 0$. By letting $\varepsilon \to 0$, we verify that the definition of outer volume does not depend on the rectangle $R$. A similar argument shows that the same is true for inner volume.

Having established that inner and outer volume are well-defined, we continue our exposition. The sums $V(E; \mathcal{G})$ (respectively, $v(E; \mathcal{G})$) overestimate (respectively, underestimate) the "volume" of $E$. If $E$ has a well-defined volume, we expect $V(E; \mathcal{G}) - v(E; \mathcal{G})$ to approach zero as $\mathcal{G}$ gets finer, i.e., $\overline{\mathrm{Vol}}\,(E) = \underline{\mathrm{Vol}}\,(E)$. (In Figure 12.2, the darker rectangles represent the difference $V(E; \mathcal{G}) - v(E; \mathcal{G})$.) Thus we make the following definition.

**12.5 DEFINITION.** Let $E$ be a bounded subset of $\mathbf{R}^n$.

    i) $E$ is called a *Jordan region* if $\overline{\text{Vol}}\,(E) = \underline{\text{Vol}}\,(E)$, in which case we shall denote this common value by $\text{Vol}\,(E)$.

    ii) $E$ is said to be *of volume zero* if $E$ is a Jordan region and $\text{Vol}\,(E) = 0$.

If $E$ is a Jordan region, we shall call $\text{Vol}\,(E)$ the *volume* (or *Jordan content*) of $E$. When $n = 1$, $\ell(E) := \text{Vol}\,(E)$ is frequently called the *length* of $E$; when $n = 2$, $\text{Area}\,(E) := \text{Vol}\,(E)$ is frequently called the *area* of $E$.

Notice that $\emptyset$ is of volume zero, since the empty sum is interpreted to be zero.

Also notice by definition that the volume of a Jordan region is approximately the volume of a finite collection of rectangles. This can be used to transfer properties about rectangles to properties about Jordan regions (see Exercise 3 below, Exercise 7 in Section 12.4, and Lemma 12.44).

Sets of volume zero are of special interest because they can be used to identify Jordan regions and can be ignored when evaluating integrals of multivariable functions (see Theorems 12.12 and 12.24 below). The following result is the first of several which enable us to identify such sets.

**12.6 Remark.** *Let $E$ be a bounded subset of $\mathbf{R}^n$. Then $E$ is a Jordan region of volume zero if and only if $\overline{\text{Vol}}\,(E) = 0$.*

PROOF. Let $R$ be a rectangle containing $E$, and suppose $\mathcal{G}$ and $\mathcal{H}$ are arbitrary grids on $R$. By Remark 12.3, $0 \leq v(E; \mathcal{G}) \leq V(E; \mathcal{H})$. Taking the supremum of this inequality over all grids $\mathcal{G}$, and then the infimum all grids $\mathcal{H}$, we obtain

$$(4) \qquad\qquad 0 \leq \underline{\text{Vol}}\,(E) \leq \overline{\text{Vol}}\,(E)$$

for any bounded subset $E$ of $\mathbf{R}^n$.

Suppose $E$ is of volume zero. By Definition 12.5, then, $\overline{\text{Vol}}\,(E) = \text{Vol}\,(E) = 0$. Conversely, if $\overline{\text{Vol}}\,(E) = 0$, then it follows from (4) that $\underline{\text{Vol}}\,(E) = \overline{\text{Vol}}\,(E) = 0$. Thus $E$ is a Jordan region of volume zero. ∎

Inequality (4) also contains the following result.

**12.7 Remark.** *A bounded subset $E$ of $\mathbf{R}^n$ is a Jordan region if and only if $\overline{\text{Vol}}\,(E) \leq \underline{\text{Vol}}\,(E)$.*

To evaluate integrals of multivariable functions over unions of sets, we need to introduce the following concept.

**12.8 DEFINITION.** Let $\mathcal{E} := \{E_\ell\}_{\ell \in \mathbf{N}}$ be a collection of subsets of $\mathbf{R}^n$.

    i) $\mathcal{E}$ is said to be *nonoverlapping* if $E_j \cap E_k$ is of volume zero for $j \neq k$.

    iv) $\mathcal{E}$ is said to be *pairwise disjoint* if $E_j \cap E_k = \emptyset$ for $j \neq k$.

Notice that since $\emptyset$ is of volume zero, every collection of pairwise disjoint sets is nonoverlapping. (The converse of this statement is false–see Exercise 6 below).

We have defined Jordan regions, but the only example we have given is the empty set. The following result shows all rectangles are Jordan regions and that $\text{Vol}\,(R)$ agrees with the usual definition of the volume of a rectangle.

**12.9 Remark.** *If $R$ is a rectangle, then $R$ is a Jordan region and* $\mathrm{Vol}\,(R) = |R|$.

PROOF. By Remark 12.7, it suffices to show that $\underline{\mathrm{Vol}}\,(R) \geq |R| \geq \overline{\mathrm{Vol}}\,(R)$. Since $\mathcal{G} = \{R\}$ is a grid on $R$, it is clear by definition that $|R| = V(R;\mathcal{G}) \geq \overline{\mathrm{Vol}}\,(R)$. On the other hand, let $\varepsilon > 0$ and suppose

$$R = [a_1, b_1] \times \cdots \times [a_n, b_n].$$

Since $b_j - a_j - 2\delta \to b_j - a_j$ as $\delta \to 0$, we can choose $\delta > 0$ so small that if

$$Q = [a_1 + \delta, b_1 - \delta] \times \cdots \times [a_n + \delta, b_n - \delta],$$

then $|R| - |Q| < \varepsilon$. Let $\mathcal{G}$ be the grid on $R$ determined by

$$\mathcal{P}_j(\mathcal{G}) = \{a_j, a_j + \delta, b_j - \delta, b_j\}.$$

Then $Q$ is the only rectangle in $\mathcal{G}$ which satisfies $Q \subset R^o$. Hence, by definition,

$$\underline{\mathrm{Vol}}\,(R) \geq v(R;\mathcal{G}) = |Q| > |R| - \varepsilon,$$

i.e., $\underline{\mathrm{Vol}}\,(R) > |R| - \varepsilon$. Taking the limit of this last inequality as $\varepsilon \to 0$, we conclude that $\underline{\mathrm{Vol}}\,(R) \geq |R|$. ∎

Thus $\mathbf{R}^n$ contains plenty of Jordan regions. Not every subset of $\mathbf{R}^n$, however, is a Jordan region.

**12.10 Example.** The set of ordered pairs

$$E = \{(x,y) : x,y \in \mathbf{Q} \cap [0,1]\}$$

is not a Jordan region.

PROOF. Let $R := [0,1] \times [0,1]$. Since no rectangle satisfies $R_j \subseteq E$ (much less $R_j \subset E^o$), $v(E;\mathcal{G}) = 0$ for all grids $\mathcal{G}$ on $R$. On the other hand, if $R_j$ is any rectangle contained in $R$, then $R_j \cap E \neq \emptyset$; hence, $R_j \cap \overline{E} \neq \emptyset$. Thus $V(E;\mathcal{G}) = 1$ for all grids $\mathcal{G}$ on $R$. It follows that

$$\underline{\mathrm{Vol}}\,(E) = 0 < 1 = \overline{\mathrm{Vol}}\,(E). \quad ∎$$

Nevertheless, every subset of a set of volume zero must be a Jordan region.

**12.11 Remark.** *If $E_0$ is a Jordan region of volume zero and $E \subseteq E_0$, then $E$ is a Jordan region of volume zero.*

PROOF. Suppose $E_0$ is a Jordan region of volume zero and $E \subseteq E_0$. By hypothesis and the Monotone Property of Infima, $0 \leq \overline{\mathrm{Vol}}\,(E) \leq \overline{\mathrm{Vol}}\,(E_0) = 0$. Hence, by Remark 12.6, $E$ is a Jordan region and $\mathrm{Vol}\,(E) = 0$. ∎

In general, it is not easy to decide whether or not a given set is a Jordan region. Topology alone cannot resolve this problem since there are open sets in $\mathbf{R}^n$ which are not Jordan regions (see Spivak [12], p. 56). Hence, we need to approach the definition of Jordan regions from a different point of view.

Notice that by definition, if $E$ is a Jordan region, then given $\varepsilon > 0$ there is a grid $\mathcal{G}$ such that $V(E;\mathcal{G}) - v(E;\mathcal{G}) < \varepsilon$. Since by Remark 12.1, $V(E;\mathcal{G}) - v(E;\mathcal{G}) = V(\partial E;\mathcal{G})$ (see Figure 12.2), we guess that a set is a Jordan region if and only if its boundary is of volume zero. The following result shows this guess is correct.

**12.12 THEOREM.** *Let $E \subset \mathbf{R}^n$ be bounded. Then $E$ is a Jordan region if and only if its boundary $\partial E$ is of volume zero.*

PROOF. Let $R$ be a rectangle which contains $E$ and $\mathcal{G}$ be a grid on $R$. By Remark 12.1 and Definition 12.4, $V(\partial E; \mathcal{G}) = V(E; \mathcal{G}) - v(E; \mathcal{G}) \geq \overline{\mathrm{Vol}}(E) - \underline{\mathrm{Vol}}(E)$. Taking the infimum of this inequality over all grids $\mathcal{G}$, we have

(5)                      $$\overline{\mathrm{Vol}}(\partial E) \geq \overline{\mathrm{Vol}}(E) - \underline{\mathrm{Vol}}(E).$$

On the other hand, given $\varepsilon > 0$, choose grids $\mathcal{H}_1$ and $\mathcal{H}_2$ such that

$$\overline{\mathrm{Vol}}(E) + \varepsilon > V(E; \mathcal{H}_1) \quad \text{and} \quad \underline{\mathrm{Vol}}(E) - \varepsilon < v(E; \mathcal{H}_2).$$

If $\mathcal{G}$ is a grid on $R$ which is finer than both $\mathcal{H}_1$ and $\mathcal{H}_2$, it follows from Remark 12.2 that

$$\overline{\mathrm{Vol}}(E) + \varepsilon > V(E; \mathcal{G}) \quad \text{and} \quad \underline{\mathrm{Vol}}(E) - \varepsilon < v(E; \mathcal{G}).$$

Subtracting these inequalities, we see by Remark 12.1 that

$$V(\partial E; \mathcal{G}) = V(E; \mathcal{G}) - v(E; \mathcal{G}) < \overline{\mathrm{Vol}}(E) - \underline{\mathrm{Vol}}(E) + 2\varepsilon.$$

Therefore, $\overline{\mathrm{Vol}}(\partial E) < \overline{\mathrm{Vol}}(E) - \underline{\mathrm{Vol}}(E) + 2\varepsilon$. Taking the limit of this inequality as $\varepsilon \to 0$, we obtain $\overline{\mathrm{Vol}}(\partial E) \leq \overline{\mathrm{Vol}}(E) - \underline{\mathrm{Vol}}(E)$. This inequality, together with (5), proves $\overline{\mathrm{Vol}}(\partial E) = \overline{\mathrm{Vol}}(E) - \underline{\mathrm{Vol}}(E)$. Hence, by Definition 12.5, $E$ is a Jordan region if and only if $\overline{\mathrm{Vol}}(\partial E) = 0$. We conclude by Remark 12.6 that $E$ is a Jordan region if and only if $\partial E$ is a Jordan region of volume zero. ∎

This result can be used to show all balls in $\mathbf{R}^n$ are Jordan regions–see Exercise 9 below. (For a formula of the volume of a ball in $\mathbf{R}^n$, see Theorem 12.70.) We shall also use it to show that the collection of Jordan regions is closed under finite unions. (For intersections and set differences, see Exercise 4 below.)

**12.13 Remark.** *If $E_1$ and $E_2$ are Jordan regions, then $E_1 \cup E_2$ is a Jordan region and*

$$\mathrm{Vol}(E_1 \cup E_2) \leq \mathrm{Vol}(E_1) + \mathrm{Vol}(E_2).$$

PROOF. To estimate the outer volume of $E_1 \cup E_2$, let $\mathcal{G}$ be a grid on a rectangle which contains $E_1 \cup E_2$. If $R_j$ intersects $\overline{E_1 \cup E_2}$, then by Theorem 9.16 (or 10.40) $R_j$ intersects $\overline{E}_1$ or $\overline{E}_2$ (or both). Hence, $V(E_1 \cup E_2; \mathcal{G}) \leq V(E_1; \mathcal{G}) + V(E_2; \mathcal{G})$. Taking the infimum of this inequality over all grids $\mathcal{G}$, we obtain

(6)       $$\overline{\mathrm{Vol}}(E_1 \cup E_2) \leq \overline{\mathrm{Vol}}(E_1) + \overline{\mathrm{Vol}}(E_2) = \mathrm{Vol}(E_1) + \mathrm{Vol}(E_2).$$

It remains to see that $E_1 \cup E_2$ is a Jordan region.

By Remark 12.6 and (6), the union of two sets of volume zero is a set of volume zero. Hence, by Theorem 12.12, $\partial E_1 \cup \partial E_2$ is of volume zero. Since (by Theorem 9.16 or 10.40) $\partial(E_1 \cup E_2) \subseteq \partial E_1 \cup \partial E_2$, it follows from Remark 12.11 that the boundary of $E_1 \cup E_2$ is of volume zero. Hence, by Theorem 12.12, $E_1 \cup E_2$ is a Jordan region. ∎

We shall now obtain a quantitative characterization of sets of volume zero. (This result can be used to prove that the image under $\phi$ of nonoverlapping Jordan regions is nonoverlapping–see Exercise 7 below. This will form a crucial step in the proof of the change of variables formula in Section 12.4.)

**12.14 THEOREM.** *Let $E$ be a bounded subset of $\mathbf{R}^n$. Then* $\text{Vol}(E) = 0$ *if and only if for every $\varepsilon > 0$ there is a finite collection of cubes $Q_k$ of the same size, i.e., all with sides of length $s$, such that*

$$\overline{E} \subset \bigcup_{k=1}^{p} Q_k \quad \text{and} \quad \sum_{k=1}^{p} |Q_k| < \varepsilon.$$

PROOF. If $\text{Vol}(E) = 0$, then by definition there exists a grid $\mathcal{G}$ such that if $\{R_1, \ldots, R_q\}$ represents all rectangles in $\mathcal{G}$ which intersect $\overline{E}$, then

$$\overline{E} \subset \bigcup_{j=1}^{q} R_j \quad \text{and} \quad \sum_{j=1}^{q} |R_j| < \frac{\varepsilon}{2}.$$

By increasing the size of the $R_j$'s slightly, we may suppose that the sides of each $R_j$ have rational lengths, and $\sum_{j=1}^{q} |R_j| < \varepsilon$. (These rectangles may no longer be nonoverlapping.) The lengths of the sides of the $R_j$'s have a common denominator, say $d$. By using a grid fine enough, we can divide each $R_j$ into cubes $Q_k^{(j)}$, for $k = 1, 2, \ldots \nu_j$ and some choice of $\nu_j \in \mathbf{N}$, such that each $Q_k^{(j)}$ has sides of common length $s = 1/d$. Since $|R_j| = \sum_{k=1}^{\nu_j} |Q_k^{(j)}|$, it follows that

$$\sum_{j=1}^{q} \sum_{k=1}^{\nu_j} |Q_k^{(j)}| = \sum_{j=1}^{q} |R_j| < \varepsilon.$$

Conversely, if such cubes exist let $R$ be a rectangle which contains the union of the $Q_k$'s and suppose

$$Q_k = [a_1^{(k)}, b_1^{(k)}] \times \cdots \times [a_n^{(k)}, b_n^{(k)}].$$

For each $j = 1, 2, \ldots, n$, the endpoints $\{a_j^{(1)}, b_j^{(1)}, \ldots, a_j^{(p)}, b_j^{(p)}\}$ can be arranged in increasing order to form a partition of the $j$th side of $R$. Thus there is a grid $\mathcal{G} = \{R_1, \ldots, R_q\}$ so fine that each $Q_k$ is a union of the $R_j$'s (see Figure 12.4). Since $V(E; \mathcal{G}) \leq \sum_{k=1}^{p} |Q_k| < \varepsilon$, it follows that $\overline{\text{Vol}}(E) < \varepsilon$ for all $\varepsilon > 0$. We conclude that $\text{Vol}(E) = 0$. ∎

**EXERCISES**

**1.** For $m = 1, 2, 3$, let $\mathcal{G}_m$ be the grid on $[0, 1] \times [0, 1]$ generated by

$$\mathcal{P}_j(\mathcal{G}_m) = \{k/2^m : k = 0, 1, \ldots, 2^m\},$$

where $j = 1, 2$. For each of the following sets, compute $V(E; \mathcal{G}_m)$ and $v(E; \mathcal{G}_m)$.

a) $E = \{(x, y) \in [0, 1] \times [0, 1] : x = 0 \quad \text{or} \quad y = 0\}$.
b) $E = \{(x, y) \in [0, 1] \times [0, 1] : y \leq x\}$.
c) $E = \{(x, y) \in [0, 1] \times [0, 1] : (2x - 1)^2 + (2y - 1)^2 \leq 1\}$.

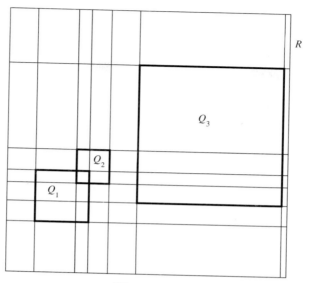

**Figure 12.4**

**2.** a) Prove that every finite subset of $\mathbf{R}^n$ is a Jordan region of volume zero.

b) Show that, even in $\mathbf{R}^2$, part a) is not true if "finite" is replaced by "countable."

c) By an interval in $\mathbf{R}^2$ we mean a set of the form

$$\{(x,c) : a \leq x \leq b\} \quad \text{or} \quad \{(c,y) : a \leq y \leq b\}$$

for some $a, b, c \in \mathbf{R}$. Prove that every interval in $\mathbf{R}^2$ is a Jordan region.

**3** . **This exercise is used in Section** $^e$**12.6.** Let $E \subset \mathbf{R}^n$. The *translation* of $E$ by an $\boldsymbol{x} \in \mathbf{R}^n$ is the set

$$\boldsymbol{x} + E = \{\boldsymbol{y} \in \mathbf{R}^n : \boldsymbol{y} = \boldsymbol{x} + \boldsymbol{z} \text{ for some } \boldsymbol{z} \in E\},$$

and the *dilation* of $E$ by a scalar $\alpha > 0$ is the set

$$\alpha E = \{\boldsymbol{y} \in \mathbf{R}^n : \boldsymbol{y} = \alpha \boldsymbol{z} \text{ for some } \boldsymbol{z} \in E\}.$$

a) Prove that $E$ is a Jordan region if and only if $\boldsymbol{x} + E$ is a Jordan region, in which case $\text{Vol}\,(\boldsymbol{x} + E) = \text{Vol}\,(E)$.

b) Prove that $E$ is a Jordan region if and only if $\alpha E$ is a Jordan region, in which case $\text{Vol}\,(\alpha E) = \alpha^n \, \text{Vol}\,(E)$.

**4** . **This exercise is used in Section** $^e$**12.5.** Suppose $E_1$, $E_2$ are Jordan regions in $\mathbf{R}^n$.

a) Prove that if $E_1 \subseteq E_2$, then $\text{Vol}\,(E_1) \leq \text{Vol}\,(E_2)$.

b) Prove that $E_1 \cap E_2$ and $E_1 \setminus E_2$ are Jordan regions.

c) Prove that if $E_1$, $E_2$ are nonoverlapping, then
$$\text{Vol}\,(E_1 \cup E_2) = \text{Vol}\,(E_1) + \text{Vol}\,(E_2).$$

d) If $E_2 \subseteq E_1$, prove $\text{Vol}\,(E_1 \setminus E_2) = \text{Vol}\,(E_1) - \text{Vol}\,(E_2)$.

e) Prove that $\text{Vol}\,(E_1 \cup E_2) = \text{Vol}\,(E_1) + \text{Vol}\,(E_2) - \text{Vol}\,(E_1 \cap E_2)$.

**5.** Let $E \subset \mathbf{R}^n$ be a Jordan region.

a) Prove $E^o$ and $\overline{E}$ are Jordan regions.

b) Prove $\text{Vol}\,(E^o) = \text{Vol}\,(\overline{E}) = \text{Vol}\,(E)$.

c) Prove that $\text{Vol}\,(E) > 0$ if and only if $E^o \neq \emptyset$.

d) Let $f : [a,b] \to \mathbf{R}$ be continuous on $[a,b]$. Prove that the graph of $y = f(x)$, $x \in [a,b]$, is a Jordan region in $\mathbf{R}^2$.

e) Does part d) hold if "continuous" is replaced by "integrable"? How about "bounded"?

**6.** Prove that every grid is a nonoverlapping collection of Jordan regions.

**$\boxed{7}$. This exercise is used in Sections 12.4 and $^e$12.5.** Suppose $V$ is a bounded, nonempty, open set in $\mathbf{R}^n$ and $\phi : V \to \mathbf{R}^n$ is continuously differentiable on $V$.

a) Let $H$ be a compact subset of $V$. Show that there is a constant $C > 0$ (depending only on $H$, $f$, and $n$) such that for all cubes $Q \subset H$, $\overline{\text{Vol}}\,(\phi(Q)) \leq C\,\text{Vol}\,(Q)$.

b) Show that if $E$ is of volume zero and $E \subset H$, then $\phi(E)$ is of volume zero.

c) If in addition $\phi$ is 1–1 and $\{E_k\}_{k \in \mathbf{N}}$ is a nonoverlapping collection of sets in $\mathbf{R}^n$ with $E_k \subset H$ for all $k \in \mathbf{N}$, prove that $\{\phi(E_k)\}_{k \in \mathbf{N}}$ is a nonoverlapping collection of sets in $\mathbf{R}^n$.

**$^*$8.** Show that if $E \subset \mathbf{R}^n$ is bounded and has only finitely many cluster points, then $E$ is a Jordan region.

**9.** a) Prove that the boundary of an open ball $B_r(\boldsymbol{a})$ is given by

$$\partial B_r(\boldsymbol{a}) = \{\boldsymbol{x} : \|\boldsymbol{x} - \boldsymbol{a}\| = r\}.$$

b) Prove that $B_r(\boldsymbol{a})$ is a Jordan region for all $\boldsymbol{a} \in \mathbf{R}^n$ and all $r \geq 0$.

**$^*$10.** A set $E \subset \mathbf{R}^n$ is said to be of *measure zero* if given $\varepsilon > 0$ there is a countable collection of rectangles $R_1, R_2, \ldots$ which covers $E$ such that $\sum_{k=1}^{\infty} |R_k| < \varepsilon$.

a) Prove that if $E \subset \mathbf{R}^n$ is of volume zero, then $E$ is of measure zero.

b) Prove that if $E \subset \mathbf{R}^n$ is countable, then $E$ is of measure zero.

c) Prove that there is a set $E \subset \mathbf{R}^2$ of measure zero which does not have zero area, in fact, is not even a Jordan region.

## 12.2   RIEMANN INTEGRATION ON JORDAN REGIONS

By analogy with the one-variable case, the integral of a nonnegative function $f$ over a Jordan region $E$ should be the volume of the set $\{(\boldsymbol{x}, t) : \boldsymbol{x} \in E, 0 \leq t \leq f(\boldsymbol{x})\}$. We should be able to approximate this volume by using $(n+1)$-dimensional rectangles whose heights approximate $t = f(\boldsymbol{x})$ and whose bases belong to some grid on $E$

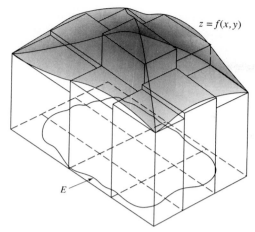

**Figure 12.5**

(see Figure 12.5). This leads us to the following definition (compare with Definition 5.13).

**12.15 DEFINITION.** Let $E$ be a Jordan region in $\mathbf{R}^n$, $f : E \to \mathbf{R}$ be a bounded function, $R$ be an $n$-dimensional rectangle such that $E \subseteq R$, and $\mathcal{G} = \{R_1, \ldots, R_p\}$ be a grid on $R$. Extend $f$ to $\mathbf{R}^n$ by setting $f(\boldsymbol{x}) = 0$ for $\boldsymbol{x} \in \mathbf{R}^n \setminus E$.

  i) The *upper sum* of $f$ on $E$ with respect to $\mathcal{G}$ is

$$U(f, \mathcal{G}) := \sum_{R_j \cap E \neq \emptyset} M_j |R_j|,$$

  where $M_j = \sup_{\boldsymbol{x} \in R_j} f(\boldsymbol{x})$.

 ii) The *lower sum* of $f$ on $E$ with respect to $\mathcal{G}$ is

$$L(f, \mathcal{G}) := \sum_{R_j \cap E \neq \emptyset} m_j |R_j|,$$

  where $m_j = \inf_{\boldsymbol{x} \in R_j} f(\boldsymbol{x})$.

iii) The *upper* and *lower integrals* of $f$ on $E$ are defined by

$$(L) \int_E f(\boldsymbol{x}) \, d\boldsymbol{x} := (L) \int_E f \, dV := \sup_{\mathcal{G}} L(f, \mathcal{G})$$

and

$$(U) \int_E f(\boldsymbol{x}) \, d\boldsymbol{x} := (U) \int_E f \, dV := \inf_{\mathcal{G}} U(f, \mathcal{G}),$$

where the supremum and infimum are taken over all grids $\mathcal{G}$ on $R$.

Modifying the proofs of Remarks 5.7, 5.8, and 5.14, we can prove the following two results.

**12.16 Remark.** *Let $E$ be a nonempty Jordan region in* $\mathbf{R}^n$, $f : E \to \mathbf{R}$ *be bounded, and $E$ be a subset of an $n$-dimensional rectangle $R$. If $\mathcal{G}$ and $\mathcal{H}$ are grids on $R$, then*

$$L(f, \mathcal{G}) \le U(f, \mathcal{H}).$$

**12.17 Remark.** *Let $E$ be a nonempty Jordan region in* $\mathbf{R}^n$ *and $f : E \to \mathbf{R}$ be bounded. Then the upper and lower integrals of $f$ over $E$ exist, do not depend on the choice of $R$, and satisfy*

$$(L) \int_E f(\boldsymbol{x}) \, d\boldsymbol{x} \le (U) \int_E f(\boldsymbol{x}) \, d\boldsymbol{x}.$$

**12.18 DEFINITION.** A real–valued bounded function $f$ defined on a Jordan region $E$ is said to be (*Riemann*) *integrable* on $E$ if for every $\varepsilon > 0$ there is a grid $\mathcal{G}$ such that

$$U(f, \mathcal{G}) - L(f, \mathcal{G}) < \varepsilon.$$

By modifying the proof of Theorem 5.15, we can establish the following result.

**12.19 Remark.** *Let $E$ be a Jordan region in* $\mathbf{R}^n$ *and $f : E \to \mathbf{R}$ be bounded. Then $f$ is integrable on $E$ if and only if*

$$(7) \qquad (L) \int_E f(\boldsymbol{x}) \, d\boldsymbol{x} = (U) \int_E f(\boldsymbol{x}) \, d\boldsymbol{x}.$$

When $f$ is integrable on $E$, we denote the common value in (7) by

$$\int_E f(\boldsymbol{x}) \, d\boldsymbol{x} \quad \text{or} \quad \int_E f \, dV$$

and call it the *integral* of $f$ over $E$. For $n = 2$ (respectively, $n = 3$) we shall frequently denote the integral $\int_E f \, dV$ by $\iint_E f \, dA$ (respectively, by $\iiint_E f \, dV$).

Notice once and for all (since the empty sum is by definition zero) that the integral of any function on $\emptyset$ is zero. To keep from continually treating the empty set as a special case, we shall tacitly assume that $E$ is a nonempty Jordan region.

The following result shows that evaluation of Riemann integrals over Jordan regions reduces to evaluation of Riemann integrals over rectangles.

**12.20 THEOREM.** *Let $E$ be a Jordan region in* $\mathbf{R}^n$, *$R$ be an $n$-dimensional rectangle which contains $E$, and $f : E \to \mathbf{R}$ be integrable on $E$. If*

$$g(\boldsymbol{x}) = \begin{cases} f(\boldsymbol{x}) & \boldsymbol{x} \in E \\ 0 & \boldsymbol{x} \notin E, \end{cases}$$

*then $g$ is integrable on $R$ and*

$$(8) \qquad \int_E f(\boldsymbol{x}) \, d\boldsymbol{x} = \int_R g(\boldsymbol{x}) \, d\boldsymbol{x}.$$

PROOF. By Definition 12.15, the upper and lower sums of $f$ and $g$ are identical; hence, they have the same upper and lower integrals. It follows from Remark 12.19 that they have the same integrals. ∎

The following result, a multidimensional analogue of Theorem 5.10, shows that if $f$ is continuous on a compact Jordan region $H$, then $f$ is integrable on $H$.

**12.21 THEOREM.** *If $E$ is a Jordan region in $\mathbf{R}^n$ and $f : E \to \mathbf{R}$ is uniformly continuous on $E$, then $f$ is integrable on $E$.*

PROOF. Let $\varepsilon > 0$ and $R$ be a rectangle containing $E$. Since $f$ is uniformly continuous on $E$, choose $\delta > 0$ such that

$$(9) \qquad \|\boldsymbol{x} - \boldsymbol{y}\| < \delta \quad \text{and} \quad \boldsymbol{x}, \boldsymbol{y} \in E \quad \text{imply} \quad |f(\boldsymbol{x}) - f(\boldsymbol{y})| < \frac{\varepsilon}{2|R|}.$$

By hypothesis, $f$ is bounded, say by some $M > 1$, and $\partial E$ is of volume zero (see Theorem 12.12). Hence, we can choose a grid $\mathcal{G} = \{R_1, \ldots, R_p\}$ on $R$ so fine that $\boldsymbol{x}, \boldsymbol{y} \in R_j$ implies $\|\boldsymbol{x} - \boldsymbol{y}\| < \delta$ and

$$\sum_{R_j \cap \partial E \neq \emptyset} |R_j| < \frac{\varepsilon}{4M}.$$

Now $M_j - m_j \leq 2M$ for all $j$ and, by (9), $M_j - m_j \leq \varepsilon/2|R|$ for each $j$ which satisfies $R_j \subseteq E$. It follows that

$$U(f, \mathcal{G}) - L(f, \mathcal{G}) = \sum_{R_j \cap E \neq \emptyset} (M_j - m_j)|R_j|$$

$$\leq \sum_{R_j \subseteq E} (M_j - m_j)|R_j| + \sum_{R_j \cap \partial E \neq \emptyset} (M_j - m_j)|R_j|$$

$$\leq \frac{\varepsilon}{2|R|} \sum_{R_j \subseteq E} |R_j| + 2M \sum_{R_j \cap \partial E \neq \emptyset} |R_j|$$

$$< \frac{\varepsilon}{2} + \frac{\varepsilon}{2} = \varepsilon.$$

We conclude by Definition 12.18 that $f$ is integrable on $E$. ∎

The following result shows that the volume of a Jordan region can be computed by integration.

**12.22 THEOREM.** *If $E$ is a Jordan region, then*

$$\text{Vol}(E) = \int_E 1 \, d\boldsymbol{x}.$$

PROOF. Let $f(\boldsymbol{x}) = 1$ for $\boldsymbol{x} \in E$ and $f(\boldsymbol{x}) = 0$ for $\boldsymbol{x} \notin E$. Since $R_j \cap E \neq \emptyset$ implies $R_j \cap \overline{E} \neq \emptyset$, and $M_j(f) = 1$ when $R_j \cap E \neq \emptyset$, it is clear, by the definition of upper sums and outer sums, that

$$U(f, \mathcal{G}) \leq V(E; \mathcal{G})$$

for any grid $\mathcal{G}$ on a rectangle $R \supset E$. Since $E$ is a Jordan set and $f$ is integrable on $E$ (by Theorem 12.21), it follows that

$$\int_E 1 \, d\boldsymbol{x} = \inf_{\mathcal{G}} U(f, \mathcal{G}) \leq \inf_{\mathcal{G}} V(E; \mathcal{G}) = \text{Vol}(E).$$

A similar argument, using lower sums and inner sums, proves the reverse inequality. ∎

As in the one-dimensional case, the integral of a sum of functions over a union of regions can be broken into simpler pieces.

**12.23 THEOREM** [LINEAR PROPERTIES]. *Let $E$ be a Jordan region in $\mathbf{R}^n$, $f, g : E \to \mathbf{R}$, and $\alpha$ be a scalar.*

i) *If $f, g$ are integrable on $E$, then so are $\alpha f$ and $f + g$. In fact,*

$$(10) \qquad \int_E \alpha f(\boldsymbol{x}) \, d\boldsymbol{x} = \alpha \int_E f(\boldsymbol{x}) \, d\boldsymbol{x}$$

*and*

$$(11) \qquad \int_E (f(\boldsymbol{x}) + g(\boldsymbol{x})) \, d\boldsymbol{x} = \int_E f(\boldsymbol{x}) \, d\boldsymbol{x} + \int_E g(\boldsymbol{x}) \, d\boldsymbol{x}.$$

ii) *If $E_1, E_2 \subseteq E$ are nonoverlapping Jordan regions and $f$ is integrable on $E_1$ and $E_2$, then $f$ is integrable on $E_1 \cup E_2$ and*

$$(12) \qquad \int_{E_1 \cup E_2} f(\boldsymbol{x}) \, d\boldsymbol{x} = \int_{E_1} f(\boldsymbol{x}) \, d\boldsymbol{x} + \int_{E_2} f(\boldsymbol{x}) \, d\boldsymbol{x}.$$

PROOF. We suppose for simplicity that $\alpha > 0$. Let $\varepsilon > 0$ and choose a grid $\mathcal{G}$ such that

$$(13) \qquad U(f, \mathcal{G}) - \varepsilon < \int_E f(\boldsymbol{x}) \, d\boldsymbol{x} < L(f, \mathcal{G}) + \varepsilon.$$

Notice that $U(\alpha f, \mathcal{G}) = \alpha U(f, \mathcal{G})$ and $L(\alpha f, \mathcal{G}) = \alpha L(f, \mathcal{G})$. Multiplying (13) by $\alpha$ we obtain

$$U(\alpha f, \mathcal{G}) - \alpha \varepsilon < \alpha \int_E f(\boldsymbol{x}) \, d\boldsymbol{x} < L(\alpha f, \mathcal{G}) + \alpha \varepsilon.$$

In particular,

$$\inf_{\mathcal{G}} U(\alpha f, \mathcal{G}) < \alpha \int_E f(\boldsymbol{x}) \, d\boldsymbol{x} + \alpha \varepsilon$$

and

$$\sup_{\mathcal{G}} L(\alpha f, \mathcal{G}) > \alpha \int_E f(\boldsymbol{x}) \, d\boldsymbol{x} - \alpha \varepsilon.$$

Taking the limit of these inequalities as $\varepsilon \to 0$, we conclude that

$$\inf_{\mathcal{G}} U(\alpha f, \mathcal{G}) \leq \alpha \int_E f(\boldsymbol{x}) \, d\boldsymbol{x} \leq \sup_{\mathcal{G}} L(\alpha f, \mathcal{G}).$$

This proves (10).

To prove (11), choose a grid $\mathcal{G}$ such that

$$U(f, \mathcal{G}) - \varepsilon < \int_E f(\boldsymbol{x}) \, d\boldsymbol{x} < L(f, \mathcal{G}) + \varepsilon$$

and

$$U(g, \mathcal{G}) - \varepsilon < \int_E g(\boldsymbol{x}) \, d\boldsymbol{x} < L(g, \mathcal{G}) + \varepsilon.$$

Adding these inequalities, we have

$$U(f, \mathcal{G}) + U(g, \mathcal{G}) - 2\varepsilon < \int_E f(\boldsymbol{x}) \, d\boldsymbol{x} + \int_E g(\boldsymbol{x}) \, d\boldsymbol{x} < L(f, \mathcal{G}) + L(g, \mathcal{G}) + 2\varepsilon.$$

By definition, $U(f + g, \mathcal{G}) \le U(f, \mathcal{G}) + U(g, \mathcal{G})$ and $L(f + g, \mathcal{G}) \ge L(f, \mathcal{G}) + L(g, \mathcal{G})$. Therefore,

$$U(f + g, \mathcal{G}) - 2\varepsilon < \int_E f(\boldsymbol{x}) \, d\boldsymbol{x} + \int_E g(\boldsymbol{x}) \, d\boldsymbol{x} < L(f + g, \mathcal{G}) + 2\varepsilon,$$

i.e.,

$$\inf_{\mathcal{G}} U(f + g, \mathcal{G}) \le \int_E f(\boldsymbol{x}) \, d\boldsymbol{x} + \int_E g(\boldsymbol{x}) \, d\boldsymbol{x} \le \sup_{\mathcal{G}} L(f + g, \mathcal{G}).$$

This proves (11).

To prove (12), set $\Omega = \partial E_1 \cup \partial E_2 \cup (E_1 \cap E_2)$ and notice that $\mathrm{Vol}\,(\Omega) = 0$. Choose grids $\mathcal{G}_i$, $i = 1, 2, 3$, such that

$$(14) \qquad\qquad U(f, \mathcal{G}_i) - \varepsilon < \int_{E_i} f(\boldsymbol{x}) \, d\boldsymbol{x} < L(f, \mathcal{G}_i) + \varepsilon$$

for $i = 1, 2$, and

$$(15) \qquad\qquad\qquad\qquad V(\Omega; \mathcal{G}_3) < \varepsilon.$$

Let $\mathcal{G} = \{R_1, \dots, R_p\}$ be a grid finer than $\mathcal{G}_1$, $\mathcal{G}_2$, and $\mathcal{G}_3$. Extend $f$ outside $E_1 \cup E_2$ to be zero, and let

$$M = \max\{|M_1|, \dots, |M_p|\}, \text{ where } M_j = \sup_{\boldsymbol{x} \in R_j} f(\boldsymbol{x}) \text{ for } 1 \le j \le p,$$

$\mathcal{I}_1 = \{j : R_j \subseteq E_1\}$, $\mathcal{I}_2 = \{j : R_j \subseteq E_2\}$, and $\mathcal{I}_3 = \{j \notin \mathcal{I}_1 \cup \mathcal{I}_2 : R_j \cap (E_1 \cup E_2) \neq \emptyset\}$. Notice that if $j \in \mathcal{I}_3$ and $R_j \cap (E_1 \cap E_2) = \emptyset$, then $R_j \cap \partial E_1 \neq \emptyset$ or $R_j \cap \partial E_2 \neq \emptyset$. Consequently,

$$(16) \qquad\qquad\qquad \sum_{j \in \mathcal{I}_3} M_j |R_j| \le M\, V(\Omega; \mathcal{G}).$$

Moreover, if $R_j \cap E_i \neq \emptyset$ and $j \notin \mathcal{I}_i$ for some $i = 1$ or $2$, then $M_j \ge 0$ (since $f$ is zero on $(E_1 \cup E_2)^c$). Since $\mathcal{G}$ is finer than $\mathcal{G}_i$, $i = 1, 2, 3$, it follows from (14), (15), and (16) that

$$U(f, \mathcal{G}) = \sum_{j \in \mathcal{I}_1} M_j |R_j| + \sum_{j \in \mathcal{I}_2} M_j |R_j| + \sum_{j \in \mathcal{I}_3} M_j |R_j|$$

$$\le U(f, \mathcal{G}_1) + U(f, \mathcal{G}_2) + M\, V(\Omega; \mathcal{G}_3)$$

$$< \int_{E_1} f(\boldsymbol{x}) \, d\boldsymbol{x} + \int_{E_2} f(\boldsymbol{x}) \, d\boldsymbol{x} + (2 + M)\varepsilon.$$

Consequently,

$$\inf_{\mathcal{G}} U(f, \mathcal{G}) \leq \int_{E_1} f(\boldsymbol{x}) \, d\boldsymbol{x} + \int_{E_2} f(\boldsymbol{x}) \, d\boldsymbol{x}.$$

A similar argument establishes

$$\sup_{\mathcal{G}} L(f, \mathcal{G}) \geq \int_{E_1} f(\boldsymbol{x}) \, d\boldsymbol{x} + \int_{E_2} f(\boldsymbol{x}) \, d\boldsymbol{x}.$$

Thus (12) holds. ∎

The following result shows that the value of an integral remains the same when the integrand is changed on a set of volume zero (compare with Exercise 6 in Section 5.1).

**12.24 THEOREM.** *Let $E$ be a Jordan region in* **R**$^n$, *$f : E \to$ **R** be integrable on $E$, and $g : E \to$ **R** be bounded. If*

$$E_0 := \{\boldsymbol{x} \in E : f(\boldsymbol{x}) \neq g(\boldsymbol{x})\}$$

*is a Jordan region of volume zero, then $g$ is integrable on $E$ and*

$$\int_E g(\boldsymbol{x}) \, d\boldsymbol{x} = \int_E f(\boldsymbol{x}) \, d\boldsymbol{x}.$$

PROOF. We begin by showing that if $A$ is a set of volume zero and $h : A \to$ **R** is bounded, then $h$ is integrable on $A$ and $\int_A h(\boldsymbol{x}) \, d\boldsymbol{x} = 0$. Let $\varepsilon > 0$ and let $M > \sup_{\boldsymbol{x} \in A} |h(\boldsymbol{x})|$. Since $A$ is of volume zero, choose a grid $\mathcal{G}$ such that $V(A; \mathcal{G}) < \varepsilon/M$. Extend $h$ to be zero off $A$ and observe that $U(h, \mathcal{G}) \leq M V(A; \mathcal{G}) < \varepsilon$, i.e., $(U) \int_A h(\boldsymbol{x}) \, d\boldsymbol{x} \leq 0$. A similar argument shows that $(L) \int_A h(\boldsymbol{x}) \, d\boldsymbol{x} \geq 0$. Therefore, $h$ is integrable on $A$ and $\int_A h(\boldsymbol{x}) \, d\boldsymbol{x} = 0$.

Applying this observation to $f$ and $g$, we see that $g$ is integrable on $E_0$ and

$$\int_{E_0} g(\boldsymbol{x}) \, d\boldsymbol{x} = \int_{E_0} f(\boldsymbol{x}) \, d\boldsymbol{x} = 0.$$

Since $g = f$ on $E \setminus E_0$, $g$ is also integrable on $E \setminus E_0$. Hence, by Theorem 12.23ii), $g$ is integrable on $E$ and

$$\int_E g(\boldsymbol{x}) \, d\boldsymbol{x} = \int_{E \setminus E_0} g(\boldsymbol{x}) \, d\boldsymbol{x} + \int_{E_0} g(\boldsymbol{x}) \, d\boldsymbol{x}$$
$$= \int_{E \setminus E_0} f(\boldsymbol{x}) \, d\boldsymbol{x} + \int_{E_0} f(\boldsymbol{x}) \, d\boldsymbol{x} = \int_E f(\boldsymbol{x}) \, d\boldsymbol{x}. \quad \blacksquare$$

This suggests a way to define the integral of $f$ on $E$ when $f$ is not defined on all of $E$. Indeed, if $f$ is defined on $E \setminus E_0$, where $E$ is a Jordan region and $E_0$ is of volume zero, and the function

$$g(\boldsymbol{x}) := \begin{cases} f(\boldsymbol{x}) & \boldsymbol{x} \in E \setminus E_0 \\ 0 & \boldsymbol{x} \in E_0 \end{cases}$$

is integrable on $E$, then we shall define

$$\int_E f(\boldsymbol{x})\, d\boldsymbol{x} := \int_E g(\boldsymbol{x})\, d\boldsymbol{x}.$$

For example,

$$\int_0^2 \frac{x^2 - 1}{x - 1}\, dx = \int_0^2 (x + 1)\, dx = 4.$$

Henceforth, the phrase "$f : E \to \mathbf{R}$ is integrable" includes the possibility that $f$ may not be defined on a subset of $E$ of volume zero.

The following result is a multidimensional analogue of Theorems 5.21 and 5.22.

**12.25 THEOREM** [COMPARISON THEOREM FOR MULTIPLE INTEGRALS]. *Let $E$ be a Jordan region in $\mathbf{R}^n$ and $f, g : E \to \mathbf{R}$ be integrable on $E$.*

i) *If $f(\boldsymbol{x}) \le g(\boldsymbol{x})$ for $\boldsymbol{x} \in E$, then*

$$\int_E f(\boldsymbol{x})\, d\boldsymbol{x} \le \int_E g(\boldsymbol{x})\, d\boldsymbol{x}.$$

ii) *If $m, M$ are scalars which satisfy $m \le f(\boldsymbol{x}) \le M$ for $\boldsymbol{x} \in E$, then*

$$m \operatorname{Vol}(E) \le \int_E f(\boldsymbol{x})\, d\boldsymbol{x} \le M \operatorname{Vol}(E).$$

iii) *The function $|f|$ is integrable on $E$ and*

(17)
$$\left| \int_E f(\boldsymbol{x})\, d\boldsymbol{x} \right| \le \int_E |f(\boldsymbol{x})|\, d\boldsymbol{x}.$$

PROOF. i) If $f \le g$ on $E$, then $L(f, \mathcal{G}) \le L(g, \mathcal{G})$ for any grid $\mathcal{G}$. Taking the supremum of this inequality over all grids $\mathcal{G}$ verifies part i).

ii) By Theorem 12.22, (10), and part i),

$$m \operatorname{Vol}(E) = \int_E m\, d\boldsymbol{x} \le \int_E f(\boldsymbol{x})\, d\boldsymbol{x} \le \int_E M\, d\boldsymbol{x} = M \operatorname{Vol}(E).$$

iii) Let $\varepsilon > 0$ and choose by Definition 12.18 a grid $\mathcal{G} = \{R_1, \ldots, R_p\}$ such that

(18)
$$U(f, \mathcal{G}) - L(f, \mathcal{G}) < \varepsilon.$$

By repeating the argument which verified (10) in Section 5.2, we have

$$\sup_{\boldsymbol{x} \in R_j} |f(\boldsymbol{x})| - \inf_{\boldsymbol{x} \in R_j} |f(\boldsymbol{x})| \le \sup_{\boldsymbol{x} \in R_j} f(\boldsymbol{x}) - \inf_{\boldsymbol{x} \in R_j} f(\boldsymbol{x}).$$

Hence, it follows from (18) that

$$U(|f|, \mathcal{G}) - L(|f|, \mathcal{G}) \le U(f, \mathcal{G}) - L(f, \mathcal{G}) < \varepsilon.$$

Thus $|f|$ is integrable on $E$. Since $-|f| \le f \le |f|$, we conclude by part i) that

$$-\int_E |f(\boldsymbol{x})|\, d\boldsymbol{x} \le \int_E f(\boldsymbol{x})\, d\boldsymbol{x} \le \int_E |f(\boldsymbol{x})|\, d\boldsymbol{x}. \quad \blacksquare$$

**12.26 THEOREM** [MEAN VALUE THEOREM FOR MULTIPLE INTEGRALS]. *Let $E$ be a Jordan region in $\mathbf{R}^n$ and $f, g : E \to \mathbf{R}$ be integrable on $E$ with $g(\boldsymbol{x}) \geq 0$ for all $\boldsymbol{x} \in E$.*

i) *There is a number $c$ satisfying*

$$(19) \qquad\qquad \inf_{\boldsymbol{x} \in E} f(\boldsymbol{x}) \leq c \leq \sup_{\boldsymbol{x} \in E} f(\boldsymbol{x})$$

*such that*

$$(20) \qquad\qquad c \int_E g(\boldsymbol{x}) \, d\boldsymbol{x} = \int_E f(\boldsymbol{x}) g(\boldsymbol{x}) \, d\boldsymbol{x}.$$

ii) *There is a number $c$ satisfying (19) such that*

$$c \, \mathrm{Vol}\,(E) = \int_E f(\boldsymbol{x}) \, d\boldsymbol{x}.$$

PROOF. i) By hypothesis, the product $fg$ is integrable on $E$ (see Exercise 7 below). Let $m = \inf_{\boldsymbol{x} \in E} f(\boldsymbol{x})$ and $M = \sup_{\boldsymbol{x} \in E} f(\boldsymbol{x})$. Since $g \geq 0$, Theorem 12.25 implies

$$(21) \qquad\qquad m \int_E g(\boldsymbol{x}) \, d\boldsymbol{x} \leq \int_E f(\boldsymbol{x}) g(\boldsymbol{x}) \, d\boldsymbol{x} \leq M \int_E g(\boldsymbol{x}) \, d\boldsymbol{x}.$$

If $\int_E g(\boldsymbol{x}) \, d\boldsymbol{x} = 0$, then (21) implies $\int_E f(\boldsymbol{x}) g(\boldsymbol{x}) \, d\boldsymbol{x} = 0$ and (20) holds for any $c$. If $\int_E g(\boldsymbol{x}) \, d\boldsymbol{x} \neq 0$, then (20) holds for

$$c = \frac{\int_E f(\boldsymbol{x}) g(\boldsymbol{x}) \, d\boldsymbol{x}}{\int_E g(\boldsymbol{x}) \, d\boldsymbol{x}}.$$

ii) Apply part i) to $g(\boldsymbol{x}) = 1$. ∎

We close this section with some optional material which generalizes a concept introduced in Section 9.5.

*\*12.27 DEFINITION.* A set $E \subset \mathbf{R}^n$ is said to be of *measure zero* if for every $\varepsilon > 0$ there is a countable collection of rectangles $\{R_j\}_{j \in \mathbf{N}}$ such that

$$E \subset \bigcup_{j=1}^{\infty} R_j \quad \text{and} \quad \sum_{j=1}^{\infty} |R_j| < \varepsilon.$$

*\*12.28 Remark.* If $E_1, E_2, \dots$ is a sequence of subsets of $\mathbf{R}^n$ and each $E_k$ is of measure zero, then

$$E = \bigcup_{k=1}^{\infty} E_k$$

*is also of measure zero.*

PROOF. Let $\varepsilon > 0$. For each $k \in \mathbf{N}$, choose a collection of rectangles $\{R_j^{(k)}\}_{j\in\mathbf{N}}$ which covers $E_k$ such that

$$\sum_{j=1}^{\infty} |R_j^{(k)}| < \frac{\varepsilon}{2^k}.$$

Clearly, the collection $\{R_j^{(k)}\}_{j,k\in\mathbf{N}}$ is countable, covers $E$, and

$$\sum_{k=1}^{\infty} \sum_{j=1}^{\infty} |R_j^{(k)}| \leq \sum_{k=1}^{\infty} \frac{\varepsilon}{2^k} = \varepsilon.$$

Consequently, $E$ is of measure zero. ∎

Every singleton $E = \{\boldsymbol{a}\}$ in $\mathbf{R}^n$ is of measure zero. In fact, by comparing Definition 12.27 with Theorem 12.14, it is clear that every set of volume zero is a set of measure zero. The converse of this statement is false. Indeed, for each $a \in \mathbf{R}$ the set $\{(a, y) : y \in [0, 1]\}$ is of volume zero, hence, is of measure zero. Thus, by Remark 12.28, $E = \mathbf{Q} \times [0, 1]$ is a set of measure zero. On the other hand, it is clear that $\underline{\mathrm{Vol}}(E) = 0 < 1 \leq \overline{\mathrm{Vol}}(E)$, so $E$ is not a set of volume zero; in fact, $E$ is not even a Jordan region.

An analogue of Lebesgue's Theorem holds for multiple integrals.

*12.29 Remark. *A bounded function $f$ is Riemann integrable on a Jordan region $E$ if and only if the set of points of discontinuity of $f$ on $E$ is of measure zero.*

The proof is similar to the proof of Theorem 9.50 (see Spivak [12], p. 53).

## EXERCISES

**1.** Using Exercise 1 in Section 1.2, compute the upper and lower sums of $f(x, y) = xy$ generated by the grid $\mathcal{G}_m$ (on $[0, 1] \times [0, 1]$) which satisfies

$$\mathcal{P}_j(\mathcal{G}_m) = \{k/2^m : k = 0, 1, \ldots, 2^m\}$$

for $j = 1, 2$. Prove

$$\lim_{m \to \infty} U(f, \mathcal{G}_m) - L(f, \mathcal{G}_m) = 0.$$

**2.** Let $E$ be a Jordan region in $\mathbf{R}^n$ and suppose $f, g$ are integrable on $E$ with

$$\int_E f(\boldsymbol{x}) \, d\boldsymbol{x} = 5 \quad \text{and} \quad \int_E g(\boldsymbol{x}) \, d\boldsymbol{x} = 2.$$

a) Find

$$\int_{E^o} (2f(\boldsymbol{x}) - 3g(\boldsymbol{x})) \, d\boldsymbol{x}, \quad \int_E (2f(\boldsymbol{x}) - 3g(\boldsymbol{x})) \, d\boldsymbol{x}, \quad \text{and} \quad \int_{\overline{E}} (2f(\boldsymbol{x}) - 3g(\boldsymbol{x})) \, d\boldsymbol{x}.$$

b) If $h$ is integrable on $E$ and $g(\boldsymbol{x}) \leq h(\boldsymbol{x}) \leq f(\boldsymbol{x})$, prove that $\int_E h(\boldsymbol{x})\,d\boldsymbol{x} \neq \pi/2$.

c) Suppose $n = 2$ and $E \subseteq [0,1] \times [0,1]$. If $g(x,y) \leq f(x,y)$ for all $(x,y) \in E$, prove there is a $0 \leq t_0 \leq 1$ such that

$$\int_E x^2(f(x,y) - g(x,y))\,dA = 3t_0.$$

**3.** Let $Q := [0,1] \times [0,1]$, $A := \{(x,y) \in Q : y \leq x\}$, $B := \{(x,y) \in Q : y \geq x\}$, and suppose $f$ is integrable on $Q$ (hence, on $A$) with $\iint_A f\,dA = 4$.

a) If $\iint_Q f\,dA = 3$, find $\iint_B f\,dA$.

b) Compute the value of $\iint_B (2f + 5)\,dA$.

c) If $f \geq 0$ on $A$ and $E := \{(x,y) \in Q : y \leq \sqrt[3]{x^4}\}$, prove that

$$\iint_E (2 + f)\,dA \leq \frac{34}{7}.$$

**4. a)** Let $E_1 \subset E$ be Jordan regions in $\mathbf{R}^n$. Prove that if $f : E \to \mathbf{R}$ is integrable on $E$, then $f$ is integrable on $E_1$.

b) Prove that if $f : \mathbf{R}^n \to \mathbf{R}$ is continuous on $\mathbf{R}^n$, then $f$ is integrable on any Jordan region in $\mathbf{R}^n$.

**5 . This exercise is used in Sections 12.4, 13.5, and 13.6.**
Let $E$ be an open Jordan region in $\mathbf{R}^n$ and $\boldsymbol{x}_0 \in E$. If $f : E \to \mathbf{R}$ is integrable on $E$ and continuous at $\boldsymbol{x}_0$, prove

$$\lim_{r \to 0+} \frac{1}{\text{Vol}\,(B_r(\boldsymbol{x}_0))} \int_{B_r(\boldsymbol{x}_0)} f(\boldsymbol{x})\,d\boldsymbol{x} = f(\boldsymbol{x}_0).$$

**6. a)** Suppose $E$ is a Jordan region in $\mathbf{R}^n$ and $f, f_k : E \to \mathbf{R}$ are integrable on $E$. If $f_k \to f$ uniformly on $E$ as $k \to \infty$, prove

$$\lim_{k \to \infty} \int_E f_k(\boldsymbol{x})\,d\boldsymbol{x} = \int_E f(\boldsymbol{x})\,d\boldsymbol{x}.$$

b) Prove that

$$\lim_{k \to \infty} \iint_E \cos(x/k)e^{y/k}\,dA$$

exists, and find its value for any Jordan region $E$ in $\mathbf{R}^2$.

**7.** Let $E$ be a Jordan region in $\mathbf{R}^n$ and $f, g : E \to \mathbf{R}$ be integrable on $E$.

a) Modifying the proof of Corollary 5.23, prove that $fg$ is integrable on $E$.

b) Prove that $f \vee g$ and $f \wedge g$ are integrable on $E$ (see Exercise 9 in Section 3.1).

8. Let $H$ be a compact, connected, nonempty Jordan region and $f : H \to \mathbf{R}$ be continuous on $H$.

   a) If $g : H \to \mathbf{R}$ is integrable and nonnegative on $H$, prove there is an $\boldsymbol{x}_0 \in H$ such that

   $$f(\boldsymbol{x}_0) \int_H g(\boldsymbol{x})\, d\boldsymbol{x} = \int_H f(\boldsymbol{x}) g(\boldsymbol{x})\, d\boldsymbol{x}.$$

   b) If $H^o \neq \emptyset$, prove there is a point $\boldsymbol{x}_0 \in H^o$ such that

   $$\int_H f(\boldsymbol{x})\, d\boldsymbol{x} = f(\boldsymbol{x}_0)\, \mathrm{Vol}\,(H).$$

9. Prove the following special case of Remark 12.29. If $E$ is a compact nonempty Jordan region in $\mathbf{R}^n$, $E_0$ is a nonempty Jordan region of volume zero, and $f : E \to \mathbf{R}$ is a bounded function which is continuous on $E \setminus E_0$, then $f$ is integrable on $E$.

10. Suppose $V$ is open in $\mathbf{R}^n$ and $f : V \to \mathbf{R}$ is continuous. Prove that if

   $$\int_E f(\boldsymbol{x})\, d\boldsymbol{x} = 0$$

   for all nonempty Jordan regions $E \subset V$, then $f = 0$ on $V$.


## 12.3   ITERATED INTEGRALS

If $f(x_1, \ldots, x_k, \ldots, x_j, \ldots, x_n)$ is defined for $x_k \in [c, d]$ and $x_j \in [a, b]$, $j \neq k$, then we shall call

$$\int_c^d \int_a^b f(x_1, \ldots, x_n)\, dx_j\, dx_k := \int_c^d \left( \int_a^b f(x_1, \ldots, x_n)\, dx_j \right) dx_k$$

an *iterated integral*, when the integrals on the right side exist. In a similar way, we define higher-order iterated integrals.

In the preceding section we defined the Riemann integral of a multivariable function but developed no practical way to evaluate it. In this section we show that, for a large collection of Jordan regions $E$, an integral over $E$ can be evaluated using iterated integrals.

For simplicity, we begin with the two-dimensional case. Recall that for each $\phi : [a, b] \to \mathbf{R}$, $(U) \int_a^b \phi(x)\, dx$ represents the upper Riemann integral of $\phi$, and $(L) \int_a^b \phi(x)\, dx$ represents the lower Riemann integral of $\phi$.

**12.30 Lemma.** *Let $R = [a, b] \times [c, d]$ be a two-dimensional rectangle and $f : R \to \mathbf{R}$ be bounded. If $f(x, \cdot)$ is integrable on $[c, d]$ for each $x \in [a, b]$, then*

$$(22) \qquad (L) \iint_R f\, dA \leq (L) \int_a^b \left( \int_c^d f(x, y)\, dy \right) dx$$

$$\leq (U) \int_a^b \left( \int_c^d f(x, y)\, dy \right) dx \leq (U) \iint_R f\, dA.$$

PROOF. Let $R_{ij} = [x_{i-1}, x_i] \times [y_{j-1}, y_j]$, where $\{x_0, \ldots, x_k\}$ is a partition of $[a, b]$ and $\{y_0, \ldots, y_\ell\}$ is a partition of $[c, d]$. Then $\mathcal{G} = \{R_{ij} : i = 1, 2, \ldots, k, j = 1, 2 \ldots, \ell\}$ is a grid on $R$.

Let $\varepsilon > 0$, choose $\mathcal{G}$ so that

$$(23) \qquad\qquad U(f, \mathcal{G}) - \varepsilon < (U) \iint_R f \, dA,$$

and set

$$(24) \qquad\qquad M_{ij} = \sup_{(x,y) \in R_{ij}} f(x, y).$$

Since $(U) \int_a^b \phi(x)\, dx = \sum_{i=1}^k (U) \int_{x_{i-1}}^{x_i} \phi(x)\, dx$ and

$$(U) \int_a^b (\phi(x) + \psi(x))\, dx \leq (U) \int_a^b \phi(x)\, dx + (U) \int_a^b \psi(x)\, dx$$

for any bounded functions $\phi$ and $\psi$ defined on $[a, b]$ (see Exercise 7 in Section 5.1), we can write

$$\begin{aligned}
(U) \int_a^b \left( \int_c^d f(x,y) dy \right) dx &= \sum_{i=1}^k (U) \int_{x_{i-1}}^{x_i} \left( \sum_{j=1}^\ell \int_{y_{j-1}}^{y_j} f(x,y) dy \right) dx \\
&\leq \sum_{i=1}^k \sum_{j=1}^\ell (U) \int_{x_{i-1}}^{x_i} \left( \int_{y_{j-1}}^{y_j} f(x,y) dy \right) dx \\
&\leq \sum_{i=1}^k \sum_{j=1}^\ell M_{ij}(x_i - x_{i-1})(y_j - y_{j-1}) = U(f, \mathcal{G}).
\end{aligned}$$

It follows from (23) that

$$(U) \int_a^b \left( \int_c^d f(x,y) dy \right) dx < (U) \iint_R f \, dA + \varepsilon.$$

Taking the limit of this inequality as $\varepsilon \to 0$, we obtain

$$(U) \int_a^b \left( \int_c^d f(x,y) dy \right) dx \leq (U) \iint_R f \, dA.$$

Similarly,

$$(L) \int_a^b \left( \int_c^d f(x,y) dy \right) dx \geq (L) \iint_R f \, dA. \quad \blacksquare$$

We are now prepared to show that, under reasonable conditions, a double integral over a rectangle reduces to an iterated integral.

**12.31 THEOREM** [FUBINI'S THEOREM]. *Let $R = [a, b] \times [c, d]$ be a two-dimensional rectangle and $f : R \to \mathbf{R}$. Suppose that $f(x, \cdot)$ is integrable on $[c, d]$ for each $x \in [a, b]$, $f(\cdot, y)$ is integrable on $[a, b]$ for each $y \in [c, d]$, and $f$ is integrable on $R$ (as a function of two variables). Then*

$$(25) \qquad \iint_R f \, dA = \int_a^b \int_c^d f(x, y) \, dy \, dx = \int_c^d \int_a^b f(x, y) \, dx \, dy.$$

PROOF. For each $x \in [a, b]$, set $g(x) = \int_c^d f(x, y) \, dy$. Since $f$ is integrable on $R$, Lemma 12.30 implies

$$\iint_R f \, dA = (U) \int_a^b g(x) \, dx = (L) \int_a^b g(x) \, dx.$$

Hence, $g$ is integrable on $[a, b]$ and the first identity in (25) holds. Reversing the roles of $x$ and $y$, we obtain

$$\iint_R f \, dA = \int_c^d \int_a^b f(x, y) \, dx \, dy.$$

Hence, the second identity in (25) holds. ∎

Notice that the hypotheses of Fubini's Theorem hold if $f$ is continuous on the rectangle $[a, b] \times [c, d]$.

The second identity in Fubini's Theorem is as important as the first. It tells us that, under certain conditions, the order of integration in an iterated integral can be reversed. Frequently, one of these iterated integrals is easier to evaluate than the other.

**12.32 Example.** Find

$$\int_0^1 \int_0^1 y^3 e^{xy^2} \, dy \, dx.$$

SOLUTION. This iterated integral looks tough to integrate. However, if we change the order of integration, using Fubini's Theorem, we obtain

$$\int_0^1 \int_0^1 y^3 e^{xy^2} \, dx \, dy = \int_0^1 y(e^{y^2} - 1) \, dy = \frac{e - 2}{2}. \quad \blacksquare$$

The following three remarks show that the hypotheses of Fubini's Theorem cannot be relaxed.

**12.33 Remark.** *There exists a function $f : \mathbf{R}^2 \to \mathbf{R}$ such that $f(x, \cdot)$ and $f(\cdot, y)$ are both integrable on $[0, 1]$, but the iterated integrals are not equal.*

PROOF. Set

$$f(x, y) = \begin{cases} 2^{2n} & (x, y) \in [2^{-n}, 2^{-n+1}) \times [2^{-n}, 2^{-n+1}), \ n \in \mathbf{N}, \\ -2^{2n+1} & (x, y) \in [2^{-n-1}, 2^{-n}) \times [2^{-n}, 2^{-n+1}), \ n \in \mathbf{N}, \\ 0 & \text{otherwise.} \end{cases}$$

Notice that for each fixed $y_0 \in [0, 1)$, $f(x, y_0)$ takes on only two nonzero values and is integrable on $[0, 1)$ in $x$. For example, if $y_0 \in [2^{-n}, 2^{-n+1})$, then $f(x, y_0) = 2^{2n}$ for $x \in [2^{-n}, 2^{-n+1})$, and $f(x, y_0) = -2^{2n+1}$ for $x \in [2^{-n-1}, 2^{-n})$; hence, $f(x, y_0)$ is bounded on $[0, 1)$, and

$$(26) \qquad \int_0^1 f(x, y_0) \, dx = \int_{2^{-n}}^{2^{-n+1}} 2^{2n} \, dx - \int_{2^{-n-1}}^{2^{-n}} 2^{2n+1} \, dx = 2^n - 2^n = 0.$$

The same is true for $f(x_0, y)$ when $x_0 \in [0, 1/2)$, but when $x_0 \in [1/2, 1)$, $f(x_0, y)$ takes on only one nonzero value, namely, $f(x_0, y) = 4$ when $y \in [1/2, 1)$, and equals zero otherwise. It follows that

$$\int_0^1 \int_0^1 f(x, y) \, dy \, dx = \int_{1/2}^1 \int_{1/2}^1 4 \, dy \, dx = 1.$$

On the other hand, by (26) we have

$$\int_0^1 \int_0^1 f(x, y) \, dx \, dy = 0.$$

Thus the iterated integrals of $f$ are not equal. (Of course, by Fubini's Theorem, $f$ itself cannot be Riemann integrable on $[0, 1) \times [0, 1)$. And, in fact, $f$ is not even bounded.) ∎

Thus the right–most equality of (25) need not hold when the double integral of $f$ does not exist.

The next two results are starred because they use Lebesgue's characterization of Riemann integrability (see Theorem 9.50 and Remark 12.29).

**\*12.34 Remark.** *There exists a bounded function* $f : \mathbf{R}^2 \to \mathbf{R}$ *such that* $f(x, \cdot)$ *and* $f(\cdot, y)$ *are both integrable on* $[0, 1]$, *but* $f$ *is not integrable on* $[0, 1] \times [0, 1]$.

PROOF. Set

$$f(x, y) = \begin{cases} 1 & (x, y) = \left( \dfrac{p}{2^n}, \dfrac{q}{2^n} \right), \quad 0 < p, q < 2^n, n \in \mathbf{N}, \\ 0 & \text{otherwise.} \end{cases}$$

Notice that if $x_0 = p/2^n$, then $f(x_0, y) = 1$ only when $y = q/2^n$ for $q = 1, 2, \ldots, 2^n - 1$. Hence, for each fixed $x_0 \in [0, 1]$, $f(x_0, y) = 0$ except for finitely many $y$'s. It follows from Exercise 6 in Section 5.1 that

$$\int_0^1 f(x, y) \, dy = 0$$

for all $x \in [0, 1]$. A similar statement holds for the $dx$ integral. Consequently,

$$\int_0^1 \int_0^1 f(x, y) \, dy \, dx = \int_0^1 \int_0^1 f(x, y) \, dx \, dy = 0.$$

To see that the double integral of $f$ does not exist, let $\varepsilon > 0$. Let $E :=$ $\{(p/2^n, q/2^n) : 0 < p, q < 2^n\}$, and choose $\mathbf{z}_0 \in [0,1] \times [0,1] \setminus E$ and $\mathbf{z}_1 \in E$ such that $\|\mathbf{z}_1 - \mathbf{z}_0\| < \varepsilon$ (just repeat the proof of Theorem 1.24 with $2^n$ in place of $n$). Since $\|\mathbf{z}_0 - \mathbf{z}_1\| < \varepsilon$ but $f(\mathbf{z}_0) = 0 \neq 1 = f(\mathbf{z}_1)$, it follows that $f$ is nowhere continuous on $[0,1] \times [0,1]$. We conclude by Lebesgue's Theorem that $f$ is not integrable on $[0,1] \times [0,1]$. ∎

Thus we cannot be sure a function of several variables is integrable just because its iterated integrals exist and are equal. (See also Exercises 5 and 9 below.)

**\*12.35 Remark.** *There exists a function $f : \mathbf{R}^2 \to \mathbf{R}$ such that $f$ integrable on $[0,1] \times [0,1]$, $f(\cdot, y)$ is integrable on $[0,1]$ for all $y \in [0,1]$, but $f(x, \cdot)$ is not integrable on $[0,1]$ for infinitely many $x \in [0,1]$.*

Proof. Let

$$f(x,y) = \begin{cases} 0 & \text{when } x = 0 \text{ or when } x \text{ or } y \text{ is irrational} \\ 1/q & \text{when } x, y \in \mathbf{Q} \text{ and } x = p/q \text{ is in reduced form.} \end{cases}$$

By the argument of Example 3.33, the function $f$ is continuous and zero on the set $([0,1] \setminus \mathbf{Q}) \times [0,1]$. Hence, by Lebesgue's Theorem, $f$ is integrable on the square $R = [0,1] \times [0,1]$. By computing its lower sums, it follows that $\iint_R f \, dA = 0$.

Similarly, for each $y \in [0,1]$, $f(\cdot, y)$ is integrable on $[0,1]$ with $\int_0^1 f(x,y) \, dx = 0$. Thus

$$\int_0^1 \left( \int_0^1 f(x,y) \, dx \right) dy = \iint_R f \, dA = 0.$$

On the other hand, since for each nonzero $x \in \mathbf{Q}$ the function $f(x, \cdot)$ is nowhere continuous, it cannot be integrable on $[0,1]$. Therefore, the other iterated integral in Fubini's Theorem does not exist. ∎

Fubini's Theorem shows us how to evaluate a double integral over a rectangle by means of iterated integrals. The following result shows that the integral of a continuous function over a rectangle in $\mathbf{R}^n$ can be evaluated using $n$ partial integrals.

**12.36 Lemma.** *Let $R = [a_1, b_1] \times \cdots \times [a_n, b_n]$ be an $n$-dimensional rectangle and $f : R \to \mathbf{R}$ be integrable on $R$. If, for each $\mathbf{x} := (x_1, \ldots, x_{n-1}) \in R_n := [a_1, b_1] \times \cdots \times [a_{n-1}, b_{n-1}]$, the function $f(\mathbf{x}, \cdot)$ is integrable on $[a_n, b_n]$, then*

$$\int_{a_n}^{b_n} f(\mathbf{x}, t) \, dt$$

*is integrable on $R_n$, and*

(27) $$\int_R f(\mathbf{x}, t) \, d(\mathbf{x}, t) = \int_{R_n} \int_{a_n}^{b_n} f(\mathbf{x}, t) \, dt \, d\mathbf{x}.$$

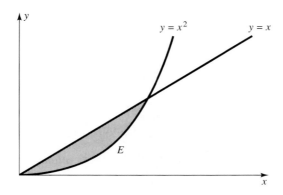

**Figure 12.6**

PROOF. By repeating the argument of Lemma 12.30, we have

$$(L) \int_R f(\boldsymbol{x}, t) \, d(\boldsymbol{x}, t) \le (L) \int_{R_n} \int_{a_n}^{b_n} f(\boldsymbol{x}, t) \, dt \, d\boldsymbol{x}$$

$$\le (U) \int_{R_n} \int_{a_n}^{b_n} f(\boldsymbol{x}, t) \, dt \, d\boldsymbol{x}$$

$$\le (U) \int_R f(\boldsymbol{x}, t) \, d(\boldsymbol{x}, t)$$

for any bounded $f$. Since $f$ is integrable on $R$, it follows that (27) holds. ∎

Using this result in conjunction with Theorem 12.20, we can evaluate integrals over a large collection of nonrectangular Jordan regions. To this end, we shall call a nonempty set $E \subset \mathbf{R}^n$ a *projectable region* if there is a compact Jordan region $H \subset \mathbf{R}^{n-1}$, an index $j \in \{1, \dots, n\}$, and continuous functions $\phi, \psi : H \to \mathbf{R}$ such that

$$E = \{(x_1, \dots, x_n) \in \mathbf{R}^n : (x_1, \dots, \widehat{x}_j, \dots, x_n) \in H$$
$$\text{and} \quad \phi(x_1, \dots, \widehat{x}_j, \dots, x_n) \le x_j \le \psi(x_1, \dots, \widehat{x}_j, \dots, x_n)\}.$$

(The notation $\widehat{x}_j$ means the variable $x_j$ is missing; hence, $(x_1, \dots, \widehat{x}_j, \dots, x_n)$ is a point in $\mathbf{R}^{n-1}$.) In this case, we say that $E$ is *generated* by $j$, $H$, $\phi$, and $\psi$.

We are more specific for regions in $\mathbf{R}^2$ and $\mathbf{R}^3$. A set $E \subset \mathbf{R}^2$ is called a *region of type I* if $E = \{(x, y) : x \in [a, b], \; \phi(x) \le y \le \psi(x)\}$ and a *region of type II* if $E = \{(x, y) : y \in [a, b], \; \phi(y) \le x \le \psi(y)\}$, where $\phi, \psi : [a, b] \to \mathbf{R}$ are continuous functions. Similarly, a set $E \subset \mathbf{R}^3$ is called a *region of type I* if $E = \{(x, y, z) : (x, y) \in H, \; \phi(x, y) \le z \le \psi(x, y)\}$, a *region of type II* if $E = \{(x, y, z) : (x, z) \in H, \; \phi(x, z) \le y \le \psi(x, z)\}$, and a *region of type III* if $E = \{(x, y, z) : (y, z) \in H, \; \phi(y, z) \le x \le \psi(y, z)\}$, where $\phi, \psi : H \to \mathbf{R}$ are continuous functions and $H$ is a compact Jordan region in $\mathbf{R}^2$.

**12.37 Example.** Prove that the set $E$ in $\mathbf{R}^2$ bounded by $y = x$ and $y = x^2$ is a region of types I and II.

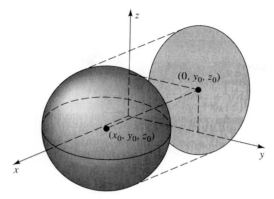

**Figure 12.7**

PROOF. The set $E$ can be described by

$$\{(x,y) : x^2 \leq y \leq x, \ x \in [0,1]\} \quad \text{or} \quad \{(x,y) : y \leq x \leq \sqrt{y}, \ y \in [0,1]\}$$

(see Figure 12.6). ∎

**12.38 Example.** Prove that the set $E$ of points $(x,y,z)$ which satisfy $4x^2+y^2+z^2 \leq 1$ is a region of types I, II, and III.

PROOF. The set $E$, an ellipsoid, can be described by

$$E = \{(x,y,z) : -\sqrt{1 - 4x^2 - y^2} \leq z \leq \sqrt{1 - 4x^2 - y^2}, \ (x,y) \in H\},$$

where $H = \{(x,y) : 4x^2 + y^2 \leq 1\}$. A similar argument shows $E$ is of types II and III. ∎

Before we show how to evaluate multiple integrals over projectable regions, we introduce additional terminology. For each $k = 1, \ldots, n$ the set

$$\Pi_k = \{\mathbf{x} \in \mathbf{R}^n : x_k = 0\}$$

will be called a *coordinate hyperplane*. Given a set $E \subseteq \mathbf{R}^n$, the *projection* of $E$ onto the coordinate hyperplane $\Pi_k$ is the set $E_k$ of points $(x_1, \ldots, x_{k-1}, 0, x_{k+1}, \ldots, x_n)$ such that $(x_1, \ldots, x_k, \ldots, x_n) \in E$ for some $x_k \in \mathbf{R}$. For example, in $\mathbf{R}^3$ the coordinate hyperplane $\Pi_1$ corresponds to the $yz$ plane, and the projection of the three-dimensional ball $B_r(x_0, y_0, z_0)$ onto $\Pi_1$ is essentially the two-dimensional ball $B_r(y_0, z_0)$ (see Figure 12.7).

The following result shows that multiple integrals over some projectable regions can be evaluated using iterated integrals.

**12.39 THEOREM.** *Let $E$ be a projectable region in $\mathbf{R}^n$ generated by $j$, $H$, $\phi$, and $\psi$. Then $E$ is a Jordan region in $\mathbf{R}^n$. Moreover, if $f : E \to \mathbf{R}$ is continuous on $E$, then*

$$(28) \quad \int_E f(\mathbf{x}) \, d\mathbf{x} = \int_H \left( \int_{\phi(x_1, \ldots \widehat{x}_j, \ldots x_n)}^{\psi(x_1, \ldots, \widehat{x}_j, \ldots x_n)} f(x_1, \ldots, x_n) \, dx_j \right) d(x_1, \ldots, \widehat{x}_j, \ldots, x_n).$$

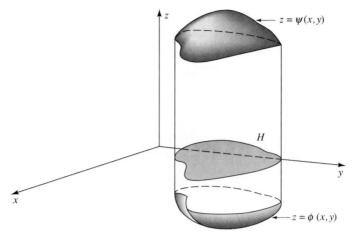

**Figure 12.8**

PROOF. By symmetry, we may suppose $j = n$. Thus

$$E = \{(\boldsymbol{x}, t) : \boldsymbol{x} = (x_1, \ldots, x_{n-1}) \in H \quad \text{and} \quad \phi(\boldsymbol{x}) \le t \le \psi(\boldsymbol{x})\}.$$

To show $E$ is a Jordan region, we must show that the volume of $\partial E$ is zero. Now $\partial E$ is made up of "lower-dimensional pieces," a bottom $B = \{(\boldsymbol{x}, t) : \boldsymbol{x} \in H \text{ and } t = \phi(\boldsymbol{x})\}$, a top $T = \{(\boldsymbol{x}, t) : \boldsymbol{x} \in H \text{ and } t = \psi(\boldsymbol{x})\}$, and a side $S = \{(\boldsymbol{x}, t) : \boldsymbol{x} \in \partial H \text{ and } \phi(\boldsymbol{x}) \le t \le \psi(\boldsymbol{x})\}$. (Figure 12.8 illustrates the situation for the case $n = 3$.) Hence, we must show that $B$, $T$, and $S$ are of volume zero.

To estimate the volume of $B$, notice that since $H$ is compact, $\phi$ is uniformly continuous on $H$. Thus, given $\varepsilon > 0$, there is a $\delta > 0$ such that

$$(29) \qquad \boldsymbol{x}, \boldsymbol{y} \in H \quad \text{and} \quad \|\boldsymbol{x} - \boldsymbol{y}\| < \delta \quad \text{imply} \quad \|\phi(\boldsymbol{x}) - \phi(\boldsymbol{y})\| < \varepsilon.$$

Since compact sets are bounded, $H$ is contained in some $(n-1)$-dimensional cube $Q$. Divide $Q$ into subcubes $Q_1, \ldots, Q_p$ such that $\boldsymbol{x}, \boldsymbol{y} \in Q_k$ implies $\|\boldsymbol{x} - \boldsymbol{y}\| < \delta$, and let $R_k = Q_k \times [\phi(\boldsymbol{a}_k) - 2\varepsilon, \phi(\boldsymbol{a}_k) + 2\varepsilon]$ for some $\boldsymbol{a}_k \in Q_k$, $k = 1, 2, \ldots, p$. Then each $R_k$ is an $n$-dimensional rectangle and

$$\sum_{k=1}^{p} |R_k| = 4\varepsilon \sum_{k=1}^{p} |Q_k| = 4\varepsilon |Q|.$$

Moreover, by (29), $B$ is covered by the rectangles $R_k$. Hence, $\overline{\text{Vol}}\,(B) < 4\varepsilon |Q|$. Taking the limit of this inequality as $\varepsilon \to 0$, we see that $B$ is of volume zero. A similar argument shows that $T$ is of volume zero.

To estimate the volume of $S$, set

$$M = \sup_{\boldsymbol{x} \in H} \psi(\boldsymbol{x}) \quad \text{and} \quad m = \inf_{\boldsymbol{x} \in H} \phi(\boldsymbol{x}).$$

Since $H$ is a Jordan region, choose $(n-1)$-dimensional cubes $Q_1, \ldots, Q_p$ which cover $\partial H$ such that

$$\sum_{k=1}^{p} |Q_k| < \varepsilon.$$

Set $R_k = Q_k \times [m, M]$ and observe that

$$S \subseteq \bigcup_{k=1}^{p} R_k \quad \text{and} \quad \sum_{k=1}^{p} |R_k| < (M - m)\varepsilon.$$

Hence, $\overline{\mathrm{Vol}}\,(S) \leq (M-m)\varepsilon$ and it follows that $S$ is of volume zero. We conclude that $\partial E$ is of volume zero; i.e., $E$ is a Jordan region.

To prove (28), let $R = [a_1, b_1] \times \cdots \times [a_n, b_n]$ be an $n$-dimensional rectangle which contains $E$, and define $g$ on $R$ by $g(\boldsymbol{x}, t) = f(\boldsymbol{x}, t)$ when $(\boldsymbol{x}, t) \in E$, and $g(\boldsymbol{x}, t) = 0$ otherwise. By Theorem 12.20 and Lemma 12.36,

$$\int_{E} f(\boldsymbol{x}, t)\, d(\boldsymbol{x}, t) = \int_{a_1}^{b_1} \cdots \int_{a_n}^{b_n} g(x_1, \ldots, x_n)\, dx_n \ldots dx_1$$

$$= \int_{H} \left( \int_{a_n}^{b_n} g(\boldsymbol{x}, t)\, dt \right) d\boldsymbol{x}.$$

But for each $\boldsymbol{x} = (x_1, \ldots, x_{n-1}) \in H$, we have

$$g(\boldsymbol{x}, t) = \begin{cases} f(\boldsymbol{x}, t) & \phi(\boldsymbol{x}) \leq t \leq \psi(\boldsymbol{x}) \\ 0 & \text{otherwise.} \end{cases}$$

Therefore,

$$\int_{a_n}^{b_n} g(\boldsymbol{x}, t)\, dt = \int_{\phi(\boldsymbol{x})}^{\psi(\boldsymbol{x})} f(\boldsymbol{x}, t)\, dt. \quad \blacksquare$$

Although we have stated Theorem 12.39 for continuous $f$, the result is evidently true whenever Lemma 12.36 applies, e.g., if $f$ is integrable on $E$ and $f(\boldsymbol{x}, \cdot)$ is integrable on $[a_n, b_n]$ for each fixed $\boldsymbol{x} \in H$.

If the set $H$ is itself projectable, then Theorem 12.39 can be applied again to $H$. Thus if $E$ is nice enough, an integral over $E$ can be evaluated using $n$ partial integrals. We close this section with several examples which illustrate this principle for the cases $n = 2$ and $n = 3$.

**12.40 Example.** Find the integral of $f(x, y, z) = x$ over the region $E$ bounded by $z = 1 - x - y$, $x = 0$, $y = 0$, and $z = 0$.

SOLUTION. The surfaces $z = 0$ and $z = 1 - x - y$ intersect when $y = 1 - x$. The projection $E_3$ is bounded by the curves $x = 0$, $y = 0$, and $y = 1 - x$. These last two curves intersect when $x = 1$. Thus $E$ is a region of the type I:

$$E = \{(x, y, z) : 0 \leq x \leq 1,\ 0 \leq y \leq 1 - x,\ 0 \leq z \leq 1 - x - y\}$$

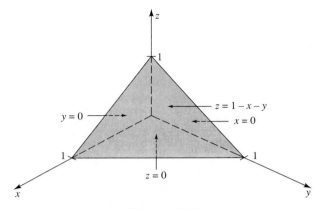

**Figure 12.9**

(see Figure 12.9). It follows that

$$\iiint_E f\, dV = \int_0^1 \int_0^{1-x} \int_0^{1-x-y} x\, dz\, dy\, dx$$

$$= \int_0^1 \int_0^{1-x} (x - x^2 - xy)\, dy\, dx$$

$$= \frac{1}{2} \int_0^1 (x - 2x^2 + x^3)\, dx = \frac{1}{24}. \ \blacksquare$$

**12.41 Example.** Find the integral of $f(x, y, z) = x^2$ over the region $E$ bounded by $|x| = 1$, $z = x^2 - y^2$, where $z \geq 0$.

SOLUTION. The surfaces $z = 0$ and $z = x^2 - y^2$ intersect when $x^2 - y^2 = 0$, i.e., $y = \pm x$. The curves $y = \pm x$ and $|x| = 1$ intersect when $x = \pm 1$. Thus the region $E$ is of type I:

$$E = \{(x, y, z) : -1 \leq x \leq 1, \ -|x| \leq y \leq |x|, \ 0 \leq z \leq x^2 - y^2\}$$

(see Figure 12.10). It follows that

$$\iiint_E f\, dV = \int_{-1}^1 \int_{-|x|}^{|x|} \int_0^{x^2 - y^2} x^2\, dz\, dy\, dx$$

$$= \int_{-1}^1 \int_{-|x|}^{|x|} (x^2 - y^2) x^2\, dy\, dx$$

$$= 4 \int_0^1 \int_0^x (x^2 - y^2) x^2\, dy\, dx = \frac{8}{3} \int_0^1 x^5\, dx = \frac{4}{9}. \ \blacksquare$$

Although Theorem 12.39 can be used in conjunction with Theorem 12.23 to handle the case when $E$ is a finite union of projectable subregions, we can sometimes

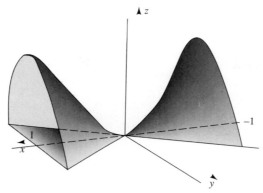

**Figure 12.10**

avoid breaking $E$ into subregions by changing our point of view. Here is a typical example.

**12.42 Example.** Find the integral of $f(x, y, z) = x - z$ over the region bounded by $z = y^2$, $z = 1$, $z = x$, and $x = 0$.

SOLUTION. The region $E$ is a union of two regions of type I (see Figure 12.11, where the "back" of $E$ is that portion of the plane $x = 0$ which is bounded by $z = y^2$ and $z = 1$, here represented by a dashed line). Therefore, we must use two integrals if we integrate $dz$ first, the integral where $z$ varies between $y^2$ and 1, and the integral where $z$ varies from $x$ to 1. It looks complicated to set up. The solution is simpler if we integrate $dx$ first. Indeed, $E$ is a single region of type III since

$$E = \{(x, y, z) : -1 \le y \le 1, \ y^2 \le z \le 1, \ 0 \le x \le z\}.$$

Thus,

$$\iiint_E f \, dV = \int_{-1}^{1} \int_{y^2}^{1} \int_0^z (x - z) \, dx \, dz \, dy$$

$$= -\frac{1}{2} \int_{-1}^{1} \int_{y^2}^{1} z^2 \, dz \, dy = \frac{1}{6} \int_{-1}^{1} (y^6 - 1) \, dy = -\frac{2}{7}. \ \blacksquare$$

**EXERCISES**

**1.** Evaluate each of the following iterated integrals.

a) $\displaystyle\int_0^1 \int_0^1 (x^2 + y) \, dx \, dy$.   b) $\displaystyle\int_0^1 \int_0^1 \sqrt{xy + x} \, dx \, dy$.   c) $\displaystyle\int_0^{\pi/2} \int_0^{\pi/2} y \cos(xy) \, dy \, dx$.

**2.** Evaluate each of the following iterated integrals. Write each as an integral over

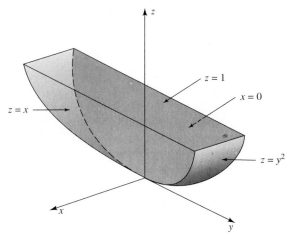

**Figure 12.11**

a region $E$, and sketch $E$ in each case.

a) $\displaystyle\int_0^1 \int_x^{x^2+1} (x+y)\, dy\, dx.$     b) $\displaystyle\int_0^1 \int_{\sqrt{y}}^1 \int_0^{x^2+y^2} 3\, dz\, dx\, dy.$

c) $\displaystyle\int_0^1 \int_y^1 \sin(x^2)\, dx\, dy.$     d) $\displaystyle\int_0^1 \int_{\sqrt{y}}^1 \int_{x^3}^1 \sqrt{x^3+z}\, dz\, dx\, dy.$

**3.** For each of the following, evaluate $\int_E f(\mathbf{x})\, d\mathbf{x}$.

   a) $f(x,y) = x\sqrt{y}$ and $E$ is bounded by $y = x$ and $y = x^2$.

   b) $f(x,y) = x+y$ and $E$ is the triangle with vertices $(0,0)$, $(0,1)$, $(2,0)$.

   c) $f(x,y) = x$ and $E$ is bounded by $y = \sqrt{x}$, $x = -\sqrt{y}$, and $y = 4$.

   d) $f(x,y,z) = x$ and $E$ is the set of points $(x,y,z)$ such that $0 \le z \le 1 - x^2$, $0 \le y \le x^2 + z^2$, and $x \ge 0$.

**4.** Compute the volume of each of the following regions.

   a) $E$ is bounded by the surfaces $x+y+z = 3$, $z = 0$, and $x^2 + y^2 = 1$.

   b) $E$ lies under the plane $z = x+y$ and over the region in the $xy$ plane bounded by the curves $x = \sqrt{y/2}$, $x = 2\sqrt{y}$, $x + y = 3$.

   c) $E$ is bounded by $z = y^2$, $x = y^2 + z^2$, $x = 0$, $z = 1$.

   d) $E$ is bounded by $y = x^3$, $x = z^2$, $z = x^2$, and $y = 0$.

**5.** a) Verify that the hypotheses of Fubini's Theorem hold when $f$ is continuous on $R$.

   b) Modify the proof of Remark 12.34 to show that Fubini's Theorem might not hold for a nonintegrable $f$, even if $f(x,y)$ is continuous in each variable separately, i.e., if $f(x,\cdot)$ is continuous for each $x \in [a,b]$ and $f(\cdot,y)$ is continuous for each $y \in [c,d]$.

**6. a)** Suppose $f_k$ is integrable on $[a_k, b_k]$ for $k = 1, \ldots, n$, and set $R = [a_1, b_1] \times \cdots \times [a_n, b_n]$. Prove that

$$\int_R f_1(x_1) \ldots f_n(x_n) \, d(x_1, \ldots, x_n) = \left( \int_{a_1}^{b_1} f_1(x_1) \, dx_1 \right) \ldots \left( \int_{a_n}^{b_n} f_n(x_n) \, dx_n \right).$$

**b)** If $Q = [0, 1]^n$ and $\boldsymbol{y} := (1, 1, \ldots, 1)$, prove

$$\int_Q e^{-\boldsymbol{x} \cdot \boldsymbol{y}} \, d\boldsymbol{x} = \left( \frac{e - 1}{e} \right)^n.$$

**7.** Let $R$ be a two-dimensional rectangle, $\phi : \mathbf{R} \to \mathbf{R}$ be defined by $\phi(x) = x - n - 1/2$ for $n \leq x < n + 1$, $n = 0, \pm 1, \pm 2, \ldots$, and set $\psi(x, y) = \phi(x)\phi(y)$.

**a)** Prove that $\iint_R \psi \, dA = 0$ if and only if at least one side of $R$ has integer length.

**b)** Suppose $R$ is tiled by rectangles $R_1 \ldots, R_N$; i.e., the $R_j$'s are nonoverlapping and $R = \cup_{j=1}^N R_j$. Prove that if each $R_j$ has at least one side of integer length, then $R$ has at least one side of integer length.

**8.** Let $E$ be a nonempty Jordan region in $\mathbf{R}^2$ and $f : E \to [0, \infty)$ be integrable on $E$. Prove that the volume of $\Omega = \{(x, y, z) : (x, y) \in E, \ 0 \leq z \leq f(x, y)\}$ (as given by Definition 12.5) satisfies

$$\mathrm{Vol}\,(\Omega) = \iint_E f \, dA.$$

**9.** Let $R = [a, b] \times [c, d]$ be a two-dimensional rectangle and $f : R \to \mathbf{R}$ be bounded.

**a)** Prove

$$(L) \iint_R f \, dA \leq (L) \int_a^b \left( (X) \int_c^d f(x, y) \, dy \right) dx$$

$$\leq (U) \int_a^b \left( (X) \int_c^d f(x, y) \, dy \right) dx$$

$$\leq (U) \iint_R f \, dA$$

for $X = U$ or $X = L$.

**b)** Prove that if $f$ is integrable on $R$, then

$$\iint_R f \, dA = \int_a^b \left( (L) \int_c^d f(x, y) \, dy \right) dx = \int_a^b \left( (U) \int_c^d f(x, y) \, dy \right) dx.$$

**c)** Compute the two iterated integrals in part b) for

$$f(x, y) = \begin{cases} 1 & y \in \mathbf{Q} \\ x & y \notin \mathbf{Q} \end{cases}$$

and $R = [0, 1] \times [0, 1]$. Prove that $f$ is not integrable on $R$.

**\*10.** [FUBINI'S THEOREM FOR IMPROPER INTEGRALS]. If $a < b$ are extended
real numbers, $c < d$ are finite real numbers, $f : (a, b) \times [c, d] \to \mathbf{R}$ is continuous,
and

$$F(y) = \int_a^b f(x, y) \, dx$$

converges uniformly on $[c, d]$, prove

$$\int_c^d f(x, y) \, dy$$

is improperly integrable on $(a, b)$ and

$$\int_c^d \int_a^b f(x, y) \, dx \, dy = \int_a^b \int_c^d f(x, y) \, dy \, dx.$$

## 12.4  CHANGE OF VARIABLES

Recall (Exercise 11 in Section 5.3) that if $\phi : [a, b] \to \mathbf{R}$ is continuously differentiable
and $\phi' \neq 0$ on $[a, b]$, then

$$\int_{\phi([a,b])} f(t) \, dt = \int_{[a,b]} f(\phi(x)) \, |\phi'(x)| \, dx$$

for all $f$ integrable on $\phi([a, b])$. We shall generalize this result to functions of several
variables; namely, we shall identify conditions under which

$$(30) \qquad \int_{\phi(E)} f(\mathbf{u}) \, d\mathbf{u} = \int_E f(\phi(\mathbf{x})) |\Delta_\phi(\mathbf{x})| \, d\mathbf{x}$$

holds. (At this point you may wish to read the discussion following the proof of
Theorem 12.46 below to see that $\Delta_\phi$ takes on a familiar form when $\phi$ is the change
from polar to rectangular coordinates.)

It takes 6 or 7 hypotheses to establish (30). These hypotheses fall into two
categories. 1) *Hypotheses made so the change of variables is possible.* Since the one-
dimensional result required $\phi$ to be continuously differentiable and $\phi' \neq 0$ (which
together imply that $\phi$ is 1–1), we expect the corresponding hypotheses for (30) to
be: $\phi$ is 1–1, continuously differentiable, and $\Delta_\phi \neq 0$. 2) *Hypotheses made so the
integrals in (30) exist.* There are four of these: $E$ and $\phi(E)$ are Jordan regions, $f$
is integrable on $\phi(E)$, and $f \circ \phi |\Delta_\phi|$ is integrable on $E$. To make the statements
below less cumbersome, we shall abbreviate hypotheses in category 2) by the phrase
"provided these integrals exist."

In practice, the hypotheses in category 2) can usually be verified by inspection.
The reason for this is threefold. Most functions are continuous (or nearly so), hence,
integrable. $E$ is frequently projectable, hence, a Jordan region. And finally, if $E$ is

a Jordan region and $\phi$ satisfies the hypotheses in category 1), then $\phi(E)$ must be a Jordan region (see Theorem 12.52 below). In particular, the essential hypotheses for (30) are those in category 1), namely, that $\phi$ be 1–1, continuously differentiable, and $\Delta_\phi \neq 0$.

To give an outline of a proof of (30), we introduce the following terminology. A function $f$ is said to satisfy a certain property $\mathcal{P}$ "locally" on a set $E$ if given $\boldsymbol{a} \in E$ there is an open ball $B$ containing $\boldsymbol{a}$ such that $f$ satisfies $\mathcal{P}$ on $B \cap E$. $f$ is said to satisfy the property $\mathcal{P}$ "globally" on $E$ if $f$ satisfies $\mathcal{P}$ for all points in $E$. To prove (30), we first obtain several preliminary results which culminate in a "local" change of variables formula (see Theorem 12.45) and then use this to obtain a "global" change of variables formula for functions $\phi$ whose Jacobians are never zero (see Theorem 12.46). In Section 12.5, we work much harder to show that the condition that $\Delta_\phi \neq 0$ can be relaxed on a set of volume zero (see Theorem 12.66).

Our first preliminary result shows that (30) holds when $f$ is identically 1 and $\phi(E)$ is a rectangle. (Since every Jordan region can be approximated by rectangles, and every integrable function is almost continuous, hence, locally nearly constant, this is the crucial case to consider. Once this has been dealt with, (30) should follow by an approximation argument.)

**12.43 Lemma.** *Let $V$ be open in $\mathbf{R}^n$, and $\phi : V \to \mathbf{R}^n$ be 1–1 and continuously differentiable on $V$. If $\Delta_\phi(\boldsymbol{a}) \neq 0$ for some $\boldsymbol{a} \in V$, then there exists an open rectangle $W \subset V$, which contains $\boldsymbol{a}$, such that*

$$(31) \qquad |R| = \int_{\phi^{-1}(R)} |\Delta_\phi(\boldsymbol{x})|\, d\boldsymbol{x}$$

*for all $n$-dimensional rectangles $R \subset \phi(W)$.*

STRATEGY: The idea behind the proof is simple. The proof is by induction on $n$. If $n = 1$ and $\phi'(a) \neq 0$, then $\phi'$ is nonzero in some open interval $I$ containing $a$. Hence, by Exercise 11 in Section 5.3, (31) holds for "rectangles" (i.e., intervals) in $\phi(I)$. Assuming the result holds for rectangles in $\mathbf{R}^{n-1}$, write $\phi = \sigma \circ \psi$, where $\sigma$ is essentially a one-dimensional function and $\psi$ is essentially an $(n-1)$-dimensional function. Apply the inductive hypothesis to $\sigma$ and $\psi$, and show (using the Chain Rule) that (31) holds for $\phi$. Here are the details.

PROOF. Suppose the theorem holds on $\mathbf{R}^{n-1}$ for some $n > 1$. Let $\phi : V \to \mathbf{R}^n$ be 1–1 and $\mathcal{C}^1$ on $V$ with $\Delta_\phi(\boldsymbol{a}) \neq 0$. Expanding this determinant along the first column, we see that $\partial \phi_i / \partial x_1 \cdot \Delta_i$ must be nonzero at $\boldsymbol{a}$ for some $i$, where $\Delta_i$ represents the cofactor of $\partial \phi_i / \partial x_1$ in $D\phi(\boldsymbol{a})$. By interchanging the roles of $\phi_i$ and $\phi_1$, we may suppose that $\partial \phi_1 / \partial x_1(\boldsymbol{a}) \neq 0$ and $\Delta_1(\boldsymbol{a}) \neq 0$. Since all these functions are continuous, choose an open set $W \subset V$ such that $\boldsymbol{a} \in W$, $\partial \phi_1 / \partial x_1(\boldsymbol{x}) \neq 0$, and $\Delta_1(\boldsymbol{x}) \neq 0$ for all $\boldsymbol{x} \in W$. By making $W$ smaller, if necessary, we may suppose that $W$ is an open rectangle; i.e., there exist open intervals $I_j$, $j = 1, \ldots, n$ such that $W = I_1 \times \cdots \times I_n$ and $a_j \in I_j$ for $j = 1, 2, \ldots, n$.

For each $\boldsymbol{x} = (x_1, \ldots, x_n) \in W$, set

$$\psi(\boldsymbol{x}) = (x_1, \phi_2(\boldsymbol{x}), \ldots, \phi_n(\boldsymbol{x}))$$

and

$$\sigma(\boldsymbol{x}) = (\phi_1(\psi^{-1}(\boldsymbol{x})), x_2, \ldots, x_n).$$

Then $\phi = \sigma \circ \psi$. Moreover, by construction and the choice of $W$,

(32) $$\Delta_\psi(\boldsymbol{x}) = \Delta_1(\boldsymbol{x}) \neq 0$$

and

(33) $$\Delta_\sigma(\psi(\boldsymbol{x})) = \frac{\partial \phi_1}{\partial x_1}(\psi^{-1} \circ \psi)(\boldsymbol{x}) = \frac{\partial \phi_1}{\partial x_1}(\boldsymbol{x}) \neq 0$$

for all $\boldsymbol{x} \in W$.

Set $W_0 = I_2 \times \cdots \times I_n$, fix $\xi \in I_1$, and for each $\boldsymbol{y} = (x_2, \ldots, x_n)$ which satisfies $(\xi, \boldsymbol{y}) \in W$ set

$$\phi_0(\boldsymbol{y}) = (\phi_2(\xi, \boldsymbol{y}), \ldots, \phi_n(\xi, \boldsymbol{y})).$$

Clearly, $\phi_0 : W_0 \to \mathbf{R}^{n-1}$ satisfies $\psi(\xi, \boldsymbol{y}) = (\xi, \phi_0(\boldsymbol{y}))$. Hence, $\phi_0$ is continuously differentiable and, by (32), $\Delta_{\phi_0}(\boldsymbol{y}) = \Delta_\psi(\xi, \boldsymbol{y}) \neq 0$ for all $\boldsymbol{y} \in W_0$. It follows from the inductive hypothesis that

(34) $$|R_0| = \int_{\phi_0^{-1}(R_0)} |\Delta_{\phi_0}(\boldsymbol{y})| \, d\boldsymbol{y}$$

for every $(n-1)$-dimensional rectangle $R_0 \subset \phi_0(W_0)$. ($W_0$, hence, $W$, may have gotten smaller again.)

Suppose $R$ is an $n$-dimensional rectangle which satisfies $R \subset \phi(W)$. Then $R = [a, b] \times R_0$, where $R_0$ is an $(n-1)$-dimensional rectangle contained in $\phi_0(W_0)$; i.e., (34) holds. For each $\boldsymbol{y} = (x_2, \ldots, x_n) \in W_0$, set $\sigma_0(t) = \sigma_1(t, \boldsymbol{y})$, the first component of the vector $\sigma(t, \boldsymbol{y})$. Then by the construction of $\sigma$ and (33), $\sigma_0 : I_1 \to \mathbf{R}$,

$$\sigma_0'(t) = \Delta_\sigma(\psi(t, \boldsymbol{y})) \neq 0, \qquad t \in [a, b],$$

and $\sigma^{-1}([a, b] \times \{\boldsymbol{y}\}) = \sigma_0^{-1}[a, b] \times \{\boldsymbol{y}\}$. Similarly, $\psi^{-1}(\{\xi\} \times R_0) = \{\xi\} \times \phi_0^{-1}(R_0)$. Since

$$\phi^{-1} = \psi^{-1} \circ \sigma^{-1}(R) = \sigma_0^{-1}[a, b] \times \phi_0^{-1}(R_0),$$

it follows, from the Chain Rule, (34), Theorem 12.39, and a one-dimensional change of variables involving $\sigma_0$, that

$$\int_{\phi^{-1}(R)} |\Delta_\phi(\boldsymbol{x})| \, d\boldsymbol{x} = \int_{\psi^{-1} \circ \sigma^{-1}(R)} |\Delta_\psi(\boldsymbol{x})| \, |\Delta_\sigma(\psi(\boldsymbol{x}))| \, d\boldsymbol{x}$$

$$= \int_{\phi_0^{-1}(R_0)} |\Delta_{\phi_0}(\boldsymbol{y})| \int_{\sigma_0^{-1}[a,b]} |\sigma_0'(t)| \, dt \, d\boldsymbol{y}$$

$$= |R_0| \cdot (b - a) = |R|. \ \blacksquare$$

Next, we show that if (31) holds and $f$ is nonnegative, then the left-side of (30) is less than or equal to the right-side of (30).

**12.44 Lemma.** *Let $V$ be open in $\mathbf{R}^n$, $\phi : V \to \mathbf{R}^n$ be 1–1 and continuously differentiable on $V$, and $\Delta_\phi \neq 0$ on $V$. If*

$$(35) \qquad |R| = \int_{\phi^{-1}(R)} |\Delta_\phi(\boldsymbol{x})| \, d\boldsymbol{x}$$

*for every $n$-dimensional rectangle $R \subset \phi(V)$, then*

$$\int_{\phi(E)} f(\boldsymbol{u}) \, d\boldsymbol{u} \leq \int_E (f \circ \phi)(\boldsymbol{x}) |\Delta_\phi(\boldsymbol{x})| \, d\boldsymbol{x}$$

*for all Jordan regions $\overline{E} \subset V$ and all $f : \phi(E) \to [0, \infty)$, provided all these integrals exist.*

PROOF. We may suppose that $V$ is nonempty. Let $E$ be a fixed Jordan region which satisfies $\overline{E} \subset V$ and let $f : \phi(E) \to [0, \infty)$ be integrable on $\phi(E)$. Let $H$ be a rectangle containing $\phi(E)$ and $\mathcal{G}$ be a grid on $H$. Given $\varepsilon > 0$, choose a grid $\mathcal{H} = \{R_1, \ldots, R_p\}$ finer than $\mathcal{G}$ such that $V(\partial(\phi(E)); \mathcal{H}) < \varepsilon$. Let $\mathcal{I}_1 = \{j : R_j \subseteq \phi(E)\}$, $\mathcal{I}_2 = \{j \notin \mathcal{I}_1 : R_j \cap \phi(E) \neq \emptyset\}$, extend $f$ to $\mathbf{R}^n$ by $f(\boldsymbol{x}) = 0$ for $\boldsymbol{x} \in \phi(E)^c$, and set $M = \sup\{m_1, \ldots, m_p\}$, where for each $j \in \mathcal{I}_1 \cup \mathcal{I}_2$,

$$m_j = \inf_{\boldsymbol{u} \in R_j} f(\boldsymbol{u}) := \inf_{\boldsymbol{x} \in \phi^{-1}(R_j)} f(\phi(\boldsymbol{x})).$$

Notice that $\{\phi^{-1}(R_j)\}_{j \in \mathcal{I}_1}$ is a nonoverlapping collection of Jordan regions (see Exercise 7 in Section 12.1) and

$$\Omega := \bigcup_{j \in \mathcal{I}_1} \phi^{-1}(R_j) \subseteq E.$$

It follows from (35), Theorem 12.25, and Theorem 12.23 that

$$
\begin{aligned}
L(f, \mathcal{G}) \leq L(f, \mathcal{H}) &= \sum_{j \in \mathcal{I}_1} m_j |R_j| + \sum_{j \in \mathcal{I}_2} m_j |R_j| \\
&\leq \sum_{j \in \mathcal{I}_1} m_j |R_j| + MV(\partial(\phi(E)); \mathcal{H}) \\
&< \sum_{j \in \mathcal{I}_1} m_j |R_j| + M\varepsilon \\
&= \sum_{j \in \mathcal{I}_1} m_j \int_{\phi^{-1}(R_j)} |\Delta_\phi(\boldsymbol{x})| \, d\boldsymbol{x} + M\varepsilon \\
&\leq \sum_{j \in \mathcal{I}_1} \int_{\phi^{-1}(R_j)} f(\phi(\boldsymbol{x})) |\Delta_\phi(\boldsymbol{x})| \, d\boldsymbol{x} + M\varepsilon \\
&= \int_\Omega f(\phi(\boldsymbol{x})) |\Delta_\phi(\boldsymbol{x})| \, d\boldsymbol{x} + M\varepsilon \\
&\leq \int_E f(\phi(\boldsymbol{x})) |\Delta_\phi(\boldsymbol{x})| \, d\boldsymbol{x} + M\varepsilon.
\end{aligned}
$$

(This is the only place where $f \geq 0$ is used.) Taking the supremum of this inequality over all grids $\mathcal{G}$ and the limit as $\varepsilon \to 0$, we obtain

$$\int_{\phi(E)} f(\boldsymbol{u}) \, d\boldsymbol{u} \leq \int_E f(\phi(\boldsymbol{x})) |\Delta_\phi(\boldsymbol{x})| \, d\boldsymbol{x}. \quad \blacksquare$$

Here is a local version of the change of variables formula we want.

**12.45 THEOREM.** *Suppose $V$ is open in $\mathbf{R}^n$, $\boldsymbol{a} \in V$, and $\phi : V \to \mathbf{R}^n$ is continuously differentiable on $V$. If $\Delta_\phi(\boldsymbol{a}) \neq 0$, then there exists an open rectangle $W \subset V$ containing $\boldsymbol{a}$ such that*

$$\int_{\phi(E)} f(\boldsymbol{u}) \, d\boldsymbol{u} = \int_E f(\phi(\boldsymbol{x})) |\Delta_\phi(\boldsymbol{x})| \, d\boldsymbol{x},$$

*provided $\overline{E} \subset W$ and all these integrals exist.*

PROOF. Since $f = f^+ - f^-$, where both $f^+$ and $f^-$ are nonnegative and are integrable when $f$ is (see Exercise 2 in Section 5.2), we may suppose that $f \geq 0$. Hence by Lemmas 12.43 and 12.44, we are half way toward the desired formula on any open set $W$ where $\phi$ is 1–1 and has a nonzero Jacobian. Such an open set can be chosen, by the Inverse Function Theorem. In fact, since $\Delta_\phi(\boldsymbol{a}) \neq 0$, there is an open set $W \subset V$, containing $\boldsymbol{a}$, such that $\phi$ is 1–1 and $\Delta_\phi \neq 0$ on $W$, and $\phi^{-1}$ is 1–1, $\mathcal{C}^1$ and $\Delta_{\phi^{-1}} \neq 0$ on $\phi(W)$.

Suppose $E$ is a Jordan region which satisfies $\overline{E} \subset W$ and $f$ is a nonnegative function integrable on $\phi(E)$. By Lemmas 12.43 and 12.44,

$$(36) \qquad \int_{\phi(E)} f(\boldsymbol{u}) \, d\boldsymbol{u} \leq \int_E f(\phi(\boldsymbol{x})) |\Delta_\phi(\boldsymbol{x})| \, d\boldsymbol{x}.$$

But $\phi^{-1}$ satisfies the same hypotheses that $\phi$ does. Hence, we can apply (36) with $\phi^{-1}$ in place of $\phi$, $f \circ \phi |\Delta_\phi|$ in place of $f$, and $E = \phi^{-1}(\phi(E))$ in place of $\phi(E)$. We obtain

$$(37) \qquad \int_E f(\phi(\boldsymbol{x})) |\Delta_\phi(\boldsymbol{x})| \, d\boldsymbol{x} \leq \int_{\phi(E)} f(\phi(\phi^{-1}(\boldsymbol{u}))) |\Delta_\phi(\phi^{-1}(\boldsymbol{u}))| \, |\Delta_{\phi^{-1}}(\boldsymbol{u})| \, d\boldsymbol{u}.$$

Since $\phi \circ \phi^{-1} = I$ on $\phi(V)$, it follows from the Chain Rule that

$$|\Delta_\phi(\phi^{-1}(\boldsymbol{u}))| \, |\Delta_{\phi^{-1}}(\boldsymbol{u})| = |\Delta_I(\boldsymbol{u})| = 1$$

for all $\boldsymbol{u} \in \phi(V) \supset \phi(E)$. Substituting this back into (37), we obtain

$$\int_E f(\phi(\boldsymbol{x})) |\Delta_\phi(\boldsymbol{x})| \, d\boldsymbol{x} \leq \int_{\phi(E)} f(\boldsymbol{u}) \, d\boldsymbol{u}.$$

We conclude by (36) that equality holds. $\blacksquare$

This local change of variables formula contains the following global result.

**12.46 THEOREM.** *Suppose $V$ is open in $\mathbf{R}^n$ and $\phi : V \to \mathbf{R}^n$ is 1–1 and continuously differentiable on $V$. If $\Delta_\phi \neq 0$ on $V$ and $\overline{E} \subset V$, then*

$$(38) \qquad \int_{\phi(E)} f(\boldsymbol{u})\, d\boldsymbol{u} = \int_E f(\phi(\boldsymbol{x}))|\Delta_\phi(\boldsymbol{x})|\, d\boldsymbol{x},$$

*provided both these integrals exist.*

PROOF. Let $f : \phi(E) \to \mathbf{R}$, $H = \overline{E}$, suppose $f$ is integrable on $\phi(E)$, and $(f \circ \phi)|\Delta_\phi|$ is integrable on $E$. By Theorem 12.45, given $\boldsymbol{a} \in H$ there is an open rectangle $V_{\boldsymbol{a}}$ such that $\boldsymbol{a} \in V_{\boldsymbol{a}} \subset V$ and

$$(39) \qquad \int_{\phi(E_i)} f(\boldsymbol{u})\, d\boldsymbol{u} = \int_{E_i} f(\phi(\boldsymbol{x}))|\Delta_\phi(\boldsymbol{x})|\, d\boldsymbol{x}$$

for every Jordan region $E_i$ which satisfies $\overline{E_i} \subset V_{\boldsymbol{a}}$. Let $W_{\boldsymbol{a}}$ be an open rectangle which satisfies $\boldsymbol{a} \in W_{\boldsymbol{a}} \subset \overline{W_{\boldsymbol{a}}} \subset V_{\boldsymbol{a}}$. Clearly,

$$H \subset \bigcup_{\boldsymbol{a} \in H} W_{\boldsymbol{a}}.$$

Since the Jordan region $E$ is bounded, $H$ is compact. Thus choose open rectangles $W_j := W_{\boldsymbol{a}_j}$ such that

$$H \subset \bigcup_{j=1}^{k} W_j.$$

Let $R$ be a huge rectangle which contains $H$ and $\mathcal{G} = \{R_i : i = 1, \ldots, N\}$ be a grid on $R$ so fine that each rectangle in $\mathcal{G}$ which intersects $H$ is a subset of some $\overline{W_j}$. (This is possible since there are only finitely many $W_j$'s.) Let $E_i = R_i \cap E$. Then $\overline{E_i} \subseteq R_i \cap H \subseteq \overline{W_j} \subset V_{\boldsymbol{a}_j}$ for some $j \in \{1, \ldots, k\}$, i.e., (39) holds. Moreover, the collection $\{E_i : i = 1, \ldots, N\}$ is a nonoverlapping family of nonempty Jordan regions whose union is $E$; hence, by Theorem 1.43 and Exercise 7 in Section 12.1, the collection $\{\phi(E_i) : i = 1, \ldots, N\}$ is a nonoverlapping family of nonempty Jordan regions whose union is $\phi(E)$. It follows from Theorem 12.23 and (39) that

$$\int_{\phi(E)} f(\boldsymbol{u})\, d\boldsymbol{u} = \sum_{i=1}^{N} \int_{\phi(E_i)} f(\boldsymbol{u})\, d\boldsymbol{u}$$

$$= \sum_{i=1}^{N} \int_{E_i} f(\phi(\boldsymbol{x}))|\Delta_\phi(\boldsymbol{x})|\, d\boldsymbol{x} = \int_E f(\phi(\boldsymbol{x}))|\Delta_\phi(\boldsymbol{x})|\, d\boldsymbol{x}. \ \blacksquare$$

To see how all this works out in practice, we begin with a familiar change of variables in $\mathbf{R}^2$. Recall that *polar coordinates* in $\mathbf{R}^2$ have the form

$$x = r\cos\theta, \qquad y = r\sin\theta,$$

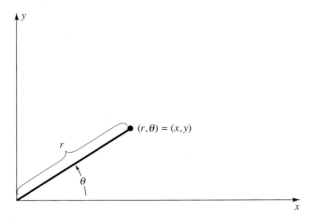

**Figure 12.12**

where $r = \|(x, y)\|$ and $\theta$ is the angle measured counterclockwise from the positive $x$ axis to the line segment $L((0,0);(x,y))$ (see Figure 12.12). Set $\phi(r, \theta) = (r\cos\theta, r\sin\theta)$ and observe that

$$(40) \qquad \Delta_\phi = \det \begin{bmatrix} \cos\theta & \sin\theta \\ -r\sin\theta & r\cos\theta \end{bmatrix} = r(\cos^2\theta + \sin^2\theta) = r.$$

Thus we abbreviate the change of variables formula from polar coordinates to rectangular coordinates by $dx\, dy = r\, dr\, d\theta$.

Although $\phi$ is not 1–1 (for example, $\phi(0, \theta) = (0, 0)$ for all $\theta \in \mathbf{R}$) and its Jacobian is not nonzero, this does not prevent us from applying Theorem 12.46 (i.e., changing variables from polar coordinates to rectangular coordinates and vice versa). Indeed, since $\phi$ is 1–1 on $\Omega := \{(r, \theta) : r > 0, 0 \le \theta < 2\pi\}$ and its Jacobian is nonzero off the set $Z := \{(r, \theta) : r = 0\}$, we can apply Theorem 12.46 to $E \cap \{(r, \theta) : r > 0\}$ and let $r \downarrow 0$. Since the end result is the same as if we applied Theorem 12.46 directly without this intermediate step, we shall do so below without any further comments. This works in part because the set $Z$, where the hypotheses of category 1) fail, is a set of volume zero (see Theorem 12.66 below).

The next two examples show that polar coordinates can be used to evaluate integrals which cannot be easily computed using rectangular coordinates.

**12.47 Example.** Find the volume of the region $E$ bounded by $z = x^2 + y^2$, $x^2 + y^2 = 4$, and $z = 0$.

SOLUTION. Clearly, $E$ lies under the function $f(x, y) = x^2 + y^2$ over the region $B = B_2(0, 0)$ (see Figure 12.13). Using polar coordinates, we obtain

$$\text{Vol}\,(E) = \iint_B (x^2 + y^2)\, dA = \int_0^{2\pi} \int_0^2 r^3\, dr\, d\theta = 8\pi. \quad \blacksquare$$

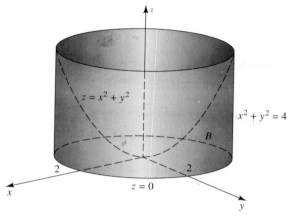

**Figure 12.13**

**12.48 Example.** Evaluate

$$\iint_E \frac{x^2 + y^2}{x} \, dA,$$

where $E = \{(x,y) : a^2 \leq x^2 + y^2 \leq 1 \text{ and } 0 \leq y \leq x\}$ for some $0 < a < 1$.

SOLUTION. Changing to polar coordinates, we see that

$$\iint_E \frac{x^2 + y^2}{x} \, dA = \int_0^{\pi/4} \int_a^1 \frac{r^3}{r \cos \theta} \, dr \, d\theta = \frac{1-a^3}{3} \int_0^{\pi/4} \sec \theta \, d\theta.$$

To integrate $\sec \theta$, multiply and divide by $\sec \theta + \tan \theta$. Using the change of variables $u = \sec \theta + \tan \theta$, we obtain

$$\int_0^{\pi/4} \sec \theta \, d\theta = \int_0^{\pi/4} \frac{\sec \theta \tan \theta + \sec^2 \theta}{\sec \theta + \tan \theta} \, d\theta$$

$$= \int_1^{1+\sqrt{2}} \frac{du}{u} = \log(1 + \sqrt{2}).$$

Consequently,

$$\iint_E \frac{x^2 + y^2}{x} \, dA = \frac{(1-a^3)\log(1+\sqrt{2})}{3}. \quad \blacksquare$$

Recall that *cylindrical coordinates* in $\mathbf{R}^3$ have the form

$$x = r \cos \theta, \qquad y = r \sin \theta, \qquad z = z,$$

where $r = \|(x,y,0)\|$ and $\theta$ is the angle measured counterclockwise from the positive $x$ axis to the line segment $L((0,0,0);(x,y,0))$. It is easy to see that this change of variables is 1–1 on $\Omega := \{(r,\theta,z) : r > 0, 0 \leq \theta < 2\pi, z \in \mathbf{R}\}$, and its Jacobian, $r$, is nonzero off $Z := \{(r,\theta,z) : r = 0\}$. We shall abbreviate the

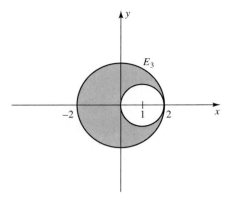

**Figure 12.14**

change of variables formula from cylindrical coordinates to rectangular coordinates by $dx\,dy\,dz = r\,dz\,dr\,d\theta$. (Note that $Z$ is a set of volume zero. As with polar coordinates, application of Theorem 12.46 can justified by applying it first for $r > 0$, and then taking the limit as $r \downarrow 0$.)

**12.49 Example.** Find the volume of the region $E$ which lies inside the paraboloid $x^2 + y^2 + z = 4$, outside the cylinder $x^2 - 2x + y^2 = 0$, and above the plane $z = 0$.

SOLUTION. The paraboloid $z = 4 - x^2 - y^2$ has vertex $(0,0,4)$ and opens downward about the $z$ axis. The cylinder $x^2 - 2x + y^2 = (x-1)^2 + y^2 - 1 = 0$ has base centered at $(1,0)$ with radius 1. Hence, the projection $E_3$ lies inside the circle $x^2 + y^2 = 4$ and outside the circle $x^2 + y^2 = 2x$ (see Figure 12.14). This last circle can be described in polar coordinates by $r^2 = 2r\cos\theta$, i.e., $r = 2\cos\theta$. Thus

$$
\mathrm{Vol}\,(E) = \iiint_E 1\,dV = \iint_{E_3} \int_0^{4-r^2} dz\,dA
$$
$$
= \int_{-\pi/2}^{\pi/2} \int_{2\cos\theta}^{2} (4 - r^2)r\,dr\,d\theta + \int_{\pi/2}^{3\pi/2} \int_0^2 (4 - r^2)r\,dr\,d\theta = 3\pi. \quad \blacksquare
$$

Recall that *spherical coordinates* in $\mathbf{R}^3$ have the form

$$
x = \rho \sin\varphi \cos\theta, \qquad y = \rho \sin\varphi \sin\theta, \qquad z = \rho \cos\varphi,
$$

where $\rho = \|(x,y,z)\|$, $\theta$ is the angle measured counterclockwise from the positive $x$ axis to the line segment $L((0,0,0);(x,y,0))$, and $\phi$ is the angle measured from the positive $z$ axis to the vector $(x,y,z)$ (see Figure 12.15). Notice that this change of variables is 1-1 on $\{(\rho,\varphi,\theta) : \rho > 0, 0 < \varphi < \pi, 0 \le \theta < 2\pi\}$ and its Jacobian, $\rho^2 \sin\varphi$ (see Exercise 8), is nonzero off $Z := \{(\rho,\varphi,\theta) : \varphi = 0, \pi, \rho = 0\}$, a Jordan region of volume zero. Hence, application of Theorem 12.46 can justified by applying it first for $\rho > 0$ and $0 < \varphi < \pi$, and then taking the limit as $\rho, \varphi \downarrow 0$ and $\varphi \uparrow \pi$. Since the end result is the same as applying Theorem 12.46 directly to any projectable region in $\mathbf{R}^3$, we shall do so, without further comments,

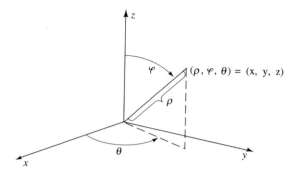

**Figure 12.15**

when changing variables to or from spherical coordinates. We shall abbreviate the change of variables formula from spherical coordinates to rectangular coordinates by $dx\,dy\,dz = \rho^2 \sin\varphi\,d\rho\,d\varphi\,d\theta$. (For spherical coordinates in $\mathbf{R}^n$, see the proof of Theorem 12.70.)

**12.50 Example.** Find

$$\iiint_Q x\,dV,$$

where $Q = B_3(0,0,0) \setminus B_2(0,0,0)$.

SOLUTION. Using spherical coordinates, we have

$$\iiint_Q x\,dV = \int_0^{2\pi}\int_0^{\pi}\int_2^3 \rho\sin\varphi\cos\theta(\rho^2\sin\varphi)\,d\rho\,d\varphi\,d\theta = 0. \quad\blacksquare$$

Theorem 12.46 can be used for other changes of variables besides polar, cylindrical, and spherical coordinates.

**12.51 Example.** Find

$$\iint_E \sin(x+y)\cos(2x-y)\,dA,$$

where $E$ is the region bounded by $y = 2x - 1$, $y = 2x + 3$, $y = -x$, and $y = -x + 1$.

SOLUTION. Let $\phi(x,y) = (2x - y, x + y)$ and observe that the integral in question looks like the right side of (38) except the Jacobian is missing. By Cramer's Rule, for each fixed $u, v \in \mathbf{R}$, the system $u = 2x - y$, $v = x + y$ has a unique solution in $x, y$. Hence, $\phi$ is 1–1 on $\mathbf{R}^2$. It is obviously continuously differentiable, and its Jacobian,

$$\Delta_\phi(x,y) = \frac{\partial(u,v)}{\partial(x,y)} = \det\begin{bmatrix} 2 & -1 \\ 1 & 1 \end{bmatrix} = 3,$$

is a nonzero constant. Hence, we can make adjustments to the integral in question so that it is precisely the right side of (38):

$$\iint_E \sin(x+y)\cos(2x-y)\,dA = \frac{1}{3}\iint_E f\circ\phi(x,y)\Delta_\phi(x,y)\,d(x,y),$$

where $f(u,v) = \cos u \sin v$. It remains to compute the left side of (38), i.e., to find what happens to $E$ under $\phi$.

Notice that $y = 2x - 1$ implies $u = 1$, $y = 2x + 3$ implies $u = -3$, $y = -x$ implies $v = 0$, and $y = -x + 1$ implies $v = 1$. Thus $\phi(E) = [-3, 1] \times [0, 1]$. Applying Theorem 12.46 and the preliminary step taken above, we find

$$\iint_E \sin(x+y)\cos(2x-y)\, dA = \frac{1}{3} \int_0^1 \int_{-3}^1 \sin v \cos u \, du \, dv$$

$$= \frac{1}{3}(\sin(1) + \sin(3))(1 - \cos(1)). \ \blacksquare$$

*We close this section with an optional result, used in Sections $^e$ 12.5 and $^e$ 15.2, which shows that under the hypotheses of category 1), $\phi(E)$ is Jordan region when $E$ is.*

$^*$**12.52 THEOREM.** *Let $V$ be a bounded, open set in $\mathbf{R}^n$, $E$ be a Jordan region, and $\phi : V \to \mathbf{R}^n$ be 1–1 and continuously differentiable on $V$. If $\overline{E} \subset V$ and $\Delta_\phi \neq 0$ on $V$, then $\phi(E)$ is a Jordan region.*

STRATEGY: The idea behind the proof is the following. By Theorem 12.14, a closed set is of volume zero if and only if it can be covered by cubes of small volume. If $E$ is a Jordan region, then $\partial E$ is a set of volume zero by Theorem 12.12, hence, can be covered by cubes of small volume. Since $\|\phi(\boldsymbol{x}) - \phi(\boldsymbol{y})\|$ is small when $\|\boldsymbol{x} - \boldsymbol{y}\|$ is, $\phi$ takes a set of small diameter to a set of small diameter, i.e., a cube of small volume to a subset of a cube of small volume. Hence, it seems likely that $\phi(\partial E)$ can be covered by cubes of small volume, i.e., is also of volume zero. Since $\phi(E)$ is a Jordan region if and only if $\partial(\phi(E))$ is of volume zero, it would remain to prove that

$$(41) \qquad\qquad \partial(\phi(E)) \subseteq \phi(\partial E).$$

(Here alone the condition $\Delta_\phi \neq 0$ enters the picture.) We now fill in the details.

PROOF. By Exercise 9 in Section 11.5, the set $\phi(E^o)$ is open and by Theorem 9.35 (or 10.61), the set $\phi(\overline{E})$ is compact, hence closed. It follows from Theorem 9.10 (or 10.34) that $\phi(E^o) \subseteq (\phi(E))^o$ and $\phi(\overline{E}) \supseteq \overline{\phi(E)}$. Therefore,

$$\partial(\phi(E)) = \overline{\phi(E)} \setminus (\phi(E))^o \subseteq \phi(\overline{E}) \setminus \phi(E^o) = \phi(\overline{E} \setminus E^o) = \phi(\partial E),$$

i.e., (41) holds. Hence, by Theorem 12.12, it suffices to show $\text{Vol}(\phi(\partial E)) = 0$.

We may suppose that $E$ is nonempty. For each $\boldsymbol{x} \in \overline{E}$, choose $r = r(\boldsymbol{x}) > 0$ such that $\overline{B_r(\boldsymbol{x})} \subset V$. Since $E$ is bounded, $\overline{E}$ is compact. Hence, we can choose points $\boldsymbol{x}_j \in \overline{E}$ and radii $r_j = r(\boldsymbol{x}_j)$ such that

$$\overline{E} \subset \bigcup_{j=1}^N B_{r_j}(\boldsymbol{x}_j) \subset H := \bigcup_{j=1}^N \overline{B_{r_j}(\boldsymbol{x}_j)}.$$

Evidently, $H$ is compact and $\overline{E} \subset H^o \subset H \subset V$.

Since rectangles are compact and convex, apply Corollary 11.30, choosing a constant $M > 0$ (which depends only on $\phi$ and $H$) such that

$$\|\phi(\boldsymbol{x}) - \phi(\boldsymbol{y})\| \le M \|\boldsymbol{x} - \boldsymbol{y}\|$$

for all $\boldsymbol{x}, \boldsymbol{y} \in R$ and any rectangle $R \subseteq H$. Now, if $Q$ is a cube with side $s$, which is contained in $H$, it follows from Remark 8.6 that $\|\phi(\boldsymbol{x}) - \phi(\boldsymbol{y})\| \le Ms\sqrt{n}$ for all $\boldsymbol{x}, \boldsymbol{y} \in Q$. Hence, $\phi(Q)$ is contained in a cube with sides of length $\le Ms\sqrt{n}$. In particular, by the definition of outer volume,

(42) $$\overline{\mathrm{Vol}}(\phi(Q)) \le (Ms)^n n^{n/2} =: C |Q|.$$

Notice that $C := M^n n^{n/2}$ depends only on $f$, $H$, and $n$, but not on $Q$.

Let $\varepsilon > 0$. Since $\partial E$ is a closed subset of $H^o$ of volume zero, use Theorem 12.14 to choose cubes $Q_1, \dots, Q_p$ such that $Q_j \subset H$,

(43) $$\partial E \subset \bigcup_{j=1}^{p} Q_j, \quad \text{and} \quad \sum_{j=1}^{p} |Q_j| < \frac{\varepsilon}{C},$$

where $C$ satisfies (42). Applying Theorem 1.43 and (42), we obtain

$$\phi(\partial E) \subset \bigcup_{j=1}^{p} \phi(Q_j) \quad \text{and} \quad \sum_{j=1}^{p} \overline{\mathrm{Vol}}(\phi(Q_j)) < \varepsilon.$$

It follows from the Monotone Property for Infima that $\overline{\mathrm{Vol}}(\phi(\partial E)) < \epsilon$. We conclude by Remark 12.6 that $\phi(\partial E)$ is of volume zero. ∎

**EXERCISES**

**1.** Evaluate each of the following integrals.

a)
$$\int_0^2 \int_0^{\sqrt{4-x^2}} \sin(x^2 + y^2) \, dy \, dx.$$

b)
$$\int_0^1 \int_0^x \sqrt[3]{(2y - y^2)^2} \, dy \, dx.$$

c)
$$\int_a^b \int_0^x \sqrt{x^2 + y^2} \, dy \, dx, \qquad 0 \le a < b.$$

**2.** For each of the following, find $\iint_E f \, dA$.

a) $f(x,y) = \cos(3x^2 + y^2)$ and $E$ is the set of points satisfying $x^2 + y^2/3 \le 1$.

b) $f(x,y) = y\sqrt{x - 2y}$ and $E$ is bounded by the triangle with vertices $(0,0)$, $(4,0)$, and $(4,2)$.

**3.** For each of the following, find $\iiint_E f \, dV$.

a) $f(x, y, z) = z^2$ and $E$ is the set of points satisfying $x^2 + y^2 + z^2 \leq 6$ and $z \geq x^2 + y^2$.

b) $f(x, y, z) = e^z$ and $E$ is the set of points satisfying $x^2 + y^2 + z^2 \leq 9$, $x^2 + y^2 \leq 1$, and $z \geq 0$.

c) $f(x, y, z) = (x - y)z$ and $E$ is the set of points satisfying $x^2 + y^2 + z^2 \leq 4$, $z \geq \sqrt{x^2 + y^2}$, and $x \geq 0$.

**4.** a) Prove that the volume bounded by the ellipsoid

$$\frac{x^2}{a^2} + \frac{y^2}{b^2} + \frac{z^2}{c^2} = 1$$

is $4\pi abc/3$.

b) Let $a, b, c, d$ be positive numbers and $r^2 < d^2/(b^2 + c^2)$. Find the volume of the region bounded by $y^2 + z^2 = r^2$ and $ax + by + cz = d$.

c) Show that for any $a \geq 0$, the volume of the region bounded by the cylinders $x^2 + z^2 = a^2$ and $y^2 + z^2 = a^2$ is $16a^3/3$.

**5.** a) Compute $\iint_E \sqrt{x - y}\sqrt{x + 2y} \, dA$, where $E$ is the parallelogram with vertices $(0, 0)$, $(2/3, -1/3)$, $(1, 0)$, $(1/3, 1/3)$.

b) Compute $\iint_E \sqrt[3]{2x^2 - 5xy - 3y^2} \, dA$, where $E$ is the parallelogram bounded by the lines $y = x/3$, $y = (x - 1)/3$, $y = -2x$, $y = 1 - 2x$.

c) Find

$$\iint_E e^{(y-x)/(y+x)} \, dA,$$

where $E$ is the trapezoid with vertices $(1, 1)$, $(2, 2)$, $(2, 0)$, $(4, 0)$.

d) Given $\int_0^1 (1 - x)f(x) \, dx = 5$, find

$$\int_0^1 \int_0^x f(x - y) \, dy \, dx.$$

**6.** Suppose $V$ is nonempty and open in $\mathbf{R}^n$ and $f : V \to \mathbf{R}^n$ is continuously differentiable with $\Delta_f \neq 0$ on $V$. Prove that

$$\lim_{r \to 0+} \frac{\text{Vol}\,(f(B_r(\mathbf{x}_0)))}{\text{Vol}\,(B_r(\mathbf{x}_0))} = |\Delta_f(\mathbf{x}_0)|$$

for every $\mathbf{x}_0 \in V$.

**7.** Show that Vol is *rotation invariant* in $\mathbf{R}^2$; i.e., if $\phi$ is a rotation on $\mathbf{R}^2$ (see Exercise 9 in Section 8.4) and $E$ is a Jordan region in $\mathbf{R}^2$, then

$$\text{Vol}\,(\phi(E)) = \text{Vol}\,(E).$$

**8.** a) Compute the Jacobian of the change of variables from spherical coordinates to rectangular coordinates.

b) Assuming Vol is translation and rotation invariant (see Exercise 3 in Section 12.1 and Exercise 7 above), verify the following classical formulas: the volume of a sphere of radius $r$ is $\frac{4}{3}\pi r^3$, and the volume of a right circular cone of altitude $h$ and radius $r$ is $\pi r^2 h/3$.

**9.** Let $\boldsymbol{v}_j = (v_{j1}, \ldots, v_{jn}) \in \mathbf{R}^n$, $j = 1, \ldots, n$, be fixed. The parallelepiped determined by the vectors $\boldsymbol{v}_j$ is the set

$$\mathcal{P}(\boldsymbol{v}_1, \ldots, \boldsymbol{v}_n) := \{t_1 \boldsymbol{v}_1 + \cdots + t_n \boldsymbol{v}_n : t_j \in [0, 1]\},$$

and the determinant of the $\boldsymbol{v}_j$'s is the number

$$\det(\boldsymbol{v}_1, \ldots, \boldsymbol{v}_n) := \det [v_{jk}]_{n \times n}.$$

Prove that

$$\mathrm{Vol}\,(\mathcal{P}(\boldsymbol{v}_1, \ldots, \boldsymbol{v}_n)) = |\det(\boldsymbol{v}_1, \ldots, \boldsymbol{v}_n)|.$$

Check this formula for $n = 2$ and $n = 3$ to see that it agrees with the classical formulas for the area of a parallelogram and the volume of a parallelepiped.

**10**. **This exercise is used in Section $^e$12.6.**

a) Prove that the improper integral $\int_0^\infty e^{-x^2}\, dx$ converges to a finite real number.

b) Prove that if $I$ is the value of the integral in part a), then

$$I^2 = \lim_{N \to \infty} \int_0^{\pi/2} \int_0^N e^{-r^2} r\, dr\, d\theta.$$

c) Show

$$\int_0^\infty e^{-x^2}\, dx = \frac{\sqrt{\pi}}{2}.$$

d) Let $Q_k$ represent the $n$-dimensional cube $[-k, k] \times \cdots \times [-k, k]$. Find

$$\lim_{k \to \infty} \int_{Q_k} e^{-\|\boldsymbol{x}\|^2}\, d\boldsymbol{x}.$$

$^e$**12.5 PARTITIONS OF UNITY** *This section uses Theorem 12.52 from the previous section.*

In this section we show that a smooth function can be broken into a sum of smooth functions, each of which is zero except on a small set, and use this to prove a global change of variables formula when the Jacobian is nonzero off a set volume zero. Later, this same technique will be used to prove the Fundamental Theorem of Calculus on manifolds (see Theorem 15.44).

**12.53 DEFINITION.** Let $f : \mathbf{R}^n \to \mathbf{R}$.

i) The *support* of $f$ is the closure of the set of points where $f$ is nonzero, i.e.,

$$\mathrm{spt}\, f := \overline{\{\boldsymbol{x} \in \mathbf{R}^n : f(\boldsymbol{x}) \neq 0\}}.$$

ii) A function $f$ is said to have *compact support* if $\mathrm{spt}\, f$ is a compact set.

**12.54 Example.** If

$$f(x) = \begin{cases} 1 & x \in \mathbf{Q} \\ 0 & x \notin \mathbf{Q}, \end{cases}$$

then spt $f = \mathbf{R}$.

**12.55 Example.** If

$$f(x) = \begin{cases} 1 & x \in (0,1) \\ 2 & x \in (1,2) \\ 0 & \text{otherwise}, \end{cases}$$

then spt $f = [0,2]$.

Since the support of a function is always closed, a function $f$ on $\mathbf{R}^n$ has compact support if and only if spt $f$ is bounded.

The following result shows that if two functions have compact support, then so does their sum (see also Exercises 1 and 2).

**12.56 Remark.** *If $f, g : \mathbf{R}^n \to \mathbf{R}$, then*

$$\operatorname{spt}(f+g) \subseteq \operatorname{spt} f \cup \operatorname{spt} g.$$

Proof. If $(f+g)(\boldsymbol{x}) \neq 0$, then $f(\boldsymbol{x}) \neq 0$ or $g(\boldsymbol{x}) \neq 0$. Thus

$$\{\boldsymbol{x} \in \mathbf{R}^n : (f+g)(\boldsymbol{x}) \neq 0\} \subseteq \{\boldsymbol{x} \in \mathbf{R}^n : f(\boldsymbol{x}) \neq 0\} \cup \{\boldsymbol{x} \in \mathbf{R}^n : g(\boldsymbol{x}) \neq 0\}.$$

Since the closure of a union equals the union of its closures (see Theorem 9.16 or 10.40), it follows that spt $(f+g) \subseteq$ spt $f \cup$ spt $g$. ∎

Let $p \in \mathbf{N}$ or $p = \infty$. The symbol $\mathcal{C}_c^p(\mathbf{R}^n)$ will denote the collection of functions $f : \mathbf{R}^n \to \mathbf{R}$ which are $\mathcal{C}^p$ on $\mathbf{R}^n$ and have compact support. In particular, it follows from Remark 12.56 that if $f_j \in \mathcal{C}_c^p(\mathbf{R}^n)$ for $j = 1, \ldots, N$, then

$$\sum_{j=1}^N f_j \in \mathcal{C}_c^p(\mathbf{R}^n).$$

We will use this observation several times below.

If $f$ is analytic (a condition stronger than $\mathcal{C}^\infty$) and has compact support, then $f$ is identically zero (see Exercise 3 below). Thus it is not at all obvious that $\mathcal{C}_c^\infty(\mathbf{R}^n)$ contains anything but the zero function. Nevertheless, we shall show that $\mathcal{C}_c^\infty(\mathbf{R}^n)$ not only contains nonzero functions, but has enough functions to "approximate" any compact set (see Theorem 12.59 and Exercise 6 below).

First, we deal with the one-dimensional case.

**12.57 Lemma.** *For every $a < b$ there is a function $\phi \in \mathcal{C}_c^\infty(\mathbf{R})$ such that $\phi(t) > 0$ for $t \in (a, b)$ and $\phi(t) = 0$ for $t \notin (a, b)$.*

PROOF. The function

$$f(t) = \begin{cases} e^{-1/t^2} & t \neq 0 \\ 0 & t = 0 \end{cases}$$

belongs to $\mathcal{C}^\infty(\mathbf{R})$ and $f^{(j)}(0) = 0$ for all $j \in \mathbf{N}$ (see Exercise 3 in Section 4.3). Hence,

$$\phi(t) = \begin{cases} e^{-1/(t-a)^2} e^{-1/(t-b)^2} & t \in (a, b) \\ 0 & \text{otherwise} \end{cases}$$

belongs to $\mathcal{C}^\infty(\mathbf{R})$, satisfies $\phi(t) > 0$ for $t \in (a, b)$, and spt $\phi = [a, b]$. ∎

Next, we show that there exists a nonzero $\mathcal{C}^\infty$ function which is constant everywhere except on a small interval.

**12.58 Lemma.** *For each $\delta > 0$ there is a function $\psi$, $\mathcal{C}^\infty$ on $\mathbf{R}$, such that $0 \leq \psi \leq 1$ on $\mathbf{R}$, $\psi(t) = 0$ for $t \leq 0$, and $\psi(t) = 1$ for $t > \delta$.*

PROOF. By Lemma 12.57, choose $\phi \in \mathcal{C}_c^\infty(\mathbf{R})$ such that $\phi(t) > 0$ for $t \in (0, \delta)$ and $\phi(t) = 0$ for $t \notin (0, \delta)$. Set

$$\psi(t) = \frac{\int_0^t \phi(u)\, du}{\int_0^\delta \phi(u)\, du}.$$

By the Fundamental Theorem of Calculus, $\psi \in \mathcal{C}^\infty(\mathbf{R})$, by construction $0 \leq \psi \leq 1$, and

$$\psi(t) = \begin{cases} 0 & t \leq 0 \\ 1 & t > \delta. \end{cases} \quad \blacksquare$$

Finally, we use these one-dimensional $\mathcal{C}^\infty$ functions to construct nonzero functions in $\mathcal{C}_c^\infty(\mathbf{R}^n)$.

**12.59 THEOREM** [$\mathcal{C}^\infty$ VERSION OF URYSOHN'S LEMMA]. *Let $H$ be compact and nonempty, $V$ be open in $\mathbf{R}^n$, and $H \subset V$. Then there is an $h \in \mathcal{C}_c^\infty(\mathbf{R}^n)$ such that $0 \leq h(\mathbf{x}) \leq 1$ for all $\mathbf{x} \in \mathbf{R}^n$, $h(\mathbf{x}) = 1$ for all $\mathbf{x} \in H$, and spt $h \subset V$.*

PROOF. Let $\phi \in \mathcal{C}_c^\infty(\mathbf{R})$ satisfy $\phi(t) > 0$ for $t \in (-1, 1)$ and $\phi(t) = 0$ for $t \notin (-1, 1)$. For each $\varepsilon > 0$ and each $\mathbf{x} \in \mathbf{R}^n$, let $Q_\varepsilon(\mathbf{x})$ represent the $n$-dimensional cube

$$Q_\varepsilon(\mathbf{x}) = \{\mathbf{y} \in \mathbf{R}^n : |y_j - x_j| \leq \varepsilon \text{ for all } j = 1, \ldots, n\}.$$

Set

(44) $$g_\varepsilon(\mathbf{y}) = \phi\left(\frac{y_1}{\varepsilon}\right) \ldots \phi\left(\frac{y_n}{\varepsilon}\right),$$

and observe by Theorem 4.10 (the Product Rule) that $g_\varepsilon$ is $\mathcal{C}^\infty$ on $\mathbf{R}^n$. By construction, $g_\varepsilon(\mathbf{y}) \geq 0$ on $\mathbf{R}^n$, $g_\varepsilon(\mathbf{y}) > 0$ for all $\mathbf{y}$ in the open ball $B_\varepsilon(\mathbf{0})$, and the support of $g_\varepsilon$ is a subset of the cube $Q_\varepsilon(\mathbf{0})$. In particular, $g_\varepsilon \in \mathcal{C}_c^\infty(\mathbf{R}^n)$.

We will use sums of translates of these $g_\varepsilon$'s to construct a $\mathcal{C}^\infty$ function, supported on $V$, which is strictly positive on $H$. It is here that the compactness of $H$ enters in a crucial way.

For each $\boldsymbol{x} \in H$, choose $\varepsilon := \varepsilon(\boldsymbol{x}) > 0$ such that $Q_\varepsilon(\boldsymbol{x}) \subset V$. Set

$$h_{\boldsymbol{x}}(\boldsymbol{y}) = g_\varepsilon(\boldsymbol{y} - \boldsymbol{x}), \qquad \boldsymbol{y} \in \mathbf{R}^n,$$

and notice that $h_{\boldsymbol{x}} \geq 0$ on $\mathbf{R}^n$, $h_{\boldsymbol{x}}(\boldsymbol{y}) > 0$ for all $\boldsymbol{y} \in B_\varepsilon(\boldsymbol{x})$, $h_{\boldsymbol{x}}(\boldsymbol{y}) = 0$ for all $\boldsymbol{y} \notin Q_\varepsilon(\boldsymbol{x})$, and $h_{\boldsymbol{x}} \in \mathcal{C}_c^\infty(\mathbf{R}^n)$. Since $H$ is compact and

$$H \subset \bigcup_{\boldsymbol{x} \in H} B_\varepsilon(\boldsymbol{x}),$$

choose points $\boldsymbol{x}_j \in H$ and positive numbers $\varepsilon_j = \varepsilon(\boldsymbol{x}_j)$, $j = 1, \ldots, N$, such that

$$H \subset B_{\varepsilon_1}(\boldsymbol{x}_1) \cup \cdots \cup B_{\varepsilon_N}(\boldsymbol{x}_N).$$

Set $Q = Q_{\varepsilon_1}(\boldsymbol{x}_1) \cup \cdots \cup Q_{\varepsilon_N}(\boldsymbol{x}_N)$ and $f = h_{\boldsymbol{x}_1} + \cdots + h_{\boldsymbol{x}_N}$. Clearly, $Q$ is compact, $Q \subset V$, and $f$ is $\mathcal{C}^\infty$ on $\mathbf{R}^n$. If $\boldsymbol{x} \notin Q$, then $\boldsymbol{x} \notin Q_{\varepsilon_j}(\boldsymbol{x}_j)$ for all $j$, hence, $f(\boldsymbol{x}) = 0$. Thus $\operatorname{spt} f \subseteq Q$. If $\boldsymbol{x} \in H$, then $\boldsymbol{x} \in B_{\varepsilon_j}(\boldsymbol{x}_j)$ for some $j$, hence, $f(\boldsymbol{x}) > 0$. It remains to flatten $f$ so that it is identically 1 on $H$. This is where Lemma 12.58 comes in.

Since $f > 0$ on the compact set $H$, $f$ has a nonzero minimum on $H$. Thus there is a $\delta > 0$ such that $f(\boldsymbol{x}) > \delta$ for $\boldsymbol{x} \in H$. By Lemma 12.58, choose $\psi \in \mathcal{C}^\infty(\mathbf{R})$ such that $\psi(t) = 0$ when $t \leq 0$, and $\psi(t) = 1$ when $t > \delta$. Set $h = \psi \circ f$. Clearly, $h \in \mathcal{C}_c^\infty(\mathbf{R}^n)$, $\operatorname{spt} h \subseteq Q \subset V$ and, since $f > \delta$ on $H$, $h = 1$ on $H$. Finally, since $0 \leq \psi \leq 1$, the same is true of $h$. ∎

This result leads directly to a decomposition theorem for $\mathcal{C}^\infty$ functions.

**12.60 THEOREM** [$\mathcal{C}^\infty$ PARTITIONS OF UNITY]. *Let $\Omega \subset \mathbf{R}^n$ be nonempty and $\{V_\alpha\}_{\alpha \in A}$ be an open covering of $\Omega$. Then there exist functions $\phi_j \in \mathcal{C}_c^\infty(\mathbf{R}^n)$ and indices $\alpha_j \in A$, $j \in \mathbf{N}$, such that the following properties hold.*

i) 
$$\phi_j \geq 0 \text{ for all } j \in \mathbf{N}.$$

ii) 
$$\operatorname{spt} \phi_j \subset V_{\alpha_j} \text{ for all } j \in \mathbf{N}.$$

iii) 
$$\sum_{j=1}^\infty \phi_j(\boldsymbol{x}) = 1 \text{ for all } \boldsymbol{x} \in \Omega.$$

iv) *If $H$ is a nonempty compact subset of $\Omega$, then there is a nonempty open set $W \supset H$ and an integer $N$ such that $\phi_j(\boldsymbol{x}) = 0$ for all $j \geq N$ and $\boldsymbol{x} \in W$. In particular,*

$$\sum_{j=1}^N \phi_j(\boldsymbol{x}) = 1 \quad \text{for all } \boldsymbol{x} \in W.$$

PROOF. For each $\boldsymbol{x} \in \Omega$, choose a bounded open set $W(\boldsymbol{x})$ and an index $\alpha \in A$ such that

$$\boldsymbol{x} \in W(\boldsymbol{x}) \subset \overline{W}(\boldsymbol{x}) \subset V_\alpha.$$

Then $\mathcal{W} = \{W(\boldsymbol{x}) : \boldsymbol{x} \in \Omega\}$ is an open covering of $\Omega$ and, by Lindelöf's Theorem, we may suppose $\mathcal{W}$ is countable, i.e., $\mathcal{W} = \{W_j\}_{j \in \mathbf{N}}$.

By construction, given $j \in \mathbf{N}$, there is an index $\alpha_j \in A$ such that

$$W_j \subset \overline{W}_j \subset V_{\alpha_j}.$$

Choose by Theorem 12.59 functions $h_j \in \mathcal{C}_c^\infty(\mathbf{R}^n)$ such that $0 \le h_j \le 1$ on $\mathbf{R}^n$, $h_j = 1$ on $\overline{W}_j$, and spt $h_j \subset V_{\alpha_j}$ for $j \in \mathbf{N}$. Set $\phi_1 = h_1$ and for $j > 1$, set

$$\phi_j = (1 - h_1) \dots (1 - h_{j-1}) h_j.$$

Then $\phi_j \ge 0$ on $\mathbf{R}^n$, and $\phi_j \in \mathcal{C}_c^\infty(\mathbf{R}^n)$ with spt $\phi_j \subseteq$ spt $h_j \subset V_{\alpha_j}$ for $j \in \mathbf{N}$. This proves parts i) and ii).

An easy induction argument establishes

$$\sum_{j=1}^{k} \phi_j = 1 - (1 - h_1) \dots (1 - h_k)$$

for $k \in \mathbf{N}$. If $\boldsymbol{x} \in \Omega$, then $\boldsymbol{x} \in W_{j_0}$ for some $j_0$ so $1 - h_{j_0}(\boldsymbol{x}) = 0$. Thus

$$\sum_{j=1}^{k} \phi_j(\boldsymbol{x}) = 1 - 0 = 1$$

for $k \ge j_0$. If $H$ is a compact subset of $\Omega$, then $H \subset W_1 \cup \dots \cup W_N$ for some $N \in \mathbf{N}$. If $W = W_1 \cup \dots \cup W_N$, then $\boldsymbol{x} \in W$ implies $h_k(\boldsymbol{x}) = 1$ for some $1 \le k \le N$, i.e., $\phi_j(\boldsymbol{x}) = 0$ for all $j > N$. Hence,

$$\sum_{j=1}^{N} \phi_j(\boldsymbol{x}) = \sum_{j=1}^{\infty} \phi_j(\boldsymbol{x}) = 1$$

for all $\boldsymbol{x} \in W$. ∎

A sequence of functions $\{\phi_j\}_{j \in \mathbf{N}}$ is called a ($\mathcal{C}^0$) *partition of unity on $\Omega$ subordinate to* a covering $\{V_\alpha\}_{\alpha \in A}$ if $\Omega$ and all the $V_\alpha$'s are open and nonempty, the $\phi_j$'s are all continuous with compact support and satisfy statements i) through iv) of Theorem 12.60. By a $\mathcal{C}^p$ *partition of unity* on $\Omega$ we shall mean a partition of unity on $\Omega$ whose functions $\phi_j$ are also $\mathcal{C}^p$ on $\Omega$. By Theorem 12.60, given any open covering $\mathcal{V}$ of any nonempty set $\Omega \subseteq \mathbf{R}^n$ and any extended real number $p \ge 0$, there exists a $\mathcal{C}^p$ partition of unity on $\Omega$ subordinate to $\mathcal{V}$.

$\mathcal{C}^p$ partitions of unity can be used to decompose a function $f$ into a sum of functions $f_j$ which have small support and are as smooth as $f$. For example, let $f$

be defined on a set $\Omega$, $\{\phi_j\}_{j \in \mathbf{N}}$ be a $\mathcal{C}^p$ partition of unity on $\Omega$ subordinate to a covering $\{V_j\}_{j \in \mathbf{N}}$, and $f_j = f\phi_j$. Then

$$f(\boldsymbol{x}) = f(\boldsymbol{x}) \sum_{j=1}^{\infty} \phi_j(\boldsymbol{x}) = \sum_{j=1}^{\infty} f(\boldsymbol{x})\phi_j(\boldsymbol{x}) = \sum_{j=1}^{\infty} f_j(\boldsymbol{x})$$

for all $\boldsymbol{x} \in \Omega$. If $f$ is continuous on $\Omega$ and $p \geq 0$, then each $f_j$ is continuous on $\Omega$; if $f$ is continuously differentiable on $\Omega$ and $p \geq 1$, then each $f_j$ is continuously differentiable on $\Omega$. Thus, $f$ can be written as a sum of functions $f_j$ which are as smooth as $f$. This allows us to pass from local results to global ones; e.g., if we know that a certain property holds on small open sets in $\Omega$, then we can show that a similar property holds on all of $\Omega$ by using a partition of unity subordinate to a covering of $\Omega$ which consists of small open sets.

To illustrate the power of this point of view, we now show that the integral can be extended from Jordan regions to open bounded sets, even though such sets are not always Jordan regions. This extension is a multidimensional version of the improper integral. (The proofs Theorems 12.64 and 12.65 are based on Spivak [12].[1])

STRATEGY: The idea behind this extension is fairly simple. Let $V$ be a bounded open set and let $f$ be *locally integrable* on $V$; i.e., $f : V \to \mathbf{R}$ is integrable on every closed Jordan region $H \subset V$. For each $\boldsymbol{x} \in V$, choose an open Jordan region $V(\boldsymbol{x})$ so small that $\boldsymbol{x} \in V(\boldsymbol{x}) \subset V$. (For example, $V(\boldsymbol{x})$ could be an open ball.) Then $\{V(\boldsymbol{x})\}_{\boldsymbol{x} \in V}$ is an open covering of $V$, and by Lindelöf's Theorem it has a countable subcover, say $\mathcal{V} = \{V_j\}_{j \in \mathbf{N}}$. Let $\{\phi_j\}_{j \in \mathbf{N}}$ be a partition of unity on $V$ subordinate to $\mathcal{V}$. Since $f$ is locally integrable on $V$, each $f\phi_j$ is integrable. Since $f = \sum_{j=1}^{\infty} f\phi_j$, it seems reasonable to define

$$\int_V f(\boldsymbol{x}) \, d\boldsymbol{x} = \sum_{j=1}^{\infty} \int_{V_j} f(\boldsymbol{x})\phi_j(\boldsymbol{x}) \, d\boldsymbol{x}.$$

Before we can proceed, we must answer two questions. Does this series converge? And if it does, will its value change when the partition of unity changes? The next two results answer these questions.

**12.61 Lemma.** *Let $V$ be a bounded open set in $\mathbf{R}^n$ and $\mathcal{V} = \{V_j\}_{j \in \mathbf{N}}$ be a sequence of nonempty open Jordan regions in $V$ which satisfies*

$$V = \bigcup_{j=1}^{\infty} V_j.$$

*Suppose $f : V \to \mathbf{R}$ is bounded on $V$ and integrable on each $V_j$. If $\{\phi_j\}_{j \in \mathbf{N}}$ is any partition of unity on $V$ subordinate to the covering $\mathcal{V}$, then*

(45)
$$\sum_{j=1}^{\infty} \int_{V_j} \phi_j(\boldsymbol{x}) f(\boldsymbol{x}) \, d\boldsymbol{x}$$

---

[1]M. Spivak, *Calculus on Manifolds*, (New York: W. A. Benjamin, Inc., 1965). Reprinted with permission of Addison-Wesley Publishing Company.

*converges absolutely.*

PROOF. Let $R$ be an $n$-dimensional rectangle containing $V$ and $M = \sup_{\boldsymbol{x} \in V} |f(\boldsymbol{x})|$. Since $\phi_j$ is supported on $V_j$, the function $\phi_j f$ is integrable on $V_j$. Moreover, if $E = \cup_{j=1}^{N} V_j$ we have

$$\sum_{j=1}^{N} \left| \int_{V_j} \phi_j(\boldsymbol{x}) f(\boldsymbol{x}) \, d\boldsymbol{x} \right| \le \sum_{j=1}^{N} \int_{E} |\phi_j(\boldsymbol{x}) f(\boldsymbol{x})| \, d\boldsymbol{x}$$

$$= \int_{E} \sum_{j=1}^{N} |\phi_j(\boldsymbol{x}) f(\boldsymbol{x})| \, d\boldsymbol{x}$$

$$\le M \int_{E} \sum_{j=1}^{N} |\phi_j(\boldsymbol{x})| \, d\boldsymbol{x} \le M \operatorname{Vol}(E) \le M|R| < \infty.$$

Therefore, the series in (45) converges absolutely. ∎

The value of the series in (45) depends neither on the partition of unity chosen nor the covering $\mathcal{V}$.

**12.62 Lemma.** *Let $V$ be a bounded open set in $\mathbf{R}^n$. Suppose that $\mathcal{V} = \{V_j\}_{j \in \mathbf{N}}$ and $\mathcal{W} = \{W_k\}_{k \in \mathbf{N}}$ are sequences of nonempty open Jordan regions in $\mathbf{R}^n$ such that*

$$V = \bigcup_{j=1}^{\infty} V_j = \bigcup_{k=1}^{\infty} W_k.$$

*Suppose further that $f : V \to \mathbf{R}$ is bounded and locally integrable on $V$. If $\{\phi_j\}_{j \in \mathbf{N}}$ is a partition of unity on $V$ subordinate to $\mathcal{V}$ and $\{\psi_k\}_{k \in \mathbf{N}}$ is a partition of unity on $V$ subordinate to $\mathcal{W}$, then*

(46)
$$\sum_{j=1}^{\infty} \int_{V_j} \phi_j(\boldsymbol{x}) f(\boldsymbol{x}) \, d\boldsymbol{x} = \sum_{k=1}^{\infty} \int_{W_k} \psi_k(\boldsymbol{x}) f(\boldsymbol{x}) \, d\boldsymbol{x}.$$

PROOF. By Lemma 12.61, both sums in (46) converge absolutely. By Exercise 5, $\{\phi_j \psi_k\}_{j,k \in \mathbf{N}}$ is a partition of unity on $V$ subordinate to the covering $\{V_j \cap W_k\}_{j,k \in \mathbf{N}}$. Thus

$$\sum_{j=1}^{\infty} \sum_{k=1}^{\infty} \int_{V} \phi_j(\boldsymbol{x}) \psi_k(\boldsymbol{x}) f(\boldsymbol{x}) \, d\boldsymbol{x}$$

also converges absolutely. Fix $j \in \mathbf{N}$. Since $\operatorname{spt} \phi_j$ is compact, choose $N \in \mathbf{N}$ so large that $\psi_k(\boldsymbol{x}) = 0$ for $k > N$ and $\boldsymbol{x} \in \operatorname{spt} \phi_j$. Hence,

$$\int_{V_j} \phi_j(\boldsymbol{x}) f(\boldsymbol{x}) \, d\boldsymbol{x} = \int_{V_j} \phi_j(\boldsymbol{x}) \sum_{k=1}^{N} \psi_k(\boldsymbol{x}) f(\boldsymbol{x}) \, d\boldsymbol{x}$$

$$= \sum_{k=1}^{N} \int_{V_j \cap W_k} \phi_j(\boldsymbol{x}) \psi_k(\boldsymbol{x}) f(\boldsymbol{x}) \, d\boldsymbol{x}$$

$$= \sum_{k=1}^{\infty} \int_{V_j \cap W_k} \phi_j(\boldsymbol{x}) \psi_k(\boldsymbol{x}) f(\boldsymbol{x}) \, d\boldsymbol{x}.$$

Thus

$$\sum_{j=1}^{\infty} \int_{V_j} \phi_j(\boldsymbol{x}) f(\boldsymbol{x}) \, d\boldsymbol{x} = \sum_{j=1}^{\infty} \sum_{k=1}^{\infty} \int_{V_j \cap W_k} \phi_j(\boldsymbol{x}) \psi_k(\boldsymbol{x}) f(\boldsymbol{x}) \, d\boldsymbol{x}.$$

Reversing the roles of $j$ and $k$ we also have

$$\sum_{k=1}^{\infty} \int_{W_k} \psi_k(\boldsymbol{x}) f(\boldsymbol{x}) \, d\boldsymbol{x} = \sum_{k=1}^{\infty} \sum_{j=1}^{\infty} \int_{V_j \cap W_k} \phi_j(\boldsymbol{x}) \psi_k(\boldsymbol{x}) f(\boldsymbol{x}) \, d\boldsymbol{x}.$$

Since these series are absolutely convergent, we may reverse the order of summation in the last double series. ∎

Using Lemma 12.62, we define the integral of a locally integrable function $f$ over a bounded open set $V$ as follows.

**12.63 DEFINITION.** Let $V$ be a bounded, nonempty, open set in $\mathbf{R}^n$ and $f : V \to \mathbf{R}$ be bounded and locally integrable on $V$. The *integral* of $f$ on $V$ is defined to be

$$I_V(f) := \sum_{j=1}^{\infty} \int_{V_j} \phi_j(\boldsymbol{x}) f(\boldsymbol{x}) \, d\boldsymbol{x},$$

where $\{\phi_j\}_{j \in \mathbf{N}}$ is any partition of unity on $V$ subordinate to an open covering $\mathcal{V} = \{V_j\}_{j \in \mathbf{N}}$ such that each $V_j$ is a nonempty Jordan region and

$$V = \bigcup_{j=1}^{\infty} V_j.$$

The following result shows that this definition agrees with the old one when $V$ is a Jordan region. Thus, we shall use the notation $\int_V f(\boldsymbol{x}) \, d\boldsymbol{x}$ for $I_V(f)$.

**12.64 THEOREM.** *If $E$ is a nonempty, open Jordan region in $\mathbf{R}^n$ and $f : E \to \mathbf{R}$ is integrable on $E$, then*

$$\int_E f(\boldsymbol{x}) \, d\boldsymbol{x} = I_E(f).$$

PROOF. Let $\varepsilon > 0$. Since $E$ is a Jordan region, choose a grid $\mathcal{G} = \{Q_1, \ldots, Q_p\}$ of some $n$-dimensional rectangle $R \supset E$ such that

(47) $$\sum_{Q_\ell \cap \partial E \neq \emptyset} |Q_\ell| < \varepsilon.$$

Let

$$H = \bigcup_{Q_\ell \subset E} Q_\ell.$$

Clearly, $H$ is compact and by (47), $\text{Vol}\,(E \setminus H) < \varepsilon$ (see Exercise 4d) in Section 12.1).

Set $M = \sup_{\boldsymbol{x} \in E} |f(\boldsymbol{x})|$. Let $\{R_j\}_{j \in \mathbf{N}}$ be a sequence of rectangles such that $R_j \subset E$ and $E = \bigcup_{j=1}^{\infty} R_j^o$, and let $\{\phi_j\}_{j \in \mathbf{N}}$ be a partition of unity on $E$ subordinate to $\mathcal{V} = \{R_j^o\}_{j \in \mathbf{N}}$. Since $H$ is compact, choose $N_1 \in \mathbf{N}$ such that $\phi_j(\boldsymbol{x}) = 0$ for $j > N_1$ and $\boldsymbol{x} \in H$. Then, for any $N \geq N_1$, we have

$$\left| \int_E f(\boldsymbol{x}) \, d\boldsymbol{x} - \sum_{j=1}^{N} \int_{R_j} \phi_j(\boldsymbol{x}) f(\boldsymbol{x}) \, d\boldsymbol{x} \right| = \left| \int_E f(\boldsymbol{x}) \, d\boldsymbol{x} - \sum_{j=1}^{N} \int_E \phi_j(\boldsymbol{x}) f(\boldsymbol{x}) \, d\boldsymbol{x} \right|$$

$$\leq \int_E \left| f(\boldsymbol{x}) - \sum_{j=1}^{N} \phi_j(\boldsymbol{x}) f(\boldsymbol{x}) \right| \, d\boldsymbol{x}$$

$$\leq M \int_E \left| 1 - \sum_{j=1}^{N} \phi_j(\boldsymbol{x}) \right| \, d\boldsymbol{x}$$

$$\leq M \operatorname{Vol}(E \setminus H) < M\varepsilon.$$

We conclude that $I_E(f)$ exists and equals $\int_E f(\boldsymbol{x}) \, d\boldsymbol{x}$. ∎

We now prove a change of variables formula valid for all open bounded sets.

**12.65 THEOREM.** *Suppose $V$ is a bounded, nonempty, open set in $\mathbf{R}^n$, $\phi : V \to \mathbf{R}^n$ is 1–1 and continuously differentiable on $V$, and $\phi(V)$ is bounded. If $\Delta_\phi \neq 0$ on $V$, then*

$$\int_{\phi(V)} f(\boldsymbol{u}) \, d\boldsymbol{u} = \int_V f(\phi(\boldsymbol{x})) |\Delta_\phi(\boldsymbol{x})| \, d\boldsymbol{x},$$

*for all bounded $f : \phi(V) \to \mathbf{R}$, provided $f$ is locally integrable on $\phi(V)$ and $(f \circ \phi) |\Delta_\phi|$ is locally integrable on $V$.*

PROOF. For each $\boldsymbol{a} \in V$, choose by Theorem 12.45 an open rectangle $W(\boldsymbol{a})$ such that $\overline{W}(\boldsymbol{a}) \subset V$ and

$$(48) \qquad \int_{\phi(W(\boldsymbol{a}))} f(\boldsymbol{u}) \, d\boldsymbol{u} = \int_{W(\boldsymbol{a})} f(\phi(\boldsymbol{x})) |\Delta_\phi(\boldsymbol{x})| \, d\boldsymbol{x}.$$

Set $\mathcal{W} = \{W(\boldsymbol{a})\}_{\boldsymbol{a} \in V}$. Then $\mathcal{W}$ is an open covering of $V$. By Lindelöf's Theorem, we may assume that $\mathcal{W} = \{W_j\}_{j \in \mathbf{N}}$. Let $\{\phi_j\}_{j \in \mathbf{N}}$ be a partition of unity on $V$ subordinate to $\mathcal{W}$, i.e., a sequence of $\mathcal{C}^\infty$ functions such that

$$\operatorname{spt} \phi_j \subset W_j \subset V, \quad j \in \mathbf{N}, \quad \text{and} \quad \sum_{j=1}^{\infty} \phi_j(\boldsymbol{x}) = 1$$

for all $\boldsymbol{x} \in V$. By Theorem 12.52, each $\phi(W_j)$ is a Jordan region. By Exercise 9 in Section 11.5, each $\phi(W_j)$ is open. And by Exercise 4 below, $\{\phi_j \circ \phi^{-1}\}_{j \in \mathbf{N}}$ is a partition of unity on $\phi(V)$ subordinate to the open covering $\{\phi(W_j)\}_{j \in \mathbf{N}}$. Hence, by Definition 12.63 and (48),

$$\int_{\phi(V)} f(\boldsymbol{u})\,d\boldsymbol{u} = \sum_{j=1}^{\infty} \int_{\phi(W_j)} (\phi_j \circ \phi^{-1})(\boldsymbol{u}) f(\boldsymbol{u})\,d\boldsymbol{u}$$

$$= \sum_{j=1}^{\infty} \int_{W_j} \phi_j(\boldsymbol{x}) f(\phi(\boldsymbol{x})) |\Delta_\phi(\boldsymbol{x})|\,d\boldsymbol{x}$$

$$= \int_V f(\phi(\boldsymbol{x})) |\Delta_\phi(\boldsymbol{x})|\,d\boldsymbol{x}. \quad \blacksquare$$

Finally, we are prepared to prove a change of variables formula for functions whose Jacobians are zero on a set of volume zero.

**12.66 THEOREM** [CHANGE OF VARIABLES FOR MULTIPLE INTEGRALS]. *Suppose $W$ is open in $\mathbf{R}^n$, $\phi : W \to \mathbf{R}^n$ is continuously differentiable, and $E$ is a Jordan region such that $\overline{E} \subset W$. If $\phi(E)$ is a Jordan region and there exists a closed set $Z$ such that $E \cap Z$ is of volume zero, and $\phi$ is 1–1 and $\Delta_\phi(\boldsymbol{x}) \neq 0$ for all $\boldsymbol{x} \in E^o \setminus Z$, then*

$$\int_{\phi(E)} f(\boldsymbol{u})\,d\boldsymbol{u} = \int_E f(\phi(\boldsymbol{x})) |\Delta_\phi(\boldsymbol{x})|\,d\boldsymbol{x}$$

*provided $f$ is integrable on $\phi(E)$ and $f \circ \phi |\Delta_\phi|$ is integrable on $E$.*

PROOF. Set $V := E^o \setminus Z$ and observe that $V$ is open and bounded. Since $\phi(E) \supseteq \phi(V)$, $\phi(V)$ is also bounded. By hypothesis, $E \setminus E^o \subseteq \partial E$ and $E \cap Z$ are of volume zero. Moreover, by Exercise 7a) in Section 12.1, $\phi(E \setminus E^o)$ and $\phi(E \cap Z)$ are of volume zero. Since $E = V \cup (E \cap Z) \cup (E \setminus E^o)$ and $\phi(E) = \phi(V) \cup \phi(E \cap Z) \cup \phi(E \setminus E^o)$, it follows from Theorems 12.23, 12.24, and 12.65 that

$$\int_{\phi(E)} f(\boldsymbol{u})\,d\boldsymbol{u} = \int_{\phi(V)} f(\boldsymbol{u})\,d\boldsymbol{u}$$

$$= \int_V f(\phi(\boldsymbol{x})) |\Delta_\phi(\boldsymbol{x})|\,d\boldsymbol{x} = \int_E f(\phi(\boldsymbol{x})) |\Delta_\phi(\boldsymbol{x})|\,d\boldsymbol{x}. \quad \blacksquare$$

We close by noting that, as general as it is, even this result can be improved. If something called the Lebesgue integral is used instead of the Riemann integral, the integrability condition concerning $(f \circ \phi)|\Delta_\phi|$ and the condition $\Delta_\phi \neq 0$ can be dropped altogether (see Apostol [1], p. 421, and Spivak [12], p. 72, respectively.)

## EXERCISES

**1.** If $f, g : \mathbf{R}^n \to \mathbf{R}$, prove spt $(fg) \subseteq$ spt $f \cap$ spt $g$.

**2.** Prove that if $f, g \in \mathcal{C}_c^\infty(\mathbf{R}^n)$, then so are $fg$ and $\alpha f$ for any scalar $\alpha$.

**\*3.** Prove that if $f$ is analytic on $\mathbf{R}$ and $f(x_0) \neq 0$ for some $x_0 \in \mathbf{R}$, then $f \notin \mathcal{C}_c^\infty(\mathbf{R})$.

**4.** Suppose $V$ is a bounded, nonempty, open set in $\mathbf{R}^n$ and $\phi : V \to \mathbf{R}^n$ is 1–1 and continuously differentiable on $V$ with $\Delta_\phi \neq 0$ on $V$. Let $\mathcal{W} = \{W_j\}_{j\in\mathbf{N}}$ be an open covering of $V$ and $\{\phi_j\}_{j\in\mathbf{N}}$ be a $\mathcal{C}^p$ partition of unity on $V$ subordinate to $\mathcal{W}$, where $p \geq 1$. Prove $\{\phi_j \circ \phi^{-1}\}_{j\in\mathbf{N}}$ is a $\mathcal{C}^1$ partition of unity on $\phi(V)$ subordinate to the open covering $\{\phi(W_j)\}_{j\in\mathbf{N}}$.

**5.** Let $V$ be open in $\mathbf{R}^n$ and $\mathcal{V} = \{V_j\}_{j\in\mathbf{N}}$, $\mathcal{W} = \{W_k\}_{k\in\mathbf{N}}$ be coverings of $V$. If $\{\phi_j\}_{j\in\mathbf{N}}$ is a $\mathcal{C}^p$ partition of unity on $V$ subordinate to $\mathcal{V}$ and $\{\psi_k\}_{k\in\mathbf{N}}$ is a $\mathcal{C}^p$ partition of unity on $V$ subordinate to $\mathcal{W}$, prove that $\{\phi_j\psi_k\}_{j,k\in\mathbf{N}}$ is a $\mathcal{C}^p$ partition of unity on $V$ subordinate to the covering $\{V_j \cap W_k\}_{j,k\in\mathbf{N}}$.

**6.** Show that, given any compact Jordan region $H \subset \mathbf{R}^n$, there is a sequence of $\mathcal{C}^\infty$ functions $\phi_j$ such that

$$\lim_{j\to\infty} \int_{\mathbf{R}^n} \phi_j(\boldsymbol{x})\, d\boldsymbol{x} = \text{Vol}\,(H).$$

$^e$**12.6 THE GAMMA FUNCTION AND VOLUME** *The last result of this section uses Dini's Theorem from Section 9.5.*

In this section we introduce the gamma function and use it to find a formula for the volume of any $n$-dimensional ball and an asymptotic estimate of $n!$.

Recall that if $f : (0, \infty) \to \mathbf{R}$ is locally integrable on $(0, \infty)$, then

$$\int_0^\infty f(t)\, dt = \lim_{\substack{x\to 0+ \\ y\to\infty}} \int_x^y f(t)\, dt.$$

In particular, it is easy to check that $\int_0^\infty e^{-\alpha t}\, dt$ is finite for all $\alpha > 0$.

**12.67 DEFINITION.** The *gamma function* is defined by

$$\Gamma(x) = \int_0^\infty t^{x-1} e^{-t}\, dt, \qquad x \in (0, \infty),$$

when this (improper) integral converges.

By definition,

$$\Gamma(1) = \int_0^\infty e^{-t}\, dt = 1,$$

and

$$\Gamma(1/2) = \int_0^\infty t^{-1/2} e^{-t}\, dt = 2 \int_0^\infty e^{-u^2}\, du = \sqrt{\pi}.$$

(We used the change of variables $t = u^2$ and Exercise 10 in Section 12.4.) It turns out that $\Gamma(x)$ is defined for all $x \in (0, \infty)$.

**12.68 THEOREM.** *For each* $x \in (0, \infty)$, $\Gamma(x)$ *exists and is finite,* $\Gamma(x + 1) = x\Gamma(x)$, *and* $\Gamma(n) = (n - 1)!$ *for* $n \in \mathbf{N}$.

PROOF. Write

$$\Gamma(x) = \int_0^1 t^{x-1} e^{-t} \, dt + \int_1^\infty t^{x-1} e^{-t} \, dt =: I_1 + I_2.$$

By l'Hôpital's Rule,

$$\lim_{t \to \infty} e^{-t/2} t^y = 0$$

for all $y \in \mathbf{R}$. Hence, $e^{-t} t^{x-1} \le e^{-t/2}$ for $t$ large and it follows, from Theorem 5.43 (the Comparison Theorem), that $I_2$ is finite for all $x \in \mathbf{R}$.

To show $I_1$ is finite for $x > 0$, suppose first that $x \ge 1$. Then $t^{x-1} \le 1$ for all $t \in [0, 1]$ and

$$I_1 = \int_0^1 t^{x-1} e^{-t} \, dt \le \int_0^1 e^{-t} \, dt = 1 - \frac{1}{e} < \infty.$$

Therefore, $\Gamma(x)$ is finite for all $x \ge 1$. Next, suppose $0 < x < 1$. Then $x + 1 \ge 1$, so $\Gamma(x + 1)$ is finite. Integration by parts yields

$$\Gamma(x) = \int_0^\infty t^{x-1} e^{-t} \, dt = \frac{t^x e^{-t}}{x} \Big|_{t=0}^\infty + \frac{1}{x} \int_0^\infty t^x e^{-t} \, dt = \frac{1}{x} \Gamma(x + 1).$$

Therefore, $\Gamma(x)$ is finite when $0 < x < 1$.

This argument also verifies $x\Gamma(x) = \Gamma(x + 1)$ for $x \in (0, \infty)$. Since $\Gamma(1) = 1$, it follows that $\Gamma(n) = (n - 1)!$ for all $n \in \mathbf{N}$. ∎

The gamma function can be used to evaluate certain integrals which cannot be evaluated by using elementary techniques of integration.

**12.69 THEOREM.** *If* $x, y \in (0, \infty)$, *then*

i)
$$\int_0^1 v^{y-1} (1 - v)^{x-1} \, dv = \frac{\Gamma(x)\Gamma(y)}{\Gamma(x + y)},$$

*and*

ii)
$$\int_0^{\pi/2} \cos^{2x-1} \varphi \sin^{2y-1} \varphi \, d\varphi = \frac{\Gamma(x)\Gamma(y)}{2\Gamma(x + y)}.$$

*In particular,*

iii)
$$\int_0^\pi \sin^{k-2} \varphi \, d\varphi = \frac{\Gamma((k - 1)/2)\Gamma(1/2)}{\Gamma(k/2)}$$

*holds for all integers $k > 2$.*

PROOF. To prove part i), make the change of variables $v = u/(1 + u)$ and write

$$\int_0^1 v^{y-1}(1 - v)^{x-1}\, dv = \int_0^\infty \left(\frac{u}{1+u}\right)^{y-1}\left(1 - \frac{u}{1+u}\right)^{x-1}\frac{du}{(1+u)^2}$$

$$= \int_0^\infty u^{y-1}\left(\frac{1}{1+u}\right)^{x+y} du.$$

It follows from two more changes of variables ($s = t/(1 + u)$ and $w = su$) and Fubini's Theorem that

$$\Gamma(x + y)\int_0^1 v^{y-1}(1 - v)^{x-1}\, dv$$

$$= \int_0^\infty \int_0^\infty u^{y-1}\left(\frac{1}{1+u}\right)^{x+y} t^{x+y-1} e^{-t}\, dt\, du$$

$$= \int_0^\infty \int_0^\infty u^{y-1} s^{x+y-1} e^{-s(u+1)}\, ds\, du$$

$$= \int_0^\infty s^{x-1} e^{-s}\left(\int_0^\infty u^{y-1} s^y e^{-su}\, du\right) ds$$

$$= \int_0^\infty s^{x-1} e^{-s}\left(\int_0^\infty w^{y-1} e^{-w}\, dw\right) ds = \Gamma(x)\Gamma(y).$$

To prove part ii) use the change of variables $v = \sin^2\varphi$ and part i) to verify

$$\int_0^{\pi/2} \cos^{2x-1}\varphi \sin^{2y-1}\varphi\, d\varphi = \frac{1}{2}\int_0^1 v^{y-1}(1 - v)^{x-1}\, dv = \frac{\Gamma(x)\Gamma(y)}{2\Gamma(x + y)}.$$

Specializing to the case $y = (k - 1)/2$ and $x = 1/2$, we obtain part iii). ∎

The connection between the gamma function and volume is contained in the following result.

**12.70 THEOREM.** *If $r > 0$ and $\mathbf{a} \in \mathbf{R}^n$, then*

$$\text{Vol}(B_r(\mathbf{a})) = \frac{2r^n \pi^{n/2}}{n\Gamma(n/2)}.$$

PROOF. By translation invariance (see Exercise 3 in Section 12.1) and Theorem 12.22, $\text{Vol}(B_r(\mathbf{a})) = \int_B 1\, d\mathbf{x}$ for $B = B_r(\mathbf{0})$. We suppose for simplicity that $n \geq 2$, and introduce a change of variables in $\mathbf{R}^n$ analogous to spherical coordinates. Namely, let

$$x_1 = \rho\cos\varphi_1, \quad x_2 = \rho\sin\varphi_1\cos\varphi_2, \quad x_3 = \rho\sin\varphi_1\sin\varphi_2\cos\varphi_3, \quad \dots,$$

$$x_{n-1} = \rho\sin\varphi_1 \dots \sin\varphi_{n-2}\cos\theta, \quad \text{and} \quad x_n = \rho\sin\varphi_1 \dots \sin\varphi_{n-2}\sin\theta,$$

where $0 \leq \rho \leq r$, $0 \leq \theta \leq 2\pi$, and $0 \leq \varphi_j \leq \pi$ for $j = 1, \ldots, n - 2$. An easy induction argument shows this change of variables has Jacobian

$$(49) \qquad \Delta := \rho^{n-1} \sin^{n-2} \varphi_1 \sin^{n-3} \varphi_2 \ldots \sin^2 \varphi_{n-3} \sin \varphi_{n-2}.$$

Hence, by Theorems 12.66 and 12.69iii),

$$
\begin{aligned}
\mathrm{Vol}\,(B_r(\boldsymbol{a})) &= \int_B 1 \, d\boldsymbol{x} \\
&= \int_0^r \int_0^\pi \cdots \int_0^\pi \int_0^{2\pi} \rho^{n-1} \sin^{n-2} \varphi_1 \ldots \sin \varphi_{n-2} \, d\theta \, d\varphi_1 \ldots d\varphi_{n-2} \, d\rho \\
&= \frac{2\pi r^n}{n} \left( \int_0^\pi \sin^{n-2} \varphi \, d\varphi \right) \cdots \left( \int_0^\pi \sin \varphi \, d\varphi \right) \\
&= \frac{2\pi r^n}{n} \cdot \frac{\Gamma((n-1)/2)\Gamma(1/2)}{\Gamma(n/2)} \cdot \frac{\Gamma((n-2)/2)\Gamma(1/2)}{\Gamma((n-1)/2)} \cdots \frac{\Gamma(1)\Gamma(1/2)}{\Gamma(3/2)}.
\end{aligned}
$$

Canceling all superfluous factors and substituting the value $\sqrt{\pi}$ for $\Gamma(1/2)$, we conclude that

$$\mathrm{Vol}\,(B_r(\boldsymbol{a})) = \frac{2\pi r^n}{n} \left( \frac{\Gamma^{n-2}(1/2)}{\Gamma(n/2)} \right) = \frac{2r^n \pi^{n/2}}{n\Gamma(n/2)}. \quad \blacksquare$$

This formula agrees with what we already know. For $n = 1$ we have

$$\mathrm{Vol}\,(B_r(0)) = \frac{2r\pi^{1/2}}{\Gamma(1/2)} = 2r,$$

for $n = 2$ we have

$$\mathrm{Vol}\,(B_r(0,0)) = \frac{2r^2 \pi}{2\Gamma(1)} = \pi r^2,$$

and for $n = 3$ we have

$$\mathrm{Vol}\,(B_r(0,0,0)) = \frac{2r^3 \pi^{3/2}}{3\Gamma(3/2)} = \frac{2r^3 \pi^{3/2}}{(3/2)\Gamma(1/2)} = \frac{4}{3}\pi r^3.$$

We close this section with an asymptotic estimate of $n!$. First, we obtain an integral representation for $n!/(n^{n+1/2}e^{-n})$.

**12.71 Lemma.** *If* $\phi(x) = x - \log x - 1$, $x > 0$, *then*

$$\frac{n!}{n^{n+1/2}e^{-n}} = \int_{-\sqrt{n}}^\infty e^{-n\phi(1+t/\sqrt{n})} \, dt.$$

PROOF. By Definition 12.67 and Theorem 12.68, we can write

$$n! = \Gamma(n+1) = \int_0^\infty x^n e^{-x} \, dx.$$

Making two changes of variables (first $x = ny$, then $y = 1 + t/\sqrt{n}$), we conclude that

$$
\frac{n!}{n^{n+1/2}e^{-n}} = \frac{1}{\sqrt{n}} \int_0^\infty \left(\frac{x}{n}\right)^n e^{-x+n}\, dx
$$

$$
= \sqrt{n} \int_0^\infty y^n e^{-n(y-1)}\, dy
$$

$$
= \sqrt{n} \int_0^\infty e^{-n\phi(y)}\, dy = \int_{-\sqrt{n}}^\infty e^{-n\phi(1+t/\sqrt{n})}\, dt. \quad \blacksquare
$$

Next, we derive some inequalities which will be used, in conjunction with Dini's Theorem, to evaluate the limit of the integral which appears in Lemma 12.71.

**12.72 Lemma.** *If $\phi(x) = x - \log x - 1$, $x > 0$, then*

$$
(x-1)\phi'(x) - 2\phi(x) > 0 \quad \text{for } 0 < x < 1,
$$

*and*

$$
(x-1)\phi'(x) - 2\phi(x) < 0 \quad \text{for } x > 1.
$$

*Moreover, there is an absolute constant $M > 0$ such that*

$$
(50) \qquad \phi(x) \geq M(x-1)^2 \quad \text{for } 0 < x < 2,
$$

*and*

$$
(51) \qquad \phi(x) \geq M(x-1) \quad \text{for } x \geq 2.
$$

PROOF. Let $\psi(x) = 2\log x - x + 1/x$ and observe that $(x-1)\phi'(x) - 2\phi(x) = \psi(x)$. Since $\psi'(x) = -(x-1)^2/x^2 < 0$ for all $x \neq 1$, $\psi$ is decreasing on $(0, \infty)$. Since $\psi(1) = 0$, it follows that $\psi > 0$ on $(0, 1)$ and $\psi < 0$ on $(1, \infty)$. This proves the first pair of inequalities.

To prove the second pair of inequalities, observe first that, by Taylor's Formula,

$$
\phi(x) = \phi(1) + \phi'(1)(x-1) + \phi''(c)\frac{(x-1)^2}{2!} = \frac{(x-1)^2}{2c^2}
$$

for some $c$ between $x$ and 1. Thus $\phi(x) \geq (x-1)^2/8$ for all $0 < x < 2$. Next, observe, since $\phi(x) > 0$ for $x > 1$ and $\phi(x)/(x-1) \to 1$ as $x \to \infty$, that $\phi(x)/(x-1)$ has a positive minimum, say $m$, on $[2, \infty)$. Thus (50) and (51) hold for $M := \min\{m, 1/8\}$. $\blacksquare$

Our final preliminary result evaluates the limit of the integral which appears in Lemma 12.71.

**12.73 Lemma.** If $\phi(x) = x - \log x - 1$ for $x > 0$, and $F_n(t) = e^{-n\phi(1+t/\sqrt{n})}$ for $n \in \mathbf{N}$ and $t > -\sqrt{n}$, then

$$\lim_{n \to \infty} \int_{-\sqrt{n}}^{\infty} F_n(t)\, dt = \int_{-\infty}^{\infty} e^{-t^2/2}\, dt.$$

STRATEGY: The idea behind the proof is simple. By l'Hôpital's Rule,

$$\lim_{n \to \infty} n\phi\left(1 + \frac{t}{\sqrt{n}}\right) = \lim_{n \to \infty} \frac{t}{2} \frac{\phi'(1 + t/\sqrt{n})}{1/\sqrt{n}} = \frac{t^2}{2} \lim_{n \to \infty} \phi''\left(1 + \frac{t}{\sqrt{n}}\right) = \frac{t^2}{2},$$

so $F_n(t) \to e^{-t^2/2}$, as $n \to \infty$, for every $t \in \mathbf{R}$. Thus $\int_{-\sqrt{n}}^{\infty} F_n(t)\, dt$ should converge to $\int_{-\infty}^{\infty} e^{-t^2/2}\, dt$ as $n \to \infty$. To prove this, we break the integral over $(-\sqrt{n}, \infty)$ into three pieces: one over $(-\sqrt{n}, -\sqrt{a})$, one over $(\sqrt{a}, \infty)$, and one over $(-\sqrt{a}, \sqrt{a})$. Since $e^{-t^2/2}$ is integrable on $(-\infty, \infty)$, the first two integrals should be small for $a$ sufficiently large. Once $a$ is fixed, we shall use Dini's Theorem on the third integral. Here are the details.

PROOF. Let $\varepsilon > 0$ and observe that

$$\left| \int_{-\sqrt{n}}^{\infty} F_n(t)\, dt - \int_{-\sqrt{n}}^{\infty} e^{-t^2/2}\, dt \right|$$

$$\leq I_1 + I_2 + I_3 + I_4$$

$$:= \left| \int_{-\sqrt{a}}^{\sqrt{a}} \left( F_n(t) - e^{-t^2/2} \right) dt \right| + \int_{|t| \geq \sqrt{a}} e^{-t^2/2}\, dt$$

$$\int_{\sqrt{a}}^{\infty} |F_n(t)|\, dt + \int_{-\sqrt{n}}^{-\sqrt{a}} |F_n(t)|\, dt$$

for any $a > 0$ and $n \in \mathbf{N}$, provided $n > a$. Hence, it suffices to prove that $|I_j| \leq \varepsilon/4$ for $j = 1, 2, 3, 4$, and $n, a$ sufficiently large.

Let $M$ be the constant given in Lemma 12.72, and choose $a > 0$ so large that

(52)
$$\int_{|t| \geq \sqrt{a}} e^{-Mt^2}\, dt < \frac{\varepsilon}{4}, \qquad \int_{\sqrt{a}}^{\infty} e^{-Mt}\, dt < \frac{\varepsilon}{4},$$

and

(53)
$$\int_{|t| \geq \sqrt{a}} e^{-t^2/2}\, dt < \frac{\varepsilon}{4}.$$

By (53), $|I_2| < \varepsilon/4$.

To estimate $|I_j|$ for $j \neq 2$, fix $t > -\sqrt{a}$ and consider the function $G(x) = e^{-x\phi(1+t/\sqrt{x})}$, $x > 0$. By the Product Rule,

$$G'(x) = e^{-x\phi(1+t/\sqrt{x})} \left( \frac{t}{2\sqrt{x}} \phi'\left(1 + \frac{t}{\sqrt{x}}\right) - \phi\left(1 + \frac{t}{\sqrt{x}}\right) \right)$$

$$= \frac{e^{-x\phi(y)}}{2} ((y-1)\phi'(y) - 2\phi(y)),$$

where $y = 1 + t/\sqrt{x}$. Thus by Lemma 12.72, $G'(x) > 0$ for $x > a$, $-\sqrt{a} < t < 0$, and $G'(x) < 0$ for $x > 0$, $t > 0$. It follows that for each $t \in (-\sqrt{a}, 0)$, $F_n(t) \uparrow e^{-t^2/2}$ as $n \to \infty$, and for each $t \in (0, \infty)$, $F_n(t) \downarrow e^{-t^2/2}$ as $n \to \infty$. Hence, by Dini's Theorem (Theorem 9.41),

$$\int_{-\sqrt{a}}^{\sqrt{a}} F_n(t) \, dt \to \int_{-\sqrt{a}}^{\sqrt{a}} e^{-t^2/2} \, dt$$

as $n \to \infty$. Thus, we can choose an $N \in \mathbf{N}$ so large that $n \geq N$ implies $|I_1| < \varepsilon$. It remains to estimate $|I_j|$ for $j = 3, 4$.

To this end, let $n > \max\{N, a\}$. By (50) and (51),

$$n\phi\left(1 + \frac{t}{\sqrt{n}}\right) \geq nM\frac{t^2}{n} = Mt^2 \quad \text{for } -\sqrt{n} < t < \sqrt{n},$$

and

$$n\phi\left(1 + \frac{t}{\sqrt{n}}\right) \geq nM\frac{t}{\sqrt{n}} \geq Mt \quad \text{for } t \geq \sqrt{n}.$$

Since $n > a$, it follows that

$$|I_3| + |I_4| = \int_{\sqrt{a}}^{\infty} |F_n(t)| \, dt + \int_{-\sqrt{n}}^{-\sqrt{a}} |F_n(t)| \, dt$$

$$\leq \int_{\sqrt{a} \leq |t| \leq \sqrt{n}} e^{-Mt^2} \, dt + \int_{\sqrt{n}}^{\infty} e^{-Mt} \, dt$$

$$< \int_{|t| \geq \sqrt{a}} e^{-Mt^2} \, dt + \int_{\sqrt{a}}^{\infty} e^{-Mt} \, dt.$$

We conclude by (52) that $|I_3| + |I_4| < \varepsilon/2$ as required. ∎

**12.74 THEOREM** [STIRLING'S FORMULA]. *For $n \in \mathbf{N}$ sufficiently large, $n! \approx \sqrt{2\pi}(n^{n+1/2})e^{-n}$, i.e.,*

$$\lim_{n \to \infty} \frac{n!}{\sqrt{2\pi}(n^{n+1/2})e^{-n}} = 1.$$

PROOF. By Exercise 10 in Section 12.4 and the change of variables $t = \sqrt{2}u$, we have

$$\int_{-\infty}^{\infty} e^{-t^2/2} \, dt = \sqrt{2} \int_{-\infty}^{\infty} e^{-u^2} \, du = 2\sqrt{2} \int_{0}^{\infty} e^{-u^2} \, du = \sqrt{2\pi}.$$

Therefore, it follows from Lemmas 12.71 and 12.73 that

$$\lim_{n \to \infty} \frac{n!}{\sqrt{2\pi}n^{n+1/2}e^{-n}} = \lim_{n \to \infty} \frac{1}{\sqrt{2\pi}} \int_{-\sqrt{n}}^{\infty} e^{-n\phi(1+t/\sqrt{n})} \, dt$$

$$= \frac{1}{\sqrt{2\pi}} \int_{-\infty}^{\infty} e^{-t^2/2} \, dt = 1. \quad \blacksquare$$

**EXERCISES**

**1.** Show
$$\int_0^\infty t^2 e^{-t^2}\, dt = \frac{\sqrt{\pi}}{4}.$$

**2.** Show
$$\int_0^1 \frac{dx}{\sqrt{-\log x}} = \sqrt{\pi}$$

**3.** Show
$$\int_{-\infty}^\infty e^{\pi t - e^t}\, dt = \Gamma(\pi).$$

**4.** Show that the volume of a four-dimensional ball of radius $r$ is $\pi^2 r^4 / 2$, and the volume of a five-dimensional ball of radius $r$ is $8\pi^2 r^5 / 15$.

**5.** Verify (49).

**6.** For $n > 2$, prove that the volume of the $n$-dimensional ellipsoid
$$E = \left\{ (x_1, \ldots, x_n) : \frac{x_1^2}{a_1^2} + \frac{x_2^2}{a_2^2} + \cdots + \frac{x_n^2}{a_n^2} \leq 1 \right\}$$
is
$$\mathrm{Vol}\,(E) = \frac{2a_1 \ldots a_n \pi^{n/2}}{n\Gamma(n/2)}.$$

**7.** For $n > 2$, prove that the volume of the $n$-dimensional cone
$$C = \{ (x_1, \ldots, x_n) : (h/r)\sqrt{x_2^2 + \ldots x_n^2} \leq x_1 \leq h \}$$
is
$$\mathrm{Vol}\,(C) = \frac{2hr^{n-1}\pi^{(n-1)/2}}{n(h-1)\Gamma((n-1)/2)}.$$

**8.** Find the value of
$$\int_{B_r(\mathbf{0})} x_k^2 \, d(x_1, \ldots, x_n)$$
for each $k \in \mathbf{N}$.

**9.** If $f : B_1(\mathbf{0}) \to \mathbf{R}$ is differentiable with
$$f(\mathbf{0}) = 0 \quad \text{and} \quad \|\nabla f(\boldsymbol{x})\| \leq 1$$
for $\boldsymbol{x} \in B_1(\mathbf{0})$, prove that the following exists and equals 0.
$$\lim_{k \to \infty} \int_{B_1(\mathbf{0})} |f(\boldsymbol{x})|^k \, d\boldsymbol{x}.$$

**10.** a) Prove that $\Gamma$ is differentiable on $(0, \infty)$ with
$$\Gamma'(x) = \int_0^\infty e^{-t} t^{x-1} \log t \, dt.$$
*b) Prove that $\Gamma$ is $\mathcal{C}^\infty$ and convex on $(0, \infty)$.

# Chapter 13

# Fundamental Theorems of Vector Calculus

This chapter is more descriptive and less rigorous than its predecessors. Our goal is to lay a practical foundation for an abstract Chapter 15.

## 13.1  CURVES

According to the dictionary, a curve is a smooth line which bends, without corners; a one-dimensional object with length but no breadth. Of course, this definition is too imprecise. It is also too restrictive. Our concept of a curve will include not only "smooth" objects like the graphs of the function $y = x^2$ and the relation $x^2 + y^2 = 1$, but also objects with "corners" like the graph of $y = |x|$.

Recall that if $I \subseteq \mathbf{R}$ and $\phi : I \to \mathbf{R}^m$, then the image of $I$ under $\phi$ is the set

$$\phi(I) = \{\boldsymbol{x} \in \mathbf{R}^m : \boldsymbol{x} = \phi(t) \quad \text{for some } t \in I\}.$$

Let $\boldsymbol{x}_0, \boldsymbol{a} \in \mathbf{R}^m$. It is easy to see that the image of $\mathbf{R}$ under $\phi(t) := t\boldsymbol{a} + \boldsymbol{x}_0$ is a straight line (see Exercise 2). We shall call it the *straight line* through $\boldsymbol{x}_0$ *in the direction* of $\boldsymbol{a}$. This is the simplest type of curve in $\mathbf{R}^m$.

A naive attempt to define a general curve in $\mathbf{R}^m$ is to insist that it be the image of an interval under some continuous function $\phi : \mathbf{R} \to \mathbf{R}^m$. It turns out that this definition is too broad. There are continuous functions (called "space-filling curves") which take the unit interval $[0, 1]$ onto the unit square $[0, 1] \times [0, 1]$ (see Boas [2]). One way to fix this definition is to use homeomorphisms, i.e., continuous functions whose inverses are also continuous. Since we are primarily interested in the differential structure of curves, we take a different approach, using differentiable functions to define curves (see Definition 13.1 below).

We begin by extending the definition of partial differentiation to include functions defined on nonopen domains. Let $m, n, p \in \mathbf{N}$, and $E$ be a nonempty subset of $\mathbf{R}^n$. A function $f : E \to \mathbf{R}^m$ is said to be $\mathcal{C}^p$ (on $E$) if there is an open set $V \supseteq E$ and a function $g : V \to \mathbf{R}^m$ whose partial derivatives of orders $j \leq p$ exist and are continuous on $V$ such that $f(\boldsymbol{x}) = g(\boldsymbol{x})$ for all $\boldsymbol{x} \in E$. In this case we define the *partial derivatives* of $f$ to be equal to the partial derivatives of $g$; e.g., $\partial f_j / \partial x_k (\boldsymbol{x}) = \partial g_j / \partial x_k (\boldsymbol{x})$ for $k = 1, 2, \ldots, n$, $j = 1, 2, \ldots, m$, and $\boldsymbol{x} \in E$. A function $f : E \to \mathbf{R}^m$ is said to be $\mathcal{C}^\infty$ (on $E$) if $f$ is $\mathcal{C}^p$ on $E$ for all $p \in \mathbf{N}$. (Notice that this agrees with Definition 4.6 when $n = 1$.)

Henceforth, $p$ will denote an element of $\mathbf{N}$ or the extended real number $\infty$.

For technical reasons (orientation is the key word), we want a curve to be more than just a set of points. Hence, we define a general curve in the following way.

### DEFINITION 13.1.

   i) A $(\mathcal{C}^p)$ *curve* in $\mathbf{R}^m$ is a pair $C = (\phi, I)$, where $I$ is a nondegenerate interval (bounded or unbounded) and $\phi : I \to \mathbf{R}^m$ is $\mathcal{C}^p$ on $I$. The set $\phi(I)$ will be called the *trace* of $C$. A point $\boldsymbol{x} \in \mathbf{R}^m$ is said to *lie on the curve* $C$ if it belongs to the trace of $C$.

   ii) A curve $C = (\phi, I)$ is called an *arc* if $I = [a, b]$ is closed and bounded, in which case $\phi(a), \phi(b)$ are called the *endpoints* of $C$.

   iii) An arc $(\phi, [a, b])$ is called *closed* if $\phi(a) = \phi(b)$.

   iv) A nonclosed curve $(\phi, I)$ is called *simple* if $\phi$ is 1–1 on $I$. A closed curve $(\phi, [a, b])$ is called *simple* if $\phi$ is 1–1 on $[a, b)$ and $\phi(a) = \phi(b)$.

Simple closed curves are also called *Jordan curves* because of the Jordan Curve Theorem. This theorem states that the trace of every simple closed curve $(\phi, I)$ in $\mathbf{R}^2$ separates $\mathbf{R}^2$ into two pieces, a bounded connected set $E$ and an unbounded connected set $\Omega$, where $\partial E = \partial \Omega = \phi(I)$ (see Griffiths [3]). (It is interesting to note that the set $E$ is not necessarily a Jordan region. This fact was discovered by W.F. Osgood[1].)

It is important to realize that a given set of points can be the trace of many different curves. For example, the line segment $\{(x, y) \in \mathbf{R}^2 : y = x, 0 < x \leq 1\}$ is the trace of $\phi(t) = (t, t)$ on $(0, 1]$, of $\psi(t) = (t/2, t/2)$ on $(0, 2]$, and of $\sigma(t) = (1/t, 1/t)$ on $[1, \infty)$. Although these functions trace the same line segment, each of them traces it differently. The function $\psi$ traces the line "twice as slowly" as $\phi$, and $\sigma$ traces the line "backwards" from $\phi$. Therefore, a curve $(\phi, I)$ is not merely a set of points, but a set of points $\phi(I)$ together with a specific way $\phi : I \to \phi(I)$ of tracing those points.

To discuss different ways of tracing out the same set of points, we introduce the following terminology. By a $\mathcal{C}^p$ *parametrization* of a curve $(\phi, I)$ we mean a simple $\mathcal{C}^p$ curve $(\psi, J)$ which satisfies $\psi(J) = \phi(I)$. The equations

$$x_j = \psi_j(t), \qquad t \in J, \quad j = 1, \ldots, m,$$

are called the *parametric equations* of $(\phi, I)$ induced by the parametrization $(\psi, J)$. (For a physical interpretation of parametrization, see Remark 13.6 below.)

---

[1] "A Jordan Curve of Positive Area," *Transactions of the American Mathematical Society*, vol. 4 (1903), pp. 107–112.

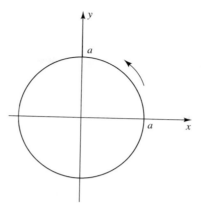

**Figure 13.1**

The following result shows that the graph of a $\mathcal{C}^p$ real-valued function on an interval is the trace of a simple $\mathcal{C}^p$ curve.

**13.2 Example.** Let $I$ be an interval and $f : I \to \mathbf{R}$ be a $\mathcal{C}^p$ function. Prove that there is a 1–1 $\mathcal{C}^p$ function $\phi : I \to \mathbf{R}^2$ whose trace coincides with the graph of $y = f(x)$ on $I$.

PROOF. If $\phi(t) = (t, f(t))$, then $\phi$ is $\mathcal{C}^p$ and 1–1 on $I$, and $\phi(I)$ is the graph of $y = f(x)$ as $x$ varies over $I$. (We shall call $(\phi, I)$ the *trivial parametrization* of $y = f(x)$.) ∎

By an *explicit curve* we mean a curve of the form $(\phi, I)$, where $\phi(t) = (t, f(t))$ (respectively, $\phi(t) = (f(t), t)$) for some $\mathcal{C}^p$ function $f : I \to \mathbf{R}$. Notice that the trace of an explicit curve consists of points $(x, y)$ which satisfy $y = f(x)$ (respectively, $x = f(y)$).

By the proof of Example 13.2, every explicit curve is a simple curve in $\mathbf{R}^2$. The following result shows that the converse of this statement is false.

**13.3 Example.** Prove that the circle $x^2 + y^2 = a^2$ is the trace of a simple closed $\mathcal{C}^\infty$ curve in $\mathbf{R}^2$.

PROOF. This circle can be described in polar coordinates by $r = a$, i.e., in rectangular coordinates by $x = a \cos\theta$, $y = a \sin\theta$. Thus, the set of points $(x, y) \in \mathbf{R}^2$ such that $x^2 + y^2 = a^2$ is the trace of the simple closed $\mathcal{C}^\infty$ curve $(\phi, I)$ defined by $\phi(t) = (a \cos t, a \sin t)$ and $I = [0, 2\pi]$. (The trace of this circle is sketched in Figure 13.1. The arrow shows the direction the point $\phi(t)$ moves as $t$ grows larger. For example, $\phi(0) = (a, 0)$ and $\phi(\pi/2) = (0, a)$.) ∎

We shall define the "length" of a curve as follows.

**13.4 DEFINITION.** The *arc length* of a simple $\mathcal{C}^1$ arc $C = (\phi, I)$ is defined to be

$$L(C) := \int_I \|\phi'(t)\| \, dt.$$

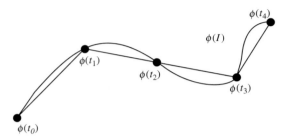

**Figure 13.2**

Notice, since $\phi'$ is continuous on the compact set $I$, that $\|\phi'\|$ is integrable; hence, $L(C)$ is finite. This is not necessarily the case if $p = 0$ (the space-filling curve is continuous but its length is infinite) or if $C$ is not an arc (see Exercise 4 below). Also notice that if the curve were not simple, some part of it could be traced more than once, giving an inflated value of its arc length.

In the case that $C$ is an explicit curve, say $y = f(x)$ on $[a, b]$, Definition 13.4 becomes

$$L(C) = \int_a^b \sqrt{1 + (f'(x))^2}\, dx,$$

which agrees with the formula for arc length introduced in elementary calculus texts.

*The following justification of Definition 13.4 is optional since we do not use it elsewhere.*

The arc length of some non–$\mathcal{C}^1$ curves can be defined by using line segments for approximation (see Figure 13.2). Namely, we say that a simple curve $C = (\phi, I)$ is *rectifiable* if

$$\|C\| := \sup\left\{ \sum_{j=1}^{k} \|\phi(t_j) - \phi(t_{j-1})\| : \{t_0, t_1, \ldots, t_k\} \text{ is a partition of } I \right\}$$

is finite, in which case we call $\|C\|$ the *arc length* of $C$.

The following result shows that every $\mathcal{C}^p$ arc, $p \geq 1$, is rectifiable, in which case these two definitions of arc length agree.

**\*13.5 THEOREM.** *If $C = (\phi, I)$ is a simple $\mathcal{C}^1$ arc, then $L(C) = \|C\|$.*

STRATEGY: The idea behind the proof is simple. By the Mean Value Theorem, each term $\|\phi(t_j) - \phi(t_{j-1})\|$ which appears in the definition of $\|C\|$ is approximately $\|\phi'(t_j)\|(t_j - t_{j-1})$, a term of a Riemann sum of the integral of $\|\phi'(t)\|$. Thus, we begin by controlling the size of $\|\phi'(t_j)\|$.

PROOF. Let $\varepsilon > 0$, write $\phi = (\phi_1, \phi_2, \ldots, \phi_m)$, and set

$$F(x_1, \ldots, x_m) = \left( \sum_{\ell=1}^{m} |\phi'_\ell(x_\ell)|^2 \right)^{1/2}$$

for $(x_1, \ldots, x_m)$ in the cube $I^m := I \times \cdots \times I$. By hypothesis, $F$ is continuous on $I^m$, and $I^m$ is evidently compact. Thus, $F$ is uniformly continuous on $I^m$; i.e., there is a $\delta > 0$ such that

$$\boldsymbol{x}, \boldsymbol{y} \in I^m \quad \text{and} \quad \|\boldsymbol{x} - \boldsymbol{y}\| < \delta \quad \text{imply} \quad |F(\boldsymbol{x}) - F(\boldsymbol{y})| < \frac{\varepsilon}{2|I|}.$$

Let $\mathcal{P} = \{u_0, \ldots, u_N\}$ be any partition of $I$. By Theorem 5.18, choose a partition $\mathcal{P}_0 = \{t_0, t_1, \ldots, t_k\}$ of $I$ finer than $\mathcal{P}$ such that $\|\mathcal{P}_0\| < \delta/\sqrt{m}$ and

$$\int_I \|\phi'(t)\| \, dt - \frac{\varepsilon}{2} < \sum_{j=1}^k \|\phi'(t_j)\|(t_j - t_{j-1}) < \int_I \|\phi'(t)\| \, dt + \frac{\varepsilon}{2}.$$

Fix $\ell \in \{1, \ldots, m\}$ and $j \in \{1, \ldots, k\}$. By Theorem 4.15 (the one-dimensional Mean Value Theorem), choose a point $c_j(\ell) \in [t_{j-1}, t_j]$ such that

$$\phi_\ell(t_j) - \phi_\ell(t_{j-1}) = \phi'_\ell(c_j(\ell))(t_j - t_{j-1}).$$

Since $\|\mathcal{P}_0\| < \delta/\sqrt{m}$, we have $|F(t_j, \ldots, t_j) - F(c_j(1), \ldots, c_j(m))| < \varepsilon/(2|I|)$. Since $\phi'(t) = (\phi'_1(t), \ldots, \phi'_m(t))$, we also have $F(t_j, \ldots, t_j) = \|\phi'(t_j)\|$ and

$$F(c_j(1), \ldots, c_j(m))(t_j - t_{j-1}) = \left( \sum_{\ell=1}^m |\phi'_\ell(c_j(\ell))|^2 \right)^{1/2} (t_j - t_{j-1})$$

$$= \|\phi(t_j) - \phi(t_{j-1})\|.$$

It follows that

$$\sum_{j=1}^k \|\phi'(t_j)\|(t_j - t_{j-1}) - \frac{\varepsilon}{2} < \sum_{j=1}^k \|\phi(t_j) - \phi(t_{j-1})\| < \sum_{j=1}^k \|\phi'(t_j)\|(t_j - t_{j-1}) + \frac{\varepsilon}{2}.$$

Combining this double inequality with the previous one, we obtain

$$\int_I \|\phi'(t)\| \, dt - \varepsilon < \sum_{j=1}^k \|\phi(t_j) - \phi(t_{j-1})\| < \int_I \|\phi'(t)\| \, dt + \varepsilon.$$

Using the left-hand inequality and the definition of $\|C\|$, we have

$$\|C\| \geq \sum_{j=1}^k \|\phi(t_j) - \phi(t_{j-1})\| > \int_I \|\phi'(t)\| \, dt - \varepsilon.$$

It follows from Definition 13.4 that $\|C\| \geq L(C)$. On the other hand, since $\mathcal{P}_0 = \{t_0, t_1, \ldots, t_k\}$ is finer than $\mathcal{P}$, it follows from the right-hand inequality that

$$\sum_{i=1}^N \|\phi(u_i) - \phi(u_{i-1})\| \leq \sum_{j=1}^k \|\phi(t_j) - \phi(t_{j-1})\| < \int_I \|\phi'(t)\| \, dt + \varepsilon.$$

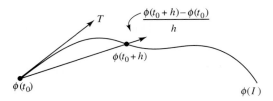

**Figure 13.3**

Taking the supremum over all partitions $\{u_0, \ldots, u_N\}$ of $I$, we have

$$\|C\| \le \int_I \|\phi'(t)\| \, dt + \varepsilon,$$

i.e., $\|C\| \le L(C)$. ∎

It is useful to interpret the trace of a curve as the path of a particle $q$ moving through space and each of its parametrizations $(\phi, I)$ as a record of a particular flight plan. (The independent variable $t$ is used to suggest "time.")

**13.6 Remark.** *Under this interpretation, if $t_0 \in I^\circ$ and $\mathbf{x}_0 = \phi(t_0)$, then $\|\phi'(t_0)\|$ is the speed of the particle $q$ at $\mathbf{x}_0$ and $\phi'(t_0)$, when nonzero, is a vector "tangent" to the curve $C$ at $\mathbf{x}_0$ which points in the direction of flight.*

PROOF. Let $t_0 \in I^\circ$ and notice that, for each sufficiently small $h > 0$, the quotient

$$\frac{\phi(t_0 + h) - \phi(t_0)}{h}$$

is a vector which points in the direction of flight along the curve $C$ and approximates the direction of the line tangent to $C$ at $\mathbf{x}_0$ (see Figure 13.3). To calculate the speed of $q$, define the *natural parameter* of the curve $C$ by

$$(1) \qquad\qquad s := \ell(t) := \int_a^t \|\phi'(u)\| \, du, \qquad t \in [a, b].$$

By the Fundamental Theorem of Calculus, $ds/dt = \ell'(t) = \|\phi'(t)\|$. Thus, the change of arc length $s$ with respect to time $t$, i.e., the speed of the particle $q$ at $\mathbf{x}_0$, is precisely $\|\phi'(t_0)\|$. ∎

Recall that the graph of a $C^p$ function on an open interval is "smooth," i.e., has a unique tangent line at each of its points. The following example shows that this is not the case for the trace of a $C^p$ *curve*.

**13.7 Example.** Let $\phi(t) = (\cos^3 t, \sin^3 t)$ and $I = [0, 2\pi]$. Show that $(\phi, I)$ is a simple closed $C^\infty$ curve in $\mathbf{R}^2$ which has "corners." (This curve is called an *astroid*.)

PROOF. Clearly, $\phi$ is $C^\infty$ on $I$ and 1–1 on $[0, 2\pi)$. Let $x = \cos^3 t$ and $y = \sin^3 t$ and observe by a double-angle formula that

$$x^2 + y^2 = \frac{3}{4}\cos^2(2t) + \frac{1}{4}.$$

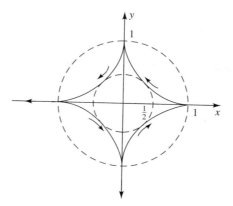

**Figure 13.4**

Hence, $\sqrt{x^2 + y^2}$ varies from a maximum of 1 (attained when $t = 0, \pi/2, \pi, 3\pi/2,$ $2\pi$) to a minimum of $1/2$ (attained when $t = \pi/4, 3\pi/4, 5\pi/4, 7\pi/4$). Since $I$ is connected and $\phi$ is differentiable, hence, continuous, the set $\phi(I)$ must also be connected. Plotting a few points, we see that $\phi(I)$ is a four-cornered star, starting at $(1, 0)$ and moving in a counterclockwise direction from $\partial B_1(0, 0)$ to $\partial B_{1/2}(0, 0)$ and back again (see Figure 13.4). As $t$ runs from 0 to $2\pi$, this curve makes one complete circuit. ∎

We shall say that a curve $(\phi, I)$ in $\mathbf{R}^2$ has a *unique tangent line* at a point $(x_0, y_0) = \phi(t_0)$ if the trace $\phi(I)$ looks like the graph of a differentiable function near $(x_0, y_0)$. When $t_0 \in I^\circ$, this means that there exist open intervals $I_0$ and $J_0$ and a differentiable function $f : J_0 \to \mathbf{R}$ such that $t_0 \in I_0 \subset I$, and either $x_0 \in J_0$ and the graph of $y = f(x)$ on $J_0$ coincides with $\phi(I_0)$, or $y_0 \in J_0$ and the graph of $x = f(y)$ on $J_0$ coincides with $\phi(I_0)$.

Is there an easy way to recognize when a curve has a unique tangent line at each of its points? To answer this question, let $(\phi, I)$ be the astroid in Example 13.7, and notice that $\phi'(t) = (0, 0)$ when $t = 0, \pi/2, \pi, 3\pi/2, 2\pi$, i.e., exactly at the points where the curve $(\phi, I)$ fails to have a unique tangent line. This agrees with Remark 13.6 since it is impossible to draw the trace of the astroid without pausing at the corners. Thus, we guess that a curve $(\phi, I)$ has a unique tangent line at some point $(x_0, y_0) = \phi(t_0)$ when $\phi'(t_0) \neq \mathbf{0}$. The following result shows that this guess is correct, except possibly at the endpoints.

**13.8 Remark.** *Suppose $I$ is an open interval. If $(\phi, I)$ is a $\mathcal{C}^1$ curve in $\mathbf{R}^2$ which has no tangent line at some point $(x_0, y_0) = \phi(t_0)$, $t_0 \in I$, then $\phi'(t_0) = \mathbf{0}$.*

PROOF. Let $(\phi_1, \phi_2)$ represent the components of $\phi$. Suppose $\phi'(t_0) \neq \mathbf{0}$. We may suppose that $\phi_1'(t_0) \neq 0$. By the Implicit Function Theorem, applied to $F(x, t) = \phi_1(t) - x$, there is an open interval $J_0$ containing $x_0$ and a continuously differentiable function $g : J_0 \to I$ such that $\phi_1(g(x)) = x$ for all $x \in J_0$ and $g(x_0) = t_0$. Set $f = \phi_2 \circ g$. Then the graph of $y = f(x)$, $x \in J_0$, coincides with the trace of $\phi$ on $g(J_0)$, i.e., near $(x_0, y_0)$. Since $f$ is differentiable at $x_0$, it follows from elementary calculus that the graph of $y = f(x)$ has a tangent line at $x_0$, a contradiction. ∎

Accordingly, we make the following definition.

**13.9 DEFINITION.** Let $(\phi, I)$ be a $\mathcal{C}^1$ curve or parametrization.

i) $(\phi, I)$ is said to be *smooth* at $t_0 \in I$ if $\phi'(t_0) \neq \mathbf{0}$.

ii) A nonclosed $(\phi, I)$ is called *smooth* if it is smooth at each point of $I$. A closed $(\phi, [a, b])$ is called *smooth* if $(\phi, [a, b))$ is smooth and $\phi'(a) = \phi'(b)$.

iii) If $C = (\phi, I)$ is simple and smooth, then the *unit tangent vector* of $C$ at $\mathbf{x}_0 = \phi(t_0)$ is defined by $T(\mathbf{x}_0) := \phi'(t_0)/\|\phi'(t_0)\|$.

By Remark 13.8, a curve which has no tangent line at some interior point cannot be smoothly parametrized. The converse of this statement is false.

**13.10 Remark.** *A curve which has a unique tangent line at each of its points can have a nonsmooth parametrization.*

PROOF. Let $\psi(t) = (t^3, 3t^3)$. The trace of $\psi$ is the line $y = 3x$ which has a unique tangent line at each of its points. Nevertheless, $\psi'(0) = \mathbf{0}$. (Of course, there are other parametrizations of this line which satisfy $\phi'(t) \neq \mathbf{0}$ for all $t \in \mathbf{R}$, for example, the trivial parametrization $\phi(t) = (t, 3t)$.) ∎

In view of this, we must be careful to distinguish between those properties which are the same for all parametrizations of a given curve, and those which are not. For example, is the arc length of a curve (see Definition 13.4 above) independent of parametrization?

To answer this question, we begin by showing that any two parametrizations of the same arc are related by a one-dimensional change of variables $\tau$.

**13.11 Remark.** *Let $I, J$ be closed bounded intervals and $\phi : I \to \mathbf{R}^m$ be 1–1 and continuous. Then $\phi(I) = \psi(J)$ for some continuous $\psi : J \to \mathbf{R}^m$ if and only if there is a continuous function $\tau$ from $J$ onto $I$ such that $\psi = \phi \circ \tau$.*

PROOF. Since $I$ is compact and $\phi$ is 1–1 and continuous on $I$, $\phi^{-1}$ is continuous from $\phi(I)$ onto $I$ (see Theorem 9.39 or 10.64). Since $\psi(J) = \phi(I)$, it follows that $\tau = \phi^{-1} \circ \psi$ is continuous from $J$ onto $I$.

Conversely, if $\tau$ is continuous from $J$ onto $I$, then $\psi = \phi \circ \tau$ is continuous from $J$ onto $\phi(I)$, i.e., $\psi(J) = \phi(I)$. ∎

Notice that if $(\psi, J)$, $(\phi, I)$ are parametrizations of the same simple $\mathcal{C}^1$ curve and $\tau = \phi^{-1} \circ \psi$ is differentiable, then by the Chain Rule,

$$(2) \qquad\qquad \psi'(u) = \phi'(\tau(u))\tau'(u), \qquad u \in J.$$

In particular, if $(\phi, I)$ is smooth, then $(\psi, J)$ is smooth if and only if $\tau'(u) \neq 0$ for all $u \in J$. This leads us to the following definition.

**13.12 DEFINITION.** Let $p \geq 1$. Two smooth $\mathcal{C}^p$ curves (or parametrizations) $(\phi, I)$, $(\psi, J)$ are said to be *smoothly equivalent* if $\phi(I) = \psi(J)$ and there is a $\mathcal{C}^p$

function $\tau : J \to \mathbf{R}$, which takes $J$ onto $I$, such that $\psi = \phi \circ \tau$ and $\tau'(u) \neq 0$ for all $u \in J$. The function $\tau$ is called the *transition* from $\psi(J)$ to $\phi(I)$.

Notice that since $\tau'$ is continuous and nonzero, either $\tau'$ is positive on $J$ or $\tau'$ is negative on $J$. Hence, by Theorem 4.24, a transition $\tau$ is always 1–1. Also notice that if two curves are smoothly equivalent, then one is simple if and only if the other one is simple (see Exercise 5 in Section 13.4).

We now answer the question posed above concerning the definition of arc length.

**13.13 Remark.** *If $(\phi, I)$ and $(\psi, J)$ are smoothly equivalent, then*

$$\int_I \|\phi'(t)\| \, dt = \int_J \|\psi'(u)\| \, du.$$

PROOF. If $\tau$ is the transition from $\psi(J)$ to $\phi(I)$, then $\tau(J) = I$. Hence, it follows from (2) and the Change of Variables Formula (Theorem 12.46) that

$$\int_I \|\phi'(t)\| \, dt = \int_{\tau(J)} \|\phi'(t)\| \, dt = \int_J \|\phi'(\tau(u))\| \, |\tau'(u)| \, du = \int_J \|\psi'(u)\| \, du. \quad \blacksquare$$

We note that the condition $\tau'(u) \neq 0$ can be relaxed at finitely many points in $J$ (see Exercise 8).

The following integral can be interpreted as the mass of a wire on a curve with density $g$ (see Appendix E).

**13.14 DEFINITION.** Let $C = (\phi, I)$ be a smooth simple arc in $\mathbf{R}^m$ and $g : \phi(I) \to \mathbf{R}$ be continuous. Then the *line integral* of $g$ on $C$ is

(3)
$$\int_C g \, ds := \int_I g(\phi(t)) \|\phi'(t)\| \, dt.$$

For an explicit curve $C$ given by $y = f(x)$, $x \in [a, b]$, this integral becomes

$$\int_C g \, ds = \int_a^b g(x, f(x)) \sqrt{1 + |f'(x)|^2} \, dx.$$

We note that by Definition 13.4, the line integral (3) equals the arc length of $C$ when $g = 1$. This explains the notation $ds$. Indeed, the parameter $s$ represents arc length (see (1) above) and, by the Fundamental Theorem of Calculus, $ds/dt = \|\phi'(t)\|$. Hence, the Leibnizian differential of $s$ satisfies $ds = \|\phi'(t)\| \, dt$. We also note that the line integral of a function $g$ on a curve is the same under smoothly equivalent parametrizations (see Exercise 8).

Since a line integral is a one-dimensional integral, it can often be evaluated by the techniques discussed in Chapter 5.

**13.15 Example.** Find $\int_C g \, ds$ where $g(x, y) = 2x + y$, $C = (\phi, I)$, $\phi(t) = (\cos t, \sin t)$, and $I = [0, \pi/2]$.

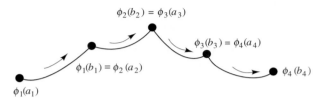

**Figure 13.5**

SOLUTION. Since $\|\phi'(t)\| = \|(-\sin t, \cos t)\| = 1$ we have

$$\int_C g \, ds = \int_0^{\pi/2} (2\cos t + \sin t) \, dt = 3. \quad \blacksquare$$

For even the simplest applications, we must have a theory rich enough to handle curves, like the boundary of the unit square $\partial([0, 1] \times [0, 1])$, which are not smooth but a union of smooth pieces. Consequently, we extend the theory developed above to finite unions of smooth curves as follows. Let $C = \cup_{j=1}^N C_j$, where each $C_j$ is a smooth simple curve with endpoints. We shall call $C$ a *piecewise smooth curve* if either of the following conditions holds. The first condition is that the traces of $C_j$ and $C_k$ are disjoint for each $j \neq k$. (This allows for nonconnected smooth curves, e.g., the boundary of an annulus $0 < a^2 < x^2 + y^2 < b^2$.) The second condition is that the $C_j$'s are joined end to end; i.e., if $C_j = (\phi_j, [a_j, b_j])$ then $\phi_{j-1}(b_{j-1}) = \phi_j(a_j)$ for $j = 2, \ldots, N$ (see Figure 13.5). (This allows for curves with finitely many corners which cannot be parametrized by a single function like the astroid. Important examples are the perimeter of a triangle and the boundary of a rectangle.) If $C$ is a piecewise smooth curve which satisfies the second condition, then the points $\phi_1(a_1)$ and $\phi_N(b_N)$ are called the *endpoints* of $C$.

The *trace* of a piecewise smooth curve $C := \cup_{j=1}^N C_j$ is defined to be the union of the traces of the $C_j$'s. $C$ is said to be *simple* if each $C_j$ is simple and the trace of $C$ does not intersect itself except possibly at the endpoints. $C$ is said to be *closed* if it is connected and its endpoints are identical. $C$ is called an *arc* if each $C_j$ is an arc. The *arc length* of a piecewise smooth arc $C$ is defined by

$$L(C) := \sum_{j=1}^N L(C_j).$$

If $C$ is a piecewise smooth, simple arc, and $g$ is continuous and real-valued on the trace of $C$, then the *line integral* of $g$ on $C$ is defined by

$$\int_C g \, ds = \sum_{j=1}^N \int_{C_j} g \, ds.$$

By a *parametrization* of $C$ we mean a collection of smooth parametrizations $(\phi_j, I_j)$ of $C_j$. Two parametrizations $\cup_{j=1}^N (\phi_j, I_j)$ and $\cup_{j=1}^N (\psi_j, J_j)$ of $C$ are said to to be smoothly equivalent if $(\phi_j, I_j)$ and $(\psi_j, J_j)$ are smoothly equivalent for each $j \in \{1, \dots, N\}$.

**13.16 Example.** Parametrize the boundary $C$ of the unit square $[0,1] \times [0,1]$ and compute $\int_C g\, ds$, where $g(x, y) = x^2 + y^3$.

SOLUTION. $C$ has four smooth pieces which can be parametrized by

$$\phi_1(t) = (t, 0), \quad \phi_2(t) = (1, t), \quad \phi_3(t) = (1 - t, 1), \quad \phi_4(t) = (0, 1 - t),$$

for $t \in [0, 1]$. Since $\|\phi_j'(t)\| = 1$, we have by definition,

$$\int_C g\, ds = \int_0^1 t^2\, dt + \int_0^1 (1 + t^3)\, dt + \int_0^1 (t^2 + 1)\, dt + \int_0^1 t^3\, dt = \frac{19}{6}. \quad \blacksquare$$

## EXERCISES

**1.** Let $\psi(t) = (a \sin t, a \cos t)$, $\sigma(t) = (a \cos 2t, a \sin 2t)$, $I = [0, 2\pi)$, and $J = [0, \pi)$. Sketch the traces of $(\psi, I)$ and $(\sigma, J)$. Note the "direction of flight" and the "speed" of each parametrization. Compare these parametrizations with the one given in Example 13.3.

**2.** Let $\boldsymbol{x}_0, \boldsymbol{a} \in \mathbf{R}^m$, $\boldsymbol{a} \neq \boldsymbol{0}$, and set $\phi(t) = t\boldsymbol{a} + \boldsymbol{x}_0$. Show that $C = (\phi, \mathbf{R})$ is a smooth simple unbounded curve which contains $\boldsymbol{x}_0$ and $\boldsymbol{x}_0 + \boldsymbol{a}$. Prove that the angle between $\phi(t_1) - \phi(0)$ and $\phi(t_2) - \phi(0)$ for any $t_1, t_2 \neq 0$ is 0 or $\pi$.

**3.** Let $I$ be an interval and $f : I \to \mathbf{R}$ be continuously differentiable with

$$|f(\theta)|^2 + |f'(\theta)|^2 \neq 0$$

for all $\theta \in I$. Prove that the graph of $r = f(\theta)$ (in polar coordinates) is the trace of a smooth curve in $\mathbf{R}^2$.

**\*4.** Show that the curve $y = \sin(1/x)$, $0 < x \leq 1$, is not rectifiable. Thus show that Theorem 13.5 can be false if $C$ is not an arc.

**5.** Sketch the trace and compute the arc length of each of the following curves.

 a) $\phi(t) = (e^t \sin t, e^t \cos t, e^t)$, $t \in [0, 2\pi]$.
 b) $y^3 = x^2$ from $(-1, 1)$ to $(1, 1)$.
 c) $\phi(t) = (t^2, t^2, t^2)$, $t \in [0, 2]$.
 d) The astroid of Example 13.7.

**6.** For each of the following, find a (piecewise) smooth parametrization of $C$ and compute $\int_C g\, ds$.

 a) $C$ is the curve $y = \sqrt{9 - x^2}$, $x \geq 0$, and $g(x, y) = xy^2$.
 b) $C$ is the portion of the ellipse $x^2/a^2 + y^2/b^2 = 1$, $a, b > 0$, which lies in the first quadrant and $g(x, y) = xy$.
 c) $C$ is the intersection of the surfaces $x^2 + z^2 = 4$ and $y = x^2$, and $g(x, y, z) = \sqrt{1 + yz^2}$.
 d) $C$ is the triangle with vertices $(0, 0, 0)$, $(1, 0, 0)$ and $(0, 2, 0)$, and $g(x, y, z) = x + y + z^3$.

**7.** Let $C = (\phi, I)$ be a smooth arc and $g_k : \phi(I) \to \mathbf{R}$ be continuous for $k \in \mathbf{N}$.

    a) If $g_k \to g$ uniformly on $\phi(I)$, prove $\int_C g_k \, ds \to \int_C g \, ds$ as $k \to \infty$.

    *b) Suppose $\{g_k\}$ is pointwise monotone and $g_k \to g$ pointwise on $\phi(I)$ as $k \to \infty$. If $g$ is continuous on $\phi(I)$, prove $\int_C g_k \, ds \to \int_C g \, ds$ as $k \to \infty$.

**8.** Suppose $(\phi, I)$ is a smooth simple arc in $\mathbf{R}^m$ and $\tau : J \to \mathbf{R}$ is a $\mathcal{C}^1$ function, 1–1 from $J$ onto $I$. If $\tau'(u) \neq 0$ for all but finitely many $u \in J$, $\psi = \phi \circ \tau$, and $g : \phi(I) \to \mathbf{R}$ is continuous, prove

$$\int_I g(\phi(t)) \|\phi'(t)\| \, dt = \int_J g(\psi(u)) \|\psi'(u)\| \, du.$$

**9.** [THE FOLIUM OF DESCARTES]. Let $C$ be the piecewise smooth curve $(\phi, I_1)$ $\cup (\phi, I_2)$, where $I_1 = (-\infty, -1)$, $I_2 = (-1, \infty)$, and

$$\phi(t) = \left( \frac{3t}{1 + t^3}, \frac{3t^2}{1 + t^3} \right).$$

Show that if $(x, y) = \phi(t)$, then $x^3 + y^3 = 3xy$. Sketch the trace of $C$.

**10.** The *absolute curvature* of a smooth curve $(\psi, I)$ at a point $\mathbf{x}_0 = \psi(t_0)$ is the number

$$\kappa(\mathbf{x}_0) = \lim_{t \to t_0} \frac{\theta(t)}{\ell(t)},$$

when this limit exists, where $\theta(t)$ is the angle between $\psi'(t)$ and $\psi'(t_0)$, and $\ell(t)$ is the arc length of $(\psi, I)$ from $\psi(t)$ to $\psi(t_0)$. (Thus $\kappa$ measures how rapidly $\theta(t)$ changes with respect to arc length.)

    a) Given $\mathbf{a}, \mathbf{b} \in \mathbf{R}^n$, prove that the absolute curvature of the line $\Lambda = (\psi, I)$, where $\psi(t) := t\mathbf{a} + \mathbf{b}$ and $I := (-\infty, \infty)$, is zero at each point $\mathbf{x}_0$ on $\Lambda$.

    b) Prove that the absolute curvature of the circle of radius $r$ (namely, $C = (\psi, I)$, where $\psi(t) = (r \cos t, r \sin t)$ and $I = [0, 2\pi)$) is $1/r$ at each point $\mathbf{x}_0$ on $C$.

**11.** Let $C = (\phi, [a, b])$ be a smooth simple arc in $\mathbf{R}^m$ and $s = \ell(t)$ be given by (1). The *natural parametrization* of $C$ is the pair $(\nu, [0, L])$, where

$$\nu(s) = (\phi \circ \ell^{-1})(s) \quad \text{and} \quad L = L(C).$$

    a) Prove that $\|\nu'(s)\| = 1$ for all $s \in [0, L]$ and the arc length of a subcurve $(\nu, [c, d])$ of $C$ is $d - c$. (This is why $(\nu, [0, L])$ is called the natural parametrization.)

    b) Show $\nu'(s)$ and $\nu''(s)$ are orthogonal for each $s \in [0, L]$.

    c) Prove that the absolute curvature (see Exercise 10 above) of $(\nu, [0, L])$ at $\mathbf{x}_0 = \nu(s_0)$ is $\kappa(\mathbf{x}_0) = \|\nu''(s_0)\|$.

    d) Show if $\mathbf{x}_0 = \phi(t_0) = \nu(s_0)$ and $m = 3$, then

$$\kappa(\mathbf{x}_0) = \|\nu'(s_0) \times \nu''(s_0)\| = \frac{\|\phi'(t_0) \times \phi''(t_0)\|}{\|\phi'(t_0)\|^3}.$$

e) Let $p \geq 1$. Prove that the absolute curvature of an explicit $\mathcal{C}^p$ curve $y = f(x)$ at $(x_0, y_0)$ is

$$\kappa = \frac{|y''(x_0)|}{(1 + (y'(x_0))^2)^{3/2}}.$$

## 13.2 ORIENTED CURVES

Every parametrization $(\phi, I)$ of a smooth curve $C$ determines a "direction of flight" along $C$, i.e., determines the direction $\phi(t)$ moves as $t$ increases on $I$, equivalently, the direction $\phi'(t)$ points. This direction is called the *orientation* of $C$ *induced by* $(\phi, I)$. (The arrows in Figures 13.1, 13.4, and 13.5 above represent the orientation of the given parametrization.) Suppose $(\phi, I)$ and $(\psi, J)$ are smoothly equivalent parametrizations of the same curve with transition $\tau$. Since $\tau$ is continuously differentiable and nonzero, either $\tau'(u) > 0$ for all $u \in J$ or $\tau'(u) < 0$ for all $u \in J$. In the first case, the vectors $\phi'(\tau(u))$ and $\psi'(u)$ point in the same direction (see (2) in Section 13.1); hence, these parametrizations determine the same orientation. In the second case, the vectors $\phi'(\tau(u))$ and $\psi'(u)$ point in opposite directions, hence, determine different orientations. Accordingly, we make the following definition.

**13.17 DEFINITION.** Two smooth curves (or parametrizations) $(\phi, I)$ and $(\psi, J)$ are said to be *orientation equivalent* if they are smoothly equivalent and the transition $\tau$ from $\psi(J)$ to $\phi(I)$ satisfies $\tau'(u) > 0$ for all $u \in J$.

In practice, a curve and its orientation are often described geometrically. The reader must provide a parametrization which traces the curve in the prescribed orientation. Here are two typical examples.

**13.18 Example.** Find a smooth parametrization of the curve $C$ in $\mathbf{R}^3$, oriented in the clockwise direction when viewed from high up on the positive $z$ axis, formed by intersecting the surfaces $x^2 + 5y^2 = 5$ and $z = x^2$.

SOLUTION. The elliptical cylinder $x^2 + 5y^2 = 5$ intersects the parabolic cylinder $z = x^2$ to form a "sagging ellipse" (the shaded region in Figure 13.6 represents that part of $z = x^2$ which lies inside the cylinder $x^2 + 5y^2 = 5$). Using $x = \sqrt{5}\sin t$, $y = \cos t$ to incorporate clockwise motion around the ellipse $x^2 + 5y^2 = 5$, we see that $z = x^2 = 5\sin^2 t$. Thus a smooth parametrization of $C$ with clockwise orientation is $\phi(t) = (\sqrt{5}\sin t, \cos t, 5\sin^2 t)$ on $I = [0, 2\pi]$. ∎

**13.19 Example.** Find a smooth parametrization of the curve $C$ in $\mathbf{R}^3$, oriented from right to left when viewed from far out on the line $y = x$ in the $xy$ plane, formed by intersecting the surfaces $z = x^2 - y^2$ and $x + y = 1$.

SOLUTION. The saddle surface $z = x^2 - y^2$ intersects the plane $x + y = 1$ to form a curve which cuts across the surface. Using $x = t$ as a parameter to incorporate right to left orientation, we see that $y = 1 - t$ and $z = t^2 - (1 - t)^2 = 2t - 1$. Thus a smooth parametrization of $C$ is $\phi(t) = (t, 1 - t, 2t - 1)$ on $I = \mathbf{R}$. In particular, $C$ is a line in the direction $(1, -1, 2)$ passing though the point $(0, 1, -1)$. ∎

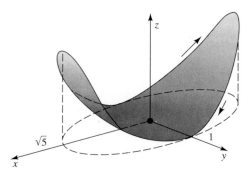

**Figure 13.6**

The following integral arises naturally in the study of fluids, electricity, and magnetism (e.g., see the discussion below).

**13.20 DEFINITION.** Let $C = (\phi, I)$ be a smooth simple arc in $\mathbf{R}^m$ and $F : \phi(I) \to \mathbf{R}^m$ be continuous. Then the *oriented line integral* of $F$ *along* $C$ is

$$(4) \qquad \int_C F \cdot T \, ds := \int_C F \cdot d\phi := \int_I F(\phi(t)) \cdot \phi'(t) \, dt.$$

The notation $F \cdot d\phi$ is self-explanatory. The notation $F \cdot T \, ds$ is consistent with (3) in Section 13.1. Indeed, $T = \phi'(t)/\|\phi'(t)\|$ is the unit tangent vector of $C$ and $ds = \|\phi'(t)\| \, dt$ is the arc length differential associated with $C$. Thus, when evaluating $F \cdot T \, ds$, the scalars $\|\phi'(t)\|$ cancel each other out.

What does this number represent? If $F$ represents the flow of a fluid, then $F \cdot T$ is the tangential component of $F$, i.e., a measure of fluid flow in the direction which the tangent $T$ points (see Appendix E). For example, suppose $C$ is the unit circle oriented in the counterclockwise direction and $F(x, y) = (-y, x)$. The unit tangent to $C$ at a point $(x, y)$ is $(-y, x)$, so $F$ points in the same direction that $T$ does. Hence, $F \cdot T = 1$ is an indication that the fluid is flowing "with the tangent" rather than against it. On the other hand, if $G(x, y) = (y, -x)$ or $H(x, y) = (x, y)$, then $G \cdot T = -1$ because the fluid is flowing against the tangent, and $H \cdot T = 0$ because the fluid is flowing orthogonally to $T$ (e.g., neither with nor against it). Therefore, the integral of $F \cdot T \, ds$ over $C$ is a measure of the circulation of $F$ around $C$ in the direction of the tangent vector. If this integral is positive, it means that the net flow of the fluid is with $T$ rather than against $T$.

Since an oriented line integral is a one–dimensional integral, it can often be evaluated by techniques introduced in Chapter 5. Here is a typical example.

**13.21 Example.** Describe the trace of $\phi(t) = (\cos t, \sin t, t)$, $t \in I = [0, 4\pi]$, and compute

$$\int_C F \cdot T \, ds,$$

where $F(x, y, z) = (1, \cos z, xy)$ and $C = (\phi, I)$.

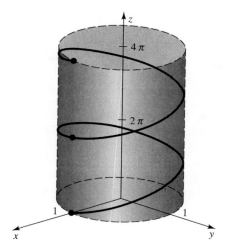

**Figure 13.7**

SOLUTION. Let $(x, y, z) = \phi(t)$. Since $x^2 + y^2 = 1$, the trace of $C$ lies on the cylinder $x^2 + y^2 = 1$, $0 \le z \le 4\pi$. As $t$ increases, the point $(x, y)$ winds around the unit circle $x^2 + y^2 = 1$ in a counterclockwise direction. Thus the trace of $\phi$ is a spiral (called the *circular helix*) which winds around the cylinder $x^2 + y^2 = 1$ (see Figure 13.7). As $t$ runs from 0 to $4\pi$ this spiral winds around the cylinder twice, and $z$ runs from 0 to $4\pi$. Since $\phi'(t) = (-\sin t, \cos t, 1)$, we have

$$\int_C F \cdot T \, ds = \int_0^{4\pi} (1, \cos t, \cos t \sin t) \cdot (-\sin t, \cos t, 1) \, dt$$

$$= \int_0^{4\pi} (-\sin t + \cos^2 t + \sin t \cos t) \, dt = 2\pi. \ \blacksquare$$

The following result shows that, unlike the line integral $\int_C g \, ds$, the oriented line integral $\int_C F \cdot T \, ds$ can give different values for different smoothly equivalent parametrizations of the same curve.

**13.22 Remark.** *If $(\phi, I)$ and $(\psi, J)$ are simple arcs which are smoothly equivalent but not orientation equivalent, then*

$$\int_I F(\phi(t)) \cdot \phi'(t) \, dt = -\int_J F(\psi(u)) \cdot \psi'(u) \, du.$$

PROOF. Let $\tau$ be the transition from $\psi(J)$ to $\phi(I)$. Since $\tau'$ is continuous and nonzero, it is either positive on $J$ or negative on $J$. Since $(\phi, I)$ and $(\psi, J)$ are not orientation equivalent, it follows that $\tau'$ is negative on $J$, i.e., $|\tau'(u)| = -\tau'(u)$ for $u \in J$. Combining this observation with the Change of Variables Formula (Theorem

12.46) and (2) in Section 13.1, we conclude that

$$\int_I F(\phi(t)) \cdot \phi'(t) \, dt = \int_J F(\phi(\tau(u))) \cdot \phi'(\tau(u)) \, |\tau'(u)| \, du$$

$$= -\int_J F(\psi(u)) \cdot \psi'(u) \, du. \quad \blacksquare$$

By the same method, one can show that the oriented integral (4) gives identical values for orientation equivalent parametrizations of the same curve (see Exercise 5). Therefore, to evaluate an oriented integral over a curve $C$ whose orientation has been described geometrically, we can use any simple, smooth parametrization of $C$ and adjust the sign of the integral to reflect the prescribed orientation. Here is a typical example.

**13.23 Example.** Find

$$\int_C F \cdot T \, ds,$$

where $F(x, y) = (y, xy)$ and $C$ is the unit circle $x^2 + y^2 = 1$ oriented in the clockwise direction.

SOLUTION. The parametrization $\phi(t) = (\cos t, \sin t)$, $t \in [0, 2\pi]$, of $C$ has counterclockwise orientation (see Example 13.3). Thus, by Remark 13.22,

$$\int_C F \cdot T \, ds = -\int_0^{2\pi} (\sin t, \sin t \cos t) \cdot (-\sin t, \cos t) \, dt$$

$$= \int_0^{2\pi} (\sin^2 t - \sin t \cos^2 t) \, dt = \pi. \quad \blacksquare$$

There is another way to represent the oriented integral (4) which uses differential notation. Recall that if $x_j = \phi_j(t)$, then $dx_j = \phi'_j(t)dt$. Hence, formally, $F(\phi(t)) \cdot \phi'(t) \, dt$ looks like

$$(F_1(\phi(t))\phi'_1(t) + \cdots + F_m(\phi(t))\phi'_m(t)) \, dt = F_1 \, dx_1 + \cdots + F_m \, dx_m.$$

This last expression is called a *differential form of degree 1* on $\mathbf{R}^m$ (more briefly, a *1–form*) and the functions $F_j$ are called its *coefficients*. A 1–form is said to be *continuous* on a set $E$ if each of its coefficients is continuous on $E$. The *oriented integral* of a continuous 1–form on a smooth simple arc $C = (\phi, I)$ in $\mathbf{R}^m$ is defined by

$$\int_C F_1 \, dx_1 + \cdots + F_m \, dx_m := \int_C F \cdot T \, ds,$$

where $F = (F_1, \ldots, F_m)$.

The following example illustrates the fact that differential forms provide a shorthand for the way an oriented line integral is computed (so we can avoid parametrizing explicit curves).

**13.24 Example.** Find

$$\int_C y\,dx + \cos x\,dy,$$

where $C$ is the explicit curve $y = x^2 + \sin x$ oriented from $(0,0)$ to $(\pi, \pi^2)$.

SOLUTION. Since $y = x^2 + \sin x$ and $dy = (2x + \cos x)\,dx$, we have

$$\int_C y\,dx + \cos x\,dy = \int_0^\pi (x^2 + \sin x)\,dx + \int_0^\pi \cos x\,(2x + \cos x)\,dx$$

$$= \frac{\pi^3}{3} + \frac{\pi}{2} - 2. \quad \blacksquare$$

Let $C = \bigcup_{j=1}^N C_j$ be a piecewise smooth curve in $\mathbf{R}^m$ (see the discussion preceding Example 13.16) with trace $E$. If $F : E \to \mathbf{R}^m$ is continuous, then the *oriented line integral* of $F$ *along* $C$ is defined to be

$$\int_C F \cdot T\,ds = \sum_{j=1}^N \int_{C_j} F \cdot T\,ds.$$

If $\omega$ is a 1–form continuous on $E$, then the *oriented integral* of $\omega$ *along* $C$ is defined to be

$$\int_C \omega = \sum_{j=1}^N \int_{C_j} \omega.$$

**13.25 Example.** Find

$$\int_C xy\,dx + (x^2 + y^2)\,dy,$$

where $C$ is the boundary of $Q = [0,1] \times [0,1]$ oriented in the counterclockwise direction.

SOLUTION. The boundary $C = \partial Q$ consists of four smooth pieces (see Figure 13.8): $C_1$ (which lies in the line $x = 0$), $C_2$ (in $y = 0$), $C_3$ (in $x = 1$), and $C_4$ (in $y = 1$). For $C_1$, let $x = 0$ and $y$ run from 1 to 0 (to maintain counterclockwise orientation on $C$). Then

$$\int_{C_1} xy\,dx + (x^2 + y^2)\,dy = \int_1^0 y^2\,dy = -\frac{1}{3}.$$

Similarly, the integrals over $C_2$, $C_3$, and $C_4$ are 0, 4/3, and $-1/2$. Hence,

$$\int_C F \cdot T\,ds = -\frac{1}{3} + 0 + \frac{4}{3} - \frac{1}{2} = \frac{1}{2}. \quad \blacksquare$$

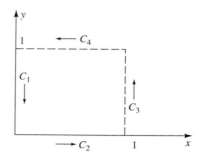

**Figure 13.8**

## EXERCISES

**1.** For each of the following, sketch the trace of $(\phi, \mathbf{R})$, describe its orientation, and verify that it is a subset of the set $S$.

a) $\phi(t) = (3t, 3\sin t, \cos t)$, $S = \{(x, y, z) : y^2 + 9z^2 = 9\}$.
b) $\phi(t) = (t^2, t^3, t^2)$, $S = \{(x, y, z) : z = x\}$.
c) $\phi(t) = (t, t^2, \sin t)$, $S = \{(x, y, z) : y = x^2\}$.
d) $\phi(t) = (\cos t, \sin t, \cos t)$, $S = \{(x, y, z) : y^2 + z^2 = 1\}$.
e) $\phi(t) = (\sin t, \sin t, t)$, $S = \{(x, y, z) : y = x\}$.

**2.** For each of the following, find a (piecewise) smooth parametrization of $C$ and compute $\int_C F \cdot T \, ds$.

a) $C$ is the curve $y = x^2$ from $(1, 1)$ to $(3, 9)$, and $F(x, y) = (xy, y - x)$.
b) $C$ is the intersection of the elliptical cylinder $y^2 + 2z^2 = 1$ with the plane $x = -1$, oriented in the counterclockwise direction when viewed from far out the positive $x$ axis, and $F(x, y, z) = (\sqrt{x^3 + y^3 + 5}, z, x^2)$.
c) $C$ is the intersection of the bent plane $y = |x|$ with the elliptical cylinder $x^2 + 3z^2 = 1$, oriented in the clockwise direction when viewed from far out the positive $y$ axis, and $F(x, y, z) = (z, -z, x + y)$.

**3.** For each of the following, compute $\int_C \omega$.

a) $C$ is the polygonal path consisting of the line segment from $(1, 1)$ to $(2, 1)$ followed by the line segment from $(2, 1)$ to $(2, 3)$, and $\omega = y \, dx + x \, dy$.
b) $C$ is the intersection of $z = x^2 + y^2$ and $x^2 + y^2 + z^2 = 1$, oriented in the counterclockwise direction when viewed from high up the positive $z$ axis, and $\omega = dx + (x + y) \, dy + (x^2 + xy + y^2) \, dz$.
c) $C$ is the boundary of the rectangle $R = [a, b] \times [c, d]$, oriented in the counterclockwise direction, and $\omega = xy \, dx + (x + y) \, dy$.
d) $C$ is the intersection of $y = x$ and $y = z^2$, $0 \le z \le 1$, oriented from left to right when viewed from far out the $y$ axis, and $\omega = \sqrt{x} \, dx + \cos y \, dy - dz$.

**4. a)** Let $c \in \mathbf{R}$, $\delta > 0$, and set $\tau(u) = \delta u + c$ for $u \in \mathbf{R}$. Prove that if $(\phi, I)$ is a smooth curve, $J = \tau^{-1}(I)$, and $\psi = \phi \circ \tau$ for $\tau(u) = \delta u + c$, then $(\psi, J)$ is orientation equivalent to $(\phi, I)$.

b) Prove that if $(\phi, I)$ is a smooth arc, then it has an orientation equivalent parametrization of the form $(\psi, [0, 1])$.

c) Obtain an analogue of b) for piecewise smooth curves.

5. Suppose $(\phi, I)$ is a smooth simple arc and $\tau$ is a $\mathcal{C}^1$ function, 1–1 from $J$ onto $I$, which satisfies $\tau'(u) > 0$ for all but finitely many $u \in J$. If $\psi = \phi \circ \tau$ prove that

$$\int_I F(\phi(t)) \cdot \phi'(t)\, dt = \int_J F(\psi(u)) \cdot \psi'(u)\, du$$

for any continuous $F : \phi(I) \to \mathbf{R}^m$.

6. **This exercise is used in Section 13.5.** Let $f : [a, b] \to \mathbf{R}$ be $\mathcal{C}^1$ on $[a, b]$ with $f'(t) \neq 0$ for $t \in [a, b]$. Prove that the explicit curve $x = f^{-1}(y)$, as $y$ runs from $f(a)$ to $f(b)$, is orientation equivalent to the explicit curve $y = f(x)$, as $x$ runs from $a$ to $b$.

7. Let $V \neq \emptyset$ be open in $\mathbf{R}^2$. A function $F : V \to \mathbf{R}^2$ is said to be *conservative* on $V$ if there is a function $f : V \to \mathbf{R}$ such that $F = \nabla f$ on $V$. Let $(x, y) \in V$ and $F = (P, Q) : V \to \mathbf{R}^2$ be continuous on $V$.

a) Suppose $C(x)$ is a horizontal line segment in $V$ terminating at $(x, y)$, i.e., a line segment of the form $L((x_1, y); (x, y))$, oriented from $(x_1, y)$ to $(x, y)$, whose trace is a subset of $V$. Prove

$$\frac{\partial}{\partial x} \int_{C(x)} F \cdot T\, ds = P(x, y).$$

Make and prove a similar statement for $\partial/\partial y$ and vertical line segments in $V$ terminating at $(x, y)$.

b) Let $(x_0, y_0) \in V$. Prove that

(*)                               $$\int_C F \cdot T\, ds = 0$$

for all closed piecewise smooth curves $C \subset V$ if and only if for all $(x, y) \in V$, the integrals

$$f(x, y) := \int_{C(x,y)} F \cdot T\, ds$$

give the same value for all piecewise smooth curves $C(x, y)$ whose traces are subsets of $V$ which start at $(x_0, y_0)$ and end at $(x, y)$.

c) Prove that $F$ is conservative on $V$ if and only if (*) holds for all closed piecewise smooth curves $C$ whose traces are subsets of $V$.

d) Prove that if $F$ satisfies (*) for all closed piecewise smooth curves $C$ whose traces are subsets of $V$, then

$$\frac{\partial P}{\partial y} = \frac{\partial Q}{\partial x}.$$

(Note: If $V$ is nice enough, the converse of this statement also holds (see Exercise 8 in Section 13.6 or Theorem 15.45).)

*8. Suppose $f : [0, 1] \to \mathbf{R}$ is increasing and continuously differentiable on $[0, 1]$. Let $T$ be the right triangle whose vertices are $(0, f(0))$, $(1, f(0))$, and $(1, f(1))$. If $c$ represents the hypotenuse of $T$, $a$ and $b$ represent the legs of $T$, and $L$ represents the arc length of the explicit curve $y = f(x)$, $x \in [0, 1]$, prove that $c \leq L \leq a + b$.

## 13.3  SURFACES

In this section we define surfaces and unoriented surface integrals, concepts which are two-dimensional analogues of curves and the line integrals discussed in Section 13.1. Recall that a smooth arc is parametrized on a closed bounded interval. On what shall we parametrize a smooth surface? Evidently, we need to use some type of closed bounded set in $\mathbf{R}^2$. Although we could use rectangles, such a restriction would be awkward when dealing with explicit surfaces with curved projections, or with surfaces described by cylindrical or spherical coordinates. It is much more efficient to build greater generality into the definition of surface, using *two-dimensional regions* instead of rectangles, i.e., using sets of the following type.

**13.26 DEFINITION.** An *m-dimensional region* is a set $E \subset \mathbf{R}^m$ such that $E = \overline{V}$ for some nonempty, open, connected Jordan region $V$ in $\mathbf{R}^m$.

Notice that every closed bounded interval is a one-dimensional region, every two-dimensional rectangle and the closure of every two-dimensional ball or ellipse is a two-dimensional region, and every three-dimensional rectangle and the closure of every three-dimensional ball or ellipsoid is a three-dimensional region.

**13.27 DEFINITION.**

i) A $(\mathcal{C}^p)$ *surface* (in $\mathbf{R}^3$) is a pair $S = (\phi, E)$, where $E$ is a two-dimensional region and $\phi : E \to \mathbf{R}^3$ is $\mathcal{C}^p$ on $E$. The *trace* of $S$ is the set $\phi(E)$. A point $(x, y, z) \in \mathbf{R}^3$ is said to *lie on* (or *belong to*) $S$ if it belongs to the trace of $S$.

ii) A surface $(\phi, E)$ is said to be *simple* if $\phi$ is 1–1 on $E$.

iii) A pair $(\psi, B)$ is called a $\mathcal{C}^p$ *parametrization* of a surface $(\phi, E)$ if $(\psi, B)$ is a simple $\mathcal{C}^p$ surface and $\psi(B) = \phi(E)$. The equations

$$x = \psi_1(u, v), \quad y = \psi_2(u, v), \quad z = \psi_3(u, v), \qquad (u, v) \in B,$$

are called *parametric equations* of $(\phi, E)$ induced by the parametrization $(\psi, B)$.

Earlier, we called the graph of a function $z = f(x, y)$ a surface. The following result shows that this designation is compatible with Definition 13.27 when $f$ is $\mathcal{C}^p$.

**13.28 Example.** Let $E$ be a two-dimensional region and $f : E \to \mathbf{R}$ be a $\mathcal{C}^p$ function. Prove that the graph of $z = f(x, y)$ is the trace of a simple $\mathcal{C}^p$ surface.

**PROOF.** If $\phi(u, v) = (u, v, f(u, v))$, then $\phi$ is $\mathcal{C}^p$ and 1–1 on $E$, and $\phi(E)$ is the graph of $z = f(x, y)$. (This is called the *trivial parametrization* of $z = f(x, y)$.) ∎

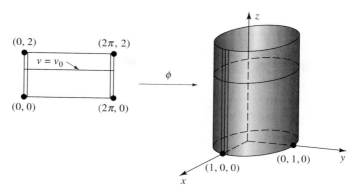

**Figure 13.9**

In a similar way we define trivial parametrizations of surfaces of the form $x = f(y, z)$ and $y = f(x, z)$. For example, the trivial parametrization of the surface $x = f(y, z)$, $(y, z) \in E$, is given by $(\phi, E)$, where $\phi(u, v) = (f(u, v), u, v)$. By an *explicit surface over* $E$ we shall mean the trivial parametrization of a surface of the form $x = f(y, z)$, $y = f(x, z)$, or $z = f(x, y)$, where $f : E \to \mathbf{R}$ is a $\mathcal{C}^p$ function and $E$ is a two-dimensional region. By the proof of Example 13.28, every explicit surface is a simple surface.

The next four examples, which show that not every surface is an explicit surface, provide model parametrizations for certain kinds of surfaces.

**13.29 Example.** Show that the truncated cylinder $x^2 + y^2 = 1$, $0 \leq z \leq 2$, is the trace of a $\mathcal{C}^\infty$ surface.

PROOF. Let $\phi(u, v) = (\cos u, \sin u, v)$ and $E = [0, 2\pi] \times [0, 2]$, and notice that $\phi$ is $\mathcal{C}^\infty$ on $E$. The corresponding parametric equations are $x = \cos u$, $y = \sin u$, $z = v$. Clearly, $x^2 + y^2 = 1$. Thus $\phi(E)$ is a subset of the cylinder $x^2 + y^2 = 1$, $0 \leq z \leq 2$. Since $E$ is connected, so is $\phi(E)$. To see that $\phi(E)$ is the entire cylinder, look at the images of horizontal line segments in $E$. The image of the line segment $v = v_0$ is a circle lying in the plane $z = v_0$, centered at $(0, 0, v_0)$, of radius 1 (see Figure 13.9). Thus, as $v_0$ ranges from 0 to 2, the images of horizontal lines $v = v_0$ cover the entire cylinder $x^2 + y^2 = 1$, $0 \leq z \leq 2$. ∎

**13.30 Example.** Show that the sphere $x^2 + y^2 + z^2 = a^2$ is the trace of a $\mathcal{C}^\infty$ surface.

PROOF. Let $\phi(u, v) = (a \cos u \cos v, a \sin u \cos v, a \sin v)$ and $E = [0, 2\pi] \times [-\pi/2, \pi/2]$. Clearly, $\phi$ is $\mathcal{C}^\infty$ on $E$. The corresponding parametric equations are $x = a \cos u \cos v$, $y = a \sin u \cos v$, $z = a \sin v$. Since $x^2 + y^2 = a^2 \cos^2 v$, we have $x^2 + y^2 + z^2 = a^2$. Thus $\phi(E)$ is a subset of the sphere centered at the origin of radius $a$. The image of the horizontal line segment $v = v_0$ is a circle, lying in the plane $z = a \sin v_0$, centered at $(0, 0, a \sin v_0)$ of radius $a \cos v_0$ (see Figure 13.10). The image of the top edge (respectively, bottom edge) of $E$, i.e., of the horizontal line $v = \pi/2$ (respectively, $v = -\pi/2$), is the north pole $(0, 0, a)$ (respectively, the south

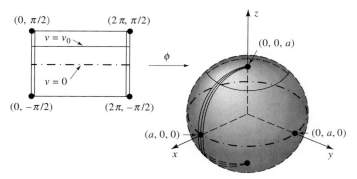

**Figure 13.10**

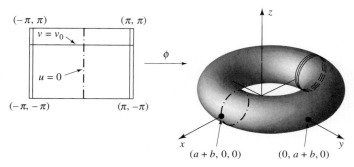

**Figure 13.11**

pole $(0, 0, -a)$). Thus, as $v_0$ ranges from $-\pi/2$ to $\pi/2$, the images of horizontal lines $v = v_0$ cover the entire sphere $x^2 + y^2 + z^2 = a^2$. ∎

Let $C$ represent the circle in the $xz$ plane centered at $(a, 0, 0)$ of radius $b$, where $a > b$. The *torus* centered at the origin with radii $a > b$ is the donut-shaped surface obtained by revolving $C$ about the $z$ axis (see Figure 13.11).

**13.31 Example.** Show that the torus centered at the origin with radii $a > b$ is the trace of a $\mathcal{C}^\infty$ surface.

PROOF.   Let $\phi(u, v) = ((a + b\cos v)\cos u, (a + b\cos v)\sin u, b\sin v)$ and $E = [-\pi, \pi] \times [-\pi, \pi]$, and notice that $\phi$ is $\mathcal{C}^\infty$ on $E$. The image of $u = 0$ is a circle in the $xz$ plane centered at $(a, 0, 0)$ of radius $b$. The images of horizontal lines $v = v_0$ are circles, parallel to the $xy$ plane, centered at $(0, 0, b\sin v_0)$ of radius $(a + b\cos v_0)$. The image of the lines $v = \pm\pi$ is a circle in the $xy$ plane centered at $(0, 0, 0)$ of radius $a - b$. Thus $\phi(E)$ covers the entire torus. ∎

**13.32 Example.** Show that the truncated cone $z = \sqrt{x^2 + y^2}$, $0 \le z \le b$, is the trace of a $\mathcal{C}^\infty$ surface.

PROOF.   Let $(x, y, z) = \phi(u, v) = (v\cos u, v\sin u, v)$ and $E = [0, 2\pi] \times [0, b]$, and notice that $\phi$ is $\mathcal{C}^\infty$ on $E$. Clearly, $x^2 + y^2 = z^2$ and $0 \le z \le b$. Thus $\phi(E)$ is a

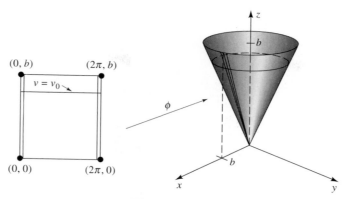

**Figure 13.12**

subset of the given cone. The image of a horizontal line $v = v_0$, $0 < v_0 \le b$, is a circle in the plane $z = v_0$ centered at $(0, 0, v_0)$ of radius $v_0$ (see Figure 13.12). Thus $\phi(E)$ is the cone $z = \sqrt{x^2 + y^2}$, $0 \le z \le b$. Notice that the image of the line $v = 0$ is the vertex $(0, 0, 0)$. ∎

We shall say that a surface $S = (\phi, E)$ has a unique tangent plane at a point $(x_0, y_0, z_0) = \phi(u_0, v_0)$ if the trace of $S$ near $(x_0, y_0, z_0)$ looks like the graph of a differentiable function. When $(u_0, v_0) \in E^o$, this means that there exist open sets $E_0$ and $V_0$, and a differentiable function $f : V_0 \to \mathbf{R}$, such that $(u_0, v_0) \in E_0 \subset E$, and one of the following conditions holds: $(x_0, y_0) \in V_0$ and the graph of $z = f(x, y)$ coincides with $\phi(E_0)$; $(x_0, z_0) \in V_0$ and the graph of $y = f(x, z)$ coincides with $\phi(E_0)$; or $(y_0, z_0) \in V_0$ and the graph of $x = f(y, z)$ coincides with $\phi(E_0)$. In this case, the tangent plane of the explicit surface generated by $f$ is called the *tangent plane* of $S$ at $(x_0, y_0, z_0)$, and its *normal (vector)* can be computed by applying Remark 11.17 or Theorem 11.20. For example, if the trace of $S$ looks like the graph of $x = f(y, z)$, then a normal to the tangent plane is given by $(1, -f_y(y_0, z_0), -f_z(y_0, z_0))$.

Normal vectors play the same role for surfaces that tangent vectors played for curves. (For example, we shall use normal vectors to define area of surfaces, smooth surfaces, and orientation of surfaces.) Thus we need to know how to compute normal vectors to surfaces given in parametric form.

**13.33 Remark.** *Let $S = (\phi, E)$ be a $\mathcal{C}^p$ surface for some $p \ge 1$ and $\phi = (\phi_1, \phi_2, \phi_3)$. If $S$ has a unique tangent plane at a point $(x_0, y_0, z_0) = \phi(u_0, v_0)$, where $(u_0, v_0) \in E^o$, then a normal to $S$ at $(x_0, y_0, z_0)$ is given by*

$$N_\phi := \left( \frac{\partial(\phi_2, \phi_3)}{\partial(u, v)}(u_0, v_0), \frac{\partial(\phi_3, \phi_1)}{\partial(u, v)}(u_0, v_0), \frac{\partial(\phi_1, \phi_2)}{\partial(u, v)}(u_0, v_0) \right).$$

PROOF. By Remark 13.6, $\phi_u(u_0, v_0)$ is tangent, at $(x_0, y_0, z_0)$, to the curve $\phi(u, v_0)$ and $\phi_v(u_0, v_0)$ is tangent, at $(x_0, y_0, z_0)$, to the curve $\phi(u_0, v)$. Hence, a normal to the tangent plane to $S$ at $(x_0, y_0, z_0)$ is given by the cross product

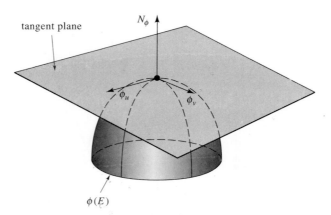

tangent plane

$N_\phi$

$\phi_u$    $\phi_v$

$\phi(E)$

**Figure 13.13**

$\phi_u(u_0, v_0) \times \phi_v(u_0, v_0) =: (\Delta_{\phi_2, \phi_3}(u_0, v_0), \Delta_{\phi_3, \phi_1}(u_0, v_0), \Delta_{\phi_1, \phi_2}(u_0, v_0))$ (see Figure 13.13). ∎

If $z = f(x, y)$ is an explicit surface and $\phi$ is its trivial parametrization, an easy calculation verifies $N_\phi = (-f_x, -f_y, 1)$. Notice that this agrees with Remark 11.17 and Theorem 11.20.

By replacing $\phi'$ by $N_\phi$, we develop a theory of surfaces which is analogous to the theory of curves presented in the previous two sections. For example, compare the following definition with Definition 13.9.

**13.34 DEFINITION.**  Let $p \geq 1$ and suppose $(\phi, E)$ is a $\mathcal{C}^p$ surface or a parametrization of a $\mathcal{C}^p$ surface.

   i) $(\phi, E)$ is said to be *smooth* at a point $(u_0, v_0) \in E$ if $N_\phi(u_0, v_0) \neq \mathbf{0}$ (equivalently, if $\|N_\phi(u_0, v_0)\| > 0$).

   ii) $(\phi, E)$ is said to be *smooth* if it is smooth at each point in $E$.

   iii) $(\phi, E)$ is said to be smooth off a set $E_0 \subset E$ if $(\phi, E)$ is smooth at each point in $E \setminus E_0$.

Notice that an explicit surface, under the trivial parametrization, is always smooth.

Analogous to the situation for curves, a surface $(\phi, E)$ which has no tangent plane at $\phi(u_0, v_0)$ cannot be smooth at $(u_0, v_0)$ (see Exercise 7). On the other hand, a surface with tangent planes at each point can have a nonsmooth parametrization. (This is consistent with our position that a surface is not just a set of points, but a set of points together with a particular parametrization.) For example, the parametrization $\phi$ of the sphere given in Example 13.30 satisfies

$$\|N_\phi\| = \|(a^2 \cos u \cos^2 v, a^2 \sin u \cos^2 v, a^2 \sin v \cos v)\| = a^2 |\cos v|.$$

Thus, although the sphere has a tangent plane at each point in its trace, the parametrization $(\phi, E)$ is not smooth at $v = \pm\pi/2$. (This happens because this parametrization takes the lines $v = \pm\pi/2$ to the north and south pole, hence, is not 1–1 there.) In particular, smoothness, as defined above, depends on the parametrization, not on just the shape of the surface.

The following result shows what happens to the normal vector $N_\phi$ under a change of parameter.

**13.35 THEOREM.** *Let $p \geq 1$ and suppose $(\phi, E)$ and $(\psi, B)$ are $\mathcal{C}^p$ surfaces in $\mathbf{R}^3$. If $\tau$ is a $\mathcal{C}^p$ function which takes $B$ into $E$ such that $\psi = \phi \circ \tau$, then*

$$N_\psi(u, v) = \Delta_\tau(u, v) N_\phi(\tau(u, v))$$

*for each $u, v \in B$.*

PROOF. Let $\phi = (\phi_1, \phi_2, \phi_3)$ and $\psi = (\psi_1, \psi_2, \psi_3)$. By Remark 13.33,

$$N_\psi = (\Delta_{(\psi_2, \psi_3)}, \Delta_{(\psi_3, \psi_1)}, \Delta_{(\psi_1, \psi_2)}).$$

Since, by hypothesis, $(\psi_i, \psi_j) = (\phi_i, \phi_j) \circ \tau$ for $i, j = 1, 2, 3$, it follows from the Chain Rule that

$$\Delta_{(\psi_i, \psi_j)}(u, v) = \Delta_\tau(u, v) \Delta_{(\phi_i, \phi_j)}(\tau(u, v))$$

for any $u, v \in B$. Therefore, $N_\psi = \Delta_\tau \cdot (N_\phi \circ \tau)$ on $B$. ∎

This leads us to the following definition (compare with Definition 13.12).

**13.36 DEFINITION.** Let $p \geq 1$. Two smooth $\mathcal{C}^p$ surfaces (or parametrizations) $(\phi, E)$, $(\psi, B)$ are said to be *smoothly equivalent* if $\psi(B) = \phi(E)$ and there is a 1–1 $\mathcal{C}^p$ function $\tau$, which takes $B$ onto $E$, such that $\psi = \phi \circ \tau$ and $\Delta_\tau(u, v) \neq 0$ for all $(u, v) \in B$. The function $\tau$ is called the *transition* from $\psi(B)$ to $\phi(E)$.

Notice that, if two surfaces are smoothly equivalent, then one is simple if and only if the other one is simple (see Exercise 5 in Section 13.4).

Analogous to Definitions 13.4 and 13.14, we define surface area and the surface integral as follows.

**13.37 DEFINITION.** Let $S = (\phi, E)$ be a smooth simple surface.

i) The *surface area* of $S$ is defined to be

$$\sigma(S) := \int_E \|N_\phi(u, v)\| \, d(u, v).$$

ii) If $g : \phi(E) \to \mathbf{R}$ is continuous, then the *surface integral* of $g$ on $S$ is defined to be

(5)
$$\iint_S g \, d\sigma := \int_E g(\phi(u, v)) \|N_\phi(u, v)\| \, d(u, v).$$

The surface integral (5) can be interpreted as the mass of a membrane with shape $\phi(E)$ and density $g$ (see Appendix E). For an explicit $\mathcal{C}^p$ surface $S$ given by $z = f(x, y)$, $(x, y) \in E$, this integral looks like

$$(6) \qquad \iint_S g\,d\sigma = \int_E g(x, y, f(x, y))\sqrt{1 + f_x^2(x, y) + f_y^2(x, y)}\; d(x, y).$$

It can be argued on heuristic grounds that this is the right definition for surface area (see Appendix E). In fact, we could have defined the surface area of $S$ by approximating it with planar regions, as we defined $\|C\|$ below Definition 13.4, by approximating it by line segments (see Price [10], p. 360). This approach, however, works only under suitable restrictions. Indeed, even when using triangular regions to approximate a bounded cylinder, the total area of the approximating regions may become infinite (see Spivak [12], p. 130). We prefer Definition 13.37i) because it is both direct and easy to use.

Notice that by Theorem 12.24, (5) makes sense when $\phi$ is 1–1 off a set of area zero, or $N_\phi(u, v)$ is undefined on a set of area zero, in particular, if $(\phi, E)$ is smooth and simple off a set $Z$ of area zero. Thus these concepts can be extended to some nonsmooth nonsimple parametrizations including those of cones and spheres.

It is easy to see that surface area and the surface integral are invariant under smoothly equivalent parametrizations, even when the condition $\Delta_\tau \neq 0$ is relaxed on a closed set of area zero (see Exercise 5). It is also easy to see that if the trace of $S$ is a subset of $\mathbf{R}^2$, then its surface area, as defined by Definition 13.37, is the same as the area of the trace of $S$, as defined by Definition 12.5 (see Exercise 4).

To compute a surface integral, one must find a suitable parametrization of the given surface and apply Definition 13.37.

**13.38 Example.** Find $\iint_S g\,d\sigma$, where $S$ is the hemisphere $z = \sqrt{a^2 - x^2 - y^2}$ and $g(x, y, z) = \sqrt{z}$.

SOLUTION. Let $\phi$ be the function defined in Example 13.30 and $E = [0, 2\pi] \times [0, \pi/2]$. Then $(\phi, E)$ is a parametrization of the hemisphere $S$ and $\|N_\phi\| = a^2 \cos v$. Since $\cos v \neq 0$ for $v \in [0, \pi/2)$, $\phi(E)$ is smooth off $[0, 2\pi] \times \{\pi/2\}$, a set of area zero. Hence,

$$\iint_S g\,d\sigma = \iint_{E_0} a^2 \cos v \sqrt{a \sin v}\, du\, dv = 2\pi a^{5/2} \int_0^{\pi/2} \cos v \sqrt{\sin v}\, dv = \frac{4\pi}{3} a^{5/2}. \quad \blacksquare$$

Continuity of $g$ is assumed in Definition 13.37 only so that the integral on the right-hand side of (5) makes sense. If one of the iterated integrals is a convergent improper integral, we can extend the definition of the surface integral in the obvious way. Using this observation, we now offer a second solution to Example 13.38 using the trivial parametrization.

ALTERNATE SOLUTION. The explicit surface $z = \sqrt{a^2 - x^2 - y^2}$ has normal $N = (-z_x, -z_y, 1) = (x/z, y/z, 1)$. (This normal does not exist on $\partial B_a(0, 0)$, but since $\partial B_a(0,0)$ is of area zero, we can ignore it when integrating over $B_a(0,0)$.) On $S$, $\|N\| = a/z$. Thus, by (6) and polar coordinates,

$$\iint_S g \, d\sigma = \int_{B_a(0,0)} \frac{a\sqrt{z}}{z} \, d(x,y) = a \int_0^{2\pi} \int_0^a r(a^2 - r^2)^{-1/4} \, dr \, d\theta = \frac{4\pi}{3} a^{5/2}.$$

(The inner integral (with respect to $r$) is an improper integral.) ∎

For even the simplest applications, we must have a theory rich enough to handle surfaces, like the boundary of the unit cube $\partial([0,1] \times [0,1] \times [0,1])$, which are not smooth but a union of smooth pieces. Consequently, we shall extend the theory developed above to finite unions of smooth surfaces. This expanded theory will be introduced using informal geometric descriptions instead of formal statements. For now, these vague descriptions will suffice because the concrete surfaces which arise in practice are easy to visualize. (Chapter 15 contains a rigorous and more mathematically satisfying treatment of these ideas.)

Before describing piecewise smooth surfaces, we must distinguish between interior points (points which lie "inside" a surface) and boundary points (points which lie on the "edge" of a surface). To illustrate the difference, consider the truncated cylinder $S$ parametrized by $(\phi, E)$ in Example 13.29. A point $(x,y,z) \in S$ lies inside $S$ if $0 < z < 2$, and on its edge if $z = 0$ or $z = 2$. (Look at Figure 13.9 to see why this terminology is appropriate.) Naively, we might guess that $(x,y,z)$ lies inside $(\phi, E)$ if $(x,y,z) \in \phi(E^o)$. This guess is incorrect, even for the cylinder; for example, $(-1,0,1) = \phi(\pi, 1) \in \phi(E^o)$ lies on a "seam" of $S$ but not an edge. Evidently, to define the interior and boundary of a general surface $S$, we must use the trace of $S$ and retreat from our earlier position that a surface is fully described by a particular parametrization $(\phi, E)$.

Accordingly, let $S$ be (the trace of) a smooth $\mathcal{C}^p$ surface in $\mathbf{R}^3$. Imagine yourself standing on a point $(x,y,z) \in S$. We shall say that $(x,y,z)$ is *interior* to $S$ if you are surrounded on all sides by points in $S$; i.e., if you take a sufficiently small step in any direction you remain on $S$. We shall denote the set of interior points of a surface $S$ by Int $(S)$ and shall define the (*manifold*) *boundary* of a surface $S$ by $\partial S := S \setminus \text{Int}(S)$.

We have used the same notation to denote the boundary of a surface as we did to denote the boundary of a set (see Definition 9.13 or 10.37) even though these concepts are not the same. We made this choice because it homogenizes the statements of all the fundamental theorems of multidimensional calculus. To avoid ambiguity, we shall henceforth refer to the boundary of a set $E$ (i.e., to $\overline{E} \setminus E^o$) as the *topological boundary* of $E$. No confusion will arise because the only boundary we use in connection with surfaces is the manifold boundary, and the only boundary we use in connection with $m$-dimensional regions is the topological boundary.

A surface $S$ is said to be *closed* if $\partial S = \emptyset$. For example, if $a > 0$, then the sphere $x^2 + y^2 + z^2 = a^2$ is closed, but the hemisphere $z = \sqrt{a^2 - x^2 - y^2}$ (respectively, the truncated paraboloid $z = x^2 + y^2$, $0 \le z \le 1$) is not closed, since its boundary is $x^2 + y^2 = a^2$, $z = 0$ (respectively, $x^2 + y^2 = 1$, $z = 1$).

By the Jordan Curve Theorem, a closed smooth curve $C$ divides $\mathbf{R}^2$ into two disjoint sets, the set of points "surrounded" by $C$ and the set of points which lie "outside" $C$. This is not the case for surfaces. Indeed, there are closed smooth surfaces (the Klein bottle is one example) which surround no points, hence, do not divide $\mathbf{R}^3$ into two disjoint sets (see Griffiths [3], p. 22, or Hocking and Young [4], p. 237).

Let $S = \cup_{j=1}^N S_j$, where each $S_j = (\phi_j, E_j)$ is a smooth simple surface. $S$ will be called *piecewise smooth* if either of the following conditions holds. The first condition is that the traces of $S_j$ and $S_k$ are disjoint for each $j \neq k$. (For example, the topological boundary of the *corona* $0 < a \leq \|(x, y, z)\| \leq b$ is such a surface.) The second condition is that the $S_j$'s are nonoverlapping (specifically, that $\phi_j(E_j^o) \cap \phi_k(E_k^o) = \emptyset$), and a portion of the boundary of each $S_j$ is matched to a portion of the boundary of some $S_k$. (For example, the topological boundary of a three-dimensional rectangle is such a surface.) We make the further restriction that the intersection of any three $S_j$'s is empty, or a finite set. This prevents a piecewise smooth surface from doubling back on itself more than once along any given edge. We shall call surfaces satisfying the first condition *nonconnected piecewise smooth surfaces* and those satisfying the second condition *connected piecewise smooth surfaces*.

The *trace* of a piecewise smooth surface $S = \cup_{j=1}^N S_j$ is defined to be the union of the traces of the $S_j$'s. If $S$ is a nonconnected piecewise smooth surface, the *boundary* $\partial S$ of $S$ is defined to be the union of the boundaries $\partial S_j$. If $S$ is a connected piecewise smooth surface, the *boundary* $\partial S$ of $S$ is defined to be the union of all points which belong to the closure of an unmatched portion of $\partial S_j$. (For example, the boundary of the box formed by removing the face $z = 1$ from the unit cube $[0, 1] \times [0, 1] \times [0, 1]$ is the unit square in the plane $z = 1$, and the boundary of the union of $x^2 + y^2 = 1$, $-3 \leq z \leq 0$, and $z = \sqrt{1 - x^2 - y^2}$ is the unit circle in the plane $z = -3$.) A *piecewise smooth parametrization* of $S = \cup_{j=1}^N S_j$ is a collection of smooth parametrizations $(\phi_j, E_j)$, where $S_j = (\phi_j, E_j)$. The surface area of $S$ is defined by

$$\sigma(S) = \sum_{j=1}^N \sigma(S_j)$$

and the *surface integral* of a real-valued function $g$ continuous on the trace of $S$ is defined by

$$\iint_S g \, d\sigma = \sum_{j=1}^N \iint_{S_j} g \, d\sigma.$$

**13.39 Example.** Let $S$ be the tetrahedron formed by taking the topological boundary of the region bounded by $x = 0$, $y = 0$, $z = 0$, and $x + y + z = 1$. Find a piecewise smooth parametrization $S$ and compute $\iint_S g \, d\sigma$, where $g(x, y, z) = x + y^2 + z^3$.

SOLUTION. The tetrahedron $S$ has four faces. They can be parametrized by $\phi_1(u, v) = (u, v, 0)$, $\phi_2(u, v) = (0, u, v)$, $\phi_3(u, v) = (u, 0, v)$, $\phi_4(u, v) = (u, v, 1 - u -$

$v$), where $(u, v)$ belongs to $E$, the triangular region with vertices $(0,0)$, $(1,0)$, and $(0,1)$. Since $\|N_{\phi_j}\| = 1$ for $j = 1, 2, 3$ and $\|N_{\phi_4}\| = \sqrt{3}$, we have

$$
\iint_S g \, d\sigma = \int_0^1 \int_0^{1-u} (u + v^2) \, dv \, du + \int_0^1 \int_0^{1-u} (u^2 + v^3) \, dv \, du
$$

$$
+ \int_0^1 \int_0^{1-u} (u + v^3) \, dv \, du
$$

$$
+ \sqrt{3} \int_0^1 \int_0^{1-u} (u + v^2 + (1 - u - v)^3)) \, dv \, du
$$

$$
= \int_0^1 \int_0^{1-u} ((2 + \sqrt{3})u + u^2 + (1 + \sqrt{3})v^2 + 2v^3
$$

$$
+ \sqrt{3}(1 - u - v)^3) \, dv \, du
$$

$$
= \int_0^1 ((2 + \sqrt{3})u - (1 + \sqrt{3})u^2 - u^3 + \frac{1 + \sqrt{3}}{3}(1 - u)^3
$$

$$
+ \frac{2 + \sqrt{3}}{4}(1 - u)^4) \, du
$$

$$
= \frac{3}{10}(2 + \sqrt{3}). \quad \blacksquare
$$

## EXERCISES

**1.** For each of the following, find the surface area of $S$.

a) $S$ is the conical shell given by $z = \sqrt{x^2 + y^2}$, where $a \le z \le b$.

b) $S$ is the sphere given in Example 13.30.

c) $S$ is the torus given in Example 13.31.

**2.** For each of the following, find a (piecewise) smooth parametrization of $S$ and of $\partial S$, and compute $\iint_S g \, d\sigma$.

a) $S$ is the portion of the surface $z = x^2 - y^2$ which lies above the $xy$ plane and between the planes $x = 1$ and $x = -1$, and $g(x, y, z) = \sqrt{1 + 4x^2 + 4y^2}$.

b) $S$ is the surface $y = x^3$, $0 \le y \le 8$, $0 \le z \le 4$, and $g(x, y, z) = x^3 z$.

c) $S$ is the portion of the hemisphere $z = \sqrt{9 - x^2 - y^2}$ which lies outside the cylinder $2x^2 + 2y^2 = 9$, and $g(x, y, z) = x + y + z$.

**3.** Find a parametrization $(\phi, E)$ of the ellipsoid

$$
\frac{x^2}{a^2} + \frac{y^2}{b^2} + \frac{z^2}{c^2} = 1
$$

which is smooth off the topological boundary $\partial E$.

**4.** a) Suppose $E$ is a two- dimensional region and $S = \{(x, y, z) \in \mathbf{R}^3 : (x, y) \in E$ and $z = 0\}$. Prove

$$
\text{Area}\,(E) = \iint_S d\sigma
$$

and

$$\iint_S g \, d\sigma = \int_E g(x, y, 0) \, d(x, y)$$

for each continuous $g : E \to \mathbf{R}$.

b) Let $f : [a, b] \to \mathbf{R}$ be a $\mathcal{C}^p$ function, $p \geq 1$, $C$ be the curve in $\mathbf{R}^2$ determined by $z = f(x)$, $a \leq x \leq b$, and $S$ be the surface in $\mathbf{R}^3$ determined by $z = f(x)$, $a \leq x \leq b$, $c \leq y \leq d$. Show $\sigma(S) = (d - c)L(C)$.

c) Let $f : [a, b] \to \mathbf{R}$ be a $\mathcal{C}^p$ function, $p \geq 1$, and $S$ be the surface obtained by revolving the curve $y = f(x)$, $a \leq x \leq b$, about the $x$ axis. Prove that the surface area of $S$ is

$$\sigma(S) = 2\pi \int_a^b |f(x)| \sqrt{1 + |f'(x)|^2} \, dx.$$

**5.** Suppose $(\psi, B)$ and $(\phi, E)$ are simple $\mathcal{C}^p$ surfaces, and $\psi = \phi \circ \tau$, where $\tau$ is a $\mathcal{C}^1$ function from $B$ onto $Z$.

a) If $(\psi, B)$ and $(\phi, E)$ are smooth and $\tau$ is 1–1 with $\Delta_\tau \neq 0$ on $B$, prove

$$\iint_E g(\phi(u, v)) \|N_\phi(u, v)\| \, du \, dv = \iint_B g(\psi(s, t)) \|N_\psi(s, t)\| \, ds \, dt$$

for all continuous $g : \phi(E) \to \mathbf{R}$.

*b) If $Z$ is a closed subset of $B$ of area zero, $(\psi, B)$ is smooth off $Z$, $\tau$ is 1–1 and $\Delta_\tau \neq 0$ on $B^o \setminus Z$, prove

$$\iint_E g(\phi(u, v)) \|N_\phi(u, v)\| \, du \, dv = \iint_B g(\psi(s, t)) \|N_\psi(s, t)\| \, ds \, dt$$

for all continuous $g : \phi(E) \to \mathbf{R}$.

**6.** Suppose $f : B_3(0, 0) \to \mathbf{R}$ is differentiable with $\|\nabla f(x, y)\| \leq 1$ for all $(x, y) \in B_3(0, 0)$. Prove that if $S$ is the paraboloid $2z = x^2 + y^2$, $0 \leq z \leq 4$, then

$$\iint_S |f(x, y) - f(0, 0)| \, d\sigma \leq 40\pi.$$

**7.** Suppose $(\phi, E)$ is a $\mathcal{C}^p$ surface and $(x_0, y_0, z_0) = \phi(u_0, v_0)$, where $(u_0, v_0) \in E^o$. If $(\phi, E)$ has no tangent plane at $(x_0, y_0, z_0)$, prove $N_\phi(u_0, v_0) = \mathbf{0}$.

**8.** Let $S = (\psi, B)$ be a smooth simple surface. Set $E = \|\psi_u\|$, $F = \psi_u \cdot \psi_v$, and $G = \|\psi_v\|$. Prove that the surface area of $S$ is $\int_B \sqrt{E^2 G^2 - F^2} \, d(u, v)$.

**9.** Suppose $S = (\phi, E)$ is a $\mathcal{C}^1$ surface smooth at $(x_0, y_0, z_0) = \phi(u_0, v_0)$. Let $C = (\psi, I)$ be a $\mathcal{C}^1$ curve whose trace is a subset of $E$ which passes through the point $(u_0, v_0)$ (i.e., there is a $t_0 \in I$ such that $\psi(t_0) = (u_0, v_0)$). Prove $(\phi \circ \psi)'(t_0) \cdot (\phi_u \times \phi_v)(u_0, v_0) = 0$.

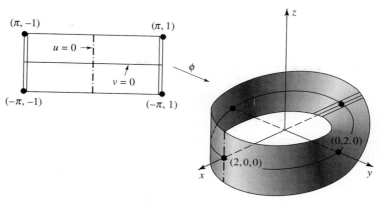

**Figure 13.14**

## 13.4   ORIENTED SURFACES

Recall that a smooth simple curve $(\phi, I)$ is oriented by using the tangent vector $\phi'(t)$ to choose a "positive direction." This works because a simple connected curve can be traversed in two, and only two, directions. Analogously, a smooth simple surface $(\phi, E)$ is oriented by using the normal vector $N_\phi$ to choose a "positive side." For a connected surface $S$, such a choice is possible if $S$ has two, and only two, sides.

A new complication arises here. There are smooth connected surfaces which have only one side. (The following example of such a surface can be made out of paper by taking a long narrow strip by the narrow edges, twisting it once, and gluing the narrow edges together.)

**13.40 Example.** [THE MÖBIUS STRIP]. Sketch the trace of the surface $(\phi, E)$, where $\phi(u, v) = ((2 + v \sin(u/2)) \cos u, (2 + v \sin(u/2)) \sin u, v \cos(u/2))$ and $E = [-\pi, \pi] \times [-1, 1]$.

SOLUTION. The image of the horizontal line $v = 0$ under $\phi$ is $(2 \cos u, 2 \sin u, 0)$, i.e., the circle in the $xy$ plane centered at the origin of radius 2. The image of each vertical line $u = u_0$ is a line segment in $\mathbf{R}^3$ which rotates through space as $u_0$ increases. For example, the image of $u = 0$ is $(2, 0, v)$, $-1 \le v \le 1$, and the image of $u = \pm\pi$ is the seam $S_0 := (-2 \mp v, 0, 0)$, $-1 \le v \le 1$, i.e., the set of points $\{(x, 0, 0) : -3 \le x \le -1\}$. Thus the trace of $(\phi, E)$ is given in Figure 13.14. ∎

To avoid such anomalies, we introduce the following concepts. The *unit normal* of a smooth surface $S = (\phi, E)$ at a point $(x_0, y_0, z_0)$ on $S$ is the vector $\boldsymbol{n}(x_0, y_0, z_0) = N_\phi(u_0, v_0)/\|N_\phi(u_0, v_0)\|$, where $\phi(u_0, v_0) = (x_0, y_0, z_0)$. Evidently, the unit normal $\boldsymbol{n}$ is well-defined only when

$$\frac{N_\phi(u_0, v_0)}{\|N_\phi(u_0, v_0)\|} = \frac{N_\phi(u_1, v_1)}{\|N_\phi(u_1, v_1)\|} \ne \boldsymbol{0}$$

for all $(u_j, v_j) \in E$ which satisfy $\phi(u_j, v_j) = (x_0, y_0, z_0)$ for $j = 0, 1$. This will surely be the case if $\phi$ is 1–1 and smooth on $E$. If $\phi$ fails to be 1–1 on $E$, however, the unit normal $\boldsymbol{n}$ might not be well-defined, even though $(\phi, E)$ is smooth on $E$ (see the Möbius strip above).

A smooth surface $S = (\phi, E)$ is said to be *orientable* if the unit normal $\boldsymbol{n}$ can be unambiguously defined at each point on the surface $S$ so that it varies continuously over $S$; i.e., if $\phi(u_0, v_0) = \phi(u_1, v_1)$, then $N_\phi(u_0, v_0)$ points in the same direction as $N_\phi(u_1, v_1)$, and if $(u_2, v_2)$ is near $(u_0, v_0)$, then $N_\phi(u_2, v_2)$ points in approximately the same direction as $N_\phi(u_0, v_0)$. (A formal definition of orientable will be given in Section 15.2.) The problem with the Möbius strip is that near the "seam" $S_0$ (defined in Example 13.40 above), $N_\phi$ points both upward and downward.

Notice that if $S$ is connected, orientable, and smooth, then $S$ has two and only two sides, a "positive" side (the side from which $\boldsymbol{n}$ points) and a "negative" side (the other side). Clearly, an explicit surface is always orientable. The Möbius strip shows that not all smooth surfaces are orientable.

**13.41 DEFINITION.** Two smooth simple surfaces (or parametrizations) $(\phi, E)$ and $(\psi, B)$ are said to be *orientation equivalent* if they are orientable, smoothly equivalent with transition $\tau$, and $\Delta_\tau(u, v) > 0$ for all $(u, v) \in B$.

By Theorem 13.35, if $(\phi, E)$ and $(\psi, B)$ are orientation equivalent, then the normal vectors they determine point in the same direction. Thus the positive side chosen by $(\phi, E)$ is the same as the positive side chosen by $(\psi, B)$.

Oriented surface integrals can be defined using the unit normal in the same way that oriented line integrals were defined using the unit tangent (compare the following definition with Definition 13.20).

**13.42 DEFINITION.** Let $S = (\phi, E)$ be a smooth simple orientable surface with unit normal $\boldsymbol{n}$, and $F : \phi(E) \to \mathbf{R}^3$ be continuous. The *oriented surface integral* of $F$ on $S$ is

$$\iint_S F \cdot \boldsymbol{n} \, d\sigma := \int_E (F \circ \phi)(u, v) \cdot N_\phi(u, v) \, d(u, v).$$

The notation of the left-most intergal is consistent with the notation in (5) since $\boldsymbol{n} = N_\phi / \|N_\phi\|$ and $d\sigma = \|N_\phi\| \, d(u, v)$.

For an explicit surface $S$ given by $z = f(x, y)$, $(x, y) \in E$, Definition 13.42 takes the form

$$(7) \qquad \iint_S F \cdot \boldsymbol{n} \, d\sigma = \int_E F(x, y, f(x, y)) \cdot (-f_x, -f_y, 1) \, d(x, y).$$

What does an oriented surface integral represent? If $F$ represents the flow of an incompressible fluid at points in the trace of $S = (\phi, E)$, then $F \cdot \boldsymbol{n}$ represents the normal component of $F$, i.e., the amount of fluid which flows in the direction of $\boldsymbol{n}$ (see Appendix E). Thus the integral of $F \cdot \boldsymbol{n} \, d\sigma$ on $S$, a measure of the flow of the fluid across the surface $S$ in the direction of $\boldsymbol{n}$, is sometimes called the *flux* of $F$ across $S$.

As was the case for the oriented line integral, the oriented surface integral can be defined on some nonsmooth surfaces (see Exercise 4). One needs to be careful, however, with the definition of orientable. If the collection of nonsmooth points cuts across the entire surface (like the peak of a pup tent or the edge of a pyramid), one has difficulty defining what it means to have a "continuously varying" normal. We shall address this problem for piecewise smooth surfaces at the end of this section. In the meantime, notice that one can define what it means for a surface $S = (\phi, E)$ to be orientable if the set of *singularities* (i.e., the set of $(x, y, z) \in \mathbf{R}^3$ such that $(x, y, z) = \phi(u, v)$ for some $(u, v) \in E$ which satisfies $N_\phi(u, v) = \mathbf{0}$) is finite. In particular, all spheres and cones are orientable.

It is easy to see that the integral of $F \cdot \boldsymbol{n}\, d\sigma$ on a surface $S$ which is smooth off a set of area zero does not change when orientation equivalent parametrizations are used (see Exercise 4). The following result shows that a change of orientation changes the value of the oriented surface integral by a minus sign.

**13.43 Remark.** *If $(\phi, E)$ and $(\psi, B)$ are smoothly equivalent but not orientation equivalent, then*

$$
\int_E F(\phi(u, v)) \cdot N_\phi(u, v)\, d(u, v) = - \int_B F(\psi(s, t)) \cdot N_\psi(s, t)\, d(s, t).
$$

PROOF. Let $\tau$ be the transition from $\psi(B)$ to $\phi(E)$. Since $\Delta_\tau$ is continuous and nonzero on the connected set $B$, and $(\phi, E)$ and $(\psi, B)$ are not orientation equivalent, we have $\Delta_\tau < 0$ on $B$. Hence, it follows from Theorem 13.35 and Theorem 12.46 (the Change of Variables Formula) that

$$
\int_B F(\psi(s, t)) \cdot N_\psi(s, t)\, d(s, t) = - \int_B |\Delta_\tau(s, t)| (F \circ \phi \circ \tau)(s, t) \cdot (N_\phi \circ \tau)(s, t)
$$

$$
= - \int_{\tau(B)} F(\phi(u, v)) \cdot N_\phi(u, v)\, d(u, v)
$$

$$
= - \int_E F(\phi(u, v)) \cdot N_\phi(u, v)\, d(u, v). \ \blacksquare
$$

Therefore, when evaluating an oriented integral on a surface $S$ whose orientation has been described geometrically, we can use any smooth, simple parametrization of $S$ and adjust the sign of the integral to reflect the prescribed orientation. Here is a typical example.

**13.44 Example.** Find the value of $\iint_S F \cdot \boldsymbol{n}\, d\sigma$, where $F(x, y, z) = (xy, x - y, z)$, $S$ is the planar region $x + y + z = 1$, $(x, y) \in [0, 1] \times [0, 1]$, and $\boldsymbol{n}$ is the downward pointing normal.

SOLUTION. The usual normal $(1, 1, 1)$ of the plane $x + y + z = 1$ points upward rather than downward. Thus, by Remark 13.43,

$$
\iint_S F \cdot \boldsymbol{n}\, d\sigma = - \int_0^1 \int_0^1 (xy, x - y, 1 - x - y) \cdot (1, 1, 1)\, dx\, dy = -\frac{1}{4}. \ \blacksquare
$$

It is convenient to have a "differential" version of oriented surface integrals. To see how to define differentials of degree 2, let $S = (\phi, E)$ be a smooth orientable surface and $x = \phi_1(u, v)$, $y = \phi_2(u, v)$, $z = \phi_3(u, v)$. By Remark 13.33,

$$N_\phi = \left( \frac{\partial(y, z)}{\partial(u, v)}, \frac{\partial(z, x)}{\partial(u, v)}, \frac{\partial(x, y)}{\partial(u, v)} \right).$$

Therefore, the oriented surface integral of a function $F = (P, Q, R) : \phi(E) \to \mathbf{R}^3$ has the form

$$\int_E \left( P \frac{\partial(y, z)}{\partial(u, v)} + Q \frac{\partial(z, x)}{\partial(u, v)} + R \frac{\partial(x, y)}{\partial(u, v)} \right) d(u, v)$$

$$=: \iint_S P \, dy \, dz + Q \, dz \, dx + R \, dx \, dy,$$

i.e.,

$$dy \, dz := \frac{\partial(y, z)}{\partial(u, v)} \, d(u, v), \quad dz \, dx := \frac{\partial(z, x)}{\partial(u, v)} \, d(u, v), \text{ and } dx \, dy := \frac{\partial(x, y)}{\partial(u, v)} \, d(u, v).$$

(These are two-dimensional analogues of the differential $dy = f'(x) \, dx$.) By a 2–*form* (or a *differential form of degree* 2) on a set $\Omega \subset \mathbf{R}^3$ we mean an expression of the form

$$P \, dy \, dz + Q \, dz \, dx + R \, dx \, dy,$$

where $P, Q, R : \Omega \to \mathbf{R}$. A 2–form is said to be continuous on $\Omega$ if its *coefficients* $P, Q, R$ are continuous on $\Omega$. The *oriented integral* of a continuous 2–form on a smooth simple surface $S$ oriented with a unit normal $\boldsymbol{n}$ is defined by

$$\iint_S P \, dy \, dz + Q \, dz \, dx + R \, dx \, dy = \iint_S (P, Q, R) \cdot \boldsymbol{n} \, d\sigma.$$

Differential forms of degree 1 were formal devices used in certain computations, e.g., to compute an oriented line integral or to estimate the increment of a function. Similarly, differential forms of degree 2 are formal devices which will be used in certain computations, e.g., to compute an oriented surface integral. They can also be used to unify the three fundamental theorems of vector calculus presented in the next two sections (see Exercise 4 in Section 15.1). (There is a less formal but time-consuming way to introduce differentials in which the differential $dx$ can be interpreted as the derivative of the projection operator $(x, y, z) \longmapsto x$ (see Spivak [12], p. 89).)

In general, the boundary of a surface is a curve. Since the boundary of the Möbius strip is a simple closed curve, the boundary of a surface may be orientable even when the surface is not.

Suppose $S$ is an oriented surface and $\partial S$ is a piecewise smooth curve. The orientation of $S$ can be used to induce an orientation on $\partial S$ in the following way. Imagine yourself standing on the positive side of $S$, close to $\partial S$. The direction of positive flow on $\partial S$ moves from right to left; i.e., as you walk around the boundary on the positive side of $S$ in the direction of positive flow, the surface lies on your left. This orientation of $\partial S$ is called the *positive orientation*, the *right-hand orientation*, or the orientation on $\partial S$ *induced* by the orientation of $S$. When $S$ is a subset of $\mathbf{R}^2$, i.e., of the $xy$ plane, we shall say that $\partial S$ is oriented *positively* if it carries the orientation induced by the upward pointing normal on $S$, i.e., the normal which points toward the upper half space $z \geq 0$. Thus if $S$ is a bounded subset of $\mathbf{R}^2$ whose boundary is a connected piecewise smooth closed curve, then the usual orientation on $S$ induces a counterclockwise orientation on $\partial S$ when viewed from high up on the positive $z$ axis. This is not the case, however, when $E$ has interior "holes." For example, if $E = \{(x,y) : a^2 < x^2 + y^2 < b^2\}$ for some $a > 0$, then the positive orientation is counterclockwise on $\{(x,y) : x^2 + y^2 = b^2\}$, but clockwise on $\{(x,y) : x^2 + y^2 = a^2\}$.

A formal definition of the positive or induced orientation will be given in Section 15.2. In the meantime, the informal geometric description given above is sufficient to identify the induced orientation in most concrete situations. Here is a typical example.

**13.45 Example.** Let $S$ be the truncated paraboloid $z = x^2 + y^2$, $0 \leq z \leq 4$, with outward pointing normal. Parametrize $\partial S$ with positive orientation.

SOLUTION. The boundary of $S$ is the circle $x^2 + y^2 = 4$ which lies in the $z = 4$ plane. The positive orientation is clockwise when viewed from high up the $z$ axis. Therefore, a parametrization of $\partial S$ is given by $\phi(t) = (2\sin t, 2\cos t, 4)$, $t \in [0, 2\pi]$. ∎

If $S = \cup S_j = \cup(\phi_j, E_j)$ is a nonconnected piecewise smooth surface, we shall call $S$ *orientable* if each $S_j$ is orientable. This definition must be modified for connected surfaces, since the Möbius strip is the union of two orientable surfaces, namely $(\phi, E_1)$ and $(\phi, E_2)$, where $\phi$ is given by Example 13.40 and $E_k = [\pi(k - 2), \pi(k - 1)] \times [-1, 1]$, $k = 1, 2$. We shall say that a connected piecewise smooth surface $S = \cup S_j = \cup(\phi_j, E_j)$ is *orientable* if it has two distinct sides. In this case, an orientation on $S$ can be given, using the normals $\pm N_{\phi_j}$ to generate a unit normal $\boldsymbol{n}_j$ on each piece $S_j$ which identifies the "positive side" in a consistent way, e.g., all normals point outward.

Oriented surface integrals can be defined for orientable piecewise smooth surfaces in the natural way. Indeed, if $S = \cup_{j=1}^N S_j$ is orientable, then the *oriented surface integral* of a vector-valued function $F$ continuous on the trace of $S$ is defined to be

$$\iint_S F \cdot \boldsymbol{n}\, d\sigma = \sum_{j=1}^N \iint_{S_j} F \cdot \boldsymbol{n}_j\, d\sigma.$$

The following three examples provide further explanation of these ideas.

**13.46 Example.** Evaluate

$$\iint_S F \cdot \boldsymbol{n}\, d\sigma,$$

where $S$ is the topological boundary of the solid bounded by the cylinder $x^2 + y^2 = 1$ and the planes $z = 0$, $z = 2$, $\boldsymbol{n}$ is the outward pointing normal, and $F(x,y,z) = (x, 0, y)$.

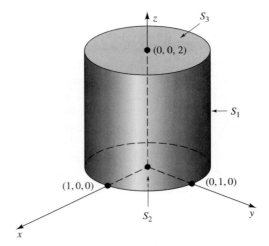

**Figure 13.15**

SOLUTION. This surface has three smooth pieces: a vertical side $S_1$, a bottom $S_2$, and a top $S_3$ (see Figure 13.15). Parametrize $S_1$ by $\phi(u,v) = (\cos u, \sin u, v)$, where $E = [0, 2\pi] \times [0, 2]$. Thus $N_\phi = (\cos u, \sin u, 0)$ and

$$\iint_{S_1} F \cdot \boldsymbol{n}\, d\sigma = \int_0^2 \int_0^{2\pi} \cos^2 u \, du \, dv = 2\pi.$$

Since the outward pointing unit normal to $S_2$ is $\boldsymbol{n} = (0, 0, -1)$, we see by Exercise 4a) in Section 13.3 that

$$\iint_{S_2} F \cdot \boldsymbol{n}\, d\sigma = -\int_{B_1(0,0)} y \, d(x, y) = -\int_0^{2\pi} \int_0^1 r^2 \sin\theta \, dr \, d\theta = 0.$$

Similarly, the integral on $S_3$ is also zero. Therefore,

$$\iint_S F \cdot \boldsymbol{n}\, d\sigma = 2\pi + 0 + 0 = 2\pi. \quad \blacksquare$$

**13.47 Example.** Find $\iint_S F \cdot \boldsymbol{n}\, d\sigma$, where $F(x, y, z) = (x + z^2, x, z)$, $S$ is the topological boundary of the solid bounded by the paraboloid $z = x^2 + y^2$ and the plane $z = 1$, and $\boldsymbol{n}$ is the outward pointing normal.

SOLUTION. The surface $S$ has two smooth pieces: the paraboloid $S_1$ given by $z = x^2 + y^2$, $0 \le z \le 1$, and the disk $S_2$ given by $x^2 + y^2 \le 1$, $z = 1$. The trivial parametrization of $S_1$ is $\phi(u, v) = (u, v, u^2 + v^2)$, $(u, v) \in B_1(0, 0)$. Note that $N_\phi = (-2u, -2v, 1)$ points inward (the wrong way). Thus, by Remark 13.43 and polar coordinates,

$$\iint_{S_1} F \cdot \boldsymbol{n}\, d\sigma = -\int_{B_1(0,0)} (-2u^2 - 2u(u^2 + v^2)^2 - 2uv + (u^2 + v^2)) \, d(u, v)$$

$$= \int_0^1 \int_0^{2\pi} (2r^2 \cos^2\theta + 2r^5 \cos\theta + 2r^2 \cos\theta \sin\theta - r^2) r \, d\theta \, dr = 0.$$

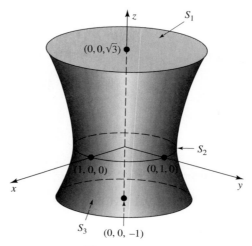

**Figure 13.16**

Since the unit outward pointing normal of $S_2$ is $\boldsymbol{n} = (0,0,1)$ and on $S_2$, $F \cdot \boldsymbol{n} = z = 1$, we see by Exercise 4a) in Section 13.3 that

$$\iint_{S_2} F \cdot \boldsymbol{n} \, d\sigma = \int_{B_1(0,0)} d(x,y) = \text{Area}\,(B_1(0,0)) = \pi.$$

Therefore,

$$\iint_{S} F \cdot \boldsymbol{n} \, d\sigma = 0 + \pi = \pi. \quad \blacksquare$$

**13.48 Example.** Compute $\iint_S F \cdot \boldsymbol{n} \, d\sigma$, where $F(x,y,z) = (x,y,z)$, $S$ is the topological boundary of the solid bounded by the hyperboloid of one sheet $x^2 + y^2 - z^2 = 1$ and the planes $z = -1$, $z = \sqrt{3}$, and $\boldsymbol{n}$ is the outward pointing normal to $S$.

SOLUTION. The surface $S$ has three smooth pieces: a top $S_1$, a side $S_2$, and a bottom $S_3$ (see Figure 13.16). Using $\boldsymbol{n} = (0,0,1)$ for $S_1$, we have

$$\iint_{S_1} F \cdot \boldsymbol{n} \, d\sigma = \int_{B_2(0,0)} \sqrt{3} \, d(x,y) = 4\sqrt{3}\pi.$$

Similarly,

$$\iint_{S_3} F \cdot \boldsymbol{n} \, d\sigma = 2\pi.$$

To integrate $F \cdot \boldsymbol{n}$ on $S_2$, let $z = u$ and note that $x^2 + y^2 = 1 + u^2$. Thus $\phi(u,v) = ((1+u^2)\cos v, (1+u^2)\sin v, u)$, $(u,v) \in [-1, \sqrt{3}] \times [0, 2\pi]$, is a parametrization of $S_2$. Since $N_\phi = (-(1+u^2)\cos v, -(1+u^2)\sin v, 2u(1+u^2))$ points inward and

$$F \cdot N_\phi = ((1+u^2)\cos v, (1+u^2)\sin v, u) \cdot$$
$$\cdot (-(1+u^2)\cos v, -(1+u^2)\sin v, 2u(1+u^2))$$
$$= -(1+u^2)^2 + 2u^2(1+u^2) = u^4 - 1,$$

we have

$$\iint_{S_2} F \cdot n \, d\sigma = - \int_{-1}^{\sqrt{3}} \int_{0}^{2\pi} (u^4 - 1) \, dv \, du$$

$$= 2\pi \int_{-1}^{\sqrt{3}} (1 - u^4) \, du = \frac{8\pi}{5}(1 - \sqrt{3}).$$

Therefore,

$$\iint_{S} F \cdot n \, d\sigma = 4\sqrt{3}\pi + 2\pi + \frac{8\pi}{5}(1 - \sqrt{3}) = \frac{6\pi}{5}(3 + 2\sqrt{3}). \quad \blacksquare$$

## EXERCISES

**1.** For each of the following, find a (piecewise) smooth parametrization of $\partial S$ which agrees with the induced orientation, and compute $\int_{\partial S} F \cdot T \, ds$.

   a) $S$ is the truncated paraboloid $y = 9 - x^2 - z^2$, $y \geq 0$, with outward pointing normal, and $F(x, y, z) = (x^2y, y^2x, x + y + z)$.

   b) $S$ is the portion of the plane $x + 2y + z = 1$ which lies in the first octant, with normal which points away from the origin, and $F(x, y, z) = (x - y, y - x, xz^2)$.

   c) $S$ is the truncated paraboloid $z = x^2 + y^2$, $1 \leq z \leq 4$, with outward pointing normal, and $F(x, y, z) = (5y + \cos z, 4x - \sin z, 3x \cos z + 2y \sin z)$.

**2.** For each of the following, compute $\iint_S F \cdot n \, d\sigma$.

   a) $S$ is the truncated paraboloid $z = x^2 + y^2$, $0 \leq z \leq 1$, $n$ is the outward pointing normal, and $F(x, y, z) = (x, y, z)$.

   b) $S$ is the truncated half cylinder $z = \sqrt{4 - y^2}$, $0 \leq x \leq 1$, $n$ is outward pointing normal, and $F(x, y, z) = (x^2 + y^2, yz, z^2)$.

   c) $S$ is the torus in Example 13.31, $n$ is the outward pointing normal, and $F(x, y, z) = (y, -x, z)$.

   d) $S$ is the portion of $z = x^2$ which lies inside the cylinder $x^2 + y^2 = 1$, $n$ is the upward pointing normal, and $F(x, y, z) = (y^2z, \cos(2 + \log(2 - x^2 - y^2)), x^2z)$.

**3.** For each of the following, compute $\iint_S \omega$.

   a) $S$ is the portion of the surface $z = x^4 + y^2$ which lies over the unit square $[0, 1] \times [0, 1]$, with upward pointing normal, and $\omega = x \, dy \, dz + y \, dz \, dx + z \, dx \, dy$.

   b) $S$ is the upper hemisphere $z = \sqrt{a^2 - x^2 - y^2}$, with outward pointing normal, and $\omega = x \, dy \, dz + y \, dz \, dx$.

   c) $S$ is the spherical cap $z = \sqrt{a^2 - x^2 - y^2}$ which lies inside the cylinder $x^2 + y^2 = b^2$, $0 < b < a$, with upward pointing normal, and $\omega = xz \, dy \, dz + dz \, dx + z \, dx \, dy$.

   d) $S$ is the truncated cone $z = 2\sqrt{x^2 + y^2}$, $0 \leq z \leq 2$, with normal which points away from the $z$ axis, and $\omega = x \, dy \, dz + y dz \, dx + z^2 \, dx \, dy$.

**4.** Suppose $(\psi, B)$ and $(\phi, E)$ are simple $\mathcal{C}^p$ surfaces, and $\psi = \phi \circ \tau$, where $\tau$ is a $\mathcal{C}^1$ function from $B$ onto $E$.

a) If $(\psi, B)$ and $(\phi, E)$ are smooth, and $\tau$ is 1–1 with $\Delta_\tau > 0$ on $B$, prove

$$\int_E F(\phi(u,v)) \cdot N_\phi(u,v)\, d(u,v) = \int_B F(\psi(s,t)) \cdot N_\psi(s,t)\, d(s,t)$$

for all continuous $F : \phi(E) \to \mathbf{R}^3$.

\*b) Suppose $Z$ is a closed subset of $B$ of area zero, $(\psi, B)$ is smooth off $Z$, and $\tau$ is 1–1 with $\Delta_\tau > 0$ on $B^\circ \setminus Z$. Prove

$$\int_E F(\phi(u,v)) \cdot N_\phi(u,v)\, d(u,v) = \int_B F(\psi(s,t)) \cdot N_\psi(s,t)\, d(s,t)$$

for all continuous $F : \phi(E) \to \mathbf{R}^3$.

**5.** Suppose $M_1$ and $M_2$ are both smooth curves or both smooth surfaces. Prove that if $M_1$ and $M_2$ are smoothly equivalent or orientation equivalent, then $M_1$ is simple if and only if $M_2$ is simple.

**6.** Let $E$ be the solid tetrahedron bounded by $x = 0$, $y = 0$, $z = 0$, and $x + y + z = 1$, and suppose its topological boundary, $T = \partial E$, is oriented with outward pointing normal. Prove

$$\iint_{\partial E} P\, dy\, dz + Q\, dz\, dx + R\, dx\, dy = \iiint_E (P_x + Q_y + R_z)\, dV$$

for all $\mathcal{C}^1$ functions $P, Q, R : E \to \mathbf{R}$.

**7.** Let $T$ be the topological boundary of the tetrahedron in Exercise 6, with outward pointing normal, and $S$ be the surface obtained by taking away the slanted face from $T$; i.e., $S$ has three triangular faces, one each in the planes $x = 0$, $y = 0$, $z = 0$. If $\partial S$ is oriented positively, prove

$$\int_{\partial S} P\, dx + Q\, dy + R\, dz = \iint_S (R_y - Q_z)\, dy\, dz + (P_z - R_x)\, dz\, dx + (Q_x - P_y)\, dx\, dy$$

for all $\mathcal{C}^1$ functions $P, Q, R : S \to \mathbf{R}$.

## 13.5  THEOREMS OF GREEN AND GAUSS

Recall by the Fundamental Theorem of Calculus that if $f$ is a $\mathcal{C}^1$ function, then

$$f(b) - f(a) = \int_a^b f'(t)\, dt.$$

Thus the integral of the derivative $f'$ on $[a, b]$ is completely determined by the values $f$ takes on the topological boundary $\{a, b\}$ of $[a, b]$.

In the next two sections we shall obtain analogues of this theorem for functions $F : \Omega \to \mathbf{R}^m$, where $\Omega$ is a surface or an $m$-dimensional region, $m = 2$ or 3. Namely, we shall show that the integral of a "derivative" of $F$ on $\Omega$ is completely determined by the values $F$ takes on the "boundary" of $\Omega$. Which "derivative" and "boundary" we use depends on whether $\Omega$ is a surface or an $m$-dimensional region and whether $m = 2$ or 3.

Our first fundamental theorem applies to two-dimensional regions in the plane.

**13.49 THEOREM** [GREEN'S THEOREM]. *Let $E$ be a two-dimensional region whose topological boundary $\partial E$ is a piecewise smooth curve oriented positively. If $P, Q : E \to \mathbf{R}$ are $\mathcal{C}^1$ and $F = (P, Q)$, then*

$$\int_{\partial E} F \cdot T \, ds = \iint_E \left( \frac{\partial Q}{\partial x} - \frac{\partial P}{\partial y} \right) dA.$$

PROOF FOR SPECIAL REGIONS. Suppose for simplicity that $E$ is of types I and II. Write the integral on the left in differential notation,

$$\int_{\partial E} P \, dx + Q \, dy = \int_{\partial E} P \, dx + \int_{\partial E} Q \, dy =: I_1 + I_2.$$

We evaluate $I_1$ first. Since $E$ is of type I, choose continuous functions $f, g : [a, b] \to \mathbf{R}$ such that

$$E = \{(x, y) \in \mathbf{R}^2 : a \le x \le b, \ f(x) \le y \le g(x)\}.$$

Thus $\partial E$ has a top $y = g(x)$, a bottom $y = f(x)$, and (possibly) one or two vertical sides (see Figure 13.17).

Since the positive orientation is counterclockwise, the trivial parametrization of the top is $y = g(x)$, where $x$ runs from $b$ to $a$, and of the bottom is $y = f(x)$, where $x$ runs from $a$ to $b$. Since $dx = 0$ on any vertical curve, the contribution of the vertical sides to $I_1$ is zero. Thus it follows from Definition 13.20 and the one-dimensional Fundamental Theorem of Calculus that

$$I_1 = \int_{\partial E} P \, dx = \int_a^b P(x, f(x)) \, dx + \int_b^a P(x, g(x)) \, dx$$

$$= - \int_a^b (P(x, g(x)) - P(x, f(x)) \, dx$$

$$= - \int_a^b \int_{f(x)}^{g(x)} \frac{\partial P}{\partial y}(x, y) \, dy \, dx = - \iint_E \frac{\partial P}{\partial y} \, dA.$$

Since $E$ is of type II, a similar argument establishes

$$I_2 = \int_{\partial E} Q \, dy = \iint_E \frac{\partial Q}{\partial x} \, dA.$$

(Here, we have changed parametrizations of $\partial E$, e.g., replaced $y = f(x)$ by $x = f^{-1}(y)$. The value of the oriented integral does not change because these parametrizations are orientation equivalent—see Exercise 6 in Section 13.2.) Adding $I_1$ and $I_2$ completes the proof. ∎

The assumption that $E$ be of types I and II was made to keep the proof simple. For a proof of Green's Theorem as stated, see Theorem 15.44 and the reference which follows it. In the meantime, it is easy to check that Green's Theorem holds for any two-dimensional region which can be divided into a finite number of regions

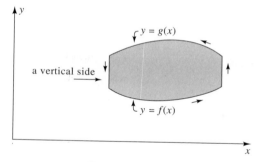

**Figure 13.17**

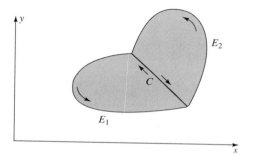

**Figure 13.18**

each of which is of types I and II. For example, consider the region $E$ illustrated in Figure 13.18. Notice that although $E$ is not of type II, it can be divided into $E_1$, $E_2$, both of which are of types I and II. Applying Theorem 13.49 to each piece, we find

$$\iint_E \left( \frac{\partial Q}{\partial x} - \frac{\partial P}{\partial y} \right) dA = \iint_{E_1} \left( \frac{\partial Q}{\partial x} - \frac{\partial P}{\partial y} \right) dA + \iint_{E_2} \left( \frac{\partial Q}{\partial x} - \frac{\partial P}{\partial y} \right) dA$$

$$= \int_{\partial E_1} F \cdot T \, ds + \int_{\partial E_2} F \cdot T \, ds$$

$$= \int_{\partial E} F \cdot T \, ds + \int_{C \cap \partial E_1} F \cdot T \, ds + \int_{C \cap \partial E_2} F \cdot T \, ds,$$

where $C$ is the common border between $E_1$ and $E_2$. Since $\partial E_1$ and $\partial E_2$ are oriented in the counterclockwise direction, the orientation of $C \cap \partial E_1$ is different from the orientation of $C \cap \partial E_2$. Since a change of orientation changes the sign of the integral, the integrals along $C$ drop out. The end result is the integral of $F \cdot T \, ds$ on $\partial E$, as promised.

Green's Theorem is often used to avoid tedious parametrizations.

**13.50 Example.** Find $\int_{\partial E} F \cdot T \, ds$, where $E = [0, 2] \times [1, 3]$, $\partial E$ has the counterclockwise orientation, and $F(x, y) = (xy, x^2 + y^2)$.

SOLUTION. Since $\partial E$ has four sides, direct evaluation requires four separate parametrizations. However, by Green's Theorem,

$$\int_{\partial E} F \cdot T \, ds = \int_0^2 \int_1^3 (2x - x) \, dy \, dx = 4. \quad \blacksquare$$

Green's Theorem is also used to avoid difficult integrals.

**13.52 Example.** Find $\int_{\partial E} F \cdot T \, ds$, where $E = B_1(0,0)$, $\partial E$ has the clockwise orientation, and $F = (xy^2, \arctan(\log(y + 3)) - x))$.

SOLUTION. The second component of $F$ looks tough to integrate. However, by Green's Theorem,

$$\int_{\partial E} F \cdot T \, ds = -\iint_{B_1(0,0)} (-1 - 2xy) \, dx \, dy$$

$$= \int_0^{2\pi} \int_0^1 (1 + 2r^2 \cos\theta \sin\theta) r \, dr \, d\theta = \pi.$$

(Note: The minus sign appears because $\partial E$ is oriented in the clockwise direction.) $\blacksquare$

By Green's Theorem, the "derivative" used to obtain a fundamental theorem of calculus for two-dimensional regions in $\mathbf{R}^2$ is $Q_x - P_y$. Here are the "derivatives" which will be used when $\Omega$ is a surface in $\mathbf{R}^3$ or a three-dimensional region.

**13.53 DEFINITION.** Let $E$ be a subset of $\mathbf{R}^3$ and $F = (P, Q, R) : E \to \mathbf{R}^3$ be $\mathcal{C}^1$ on $E$. The *curl* of $F$ is

$$\operatorname{curl} F = \left( \frac{\partial R}{\partial y} - \frac{\partial Q}{\partial z}, \frac{\partial P}{\partial z} - \frac{\partial R}{\partial x}, \frac{\partial Q}{\partial x} - \frac{\partial P}{\partial y} \right),$$

and the *divergence* of $F$ is

$$\operatorname{div} F = \frac{\partial P}{\partial x} + \frac{\partial Q}{\partial y} + \frac{\partial R}{\partial z}.$$

Notice that if $F = (P, Q, 0)$, where $P$ and $Q$ are as in Green's Theorem, then $\operatorname{curl} F = (0, 0, Q_x - P_y)$ is the derivative used for Green's Theorem.

These derivatives take on a more easily remembered form by using the notation

$$\nabla = (\frac{\partial}{\partial x}, \frac{\partial}{\partial y}, \frac{\partial}{\partial z}).$$

Indeed, $\operatorname{curl} F = \nabla \times F$   and   $\operatorname{div} F = \nabla \cdot F$.

If $E$ is a three-dimensional region whose topological boundary is a piecewise smooth orientable surface, then the *positive orientation* on $\partial E$ is determined by the unit normal which points away from $E^o$. If $E$ is convex, this means that $\mathbf{n}$ points outward. This is not the case, however, when $E$ has interior "bubbles." For example, if $E = \{\mathbf{x} : a < \|\mathbf{x}\| < b\}$ for some $a > 0$, then $\mathbf{n}$ points away from the origin on $\{\mathbf{x} : \|\mathbf{x}\| = b\}$ but toward the origin on $\{\mathbf{x} : \|\mathbf{x}\| = a\}$.

Our next fundamental theorem applies to the case when $\Omega$ is a three-dimensional region. This result is also called the *Divergence Theorem*.

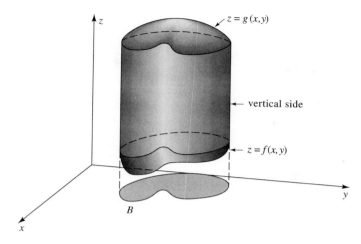

**Figure 13.19**

**13.54 THEOREM** [GAUSS'S THEOREM]. *Let $E$ be a three-dimensional region whose topological boundary $\partial E$ is a piecewise smooth surface oriented positively. If $F : E \to \mathbf{R}^3$ is $\mathcal{C}^1$ on $E$, then*

$$\iint_{\partial E} F \cdot \boldsymbol{n} \, d\sigma = \iiint_E \operatorname{div} F \, dV.$$

PROOF FOR SPECIAL REGIONS. Suppose for simplicity that $E$ is a region of types I, II, and III. Let $F = (P, Q, R)$ and write the surface integral in differential form:

$$\iint_{\partial E} F \cdot \boldsymbol{n} \, d\sigma = \iint_{\partial E} P \, dy \, dz + \iint_{\partial E} Q \, dz \, dx + \iint_{\partial E} R \, dx \, dy =: I_1 + I_2 + I_3.$$

We evaluate $I_3$ first.

Since $E$ is of type I, there exist a two-dimensional region $B \subset \mathbf{R}^2$ and continuous functions $f, g : B \to \mathbf{R}$ such that

$$E = \{(x, y, z) \in \mathbf{R}^3 : (x, y) \in B, \ f(x, y) \leq z \leq g(x, y)\}.$$

Thus $\partial E$ has a top $z = g(x, y)$, a bottom $z = f(x, y)$, and (possibly) a vertical side (see Figure 13.19). Any normal to $\partial E$ on the vertical side is parallel to the $xy$ plane. Since $dx \, dy$ is the third component of a normal to $\partial E$, it must be zero on the vertical portion. Therefore, $I_3$ can be evaluated by integrating over the top and bottom of $\partial E$. Notice that, by hypothesis, the unit normal on the bottom portion points downward and the unit normal on the top portion points upward. By using trivial parametrizations and Theorem 5.28 (the Fundamental Theorem of Calculus), we obtain

$$I_3 = \iint_{\partial E} R \, dx \, dy = \int_B (R(x, y, g(x, y)) - R(x, y, f(x, y)) \, d(x, y)$$

$$= \int_B \int_{f(x,y)}^{g(x,y)} \frac{\partial R}{\partial z}(x, y, z) \, dz \, d(x, y) = \iiint_E \frac{\partial R}{\partial z} \, dV.$$

Similarly, since $E$ is of type II,

$$I_2 = \iiint_E \frac{\partial Q}{\partial y} \, dV,$$

and since $E$ is of type III,

$$I_1 = \iiint_E \frac{\partial P}{\partial x} \, dV.$$

Adding $I_1 + I_2 + I_3$ verifies the theorem. ∎

The assumption that $E$ be of types I, II, and III was made to keep the proof simple. For a proof of Gauss's Theorem as stated, see Theorem 15.44 and the reference which follows it. In the meantime, it is easy to check that Gauss's Theorem holds for any three-dimensional region $E$ which can be divided into a finite number of regions $E_j$, each of which is of types I, II, and III. For example, if $E = E_1 \cup E_2$, then

$$\iiint_E \operatorname{div} F \, dV = \iiint_{E_1} \operatorname{div} F \, dV + \iiint_{E_2} \operatorname{div} F \, dV$$

$$= \iint_{\partial E} F \cdot \boldsymbol{n} \, d\sigma + \iint_{S \cap \partial E_1} F \cdot \boldsymbol{n} \, d\sigma + \iint_{S \cap \partial E_2} F \cdot \boldsymbol{n} \, d\sigma,$$

where $S$ is the common surface between $E_1$ and $E_2$. Since $E_1$ and $E_2$ have outward pointing normals, the orientation of $S \cap \partial E_1$ is different from the orientation of $S \cap \partial E_2$, and the integrals over $S$ cancel each other out.

The next two examples show that, like Green's Theorem, Gauss's Theorem can be used to avoid difficult integrals and tedious parametrizations.

**13.55 Example.** Use Theorem 13.54 to evaluate $\iint_S F \cdot \boldsymbol{n} \, d\sigma$, where $S$ is the topological boundary of the solid

$$E = \{(x, y, z) : x^2 + y^2 \le z \le 1\},$$

$\boldsymbol{n}$ is the outward pointing normal, and $F(x, y, z) = (2x + z^2, x^5 + z^7, \cos(x^2) + \sin(y^3) - z^2)$.

SOLUTION. Since $\operatorname{div} F = 2 - 2z$, it follows from Gauss's Theorem that

$$\iint_S F \cdot \boldsymbol{n} \, d\sigma = \iiint_E (2 - 2z) \, dV = 2 \int_0^{2\pi} \int_0^1 \int_{r^2}^1 (1 - z) r \, dz \, dr \, d\theta = \frac{\pi}{3}. \quad ∎$$

**13.56 Example.** Evaluate $\iint_{\partial Q} F \cdot \boldsymbol{n} \, d\sigma$, where $Q$ is the unit cube $[0, 1] \times [0, 1] \times [0, 1]$, $\boldsymbol{n}$ is the outward pointing normal, and $F(x, y, z) = (2x - z, x^2 y, -xz^2)$.

SOLUTION. Since $\partial Q$ has six sides, direct evaluation of this integral requires six separate integrals. However, by Gauss's Theorem,

$$\iint_{\partial Q} F \cdot \boldsymbol{n} \, d\sigma = \int_0^1 \int_0^1 \int_0^1 (2 + x^2 - 2xz) \, dx \, dy \, dz = \frac{11}{6}. \quad ∎$$

These definitions and results take on new meaning when examined in the context of fluid flow. When $F$ represents the flow of an incompressible fluid near a point $\boldsymbol{a}$, curl $F(\boldsymbol{a})$ measures the tendency of the fluid to swirl in a counterclockwise direction about $\boldsymbol{a}$ (see Exercise 6 in Section 13.6), and div $F(\boldsymbol{a})$ measures the tendency of the fluid to spread out from $\boldsymbol{a}$ (see Exercise 7 below). (This explains the etymology of the words *curl* and *divergence*.) For example, if $F(x,y,z) = (x,y,z)$, then the fluid is not swirling at all, but spreading straight out from the origin. Accordingly, curl $F = \boldsymbol{0}$ and div $F = 3$. On the other hand, if $G(x,y,z) = (y,-x,0)$, then the fluid is swirling around in a circular motion about the origin. Accordingly, curl $G = (0,0,-1)$ but div $G = 0$. Note the minus sign in the component of curl $G$. This fluid swirls about the origin in a clockwise direction, so runs against counterclockwise motion.

When the fluid flows over a two-dimensional region $E \subset \mathbf{R}^2$, the integral of $F{\cdot}T\,ds$ over $C$ represents the circulation of the fluid around $C$ in the direction of $T$ (see the comments following Definition 13.20). Thus Green's Theorem tells us that the circulation of a fluid around $\partial E$ in the direction of the tangent is determined by how strongly the fluid swirls inside $E$. When $F$ represents the flow of an incompressible fluid through a three-dimensional region $E \subset \mathbf{R}^3$ and $S = \partial E$, the integral $\iint_S F \cdot \boldsymbol{n}$ represents the flux of the fluid across the surface $S$ (see the comments following Definition 13.42). Thus Gauss's Theorem tells us that the flux of the fluid across $S = \partial E$ is determined by how strongly the fluid is spreading out inside $E$.

We close this section by admitting that the interpretations of curl and divergence given above are imperfect at best. For example, the vector field $F(x,y,z) = (0,z,0)$ has curl $(-1,0,0)$. Here the fluid is shearing in layers with flow parallel to the $xy$ plane in the direction of the positive $y$ axis when $z > 0$. Although the fluid is not swirling, it does tend to rotate a stick placed in the fluid parallel to the $z$ axis (for example, the line segment $\{(0,1,z) : 0 \le z \le 1\}$) because more force is applied to the top than the bottom. This tendency toward rotation is reflected by the value of the curl. (Notice that the rotation is clockwise and the curl has a negative first component.)

## EXERCISES

**1.** For each of the following, evaluate $\int_C F \cdot T\,ds$.

    a) $C$ is the topological boundary of the two-dimensional region in the first quadrant bounded by $x = 0$, $y = 0$, and $y = \sqrt{4 - x^2}$, oriented in the counterclockwise direction, and $F(x,y) = (\sin(\sqrt{x^3 - x^2}), xy)$.

    b) $C$ is the perimeter of the rectangle with vertices $(0,0)$, $(2,0)$, $(0,3)$, $(2,3)$, oriented in the counterclockwise direction, and $F(x,y) = (e^y, \log(x+1))$.

    c) $C = C_1 \cup C_2$, where $C_1 = \partial B_1(0,0)$ oriented in the counterclockwise direction, $C_2 = \partial B_2(0,0)$ oriented in the clockwise direction, and $F(x,y) = (f(x^2 + y^2), xy^2)$, where $f$ is a $C^1$ function on $[1,2]$.

**2.** For each of the following, evaluate $\int_C \omega$.

    a) $C$ is the topological boundary of the rectangle $[a,b] \times [c,d]$, oriented in the counterclockwise direction, and $\omega = (f(x) + y)\,dx + xy\,dy$, where $f : [0,1] \to$

**R** is any continuous function.

b) $C$ is the topological boundary of the two-dimensional region bounded by $y = x^2$ and $y = x$, oriented in the clockwise direction, and $\omega = yf(x)\,dx + (x^2 + y^2)\,dy$, where $f : [0,1] \to \mathbf{R}$ is continuously differentiable and satisfies $\int_0^1 xf(x)\,dx = \int_0^1 x^2 f(x)\,dx$.

c) $C$ is the topological boundary of a two-dimensional region $E$ which satisfies the conclusion of Green's Theorem, oriented positively, and $\omega = e^x \sin y\,dy - e^x \cos y\,dx$.

**3.** For each of the following, evaluate $\iint_S F \cdot \boldsymbol{n}\,d\sigma$, where $\boldsymbol{n}$ is the outward pointing normal.

a) $S$ is the topological boundary of the rectangle $[0,1] \times [0,2] \times [0,3]$ and $F(x,y,z) = (x + e^z, y + e^z, e^z)$.

b) $S$ is the truncated cylinder $x^2 + y^2 = 1$, $0 \le z \le 1$ together with the disks $x^2 + y^2 \le 1$, $z = 0, 1$, and $F(x,y,z) = (x^2, y^2, z^2)$.

c) $S$ is the topological boundary of $E$, where $E \subset \mathbf{R}^3$ is bounded by $z = 2 - x^2$, $z = x^2$, $y = 0$, $z = y$, and $F(x,y,z) = (x + f(y,z), y + g(x,z), z + h(x,y))$ and $f, g, h : \mathbf{R}^2 \to \mathbf{R}$ are continuously differentiable.

d) $S$ is the ellipsoid $x^2/a^2 + y^2/b^2 + z^2/c^2 = 1$ and $F(x,y,z) = (x|y|, y|z|, z|x|)$.

**4.** For each of the following, find $\iint_S \omega$, where $\boldsymbol{n}$ is the outward pointing normal.

a) $S$ is the topological boundary of the three-dimensional region enclosed by $y = x^2$, $z = 0$, $z = 1$, $y = 4$, and $\omega = xyz\,dy\,dz + (x^2 + y^2 + z^2)\,dz\,dx + (x + y + z)\,dx\,dy$.

b) $S$ is the truncated hyperboloid of one sheet $x^2 - y^2 + z^2 = 1$, $0 \le y \le 1$, together with the disks $x^2 + z^2 \le 1$, $y = 0$, and $x^2 + z^2 \le 2$, $y = 1$, and $\omega = xy|z|\,dy\,dz + x^2|z|\,dz\,dx + (x^3 + y^3)\,dx\,dy$.

c) $S$ is the topological boundary of $E$, where $E \subset \mathbf{R}^3$ is bounded by the surfaces $x^2 + y + z^2 = 4$ and $4x + y + 2z = 5$, and $\omega = (x + y^2 + z^2)\,dy\,dz + (x^2 + y + z^2)\,dz\,dx + (x^2 + y^2 + z)\,dx\,dy$.

**5. a)** Prove that if $E$ is a Jordan region whose topological boundary is a smooth curve oriented in the counterclockwise direction, then

$$\text{Area}\,(E) = \frac{1}{2} \int_{\partial E} x\,dy - y\,dx.$$

b) Find the area enclosed by the loop in the Folium of Descartes, i.e., by

$$\phi(t) = \left( \frac{3t}{1 + t^3}, \frac{3t^2}{1 + t^3} \right), \qquad t \in [0, \infty).$$

c) Find an analogue of part a) for the volume of a Jordan region $E$ in $\mathbf{R}^3$.

d) Compute the volume of the torus with radii $a > b$ (see Example 13.31).

**6. a)** Show Green's Theorem does not hold if continuity of $P$, $Q$ is relaxed at one point in $E$. (Hint: Consider $P = y/(x^2 + y^2)$, $Q = -x/(x^2 + y^2)$, and $E = B_1(0,0)$.)

b) Show that Gauss's Theorem does not hold if continuity of $F$ is relaxed at one point in $E$.

**7**. **This exercise is used in Section 13.6.** Suppose $V \neq \emptyset$ is an open set in $\mathbf{R}^3$ and $F : V \to \mathbf{R}^3$ is $\mathcal{C}^1$. Prove that

$$\operatorname{div} F(\boldsymbol{x}_0) = \lim_{r \to 0+} \frac{1}{\operatorname{Vol}(B_r(\boldsymbol{x}_0))} \iint_{\partial B_r(\boldsymbol{x}_0)} F \cdot \boldsymbol{n} \, d\sigma$$

for each $\boldsymbol{x}_0 \in V$, where $\boldsymbol{n}$ is the outward pointing normal of $B_r(\boldsymbol{x}_0)$.

**8.** Let $F, G : \mathbf{R}^3 \to \mathbf{R}^3$ and $f : \mathbf{R}^3 \to \mathbf{R}$ be differentiable. Prove the following analogues of the Sum and Product Rules for the "derivatives" curl and divergence.

a)
$$\nabla \times (F + G) = (\nabla \times F) + (\nabla \times G).$$

b)
$$\nabla \times (fF) = f(\nabla \times F) + (\nabla f \times F).$$

c)
$$\nabla \cdot (fF) = \nabla f \cdot F + f \cdot (\nabla \cdot F).$$

d)
$$\nabla \cdot (F + G) = \nabla \cdot F + \nabla \cdot G.$$

e)
$$\nabla \cdot (F \times G) = (\nabla \times F) \cdot G - (\nabla \times G) \cdot F.$$

**9**. **This exercise is used in Section 13.6.** Let $E \subset \mathbf{R}^2$. Recall that the *gradient* of a $\mathcal{C}^1$ function $f : E \to \mathbf{R}$ is defined by

$$\operatorname{grad} f := \nabla f := (f_x, f_y).$$

a) Suppose $E$ and $F$ satisfy the conclusion of Green's Theorem and $f : E \to \mathbf{R}$ is a $\mathcal{C}^2$ function. If $F = \operatorname{grad} f$ on $E$, prove

$$\int_{\partial E} F \cdot T \, ds = 0.$$

b) Prove that if $f$ and $F$ are $\mathcal{C}^2$ at $\boldsymbol{x}_0$, then $\operatorname{curl} \operatorname{grad} f(\boldsymbol{x}_0) = \boldsymbol{0}$ and $\operatorname{div} \operatorname{curl} F(\boldsymbol{x}_0) = 0$.

c) Suppose $E$ and $F$ satisfy the conclusion of Gauss's Theorem and $f : E \to \mathbf{R}$ is a $\mathcal{C}^2$ function. If $F = \operatorname{grad} f$ on $E$, prove

$$\iint_{\partial E} fF \cdot \boldsymbol{n} \, d\sigma = \iiint_E F \cdot F \, dV.$$

**10.** Let $E$ be a set in $\mathbf{R}^m$. For each $u : E \to \mathbf{R}$ which has second-order partial derivatives on $E$, *Laplace's equation* is defined by

$$\Delta u := \sum_{j=1}^{m} \frac{\partial^2 u}{\partial x_j^2}.$$

a) Show that if $u$ is $\mathcal{C}^2$ on $E$, then $\Delta u = \nabla \cdot (\nabla u)$ on $E$.

b) [GREEN'S FIRST IDENTITY]. Show that if $E \subset \mathbf{R}^3$ satisfies the conclusion of Gauss's Theorem for all $\mathcal{C}^1$ functions $F$, then

$$\iiint_E (u \Delta v + \nabla u \cdot \nabla v) \, dV = \iint_{\partial E} u \nabla v \cdot \boldsymbol{n} \, d\sigma$$

for all $\mathcal{C}^2$ functions $u, v : E \to \mathbf{R}$.

c) [GREEN'S SECOND IDENTITY]. Show that if $E \subset \mathbf{R}^3$ satisfies the conclusion of Gauss's Theorem for all $\mathcal{C}^1$ functions $F$, then

$$\iiint_E (u \Delta v - v \Delta u) \, dV = \iint_{\partial E} (u \nabla v - v \nabla u) \cdot \boldsymbol{n} \, d\sigma$$

for all $\mathcal{C}^2$ functions $u, v : E \to \mathbf{R}$.

d) A function $u : E \to \mathbf{R}$ is said to be *harmonic* on $E$ if $u$ is $\mathcal{C}^2$ on $E$ and $\Delta u(\boldsymbol{x}) = 0$ for all $\boldsymbol{x} \in E$. Suppose $E$ is a nonempty open region in $\mathbf{R}^3$ which satisfies the conclusion of Gauss's Theorem for all $\mathcal{C}^1$ functions $F$. If $u$ is harmonic on $E$, $u$ is continuous on $\overline{E}$, and $u = 0$ on $\partial E$, prove $u = 0$ on $\overline{E}$.

e) Suppose $V$ is open and nonempty in $\mathbf{R}^2$, $u$ is $\mathcal{C}^2$ on $V$, and $u$ is continuous on $\overline{V}$. Prove that $u$ is harmonic on $V$ if and only if

$$\int_{\partial E} (u_x \, dy - u_y \, dx) = 0$$

for all two-dimensional regions $E \subset V$ which satisfy the conclusion of Green's Theorem for all $\mathcal{C}^1$ functions $F = (P, Q)$.

## 13.6  STOKES'S THEOREM

Our final fundamental theorem applies to surfaces in $\mathbf{R}^3$ whose boundaries are curves.

**13.57 THEOREM** [STOKES'S THEOREM]. *Let $S$ be an oriented $\mathcal{C}^2$ surface in $\mathbf{R}^3$ with unit normal $\boldsymbol{n}$. If the boundary $\partial S$ is a piecewise smooth curve oriented positively and $F : S \to \mathbf{R}^3$ is $\mathcal{C}^1$, then*

$$\int_{\partial S} F \cdot T \, ds = \iint_S \operatorname{curl} F \cdot \boldsymbol{n} \, d\sigma.$$

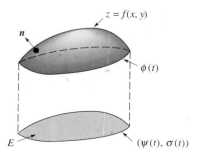

**Figure 13.20**

PROOF FOR CERTAIN EXPLICIT SURFACES. Suppose for simplicity that $S$ is an explicit $\mathcal{C}^2$ surface which lies over $E$, a two-dimensional region which satisfies the conclusion of Green's Theorem for all $\mathcal{C}^1$ functions $F$. Let $F = (P, Q, R)$ and write the line integral in differential notation:

$$\int_{\partial S} F \cdot T \, ds = \int_{\partial S} P \, dx + Q \, dy + R \, dz.$$

Without loss of generality, suppose $S$ is determined by $z = f(x, y)$, $(x, y) \in E$, where $f : E \to \mathbf{R}$ is a $\mathcal{C}^2$ function and $S$ is oriented with the upward pointing normal. Thus $\boldsymbol{n} = N/\|N\|$, where $N = (-f_x, -f_y, 1)$.

Let $(\psi(t), \sigma(t))$, $t \in [a, b]$, be a piecewise smooth parametrization of $\partial E$ oriented in the counterclockwise direction. Then

$$\phi(t) = (\psi(t), \sigma(t), f(\psi(t), \sigma(t))), \qquad t \in [a, b],$$

is a piecewise smooth parametrization of $\partial S$ which is oriented positively (see Figure 13.20). If $x = \psi(t)$, $y = \sigma(t)$, and $z = f(\psi(t), \sigma(t))$, then $dx = \psi'(t) \, dt$, $dy = \sigma'(t) \, dt$, and

$$dz = \frac{\partial z}{\partial x} \, dx + \frac{\partial z}{\partial y} \, dy.$$

Thus, by definition,

$$(8) \qquad \int_{\partial S} P \, dx + Q \, dy + R \, dz = \int_{\partial E} \left(P + R\frac{\partial z}{\partial x}\right) dx + \left(Q + R\frac{\partial z}{\partial y}\right) dy.$$

We shall apply Green's Theorem to this last integral. By the Chain Rule and the Product Rule,

$$\frac{\partial}{\partial x}\left(Q + R\frac{\partial z}{\partial y}\right) = \frac{\partial Q}{\partial x} + \frac{\partial Q}{\partial z}\frac{\partial z}{\partial x} + \frac{\partial R}{\partial x}\frac{\partial z}{\partial y} + \frac{\partial R}{\partial z}\frac{\partial z}{\partial x}\frac{\partial z}{\partial y} + R\frac{\partial^2 z}{\partial x \partial y}$$

and

$$\frac{\partial}{\partial y}\left(P + R\frac{\partial z}{\partial x}\right) = \frac{\partial P}{\partial y} + \frac{\partial P}{\partial z}\frac{\partial z}{\partial y} + \frac{\partial R}{\partial y}\frac{\partial z}{\partial x} + \frac{\partial R}{\partial z}\frac{\partial z}{\partial y}\frac{\partial z}{\partial x} + R\frac{\partial^2 z}{\partial y \partial x}.$$

Since $z = f(x, y)$ is $C^2$, the mixed second-order partial derivatives above are equal. Therefore,

$$\frac{\partial}{\partial x}(Q + R\frac{\partial z}{\partial y}) - \frac{\partial}{\partial y}(P + R\frac{\partial z}{\partial x})$$

$$= \left(\frac{\partial R}{\partial y} - \frac{\partial Q}{\partial z}\right)\left(-\frac{\partial z}{\partial x}\right) + \left(\frac{\partial P}{\partial z} - \frac{\partial R}{\partial x}\right)\left(-\frac{\partial z}{\partial y}\right) + \left(\frac{\partial Q}{\partial x} - \frac{\partial P}{\partial y}\right)$$

$$= \operatorname{curl} F \cdot N.$$

Hence, it follows from (8), Green's Theorem, and (7) that

$$\int_{\partial S} F \cdot T \, ds = \int_E \operatorname{curl} F \cdot N \, d(x, y) = \iint_S \operatorname{curl} F \cdot \boldsymbol{n} \, d\sigma. \quad \blacksquare$$

The assumption that $S$ be an explicit surface over a "Green's region" was made to keep the proof simple. For a proof of Stokes's Theorem as stated, see Theorem 15.44 and the reference which follows it. In the meantime, it is easy to check that Stokes's Theorem holds for any surface which can be divided into a finite number of such explicit surfaces. As before, the common boundaries appear twice, each time in a different orientation, hence, cancel each other out.

Stokes's Theorem can be used to replace complicated line integrals by simple surface integrals.

**13.58 Example.** Compute $\int_C F \cdot T \, ds$, where $C$ is the circle $x^2 + z^2 = 1$, $y = 0$, oriented in the counterclockwise direction when viewed from far out on the $y$ axis, and $F(x, y, z) = (x^2 z + \sqrt{x^3 + x^2 + 2}, xy, xy + \sqrt{z^3 + z^2 + 2})$.

SOLUTION. Since $\operatorname{curl} F = (x, x^2 - y, y)$, using Stokes's Theorem is considerably easier than trying to integrate $F \cdot T \, ds$ directly. Let $S$ be the disk $x^2 + z^2 \leq 1$, $y = 0$, and notice that $\partial S = C$. Since $C$ is oriented in the counterclockwise direction, the normal to $S$ must point toward the positive $y$ axis, i.e., $\boldsymbol{n} = (0, 1, 0)$. Thus $\operatorname{curl} F \cdot \boldsymbol{n} = x^2 - y = x^2$ on $S$, and Stokes's Theorem implies

$$\int_C F \cdot T \, ds = \iint_S x^2 \, dA = \int_0^{2\pi} \int_0^1 r^3 \cos^2 \theta \, dr \, d\theta = \frac{\pi}{4}. \quad \blacksquare$$

In Example 13.58, we could have chosen any surface $S$ whose boundary is $C$. Thus Stokes's Theorem can also be used to replace complicated surface integrals by simpler ones.

**13.59 Example.** Find $\iint_S \operatorname{curl} F \cdot \boldsymbol{n} \, d\sigma$, where $S$ is the semi-ellipsoid $9x^2 + 4y^2 + 36z^2 = 36$, $z \geq 0$, $\boldsymbol{n}$ is the upward pointing normal, and

$$F(x, y, z) = (\cos x \sin z + xy, x^3, e^{x^2 + z^2} - e^{y^2 + z^2} + \tan(xy)).$$

SOLUTION. Let $C = \partial S$. The integral of $\operatorname{curl} F \cdot \boldsymbol{n} \, d\sigma$ over $S$ and the integral of $F \cdot T \, ds$ over $C$ are both complicated. But, by Stokes's Theorem, the integral of $F \cdot T \, ds$ over $C$ is the same as the integral of curl $F \cdot \boldsymbol{n} \, d\sigma$ over any oriented $C^2$ surface $E$ satisfying $\partial E = C$. Let $E$ be the two-dimensional region $9x^2 + 4y^2 \leq 36$. On $E$, $\boldsymbol{n} = (0, 0, 1)$. Thus we only need the third component of curl $F$:

$$(\mathrm{curl}\, F)_3 := \frac{\partial}{\partial x}(x^3) - \frac{\partial}{\partial y}(\cos x \sin z + xy) = 3x^2 - x.$$

Therefore,

$$\iint_S \mathrm{curl}\, F \cdot \boldsymbol{n} \, d\sigma = \int_E (3x^2 - x) \, d(x, y).$$

Let $x = 2r \cos \theta$ and $y = 3r \sin \theta$. By a change of variables,

$$\int_E (3x^2 - x) \, d(x, y) = \int_0^{2\pi} \int_0^1 (12r^2 \cos^2 \theta - 2r \cos \theta) 6r \, dr \, d\theta = 18\pi. \quad \blacksquare$$

Stokes's Theorem can also be used to replace complicated surface integrals by simple line integrals.

**13.60 Example.** Let $S$ be the union of the truncated paraboloid $z = x^2 + y^2$, $0 \leq z \leq 1$, and the truncated cylinder $x^2 + y^2 = 1$, $1 \leq z \leq 3$. Compute

$$\iint_S F \cdot \boldsymbol{n} \, d\sigma,$$

where $\boldsymbol{n}$ is the outward pointing normal and $F(x, y, z) = (x + z^2, 0, -z - 3)$.

SOLUTION. The boundary of $S$ is $x^2 + y^2 = 1$, $z = 3$. To use Stokes's Theorem, we must find a function $G = (P, Q, R) : S \to \mathbf{R}^3$ such that

$$\mathrm{curl}\, G = F,$$

i.e., such that

$$(9) \qquad\qquad \frac{\partial R}{\partial y} - \frac{\partial Q}{\partial z} = x + z^2,$$

$$(10) \qquad\qquad \frac{\partial P}{\partial z} - \frac{\partial R}{\partial x} = 0,$$

and

$$(11) \qquad\qquad \frac{\partial Q}{\partial x} - \frac{\partial P}{\partial y} = -z - 3.$$

Starting with (9), set

$$(12) \qquad\qquad \frac{\partial Q}{\partial z} = -x \quad \text{and} \quad \frac{\partial R}{\partial y} = z^2.$$

The left side of (12) implies $Q = -xz + g(x, y)$ for some $g : \mathbf{R}^2 \to \mathbf{R}$. Similarly, the right side of (12) leads to $R = z^2 y + h(x, z)$ for some $h : \mathbf{R}^2 \to \mathbf{R}$. Thus $Q_x = -z + g_x$ will solve (11) if we set $g = 0$ and $P_y = 3$, i.e., $P = 3y + \sigma(x, z)$ for some $\sigma : \mathbf{R}^2 \to \mathbf{R}$. Hence, $P_z - R_x = \sigma_z - h_x$ will satisfy (10) if $\sigma = h = 0$. Therefore, $P = 3y$, $Q = -xz$ and $R = yz^2$, i.e., $G = (3y, -xz, yz^2)$.

Parametrize $\partial S$ by $\phi(t) = (\sin t, \cos t, 3)$, $t \in [0, 2\pi]$, and observe that

$$(G \circ \phi) \cdot \phi' = (3 \cos t, -3 \sin t, 9 \cos t) \cdot (\cos t, -\sin t, 0) = 3 \cos^2 t + 3 \sin^2 t = 3.$$

Consequently, Stokes's Theorem implies

$$\iint_S F \cdot \boldsymbol{n} \, d\sigma = \iint_S \operatorname{curl} G \cdot \boldsymbol{n} \, d\sigma = \int_{\partial S} G \cdot T \, ds = \int_0^{2\pi} 3 \, dt = 6\pi. \quad \blacksquare$$

The solution to Example 13.60 involved finding a function $G$ which satisfied $\operatorname{curl} G = F$. This function is not unique. Indeed, we could have begun with

$$\frac{\partial Q}{\partial z} = -z^2 \quad \text{and} \quad \frac{\partial R}{\partial y} = x$$

instead of (12). This leads to a different solution:

$$\widetilde{G}(x, y, z) = (zy, -(3x + z^3/3), xy).$$

The technique used to solve Example 13.60, however, is perfectly valid. Indeed, by Stokes's Theorem the value of the oriented line integral of $G \cdot T$ will be the same for all $\mathcal{C}^1$ functions $G$ which satisfy $\operatorname{curl} G = F$.

This technique works only when the system of partial differential equations $\operatorname{curl} G = F$ has a solution $G$. To avoid searching for something which does not exist, we must be able to discern beforehand whether such a solution exists. To discover how to do this, suppose $G$ is a $\mathcal{C}^2$ function which satisfies $\operatorname{curl} G = F$ on some set $E$. Then $\operatorname{div} F = 0$ on $E$ by Exercise 9b) in Section 13.5. Thus the condition $\operatorname{div} F = 0$ is necessary for existence of a solution $G$ to $\operatorname{curl} G = F$. The following result shows that if $E$ is nice enough, this condition is also sufficient (see also Theorem 15.45).

**13.61 THEOREM.** *Let $\Omega$ be a ball centered at $(0, 0, 0)$ or a rectangle containing $(0, 0, 0)$, and $F : \Omega \to \mathbf{R}^3$ be $\mathcal{C}^1$ on $\Omega$. Then the following three statements are equivalent.*

   i) *There is a $\mathcal{C}^2$ function $G : \Omega \to \mathbf{R}^3$ such that $\operatorname{curl} G = F$ on $\Omega$.*
   ii) *If $F$, $E$, and $S = \partial E$ satisfy the conclusion of Gauss's Theorem and $E \subset \Omega$, then*

(13)
$$\iint_S F \cdot \boldsymbol{n} \, d\sigma = 0.$$

   iii) *The identity $\operatorname{div} F = 0$ holds everywhere on $\Omega$.*

PROOF. If i) holds, then $\operatorname{div} F = \operatorname{div}(\operatorname{curl} G) = 0$ since the first-order partial derivatives of $G$ commute. Thus (13) holds by Gauss's Theorem. (This works for any set $\Omega$.)

If ii) holds, then by Gauss's Theorem and Exercise 7 in Section 13.5,

$$\operatorname{div} F(\boldsymbol{x}_0) = \lim_{r \to 0+} \frac{1}{\operatorname{Vol}(B_r(\boldsymbol{x}_0))} \iiint_{B_r(\boldsymbol{x}_0)} \operatorname{div} F \, dV$$

$$= \lim_{r \to 0+} \frac{1}{\operatorname{Vol}(B_r(\boldsymbol{x}_0))} \iint_{\partial B_r(\boldsymbol{x}_0)} F \cdot \boldsymbol{n} \, d\sigma = 0$$

for each $\boldsymbol{x}_0 \in \Omega^o$. (Recall by definition that any $n$-dimensional region has a nonempty interior.) Since $\operatorname{div} F$ is continuous on $\Omega$, it follows that $\operatorname{div} F = 0$ everywhere on $\Omega$. (This works for any three-dimensional region $\Omega$.)

Finally, suppose iii) holds. Let $F = (p, q, r)$ and suppose for simplicity that $G = (0, Q, R)$. If $\operatorname{curl} G = F$, then

(14) $$R_y - Q_z = p, \quad -R_x = q, \quad Q_x = r.$$

Integrating the last two identities, we have

$$R = -\int_0^x q(u, y, z) \, du + g(y, z) \quad \text{and} \quad Q = \int_0^x r(u, y, z) \, du + h(y, z)$$

for some $g, h : \mathbf{R}^2 \to \mathbf{R}$. (Note: By hypothesis, the line segment from $(0, y, z)$ to $(x, y, z)$ is a subset of $\Omega$, so these integrals make sense.) Differentiating under the integral sign (Theorem 11.5), and using condition iii), the first identity becomes

$$p = R_y - Q_z = -\int_0^x (q_y(u, y, z) + r_z(u, y, z)) \, du + g_y - h_z$$

$$= \int_0^x p_x(u, y, z) \, du + g_y - h_z = p(x, y, z) - p(0, y, z) + g_y - h_z.$$

Thus (14) can be solved by $g_y = p(0, y, z)$ and $h = 0$; i.e.,

$$Q = \int_0^x r(u, y, z) \, du \quad \text{and} \quad R = \int_0^y p(0, v, z) \, dv - \int_0^x q(u, y, z) \, du. \quad \blacksquare$$

We notice that Theorem 13.61 holds for any three-dimensional region $\Omega$ which satisfies the following property: given $(x, y, z) \in \Omega$, the line segments $L((0, y, z); (x, y, z))$ and $L((0, 0, z); (0, y, z))$ are both subsets of $\Omega$. However, as the following result shows, Theorem 13.61 is false without some restriction on $\Omega$.

**13.62 Remark.** *Let* $\Omega = B_1(0, 0, 0) \setminus \{(0, 0, 0)\}$ *and*

$$F(x, y, z) = \left( \frac{x}{w^{3/2}}, \frac{y}{w^{3/2}}, \frac{z}{w^{3/2}} \right),$$

*where $w = w(x, y, z) = x^2 + y^2 + z^2$. Then* $\operatorname{div} F = 0$ *on $\Omega$, but there is no $G$ which satisfies* $\operatorname{curl} G = F$.

PROOF. By definition,

$$\operatorname{div} F = \frac{-2x^2 + y^2 + z^2}{w^{5/2}} + \frac{x^2 - 2y^2 + z^2}{w^{5/2}} + \frac{x^2 + y^2 - 2z^2}{w^{5/2}} = 0.$$

Let $S$ represent the unit sphere $\partial B_1(0, 0, 0)$ oriented with the outward pointing normal, and suppose there is a $G$ such that $\operatorname{curl} G = F$. On the one hand, since $F = (x, y, z) = \boldsymbol{n}$ on $S$ implies $F \cdot \boldsymbol{n} = x^2 + y^2 + z^2 = 1$, we have

(15)
$$\iint_S F \cdot \boldsymbol{n} \, d\sigma = \iint_S 1 \, dA = \sigma(S) = 4\pi.$$

On the other hand, dividing $S$ into the upper hemisphere $S_1$ and the lower hemisphere $S_2$, we have by Stokes's Theorem that

(16)
$$\iint_S F \cdot \boldsymbol{n} \, d\sigma = \iint_{S_1} F \cdot \boldsymbol{n} \, d\sigma + \iint_{S_2} F \cdot \boldsymbol{n} \, d\sigma$$
$$= \int_{\partial S_1} G \cdot T_1 \, ds + \int_{\partial S_2} G \cdot T_2 \, ds = 0.$$

This last step follows from the fact that $\partial S_1 = \partial S_2$ and $T_1 = -T_2$. Since (15) and (16) are incompatible, we conclude that there is no $G$ which satisfies $\operatorname{curl} G = F$. ∎

## EXERCISES

**1.** For each of the following, evaluate $\int_C F \cdot T \, ds$.

    a) $C$ is the curve formed by intersecting the cylinder $x^2 + y^2 = 1$ with $z = -x$, oriented in the counterclockwise direction when viewed from high on the positive $z$ axis, and $F(x, y, z) = (xy^2, 0, xyz)$.

    b) $C$ is the intersection of the cubic cylinder $z = y^3$ and the circular cylinder $x^2 + y^2 = 3$, oriented in the clockwise direction when viewed from high up the positive $z$ axis, and $F(x, y, z) = (e^x + z, xy, ze^y)$.

**2.** For each of the following, evaluate $\iint_S \operatorname{curl} F \cdot \boldsymbol{n} \, d\sigma$.

    a) $S$ is the "bottomless" surface in the upper half space $z \geq 0$ bounded by $y = x^2$, $z = 1 - y$, $\boldsymbol{n}$ is the outward pointing normal, and $F(x, y, z) = (x \sin z^3, y \cos z^3, x^3 + y^3 + z^3)$.

    b) $S$ is the truncated paraboloid $z = 3 - x^2 - y^2$, $z \geq 0$, $\boldsymbol{n}$ is the outward pointing normal, and $F(x, y, z) = (y, xyz, y)$.

    c) $S$ is the hemisphere $z = \sqrt{10 - x^2 - y^2}$, $\boldsymbol{n}$ is the inward pointing normal, and $F(x, y, z) = (x, x, x^2 y^3 \log(z + 1))$.

    d) $S$ is the "bottomless" tetrahedron in the upper half space $z \geq 0$ bounded by $x = 0$, $y = 0$, $x + 2y + 3z = 1$, $z \geq 0$, $\boldsymbol{n}$ is the outward pointing normal, and $F(x, y, z) = (xy, yz, xz)$.

**3.** For each of the following, evaluate $\iint_S F \cdot \boldsymbol{n}\, d\sigma$ using Stokes's Theorem or Gauss's Theorem.

   a) $S$ is the sphere $x^2 + y^2 + z^2 = 1$, $\boldsymbol{n}$ is the outward pointing normal, and $F(x, y, z) = (xz^2, x^2 y - z^3, 2xy + y^2 z)$.

   b) $S$ is the portion of the plane $z = y$ which lies inside the ball $B_1(\boldsymbol{0})$, $\boldsymbol{n}$ is the upward pointing normal, and $F(x, y, z) = (xy, xz, -yz)$.

   c) $S$ is the truncated cone $y = 2\sqrt{x^2 + z^2}$, $2 \leq y \leq 4$, $\boldsymbol{n}$ is the outward pointing normal, and $F(x, y, z) = (x, -2y, z)$.

   d) $S$ is a union of truncated paraboloids $z = 4 - x^2 - y^2$, $0 \leq z \leq 4$, and $z = x^2 + y^2 - 4$, $-4 \leq z \leq 0$, $\boldsymbol{n}$ is the outward pointing normal, and

   $$F(x, y, z) = (x + y^2 + \sin z, x + y^2 + \cos z, \cos x + \sin y + z).$$

   e) $S$ is the union of three surfaces $z = x^2 + y^2$ $(0 \leq z \leq 2)$, $2 = x^2 + y^2$ $(2 \leq z \leq 5)$, and $z = 7 - x^2 - y^2$ $(5 \leq z \leq 6)$, $\boldsymbol{n}$ is the outward pointing normal, and $F(x, y, z) = (2y, 2z, 1)$.

**4.** For each of the following, evaluate $\int_S \omega$ using Stokes's Theorem or Gauss's Theorem.

   a) $S$ is topological boundary of cylindrical solid $y^2 + z^2 \leq 9$, $0 \leq x \leq 2$, with outward pointing normal, and $\omega = xy\, dy\, dz + (x^2 - z^2)\, dz\, dx + xz\, dx\, dy$.

   b) $S$ is the truncated cylinder $x^2 + z^2 = 8$, $0 \leq y \leq 1$, with outward pointing normal, and $\omega = (x - 2z)\, dy\, dz - y\, dz\, dx$.

   c) $S$ is the topological boundary of $R = [0, \pi/2] \times [0, 1] \times [0, 3]$, with outward pointing normal, and $\omega = e^y \cos x\, dy\, dz + x^2 z\, dz\, dx + (x + y + z)\, dx\, dy$.

   d) $S$ is the intersection of the elliptic cylindrical solid $2x^2 + z^2 \leq 1$ and the plane $x = y$, with normal which points toward the positive $x$ axis, and $\omega = x\, dy\, dz - y\, dz\, dx + \sin y\, dx\, dy$.

**5.** Prove Green's Theorem is a corollary of Stokes's Theorem.

**6.** Let $\Pi$ be a plane in $\mathbf{R}^3$ with unit normal $\boldsymbol{n}$ and $\boldsymbol{x}_0 \in \Pi$. For each $r > 0$ let $S_r$ be the disk in $\Pi$ centered at $\boldsymbol{x}_0$ of radius $r$, i.e., $S_r = B_r(\boldsymbol{x}_0) \cap \Pi$. Prove that if $F : B_1(\boldsymbol{x}_0) \to \mathbf{R}$ is $\mathcal{C}^1$ and $\partial S_r$ carries the orientation induced by $\boldsymbol{n}$, then

$$\operatorname{curl} F(\boldsymbol{x}_0) \cdot \boldsymbol{n} = \lim_{r \to 0+} \frac{1}{\sigma(S_r)} \int_{\partial S_r} F \cdot T\, ds.$$

**7.** Let $S$ be an orientable surface with unit normal $\boldsymbol{n}$ and nonempty boundary $\partial S$ which satisfies the conclusion of Stokes's Theorem for all $\mathcal{C}^1$ functions $F$.

   a) Suppose $F : S \to \mathbf{R}^3 \setminus \{\boldsymbol{0}\}$ is $\mathcal{C}^1$, $\partial S$ is smooth, and $T$ is the unit tangent vector on $\partial S$ induced by $\boldsymbol{n}$. If the angle between $T(\boldsymbol{x}_0)$ and $F(\boldsymbol{x}_0)$ is never obtuse for any $\boldsymbol{x}_0 \in \partial S$, and $\iint_S \operatorname{curl} F \cdot \boldsymbol{n}\, d\sigma = 0$, prove that $T(\boldsymbol{x}_0)$ and $F(\boldsymbol{x}_0)$ are orthogonal for all $\boldsymbol{x}_0 \in \partial S$.

   b) If $F, F_k : S \to \mathbf{R}^3$ are $\mathcal{C}^1$ and $F_k \to F$ uniformly on $\partial S$, prove that

$$\lim_{k \to \infty} \iint_S \operatorname{curl} F_k \cdot \boldsymbol{n}\, d\sigma = \iint_S \operatorname{curl} F \cdot \boldsymbol{n}\, d\sigma.$$

**8.** Suppose $E$ is a two-dimensional region such that if $(x, y) \in E$, then the line segments from $(0,0)$ to $(x,0)$ and from $(x,0)$ to $(x,y)$ are both subsets of $E$. If $F : E \to \mathbf{R}^2$ is $\mathcal{C}^1$, prove that the following three statements are equivalent.

  a) $F = \nabla f$ on $E$ for some $f : E \to \mathbf{R}$.
  b) $F = (P, Q)$ is *exact*, i.e., $Q_x = P_y$ on $E$.
  c) $\int_C F \cdot T \, ds = 0$ for all piecewise smooth curves $C = \partial\Omega$ oriented counter-clockwise, where $\Omega$ is a two-dimensional region which satisfies the conclusion of Green's Theorem, and $\Omega \subset E$.

**9.** Let $\Omega$ be a three-dimensional region and $F : \Omega \to \mathbf{R}^3$ be $\mathcal{C}^1$ on $\Omega$. Suppose further that, for each $(x, y, z) \in \Omega$, both the line segments $L((x, y, 0); (x, y, z))$ and $L((x, 0, 0); (x, y, 0))$ are subsets of $\Omega$. Prove the following statements are equivalent.

  a) There is a $\mathcal{C}^2$ function $G : \Omega \to \mathbf{R}^3$ such that $\operatorname{curl} G = F$ on $\Omega$.
  b) If $F$, $E$, and $S = \partial E$ satisfy the conclusion of Gauss's Theorem and $E \subset \Omega$, then

$$\iint_S F \cdot \boldsymbol{n} \, d\sigma = 0.$$

  c) The identity $\operatorname{div} F = 0$ holds everywhere on $\Omega$.

**10.** Let $F$ be $\mathcal{C}^1$ and exact on $\mathbf{R}^2 \setminus \{(0,0)\}$ (see Exercise 8b) above).

  a) Suppose $C_1$ and $C_2$ are disjoint smooth simple curves, oriented in the counterclockwise direction, and $E$ is a two-dimensional region whose topological boundary $\partial E$ is the union of the traces of $C_1$ and $C_2$. (Note: This means that $E$ has a hole with one of the $C_j$'s as the outer boundary and the other as the inner boundary.) If $(0,0) \notin E$, prove that

$$\int_{C_1} F \cdot T \, ds = \int_{C_2} F \cdot T \, ds.$$

  b) Suppose $E$ is a two-dimensional region which satisfies $(0,0) \in E^o$. If $\partial E$ is a smooth simple curve oriented in the counterclockwise direction, and

$$F(x, y) = \left( \frac{-y}{x^2 + y^2}, \frac{x}{x^2 + y^2} \right),$$

  compute $\int_{\partial E} F \cdot T \, ds$.
  c) State and prove an analogue of part a) for functions $F : \mathbf{R}^3 \setminus \{(0,0,0)\}$, three-dimensional regions, and smooth simple surfaces.

# Chapter 14

# Fourier Series

*e*14.1 **INTRODUCTION** *This section uses no material from any other enrichment section.*

In Chapter 7 we studied power series and their partial sums, *classical* polynomials. In this chapter, we shall study the following objects.

**14.1 DEFINITION.** Let $a_k, b_k \in \mathbf{R}$ and $N$ be a nonnegative integer.

i) A *trigonometric series* is a series of the form

$$\frac{a_0}{2} + \sum_{k=1}^{\infty}(a_k \cos kx + b_k \sin kx).$$

ii) A *trigonometric polynomial* of order $N$ is a function $P : \mathbf{R} \to \mathbf{R}$ of the form

$$P(x) = \frac{a_0}{2} + \sum_{k=1}^{N}(a_k \cos kx + b_k \sin kx).$$

(Here, $\cos kx$ is shorthand for $\cos(kx)$, and $\sin kx$ is shorthand for $\sin(kx)$.)

Calculus was invented with the tacit assumption that power series provided a unified function theory; i.e., every function has a power series expansion (see Klein [5]). When Cauchy showed that this assumption was false (see Remark 7.42), mathematicians began to wonder whether some other type of series would provide a unified function theory. Euler (respectively, Fourier) had shown that the position of a vibrating string (respectively, the temperature along a metal rod) can be represented by trigonometric series. Thus, it was natural to ask: Does every function have a trigonometric series expansion? In this chapter we shall examine this question, and the following calculation will help to answer it.

**14.2 Lemma** [ORTHOGONALITY]. *Let $k, j$ be nonnegative integers. Then*

i)
$$\int_{-\pi}^{\pi} \cos kx \cos jx \, dx = \begin{cases} 2\pi & k = j = 0 \\ \pi & k = j \neq 0 \\ 0 & k \neq j \end{cases}$$

ii)
$$\int_{-\pi}^{\pi} \sin kx \sin jx \, dx = \begin{cases} \pi & k = j \neq 0 \\ 0 & k \neq j \end{cases}$$

*and*

iii)
$$\int_{-\pi}^{\pi} \sin kx \cos jx \, dx = 0.$$

PROOF. Let
$$I = \int_{-\pi}^{\pi} \cos kx \cos jx \, dx.$$

If $k = j = 0$, then $I = \int_{-\pi}^{\pi} dx = 2\pi$. If $k = j \neq 0$, then by a half-angle formula and elementary integration, we have

$$I = \int_{-\pi}^{\pi} \cos^2 kx \, dx = \frac{1}{2} \int_{-\pi}^{\pi} (1 + \cos 2kx) \, dx = \pi.$$

And if $k \neq j$, then by a sum-angle formula and elementary integration, we have

$$I = \frac{1}{2} \int_{-\pi}^{\pi} (\cos(k+j)x + \cos(k-j)x) \, dx = 0.$$

This proves part i). Similar arguments prove parts ii) and iii). ∎

Notice once and for all that the question concerning representation of functions by trigonometric series has a built-in limitation. A function $f : \mathbf{R} \to \mathbf{R}$ is said to be *periodic* (of period $2\pi$) if $f(x+2\pi) = f(x)$ for all $x \in \mathbf{R}$. Since $\cos kx$ and $\sin kx$ are periodic, it is clear that every trigonometric polynomial is periodic. Therefore, any function which is the pointwise or uniform limit of a trigonometric series must also be periodic. For this reason, we will usually restrict our attention to the interval $[-\pi, \pi]$ and assume that $f(-\pi) = f(\pi)$.

The following definition, which introduces a special type of trigonometric series, plays a crucial role in the representation of periodic functions by trigonometric series.

**14.3 DEFINITION.** Let $f$ be integrable on $[-\pi, \pi]$ and $N$ be a nonnegative integer.

i) The *Fourier coefficients* of $f$ are the numbers

$$a_k(f) = \frac{1}{\pi} \int_{-\pi}^{\pi} f(x) \cos kx \, dx, \qquad k = 0, 1, \ldots,$$

and

$$b_k(f) = \frac{1}{\pi} \int_{-\pi}^{\pi} f(x) \sin kx \, dx, \qquad k = 1, 2, \cdots.$$

ii) The *Fourier series* of $f$ is the trigonometric series

$$(Sf)(x) = \frac{a_0(f)}{2} + \sum_{k=1}^{\infty} (a_k(f) \cos kx + b_k(f) \sin kx).$$

iii) The *partial sum* of $Sf$ of order $N$ is the trigonometric polynomial defined, for each $x \in \mathbf{R}$, by $(S_0 f)(x) = a_0(f)/2$ if $N = 0$, and

$$(S_N f)(x) = \frac{a_0(f)}{2} + \sum_{k=1}^{N} (a_k(f) \cos kx + b_k(f) \sin kx)$$

if $N \in \mathbf{N}$.

The following result shows why Fourier series play such an important role in the representation of periodic functions by trigonometric series.

**14.4 THEOREM** [FOURIER]. *If a trigonometric series*

$$S := \frac{a_0}{2} + \sum_{k=1}^{\infty} (a_k \cos kx + b_k \sin kx)$$

*converges uniformly on* $\mathbf{R}$ *to a function* $f$, *then* $S$ *is the Fourier series of* $f$; *i.e.*, $a_k = a_k(f)$ *for* $k = 0, 1, \ldots$, *and* $b_k = b_k(f)$ *for* $k = 1, 2, \cdots$.

PROOF. Fix an integer $k \geq 0$. Since

$$f(x) = \frac{a_0}{2} + \sum_{j=1}^{\infty} (a_j \cos jx + b_j \sin jx)$$

converges uniformly and $\cos kx$ is bounded,

$$(1) \qquad f(x) \cos kx = \frac{a_0}{2} \cos kx + \sum_{j=1}^{\infty} (a_j \cos jx \cos kx + b_j \sin jx \cos kx)$$

also converges uniformly. Since $f$ is the uniform limit of continuous functions, $f$ is continuous, hence, integrable on $[-\pi, \pi]$. Integrating (1) term by term and using orthogonality, we obtain

$$a_k(f) = \frac{1}{\pi} \int_{-\pi}^{\pi} f(x) \cos kx \, dx$$

$$= \frac{a_0}{2\pi} \int_{-\pi}^{\pi} \cos kx \, dx + \sum_{j=1}^{\infty} \left( \frac{a_j}{\pi} \int_{-\pi}^{\pi} \cos kx \cos jx \, dx + \frac{b_j}{\pi} \int_{-\pi}^{\pi} \cos kx \sin jx \, dx \right)$$

$$= a_k.$$

A similar argument establishes $b_k(f) = b_k$. ∎

There are two central questions in the study of trigonometric series.

THE CONVERGENCE QUESTION. *Given a function* $f : \mathbf{R} \to \mathbf{R}$, *periodic on* $\mathbf{R}$ *and integrable on* $[-\pi, \pi]$, *does the Fourier series of* $f$ *converge to* $f$?

THE UNIQUENESS QUESTION. *If a trigonometric series* $S$ *converges to some function* $f$ *integrable on* $[-\pi, \pi]$, *is* $S$ *the Fourier series of* $f$?

We shall answer these questions for pointwise and uniform convergence when $f$ is continuous and of bounded variation. We notice in passing that, by Theorem 14.4, the answer to the Uniqueness Question is yes if uniform convergence is used.

The following special trigonometric polynomials arise naturally in connection with the Convergence Question (see Exercise 2).

**14.5 DEFINITION.** Let $N$ be a nonnegative integer.

   i) The *Dirichlet kernel* of order $N$ is the function defined, for each $x \in \mathbf{R}$, by $D_0(x) = 1/2$ if $N = 0$, and

$$D_N(x) = \frac{1}{2} + \sum_{k=1}^{N} \cos kx$$

   if $N \in \mathbf{N}$.

   ii) The *Fejér kernel* of order $N$ is the function defined, for each $x \in \mathbf{R}$, by $K_0(x) = 1/2$ if $N = 0$, and

(2)
$$K_N(x) = \frac{1}{2} + \sum_{k=1}^{N} \left(1 - \frac{k}{N+1}\right) \cos kx$$

   if $N \in \mathbf{N}$.

The following result shows that there is a simple relationship between Fejér kernels and Dirichlet kernels.

**14.6 Remark.** *If $N$ is a nonnegative integer, then*

$$K_N(x) = \frac{D_0(x) + \cdots + D_N(x)}{N+1}$$

*for all* $x \in \mathbf{R}$.

PROOF. The identity is trivial if $N = 0$. To prove the identity for $N \in \mathbf{N}$, fix $x \in \mathbf{R}$. By definition,

$$K_N(x) = \frac{1}{N+1}\left(\frac{N+1}{2} + \sum_{k=1}^{N}(N-k+1)\cos kx\right)$$

$$= \frac{1}{N+1}\left(\frac{1}{2} + \frac{N}{2} + \sum_{k=1}^{N}\sum_{j=k}^{N} 1 \cdot \cos kx\right)$$

$$= \frac{1}{N+1}\left(\frac{1}{2} + \sum_{j=1}^{N}\left(\frac{1}{2} + \sum_{k=1}^{j}\cos kx\right)\right) = \frac{D_0(x) + \cdots + D_N(x)}{N+1}. \quad ∎$$

The next result shows that Dirichlet and Fejér kernels can be represented by quotients of trigonometric functions.

**14.7 THEOREM.** *If $x \in \mathbf{R}$ cannot be written in the form $2k\pi$ for any $k \in \mathbf{Z}$, then*

$$(3) \qquad\qquad D_N(x) = \frac{\sin\left(N + \dfrac{1}{2}\right)x}{2\sin\dfrac{x}{2}}$$

*and*

$$(4) \qquad\qquad K_N(x) = \frac{2}{N+1}\left(\frac{\sin\left(\dfrac{N+1}{2}\right)x}{2\sin\dfrac{x}{2}}\right)^2$$

*for $N = 0, 1, \cdots$.*

PROOF. The formulas are trivial for $N = 0$. Fix $N \in \mathbf{N}$. Applying a sum-angle formula and telescoping, we have

$$\left(D_N(x) - \frac{1}{2}\right)\sin\frac{x}{2} = \sum_{k=1}^{N}\cos kx \sin\frac{x}{2}$$

$$= \frac{1}{2}\sum_{k=1}^{N}\left(\sin(k + \frac{1}{2})x - \sin(k - \frac{1}{2})x\right)$$

$$= \frac{1}{2}\left(\sin(N + \frac{1}{2})x - \sin\frac{x}{2}\right).$$

Solving this equation for $D_N(x)$ verifies (3).

Let $k \in \mathbf{N}$. By (3) and another sum-angle formula,

$$D_k(x)\sin^2\frac{x}{2} = \frac{1}{2}\sin\frac{x}{2}\sin(k + \frac{1}{2})x = \frac{1}{4}(\cos kx - \cos(k+1)x).$$

This identity also holds for $k = 0$. Applying Remark 14.6 and telescoping, we have

$$(N+1)K_N(x)\sin^2\frac{x}{2} = \sum_{k=0}^{N}D_k(x)\sin^2\frac{x}{2}$$

$$= \frac{1}{4}\sum_{k=0}^{N}(\cos kx - \cos(k+1)x)$$

$$= \frac{1}{4}(1 - \cos(N+1)x) = \frac{1}{2}\sin^2\left(\frac{N+1}{2}\right)x.$$

Solving this equation for $K_N(x)$ verifies (4). ∎

These identities will be used in the next section to obtain a partial answer to the Convergence Question.

The next two examples illustrate the general principle that the Fourier coefficients of many common functions can be computed using integration by parts.

**14.8 Example.** Prove that the Fourier series of $f(x) = x$ is

$$2\sum_{k=1}^{\infty} \frac{(-1)^{k+1}}{k} \sin kx.$$

PROOF. Since $x \cos kx$ is odd and $x \sin kx$ is even, we see that $a_k(f) = 0$ for $k = 0, 1, \ldots$, and

$$b_k(f) = \frac{2}{\pi} \int_0^{\pi} x \sin kx \, dx$$

for $k = 1, 2, \cdots$. Integrating by parts, we conclude that

$$b_k(f) = \frac{2}{\pi} \left( -\frac{x \cos kx}{k} \Big|_0^{\pi} + \frac{1}{k} \int_0^{\pi} \cos kx \, dx \right) = \frac{2(-1)^{k+1}}{k}. \quad \blacksquare$$

**14.9 Example.** Prove that the Fourier series of $f(x) = |x|$ is

$$\frac{\pi}{2} - \frac{4}{\pi} \sum_{k=1}^{\infty} \frac{\cos(2k-1)x}{(2k-1)^2}.$$

PROOF. Since $|x| \cos kx$ is even and $|x| \sin kx$ is odd, we see that $b_k(f) = 0$ for $k = 1, 2 \ldots$, and

$$a_k(f) = \frac{2}{\pi} \int_0^{\pi} x \cos kx \, dx$$

for $k = 0, 1, \cdots$. If $k = 0$, then

$$a_k(f) = \frac{2}{\pi} \left( \frac{\pi^2}{2} \right) = \pi,$$

i.e., $a_0(f)/2 = \pi/2$. If $k > 0$, then integration by parts yields

$$a_k(f) = \frac{2}{\pi k^2} (\cos k\pi - 1) = \begin{cases} 0 & \text{if } k \text{ is even,} \\ -\dfrac{4}{\pi k^2} & \text{if } k \text{ is odd.} \end{cases} \quad \blacksquare$$

**EXERCISES**

**1.** Compute the Fourier series of $x^2$ and of $\cos^2 x$.

**2.** Prove that if $f : \mathbf{R} \to \mathbf{R}$ is integrable on $[-\pi, \pi]$, then

$$(S_N f)(x) = \frac{1}{\pi} \int_{-\pi}^{\pi} f(t) D_N(x - t) \, dt$$

for all $x \in [-\pi, \pi]$ and $N \in \mathbf{N}$.

**3.** Show that if $f$, $g$ are integrable on $[-\pi, \pi]$ and $\alpha \in \mathbf{R}$, then

$$a_k(f + g) = a_k(f) + a_k(g), \quad a_k(\alpha f) = \alpha a_k(f), \qquad k = 0, 1, \cdots,$$

and

$$b_k(f + g) = b_k(f) + b_k(g), \quad b_k(\alpha f) = \alpha b_k(f), \qquad k = 1, 2, \cdots.$$

**4.** Suppose $f : \mathbf{R} \to \mathbf{R}$ is differentiable and periodic and $f'$ is integrable on $[-\pi, \pi]$. Prove that

$$a_k(f') = k b_k(f) \quad \text{and} \quad b_k(f') = -k a_k(f), \qquad k \in \mathbf{N}.$$

**5.** Suppose $f_N : [-\pi, \pi] \to \mathbf{R}$ are integrable and $f_N \to f$ uniformly on $[-\pi, \pi]$ as $N \to \infty$.

a) Prove $a_k(f_N) \to a_k(f)$ and $b_k(f_N) \to b_k(f)$, as $N \to \infty$, uniformly in $k$.

b) Show that a) holds under the weaker hypothesis

$$\lim_{N \to \infty} \int_{-\pi}^{\pi} |f(x) - f_N(x)| \, dx = 0.$$

**6.** Let

$$f(x) = \begin{cases} \dfrac{x}{|x|} & x \neq 0 \\ 0 & x = 0. \end{cases}$$

a) Compute the Fourier coefficients of $f$.

b) Prove that

$$(S_{2N} f)(x) = \frac{2}{\pi} \int_0^x \frac{\sin 2Nt}{\sin t} \, dt$$

for $x \in [-\pi, \pi]$ and $N \in \mathbf{N}$.

*c) [GIBBS'S PHENOMENON]. Prove that

$$\lim_{N \to \infty} (S_{2N} f)(\frac{\pi}{2N}) = \frac{2}{\pi} \int_0^\pi \frac{\sin t}{t} \, dt \approx 1.179$$

## $^e$14.2   SUMMABILITY OF FOURIER SERIES   *This section uses material from Section 14.1.*

The Convergence Question posed in Section 14.1 is very difficult to answer, even for continuous functions. In this section we replace it with an easier question and show that the answer to this question is yes. Namely, we shall show that the Fourier series of any continuous periodic function $f$ is uniformly summable to $f$. By summable, we mean the following concept.

**14.10 DEFINITION.** A series $\sum_{k=0}^{\infty} a_k$ with partial sums $s_N = \sum_{k=0}^{N} a_k$ is said to be *Cesàro summable* to $L$ if its *Cesàro means*

$$\sigma_N := \frac{s_0 + \cdots + s_N}{N+1}$$

converge to $L$ as $N \to \infty$.

The following result shows that summability is a generalization of convergence.

**14.11 Remark.** *If $\sum_{k=0}^{\infty} a_k$ converges to a finite number $L$, then it is Cesàro summable to $L$.*

PROOF. Let $\varepsilon > 0$. Choose $N_1 \in \mathbf{N}$ such that $k \geq N_1$ implies $|s_k - L| < \varepsilon/2$. Use the Archimedean Principle to choose $N_2 \in \mathbf{N}$ such that $N_2 > N_1$ and

$$\sum_{k=0}^{N_1} |s_k - L| < \frac{\varepsilon N_2}{2}.$$

If $N > N_2$, then

$$|\sigma_N - L| \leq \frac{1}{N+1} \sum_{k=0}^{N_1} |s_k - L| + \frac{1}{N+1} \sum_{k=N_1+1}^{N} |s_k - L|$$

$$\leq \frac{\varepsilon N_2}{2(N+1)} + \frac{\varepsilon}{2} \left( \frac{N - N_1}{N+1} \right) < \frac{\varepsilon}{2} + \frac{\varepsilon}{2} = \varepsilon. \ \blacksquare$$

The converse of Remark 14.11 is false. Indeed, although the series $\sum_{k=0}^{\infty} (-1)^k$ does not converge, its Cesàro means satisfy

$$\sigma_N = \begin{cases} \dfrac{N+2}{2(N+1)} & N \text{ is even} \\[2mm] \dfrac{1}{2} & N \text{ is odd,} \end{cases}$$

whence $\sigma_N \to 1/2$ as $N \to \infty$.

It is easier to show that a series is Cesàro summable than to show it converges. Thus the following question is easier to answer than the Convergence Question.

THE SUMMABILITY QUESTION. *Given a function $f : \mathbf{R} \to \mathbf{R}$, periodic on $\mathbf{R}$ and integrable on $[-\pi, \pi]$, is $Sf$ Cesàro summable to $f$?*

The Cesàro means of a Fourier series $Sf$ are denoted by

$$(\sigma_N f)(x) := \frac{(S_0 f)(x) + \cdots + (S_N f)(x)}{N+1},$$

$N = 0, 1, \ldots$. The following result shows that the Cesàro means of a Fourier series can always be represented by an integral equation. This is important because it allows us to estimate the remainder $\sigma_N f - f$, using techniques of integration.

**14.12 Lemma.** Let $f : \mathbf{R} \to \mathbf{R}$ be periodic on $\mathbf{R}$ and integrable on $[-\pi, \pi]$. Then

$$(\sigma_N f)(x) = \frac{1}{\pi} \int_{-\pi}^{\pi} f(x - t) K_N(t)\, dt$$

for all $N = 0, 1, \ldots$, and all $x \in \mathbf{R}$.

PROOF. Fix $j, N \in \mathbf{N}$ and $x \in \mathbf{R}$. By definition and a sum-angle formula,

$$
\begin{aligned}
a_j(f) &\cos jx + b_j(f) \sin jx \\
&= \frac{1}{\pi} \int_{-\pi}^{\pi} f(u) \cos ju \cos jx\, du + \frac{1}{\pi} \int_{-\pi}^{\pi} f(u) \sin ju \sin jx\, du \\
&= \frac{1}{\pi} \int_{-\pi}^{\pi} f(u)(\cos ju \cos jx + \sin ju \sin jx)\, du \\
&= \frac{1}{\pi} \int_{-\pi}^{\pi} f(u) \cos j(x - u)\, du.
\end{aligned}
$$

Summing this identity over integers $j = 1, 2, \ldots, k$ and adding $a_0(f)/2$, we have

$$
\begin{aligned}
(S_k f)(x) &= \frac{a_0(f)}{2} + \sum_{j=1}^{k} (a_j(f) \cos jx + b_j(f) \sin jx) \\
&= \frac{1}{\pi} \int_{-\pi}^{\pi} f(u) \left( \frac{1}{2} + \sum_{j=1}^{k} \cos j(x - u) \right) du \\
&= \frac{1}{\pi} \int_{-\pi}^{\pi} f(u) D_k(x - u)\, du
\end{aligned}
$$

for $k = 0, 1, \ldots$. Making the change of variables $t = x - u$ and using the fact that both $f$ and $D_k$ are periodic, we obtain

$$(S_k f)(x) = \frac{1}{\pi} \int_{-\pi}^{\pi} f(x - t) D_k(t)\, dt, \qquad k = 0, 1, \cdots .$$

We conclude by Remark 14.11 that

$$
\begin{aligned}
(\sigma_N f)(x) &= \frac{1}{N+1} \sum_{k=0}^{N} (S_k f)(x) \\
&= \frac{1}{N+1} \sum_{k=0}^{N} \frac{1}{\pi} \int_{-\pi}^{\pi} f(x - t) D_k(t)\, dt = \frac{1}{\pi} \int_{-\pi}^{\pi} f(x - t) K_N(t)\, dt. \blacksquare
\end{aligned}
$$

To answer the Summability Question we need to know more about Fejér kernels. The following result shows that Fejér kernels satisfy some very nice properties.

**14.13 Lemma.** *For each nonnegative integer* $N$,

(5) $$K_N(t) \geq 0 \quad \text{for all } t \in \mathbf{R},$$

*and*

(6) $$\frac{1}{\pi} \int_{-\pi}^{\pi} K_N(t)\, dt = 1.$$

*Moreover, for each* $0 < \delta < \pi$,

(7) $$\lim_{N \to \infty} \int_{\delta}^{\pi} |K_N(t)|\, dt = 0.$$

PROOF. Fix $N \geq 0$. If $t = 2j\pi$ for some $j \in \mathbf{Z}$, then $D_k(t) = k + 1/2 \geq 0$ for all $k \geq 0$, whence $K_N(t) \geq 0$. If $t \neq 2j\pi$ for any $j \in \mathbf{Z}$, then by Theorem 14.7,

$$K_N(t) = \frac{2}{N+1} \left( \frac{\sin\left(\frac{N+1}{2}\right) t}{2 \sin \frac{t}{2}} \right)^2 \geq 0.$$

This proves (5). By Definition 14.5 and orthogonality,

$$\int_{-\pi}^{\pi} K_N(t)\, dt = \int_{-\pi}^{\pi} \left( \frac{1}{2} + \sum_{k=1}^{N} \left( 1 - \frac{k}{N+1} \right) \cos kt \right) dt = \pi.$$

This proves (6).

To prove (7), fix $0 < \delta < \pi$ and observe that $\sin t/2 \geq \sin \delta/2$ for $t \in [\delta, \pi]$. Hence, it follows from Theorem 14.7 that

$$\int_{\delta}^{\pi} |K_N(t)|\, dt \leq \frac{2}{N+1} \int_{\delta}^{\pi} \left( \frac{\sin\left(\frac{N+1}{2}\right) t}{2 \sin \frac{\delta}{2}} \right)^2 dt \leq \frac{\pi}{2(N+1)\sin^2 \frac{\delta}{2}}.$$

Since $\delta$ is fixed, this last expression tends to 0 as $N \to \infty$. ∎

Using these properties, we can answer the Summability Question for continuous functions (see also Exercises 6 and 8).

**14.14 THEOREM** [FEJÉR]. *Suppose* $f : \mathbf{R} \to \mathbf{R}$ *is periodic on* $\mathbf{R}$ *and integrable on* $[-\pi, \pi]$.

i) *If*
$$L = \lim_{h \to 0} \frac{f(x_0 + h) + f(x_0 - h)}{2}$$

*exists for some* $x_0 \in \mathbf{R}$, *then* $(\sigma_N f)(x_0) \to L$ *as* $N \to \infty$.

ii) *If* $f$ *is continuous on some closed interval* $I$, *then* $\sigma_N f \to f$ *uniformly on* $I$ *as* $N \to \infty$.

PROOF. Since $f$ is periodic, we may suppose that $x_0 \in [-\pi, \pi]$. Fix $N \in \mathbf{N}$. By (6), Lemma 14.12, and a change of variables,

$$(8) \qquad (\sigma_N f)(x_0) - L = \frac{1}{\pi} \int_{-\pi}^{\pi} K_N(t)(f(x_0 - t) - L)\, dt$$

$$= \frac{2}{\pi} \int_0^{\pi} K_N(t) \left( \frac{f(x_0 + t) + f(x_0 - t)}{2} - L \right) dt$$

$$=: \frac{2}{\pi} \int_0^{\pi} K_N(t) F(x_0, t)\, dt.$$

Let $\varepsilon > 0$ and choose $0 < \delta < \pi$ such that $|t| < \delta$ implies $|F(x_0, t)| < \varepsilon/3$. By (5) and (6) we have

$$(9) \qquad \frac{2}{\pi} \int_0^{\delta} K_N(t)|F(x_0, t)|\, dt < \frac{2\varepsilon}{3\pi} \int_0^{\delta} K_N(t)\, dt \leq \frac{2\varepsilon}{3}.$$

On the other hand, choose by (7) an $N_1 \in \mathbf{N}$ such that $N \geq N_1$ implies $\int_\delta^\pi K_N(t)\, dt < \varepsilon/3M$, where $M := \sup_{x \in \mathbf{R}} |F(x)|$. Then

$$\frac{2}{\pi} \int_\delta^\pi K_N(t)|F(x_0, t)|\, dt \leq M \int_\delta^\pi K_N(t)\, dt < \frac{\varepsilon}{3},$$

and it follows from (8) and (9) that

$$(10) \quad |(\sigma_N f)(x_0) - L| \leq \frac{2}{\pi} \int_0^{\delta} K_N(t)|F(x_0, t)|\, dt + \frac{2}{\pi} \int_\delta^\pi K_N(t)|F(x_0, t)|\, dt < \varepsilon$$

for all $N \geq N_1$. This proves part i).

To prove part ii), suppose $f$ is continuous on some closed interval $I$. Since $f$ is periodic, we may suppose that $I \subseteq [-\pi, \pi]$. Thus $I$ is closed and bounded, and $f$ is uniformly continuous on $I$. Repeating the estimates above, we see that (10) holds uniformly for all $x_0 \in I$. ∎

**14.15 COROLLARY.** *If $f : \mathbf{R} \to \mathbf{R}$ is continuous and periodic, then $\sigma_N f$ converges to $f$ uniformly on $\mathbf{R}$ as $N \to \infty$.*

**14.16 COROLLARY** [COMPLETENESS]. *If $f : \mathbf{R} \to \mathbf{R}$ is continuous and periodic, and $a_{k-1}(f) = b_k(f) = 0$ for $k \in \mathbf{N}$, then $f(x) = 0$ for all $x \in \mathbf{R}$.*

PROOF. By hypothesis, $(\sigma_N f)(x) = 0$ for all $N \in \mathbf{N}$ and $x \in \mathbf{R}$. Hence, by Corollary 14.15, $f(x) = \lim_{N \to \infty} (\sigma_N f)(x) = 0$ for all $x \in \mathbf{R}$. ∎

**14.17 COROLLARY.** *Let $f : \mathbf{R} \to \mathbf{R}$ be continuous and periodic. Then there is a sequence of trigonometric polynomials $T_1, T_2, \ldots,$ such that $T_N \to f$ uniformly on $\mathbf{R}$.*

PROOF. Set $T_N = \sigma_N f$ for $N \in \mathbf{N}$, and apply Corollary 14.15. ∎

This result can be used to prove the following density result for classical polynomials, i.e., polynomials of the form $P(x) = \sum_{k=0}^{n} c_k x^k$.

**14.18 THEOREM** [WEIERSTRASS APPROXIMATION THEOREM]. *Let* $[a, b]$ *be a closed bounded interval, and suppose* $f : [a, b] \to \mathbf{R}$ *is continuous on* $[a, b]$. *Given* $\varepsilon > 0$ *there is a (classical) polynomial* $P$ *on* $\mathbf{R}$ *such that*

$$|f(x) - P(x)| < \varepsilon$$

*for all* $x \in [a, b]$.

PROOF. By considering $g(t) := f(a + (b - a)t/\pi)$, which is continuous on $[0, \pi]$, we may suppose that $a = 0$ and $b = \pi$.

Let $\varepsilon > 0$. Extend $f$ from $[0, \pi]$ to $\mathbf{R}$, so that $f$ is continuous and periodic. (For example, we could insist that the graph of $y = f(x)$ on $[\pi, 2\pi]$ is the chord from $(\pi, f(\pi))$ to $(2\pi, f(0))$ and then define $f(x + 2k\pi) := f(x)$ for $k \in \mathbf{Z}$.) By Corollary 14.17, there is a trigonometric polynomial $T$ such that

$$|T(x) - f(x)| < \frac{\varepsilon}{2}$$

for $x \in \mathbf{R}$. Since each $\cos kx$ and $\sin kx$ is analytic on $\mathbf{R}$, so is $T$. Since analytic functions are uniform limits of their Taylor series, it follows that there is a polynomial $P$ on $\mathbf{R}$ such that

$$|T(x) - P(x)| < \frac{\varepsilon}{2}$$

for $x \in [-\pi, \pi] \supseteq [a, b]$. We conclude that

$$|f(x) - P(x)| \le |f(x) - T(x)| + |T(x) - P(x)| < \varepsilon$$

for all $x \in [a, b]$. ∎

**EXERCISES**

1. Let $E \subseteq \mathbf{R}$ and suppose $f, f_k : \mathbf{R} \to \mathbf{R}$ are bounded functions. Prove that if $\sum_{k=0}^{\infty} f_k(x)$ converges to $f(x)$ uniformly on $E$, then

$$\sigma_N(x) := \sum_{k=0}^{N} \left(1 - \frac{k}{N+1}\right) f_k(x)$$

   converges to $f(x)$ uniformly on $E$ as $N \to \infty$.

2. If $f : \mathbf{R} \to \mathbf{R}$ is periodic on $\mathbf{R}$ and integrable on $[-\pi, \pi]$, prove that the Cesàro means of $Sf$ are uniformly bounded; i.e., there is an $M > 0$ such that

$$|(\sigma_N f)(x)| \le M$$

   for all $x \in \mathbf{R}$ and $N \in \mathbf{N}$.

**3.** Let

$$S = \frac{a_0}{2} + \sum_{k=1}^{\infty} (a_k \cos kx + b_k \sin kx)$$

be a trigonometric series and set

$$\sigma_N(x) = \frac{a_0}{2} + \sum_{k=1}^{N} \left(1 - \frac{k}{N+1}\right)(a_k \cos kx + b_k \sin kx)$$

for $x \in \mathbf{R}$ and $N \in \mathbf{N}$. Prove that $S$ is the Fourier series of some continuous periodic function $f : \mathbf{R} \to \mathbf{R}$ if and only if $\sigma_N$ converges uniformly on $\mathbf{R}$, as $N \to \infty$.

**4.** Let $f$ be integrable on $[-\pi, \pi]$ and $L \in \mathbf{R}$.

   a) Prove that if $(\sigma_N f)(x_0) \to L$ as $N \to \infty$ and if $(Sf)(x_0)$ converges, then
   $(S_N f)(x_0) \to L$.

   b) Prove that

$$\sin \sqrt{2}\pi + \sum_{k=1}^{\infty} \frac{4(-1)^k \sin \sqrt{2}\pi}{2 - k^2} \cos kx$$

   converges to $\sqrt{2}\pi \cos \sqrt{2}x$ uniformly on $\mathbf{R}$.

**5.** Suppose $f : [a, b] \to \mathbf{R}$ is continuous and

$$\int_a^b x^n f(x)\, dx = 0$$

for all integers $n \ge 0$.

   a) Evaluate $\int_a^b P(x)f(x)\, dx$ for any polynomial $P$ on $\mathbf{R}$.

   b) Prove $\int_a^b |f(x)|^2\, dx = 0$.

   c) Show that $f(x) = 0$ for all $x \in [a, b]$.

**6.** [SUMMABILITY KERNELS]. Let $\phi_N : \mathbf{R} \to \mathbf{R}$ be a sequence of continuous, periodic functions on $\mathbf{R}$ which satisfy

$$\int_0^{2\pi} \phi_N(t)\, dt = 1 \quad \text{and} \quad \int_0^{2\pi} |\phi_N(t)|\, dt \le M < \infty$$

for all $N \in \mathbf{N}$, and

$$\lim_{N \to \infty} \int_\delta^{2\pi - \delta} |\phi_N(t)|\, dt = 0$$

for each $0 < \delta < 2\pi$. Suppose $f : \mathbf{R} \to \mathbf{R}$ is continuous and periodic. Prove that

$$\lim_{N \to \infty} \int_0^{2\pi} f(x - t)\phi_N(t)\, dt = f(x)$$

uniformly for $x \in \mathbf{R}$.

**7.** Let $[a, b]$ be a nondegenerate, closed, bounded interval.

    a) Prove that given any polynomial $P$ on $\mathbf{R}$ and any $\varepsilon > 0$, there is a polynomial $Q$ on $\mathbf{R}$, with rational coefficients, such that $|P(x) - Q(x)| < \varepsilon$ for all $x \in [a, b]$.

    *b) Prove that the space $\mathcal{C}[a, b]$ (see Example 10.6) is separable.

**\*8.** A sequence of functions $f_N : \mathbf{R} \to \mathbf{R}$ is said to converge *almost everywhere* to a function $f$ if there is a set $E$ of measure zero such that $f_N(x) \to f(x)$, as $N \to \infty$, for every $x \in \mathbf{R} \setminus E$. Suppose $f : \mathbf{R} \to \mathbf{R}$ is also periodic. Prove that if $f$ is Riemann integrable on $[-\pi, \pi]$, then $\sigma_N f \to f$ almost everywhere as $N \to \infty$.

### $^e$14.3  GROWTH OF FOURIER COEFFICIENTS    *This section uses material from Sections 5.5 and 14.2.*

By Theorem 14.14, a continuous periodic function $f$ is completely determined by its Fourier coefficients. In this section we ask to what extent does smoothness of $f$ affect the growth of these coefficients.

We begin with a computational result.

**14.19 Lemma.** *If $f : \mathbf{R} \to \mathbf{R}$ is integrable on $[-\pi, \pi]$ and $N$ is a nonnegative integer, then*

$$(11) \qquad \frac{1}{\pi} \int_{-\pi}^{\pi} f(x)(S_N f)(x)\, dx = \frac{|a_0(f)|^2}{2} + \sum_{k=1}^{N} \left( |a_k(f)|^2 + |b_k(f)|^2 \right)$$

$$= \frac{1}{\pi} \int_{-\pi}^{\pi} |(S_N f)(x)|^2\, dx.$$

PROOF. Fix $N \geq 0$. Since $f$ and $S_N f$ are integrable on $[-\pi, \pi]$, both integrals in (11) exist. By definition and orthogonality,

$$\frac{1}{\pi} \int_{-\pi}^{\pi} f(x) \frac{a_0(f)}{2}\, dx = \frac{|a_0(f)|^2}{2} = \frac{1}{\pi} \int_{-\pi}^{\pi} (S_N f)(x) \frac{a_0(f)}{2}\, dx.$$

Similarly,

$$\frac{1}{\pi} \int_{-\pi}^{\pi} f(x) a_k(f) \cos kx\, dx = |a_k(f)|^2 = \frac{1}{\pi} \int_{-\pi}^{\pi} (S_N f)(x) a_k(f) \cos kx\, dx$$

and

$$\frac{1}{\pi} \int_{-\pi}^{\pi} f(x) b_k(f) \sin kx\, dx = |b_k(f)|^2 = \frac{1}{\pi} \int_{-\pi}^{\pi} (S_N f)(x) b_k(f) \sin kx\, dx$$

for $k \in \mathbf{N}$. Adding these identities for $k = 0, \ldots, N$ verifies (11). ∎

Next, we use this result to identify a growth condition satisfied by the Fourier coefficients of any Riemann integrable function.

**14.20 THEOREM** [BESSEL'S INEQUALITY]. *If $f : \mathbf{R} \to \mathbf{R}$ is (Riemann) integrable on $[-\pi, \pi]$, then $\sum_{k=1}^{\infty} |a_k(f)|^2$ and $\sum_{k=1}^{\infty} |b_k(f)|^2$ are convergent series. In fact,*

$$
(12) \qquad \frac{|a_0(f)|^2}{2} + \sum_{k=1}^{\infty} \left( |a_k(f)|^2 + |b_k(f)|^2 \right) \leq \frac{1}{\pi} \int_{-\pi}^{\pi} |f(x)|^2 \, dx.
$$

PROOF. Fix $N \in \mathbf{N}$. By Lemma 14.19,

$$
0 \leq \frac{1}{\pi} \int_{-\pi}^{\pi} |f(x) - (S_N f)(x)|^2 \, dx
$$

$$
= \frac{1}{\pi} \int_{-\pi}^{\pi} |f(x)|^2 \, dx - \frac{2}{\pi} \int_{-\pi}^{\pi} f(x)(S_N f)(x) \, dx + \frac{1}{\pi} \int_{-\pi}^{\pi} |(S_N f)(x)|^2 \, dx
$$

$$
= \frac{1}{\pi} \int_{-\pi}^{\pi} |f(x)|^2 \, dx - \left( \frac{|a_0(f)|^2}{2} + \sum_{k=1}^{N} \left( |a_k(f)|^2 + |b_k(f)|^2 \right) \right).
$$

Therefore,

$$
\frac{|a_0(f)|^2}{2} + \sum_{k=1}^{N} \left( |a_k(f)|^2 + |b_k(f)|^2 \right) \leq \frac{1}{\pi} \int_{-\pi}^{\pi} |f(x)|^2 \, dx
$$

for all $N \in \mathbf{N}$. Taking the limit of this inequality as $N \to \infty$ verifies (12). Since $|f|^2$ is Riemann integrable when $f$ is, it follows that both $\sum_{k=1}^{\infty} |a_k(f)|^2$ and $\sum_{k=1}^{\infty} |b_k(f)|^2$ are convergent series. ∎

**14.21 COROLLARY** [RIEMANN–LEBESGUE LEMMA]. *If $f$ is integrable on $[-\pi, \pi]$, then*

$$
\lim_{k \to \infty} a_k(f) = \lim_{k \to \infty} b_k(f) = 0.
$$

PROOF. Since the terms of a convergent series converge to zero, it follows from Bessel's inequality that $a_k(f)$ and $b_k(f)$ converge to zero as $k \to \infty$. ∎

Our next major result shows that Bessel's inequality is actually an identity when $f$ is continuous and periodic. First, we show that the partial sums of the Fourier series of a function $f$ are the best approximations to $f$ in the following sense.

**14.22 Lemma.** *Let $N \in \mathbf{N}$. If $f$ is (Riemann) integrable on $[-\pi, \pi]$ and*

$$
T_N = \frac{c_0}{2} + \sum_{k=1}^{N} (c_k \cos kx + d_k \sin kx)
$$

*is any trigonometric polynomial of degree $N$, then*

$$
\int_{-\pi}^{\pi} |f(x) - (S_N f)(x)|^2 \, dx \leq \int_{-\pi}^{\pi} |f(x) - T_N(x)|^2 \, dx.
$$

PROOF. Notice by (11) that

$$\int_{-\pi}^{\pi} |f(x) - T_N(x)|^2 \, dx$$

$$= \int_{-\pi}^{\pi} |f(x) - (S_N f)(x) + (S_N f)(x) - T_N(x)|^2 \, dx$$

$$= \int_{-\pi}^{\pi} |f(x) - (S_N f)(x)|^2 \, dx$$

$$\quad + 2 \int_{-\pi}^{\pi} (f(x) - (S_N f)(x))((S_N f)(x) - T_N(x)) \, dx$$

$$\quad + \int_{-\pi}^{\pi} |(S_N f)(x) - T_N(x)|^2 \, dx$$

$$\geq \int_{-\pi}^{\pi} |f(x) - (S_N f)(x)|^2 \, dx + 2 \int_{-\pi}^{\pi} ((S_N f)(x)T_N(x) - f(x)T_N(x)) \, dx.$$

This last term is zero since, by orthogonality,

$$\frac{1}{\pi} \int_{-\pi}^{\pi} ((S_N f)(x)T_N(x) - f(x)T_N(x)) \, dx$$

$$= \frac{a_0(f)c_0}{4} + \sum_{k=1}^{N} (a_k(f)c_k + b_k(f)d_k)$$

$$\quad - \frac{c_0}{2\pi} \int_{-\pi}^{\pi} f(x) \, dx - \sum_{j=1}^{N} \frac{c_j}{\pi} \int_{-\pi}^{\pi} f(x) \cos jx \, dx$$

$$\quad - \sum_{j=1}^{N} \frac{d_j}{\pi} \int_{-\pi}^{\pi} f(x) \sin jx \, dx$$

$$= \frac{a_0(f)c_0}{4} + \sum_{k=1}^{N} (a_k(f)c_k + b_k(f)d_k)$$

$$\quad - \left( \frac{a_0(f)c_0}{4} + \sum_{k=1}^{N} a_k(f)c_k + b_k(f)d_k \right)$$

$$= 0.$$

Consequently,

$$\int_{-\pi}^{\pi} |f(x) - T_N(x)|^2 \, dx \geq \int_{-\pi}^{\pi} |f(x) - (S_N f)(x)|^2 \, dx. \quad \blacksquare$$

**14.23 THEOREM** [PARSEVAL'S IDENTITY]. *If $f : \mathbf{R} \to \mathbf{R}$ is periodic and continuous, then*

$$(13) \qquad \frac{|a_0(f)|^2}{2} + \sum_{k=1}^{\infty} \left( |a_k(f)|^2 + |b_k(f)|^2 \right) = \frac{1}{\pi} \int_{-\pi}^{\pi} |f(x)|^2 \, dx.$$

PROOF. By Bessel's inequality, we need only show that the left side of (13) is greater than or equal to the right side of (13). Since $f$ is continuous and periodic, $\sigma_N f \to f$ uniformly on $\mathbf{R}$ as $N \to \infty$ by Fejér's Theorem. Hence, it follows from Lemmas 14.19 and 14.22 that

$$\frac{1}{\pi} \int_{-\pi}^{\pi} |f(x)|^2 \, dx - \frac{|a_0(f)|^2}{2} - \sum_{k=1}^{N} \left( |a_k(f)|^2 + |b_k(f)|^2 \right)$$

$$= \frac{1}{\pi} \int_{-\pi}^{\pi} |f(x) - (S_N f)(x)|^2 \, dx \leq \frac{1}{\pi} \int_{-\pi}^{\pi} |f(x) - (\sigma_N f)(x)|^2 \, dx \to 0$$

as $N \to \infty$. In particular,

$$\frac{1}{\pi} \int_{-\pi}^{\pi} |f(x)|^2 \, dx \leq \frac{|a_0(f)|^2}{2} + \sum_{k=1}^{\infty} \left( |a_k(f)|^2 + |b_k(f)|^2 \right). \quad \blacksquare$$

The Riemann–Lebesgue Lemma can be improved if $f$ is smooth and periodic. In fact, the following result shows that, the smoother $f$ is, the more rapidly its Fourier coefficients converge to zero.

**14.24 THEOREM.** Let $f : \mathbf{R} \to \mathbf{R}$ and $j \in \mathbf{N}$. If $f^{(j)}$ exists and is integrable on $[-\pi, \pi]$ and $f^{(\ell)}$ is periodic for each $0 \leq \ell < j$, then

(14)
$$\lim_{k \to \infty} k^j a_k(f) = \lim_{k \to \infty} k^j b_k(f) = 0.$$

PROOF. Fix $k \in \mathbf{N}$. Since $f$ is periodic, integration by parts yields

$$a_k(f') = \frac{1}{\pi} \int_{-\pi}^{\pi} f'(x) \cos kx \, dx = \frac{k}{\pi} \int_{-\pi}^{\pi} f'(x) \sin kx \, dx = kb_k(f).$$

Similarly, $b_k(f') = -ka_k(f)$, hence $a_k(f'') = kb_k(f') = -k^2 a_k(f)$. Iterating, we obtain

$$|a_k(f^{(j)})| = \begin{cases} |k^j a_k(f)| & \text{when } j \text{ is even,} \\ |k^j b_k(f)| & \text{when } j \text{ is odd.} \end{cases}$$

A similar identity holds for $|b_k(f^{(j)})|$. Since the Riemann–Lebesgue Lemma implies $a_k(f^{(j)})$ and $b_k(f^{(j)}) \to 0$ as $k \to \infty$, it follows that $k^j a_k(f) \to 0$ and $k^j b_k(f) \to 0$ as $k \to \infty$. $\blacksquare$

This result shows that if $f$ is continuously differentiable and periodic, then $ka_k(f)$ and $kb_k(f)$ both converge to zero as $k \to \infty$. Recall that if $f$ is continuously differentiable on $[-\pi, \pi]$ then $f$ is of bounded variation (see Remark 5.51). Thus it is natural to ask: How rapidly do $ka_k(f)$ and $kb_k(f)$ grow when $f$ is a function of bounded variation? To answer this question, let $\{x_0, x_1, \ldots, x_n\}$ be a partition of $[-\pi, \pi]$. Using Riemann sums, the Mean Value Theorem, Abel's Formula, and $\sin kx_0 = \sin kx_n = 0$, we can convince ourselves that

$$\pi a_k(f) = \int_{-\pi}^{\pi} f(x) \cos kx \, dx \approx \sum_{j=1}^{n} f(x_j) \cos kx_j (x_j - x_{j-1})$$

$$\approx \frac{1}{k} \sum_{j=1}^{n} f(x_j)(\sin kx_j - \sin kx_{j-1})$$

$$= \frac{1}{k} \sum_{j=1}^{n-1} (f(x_j) - f(x_{j+1})) \sin kx_j.$$

Since the absolute value of this last sum is bounded by $\operatorname{Var} f$, we guess that $k|a_k(f)| \le \operatorname{Var} f / \pi$.

To prove our guess is correct, suppose for a moment that $f$ is increasing, periodic, and differentiable on $[-\pi, \pi]$, and $\phi(x) = \sin kx$. Then by Definition 14.3, periodicity, integration by parts, and the Fundamental Theorem of Calculus, we can estimate the Fourier coefficients of $f$ as follows:

$$\pi k |a_k(f)| = \left| \int_{-\pi}^{\pi} f(x)\phi'(x) \, dx \right|$$

$$= \left| f(x)\phi(x) \Big|_{-\pi}^{\pi} - \int_{-\pi}^{\pi} f'(x)\phi(x) \, dx \right|$$

$$= \left| \int_{-\pi}^{\pi} f'(x)\phi(x) \, dx \right| \le \int_{-\pi}^{\pi} f'(x) \, dx$$

$$= \sum_{j=1}^{n} \int_{x_{j-1}}^{x_j} f'(x) \, dx = \sum_{j=1}^{n} f(x_j) - f(x_{j-1}) \le \operatorname{Var} f.$$

The following result shows that this estimate is valid even when $f$ is neither differentiable nor increasing.

**14.25 Lemma.** *Suppose that $f$ and $\phi$ are periodic, where $f$ is of bounded variation on $[-\pi, \pi]$ and $\phi$ is continuously differentiable on $[-\pi, \pi]$. If $M := \sup_{x \in [-\pi, \pi]} |\phi(x)|$, then*

$$(15) \qquad \left| \int_{-\pi}^{\pi} f(x)\phi'(x) \, dx \right| \le M \operatorname{Var} f.$$

PROOF. Since $f$ is of bounded variation and $\phi'$ is continuous on $[-\pi, \pi]$, the product $f\phi'$ is integrable on $[-\pi, \pi]$ (see Corollary 5.23 and the comments following Corollary 5.57).

Let $\varepsilon > 0$ and set $C = \sup_{x \in [-\pi, \pi]} |f(x)|$. Since $\phi'$ is uniformly continuous and $f\phi'$ is integrable on $[-\pi, \pi]$, choose a partition $P = \{x_0, x_1, \ldots, x_{2n}\}$ of $[-\pi, \pi]$ such that

$$(16) \qquad w, c \in [x_{j-1}, x_j] \quad \text{implies} \quad |\phi'(w) - \phi'(c)| < \frac{\varepsilon}{4\pi C}$$

and

(17)
$$\left| \sum_{j=1}^{2n} f(w_j)\phi'(w_j)(x_j - x_{j-1}) - \int_{-\pi}^{\pi} f(x)\phi'(x)\,dx \right| < \frac{\varepsilon}{2}$$

for any choice of $w_j \in [x_{j-1}, x_j]$.

Set
$$A := \sum_{j=1}^{2n} f(w_j)(\phi(x_j) - \phi(x_{j-1})),$$

where $w_j = x_j$ when $j$ is even, $w_j = x_{j-1}$ when $j$ is odd. By the Mean Value Theorem, choose $c_j \in [x_{j-1}, x_j]$ such that $\phi(x_j) - \phi(x_{j-1}) = \phi'(c_j)(x_j - x_{j-1})$. Then
$$A = \sum_{j=1}^{2n} f(w_j)\phi'(c_j)(x_j - x_{j-1}).$$

Hence, it follows from (17) and (16) that

$$\left| A - \int_{-\pi}^{\pi} f(x)\phi'(x)\,dx \right|$$

$$\leq \left| \sum_{j=1}^{2n} f(w_j)\phi'(c_j)(x_j - x_{j-1}) - \sum_{j=1}^{2n} f(w_j)\phi'(w_j)(x_j - x_{j-1}) \right|$$

$$+ \left| \sum_{j=1}^{2n} f(w_j)\phi'(w_j)(x_j - x_{j-1}) - \int_{-\pi}^{\pi} f(x)\phi'(x)\,dx \right|$$

$$< \sum_{j=1}^{2n} |f(w_j)|\,|\phi'(c_j) - \phi'(w_j)|(x_j - x_{j-1}) + \frac{\varepsilon}{2}$$

$$\leq \frac{\varepsilon}{4\pi} \sum_{j=1}^{2n} (x_j - x_{j-1}) + \frac{\varepsilon}{2} = \frac{\varepsilon}{2} + \frac{\varepsilon}{2} = \varepsilon.$$

Combining this observation with the triangle inequality, we obtain

(18)
$$\left| \int_{-\pi}^{\pi} f(x)\phi'(x)\,dx \right| \leq |A| + \varepsilon.$$

On the other hand, by the choice of the $w_j$'s,

$$A = \sum_{j=1}^{n} f(x_{2j-2})(\phi(x_{2j-1}) - \phi(x_{2j-2})) + \sum_{j=1}^{n} f(x_{2j})(\phi(x_{2j}) - \phi(x_{2j-1}))$$

$$= \sum_{j=1}^{n} \phi(x_{2j-1})(f(x_{2j-2}) - f(x_{2j}))$$

$$+ \sum_{j=1}^{n} (f(x_{2j})\phi(x_{2j}) - f(x_{2j-2})\phi(x_{2j-2})).$$

Since $f$ and $\phi$ are periodic, this last sum telescopes to 0. Therefore,

$$|A| = \left| \sum_{j=1}^{n} \phi(x_{2j-1})(f(x_{2j-2}) - f(x_{2j})) \right|$$

$$\leq \sum_{j=1}^{n} |\phi(x_{2j-1})| \, |f(x_{2j-2}) - f(x_{2j})| \leq M \mathrm{Var} f.$$

This, together with (18), proves

$$\left| \int_{-\pi}^{\pi} f(x)\phi'(x) \, dx \right| \leq M \mathrm{Var} f + \varepsilon.$$

Taking the limit of this inequality as $\varepsilon \to 0$, we conclude that (15) holds. ∎

We now estimate the rate of growth of Fourier coefficients of functions of bounded variation.

**14.26 THEOREM.** *If $f : \mathbf{R} \to \mathbf{R}$ is periodic and of bounded variation on $[-\pi, \pi]$, then*

$$|ka_k(f)| \leq \frac{\mathrm{Var}\, f}{\pi} \quad \text{and} \quad |kb_k(f)| \leq \frac{\mathrm{Var}\, f}{\pi}$$

*for $k \in \mathbf{N}$.*

PROOF. Fix $k \in \mathbf{N}$ and set $\phi(x) = \sin kx$. Then $\phi$ is periodic and $\phi'(x) = k \cos kx$ is continuously differentiable on $[0, 2\pi]$. Hence, it follows from Lemma 14.25 that

$$|ka_k(f)| = \left| \frac{1}{\pi} \int_{-\pi}^{\pi} f(x) k \cos kx \, dx \right| = \left| \frac{1}{\pi} \int_{-\pi}^{\pi} f(x)\phi'(x) \, dx \right| \leq \frac{\mathrm{Var}\, f}{\pi}.$$

A similar argument proves $|kb_k(f)| \leq \mathrm{Var} f / \pi$. ∎

## EXERCISES

**1.** If $f$ is integrable on $[-\pi, \pi]$ and $\alpha \in \mathbf{R}$, prove

$$\lim_{k \to \infty} \int_{-\pi}^{\pi} f(x) \sin(k + \alpha)x \, dx = 0.$$

**2.** Prove that there is no continuous function whose Fourier coefficients satisfy $|a_k(f)| \geq 1/\sqrt{k}$ for $k \in \mathbf{N}$.

**3.** Prove that if $f : \mathbf{R} \to \mathbf{R}$ belongs to $\mathcal{C}^2(\mathbf{R})$ and $f$, $f'$ are both periodic, then $Sf$ converges to $f$ uniformly and absolutely on $\mathbf{R}$. (See also Exercise 5 in Section 14.4.)

**4.** If $f : \mathbf{R} \to \mathbf{R}$ belongs to $\mathcal{C}^\infty(\mathbf{R})$ and $f^{(j)}$ is periodic for all $j \geq 0$, prove that $Sf$ is term-by-term differentiable on $\mathbf{R}$. In fact, show that

$$\frac{d^j f}{dx^j}(x) = \sum_{k=1}^{\infty} \frac{d^j}{dx^j} (a_k(f) \cos kx + b_k(f) \sin kx)$$

uniformly for all $j \in \mathbf{N}$.

5. Suppose $f : \mathbf{R} \to \mathbf{R}$ is periodic on $\mathbf{R}$, integrable on $[-\pi, \pi]$, and $a_k(f) \geq 0$ for $k = 0, 1, \cdots$.

   a) Prove $(S_k f)(0) \geq (S_j f)(0)$ for all $k \geq j \geq 0$.
   b) Prove $S_N f(0) \leq 2\sigma_{2N} f(0)$ for $N \in \mathbf{N}$.
   c) Prove that $\sum_{k=1}^{\infty} |a_k(f)| < \infty$.
   d) Suppose $f$ is also even. Prove that $f$ must be continuous and $Sf$ converges uniformly and absolutely on $\mathbf{R}$.

6. Suppose $f : \mathbf{R} \to \mathbf{R}$ is continuous and periodic. The *modulus of continuity* of $f$ is defined by

$$\omega(f, \delta) = \sup_{\substack{t \in [0, 2\pi] \\ |h| \leq \delta}} |f(t + h) - f(t)|.$$

   a) Show that

$$a_k(f) = \frac{1}{2\pi} \int_{-\pi}^{\pi} \left( f(u) - f\left( u + \frac{\pi}{k} \right) \right) \cos ku \, du$$

   for $k \in \mathbf{N}$.
   b) Prove

$$|a_k(f)| \leq \omega\left( f, \frac{\pi}{k} \right) \quad \text{and} \quad |b_k(f)| \leq \omega\left( f, \frac{\pi}{k} \right)$$

   for $k \in \mathbf{N}$.
   c) Use part b) to give a different proof the Riemann–Lebesgue Lemma in the special case when $f$ is periodic and continuous.

## $^e$14.4  CONVERGENCE OF FOURIER SERIES   *This section uses material from Sections 5.5, 14.2, and 14.3.*

We shall prove that, under certain conditions, a summable series must also be convergent. Such results, called *Tauberian theorems*, will be used to obtain a partial answer to the Convergence Question posed in Section 14.1 and further results concerning the growth of Fourier coefficients.

The following result was the first Tauberian theorem discovered.

**14.27 THEOREM** [TAUBER]. *Let $a_k \geq 0$ and $L \in \mathbf{R}$. If $\sum_{k=0}^{\infty} a_k$ is Cesàro summable to $L$, then*

$$\sum_{k=0}^{\infty} a_k = L.$$

PROOF. By Remark 14.11, it suffices to prove $\sum_{k=0}^{\infty} a_k < \infty$. Suppose to the contrary that $\sum_{k=0}^{\infty} a_k = \infty$. Then given $M > 0$ there is an $n_0 \in \mathbf{N}$ such that $n \geq n_0$ implies $s_n := \sum_{k=0}^{n} a_k \geq M$. Let $N > n_0$. Then

$$\sigma_N := \frac{s_0 + s_1 + \cdots + s_{n_0}}{N + 1} + \frac{s_{n_0+1} + \cdots + s_N}{N + 1} \geq 0 + \frac{N - n_0}{N + 1} M.$$

Taking the limit of this last inequality as $N \to \infty$, we obtain $L \geq M$ for all $M > 0$. We conclude that $L = \infty$, a contradiction. ∎

This result can be used to improve the Riemann–Lebesgue Lemma for certain types of functions.

**14.28 COROLLARY.** *Let $f : \mathbf{R} \to \mathbf{R}$ be periodic on $\mathbf{R}$ and integrable on $[-\pi, \pi]$. If $a_k(f) = 0$ and $b_k(f) \geq 0$ for $k \in \mathbf{N}$, then*

$$\sum_{k=1}^{\infty} \frac{b_k(f)}{k} < \infty.$$

PROOF. By considering $g = f - a_0(f)$ we may suppose that $a_0(f) = 0$. Let

$$F(x) = \int_0^x f(t) \, dt.$$

By Theorem 5.26, $F$ is continuous on $\mathbf{R}$. Since $a_0(f) = 0$, $F$ is also periodic. Hence, by Fejér's Theorem, $(\sigma_N F)(0) \to F(0) = 0$ as $N \to \infty$. Integrating by parts, we obtain

$$a_k(F) = \frac{b_k(f)}{k} \geq 0 \quad \text{and} \quad b_k(F) = -\frac{a_k(f)}{k} = 0.$$

It follows that $\sum_{k=1}^{\infty} b_k(f)/k$ is Cesàro summable (to $-a_0(F)/2$) and has nonnegative terms. We conclude by Tauber's Theorem that $\sum_{k=1}^{\infty} b_k(f)/k$ converges. ∎

We are now in a position to see that the converse of the Riemann–Lebesgue Lemma is false. Indeed, if

$$\sum_{k=2}^{\infty} \frac{\sin kx}{\log k}$$

were the Fourier series of some integrable function, then by Corollary 14.28,

$$\sum_{k=2}^{\infty} \frac{1}{k \log k}$$

would converge, a contradiction of the Integral Test.

The following result is one of the deepest Tauberian theorems.

**14.29 THEOREM** [HARDY]. *Let $E \subseteq \mathbf{R}$ and suppose $f_k : E \to \mathbf{R}$ is a sequence of functions which satisfies*

$$(19) \qquad\qquad |k f_k(x)| \leq M$$

*for all $x \in E$, all $k \in \mathbf{N}$, and some $M > 0$. If $\sum_{k=0}^{\infty} f_k$ is uniformly Cesàro summable to $f$ on $E$, then $\sum_{k=0}^{\infty} f_k$ converges uniformly to $f$ on $E$.*

PROOF. Fix $x \in E$ and suppose without loss of generality that $M \geq 1$. For each $n = 0, 1, \ldots$, set

$$s_n(x) = \sum_{k=0}^{n} f_k(x), \qquad \sigma_n(x) = \frac{s_0(x) + \cdots + s_n(x)}{n+1},$$

and consider the delayed averages

$$\sigma_{n,k}(x) := \frac{s_n(x) + \cdots + s_{n+k}(x)}{k + 1}$$

defined for $n, k \geq 0$.

Let $0 < \varepsilon < 1$. For each $n \in \mathbf{N}$ choose $k = k(n) \in \mathbf{N}$ such that $k \leq n\varepsilon/(2M) < k + 1$. Then

(20) $$\frac{n-1}{k+1} < \frac{n}{k+1} < \frac{2M}{\varepsilon} < \infty.$$

Moreover, since

$$\sigma_{n,k}(x) - s_n(x) = \frac{(s_n(x) - s_n(x)) + \cdots + (s_{n+k}(x) - s_n(x))}{k + 1}$$

$$= \sum_{j=n+1}^{n+k} \left(1 - \frac{j-n}{k+1}\right) f_j(x),$$

it follows from (19) and the choice of $k = k(n)$ that

(21) $$|\sigma_{n,k}(x) - s_n(x)| \leq \sum_{j=n+1}^{n+k} |f_j(x)| \leq M \sum_{j=n+1}^{n+k} \frac{1}{j} < \frac{Mk}{n+1} < \frac{\varepsilon}{2}.$$

Since $\sigma_n \to f$ uniformly on $E$, choose $N \in \mathbf{N}$ such that

(22) $$n \geq N \quad \text{and} \quad x \in E \quad \text{imply} \quad |\sigma_n(x) - f(x)| < \frac{\varepsilon^2}{12M}.$$

Since

$$\sigma_{n,k}(x) = \left(1 + \frac{n-1}{k+1}\right)\sigma_{n+k} - \left(\frac{n-1}{k+1}\right)\sigma_{n-1},$$

it follows (20), (21), and (22) that

$$|s_n(x) - f(x)| \leq |s_n(x) - \sigma_{n,k}(x)| + |\sigma_{n,k} - f(x)|$$

$$< \frac{\varepsilon}{2} + \left(1 + \frac{n-1}{k+1}\right)|\sigma_{n+k}(x) - f(x)|$$

$$+ \left(\frac{n-1}{k+1}\right)|\sigma_{n-1}(x) - f(x)|$$

$$< \frac{\varepsilon}{2} + \left(1 + \frac{2M}{\varepsilon}\right)\left(\frac{\varepsilon^2}{12M}\right) + \frac{2M}{\varepsilon}\left(\frac{\varepsilon^2}{12M}\right)$$

$$= \frac{\varepsilon}{2} + \frac{\varepsilon^2}{12M} + \frac{\varepsilon}{3} < \frac{\varepsilon}{2} + \frac{\varepsilon}{12} + \frac{\varepsilon}{3} < \varepsilon$$

for any $n > N$ and $x \in E$. We conclude that $s_n \to f$ uniformly on $E$ as $n \to \infty$. ∎

We are prepared to answer the Convergence Question posed in Section 14.1 for piecewise continuous functions of bounded variation.

**14.30 THEOREM** [DIRICHLET–JORDAN]. *If $f : \mathbf{R} \to \mathbf{R}$ is periodic on $\mathbf{R}$ and of bounded variation on $[-\pi, \pi]$, then*

$$\lim_{N \to \infty} (S_N f)(x) = \frac{f(x+) + f(x-)}{2}$$

*for every $x \in \mathbf{R}$. If $f$ is also continuous on some closed interval $I$, then*

$$\lim_{N \to \infty} S_N f = f$$

*uniformly on $I$.*

PROOF. Since $f$ is periodic and of bounded variation, the one-sided limits $f(x+)$ and $f(x-)$ exist for each $x \in \mathbf{R}$, and $f$ is Riemann integrable on $[-\pi, \pi]$ (see the comments which follow the proof of Corollary 5.57). Hence, by Fejér's Theorem, both conclusions hold if $S_N$ is replaced by $\sigma_N$. Since Theorem 14.26 implies

$$|k a_k(f) \cos kx| \quad \text{and} \quad |k b_k(f) \cos kx| \le \frac{\operatorname{Var} f}{\pi}$$

for $k \in \mathbf{N}$, it follows from Hardy's Theorem that both conclusions hold as stated. ∎

We close this section with an application of Fourier series to an extremal problem. We will show that among all smooth simple closed curves in $\mathbf{R}^2$ with a given arc length, the largest area is enclosed by a circle. (The proof presented here comes from Marsden [7].)

**14.31 THEOREM** [THE ISOPERIMETRIC PROBLEM]. *Let $E$ be a region in $\mathbf{R}^2$ whose topological boundary $C = \partial E$ is a smooth closed simple curve of length $2\pi$. If $A = \operatorname{Area}(E)$, then $A \le \pi$. Moreover, $A = \pi$ if and only if $E = B_1(a, b)$ for some $a, b \in \mathbf{R}$.*

PROOF. Let $(\nu, [0, 2\pi])$ be the natural parametrization of $C$, i.e., $\|\nu'(s)\| = 1$ for all $s \in [0, 2\pi]$. Set

$$a = \frac{1}{2\pi} \int_0^{2\pi} \nu_1(s) \, ds, \qquad b = \frac{1}{2\pi} \int_0^{2\pi} \nu_2(s) \, ds,$$

$$P(s) = \nu_1(s) - a, \quad Q(s) = \nu_2(s) - b, \quad \text{and} \quad \phi(s) = (P(s), Q(s))$$

for $s \in [0, 2\pi]$, where $(\nu_1, \nu_2) := \nu$. Clearly, $(\phi, [0, 2\pi])$ is a smooth parametrization of $\partial E - (a, b)$ whose trace is a smooth closed simple curve with arc length $2\pi$ which encloses a region with area $A$. Moreover,

$$(23) \qquad |P'(s)|^2 + |Q'(s)|^2 = 1,$$

$$(24) \qquad \frac{1}{2\pi} \int_0^{2\pi} P(s) \, ds = 0, \qquad \frac{1}{2\pi} \int_0^{2\pi} Q(s) \, ds = 0,$$

and by Green's Theorem,

$$(25) \qquad A = \iint_E dA = \int_{\partial E} x\,dy = \int_0^{2\pi} P(s)Q'(s)\,ds.$$

Let $a_k, b_k$ (respectively, $c_k, d_k$) represent the Fourier coefficients of $P$ (respectively, $Q$). Since $(\phi, [0, 2\pi])$ is smooth and closed, $P$ and $Q$ are continuously differentiable and periodic. By (24) and the Dirichlet–Jordan Theorem,

$$(26) \qquad P(s) = \sum_{k=1}^\infty (a_k \cos ks + b_k \sin ks), \qquad Q(s) = \sum_{k=1}^\infty (c_k \cos ks + d_k \sin ks),$$

(27)

$$P'(s) = \sum_{k=1}^\infty (kb_k \cos ks - ka_k \sin ks), \quad \text{and} \quad Q'(s) = \sum_{k=1}^\infty (kd_k \cos ks - kc_k \sin ks)$$

uniformly on $[0, 2\pi]$. Hence, by (23) and Parseval's Identity,

$$2\pi = \int_0^{2\pi} (|P'(s)|^2 + |Q'(s)|^2)\,ds = \pi \sum_{k=1}^\infty k^2(a_k^2 + b_k^2 + c_k^2 + d_k^2).$$

Moreover, by (25) and orthogonality

$$A = \int_0^{2\pi} P(s)Q'(s)\,ds = \pi \sum_{k=1}^\infty k(a_k d_k - b_k c_k).$$

It follows that

$$\pi - A = \frac{\pi}{2} \sum_{k=2}^\infty (k^2 - k)(a_k^2 + b_k^2 + c_k^2 + d_k^2) + \frac{\pi}{2} \sum_{k=1}^\infty k((a_k - d_k)^2 + (c_k + b_k)^2) \geq 0.$$

In particular, $A \leq \pi$ and $A = \pi$ if and only if $a_1 = d_1$, $c_1 = -b_1$, and $a_k = b_k = c_k = d_k = 0$ for $k \geq 2$.

Suppose $A = \pi$. Then $P(s) = a_1 \cos s + b_1 \sin s$ and $Q(s) = -b_1 \cos s + a_1 \sin s = -P(s + \frac{\pi}{2})$. Thus $P'(s) = -Q(s)$ and $Q'(s) = -P''(s) = P(s)$ for all $s \in [0, 2\pi]$. It follows from (23) that $\phi([0, 2\pi])$ is a subset of $\partial B_1(0, 0)$. Since $\phi(0) = \phi(2\pi)$, we must have $\phi([0, 2\pi]) = \partial B_1(0, 0)$. Therefore, $C$ is the boundary of the disk $E = B_1(a, b)$. ∎

## EXERCISES

**1.** Suppose $f$ is continuous and of bounded variation on $[-\pi, \pi]$. Prove that $S_N f \to f$ pointwise on $(-\pi, \pi)$ and uniformly on any $[a, b] \subset (-\pi, \pi)$.

**2. a)** Prove that

$$x = 2 \sum_{k=1}^{\infty} \frac{(-1)^{k+1}}{k} \sin kx$$

pointwise on $(-\pi, \pi)$ and uniformly on any $[a, b] \subset (-\pi, \pi)$.

**b)** Prove that

$$|x| = \frac{\pi}{2} - \frac{4}{\pi} \sum_{k=1}^{\infty} \frac{\cos(2k-1)x}{(2k-1)^2}$$

uniformly on $[-\pi, \pi]$.

**c)** Find a value for

$$\sum_{k=1}^{\infty} \frac{1}{(2k-1)^2}.$$

**3.** Prove that if $f$ is continuous, odd, and periodic, then $\sum_{k=1}^{\infty} b_k(f)/k$ converges.

**4.** Let $L \in \mathbf{R}$. A series $\sum_{k=0}^{\infty} a_k$ is said to be *Abel summable* to $L$ if

$$\lim_{r \to 1-} \sum_{k=0}^{\infty} a_k r^k = L.$$

**a)** Let $S_k = \sum_{j=0}^{k} a_k$. Prove that

$$\sum_{k=0}^{\infty} a_k r^k = (1-r) \sum_{k=0}^{\infty} S_k r^k = (1-r)^2 \sum_{k=0}^{\infty} (k+1)\sigma_k r^k,$$

provided any one of these series converges for all $0 < r < 1$.

**b)** Prove that if $\sum_{k=0}^{\infty} a_k$ is Cesàro summable to $L$, then it is Abel summable to $L$.

**c)** Prove that if $f$ is continuous, periodic, and of bounded variation on $\mathbf{R}$, then $Sf$ is Abel summable to $f$ uniformly on $\mathbf{R}$.

**d)** Show that if $a_k \geq 0$ and $\sum_{k=0}^{\infty} a_k$ is Abel summable to $L$, then $\sum_{k=0}^{\infty} a_k$ converges to $L$.

**5. [BERNSTEIN].** Let $f : \mathbf{R} \to \mathbf{R}$ be periodic and $\alpha > 0$. Suppose $f$ is Lipschitz of order $\alpha$; i.e., there is a constant $M > 0$ such that

$$|f(x+h) - f(x)| \leq M|h|^{\alpha}$$

for all $x, h \in \mathbf{R}$.

**a)** Prove

$$\frac{1}{\pi} \int_{-\pi}^{\pi} |f(x+h) - f(x-h)|^2 \, dx = 4 \sum_{k=1}^{\infty} (a_k^2(f) + b_k^2(f)) \sin^2 kh$$

holds for each $h \in \mathbf{R}$.

b) If $h = \pi/2^{n+1}$, prove $\sin^2 kh \geq 1/2$ for all $k \in [2^{n-1}, 2^n]$.

c) Combine a) and b) to prove

$$\left\{ \sum_{k=2^{n-1}}^{2^n-1} (a_k^2(f) + b_k^2(f)) \right\}^{1/2} \leq M^2 \left( \frac{\pi}{2^{n+1}} \right)^{2\alpha}$$

for $n = 1, 2, 3, \ldots$.

d) Assuming

$$\sum_{k=2^{n-1}}^{2^n-1} (|a_k(f)| + |b_k(f)|) \leq 2^{n/2} \left( \sum_{k=2^{n-1}}^{2^n-1} (a_k^2(f) + b_k^2(f)) \right)^{1/2}$$

(see Exercise 9 in Section 11.6), prove that if $f$ is Lipschitz of order $\alpha$ for some $\alpha > 1/2$, then $Sf$ converges absolutely and uniformly on $\mathbf{R}$.

e) Prove that if $f : \mathbf{R} \to \mathbf{R}$ is periodic and continuously differentiable, then $Sf$ converges absolutely and uniformly on $\mathbf{R}$.

*6. Suppose $f : \mathbf{R} \to \mathbf{R}$ is periodic and of bounded variation on $[-\pi, \pi]$. Prove that $S_N f \to f$ almost everywhere as $N \to \infty$ (see Exercise 8 in Section 14.2).

## $^e$14.5   UNIQUENESS   *This section uses material from Section 14.4.*

In this section we examine the Uniqueness Question posed in Section 14.1. We begin with the following generalization of the second derivative.

**14.32 DEFINITION.** Let $x_0 \in \mathbf{R}$ and $I$ be an open interval containing $x_0$. A function $F : I \to \mathbf{R}$ is said to have a *second symmetric derivative* at $x_0$ if

$$D_2 F(x_0) = \lim_{h \to 0+} \frac{F(x_0 + 2h) + F(x_0 - 2h) - 2F(x_0)}{4h^2}$$

exists.

**14.33 Remark.** *Let $x_0 \in \mathbf{R}$ and $I$ be an open interval containing $x_0$. If $F$ is differentiable on $I$ and $F''(x_0)$ exists, then $F$ has a second symmetric derivative at $x_0$ and $D_2 F(x_0) = F''(x_0)$.*

PROOF. Set $G(t) = F(x_0 + 2t) + F(x_0 - 2t)$ for $t \in I$ and $H(t) = 4t^2$ and fix $t \in I$. By Theorem 4.15 (the Generalized Mean Value Theorem),

$$\frac{F(x_0 + 2t) + F(x_0 - 2t) - 2F(x_0)}{4t^2} = \frac{G(t) - G(0)}{H(t) - H(0)} = \frac{G'(c)}{H'(c)}$$

$$= \frac{F'(x_0 + 2c) - F'(x_0 - 2c)}{4c}$$

for some $c$ between $0$ and $t$. Since $c \to 0$ as $t \to 0$, it follows that

$$D_2 F(x_0) = \lim_{c \to 0} \frac{F'(x_0 + 2c) - F'(x_0 - 2c)}{4c}$$

$$= \frac{1}{2} \lim_{c \to 0} \left( \frac{F'(x_0 + 2c) - F'(x_0)}{2c} + \frac{F'(x_0) - F'(x_0 - 2c)}{2c} \right)$$

$$= \frac{1}{2} (F''(x_0) + F''(x_0)) = F''(x_0). \quad \blacksquare$$

The converse of Remark 14.33 is false. Indeed, if

$$F(x) = \begin{cases} 1 & x > 0 \\ 0 & x = 0 \\ -1 & x < 0, \end{cases}$$

then $D_2 F(0) = 0$ but $F''(0)$ does not exist.

The following result reinforces further the analogy between the second derivative and the second symmetric derivative (see also Exercises 1 and 5).

**14.34 Lemma.** *Let $[a, b]$ be a closed bounded interval. If $F : [a, b] \to \mathbf{R}$ is continuous on $[a, b]$ and $D_2 F(x) = 0$ for all $x \in (a, b)$, then $F$ is linear on $[a, b]$; i.e., there exist constants $m, \gamma$ such that $F(x) = mx + \gamma$ for all $x \in [a, b]$.*

PROOF. Let $\varepsilon > 0$. By hypothesis,

$$\phi(x) := F(x) - F(a) - \left( \frac{F(b) - F(a)}{b - a} \right)(x - a) + \varepsilon(x - a)(x - b)$$

is continuous on $[a, b]$, and by Remark 14.33,

(29) $$D_2 \phi(x) = D_2 F(x) + 2\varepsilon = 2\varepsilon$$

for $x \in (a, b)$.

We claim that $\phi(x) \le 0$ for $x \in [a, b]$. Clearly, $\phi(a) = \phi(b) = 0$. If $\phi(x) > 0$ for some $x \in (a, b)$, then $\phi$ attains its maximum at some $x_0 \in (a, b)$. By Exercise 1, $D_2 \phi(x_0) \le 0$, hence, by (29), $2\varepsilon \le 0$, a contradiction. This proves the claim.

Fix $x \in [a, b]$. We have shown that

$$F(x) - F(a) - \left( \frac{F(b) - F(a)}{b - a} \right)(x - a) \le \varepsilon(x - a)(b - x).$$

A similar argument establishes that

$$F(x) - F(a) - \left( \frac{F(b) - F(a)}{b - a} \right)(x - a) \ge -\varepsilon(x - a)(b - x).$$

Therefore,

$$\left| F(x) - F(a) - \left( \frac{F(b) - F(a)}{b - a} \right)(x - a) \right| \le \varepsilon(x - a)(b - x) \le \varepsilon(b - a)^2.$$

Taking the limit of this inequality as $\varepsilon \to 0$, we conclude that

$$F(x) = F(a) + \left( \frac{F(b) - F(a)}{b - a} \right)(x - a)$$

for all $x \in [a, b]$, i.e., $F$ is linear on $[a, b]$. $\quad \blacksquare$

**14.35 DEFINITION.** The *second formal integral* of a trigonometric series,

$$S = \frac{a_0}{2} + \sum_{k=1}^{\infty} (a_k \cos kx + b_k \sin kx),$$

is the function

$$F(x) = \frac{a_0}{4} x^2 - \sum_{k=1}^{\infty} \frac{1}{k^2} (a_k \cos kx + b_k \sin kx).$$

By the Weierstrass $M$–Test, if the coefficients of $S$ are bounded, then the second formal integral of $S$ converges uniformly on $\mathbf{R}$. In particular, the second formal integral always exists when the coefficients of $S$ converge to zero.

Notice that the second formal integral of a trigonometric series $S$ is the result of integrating $S$ twice term by term. Hence, it is not unreasonable to expect that two derivatives of the second formal integral $F$ might recapture the original series $S$. Although this statement is not quite correct, the following result shows there is a simple connection between the limit of the series $S$ and the second *symmetric* derivative of $F$.

**14.36 THEOREM** [RIEMANN]. *Suppose*

$$S = \frac{a_0}{2} + \sum_{k=1}^{\infty} (a_k \cos kx + b_k \sin kx)$$

*is a trigonometric series whose coefficients $a_k$, $b_k \to 0$ as $k \to \infty$ and let $F$ be the second formal integral of $S$. If $S(x_0)$ converges to $L$ for some $x_0 \in \mathbf{R}$, then $D_2 F(x_0) = L$.*

PROOF. Let $F_N$ denote the partial sums of $F$. After several applications of Theorem B.3, we observe that

$$\lim_{h \to 0} \frac{F_N(x_0 + 2h) + F_N(x_0 - 2h) - 2F_N(x_0)}{4h^2}$$

$$= \lim_{h \to 0} \left( \frac{a_0}{2} + \sum_{k=1}^{N} (a_k \cos kx_0 + b_k \sin kx_0) \left( \frac{\sin kh}{kh} \right)^2 \right)$$

$$= \frac{a_0}{2} + \sum_{k=1}^{N} (a_k \cos kx_0 + b_k \sin kx_0)$$

holds for any $N \in \mathbf{N}$. Therefore, it suffices to show that given $\varepsilon > 0$ there is an $N \in \mathbf{N}$ such that

$$(30) \qquad |R_N| := \left| \sum_{k=N+1}^{\infty} (a_k \cos kx_0 + b_k \sin kx_0) \left( \frac{\sin kh}{kh} \right)^2 \right| < \varepsilon$$

for all $|h| \leq 1$.

Let

$$A_k = \sum_{j=k+1}^{\infty} (a_j \cos jx_0 + b_j \sin jx_0) \quad \text{and} \quad B_k = \left( \frac{\sin kh}{kh} \right)^2$$

for $k \in \mathbf{N}$. Since $A_n \to 0$ as $n \to \infty$, we have by Abel's Formula that

$$(31) \qquad R_N := \lim_{n \to \infty} \sum_{k=N+1}^{n} (A_{k-1} - A_k) B_k$$

$$= \lim_{n \to \infty} \left( (A_N - A_n) B_n - \sum_{k=N+1}^{n-1} (A_N - A_k)(B_{k+1} - B_k) \right)$$

$$= A_N B_{N+1} + \sum_{k=N+1}^{\infty} A_k (B_{k+1} - B_k).$$

Moreover, by the Fundamental Theorem of Calculus,

$$(32) \qquad |B_{k+1} - B_k| = \left| \int_{kh}^{(k+1)h} \frac{d}{dt} \left( \frac{\sin t}{t} \right)^2 dt \right|.$$

Since

$$\frac{d}{dt} \left( \frac{\sin t}{t} \right)^2 = \frac{2 \sin t}{t} \left( \frac{t \cos t - \sin t}{t^2} \right)$$

is bounded near $t = 0$ and is bounded by $2(t+1)/t^3 < 2/t^2$ for $t \geq 2$, it is clear that the improper integral

$$C = \int_0^{\infty} \left| \frac{d}{dt} \left( \frac{\sin t}{t} \right)^2 \right| dt$$

converges. Since $\{B_k\}$ is bounded and $A_N \to 0$ as $N \to \infty$, we can choose an $N \in \mathbf{N}$ such that

$$(33) \qquad |A_N B_{N+1}| < \frac{\varepsilon}{2} \quad \text{and} \quad k \geq N \quad \text{implies} \quad |A_k| < \frac{\varepsilon}{2C}.$$

It follows from (32) that

$$\sum_{k=N+1}^{\infty} A_k (B_{k+1} - B_k) \leq \frac{\varepsilon}{2C} \sum_{k=N+1}^{\infty} \left| \int_{kh}^{(k+1)h} \frac{d}{dt} \left( \frac{\sin t}{t} \right)^2 dt \right|$$

$$\leq \frac{\varepsilon}{2C} \int_0^{\infty} \left| \frac{d}{dt} \left( \frac{\sin t}{t} \right)^2 \right| dt = \frac{\varepsilon}{2}.$$

Combining this inequality with (31) and (33), we conclude that $|R_N| < \varepsilon$. ∎

The following result shows that the hypotheses of Riemann's Theorem are satisfied by any trigonometric series which converges pointwise on a nondegenerate interval.

**14.37 THEOREM** [The Cantor–Lebesgue Lemma]. *If*

$$S = \frac{a_0}{2} + \sum_{k=1}^{\infty} (a_k \cos kx + b_k \sin kx)$$

*is a trigonometric series which converges pointwise on a nondegenerate interval* $[a, b]$, *then its coefficients satisfy* $a_k$, $b_k \to 0$ *as* $k \to \infty$.

Proof. Set $\rho_0 = a_0/2$ and $\rho_k^2 = a_k^2 + b_k^2$ for $k \in \mathbf{N}$. If the result is false, then there is a $\delta > 0$ such that $\rho_k > \delta$ for infinitely many $k \in \mathbf{N}$.

Set $\theta_0 = 0$ and for each $k \in \mathbf{N}$ define $\theta_k \in \mathbf{R}$ so that $a_k = \rho_k \cos k\theta_k$, $b_k = \rho_k \sin k\theta_k$. By a sum-angle formula,

$$\frac{a_0}{2} + \sum_{k=1}^{n} (a_k \cos kx + b_k \sin kx) = \sum_{k=0}^{n} \rho_k \cos k(x - \theta_k)$$

for each $x \in \mathbf{R}$ and $n \in \mathbf{N}$. Since $S$ converges on $[a, b]$, it follows that

(34) $$\lim_{k \to \infty} \rho_k \cos k(x - \theta_k) = 0$$

for all $x \in [a, b]$.

Set $I_0 = [a, b]$ and $k_0 = 1$. Fix $j \geq 0$ and suppose a closed interval $I_j \subseteq I_0$ and an integer $k_j > k_0$ have been chosen. Choose $k_{j+1} > k_j$ such that $k_{j+1}|I_j| > 2\pi$ and $\rho_{k_{j+1}} > \delta$. Clearly, $k_{j+1}(x - \theta_{k_{j+1}})$ runs over an interval of length $> 2\pi$ as $x$ runs over $I_j$. Hence, we can choose a closed interval $I_{j+1} \subseteq I_j$ such that

$$x \in I_{j+1} \quad \text{implies} \quad \cos k_{j+1}(x - \theta_{k_{j+1}}) \geq \frac{1}{2}.$$

By induction, then, there exist integers $1 < k_1 < k_2 < \ldots$ and a nested sequence of closed intervals $I_0 \supseteq I_1 \supseteq \ldots$ such that

(35) $$\rho_{k_j} \cos k_j(x - \theta_{k_j}) \geq \frac{\delta}{2}$$

for $x \in I_j$, $j \in \mathbf{N}$. By the Nested Interval Property, there is an $x \in I_j$ for all $j \in \mathbf{N}$. This $x$ must satisfy (35) for all $j \in \mathbf{N}$ and must belong to $[a, b]$ by construction. Since this contradicts (34), we conclude that $\rho_k \to 0$ as $k \to \infty$. ∎

We are now prepared to answer the Uniqueness Question for continuous functions of bounded variation.

**14.38 THEOREM** [Cantor]. *Suppose*

$$S = \frac{a_0}{2} + \sum_{k=1}^{\infty} (a_k \cos kx + b_k \sin kx)$$

*converges pointwise on* $[-\pi, \pi]$ *to a function* $f$ *which is periodic and continuous on* **R**, *and of bounded variation on* $[-\pi, \pi]$. *Then* $S$ *is the Fourier series of* $f$; *i.e.,* $a_k = a_k(f)$ *for* $k = 0, 1, \ldots$, *and* $b_k = b_k(f)$ *for* $k = 1, 2, \ldots$

PROOF. Suppose first that $f(x) = 0$ for all $x \in \mathbf{R}$. By the Cantor–Lebesgue Lemma, the coefficients $a_k, b_k$ tend to zero as $k \to \infty$. Thus the second formal integral $F$ of $S$ is continuous on $\mathbf{R}$ and by Riemann's Theorem has a second symmetric derivative which satisfies $D_2 F(x) = 0$ for $x \in \mathbf{R}$. It follows that $F$ is linear on $\mathbf{R}$; i.e., there exist numbers $m$ and $\gamma$ such that

$$mx + \gamma = \frac{a_0}{4}x^2 - \sum_{k=1}^{\infty} \frac{1}{k^2}(a_k \cos kx + b_k \sin kx)$$

for $x \in \mathbf{R}$. Since the series in this expression is periodic, it must be the case that $m = a_0 = 0$, i.e.,

$$\gamma + \sum_{k=1}^{\infty} \frac{1}{k^2}(a_k \cos kx + b_k \sin kx) = 0$$

for all $x \in \mathbf{R}$. Since this series converges uniformly, it follows from Theorem 14.4 that $\gamma = 0$ and $a_k = b_k = 0$ for $k \in \mathbf{N}$. This proves the theorem when $f = 0$.

If $f$ is periodic, continuous, and of bounded variation on $[-\pi, \pi]$, then $S_N f \to f$ uniformly on $\mathbf{R}$ by Theorem 14.30. Hence, the series $S - Sf$ converges pointwise on $\mathbf{R}$ to zero. It follows from the case already considered that $a_k - a_k(f) = 0$ for $k = 0, 1, \ldots$, and $b_k - b_k(f) = 0$ for $k = 1, 2, \cdots$. ∎

### EXERCISES

**1.** Suppose $F : \mathbf{R} \to \mathbf{R}$ has a second symmetric derivative at some $x_0$. Prove that if $F(x_0)$ is a local maximum, then $D_2 F(x_0) \leq 0$, and if $F(x_0)$ is a local minimum, then $D_2 F(x_0) \geq 0$.

**2.** Prove that if the coefficients of a trigonometric series are bounded, then its second formal integral converges uniformly on $\mathbf{R}$.

**3.** Prove that if $f : \mathbf{R} \to \mathbf{R}$ is periodic, then there exists at most one trigonometric series which converges to $f$ pointwise on $\mathbf{R}$.

**4.** Suppose $f : \mathbf{R} \to \mathbf{R}$ is periodic, piecewise continuous, and of bounded variation on $\mathbf{R}$. Prove that if $S$ is a trigonometric series which converges to $(f(x+) + f(x-))/2$ for all $x \in \mathbf{R}$, then $S$ is the Fourier series of $f$.

**\*5.** Suppose $F : (a, b) \to \mathbf{R}$ is continuous and $D_2 F(x) > 0$ for all $x \in (a, b)$. Prove that $F$ is convex on $(a, b)$.

# Chapter 15

# Differentiable Manifolds

This chapter is considerably more abstract than those preceding it. Our aim is to show that the theorems of Green, Gauss, and Stokes are special cases of a more general theory in which differential forms of degree 1 and 2 are replaced by differential forms of degree $n$, and curves and surfaces are replaced by $n$-dimensional manifolds. Differential forms of degree $n$ are introduced in Section 15.1, $n$-dimensional manifolds are introduced in Section 15.2, and an $n$-dimensional version of Stokes's Theorem is proved in Section 15.3.

$^e$**15.1 DIFFERENTIAL FORMS ON $\mathbf{R}^n$** *This section uses no material from any other enrichment section.*

We introduced differential forms of degree 1 and 2 in Sections 13.2 and 13.4. In this section we introduce differential forms of degree $r$. It turns out that, in so far as calculus is concerned, the actual definition of differential forms is not as important as their algebraic structure. For this reason, we begin with the following formal definition. (For a more constructive approach to differential forms which interprets $dx_i$ as a derivative of the projection operator $(x_1, \ldots, x_n) \longmapsto x_i$, see Spivak [12], p. 89.)

**15.1 DEFINITION.** Let $0 \le r \le n$ and $V$ be open in $\mathbf{R}^n$.

   i) A 0–*form* (or *differential form of degree $r = 0$*) on $V$ is a function $f : V \to \mathbf{R}$.

   ii) Let $r > 0$. An $r$–*form* (or *differential form of degree $r$*) on $V$ is an expression of the form

(1)
$$\omega = \sum f_{i_1, \ldots, i_r}\, dx_{i_1} \ldots dx_{i_r},$$

where the sum is taken over all integers $i_j$ which satisfy $1 \le i_1 < i_2 < \cdots < i_r \le n$, each *coefficient function* $f_{i_1, \ldots, i_r}$ is a 0–form on $V$, and the $dx_{i_j}$'s are symbols which (for us) will take on meaning only in the context of integration (see Definition 15.38 below). If all the coefficient functions are zero, then $\omega$ is called the *zero* $r$–form and is denoted by 0.

iii) Two $r$–forms, $\omega = \sum f_{i_1,\dots,i_r}\, dx_{i_1} \dots dx_{i_r}$ and $\eta = \sum g_{i_1,\dots,i_r}\, dx_{i_1} \dots dx_{i_r}$, are said to be *equal* on $V$ if $f_{i_1,\dots,i_r}(\boldsymbol{x}) = g_{i_1,\dots,i_r}(\boldsymbol{x})$ for all $1 \le i_1 < i_2 < \dots < i_r \le n$ and all $\boldsymbol{x} \in V$.

iv) An $r$–form $\omega$ is said to be *decomposable* on $V$ if there exist integers $1 \le i_1 < \dots < i_r \le n$ and a 0–form $f$ such that

$$\omega = f\, dx_{i_1} \dots dx_{i_r}$$

on $V$.

v) An $r$–form is said to be continuous (respectively, $\mathcal{C}^p$) on $V$ if all of its coefficient functions are continuous (respectively, $\mathcal{C}^p$) on $V$.

vi) The *support* of an $r$–form (notation: $\operatorname{spt}\omega$) is the union of the supports of its coefficient functions; i.e., if $\omega$ is given by (1), then

$$\operatorname{spt}\omega = \bigcup_{1 \le i_1 < \dots < i_r \le n} \operatorname{spt}\left(f_{i_1,\dots,i_r}\right).$$

If $\operatorname{spt}\omega \subseteq E$, then $\omega$ is said to be *supported* on $E$.

Let $V$ be open in $\mathbf{R}^n$. Since there is only one collection of indices which satisfies $1 \le i_1 < \dots < i_r \le n$ for $r = n$, an $n$–form on $V$ is an expression of the form

$$\omega = f\, dx_1 \dots dx_n$$

for some 0–form $f$ (i.e., a function) on $V$. Thus every $n$–form on $V \subseteq \mathbf{R}^n$ is decomposable. At the other extreme, a general 1–form on $V$ is an expression of the form

$$\omega = \sum_{j=1}^{n} f_j\, dx_j,$$

where each $f_j$ is a 0–form on $V$. An example of a 1–form is the total differential of a differentiable function $z = f(x,y)$, i.e., $dz = f_x\, dx + f_y\, dy$.

An $(n-1)$–form on $\mathbf{R}^n$ is an expression of the form

$$\omega = \sum_{j=1}^{n} f_j\, dx_1 \dots \widehat{dx_j} \dots dx_n.$$

The notation $\widehat{dx_j}$ indicates that the differential $dx_j$ is missing. Thus, a 2–form on $\mathbf{R}^3$ is an expression of the form

(2) $$\omega = f_1\, dy\, dz + f_2\, dx\, dz + f_3\, dx\, dy.$$

In Chapter 13, we used Jacobians to define differential forms of degree 2 on a smooth orientable surface $S = (\phi, E)$ and to associate with each 2–form an oriented integral on $S$ (see also Exercise 5 below). In the same way, we shall associate $n$–forms on $\mathbf{R}^n$ with oriented integrals over certain geometric objects called $n$-dimensional manifolds. First, we introduce an algebraic structure on the collection of differential forms which is compatible with this identification.

Addition of differential forms can be realized by grouping like terms and simplifying coefficients. For example, the sum of $x^2\,dy\,dz + y\,dx\,dy$ and $(1-x^2)\,dy\,dz$ is

$$x^2\,dy\,dz + y\,dx\,dy + (1-x^2)\,dy\,dz = dy\,dz + y\,dx\,dy.$$

In particular, if $V$ is open in $\mathbf{R}^n$ and

$$\omega = \sum f_{i_1,\dots,i_r}\,dx_{i_1}\dots dx_{i_r}, \qquad \eta = \sum g_{i_1,\dots,i_r}\,dx_{i_1}\dots dx_{i_r}$$

are $r$–forms on $V$, then

$$\omega + \eta = \sum (f_{i_1,\dots,i_r} + g_{i_1,\dots,i_r})\,dx_{i_1}\dots dx_{i_r}.$$

It is clear that addition of differential forms satisfies the usual laws of algebra, e.g., the Commutative Law and the Associative Law.

The product of a 0–form (this includes scalars) and an $r$–form can be defined by

$$g\Big(\sum f_{i_1,\dots,i_r}\,dx_{i_1}\dots dx_{i_r}\Big) = \sum g f_{i_1,\dots,i_r}\,dx_{i_1}\dots dx_{i_r}.$$

It is clear that if $\omega$, $\eta$ are $r$–forms and $f$, $g$ are 0–forms, then

$$f(\omega + \eta) = f\omega + f\eta, \quad \text{and} \quad (f+g)\omega = f\omega + g\omega.$$

Multiplication of differential forms of degrees $r, s > 0$ is somewhat more complicated to describe. To explain what happens, recall that if $S = (\phi, E)$ is a smooth orientable surface in $\mathbf{R}^3$, then differential forms of degree 2 are defined by

$$dy\,dz = \frac{\partial(y,z)}{\partial(u,v)}\,d(u,v), \quad dz\,dx = \frac{\partial(z,x)}{\partial(u,v)}\,d(u,v), \quad \text{and } dx\,dy = \frac{\partial(x,y)}{\partial(u,v)}\,d(u,v),$$

where $x = \phi_1(u,v)$, $y = \phi_2(u,v)$, $z = \phi_3(u,v)$. The notation $dx\,dy$ looks like a product of 1–forms. Does this product satisfy the usual algebraic laws? Certainly not. Since interchanging two rows of a determinant changes its sign (see Appendix C), it is clear that multiplication of 1–forms must satisfy the *Anticommutative Property*, e.g., $dx\,dy = -dy\,dx$. Since the determinant of any matrix with two identical rows is zero (see Appendix C), it is also clear that multiplication of 1–forms must satisfy the *Nilpotent Property*, e.g., $dx\,dx = 0$.

Based on these observations, we define multiplication of differential forms in the following way. First, we assume the *Anticommutative Property* and the *Nilpotent Property* hold for all decomposable 1–forms, i.e.,

$$dx_j\,dx_k = -dx_k\,dx_j \quad \text{and} \quad dx_j\,dx_j = 0,$$

for $k, j = 1, \ldots, n$. Next, we multiply two differential forms by assuming the distributive law holds, grouping like terms, and simplifying the resulting expression using the Nilpotent Property and the Anticommutative Property. For example, the product of $x^2\,dx$ and $y\,dy + z\,dz$ is

$$(x^2\,dx)(y\,dy + z\,dz) = x^2 y\,dx\,dy + x^2 z\,dx\,dz,$$

and the product of $\sin x\,dz$ and $x^2\,dx + xy\,dy + \log z\,dz$ is

$$(\sin x\,dz)(x^2\,dx + xy\,dy + \log z\,dz) = -xy \sin x\,dy\,dz - x^2 \sin x\,dx\,dz.$$

In particular, if $\omega = \sum_{j=1}^{N} \omega_j$ and $\eta = \sum_{k=1}^{L} \eta_k$ is a sum of differential forms, then

$$\omega\eta = \sum_{j=1}^{N} \sum_{k=1}^{L} \omega_j \eta_k.$$

Although the Anticommutative Property and the Nilpotent Property may seem strange, they are natural consequences of the fact that $dx\,dy$ comes not from an iterated integral but an oriented integral. For example, the Anticommutative Property reflects the fact that when orientation is changed, the sign of the integral changes.

**15.2 Example.** Find $\omega + \eta$, $\omega - \eta$ and $\omega\eta$, if $\omega = x^2\,dx\,dz + xy\,dy\,dz$ and $\eta = 2y\,dx\,dz$.

SOLUTION. By definition,

$$\omega + \eta = (x^2 + 2y)\,dx\,dz + xy\,dy\,dz = xy\,dy\,dz - (x^2 + 2y)\,dz\,dx,$$

$$\omega - \eta = (x^2 - 2y)\,dx\,dz + xy\,dy\,dz = xy\,dy\,dz + (2y - x^2)\,dz\,dx,$$

and

$$\begin{aligned}
\omega\eta &= (x^2\,dx\,dz + xy\,dy\,dz)(2y\,dx\,dz) \\
&= 2x^2 y\,dx\,dz\,dx\,dz + 2xy^2\,dy\,dz\,dx\,dz \\
&= -2x^2 y\,dx\,dx\,dz\,dz - 2xy^2\,dy\,dx\,dz\,dz = 0. \quad \blacksquare
\end{aligned}$$

Using products of 1–forms, we see that an $r$–form is an expression of the form

$$\sum f_{i_1,\ldots,i_r}\,dx_{i_1} \ldots dx_{i_r},$$

where the sum is taken over all integers $i_j \in \{1, \ldots, n\}$; i.e., it is no longer necessary that the $i_j$'s increase in $j$. Because of the connection between 2–forms and oriented surface integrals, we will frequently use

$$\omega = P\,dy\,dz + Q\,dz\,dx + R\,dx\,dy$$

to represent a generic 2–form on $\mathbf{R}^3$ rather than (2).

Here is a summary of the algebraic laws satisfied by addition and multiplication of differential forms.

**15.3 THEOREM.** Let $V$ be open in $\mathbf{R}^n$, $f$ be a 0–form on $V$, $\omega$ be an $r$–form on $V$, $\eta$ be an $s$–form on $V$, and $\theta$ be a $t$–form on $V$.

   i) If $r = s$, then $\omega + \eta$ is an $r$–form, $\omega + \eta = \eta + \omega$, and $(\omega + \eta)\theta = \omega\theta + \eta\theta$. If $r = s = t$, then $(\omega + \eta) + \theta = \omega + (\eta + \theta)$.
   ii) For any $r$ and $s$, $\omega\eta = (-1)^{rs}\eta\omega$.
   iii) For any $r, s$, and $t$, $(\omega\eta)\theta = \omega(\eta\theta)$ and $f(\omega\eta) = (f\omega)\eta = \omega(f\eta)$.

PROOF. Properties i) and iii) hold, by definition.

To prove ii), we may suppose that $\omega$ and $\eta$ are decomposable; i.e., suppose that $\omega = f\,dx_{i_1}\ldots dx_{i_r}$ and $\eta = g\,dx_{j_1}\ldots dx_{j_s}$. By definition, the product of $\omega$ and $\eta$ is the $(r + s)$–form

$$\omega\eta = fg\,dx_{i_1}\ldots dx_{i_r}\,dx_{j_1}\ldots dx_{j_s}.$$

Successive applications of the Anticommutative Property yield

$$\begin{aligned}
\omega\eta &= fg\,dx_{i_1}\ldots dx_{i_r}\,dx_{j_1}\ldots dx_{j_s}\\
&= (-1)^r fg\,dx_{j_1}\,dx_{i_1}\,dx_{i_2}\ldots dx_{i_r}\,dx_{j_2}\ldots dx_{j_s}\\
&= \cdots = (-1)^{rs}gf\,dx_{j_1}\ldots dx_{j_s}\,dx_{i_1}\ldots dx_{i_r} = (-1)^{rs}\eta\omega.
\end{aligned}$$

This completes the proof of part ii). ∎

In Section 11.4 we introduced the total differential of a function $z = f(x, y)$ of two variables as $dz = f_x\,dx + f_y\,dy$. This gives us two definitions for $dz\,dx$ and $dy\,dz$, one using Jacobians and one "multiplying" the total differential $dz$ by the 1–forms $dx$ and $dy$. These two definitions are compatible. Indeed, using the trivial parametrization of the surface $z = f(x, y)$, the Jacobian definition yields

(3)                        $dy\,dz = -f_x\,d(x, y)$    and    $dz\,dx = -f_y\,d(x, y)$.

On the other hand, multiplying the 1–form $dz = f_x\,dx + f_y\,dy$ on the left by $dy$ we have, by the Nilpotent Property and the Anticommutative Property, that

$$dy\,dz = dy(f_x\,dx + f_y\,dy) = f_x\,dy\,dx + f_y\,dy\,dy = -f_x\,dx\,dy.$$

A similar computation leads to $dz\,dx = -f_y\,dx\,dy$. Thus if we identify $d(x, y)$ with $dx\,dy$, (3) holds no matter which definition we use. (Identification of $d(x, y)$ with $dx\,dy$ is justified by using the "identity chart"—see Remark 15.41.)

The following result contains an important computation which relates the $n$–fold product of $n$–forms on $\mathbf{R}^n$ to the determinant operator.

**15.4 THEOREM.** Let $V$ be open in $\mathbf{R}^n$ and $\omega_1, \omega_2, \ldots, \omega_n$ be 1–forms on $V$. If $A = [a_{ij}]_{n\times n}$ is a real matrix, then

$$\left(\sum_{j=1}^n a_{1j}\omega_j\right)\cdots\left(\sum_{j=1}^n a_{nj}\omega_j\right) = (\det A)\omega_1\ldots\omega_n.$$

PROOF. The proof is by induction on $n$. If $n = 1$ there is nothing to prove. Suppose the theorem holds for some integer $n \geq 1$. By Theorem 15.3 and the Nilpotent Property, we have

$$\left(\sum_{j=1}^{n} a_{1j}\omega_j\right) \cdots \left(\sum_{j=1}^{n} a_{nj}\omega_j\right)$$

$$= \left(a_{11}\omega_1 + \sum_{j=2}^{n} a_{1j}\omega_j\right) \cdots \left(a_{n1}\omega_1 + \sum_{j=2}^{n} a_{nj}\omega_j\right)$$

$$= a_{11}\omega_1 \left(\sum_{j=2}^{n} a_{2j}\omega_j\right) \cdots \left(\sum_{j=2}^{n} a_{nj}\omega_j\right)$$

$$+ \left(\sum_{j=2}^{n} a_{1j}\omega_j\right) a_{21}\omega_1 \left(\sum_{j=2}^{n} a_{3j}\omega_j\right) \cdots \left(\sum_{j=2}^{n} a_{nj}\omega_j\right)$$

$$+ \cdots + \left(\sum_{j=2}^{n} a_{1j}\omega_j\right) \cdots \left(\sum_{j=2}^{n} a_{(n-1)j}\omega_j\right) a_{n1}\omega_1$$

$$+ \left(\sum_{j=2}^{n} a_{1j}\omega_j\right) \cdots \left(\sum_{j=2}^{n} a_{nj}\omega_j\right).$$

Continue this string of identities using the Anticommutative Property, the inductive hypothesis, the definition of $\det A$ in terms of cofactors of $A$, and the Nilpotent Property. We obtain

$$\left(\sum_{j=1}^{n} a_{1j}\omega_j\right) \cdots \left(\sum_{j=1}^{n} a_{nj}\omega_j\right)$$

$$= a_{11} \det A_{11}(\omega_1\omega_2 \ldots \omega_n) + (-1)^1 a_{21} \det A_{21}(\omega_1\omega_2 \ldots \omega_n)$$

$$+ \cdots + (-1)^{n-1} a_{n1} \det A_{n1}(\omega_1\omega_2 \ldots \omega_n)$$

$$+ \left(\sum_{j=2}^{n} a_{1j}\omega_j\right) \det A_{11}(\omega_2 \ldots \omega_n)$$

$$= (\det A)\omega_1 \ldots \omega_n + 0 = (\det A)\omega_1 \ldots \omega_n. \blacksquare$$

The derivative of a differential form is defined as follows. (This derivative can be used to unify the three operators *grad*, *curl*, and *div*—see Exercise 4 below.)

**15.5 DEFINITION.** Let $V$ be open in $\mathbf{R}^n$ and $\omega = \sum f_{i_1,\ldots,i_r} dx_{i_1} \ldots dx_{i_r}$ be a $C^1$ $r$–form on $V$.

   i) If $\omega = f$ is a 0–form, then the *exterior derivative* of $\omega$ is the 1–form

$$d\omega := \sum_{j=1}^{n} \frac{\partial f}{\partial x_j} dx_j.$$

ii) If $\omega = f\, dx_{i_1} \ldots dx_{i_r}$ is a decomposable $r$–form, $r > 0$, then the *exterior derivative* of $\omega$ is the $r + 1$ form

$$d\omega := df\, dx_{i_1} \ldots dx_{i_r}.$$

iii) If $\omega$ is a differential form of degree $r > 0$, i.e., $\omega = \sum_{j=1}^{N} \omega_j$, where each $\omega_j$ is a decomposable $r$–form, then the *exterior derivative* of $\omega$ is the $r + 1$ form

$$d\omega := \sum_{j=1}^{N} d\omega_j.$$

iv) If $\omega$ is $\mathcal{C}^2$ $r$–form on $V$, then the *second exterior derivative* of $\omega$ is $d^2\omega := d(d\omega)$.

**15.6 Example.** Find $d\omega$ and $d^2\omega$ if $\omega(x, y, z, t) = xy\, dx\, dy + (x + z + t)\, dz\, dt$.

SOLUTION. By definition,

$$d\omega = (y\, dx + x\, dy)\, dx\, dy + (dx + dz + dt)\, dz\, dt = dx\, dz\, dt;$$

hence, $d^2\omega = (d1)\, dx\, dy\, dz = 0.$ ∎

It is clear that for 0–forms, the exterior derivative satisfies the following rules.

**15.7 Remark.** *Let $\omega$ and $\eta$ be $\mathcal{C}^1$ 0–forms on some open set $V \subset \mathbf{R}^n$, and $\alpha$ be a scalar. Then $d(\alpha\omega)$, $d(\omega + \eta)$, and $d(\omega\eta)$ are continuous 1–forms on $V$ with*

$$d(\alpha\omega) = \alpha d\omega,$$

$$d(\omega + \eta) = d\omega + d\eta,$$

*and*

$$d(\omega\eta) = \eta d\omega + \omega d\eta.$$

Analogues of these rules hold for arbitrary $r$–forms.

**15.8 THEOREM.** *Let $V$ be open in $\mathbf{R}^n$ and $\alpha$ be a scalar. If $\omega$ is a $\mathcal{C}^1$ $r$–form on $V$ and $\eta$ is a $\mathcal{C}^1$ $s$–form on $V$, then $d(\alpha\omega)$ and $d(\omega + \eta)$ (when $r = s$) are continuous $(r + 1)$–forms on $V$, and $d(\omega\eta)$ is a continuous $(r + s + 1)$–form on $V$. Moreover,*

$$d(\alpha\omega) = \alpha d\omega,$$

$$d(\omega + \eta) = d\omega + d\eta$$

*(when $r = s$), and*

(4)                                              $$d(\omega\eta) = d\omega\, \eta + (-1)^r \omega\, d\eta.$$

PROOF. By Definition 15.5iii), we may suppose $\omega$ and $\eta$ are decomposable, i.e.,

$$\omega = f\,dx_{i_1}\ldots dx_{i_r} \quad \text{and} \quad \eta = g\,dx_{j_1}\ldots dx_{j_s}.$$

By Definition 15.5ii) and Remark 15.7,

$$d(\alpha\omega) = d(\alpha f)dx_{i_1}\ldots dx_{i_r} = \alpha df\,dx_{i_1}\ldots dx_{i_r} = \alpha d\omega.$$

Similarly, if $r = s$ and $i_\nu = j_\nu$, $\nu = 1,\ldots,r$, then

$$d(\omega + \eta) = d(f + g)\,dx_{i_1}\ldots dx_{i_r} = df\,dx_{i_1}\ldots dx_{i_r} + dg\,dx_{i_1}\ldots dx_{i_r} = d\omega + d\eta.$$

To prove (4), consider first the case $r = 0$, i.e., $\omega = f$ is a 0–form. By Remark 15.7 and Theorem 15.3ii),

$$d(\omega\eta) = d(fg)dx_{j_1}\ldots dx_{j_s} = (g\,df + f\,dg)dx_{j_1}\ldots dx_{j_s}$$
$$= df\,g\,dx_{j_1}\ldots dx_{j_s} + f\,dg\,dx_{j_1}\ldots dx_{j_s} = d\omega\,\eta + \omega\,d\eta.$$

Next, suppose $r > 0$. If $i_\nu = j_\mu$ for some indices $\nu$ and $\mu$, then the Nilpotent Property implies

$$\omega\eta = 0 = d\omega\,\eta = \omega\,d\eta.$$

On the other hand, if all the indices are distinct, then since $g$ is a 0–form and $dg$ is a 1–form, we have by Theorem 15.3ii) that

$$d(\omega\eta) = d(fg\,dx_{i_1}\ldots dx_{i_r}dx_{j_1}\ldots dx_{j_s})$$
$$= (g\,df + f\,dg)dx_{i_1}\ldots dx_{i_r}dx_{j_1}\ldots dx_{j_s}$$
$$= df\,dx_{i_1}\ldots dx_{i_r}\,g\,dx_{j_1}\ldots dx_{j_s}$$
$$\quad + (-1)^{r\cdot1}f\,dx_{i_1}\ldots dx_{i_r}\,dg\,dx_{j_1}\ldots dx_{j_s}$$
$$= d\omega\,\eta + (-1)^r\omega\,d\eta. \quad\blacksquare$$

Equation (4) is called the *Product Rule*.

The following result shows that the second exterior derivative of a $\mathcal{C}^2$ $r$–form is always zero. (By Exercise 4 below, this result generalizes Exercise 9b in Section 13.5.)

**15.9 THEOREM.** *If $\omega$ is a $\mathcal{C}^2$ $r$–form on an open set $V \subseteq \mathbf{R}^n$, then $d^2\omega = 0$.*

PROOF. We may suppose that $\omega$ is decomposable; i.e., $\omega = f\,dx_{i_1}\ldots dx_{i_r}$. The proof is by induction on $r$. Suppose $r = 0$, i.e., $\omega = f$. By the Nilpotent Property, the Anticommutative Property, and the fact that the first-order partial derivatives of $f$ commute, we have

$$d^2\omega = d\left(\sum_{j=1}^n \frac{\partial f}{\partial x_j}\,dx_j\right) = \sum_{j=1}^n\sum_{k=1}^n \frac{\partial^2 f}{\partial x_k\partial x_j}\,dx_k\,dx_j$$

$$= \sum_{j<k}\left(\frac{\partial^2 f}{\partial x_k\partial x_j} - \frac{\partial^2 f}{\partial x_j\partial x_k}\right)dx_k\,dx_j = 0.$$

Suppose $r = 1$, i.e., $\omega = f\, dx_k$. Since all first-order partial derivatives of the function 1 are zero, we have by definition that $d^2 x_k = d(1\, dx_k) = 0$. Thus, by the Product Rule and the case $r = 0$,

$$d^2\omega = d(df\,dx_k) = d^2 f\, dx_k - df\, d^2 x_k = 0.$$

Finally, suppose there is an $r > 1$ such that the theorem holds for all $s$–forms, $0 \le s < r$. By definition,

$$d\omega = d(f\,dx_{i_1}\dots dx_{i_{r-1}})\, dx_{i_r}.$$

Hence, by the Product Rule and the inductive hypothesis (for $s = 1$ and $s = r - 1$), we have

$$d^2\omega = d^2(f\,dx_{i_1}\dots dx_{i_{r-1}})\, dx_{i_r} + (-1)^r d(f\,dx_{i_1}\dots dx_{i_{r-1}})\, d^2 x_{i_r} = 0. \quad \blacksquare$$

The following definition shows how to use a continuously differentiable function $\phi : \mathbf{R}^n \to \mathbf{R}^m$ to transform differentials from $\mathbf{R}^m$ to $\mathbf{R}^n$. (This concept will be used later to define integration of $r$–forms over manifolds.)

**15.10 DEFINITION.** Let $U$ be open in $\mathbf{R}^n$, $V$ be open in $\mathbf{R}^m$, $\phi : U \to V$ be $\mathcal{C}^1$ on $U$, and

$$\omega = \sum f_{i_1,\dots,i_r}\, dx_{i_1}\dots dx_{i_r}$$

be an $r$–form on $V$. Then the *differential transform* (induced by $\phi$) of $\omega$ is the $r$–form on $U$ defined by

$$\phi^*(\omega) = \sum \phi^*(f_{i_1,\dots,i_r})\phi^*(dx_{i_1})\dots\phi^*(dx_{i_r}),$$

where $\phi^*(f) = f \circ \phi$ for every 0–form $f$ and

$$\phi^*(dx_i) = d\phi_i = \sum_{j=1}^{n} \frac{\partial \phi_i}{\partial u_j} du_j$$

for every $i = 1, 2, \dots, m$.

For the next several remarks, let $U$ be open in $\mathbf{R}^n$, $V$ be open in $\mathbf{R}^m$, and $\phi : U \to V$.

**15.11 Remark.** *If* $\omega$ *is a* $\mathcal{C}^1$ $r$–*form on* $V$ *and* $\phi$ *is* $\mathcal{C}^2$ *on* $U$, *then* $\phi^*(\omega)$ *is a* $\mathcal{C}^1$ $r$–*form on* $U$.

PROOF. By definition, $(\phi^* \circ f)(\mathbf{u}) = f(\phi(\mathbf{u}))$ is a $\mathcal{C}^1$ 0–form on $U$ for every $\mathcal{C}^1$ 0–form $f$ on $V$ and

$$\phi^*(dx_i) = \sum_{j=1}^{n} \frac{\partial \phi_i}{\partial u_j} du_j$$

is a $\mathcal{C}^1$ 1–form on $U$ for $i = 1, 2, \ldots, m$. Hence, it is clear that $\phi^*(\omega)$ is a $\mathcal{C}^1$ $r$–form on $U$ when $\omega$ is a $\mathcal{C}^1$ $r$–form on $V$. ∎

**15.12 Remark.** *The differential transform $\phi^*$ is linear; i.e., if $\omega$ and $\eta$ are $r$–forms on $V$ and $\phi$ is $\mathcal{C}^1$ on $U$, then*

$$\phi^*(\omega + \eta) = \phi^*(\omega) + \phi^*(\eta).$$

PROOF. We may suppose that $\omega = f\,dx_{i_1} \ldots dx_{i_r}$ and $\eta = g\,dx_{i_1} \ldots dx_{i_r}$. Thus,

$$\phi^*(\omega + \eta) = \phi^*((f + g)\,dx_{i_1} \ldots dx_{i_r})$$
$$= (f \circ \phi + g \circ \phi)\phi^*(dx_{i_1}) \ldots \phi^*(dx_{i_r}) = \phi^*(\omega) + \phi^*(\eta).\ \blacksquare$$

**15.13 Remark.** *The differential transform $\phi^*$ is multiplicative, i.e., if $\omega$ is an $r$–form on $V$ and $\eta$ is an $s$–form on $V$ and $\phi$ is $\mathcal{C}^1$ on $U$, then*

$$\phi^*(\omega\eta) = \phi^*(\omega)\phi^*(\eta).$$

PROOF. We may suppose that $\omega = f\,dx_{i_1} \ldots dx_{i_r}$ and $\eta = g\,dx_{j_1} \ldots dx_{j_s}$. Thus,

$$\phi^*(\omega\eta) = \phi^*((fg)\,dx_{i_1} \ldots dx_{i_r} dx_{j_1} \ldots dx_{j_s})$$
$$= (f \circ \phi)(g \circ \phi)\phi^*(dx_{i_1}) \ldots \phi^*(dx_{i_r})\phi^*(dx_{j_1}) \ldots \phi^*(dx_{j_s})$$
$$= \phi^*(\omega)\phi^*(\eta).\ \blacksquare$$

**15.14 Remark.** *The differential transform $\phi^*$ and the exterior derivative $d$ commute, i.e., if $\omega$ is a $\mathcal{C}^1$ $r$–form on $V$ and $\phi$ is $\mathcal{C}^2$ on $U$, then*

$$(5) \qquad\qquad \phi^*(d\omega) = d(\phi^*(\omega)).$$

PROOF. We may suppose $\omega$ is decomposable. The proof is by induction on $r$. Suppose $r = 0$, i.e., $\omega = f$. Then, by definition and the Chain Rule,

$$\phi^*(d\omega) = \phi^*\left(\sum_{k=1}^{m} \frac{\partial f}{\partial x_k} dx_k\right) = \sum_{k=1}^{m} \phi^*\left(\frac{\partial f}{\partial x_k}\right) \phi^*(dx_k)$$
$$= \sum_{k=1}^{m} \left(\frac{\partial f}{\partial x_k} \circ \phi\right) \sum_{j=1}^{n} \frac{\partial \phi_k}{\partial u_j} du_j$$
$$= \sum_{j=1}^{n} \frac{\partial(f \circ \phi)}{\partial u_j} du_j = d(f \circ \phi) = d(\phi^*(\omega)).$$

Suppose $r = 1$, i.e., $\omega = f\,dx_k$. Then, by definition, the multiplicative property of $\phi^*$, and the case $r = 0$, we have

$$\phi^*(d\omega) = \phi^*(df\,dx_k) = \phi^*(df)\phi^*(dx_k) = d(f \circ \phi)\,d\phi_k.$$

On the other hand, since $\phi^*(\omega) = (f \circ \phi)\phi^*(dx_k) = (f \circ \phi) \, d\phi_k$, it follows from the Product Rule, the Nilpotent Property, and Theorem 15.9, that

$$d(\phi^*(\omega)) = d(f \circ \phi) \, d\phi_k + (f \circ \phi) \, d^2\phi_k = d(f \circ \phi) \, d\phi_k.$$

Thus (5) holds when $\omega$ is a 1–form.

Finally, suppose there is an $r > 1$ such that (5) holds for all $s$–forms, $0 \le s < r$. Let $\omega$ be a decomposable $r$–form and write $\omega = \theta\eta$, where $\theta$ is a 1–form and $\eta$ is an $(r-1)$–form. By the Product Rule,

$$d\omega = (d\theta)\eta - \theta d\eta.$$

Hence, it follows from the inductive hypothesis, the Product Rule, and the multiplicative property of $\phi^*$ that

$$\begin{aligned}
\phi^*(d\omega) &= \phi^*(d\theta)\phi^*(\eta) - \phi^*(\theta)\phi^*(d\eta) \\
&= d(\phi^*\theta)\phi^*(\eta) - \phi^*(\theta)d(\phi^*\eta) \\
&= d((\phi^*\theta)(\phi^*\eta)) = d(\phi^*(\theta\eta)) = d(\phi^*(\omega)). \blacksquare
\end{aligned}$$

The following result shows that differential transforms can be used to define the oriented line and surface integrals introduced in Sections 13.2 and 13.4 (see Exercise 5 below).

**15.15 THEOREM** [FUNDAMENTAL THEOREM OF DIFFERENTIAL TRANSFORMS]. *Let $m \ge n$, $U$ be open in* **R**$^n$*, $V$ be open in* **R**$^m$*, and $\phi : U \to V$ be $\mathcal{C}^1$ on $U$. If*

$$\omega = \sum f_{i_1,\ldots,i_n} dx_{i_1} \ldots dx_{i_n}$$

*is an $n$–form on $V$, then*

$$\phi^*(\omega) = \sum (f_{i_1,\ldots,i_n} \circ \phi) \frac{\partial(\phi_{i_1},\ldots,\phi_{i_n})}{\partial(u_1,\ldots,u_n)} du_1 \ldots du_n.$$

PROOF. We may suppose that $\omega$ is decomposable. If $n = 1$, i.e., $\omega = f \, dx_j$, then by definition,

$$\phi^*(\omega) = \phi^*(f) \, \phi^*(dx_j) = (f \circ \phi)\phi' du.$$

If $n > 1$, i.e., $\omega = f \, dx_{i_1} \ldots dx_{i_n}$, then by Definition 15.10 and Theorem 15.4,

$$\begin{aligned}
\phi^*(\omega) &= \phi^*(f) \, \phi^*(dx_{i_1}) \ldots \phi^*(dx_{i_n}) \\
&= (f \circ \phi) \left( \sum_{k=1}^n \frac{\partial\phi_{i_1}}{\partial u_k} \, du_k \right) \ldots \left( \sum_{k=1}^n \frac{\partial\phi_{i_n}}{\partial u_k} \, du_k \right) \\
&= (f \circ \phi) \frac{\partial(\phi_{i_1},\ldots,\phi_{i_n})}{\partial(u_1,\ldots,u_n)} du_1 \ldots du_n. \blacksquare
\end{aligned}$$

## EXERCISES

**1.** Algebraically simplify the following differential forms.

     a) $3(dx + dy) \, dz + 2(dx + dz) \, dy$.

     b) $(x \, dy - y \, dx)(x \, dz - z \, dy)$.

     c) $(x^2 \, dx \, dy - \cos x \, dy \, dz)(y^2 \, dy + \cos x \, dw) - (x^3 \, dy \, dz - \sin x \, dy \, dw)(y^3 \, dy + \sin x \, dz)$.

**2.** Compute the exterior derivatives of the following differential forms.

     a) $x^2 \, dy - y^2 \, dx$.

     b) $\sin(xy) \, dz \, dw + \cos(zw) \, dx \, dy$.

     c) $\sqrt{x^2 + y^2} \, dy \, dz - \sqrt{x^2 + y^2} \, dx \, dz$.

     d) $(e^{xy} \, dz + e^{yz} \, dx)(\sin x \, dy + \cos y \, dx)$.

**3.** a) Prove that if $\omega$ is an $r$–form, $r$ odd, then $\omega^2 = 0$.

     b) Prove that if $\omega_j$ are decomposable $r$–forms, $r$ even, and $\omega = \sum_{j=1}^{N} \omega_j$, then

$$\omega^2 = 2 \sum_{\substack{k,j=1 \\ j<k}}^{N} \omega_j \omega_k.$$

**4.** **This exercise is used in Section 15.3.** If $f, g$ are 0–forms and $\omega, \eta$ are $r$–forms, define

$$(f, g) \cdot (\omega, \eta) = f\omega + g\eta.$$

     a) Prove that if $f : \mathbf{R}^3 \to \mathbf{R}$ is $\mathcal{C}^1$ and grad $f := (f_x, f_y, f_z)$, then the exterior derivative of the 0–form $\omega = f$ can be written in the form

$$d\omega = (\text{ grad } f) \cdot (dx, dy, dz).$$

     b) Prove that if $F = (P, Q, R) : \mathbf{R}^3 \to \mathbf{R}^3$ is $\mathcal{C}^1$, then the exterior derivative of the 1–form $\omega = P \, dx + Q \, dy + R \, dz$ can be written in the form

$$d\omega = (\text{curl } F) \cdot (dy \, dz, dz \, dx, dx \, dy).$$

     and the exterior derivative of the 2–form $\eta = P \, dy \, dz + Q \, dz \, dx + R \, dx \, dy$ can be written in the form

$$d\eta = (\text{div } F) \, dx \, dy \, dz.$$

**5.** **This exercise is used in Section 15.3.** Let $I$ be an interval and $E$ be a Jordan region. Define the integral of a continuous 1–form $\omega = f \, dt$ on $I$ and a continuous 2–form $\eta = g \, du \, dv$ on $E$ by

$$\int_I \omega = \int_I f(t) \, dt \quad \text{and} \quad \iint_E \eta = \iint_E g(u, v) \, d(u, v).$$

a) Let $C = (\phi, I)$ be a smooth simple $\mathcal{C}^1$ curve in $\mathbf{R}^2$, $F = (P, Q) : \phi(I) \to \mathbf{R}^2$ be continuous, and $\omega = P\,dx + Q\,dy$. Prove that

$$\int_C F \cdot T\,ds = \int_I \phi^*(\omega).$$

b) Let $S = (\phi, E)$ be a smooth simple orientable $\mathcal{C}^1$ surface in $\mathbf{R}^3$, $F = (P, Q, R) : \phi(E) \to \mathbf{R}^3$ be continuous, and

$$\eta = P\,dy\,dz + Q\,dz\,dx + R\,dx\,dy.$$

Prove that

$$\iint_S F \cdot \boldsymbol{n}\,d\sigma = \iint_E \phi^*(\eta).$$

## *e*15.2    DIFFERENTIABLE MANIFOLDS.    *This section uses no material from any other enrichment section.*

In Chapter 13 we introduced one-dimensional objects (curves), two-dimensional objects (surfaces), and corresponding oriented integrals. We shall extend these ideas to higher dimensions.

A problem which surfaced several times in Chapter 13 is that one parametrization by itself does not fully describe a surface. For example, the boundary of a surface could not be defined using one parametrization alone. To avoid this problem, we adopt a different point of view here. Instead of thinking of a surface as a particular parametrization $\phi : E \to S$, we will think of a surface as a set of points $S$ together with a class of functions $h_\alpha : S \to E$ related to each other in a natural way. (Each $h_\alpha^{-1}$ can be thought of as a parametrization of a piece of $S$.) This point of view has been used by map makers for centuries. The earth (a particular surface) can be described by an atlas, which is itself a collection of two-dimensional maps (or charts) which represent overlapping portions of its surface. If we know how the individual charts fit together (see (7) below), we can study the whole surface by using this atlas.

**15.16 DEFINITION.** Let $M$ be a set.

i) An *n-dimensional chart* of $M$ at a point $x \in M$ is a pair $(V, h)$, where $x \in V$, $V \subseteq M$, $h : V \to \mathbf{R}^n$ is 1–1, and $h(V)$ is open in $\mathbf{R}^n$. (We shall drop the adjective "$n$-dimensional" when no confusion arises.)

ii) An *n-dimensional $\mathcal{C}^p$ atlas* of $M$ is a collection

(6)
$$\mathcal{A} = \{(V_\alpha, h_\alpha) : \alpha \in A\}$$

of $n$-dimensional charts of $M$ such that $h_\beta(V_\alpha \cap V_\beta)$ is open in $\mathbf{R}^n$,

(7)
$$h_\alpha \circ h_\beta^{-1} \text{ is } \mathcal{C}^p \text{ on } h_\beta(V_\alpha \cap V_\beta) \text{ for all } \alpha, \beta \in A,$$

and
$$M = \bigcup_{\alpha \in A} V_\alpha.$$

The functions $h_\alpha \circ h_\beta^{-1}$ are called the *transition maps* of the atlas $\mathcal{A}$.

Notice, then, that if $(V, h)$ is a chart of $M$, then $(h^{-1}, h(V))$ is a parametrization of a portion $V$ of $M$.

**15.17 Example.** If $C = \phi(I)$, where $(\phi, I)$ is a simple curve and $I$ is an open interval, prove $\{(C, \phi^{-1})\}$ is a one-dimensional $\mathcal{C}^\infty$ atlas of $C$.

PROOF. Since $(\phi, I)$ is simple, $\phi^{-1}$ exists on $\phi(I)$. $(C, \phi^{-1})$ is evidently a one-dimensional chart of $C$, so $\{(C, \phi^{-1})\}$ is an atlas of $C$. Since the transition map $(\phi^{-1} \circ \phi)(\boldsymbol{x}) = \boldsymbol{x}$ is the identity function on $C$, this atlas is $\mathcal{C}^\infty$. ∎

A similar argument establishes the following two remarks.

**15.18 Remark.** *If $V$ is open in $\mathbf{R}^2$, $\phi : V \to \mathbf{R}^m$ is 1–1 on $V$, and $S = \phi(V)$, then $\{(S, \phi^{-1})\}$ is a two-dimensional $\mathcal{C}^\infty$ atlas of $S$.*

**15.19 Remark.** *If $V$ is open in $\mathbf{R}^n$ and $I(\boldsymbol{x}) = \boldsymbol{x}$ is the identity function on $\mathbf{R}^n$, then $\{(V, I)\}$ is an $n$-dimensional $\mathcal{C}^\infty$ atlas of $V$.* (We shall call $(V, I)$ the *identity chart*.)

Not all atlases consist of one chart.

**15.20 Example.** For each $t \in \mathbf{R}$, set $\phi(t) = (\cos t, \sin t)$, $\psi(t) = (\cos(t + \pi), \sin(t + \pi))$, $V = \phi(I)$, and $U = \psi(I)$, where $I = (0, 2\pi)$. If $h = \phi^{-1}$ on $V$ and $g = \psi^{-1}$ on $U$, prove
$$\mathcal{A} = \{(V, h), (U, g)\}$$
is a one-dimensional $\mathcal{C}^\infty$ atlas of the unit circle $x^2 + y^2 = 1$.

PROOF. Let $M$ represent the set of points $(x, y)$ such that $x^2 + y^2 = 1$. Since $\phi$ (respectively $\psi$) is 1–1 from $I$ onto $V$ (respectively, $I$ onto $U$) and $V \cup U = M$, $(V, h)$ and $(U, g)$ are charts which cover $M$. It is easy to see that the transition maps are $\mathcal{C}^\infty$. For example, $g(V \cap U) = (0, \pi) \cup (\pi, 2\pi)$ and, on the interval $(0, \pi)$, $(h \circ g^{-1})(t) = t + \pi$. Thus $\mathcal{A}$ is a $\mathcal{C}^\infty$ atlas of $M$. ∎

The following concept is a replacement for smooth equivalence of parametrizations.

**15.21 DEFINITION.** Two $n$-dimensional atlases $\mathcal{A}$, $\mathcal{B}$ of $M$ are said to be $\mathcal{C}^p$ *compatible* (notation: $\mathcal{A} \sim \mathcal{B}$) if $\mathcal{A} \cup \mathcal{B}$ is an $n$-dimensional $\mathcal{C}^p$ atlas on $M$.

Notice that $\mathcal{C}^p$ compatibility is an equivalence relation (see Exercise 2 below), i.e., any atlas $\mathcal{A}$ is $\mathcal{C}^p$ compatible with itself; if $\mathcal{A}$ is $\mathcal{C}^p$ compatible with $\mathcal{B}$, then $\mathcal{B}$ is $\mathcal{C}^p$ compatible with $\mathcal{A}$; and if $\mathcal{A}$ is $\mathcal{C}^p$ compatible with $\mathcal{B}$, and $\mathcal{B}$ is $\mathcal{C}^p$ compatible with $\mathcal{D}$, then $\mathcal{A}$ is $\mathcal{C}^p$ compatible with $\mathcal{D}$. (For some elementary remarks about equivalence relations and equivalence classes, see Appendix F.)

**15.22 DEFINITION.** An *n-dimensional $\mathcal{C}^p$ manifold* is a set $M$ together with an equivalence class $\overline{\mathcal{A}}$ of $n$-dimensional $\mathcal{C}^p$ atlases on $M$. By an *atlas* of $M$ we mean an atlas in $\overline{\mathcal{A}}$. By a *chart* of $M$ we mean a chart in some atlas of $M$.

Atlases of an $n$-dimensional manifold $M$ can be used to "pull-back" concepts from $\mathbf{R}^n$ to $M$.

**15.23 DEFINITION.** Let $M$ be an $n$-dimensional $\mathcal{C}^p$ manifold and $\mathcal{A}$ be an atlas of $M$. A set $W \subseteq M$ is said to be *open* if $h(V \cap W)$ is open in $\mathbf{R}^n$ for all charts $(V, h) \in \mathcal{A}$.

The following result shows that this definition does not depend on the atlas chosen from the manifold structure of $M$.

**15.24 Remark.** *Let $\mathcal{A}$ and $\mathcal{B}$ be $\mathcal{C}^p$ compatible atlases of $M$ and suppose $W \subseteq M$. Then $h(V \cap W)$ is open in $\mathbf{R}^n$ for all $(V, h) \in \mathcal{A}$ if and only if $g(U \cap W)$ is open in $\mathbf{R}^n$ for all $(U, g) \in \mathcal{B}$ .*

PROOF. Let $W \subseteq M$ such that $h(V \cap W)$ is open in $\mathbf{R}^n$ for all $(V, h) \in \mathcal{A}$ and suppose $(U, g) \in \mathcal{B}$. If $W \cap U = \emptyset$, then $g(W \cap U) = \emptyset$ is open in $\mathbf{R}^n$ by definition. If $W \cap U \neq \emptyset$, choose $(V, h) \in \mathcal{A}$ such that $W \cap V \cap U \neq \emptyset$. Since $h(W \cap V)$ and $g(U)$ are open in $\mathbf{R}^n$ and the transition map $h \circ g^{-1}$ is $\mathcal{C}^p$, hence continuous, it follows that

$$g(W \cap V \cap U) = (g \circ h^{-1})(h(W \cap V)) \cap g(U) = (h \circ g^{-1})^{-1}(h(W \cap V)) \cap g(U)$$

is open in $\mathbf{R}^n$. Since

$$g(W \cap U) = \bigcup_{(V,h) \in \mathcal{A}} g(W \cap V \cap U),$$

we conclude that $g(W \cap U)$ is open in $\mathbf{R}^n$. Reversing the roles of $\mathcal{A}$ and $\mathcal{B}$ proves the converse. ∎

Using open sets, we can define what we mean by continuity of a function on a manifold (compare with Theorem 9.33 or 10.58.)

**15.25 DEFINITION.** Let $M$ be an $n$-dimensional $\mathcal{C}^p$ manifold and $\mathcal{A}$ be an atlas of $M$.

  i) A function $f : M \to \mathbf{R}^k$ is said to be *continuous* on $M$ if $f^{-1}(U)$ is open in $M$ for every open set $U \subset \mathbf{R}^k$.

  ii) A function $f : \mathbf{R}^k \to M$ is said to be *continuous* on a set $E \subset \mathbf{R}^k$ if $f^{-1}(W) \cap E$ is relatively open in $E$ for every open set $W$ in $M$.

**15.26 Remark.** *If $(V, h)$ is a chart from an atlas $\mathcal{A}$ of $M$, then $h$ is a homeomorphism; i.e., $h$ is continuous on $V$ and $h^{-1}$ is continuous on $h(V)$.*

PROOF. If $W \subseteq V$ is open in $M$, then $(h^{-1})^{-1}(W) = h(W)$ is open in $\mathbf{R}^n$. Hence, $h^{-1}$ is continuous on $h(V)$ by Definition 15.25. On the other hand, suppose $\Omega \subset h(V)$ is open in $\mathbf{R}^n$ and $W = h^{-1}(\Omega)$. Let $(U, g)$ be any chart in $\mathcal{A}$. Then

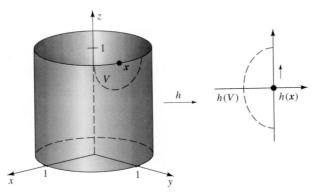

**Figure 15.1**

$g(U \cap W) = g(U) \cap g \circ h^{-1}(\Omega)$ is open in $\mathbf{R}^n$; i.e., $W$ is open in $M$ by Definition 15.23. Hence, $h$ is continuous on $V$. ∎

To define the boundary of a manifold, we introduce the following terminology. By a *half-space* of $\mathbf{R}^n$ we mean a set of the form

$$\{(x_1, \ldots, x_n) : x_j \geq \alpha\} \quad \text{or} \quad \{(x_1, \ldots, x_n) : x_j \leq \alpha\},$$

where $\alpha \in \mathbf{R}$ and $j \in \{1, 2, \ldots, n\}$. We shall refer to the special case

$$\mathcal{H}_1 := \{(x_1, \ldots, x_n) : x_1 \leq 0\}$$

as *left half-space*. If $n = 2$, we shall refer to half-spaces as *half-planes*.

A simple curve parametrized on an open interval is a one-dimensional manifold (see Example 15.17 above). What about surfaces? A smooth surface with empty boundary is a two-dimensional manifold, but the restriction in Definition 15.16i) that $h(V)$ be open prevents any surface whose boundary is nonempty from being a manifold. For example, the cylinder $1 = x^2 + y^2$, $0 \leq z \leq 1$, does not satisfy Definition 15.16 because there is no way to construct an "open" chart at points on its boundary (see Remark 15.29 below). Loosely speaking, this is because at a point on the boundary, the surface does not look like an open set, but rather like a relatively open set in a half-plane (see Figure 15.1).

Accordingly, we make the following definition.

**15.27 DEFINITION.** Let $M$ be a set.

i) An *$n$-dimensional chart-with-smooth-boundary* of $M$ at a point $x \in M$ is a pair $(V, h)$, where $x \in V$, $V \subseteq M$, $h : V \to \mathbf{R}^n$ is 1–1 on $V$, and $h(V)$ is relatively open in some half-space $\mathcal{H}$ of $\mathbf{R}^n$. If $h(V) \cap \partial\mathcal{H} = \emptyset$, then $(V, h)$ is called an *interior chart*. If $h(V) \cap \partial\mathcal{H} \neq \emptyset$, then $(V, h)$ is called a *boundary chart*.

ii) An *$n$-dimensional $C^p$ atlas-with-smooth-boundary* of $M$ is a collection

(8) $$\mathcal{A} = \{(V_\alpha, h_\alpha) : \alpha \in A\}$$

of $n$-dimensional charts-with-smooth-boundary of $M$ such that $h_\beta(V_\alpha \cap V_\beta)$ is relatively open in some half-space $\mathcal{H}$,

(9)               $h_\alpha \circ h_\beta^{-1}$ is $\mathcal{C}^p$ on $h_\beta(V_\alpha \cap V_\beta)$ for all $\alpha, \beta \in A$,

and

$$M = \bigcup_{\alpha \in A} V_\alpha.$$

The functions $h_\alpha \circ h_\beta^{-1}$ are called the *transition maps* of the atlas $\mathcal{A}$.

iii) Two $n$-dimensional atlases-with-smooth-boundary $\mathcal{A}$, $\mathcal{B}$ of $M$ are said to be $\mathcal{C}^p$ *compatible* (notation: $\mathcal{A} \sim \mathcal{B}$) if $\mathcal{A} \cup \mathcal{B}$ is an $n$-dimensional $\mathcal{C}^p$ atlas-with-smooth-boundary on $M$.

It is easy to check that $\mathcal{C}^p$ compatibility of atlases-with-smooth-boundary is an equivalence relation. We also note that, since any open subset of $\mathcal{H}$ is relatively open in $\mathcal{H}$, every atlas is an atlas-with-smooth-boundary.

We now expand the definition of manifold, using atlases-with-smooth-boundary.

### 15.28 DEFINITION.

i) An *$n$-dimensional $\mathcal{C}^p$ manifold-with-smooth-boundary* is a set $M$ together with an equivalence class of $n$-dimensional atlases-with-smooth-boundary.

ii) A point $x \in M$ is said to be a *boundary point* if it belongs to $V$ for some boundary chart $(V, h)$ of $M$, with $h(V)$ relatively open in some half-space $\mathcal{H}$, and $h(x) \in \partial\mathcal{H}$. The collection of all boundary points, called the *boundary* of $M$, is denoted by $\partial M$.

The following result shows that if the transition maps of a manifold have nonzero Jacobian, then the definition of boundary point is unambiguous.

**15.29 Remark.** *Let $M$ be an $n$-dimensional $\mathcal{C}^p$ manifold-with-smooth-boundary, $\boldsymbol{x} \in M$, and $(U, g)$, $(V, h)$ be charts of $M$ at $\boldsymbol{x}$ whose transition map $\phi = g \circ h^{-1}$ satisfies $\Delta_\phi \neq 0$ on $h(U \cap V)$. If $g(U)$ (respectively, $h(V)$) is relatively open in some half-space $\mathcal{H}$ (respectively, $\mathcal{K}$), and $g(\boldsymbol{x}) \in \partial\mathcal{H}$, then $h(\boldsymbol{x}) \in \partial\mathcal{K}$.*

PROOF. Set $\Omega = (h(U \cap V))^o$. If $h(\boldsymbol{x}) \notin \partial\mathcal{K}$, then $h(\boldsymbol{x}) \in \Omega$. Since the transition map $\phi$ is $\mathcal{C}^p$ and has nonzero Jacobian, it follows from Theorem 11.39 (the Inverse Function Theorem) that $\phi(\Omega)$ is open in $\mathbf{R}^n$. But

$$g(\boldsymbol{x}) \in \phi(\Omega) \subseteq (g \circ h^{-1})(h(U \cap V)) = g(U \cap V) \subseteq g(U).$$

Hence, $g(\boldsymbol{x})$ belongs to the interior of $g(U)$, i.e., cannot belong to $\partial\mathcal{H}$. ∎

Is Definition 15.28 general enough to include every smooth surface whose boundary is made up of smooth curves? At first glance, the answer to this question seems to be no because of the restriction that $h(V)$ be relatively open in some half-plane, i.e., part of its boundary be a straight line. Nevertheless, if $S$ is a smooth surface with smooth boundary, one can always find a smoothly equivalent parametrization

$(\psi, B)$ of $S$ such that $\partial B$ is made up of straight lines (see Munkres [8], p. 51). In particular, every smooth surface with smooth boundary is a two-dimensional manifold-with-smooth-boundary.

Our goal is to prove Stokes's Theorem for manifolds-with-smooth-boundary. We shall deal exclusively with manifolds $M$ which are subsets of $\mathbf{R}^m$ for some $m \geq n$. In this case we have two competing concepts: open sets defined by the manifold structure (Definition 15.23) and open sets defined by the relative topology (Definition 9.25 or 10.54). The purpose of the following definition is to make sure that these concepts coincide.

**15.30 DEFINITION.** An $n$-dimensional manifold $M$ is said to be *continuously embedded* in $\mathbf{R}^m$ if $m \geq n$, $M$ is a closed subset of $\mathbf{R}^m$, and if $(V, h)$ is a chart of $M$, then $V$ is relatively open in $M$ and $h$ is a *homeomorphism* from the relative topology on $V$ to the usual topology on $\mathbf{R}^n$; i.e., if $U$ is relatively open in $V$ then $h(U)$ is open in $\mathbf{R}^n$, and if $\Omega$ is open in $\mathbf{R}^n$, then $h^{-1}(\Omega) \cap V$ is relatively open in $V$.

From now on, by a manifold we mean a manifold-with-smooth-boundary (whose boundary may or may not be empty) which is continuously embedded in some $\mathbf{R}^m$. We now define what it means for a manifold to be orientable.

**15.31 DEFINITION.** Let $m \geq n$ and $M \subset \mathbf{R}^m$.

i) A $\mathcal{C}^p$ atlas (8) is said to be *oriented* if

$$(10) \qquad\qquad \Delta_{h_\alpha \circ h_\beta^{-1}}(\boldsymbol{u}) > 0$$

for all $\boldsymbol{u} \in h_\beta(V_\alpha \cap V_\beta)$ and $\alpha, \beta \in A$ (compare with Definition 13.41).

ii) An $n$-dimensional $\mathcal{C}^p$ manifold $M$ is said to be *orientable* if it has an oriented atlas.

iii) Two oriented $\mathcal{C}^p$ atlases $\mathcal{A}$, $\mathcal{B}$ of an orientable manifold $M$ are said to be *orientation compatible* if $\mathcal{A} \cup \mathcal{B}$ is an oriented atlas. (Note that orientation compatibility is an equivalence relation.)

iv) An *orientation* of an $n$-dimensional orientable $\mathcal{C}^p$ manifold $M$ is an equivalence class of oriented $\mathcal{C}^p$ atlases. If $\mathcal{A}$ is an oriented $\mathcal{C}^p$ atlas of $M$, then the *orientation* generated by $\mathcal{A}$ is the orientation of $M$ which contains $\mathcal{A}$.

An orientation of a manifold $M$ can be used to induce an orientation of $\partial M$ in the following way.

**15.32 DEFINITION.** Let $M$ be a manifold with orientation $\mathcal{O}$ and $\mathcal{A}$ be an atlas of $M$ consisting of charts $(V, h)$ from $\mathcal{O}$ which satisfy $h_1(\boldsymbol{x}) \leq 0$ for all $\boldsymbol{x} \in V$, and $h_1(\boldsymbol{x}) = 0$ if and only if $\boldsymbol{x} \in \partial M \cap V$. The *orientation induced* on $\partial M$ by $\mathcal{A}$ is the orientation of $\partial M$ generated by the atlas

$$\widetilde{\mathcal{A}} = \{(\widetilde{V}, \widetilde{h}) : (V, h) \in \mathcal{A}\},$$

where $\widetilde{V} = V \cap \partial M$ and $\widetilde{h}(\boldsymbol{x}) = (h_2(\boldsymbol{x}), \dots, h_n(\boldsymbol{x}))$.

The following result shows that $\widetilde{\mathcal{A}}$ is an oriented atlas of $\partial M$ when $\mathcal{O}$ is an orientation of $M$.

**15.33 Remark.** *Suppose $(V, h)$ and $(U, g)$ are charts from an orientation $\mathcal{O}$ of an oriented manifold $M$ which satisfy $h_1(\boldsymbol{x}) \leq 0$ (respectively, $g_1(\boldsymbol{x}) \leq 0$) for $\boldsymbol{x} \in V$ (respectively, $\boldsymbol{x} \in U$), and $h_1(\boldsymbol{x}) = 0$ (respectively, $g_1(\boldsymbol{x}) = 0$) if and only if $\boldsymbol{x} \in \partial M \cap V$ (respectively, $\boldsymbol{x} \in \partial M \cap U$). If $\widetilde{h} = (h_2, \ldots, h_n)$ and $\widetilde{g} = (g_2, \ldots, g_n)$, then*

$$\Delta_{\widetilde{h} \circ \widetilde{g}^{-1}}(\boldsymbol{u}) > 0 \text{ for all } \boldsymbol{u} \in \widetilde{g}(\widetilde{V} \cap \widetilde{U}).$$

PROOF. Let $(t, \boldsymbol{u}) = (t, u_2, \ldots, u_n)$ represent a general point in $\mathbf{R}^n$, $\phi = h \circ g^{-1}$ be the transition from $g(U)$ to $h(V)$, and $\phi_1$ be the first component of $\phi$. By Remark 15.29, $\phi$ takes boundary points to boundary points. Since $h_1(\boldsymbol{x}) = g_1(\boldsymbol{x}) = 0$ for $\boldsymbol{x} \in \partial M \cap V \cap U$, it follows that $\phi_1(0, \boldsymbol{u}) = 0$ for all $\boldsymbol{u} \in \widetilde{g}(\widetilde{V} \cap \widetilde{U})$. Consequently, the first row of the Jacobian matrix $D(h \circ g^{-1})(0, \boldsymbol{u})$ is given by

$$\frac{\partial \phi_1}{\partial t}(0, \boldsymbol{u}) \quad 0 \quad \ldots \quad 0.$$

It follows that

$$\Delta_{h \circ g^{-1}}(0, \boldsymbol{u}) = \frac{\partial \phi_1}{\partial t}(0, \boldsymbol{u}) \cdot \Delta_{\widetilde{h} \circ \widetilde{g}^{-1}}(\boldsymbol{u}).$$

Moreover, the conditions $h_1 \leq 0$ on $V$ and $g_1 \leq 0$ on $U$ imply

$$\frac{\partial \phi_1}{\partial t}(0, \boldsymbol{u}) = \lim_{t \to 0-} \frac{\phi_1(t, \boldsymbol{u}) - \phi_1(0, \boldsymbol{u})}{t} = \lim_{t \to 0-} \frac{\phi_1(t, \boldsymbol{u})}{t} \geq 0.$$

Since $\Delta_{h \circ g^{-1}} > 0$ on $g(V \cap U)$, we conclude that $\Delta_{\widetilde{h} \circ \widetilde{g}^{-1}}(\boldsymbol{u}) > 0$ for each $\boldsymbol{u} \in \widetilde{g}(\widetilde{V} \cap \widetilde{U})$. ∎

For two-dimensional manifolds, the condition $h_1(\boldsymbol{x}) \leq 0$ makes the induced orientation, as defined above, agree with the right-handed orientation introduced in Section 13.4 (see Figure 15.1)

Our definition of $n$-dimensional manifolds is quite general, but not general enough. It does not include $n$-dimensional rectangles. (There is no way to parametrize a corner of a two-dimensional rectangle by using relatively open sets in a half-plane).

One way to fix this is to extend the definition of charts to include "corner" charts. This extension is still not general enough to include all piecewise smooth curves; for example, it does not include curves with cusps (e.g., $y = x^{2/3}$). The theory can be extended once again by taking limits of "manifolds with corners." For details see Loomis and Sternberg [6].

We will take a less ambitious approach by treating the rectangular case separately. By a chart of an $n$-dimensional region $R$ in $\mathbf{R}^n$ (this includes all $n$-dimensional rectangles) we mean a pair $(E, h)$ where $R^\circ \subseteq E \subseteq R$ and $h : V \to \mathbf{R}^n$ is 1–1 and continuously differentiable on some open set $V$ which contains $R$ with $\Delta_h \neq 0$ on $V$. (Notice that, by the Inverse Function Theorem, $h$ is a homeomorphism on $V$ and that, by Theorem 12.52, $h(E)$ is a Jordan region.) Using such charts, we can define atlases and manifolds in the same way as above. The end result is that we

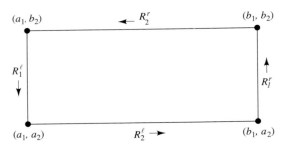

**Figure 15.2**

can consider any $n$-dimensional region in $\mathbf{R}^n$ to be a manifold. Notice that if $I$ represents the identity function on $\mathbf{R}^n$, i.e., $I(\boldsymbol{x}) = \boldsymbol{x}$ for all $\boldsymbol{x} \in \mathbf{R}^n$, and if $R$ is an $n$-dimensional rectangle, then $\{(R, I)\}$ is an atlas of $R$. The orientation generated by the identity chart is called the *usual orientation* on $R$. Also notice that, since $R$ is closed, the "manifold" boundary of $R$ is precisely its topological boundary.

When $R$ is a rectangle, what orientation is induced on $\partial R$ by the usual orientation of $R$? The following result answers this question by showing how to find an atlas of the induced orientation for arbitrary $n \in \mathbf{N}$. Notice that, for the special case $n = 2$, this orientation on $\partial R$ is counterclockwise orientation (see Figure 15.2).

**15.34 THEOREM.** Let $R = [a_1, b_1] \times \cdots \times [a_n, b_n]$ be an *$n$-dimensional rectangle. For each $j = 1, \ldots, n$, set*

$$R_j^\ell = [a_1, b_1] \times \cdots \times \{a_j\} \times \cdots \times [a_n, b_n],$$

$$R_j^r = [a_1, b_1] \times \cdots \times \{b_j\} \times \cdots \times [a_n, b_n],$$

$$h_j(x_1, \ldots, x_n) = \begin{cases} (a_1 - x_1, -x_2, x_3, \ldots, x_n) & j = 1 \\ (a_j - x_j, (-1)^j x_1, x_2, \ldots, \widehat{x}_j, \ldots, x_n) & j \neq 1, \end{cases}$$

and

$$g_j(x_1, \ldots, x_n) = \begin{cases} (x_1 - b_1, x_2, x_3, \ldots, x_n) & j = 1 \\ (x_j - b_j, (-1)^{j+1} x_1, x_2, \ldots, \widehat{x}_j, \ldots, x_n) & j \neq 1. \end{cases}$$

*(The notation $\widehat{x}_j$ indicates that this variable is missing.) If $V_j = R^o \cup R_j^\ell$ and $U_j = R^o \cup R_j^r$, then*

$$\mathcal{A} = \{(V_j, h_j), (U_j, g_j) : j = 1, \ldots, n\}$$

*is an oriented atlas of $R$ which is compatible with the usual orientation. In particular, if $\widetilde{V}_j = V_j \cap R_j^\ell$, $\widetilde{U}_j = U_j \cap R_j^r$, $\widetilde{h}_j$, and $\widetilde{g}_j$ are defined as in Definition 15.32, then*

$$\widetilde{\mathcal{A}} = \{(\widetilde{V}_j, \widetilde{h}_j), (\widetilde{U}_j, \widetilde{g}_j) : j = 1, \ldots, n\}$$

is an oriented atlas of $\partial R$ which belongs to the orientation induced by the usual orientation.

PROOF. Fix $1 \le j \le n$ and let $I(\boldsymbol{x}) = \boldsymbol{x}$ represent the identity function on $\mathbf{R}^n$. By definition, if $j = 1$, then

$$\Delta_{h_j \circ I} = \det \begin{bmatrix} -1 & 0 & 0 & \cdots & 0 \\ 0 & -1 & 0 & \cdots & 0 \\ 0 & 0 & 1 & \cdots & 0 \\ \vdots & \vdots & \vdots & \ddots & \vdots \\ 0 & 0 & 0 & \cdots & 1 \end{bmatrix} = 1 > 0.$$

If $j \ne 1$, then by factoring $-1$ out of the first row and interchanging $j - 1$ rows we have

$$\Delta_{h_j \circ I} = (-1)^j \det \begin{bmatrix} (-1)^j & 0 & 0 & \cdots & 0 \\ 0 & 1 & 0 & \cdots & 0 \\ 0 & 0 & 1 & \cdots & 0 \\ \vdots & \vdots & \vdots & \ddots & \vdots \\ 0 & 0 & 0 & \cdots & 1 \end{bmatrix} = (-1)^{2j} = 1 > 0.$$

Thus the chart $(V_j, h_j)$ is compatible with the usual orientation on $R$.

Let $h_{j1}$ represent the first component of the function $h_j$, $j = 1, \ldots, n$. Clearly, if $\boldsymbol{x} \in R$, then $h_{j1}(\boldsymbol{x}) \le 0$; and $h_{j1}(\boldsymbol{x}) = 0$ if and only if $x_j = a_j$, i.e., if and only if $\boldsymbol{x} \in R_j^\ell$. Thus $(\widetilde{V}_j, \widetilde{h}_j)$ belongs to the orientation induced on $\partial R$ by the usual orientation. A similar argument works for the "right- hand" boundaries $R_j^r$. ∎

We mentioned in Chapter 13 that a connected smooth curve or connected smooth orientable surface has only two orientations. This is a general principle shared by all connected orientable $\mathcal{C}^1$ manifolds (see Theorem 15.36 below). First, we prove the following result.

**15.35 Lemma.** *If $M$ is a connected orientable $\mathcal{C}^1$ manifold and $\mathcal{A}$, $\mathcal{B}$ are oriented atlases of $M$, then either*

$$\Delta_{h \circ g^{-1}}(\boldsymbol{u}) > 0$$

*for all $(V, h) \in \mathcal{A}$, $(U, g) \in \mathcal{B}$, and $\boldsymbol{u} \in g(V \cap U)$, or*

$$\Delta_{h \circ g^{-1}}(\boldsymbol{u}) < 0$$

*for all $(V, h) \in \mathcal{A}$, $(U, g) \in \mathcal{B}$, and $\boldsymbol{u} \in g(V \cap U)$.*

PROOF. Set

$$A = \{\boldsymbol{x} \in M : \Delta_{h \circ g^{-1}}(g(\boldsymbol{x})) > 0 \text{ for some } (V, h) \in \mathcal{A} \text{ and } (U, g) \in \mathcal{B}\},$$

and

$$B = \{\boldsymbol{x} \in M : \Delta_{h \circ g^{-1}}(g(\boldsymbol{x})) < 0 \text{ for some } (V, h) \in \mathcal{A} \text{ and } (U, g) \in \mathcal{B}\}.$$

We must show that $M = A$ or $M = B$. Since $M$ is connected, it suffices to show that $A$ and $B$ are (relatively) open in $M$, $M = A \cup B$, and $A \cap B = \emptyset$ (see Definition 9.26 or 10.53).

To show $A$ is open in $M$, let $\boldsymbol{x}_0 \in A$ and choose $(V, h) \in \mathcal{A}$, $(U, g) \in \mathcal{B}$ such that $\Delta_{h \circ g^{-1}}(g(\boldsymbol{x}_0)) > 0$. Then

$$\boldsymbol{x}_0 \in \Omega := g^{-1}(\Delta_{h \circ g^{-1}}^{-1}((0, \infty))).$$

Since $h \circ g^{-1}$ is continuously differentiable on $g(U)$, its Jacobian is continuous on $g(U)$. Hence $\Omega$, the inverse image of the open set $(0, \infty)$ under the continuous function $\Delta_{h \circ g^{-1}} \circ g$, must be open in $M$ (Theorem 9.33 or 10.58). It follows that $A$ is open in $M$. A similar argument proves $B$ is open in $M$.

To show $M = A \cup B$ we must show that $\Delta_{h \circ g^{-1}}(g(\boldsymbol{x})) \neq 0$ for all $\boldsymbol{x} \in M$. Suppose to the contrary that $\Delta_{h \circ g^{-1}}(g(\boldsymbol{x})) = 0$ for some $\boldsymbol{x} \in M$. Since $\mathcal{A}$ is an atlas of $M$, choose $(W, \sigma) \in \mathcal{A}$ such that $\boldsymbol{x} \in W$ and set $\boldsymbol{u} = \sigma(\boldsymbol{x})$. By (10) and the Chain Rule,

$$0 < \Delta_{h \circ \sigma^{-1}}(\boldsymbol{u}) = \Delta_{h \circ g^{-1}}(g \circ \sigma^{-1}(\boldsymbol{u}))\Delta_{g \circ \sigma^{-1}}(\boldsymbol{u}) = 0 \cdot \Delta_{g \circ \sigma^{-1}}(\boldsymbol{u}) = 0,$$

a contradiction. Thus $M = A \cup B$.

Finally, to show $A \cap B$ is empty, suppose to the contrary that there is an $\boldsymbol{x} \in A \cap B$. By definition, this means there exist charts $(V_i, h_i) \in \mathcal{A}$ and $(U_i, g_i) \in \mathcal{B}$ such that

$$(11) \qquad (-1)^{i+1}\Delta_{h_i \circ g_i^{-1}}(g_i(\boldsymbol{x})) > 0$$

for $i = 1, 2$. Since $\mathcal{A}$ is an orientation, we have by (10) and the Chain Rule that

$$0 < \Delta_{h_1 \circ h_2^{-1}}(h_2(\boldsymbol{x})) = \Delta_{h_1 \circ g_1^{-1}}(g_1 \circ g_2^{-1} \circ g_2 \circ h_2^{-1} \circ h_2(\boldsymbol{x})) \cdot$$
$$\cdot \Delta_{g_1 \circ g_2^{-1}}(g_2 \circ h_2^{-1} \circ h_2(\boldsymbol{x}))\Delta_{g_2 \circ h_2^{-1}}(h_2(\boldsymbol{x}))$$
$$= \Delta_{h_1 \circ g_1^{-1}}(g_1(\boldsymbol{x}))\Delta_{g_1 \circ g_2^{-1}}(g_2(\boldsymbol{x}))\Delta_{g_2 \circ h_2^{-1}}(h_2(\boldsymbol{x})).$$

By (11), the first (respectively, third) of these factors is positive (respectively, negative). Hence, the second factor must be negative. But the second factor is positive since both $g_1$ and $g_2$ come from the same oriented atlas $\mathcal{B}$. This contradiction proves the lemma. ∎

**15.36 THEOREM.** *Let $M$ be a connected orientable $C^p$ manifold. Then $M$ has exactly two orientations.*

PROOF. We first show that $M$ has at least two orientations. Let $\mathcal{A} = \{(V_\alpha, h_\alpha) : \alpha \in A\}$ be an oriented atlas of $M$ with $h_\alpha = (h_{\alpha 1}, \ldots, h_{\alpha n})$, and consider

$$\mathcal{B} = \{(V_\alpha, g_\alpha) : \alpha \in A\},$$

where $g_\alpha = (-h_{\alpha 1}, h_{\alpha 2}, \ldots, h_{\alpha n})$. Clearly, $\mathcal{B}$ is a $C^p$ atlas of $M$. Since

$$\Delta_{g_\alpha \circ g_\beta^{-1}} \circ g_\beta = \Delta_{h_\alpha \circ h_\beta^{-1}} \circ h_\beta > 0,$$

$\mathcal{B}$ is orientable. Since

$$\Delta_{g_\alpha \circ h_\beta^{-1}} \circ h_\beta = -\Delta_{h_\alpha \circ h_\beta^{-1}} \circ h_\beta < 0,$$

$\mathcal{B}$ is not orientation compatible with $\mathcal{A}$. Thus $M$ has at least two orientations.

To show that $M$ has no more than two orientations, suppose to the contrary that $M$ has three distinct orientations. Let $\mathcal{A}$, $\mathcal{B}$, and $\mathcal{O}$ be atlases from each of these orientations, and choose $(V, h) \in \mathcal{A}$, $(U, g) \in \mathcal{B}$ such that $V \cap U \neq \emptyset$. Since these orientations are distinct, there exist $(W_i, \sigma_i) \in \mathcal{O}$, $i = 1, 2$, such that

$$\Delta_{h \circ \sigma_1^{-1}}(\sigma_1(\boldsymbol{x})) < 0 \quad \text{and} \quad \Delta_{\sigma_2 \circ g^{-1}}(g(\boldsymbol{y})) < 0$$

for some $\boldsymbol{x} \in V \cap W_1$ and $\boldsymbol{y} \in U \cap W_2$. By Lemma 15.35, $\Delta_{h \circ \sigma^{-1}} \circ \sigma < 0$ and $\Delta_{\sigma \circ g^{-1}} \circ g < 0$ on $M$ for all $(W, \sigma) \in \mathcal{O}$. Let $\boldsymbol{x} \in V \cap U$ and choose $(W, \sigma) \in \mathcal{O}$ such that $\boldsymbol{x} \in W$. By the Chain Rule,

$$\Delta_{h \circ g^{-1}}(g(\boldsymbol{x})) = \Delta_{h \circ \sigma^{-1}}(\sigma(\boldsymbol{x}))\Delta_{\sigma \circ g^{-1}}(g(\boldsymbol{x})).$$

This is a product of two negative numbers, hence, positive. It follows from Lemma 15.35 that $\Delta_{h \circ g^{-1}} \circ g > 0$ on $U$ for all $(V, h) \in \mathcal{A}$ and $(U, g) \in \mathcal{B}$. Therefore, $\mathcal{A}$ is orientation compatible with $\mathcal{B}$, a contradiction. $\blacksquare$

### EXERCISES

1. Let $M$ be a $\mathcal{C}^p$ manifold (not necessarily continuously embedded in some $\mathbf{R}^m$).
   a) If $\{V_\alpha\}_{\alpha \in A}$ is a collection of open sets in $M$, prove that $\cup_{\alpha \in A} V_\alpha$ is open in $M$.
   b) If $V_1, \ldots, V_N$ are open in $M$, prove $\cap_{j=1}^N V_j$ is open in $M$.
2. Prove that $\mathcal{C}^p$ compatibility and orientation compatibility are equivalence relations.
3. Prove that the boundary of an $n$-dimensional $\mathcal{C}^p$ manifold-with-smooth-boundary is an $(n-1)$-dimensional manifold.
4. **This exercise is used in Section 15.3.** *Translation* on $\mathbf{R}^n$ by an $\boldsymbol{a} \in \mathbf{R}^n$ is defined by $\sigma(\boldsymbol{x}) = \boldsymbol{x} + \boldsymbol{a}$ for $\boldsymbol{x} \in \mathbf{R}^n$. *Dilation* on $\mathbf{R}^n$ by a $\delta > 0$ is defined by $\sigma(\boldsymbol{x}) = \delta \boldsymbol{x}$ for $\boldsymbol{x} \in \mathbf{R}^n$.
   a) Prove that if $\mathcal{A}$ is an oriented atlas of a manifold $M$ and $\sigma$ is a translation or a dilation, then $\mathcal{B} = \{(V, \sigma \circ h) : (V, h) \in \mathcal{A}\}$ is an atlas of $M$ which is orientation compatible with $\mathcal{A}$.
   b) Let $\mathcal{A}$ be an orientation of a manifold $M$ and $\boldsymbol{x} \in M$ be an interior point. Prove there is a chart $(V, h)$ at $\boldsymbol{x}$ such that $h(V) = B_1(\boldsymbol{0})$ and $h(\boldsymbol{x}) = \boldsymbol{0}$.
5. Let $\mathcal{A}$ be an $n$-dimensional $\mathcal{C}^\infty$ atlas of a manifold $M$. A function $f : M \to \mathbf{R}^k$ is said to be $\mathcal{C}^p$ on $M$ if $f \circ h^{-1} : h(V) \to \mathbf{R}^k$ is $\mathcal{C}^p$ for all charts $(V, h) \in \mathcal{A}$.
   a) Prove that this definition is independent of the atlas $\mathcal{A}$.
   b) Prove that the composition of $\mathcal{C}^p$ functions is a $\mathcal{C}^p$ function.
   c) Prove if $(V, h)$ is a chart of $M$, then $h$ is a $\mathcal{C}^\infty$ function on $V$.

**6.** Prove that the sphere $x_1^2 + \cdots + x_n^2 = a^2$ is an $(n-1)$-dimensional manifold in $\mathbf{R}^n$.

$^e$**15.3  STOKES'S THEOREM ON MANIFOLDS**    *This section uses material from Sections 12.5, 15.1, and 15.2.*

We shall define oriented integrals of $n$–forms on $n$-dimensional manifolds and obtain a fundamental theorem of calculus for these integrals. Recall that, for us, a manifold $M$ is closed and continuously embedded in $\mathbf{R}^m$ (see Definition 15.30). In particular, given $W$ open in $M$, there is an open set $\Omega \subset \mathbf{R}^m$ such that $\Omega \cap M = W$. This assumption is not essential, and all results stated in this section are valid without it. We make it to simplify the proof that partitions of unity exist on a manifold.

**15.37 Lemma** [$\mathcal{C}^\infty$ Partitions of Unity on a Compact Manifold]. *Let $M$ be a compact $n$-dimensional $\mathcal{C}^p$ manifold in $\mathbf{R}^m$ with orientation $\mathcal{O}$, and $\{U_\alpha\}_{\alpha \in A}$ be an open covering of $M$. Then there exist $\mathcal{C}^\infty$ functions $\phi_j : \mathbf{R}^m \to \mathbf{R}$, $j = 1, 2, \ldots, N$, and an atlas $\{(V_j, h_j) \in \mathcal{O} : j = 1, \ldots, N\}$ of $M$ such that*

  i) *given $j \in \{1, \ldots, N\}$ there is an $\alpha \in A$ such that $V_j \subset U_\alpha$,*
  ii) *$0 \le \phi_j(\boldsymbol{x}) \le 1$ for $\boldsymbol{x} \in \mathbf{R}^m$, $j = 1, \ldots, N$,*
  iii) *$\operatorname{spt} \phi_j \cap M \subset V_j$, for $j = 1, \ldots, N$, and*
  iv) *$\sum_{j=1}^N \phi_j(\boldsymbol{x}) = 1$ for $\boldsymbol{x} \in M$.*

PROOF. For each $\boldsymbol{x} \in M$ choose a chart $(V_{\boldsymbol{x}}, h_{\boldsymbol{x}}) \in \mathcal{O}$ such that $\boldsymbol{x} \in V_{\boldsymbol{x}} \subseteq U_\alpha$ for some $\alpha \in A$. Choose bounded open sets $\Omega_{\boldsymbol{x}}$ and $B_{\boldsymbol{x}}$ in $\mathbf{R}^m$ such that $\boldsymbol{x} \in \Omega_{\boldsymbol{x}} \subset \overline{\Omega}_{\boldsymbol{x}} \subset B_{\boldsymbol{x}}$ and $B_{\boldsymbol{x}} \cap M = V_{\boldsymbol{x}}$. By Theorem 12.59, there is a $\mathcal{C}^\infty$ function $\psi_{\boldsymbol{x}}$ on $\mathbf{R}^m$ such that $\psi_{\boldsymbol{x}} = 1$ on $\Omega_{\boldsymbol{x}}$ and $\operatorname{spt} \psi_{\boldsymbol{x}} \subset B_{\boldsymbol{x}}$. Clearly, $\{\Omega_{\boldsymbol{x}}\}_{\boldsymbol{x} \in M}$ is an open covering of $M$. Since $M$ is compact, choose a finite subcover $\{\Omega_1, \ldots, \Omega_N\}$. If $\Omega_j = \Omega_{\boldsymbol{x}}$, set $\psi_j = \psi_{\boldsymbol{x}}$ and $(V_j, h_j) = (V_{\boldsymbol{x}}, h_{\boldsymbol{x}})$. By construction, $\psi_j = 1$ on $\Omega_j$ and $\operatorname{spt} \psi_j \cap M \subset V_j$.

Set $\phi_1 = \psi_1$, $\phi_2 = (1 - \psi_1)\psi_2$, $\ldots$, $\phi_N = (1 - \psi_1) \ldots (1 - \psi_{N-1})\psi_N$. Then $\phi_j$ is $\mathcal{C}^\infty$ on $\mathbf{R}^m$ and $\operatorname{spt} \phi_j \cap M \subset \operatorname{spt} \psi_j \cap M \subset V_j$. This verifies i), ii), and iii). It is easy to see by induction that

$$\sum_{j=1}^N \phi_j = 1 - (1 - \psi_1) \ldots (1 - \psi_N).$$

Since $\{\Omega_j\}$ covers $M$, this verifies iv). ∎

We shall call the functions $\phi_1, \ldots, \phi_N$ given in Lemma 15.37 a $\mathcal{C}^\infty$ *partition of unity on $M$ subordinate to the covering* $\{U_\alpha\}_{\alpha \in A}$.

By a differential form on a manifold $M$ we mean an $r$–form on some open set $\Omega \subset \mathbf{R}^m$ such that $M \subset \Omega$. Thus a decomposable $r$–form on $M$ has the form

$$\omega = f\, dx_{i_1} \ldots dx_{i_r},$$

where $f$ is a 0–form on some open set $\Omega$ which contains $M$.

We are now prepared to define the integral of a differential form on an oriented manifold. (This definition includes oriented line integrals and oriented surface integrals—see Exercise 5 in Section 15.1.)

**15.38 DEFINITION.** Let $m \geq n$, $M$ be an $n$-dimensional oriented $\mathcal{C}^p$ manifold in $\mathbf{R}^m$, $\omega$ be a continuous $n$–form on $M$, and $\mathcal{O}$ be an orientation of $M$.

   i) If $\phi : V \to \mathbf{R}$ is continuous on $V$ for some chart $(V, h) \in \mathcal{O}$ and spt $\phi \cap M \subseteq V$, then the *oriented integral* of $\phi\omega$ on $M$ is defined by

$$\int_M \phi\omega = \int_{h(V)} (h^{-1})^*(\phi\omega).$$

   ii) If $M$ is compact, then the *oriented integral* of $\omega$ on $M$ is defined by

$$\int_M \omega = \sum_{j=1}^N \int_M \phi_j\omega,$$

where $\phi_1, \ldots, \phi_N$ is any $\mathcal{C}^\infty$ partition of unity on $M$ subordinate to the orientation $\mathcal{O}$.

The following two remarks show that these definitions make sense.

**15.39 Remark.** *The value of $\int_M \phi\omega$ does not depend on the chart chosen.*

**PROOF.** Let $(U, g) \in \mathcal{O}$ be another chart which satisfies $U \supseteq$ spt $\phi \cap M$. We may suppose that $\omega$ is decomposable, i.e., $\omega = f\,dx_{i_1} \ldots dx_{i_n}$. Let $H = h^{-1}$ and $G = g^{-1}$. Since $\mathcal{O}$ is an orientation, $\Delta_{h \circ g^{-1}} \geq 0$ on $g(U \cap V)$. Moreover, by the Chain Rule,

$$\Delta_{(G_{i_1}, \ldots, G_{i_n})} = \Delta_{(H_{i_1}, \ldots, H_{i_n})} \circ h \circ g^{-1} \Delta_{h \circ g^{-1}}.$$

Since $h(V \cap U) = (h \circ g^{-1}) \circ g(V \cap U)$ and $\phi$ is supported in $V \cap U$, it follows from the Fundamental Theorem of Differential Transforms and a change of variables in $\mathbf{R}^n$ that

$$\int_{h(V)} (h^{-1})^*(\phi\omega)(\boldsymbol{u})\, d\boldsymbol{u}$$

$$= \int_{h(V)} (\phi \circ h^{-1})(\boldsymbol{u})\,(f \circ h^{-1})(\boldsymbol{u}) \Delta_{(H_{i_1}, \ldots, H_{i_n})}(\boldsymbol{u})\, d\boldsymbol{u}$$

$$= \int_{g(U)} (\phi \circ g^{-1})(\boldsymbol{y})\,(f \circ g^{-1})(\boldsymbol{y}) \cdot$$

$$\cdot\, \Delta_{(H_{i_1}, \ldots, H_{i_n})}(h \circ g^{-1}(\boldsymbol{y})) |\Delta_{h \circ g^{-1}}(\boldsymbol{y})|\, d\boldsymbol{y}$$

$$= \int_{g(U)} (\phi \circ g^{-1})(\boldsymbol{y})\,(f \circ g^{-1})(\boldsymbol{y}) \Delta_{(G_{i_1}, \ldots, G_{i_n})}(\boldsymbol{y})\, d\boldsymbol{y}$$

$$= \int_{g(U)} (g^{-1})^*(\phi\omega)(\boldsymbol{y})\, d\boldsymbol{y}. \quad \blacksquare$$

**15.40 Remark.** *The value of $\int_M \omega$ does not depend on the $\mathcal{C}^\infty$ partition of unity $\phi_1, \ldots, \phi_N$ chosen.*

PROOF. If $\psi_1, \ldots, \psi_L$ is another $\mathcal{C}^\infty$ partition of unity subordinate to $\mathcal{O}$, then

$$\sum_{k=1}^{L} \int_M \psi_k \omega = \sum_{k=1}^{L} \int_M \psi_k (\sum_{j=1}^{N} \phi_j) \omega = \sum_{k=1}^{L} \sum_{j=1}^{N} \int_M \psi_k \phi_j \omega$$

$$= \sum_{j=1}^{N} \int_M (\sum_{k=1}^{L} \psi_k) \phi_j \omega = \sum_{j=1}^{N} \int_M \phi_j \omega. \quad \blacksquare$$

The following result justifies the identification of $d(x, y)$ with $dx\, dy$ made below (3) in Section 14.1. (See also Exercise 5 in Section 15.1.)

**15.41 Remark.** *If $R$ is an $n$-dimensional rectangle with the usual orientation, and $\omega = f\, dx_1 \ldots dx_n$ is an $n$-form on some open $\Omega \supset R$, then*

$$\int_R \omega = \int_R f(\boldsymbol{x})\, d\boldsymbol{x}.$$

PROOF. By hypothesis, $(R, I)$ is a chart of $R$. Hence, by Definition 15.38i),

$$\int_R \omega = \int_{I(R)} I^*(\omega)(\boldsymbol{x})\, d\boldsymbol{x} = \int_R f(\boldsymbol{x})\, d\boldsymbol{x}. \quad \blacksquare$$

We shall prove that the oriented integral of the exterior derivative $d\omega$ of a differential form on a manifold $M$ is determined by the behavior of $\omega$ on $\partial M$.

STRATEGY: The idea behind the proof is straightforward. First, we prove the result when $M$ is a rectangle (Lemma 15.42). (This case follows directly from the one-dimensional Fundamental Theorem of Calculus because the boundary of a rectangle only moves in one dimension at a time.) Next, we pull-back this result to sufficiently small charts on $M$ (Lemma 15.43). Finally, by using a $\mathcal{C}^\infty$ partition of unity subordinate to a covering by sufficiently small charts, we establish the general result. (The proofs of Lemma 15.43 and Theorem 15.45 presented here come from Spivak [12].[1])

**15.42 Lemma.** *Let $R = [a_1, b_1] \times \cdots \times [a_n, b_n]$ be an $n$-dimensional rectangle, $\omega$ be a $\mathcal{C}^1$ $(n-1)$-form on an open set $U$ which contains $R$, and suppose $R$ has the usual orientation. If $\partial R$ carries the induced orientation, then*

$$\int_R d\omega = \int_{\partial R} \omega.$$

PROOF. Let $\widetilde{\mathcal{A}} = \{(\widetilde{V}_j, \widetilde{h}_j), (\widetilde{U}_j, \widetilde{g}_j) : j = 1, \ldots, n\}$ be the atlas of $\partial R$ introduced in Theorem 15.34, and set $H_j = \widetilde{h}_j^{-1}$, $G_j = \widetilde{g}_j^{-1}$. We claim that

(12) $$(H_j)^* \omega = (-1)^j (f_j \circ \widetilde{h}_j^{-1}) du_1 \ldots du_{n-1}$$

---

[1]M. Spivak, *Calculus on Manifolds*, (New York: W. A. Benjamin, Inc., 1965). Reprinted with permission of Addison-Wesley Publishing Company

and

(13) $$(G_j)^*\omega = (-1)^{j+1}(f_j \circ \widetilde{g}_j^{-1})du_1 \ldots du_{n-1}$$

for any $(n-1)$–form $\omega = \sum_{i=1}^{n} f_i dx_1 \ldots \widehat{dx_i} \ldots dx_n$ on $U$.

To prove (12), fix $j$ and notice by construction that

$$H_j(u_1,\ldots,u_{n-1}) = \begin{cases} (a_1,-u_1,u_2,\ldots,u_{n-1}) & j = 1 \\ ((-1)^j u_1, u_2,\ldots,a_j,\ldots,u_{n-1}) & j > 1. \end{cases}$$

Representing the $i$th component of $H_j$ by $H_{ji}$, we have

$$\Delta_i := \frac{\partial(H_{j1},\ldots,\widehat{H_{ji}},\ldots,H_{jn})}{\partial(u_1,\ldots,u_{n-1})} = \begin{cases} (-1)^j & i = j \\ 0 & i \neq j. \end{cases}$$

Hence, by the Fundamental Theorem of Differential Transforms,

$$(H_j)^*\omega = \sum_{i=1}^{n}(f_i \circ \widetilde{h}_j^{-1})\Delta_i du_1 \ldots du_{n-1} = (-1)^j(f_j \circ \widetilde{h}_j^{-1})du_1 \ldots du_{n-1}.$$

This proves (12). A similar argument proves (13).

Using (12) and (13), a change of variables in the second variable when $j = 1$ (respectively, in the first variable when $j > 1$), the Fundamental Theorem of Calculus in the $j$th variable, and Fubini's Theorem, we see that

$$\int_{\partial R}\omega = \sum_{j=1}^{n}\left(\int_{R_j^\ell}\omega + \int_{R_j^r}\omega\right)$$

$$= \sum_{j=1}^{n}(-1)^j\left(\int_{\widetilde{h}_j(\widetilde{V}_j)}(f_j \circ \widetilde{h}_j^{-1})(\boldsymbol{u})\,d\boldsymbol{u} - \int_{\widetilde{g}_j(\widetilde{U}_j)}(f_j \circ \widetilde{g}_j^{-1})(\boldsymbol{u})\,d\boldsymbol{u}\right)$$

$$= \sum_{j=1}^{n}(-1)^j\int_{a_1}^{b_1}\ldots\widehat{\int_{a_j}^{b_j}}\ldots\int_{a_n}^{b_n}$$

$$(f_j(x_1,\ldots,a_j,\ldots,x_n) - f_j(x_1,\ldots,b_j,\ldots,x_n))\cdot$$

$$\cdot dx_n \ldots \widehat{dx_j} \ldots dx_1$$

$$= \sum_{j=1}^{n}(-1)^{j+1}\int_{a_1}^{b_1}\ldots\int_{a_n}^{b_n}\frac{\partial f_j}{\partial x_j}(x_1,\ldots,x_n)\,dx_n \ldots dx_1$$

$$= \int_R \sum_{j=1}^{n}(-1)^{j+1}\frac{\partial f_j}{\partial x_j}(\boldsymbol{x})\,d\boldsymbol{x},$$

i.e.,

(14) $$\int_{\partial R}\omega = \int_R \sum_{j=1}^{n}(-1)^{j+1}\frac{\partial f_j}{\partial x_j}(\boldsymbol{x})\,d\boldsymbol{x}.$$

On the other hand, it is clear that

$$d(gdx_1 \ldots \widehat{dx_j} \ldots dx_n) = (-1)^{j+1} \frac{\partial g}{\partial x_j} dx_1 \ldots dx_n$$

for any differentiable function $g$. Thus

$$d\omega = \sum_{j=1}^{n} (-1)^{j+1} \frac{\partial f_j}{\partial x_j} dx_1 \ldots dx_n.$$

We conclude by Remark 15.41 and (14) that

$$\int_R d\omega = \int_R \sum_{j=1}^{n} (-1)^{j+1} \frac{\partial f_j}{\partial x_j}(\boldsymbol{x}) \, d\boldsymbol{x} = \int_{\partial R} \omega. \quad \blacksquare$$

**15.43 Lemma.** *Let $M$ be an $n$-dimensional orientable $\mathcal{C}^2$ manifold-with-smooth-boundary and $\mathcal{O}$ be an orientation of $M$. For each $\boldsymbol{x} \in M$ there is a chart $(V, h) \in \mathcal{O}$ at $\boldsymbol{x}$ such that if $\eta$ is any $\mathcal{C}^1$ $(n-1)$–form supported in $V$, then*

$$\int_V d\eta = \int_{V \cap \partial M} \eta.$$

PROOF. Suppose first that $\boldsymbol{x} \notin \partial M$. Then there is a chart $(U, h)$ at $\boldsymbol{x}$ such that $h(U)$ is open in $\mathbf{R}^n$. Let $R$ be an $n$-dimensional rectangle such that

$$h(\boldsymbol{x}) \in R^o \subset R \subset h(U)$$

and set $V = h^{-1}(R^o)$ (see Figure 15.3). Let $\eta$ be an $(n-1)$–form supported in $V$. We may suppose that $\eta$ is decomposable, i.e.,

$$(h^{-1})^*(\eta) = f dx_1 \ldots \widehat{dx_j} \ldots dx_n.$$

By definition and (5),

$$(h^{-1})^*(d\eta)) = d((h^{-1})^*(\eta)) = (-1)^{j-1} \frac{\partial f}{\partial x_j} dx_1 \ldots dx_n.$$

Since spt $f \subset h(V) = R^o$, it follows from Definition 15.38 and Lemma 15.42 (using the identity chart on $R$) that

$$\int_V d\eta = \int_R (-1)^{j-1} \frac{\partial f}{\partial x_j}(\boldsymbol{x}) \, d\boldsymbol{x} = \int_R d((h^{-1})^*(\eta)) = \int_{\partial R} (h^{-1})^*(\eta).$$

Since spt $((h^{-1})^*\eta) \subset R^o$ and $R^o \cap \partial R = \emptyset$, this last integral is zero, i.e., $\int_V d\eta = 0$. On the other hand, $h(U)$ is open in $\mathbf{R}^n$ so $h(V) \subseteq h(U)$ contains no boundary points of $M$. Therefore, $V \cap \partial M = \emptyset$ and

$$\int_{V \cap \partial M} \eta = 0 = \int_V d\eta.$$

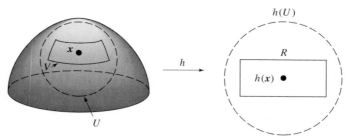

**Figure 15.3**

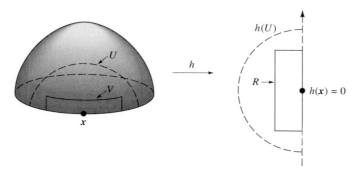

**Figure 15.4**

Next, suppose $\boldsymbol{x} \in \partial M$. Let $(U, h)$ be a chart at $\boldsymbol{x}$ such that $h(U)$ is relatively open in the left half-space $\mathcal{H}_1$ and $h(\boldsymbol{x}) = \boldsymbol{0}$. Let $R$ be a $n$-dimensional rectangle such that $R^o \subset R \subset U$ and $R \cap \partial \mathcal{H}_1 = R_1^r$ (see Figure 15.4), and set $V = h^{-1}(R^o \cup R_1^r)$. Let $\eta$ be a decomposable $\mathcal{C}^1$ $(n-1)$–form supported on $V$, with

$$(h^{-1})^*(\eta) = f\, dx_1 \ldots \widehat{dx_j} \ldots dx_n,$$

and $\boldsymbol{u} = (u_1, \ldots, u_{n-1}) \in \partial R$. Then $f$ is identically zero on $\partial R \setminus R_1^r$, and it follows from Definition 15.38 and Lemma 15.42 that

$$\int_{V \cap \partial M} \eta = \int_{R_1^r} f(\boldsymbol{u})\, d\boldsymbol{u} = \sum_{j=1}^{n} (-1)^j \left( \int_{R_j^\ell} f(\boldsymbol{u})\, d\boldsymbol{u} - \int_{R_j^r} f(\boldsymbol{u})\, d\boldsymbol{u} \right)$$

$$= \int_{\partial R} (h^{-1})^*(\eta)(\boldsymbol{u})\, d\boldsymbol{u} = \int_R (h^{-1})^*(d\eta)(\boldsymbol{x})\, d\boldsymbol{x}$$

$$= \int_{h(V)} (h^{-1})^*(d\eta)(\boldsymbol{x})\, d\boldsymbol{x} = \int_V d\eta. \;\blacksquare$$

We are now prepared to prove the general result.

**15.44 THEOREM** [STOKES'S THEOREM ON MANIFOLDS]. *Let $M$ be a compact $n$-dimensional oriented $\mathcal{C}^2$ manifold-with-smooth-boundary. If $\omega$ is a $\mathcal{C}^1$ $(n-1)$–form on $M$, then*

$$\int_M d\omega = \int_{\partial M} \omega.$$

PROOF. By Lemma 15.43, choose an open covering $\mathcal{V} = \{V_{\boldsymbol{x}}\}_{\boldsymbol{x} \in M}$ of $M$ such that $\boldsymbol{x} \in V_{\boldsymbol{x}}$ and

$$(15) \qquad \int_{V_{\boldsymbol{x}}} d\eta = \int_{V_{\boldsymbol{x}} \cap \partial M} \eta$$

for all $(n-1)$–forms $\eta$ supported in $V_{\boldsymbol{x}}$. Since $M$ is compact, choose open sets $V_j = V_{\boldsymbol{x}_j}$, $j = 1, \ldots, N$, which cover $M$ and a $\mathcal{C}^\infty$ partition of unity $\phi_1, \ldots, \phi_N$ on $M$ such that $\operatorname{spt} \phi_j \cap M \subseteq V_j$. Set $\eta_j = \phi_j \omega$ and observe that $\operatorname{spt} \eta_j \cap M \subseteq \operatorname{spt} \phi_j \cap M \subseteq V_j$ for each $j$ and $\omega = \sum_{j=1}^{N} \eta_j$. Hence, by (15),

$$\int_M d\omega = \sum_{j=1}^{N} \int_M d\eta_j = \sum_{j=1}^{N} \int_{V_j} d\eta_j$$

$$= \sum_{j=1}^{N} \int_{V_j \cap \partial M} \eta_j = \sum_{j=1}^{N} \int_{\partial M} \eta_j = \int_{\partial M} \omega. \quad \blacksquare$$

This result extends Theorems 13.49, 13.54, and 13.57 (the theorems of Green, Gauss, and Stokes) to regions with smooth boundaries (see Exercises 4 and 5 in Section 15.1). Theorem 15.44 also holds for manifolds with singularities, i.e., piecewise smooth boundaries. (For a treatment of manifolds with singularities, see Loomis and Sternberg [6].)

We close this section with an $n$-dimensional analogue of Theorem 13.61. A set $V \subset \mathbf{R}^n$ is said to be *star-shaped* (centered at $\mathbf{0}$) if for each $\boldsymbol{x} \in V$ the line segment between $\boldsymbol{x}$ and $\mathbf{0}$ lies in $V$, i.e., $t\boldsymbol{x} \in V$ for all $0 \le t \le 1$. An $r$–form $\omega$ is said to be *exact* on $V$ if there is an $(r-1)$–form $\eta$ on $V$ such that $d\eta = \omega$.

**15.45 THEOREM** [THE POINCARÉ LEMMA]. *Let $V$ be an open star-shaped set in $\mathbf{R}^n$ and $\omega$ be a $\mathcal{C}^1$ $r$–form on $V$. Then $\omega$ is exact on $V$ if and only if $d\omega = 0$ on $V$.*

PROOF. For each $r$–form

$$\omega = \sum_{1 \le i_1 < \cdots < i_r \le n} f_{i_1, \ldots, i_r}(\boldsymbol{x}) \, dx_{i_1} \ldots dx_{i_r}$$

on $V$, define an $(r-1)$–form $\Lambda(\omega)$ on $V$ by

$$(16) \qquad \Lambda(\omega) = \sum_{1 \le i_1 < \cdots < i_r \le n} \sum_{k=1}^{r} (-1)^{k-1} \left( \int_0^1 t^{r-1} f_{i_1, \ldots, i_r}(t\boldsymbol{x}) \, dt \right) x_{i_k} \cdot$$
$$\cdot \, dx_{i_1} \ldots \widehat{dx_{i_k}} \ldots dx_{i_r}.$$

Since $V$ is star-shaped and $f$ is defined on $V$, the integrals in (16) make sense for each $\boldsymbol{x} \in V$. Thus $\Lambda(\omega)$ is an $(r-1)$–form on $V$. We claim that

$$(17) \qquad \Lambda(d\omega) + d(\Lambda(\omega)) = \omega$$

for every $r$–form $\omega$ on $V$.

To prove (17) we may suppose that $\omega$ is decomposable, i.e., $\omega = f\, dx_{i_1} \ldots dx_{i_r}$. By definition,

$$d\omega = \sum_{j=1}^{n} \frac{\partial f}{\partial x_j}\, dx_j\, dx_{i_1} \ldots dx_{i_r}.$$

is an $(r+1)$–form on $V$. Letting $i_0 = j$, we have by (16) that

(18)

$$\Lambda(d\omega) = \sum_{k=0}^{r}(-1)^k \sum_{j=1}^{n}\left(\int_0^1 t^r \frac{\partial f}{\partial x_j}(t\boldsymbol{x})\, dt\right) x_{i_k}\, dx_{i_0}\, dx_{i_1} \ldots \widehat{dx_{i_k}} \ldots dx_{i_r}$$

$$= \sum_{j=1}^{n}\left(\int_0^1 t^r \frac{\partial f}{\partial x_j}(t\boldsymbol{x})\, dt\right) x_j\, dx_{i_1} \ldots dx_{i_r}$$

$$- \sum_{j=1}^{n}\sum_{k=1}^{r}(-1)^{k-1}\left(\int_0^1 t^r \frac{\partial f}{\partial x_j}(t\boldsymbol{x})\, dt\right) x_{i_k} \cdot$$

$$\cdot\, dx_j\, dx_{i_1} \ldots \widehat{dx_{i_k}} \ldots dx_{i_r}$$

On the other hand, by the Product Rule, differentiating under the integral sign (see Theorem 11.5), and the Chain Rule, we have

$$d\left(\left(\int_0^1 t^{r-1} f(t\boldsymbol{x})\, dt\right) x_{i_k}\right)$$

$$= d\left(\int_0^1 t^{r-1} f(t\boldsymbol{x})\, dt\right) x_{i_k} + \left(\int_0^1 t^{r-1} f(t\boldsymbol{x})\, dt\right) dx_{i_k}$$

$$= \sum_{j=1}^{n}\left(\int_0^1 t^r \frac{\partial f}{\partial x_j}(t\boldsymbol{x})\, dt\right) x_{i_k}\, dx_j + \left(\int_0^1 t^{r-1} f(t\boldsymbol{x})\, dt\right) dx_{i_k}.$$

Thus, by the Anticommutative Property, the exterior derivative of $\Lambda(\omega)$ is

(19)

$$d(\Lambda(\omega)) = \sum_{j=1}^{n}\sum_{k=1}^{r}(-1)^{k-1}\left(\int_0^1 t^r \frac{\partial f}{\partial x_j}(t\boldsymbol{x})\, dt\right) x_{i_k}\, dx_j\, dx_{i_1} \ldots \widehat{dx_{i_k}} \ldots dx_{i_r}$$

$$+ r\left(\int_0^1 t^{r-1} f(t\boldsymbol{x})\, dt\right) dx_{i_1} \ldots dx_{i_r}.$$

Adding (18) and (19), we obtain by the Product Rule and the one-dimensional

Fundamental Theorem of Calculus that

$$\Lambda(d\omega) + d(\Lambda(\omega)) = \sum_{j=1}^{n} \left( \int_0^1 t^r \frac{\partial f}{\partial x_j}(t\boldsymbol{x})\, dt \right) x_j \, dx_{i_1} \dots dx_{i_r}$$

$$+ r \left( \int_0^1 t^{r-1} f(t\boldsymbol{x})\, dt \right) dx_{i_1} \dots dx_{i_r}$$

$$= \left( \int_0^1 \frac{d}{dt} \left( t^r f(t\boldsymbol{x}) \right) dt \right) dx_{i_1} \dots dx_{i_r}$$

$$= f(\boldsymbol{x}) \, dx_{i_1} \dots dx_{i_r} = \omega.$$

This proves (17).

Theorem 15.45 is now easy to prove. If $\omega$ is exact and $\mathcal{C}^1$, then there is a $\mathcal{C}^2$ form $\eta$ such that $d\eta = \omega$. Thus $d\omega = d^2\eta = 0$ by Theorem 15.9. (This part works whether $V$ is star-shaped or not.)

Conversely, if $d\omega = 0$, then by (17), $d(\Lambda(\omega)) = \omega$. Thus set $\eta = \Lambda(\omega)$.  ∎

## EXERCISES

**1.** Compute $\int_{\partial B_a(0,0,0,0)} x^3 \, dy\, dz\, dw + y^2 \, dx\, dz\, dw$.

**2.** Compute $\int_M \sum_{j=1}^{n} x_j^2 dx_1 \dots \widehat{dx_j} \dots dx_n$, where $M$ is the boundary of the unit $n$-dimensional rectangle $Q = [0, a_1] \times \cdots \times [0, a_n]$.

**3.** Let $E$ be a compact $n$-dimensional Jordan region in $\mathbf{R}^n$, $n > 1$. If $\partial E$ is an $(n-1)$–dimensional manifold, prove

$$\int_{\partial E} \sum_{j=1}^{n} dx_1 \dots \widehat{dx_j} \dots dx_n = \begin{cases} \text{Vol}(E) & \text{if } n \text{ is odd} \\ 0 & \text{if } n \text{ is even.} \end{cases}$$

**4.** Let $r \in \mathbf{N}$, $m > n = 2r + 2$, $V$ be a star-shaped open set in $\mathbf{R}^m$, and $M$ be a compact $n$-dimensional $\mathcal{C}^2$ manifold-with-smooth-boundary in $\mathbf{R}^m$. If $M \subset V$ and $\omega$ is an exact $\mathcal{C}^1$ $r+1$–form on $V$ with $\omega = d\eta$, prove

$$\int_{\partial M} \eta\omega = \int_M \omega^2.$$

# Appendices

## A. ALGEBRAIC LAWS

In this section we derive several consequences of the ordered field axioms (i.e., Postulates 1 and 2 in Section 1.1).

**A.1 THEOREM.** *Let $x, a \in \mathbf{R}$.*
  i) *If $a = x + a$, then $x = 0$.*
  ii) *If $a = x \cdot a$ and $a \neq 0$, then $x = 1$.*

PROOF. i) Since the additive inverse of $a$ exists, we can add $-a$ to the equation $a = x + a$. Using the Associative Property, and the fact that 0 is the additive identity, we obtain

$$0 = a + (-a) = (x + a) + (-a) = x + (a + (-a)) = x + 0 = x.$$

ii) Since the multiplicative inverse of $a$ exists, we can multiply $a = x \cdot a$ by $a^{-1}$. Using the Associative Property, and the fact that 1 is the multiplicative identity, we obtain

$$1 = a \cdot a^{-1} = (x \cdot a) \cdot a^{-1} = x \cdot (a \cdot a^{-1}) = x \cdot 1 = x. \quad \blacksquare$$

Theorem A.1 shows that the additive and multiplicative identities are unique. The following result shows that additive and multiplicative inverses are also unique. Thus "unique" can be dropped from the statements in Postulate 1.

**A.2 THEOREM.**
  i) *If $a, b \in \mathbf{R}$ and $a + b = 0$ then $b = -a$.*
  ii) *If $a, b \in \mathbf{R}$ and $ab = 1$ then $b = a^{-1}$.*

PROOF. i) By hypothesis and the Associative Property,

$$-a = -a + (a + b) = (-a + a) + b = 0 + b = b.$$

ii) Since $1 \neq 0$, $a \neq 0$. Thus it follows from hypothesis and the Associative Property that

$$a^{-1} = a^{-1}(ab) = (a^{-1}a)b = 1 \cdot b = b. \quad \blacksquare$$

**A.3 THEOREM.** *For all $a, b \in \mathbf{R}$, $0 \cdot a = 0$, $-a = (-1) \cdot a$, $-(-a) = a$, $(-1)^2 = 1$, and $-(a - b) = b - a$.*

PROOF. Since 1 is the multiplicative identity and 0 is the additive identity, it follows from the Distributive Property that

$$a + 0 \cdot a = 1 \cdot a + 0 \cdot a = (1 + 0) \cdot a = 1 \cdot a = a.$$

$$a + x = a \implies x = 0$$

Hence, by Theorem A.1, $0 \cdot a = 0$. Similarly,

$$a + (-1) \cdot a = (1 + (-1)) \cdot a = 0 \cdot a = 0.$$

Since additive inverses are unique, it follows that $(-1) \cdot a = -a$. Since $-a + a = a + (-a) = 0$, a similar argument proves $-(-a) = a$. Substituting $a = -1$, we have

$$(-1)(-1) = -(-1) = 1.$$

Finally, for any $a, b \in \mathbf{R}$, we also have

$$-(a - b) = (-1)(a - b) = (-1)a + (-1)(-b) = -a + b = b - a. \quad \blacksquare$$

**A.4 THEOREM.** *Let $a, b, c \in \mathbf{R}$.*

    i) *If $a \cdot b = 0$, then $a = 0$ or $b = 0$.*
    ii) *If $a \cdot b = a \cdot c$ and $a \neq 0$, then $b = c$.*

PROOF. i) If $a = 0$, we are done. If $a \neq 0$, then multiplying the identity $0 = a \cdot b$ by $a^{-1}$, we have

$$0 = a^{-1} \cdot 0 = a^{-1} \cdot (a \cdot b) = (a^{-1} \cdot a) \cdot b = 1 \cdot b = b.$$

ii) If $a \cdot b = a \cdot c$, then by Theorem A.3 we have

$$a \cdot (b - c) = a \cdot (b + (-1)c) = a \cdot b + (-1)a \cdot c = a \cdot b - a \cdot c = 0.$$

Since $a \neq 0$, it follows from part i) that $b - c = 0$, i.e., $b = c$. $\quad \blacksquare$

A subset $E$ of $\mathbf{R}$ is called *inductive* if

$$(1) \hspace{10em} 1 \in E$$

and

$$(2) \hspace{6em} \text{for every } x \in E, \ x + 1 \text{ also belongs to } E.$$

Notice by Postulate 1, $\mathbf{R}$ is an inductive set.

Define $\mathbf{N}$ to be the set of elements which belong to ALL inductive sets, and set $\mathbf{Z} := \{k \in \mathbf{R} : k \in \mathbf{N}, -k \in \mathbf{N}, \text{ or } k = 0\}$. Notice that $\mathbf{N}$ is the smallest inductive set, i.e., $\mathbf{N} \subseteq E$ for any inductive set $E$. Indeed, if $k \in \mathbf{N}$ and $E$ is inductive, then by definition, $k \in E$.

We first show that $\mathbf{N}$ and $\mathbf{Z}$, as defined above, satisfy the assumptions we made in Remark 1.1.

## A.5 THEOREM.

   i) *Given $n \in \mathbf{Z}$, one and only one of the following statements holds: $n \in \mathbf{N}$, $-n \in \mathbf{N}$, or $n = 0$.*

   ii) *$n \in \mathbf{N}$ implies $n + 1 \in \mathbf{N}$ and $n \geq 1$.*

   iii) *If $n \in \mathbf{N}$ and $n \neq 1$, then $n - 1 \in \mathbf{N}$.*

   iv) *If $n \in \mathbf{Z}$ and $n > 0$, then $n \in \mathbf{N}$.*

PROOF. i) Since $(0, \infty)$ is an inductive set, all elements of $\mathbf{N}$ are positive. By the definition of $\mathbf{Z}$, given $n \in \mathbf{Z}$, one of the following statements holds: $n \in \mathbf{N}$, $-n \in \mathbf{N}$, or $n = 0$. It follows that either $n > 0$, $n < 0$, or $n = 0$. Since the Trichotomy Property implies only one of these conditions can hold for a given $n$, property i) is proved.

ii) Since $\mathbf{N}$ is inductive, $n \in \mathbf{N}$ implies $n + 1 \in \mathbf{N}$. Since $[1, \infty)$ is an inductive set, every $n \in \mathbf{N}$ satisfies $n \geq 1$.

iii) Suppose to the contrary that there is an $n_0 \in \mathbf{N}$ such that $n_0 \neq 1$ and $n_0 - 1 \notin \mathbf{N}$. Consider the set $E := \{k \in \mathbf{N} : k \neq n_0\}$. Since $1 \neq n_0$, $1 \in E$. If $x \in E$, then $x \in \mathbf{N}$; hence, $x \neq n_0 - 1$. It follows that $x + 1 \neq n_0$, i.e., $x + 1 \in E$. Thus $E$ is an inductive set. Since $E \subset \mathbf{N}$, this contradicts the fact that $\mathbf{N}$ was defined to be the smallest inductive set.

iv) Suppose to the contrary that $n \in \mathbf{Z}$, $n > 0$, but $n \notin \mathbf{N}$. Then by part i), $-n \in \mathbf{N}$, so by part ii), $-n \geq 1$. Using the second Multiplicative Property, it follows that $n \leq -1 < 0$, a contradiction. ∎

## A.6 COROLLARY. *If $x, e \in \mathbf{N}$ and $x < e$, then $x + 1 \leq e$.*

PROOF. By hypothesis, $e - x > 0$. Hence, by Theorem A.5ii) and iv), it suffices to show that $e - x \in \mathbf{Z}$. Consider the set $A = \{k \in \mathbf{N} : k - 1 \in \mathbf{Z}\}$. Clearly, $1 \in A$. Moreover, if $k \in A$, then $(k + 1) - 1 = k \in \mathbf{N} \subset \mathbf{Z}$. Thus $A$ is an inductive set, i.e., contains $\mathbf{N}$. In particular, $e \in A$. Similarly, $B = \{k \in \mathbf{N} : e - k \in \mathbf{Z}\}$ is an inductive set, hence, contains $x$. We conclude that $e - x \in \mathbf{Z}$. ∎

We close this section by proving that, under mild assumptions, the Axiom of Induction and the Well-Ordering Principle are equivalent.

## A.7 THEOREM. *Suppose the Ordered Field axioms and Theorem A.5iii) hold. Then the Axiom of Induction holds if and only if the Well-Ordering Principle holds.*

PROOF. By the proof of Theorem 1.11 (which used Theorem A.5iii) at a crucial spot), the Well-Ordering Principle implies the Axiom of Induction.

Conversely, if the Axiom of Induction holds, then the elements of $\mathbf{N}$ belong to every inductive set. This is the "definition" of $\mathbf{N}$ we made below (2) above. It follows that all results above are valid. (We will use Theorem A.5ii) and Corollary A.6.)

Suppose $E$ is a nonempty subset of $\mathbf{N}$, and consider the set

$$A := \{x \in \mathbf{N} : x \leq e \quad \text{for all} \quad e \in E\}.$$

$A$ is nonempty since $1 \in A$. $A$ is not the whole set $\mathbf{N}$ since if $e_0 \in E$ then $e_0 + 1$ cannot belong to $A$. Hence, by the Axiom of Induction, $A$ cannot be an inductive set. In particular, there is an $x \in A$ such that $x + 1 \notin A$.

We claim that this $x$ is a least element of $E$; i.e., $x$ is a lower bound of $E$ and $x \in E$. That $x$ is a lower bound of $E$ is obvious, since by construction, $x \in A$ implies $x \leq e$ for

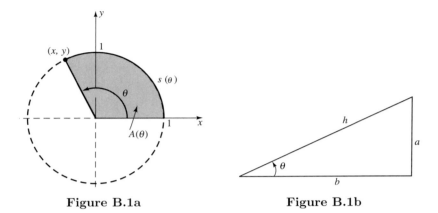

**Figure B.1a**                    **Figure B.1b**

all $e \in E$. On the other hand, if $x \notin E$, then $x < e$ for all $e \in E$. Hence, by Corollary A.6, $x + 1 \leq e$ for all $e \in E$, i.e., $x + 1 \in A$, a contradiction of the choice of $x$. ∎

## B. TRIGONOMETRY

In this section we derive some trigonometric identities, by using elementary geometry and algebra.

Let $(x, y)$ be a point on the unit circle $x^2 + y^2 = 1$ and $\theta$ be the angle measured counterclockwise from the positive $x$ axis to the line segment from $(0,0)$ to $(x,y)$ (see Figure B.1a). (We shall refer to $(x, y)$ as the point *determined* by the angle $\theta$.) Define

$$\sin\theta = y, \qquad \cos\theta = x, \quad \text{and} \quad \tan\theta = \frac{y}{x}.$$

By the Law of Similar Triangles, given a right triangle with base angle $\theta$, altitude $a$, base $b$, and hypotenuse $h$ (see Figure B.1b), $\sin\theta = a/h$, $\cos\theta = b/h$, and $\tan\theta = a/b = \sin\theta/\cos\theta$.

**B.1 THEOREM.** *Given a circle $C : x^2 + y^2 = r^2$ of radius $r$, let $s(\theta)$ represent the length of the arc on $C$ swept out by $\theta$, and $A(\theta)$ represent the area of the angular sector swept out by $\theta$ (see Figure B.1a). If the angle $\theta$ is measured in radians (not degrees), then*

$$s(\theta) = r\theta \quad \text{and} \quad A(\theta) = \frac{r^2\theta}{2}.$$

PROOF. Since there are $2\pi$ radians in a complete circle and the circumference of a circle of radius $r$ is $2\pi r$, we have

$$\frac{s(\theta)}{2\pi r} = \frac{\theta}{2\pi},$$

i.e., $s(\theta) = r\theta$. Similarly, since the area of a circle is $\pi r^2$, we have

$$\frac{A(\theta)}{\pi r^2} = \frac{\theta}{2\pi},$$

i.e., $A(\theta) = r^2\theta/2$. ∎

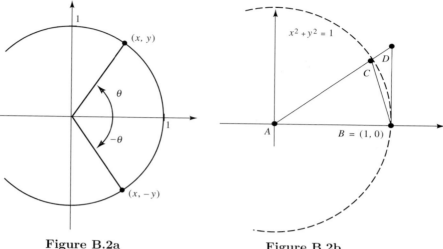

**Figure B.2a**            **Figure B.2b**

## B.2 THEOREM.

    i) $\sin(0) = 0$ *and* $\cos(0) = 1$.

    ii) *For any* $\theta \in \mathbf{R}$, $|\sin\theta| \leq 1$, $|\cos\theta| \leq 1$, $\sin(-\theta) = -\sin\theta$, $\cos(-\theta) = \cos\theta$, *and* $\sin^2\theta + \cos^2\theta = 1$.

    iii) *If* $\theta$ *is measured in radians, then* $\sin(\pi/2) = 1$, $\cos(\pi/2) = 0$, $\sin(\theta + 2\pi) = \sin\theta$, *and* $\cos(\theta + 2\pi) = \cos\theta$. *Moreover, if* $0 < \theta < \pi/2$, *then* $0 < \theta\cos\theta < \sin\theta < \theta$.

    iv) *If* $\theta \in \mathbf{R}$ *is measured in radians, then* $|\sin\theta| \leq |\theta|$.

PROOF. Let $\theta \in \mathbf{R}$ and $(x, y)$ be the point on the unit circle determined by $\theta$.

i) If $\theta = 0$, then $(x, y) = (1, 0)$ (see Figure B.1a). Hence, $\sin(0) = 0$ and $\cos(0) = 1$.

ii) Clearly, $|\sin\theta| = |y| = \sqrt{y^2} \leq \sqrt{x^2 + y^2} = 1$, and similarly, $|\cos\theta| \leq 1$. By definition (see Figure B.2a), $\sin(-\theta) = -y = -\sin\theta$ and $\cos(-\theta) = x = \cos\theta$. Moreover,

$$\sin^2\theta + \cos^2\theta = x^2 + y^2 = 1.$$

iii) If $\theta = \pi/2$, then $(x, y) = (0, 1)$, so $\sin(\pi/2) = 1$ and $\cos(\pi/2) = 0$. Fix $\theta \in (0, \pi/2)$ and consider Figure B.2b. Since $\sin\theta$ is the altitude of triangle $ABC$ and the shortest distance between two points is a straight line, we have by Theorem B.1 that

$$\sin\theta < s(\theta) = \theta.$$

On the other hand, the triangle $ABC$ is a proper subset of the angular sector swept out by $\theta$, which is a proper subset of the triangle $ABD$. Hence,

$$\text{Area}\,(ABC) < A(\theta) < \text{Area}\,(ABD).$$

Since the area of a triangle is one-half the product of its base and its altitude, it follows from Theorem B.1 that

(3)                         $$\frac{\sin\theta}{2} < \frac{\theta}{2} < \frac{\tan\theta}{2}.$$

But $0 < \cos\theta < 1$ for all $\theta \in (0, \pi/2)$. Multiplying (3) by $2\cos\theta$, we conclude that

$$(4) \qquad\qquad \sin\theta\cos\theta < \theta\cos\theta < \sin\theta.$$

iv) By part iii), $|\sin\theta| = \sin\theta \le \theta = |\theta|$ for all $0 \le \theta \le \pi/2$. Since $\sin(-\theta) = -\sin\theta$, it follows that $|\sin\theta| \le |\theta|$ for all $\theta \in [-\pi/2, \pi/2]$. But if $\theta \notin [-\pi/2, \pi/2]$, then $|\sin\theta| \le 1 < \pi/2 < |\theta|$. Therefore, $|\sin\theta| \le |\theta|$ for all $\theta \in \mathbf{R}$. $\blacksquare$

The next result shows how to compute the sine and cosine of a sum of angles.

**B.3 THEOREM.**

i) [SUM-ANGLE FORMULAS]. *If $\theta, \varphi \in \mathbf{R}$, then*

$$\cos(\theta \pm \varphi) = \cos\theta\cos\varphi \mp \sin\theta\sin\varphi,$$

*and*

$$\sin(\theta \pm \varphi) = \sin\theta\cos\varphi \pm \cos\theta\sin\varphi.$$

ii) [DOUBLE-ANGLE FORMULAS]. *If $\theta \in \mathbf{R}$, then*

$$\cos^2\theta = \frac{1 + \cos(2\theta)}{2},$$

$$\sin^2\theta = \frac{1 - \cos(2\theta)}{2},$$

*and*

$$\cos\theta = 1 - 2\sin^2(\theta/2).$$

iii) [SHIFT-FORMULAS]. *If $\varphi$ is measured in radians, then*

$$\sin\varphi = \cos\left(\frac{\pi}{2} - \varphi\right),$$

*and*

$$\cos\varphi = \sin\left(\frac{\pi}{2} - \varphi\right)$$

*for all $\varphi \in \mathbf{R}$.*

PROOF. Suppose first that $\theta > \varphi$. Consider the chord $A$ cut from the unit circle by a central angle $\theta - \varphi$, and the chord $B$ cut from the unit circle by a central angle $\varphi - \theta$ (see Figure B.3). Since $\sin^2\theta + \cos^2\theta = 1$, we have

$$(5) \qquad A^2 = (\cos\theta - \cos\varphi)^2 + (\sin\theta - \sin\varphi)^2 = 2 - 2(\cos\theta\cos\varphi + \sin\theta\sin\varphi),$$

and

$$B^2 = (\cos(\theta - \varphi) - 1)^2 + (\sin(\theta - \varphi))^2 = 2 - 2\cos(\theta - \varphi).$$

Since $|\theta - \varphi| = |\varphi - \theta|$, the lengths of these chords must be equal. Thus

$$(6) \qquad\qquad \cos(\theta - \varphi) = \cos\theta\cos\varphi + \sin\theta\sin\varphi$$

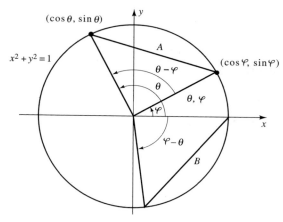

**Figure B.3**

for $\theta < \varphi$. A similar argument establishes (6) for $\varphi < \theta$. Since (6) is trivial when $\theta = \varphi$, we have proved that (6) holds for all $\theta$ and $\varphi$. Combining the identities $\sin(-\theta) = -\sin\theta$ and $\cos(-\theta) = \cos\theta$ with (6), we obtain

$$\cos(\theta + \varphi) = \cos(\theta - (-\varphi)) = \cos\theta\cos\varphi - \sin\theta\sin\varphi.$$

This and (6) verify the first identity in part i).

Applying this identity to $\theta = \pi/2$, we see by Theorem B.2ii) that

$$\cos\left(\frac{\pi}{2} - \varphi\right) = \cos\left(\frac{\pi}{2}\right)\cos\varphi + \sin\left(\frac{\pi}{2}\right)\sin\varphi = \sin\varphi;$$

i.e., the first identity in part iii) holds. Combining the first identities in parts i) and iii), we obtain

$$\sin(\theta \pm \varphi) = \cos\left(\left(\frac{\pi}{2} - \theta\right) \mp \varphi\right)$$
$$= \cos(\frac{\pi}{2} - \theta)\cos(-\varphi) \mp \sin(\frac{\pi}{2} - \theta)\sin(-\varphi)$$
$$= \sin\theta\cos\varphi \pm \cos\theta\sin\varphi.$$

This proves the second identity in part i). Specializing to the case $\theta = \pi/2$, we obtain $\sin(\pi/2 - \varphi) = \cos\varphi$. Thus parts i) and iii) have been proved.

To establish part ii), notice by part i) and Theorem B.2 that

$$\cos(2\theta) = \cos(\theta + \theta) = \cos^2\theta - \sin^2\theta = 2\cos^2\theta - 1.$$

Hence, $\cos^2\theta = (1 + \cos(2\theta))/2$. Similar arguments establish the rest of part ii). ∎

We close this section with the Law of Cosines, a generalization of the Pythagorean Theorem.

**B.4 THEOREM** [LAW OF COSINES]. *If $T$ is a triangle with sides of length $a$, $b$, $c$, and $\theta$ is the angle opposite the side of length $c$, then*

$$c^2 = a^2 + b^2 - 2ab\cos\theta.$$

PROOF. Suppose without loss of generality that $\theta$ is acute, and rotate $T$ so $b$ is its base. Let $h$ be the altitude of $T$, and notice that $h$ cuts a right triangle out of $T$ whose sides are $a$ and $h$ and the angle opposite $h$ is $\theta$. By the definition of $\sin\theta$ and $\cos\theta$, $h = a\sin\theta$ and the length $d$ of the base of this right triangle is $d = b - a\cos\theta$. Substituting these values into the equation $c^2 = h^2 + d^2$ (which follows directly from the Pythagorean Theorem), we obtain

$$c^2 = (a\sin\theta)^2 + (b - a\cos\theta)^2$$
$$= a^2\sin^2\theta + a^2\cos^2\theta + b^2 - 2ab\cos\theta = a^2 + b^2 - 2ab\cos\theta. \;\blacksquare$$

## C.  MATRICES AND DETERMINANTS

In this section we prove several elementary results about matrices and determinants. We assume the student is familiar with the concept of row and column reduction to canonical form.

Recall that an $m \times n$ matrix $B$ is a rectangular array which has $m$ rows and $n$ columns:

$$B = [b_{ij}]_{m \times n} = \begin{bmatrix} b_{11} & b_{12} & \ldots & b_{1n} \\ b_{21} & b_{22} & \ldots & b_{2n} \\ \vdots & \vdots & \ddots & \vdots \\ b_{m1} & b_{m2} & \ldots & b_{mn} \end{bmatrix}.$$

The notation $b_{ij}$ indicates the *entry* in the $i$th row and $j$th column. We shall call $B$ *real* if all its entries $b_{ij}$ belong to **R**.

**C.1 DEFINITION.** Let $B = [b_{ij}]_{m \times n}$ and $C = [c_{k\ell}]_{p \times q}$ be real matrices.

i) $B$ and $C$ are said to be *equal* if $m = p$, $n = q$, and $b_{ij} = c_{ij}$ for $i = 1, 2, \ldots, m$, and $j = 1, 2, \ldots, n$.

ii) The $m \times n$ *zero matrix* is the matrix $O = O_{m \times n} = [b_{ij}]_{m \times n}$ where $b_{ij} = 0$ for $i = 1, \ldots, m$, and $j = 1, \ldots, n$.

iii) The $n \times n$ *identity matrix* is the matrix $I = I_{n \times n} = [b_{ij}]_{n \times n}$ where $b_{ii} = 1$ for $i = 1, \ldots, n$, and $b_{ij} = 0$ for $i \neq j$, $i, j = 1, \ldots, n$.

iv) The *product* of a matrix $B$ and a scalar $\alpha$ is defined by

$$\alpha B = [\alpha b_{ij}]_{m \times n}.$$

v) The *negative* of a matrix $B$ is defined by $-B = (-1)B$.

vi) When $m = p$ and $n = q$, the *sum* of $B$ and $C$ is defined by

$$B + C = [b_{ij} + c_{ij}]_{m \times n}.$$

vii) When $n = p$, the *product* of $B$ and $C$ is defined by

$$BC = \left[ \sum_{\nu=1}^{n} b_{i\nu} c_{\nu j} \right]_{m \times q}.$$

**C.2 Example.** Compute $B + C$, $3B$, $-C$, $BC$, and $CB$, where

$$B = \begin{bmatrix} 1 & 0 \\ 2 & 3 \end{bmatrix} \quad \text{and} \quad C = \begin{bmatrix} -1 & 1 \\ -2 & 0 \end{bmatrix}.$$

SOLUTION. By definition,

$$B + C = \begin{bmatrix} 0 & 1 \\ 0 & 3 \end{bmatrix}, \qquad 3B = \begin{bmatrix} 3 & 0 \\ 6 & 9 \end{bmatrix},$$

$$-C = \begin{bmatrix} 1 & -1 \\ 2 & 0 \end{bmatrix}, \qquad BC = \begin{bmatrix} -1 & 1 \\ -8 & 2 \end{bmatrix}, \quad \text{and} \quad CB = \begin{bmatrix} 1 & 3 \\ -2 & 0 \end{bmatrix}. \quad \blacksquare$$

These operations do not satisfy all the usual laws of algebra. (For example, the last two computations show matrix multiplication is not commutative.) Here is a list of algebraic laws satisfied by real matrices.

**C.3 THEOREM.** *Let $A = [a_{ij}]$, $B = [b_{ij}]$, and $C = [c_{ij}]$ be real matrices and $\alpha, \beta$ be scalars.*

i) $(\alpha + \beta)C = \alpha C + \beta C.$

ii) *If $B + C$ is defined, then $\alpha(B + C) = \alpha B + \alpha C$, and $B + C = C + B$.*

iii) *If $BC$ is defined, then $\alpha(BC) = (\alpha B)C = B(\alpha C)$.*

iv) *If $AB$ and $AC$ are defined, then $A(B + C) = AB + AC$. If $BA$ and $CA$ are defined, then $(B + C)A = BA + CA$.*

v) *If $A + B$ and $B + C$ are defined, then $(A + B) + C = A + (B + C)$. If $AB$ and $BC$ are defined, then $(AB)C = A(BC)$.*

vi) *If $B$ is an $m \times n$ matrix, then*

$$B + O_{m\times n} = B, \qquad B - B = O_{m\times n},$$

$$BO_{n\times q} = O_{m\times q}, \qquad O_{p\times m}B = O_{p\times n}, \quad \text{and} \quad 0B = O_{m\times n}.$$

vii) *If $B$ is an $n \times n$ matrix, then*

$$I_{n\times n}B = BI_{n\times n} = B.$$

PROOF. By definition,

$$(\alpha + \beta)C = [(\alpha + \beta)c_{ij}] = [\alpha c_{ij} + \beta c_{ij}] = \alpha C + \beta C,$$

and

$$\alpha(B + C) = \alpha[b_{ij} + c_{ij}] = [\alpha(b_{ij} + c_{ij})] = [\alpha b_{ij}] + [\alpha c_{ij}] = \alpha B + \alpha C.$$

A similar argument establishes $B + C = C + B$.

Let $B$ be an $m \times n$ matrix and $C$ be an $n \times q$ matrix. By definition,

$$\alpha(BC) = \left[\alpha \sum_{\nu=1}^{n} b_{i\nu}c_{\nu j}\right] = \left[\sum_{\nu=1}^{n}(\alpha b_{i\nu})c_{\nu j}\right] = (\alpha B)C.$$

A similar argument establishes $\alpha(BC) = B(\alpha C)$.

Let $A$ be an $m \times n$ matrix and $B, C$ be $n \times q$ matrices. By definition,

$$A(B+C) = \left[\sum_{\nu=1}^{n} a_{i\nu}(b_{\nu j} + c_{\nu j})\right] = \left[\sum_{\nu=1}^{n} a_{i\nu}b_{\nu j} + \sum_{\nu=1}^{n} a_{i\nu}c_{\nu j}\right] = AB + AC.$$

A similar argument establishes $(B+C)A = BA + CA$.

Let $A$ be an $m \times n$ matrix, $B$ be an $n \times p$ matrix, and $C$ be a $p \times q$ matrix. By definition,

$$(AB)C = \left[\sum_{\nu=1}^{n} a_{i\nu}b_{\nu j}\right][c_{jk}]$$

$$= \left[\sum_{j=1}^{p}\left(\sum_{\nu=1}^{n} a_{i\nu}b_{\nu j}\right)c_{jk}\right]$$

$$= \left[\sum_{\nu=1}^{n}\left(a_{i\nu}\sum_{j=1}^{p} b_{\nu j}c_{jk}\right)\right]$$

$$= A\left[\sum_{j=1}^{p} b_{\nu j}c_{jk}\right] = A(BC).$$

A similar argument establishes $(A+B)+C = A+(B+C)$.

By definition,

$$B + O_{m \times n} = [b_{ij} + 0] = [b_{ij}] = B, \qquad B - B = [b_{ij} - b_{ij}] = O_{m \times n},$$

$$BO_{n \times q} = [\sum_{\nu=1}^{n} b_{i\nu} \cdot 0] = O_{m \times q}, \qquad O_{p \times m}B = [\sum_{\nu=1}^{m} 0 \cdot b_{\nu j}] = O_{p \times n},$$

and $0 \cdot B = [0 \cdot b_{ij}] = O_{m \times n}$. And, since $I = [\delta_{ij}]$, where

$$\delta_{ij} = \begin{cases} 1 & i = j, \\ 0 & i \neq j, \end{cases}$$

we have $I_{n \times n}B = [\sum_{\nu=1}^{n} \delta_{i\nu}b_{\nu j}] = [b_{ij}] = B = BI_{n \times n}$. ∎

A square matrix is a matrix with as many rows as columns. Clearly, if $B$ and $C$ are square real matrices of the same size, then both $B + C$ and $BC$ are defined. This gives room for more algebraic structure. An $n \times n$ real matrix $B$ is said to be *invertible* if there is an $n \times n$ matrix $B^{-1}$, called the *inverse* of $B$, which satisfies

$$BB^{-1} = B^{-1}B = I.$$

The following result shows that matrix inverses are unique.

**C.4 THEOREM.** *Let $A$, $B$ be $n \times n$ real matrices. If $B$ is invertible and $BA = I$, then $B^{-1} = A$.*

PROOF. By Theorem C.3 and definition,

$$B^{-1} = B^{-1}I = B^{-1}(BA) = (B^{-1}B)A = IA = A. \quad \blacksquare$$

If $B = [b_{ij}]_{n \times n}$ is square, recall that the *minor matrix* $B_{ij}$ of $B$ is the $(n-1) \times (n-1)$ matrix obtained by removing the $i$th row and the $j$th column from $B$. For example, if

$$B = \begin{bmatrix} 1 & 2 & 3 \\ -1 & -2 & -3 \\ 4 & 5 & 6 \end{bmatrix},$$

then

$$B_{21} = \begin{bmatrix} 2 & 3 \\ 5 & 6 \end{bmatrix}.$$

Minor matrices can be used to define an operation on square real matrices (the determinant) which makes invertible matrices easy to identify (see Theorem C.6 below).

The *determinant* can be defined recursively as follows. Let $B$ be an $n \times n$ real matrix.

  i) If $n = 1$, then the determinant of $B$ is defined by $\det[b] = b$.

  ii) If $n = 2$, then the determinant of $B$ is defined by

$$\det \begin{bmatrix} a & b \\ c & d \end{bmatrix} = ad - bc.$$

  iii) If $n > 2$, then the determinant of $B$ is defined recursively by

$$\det[b_{ij}]_{n \times n} = b_{11} \det B_{11} - b_{12} \det B_{12} + \cdots + (-1)^{n-1} b_{1n} \det B_{1n},$$

    where $B_{1j}$ are minor matrices of $B$.

The following result shows what an elementary column operation does to the determinant of a matrix.

**C.5 THEOREM.** *Let $B = [b_{ij}]$ and $C = [c_{ij}]$ be $n \times n$ real matrices, $n \geq 2$.*

  i) *If $C$ is obtained from $B$ by interchanging two columns, then $\det C = -\det B$.*

  ii) *If $C$ is obtained from $B$ by multiplying one column of $B$ by a scalar $\alpha$, then $\det C = \alpha \det B$.*

  iii) *If $C$ is obtained from $B$ by multiplying one column of $B$ by a scalar and adding it to another column of $B$, then $\det C = \det B$.*

PROOF. Since

$$\det \begin{bmatrix} a & b \\ c & d \end{bmatrix} = ad - bc = -(bc - ad) = -\det \begin{bmatrix} b & a \\ d & c \end{bmatrix},$$

part i) holds for $2 \times 2$ matrices. Suppose part i) holds for $(n-1) \times (n-1)$ matrices. Suppose further that there are indices $j_0 < j_1$ such that $b_{ij_0} = c_{ij_1}$ and $b_{ij_1} = c_{ij_0}$ for $i = 1, \ldots, n$. By the inductive hypothesis, $\det C_{1j} = -\det B_{1j}$ for $j \neq j_0$ and $j \neq j_1$,

$$\det C_{1j_0} = (-1)^{j_1 - j_0} \det B_{1j_1}, \quad \text{and} \quad \det C_{1j_1} = (-1)^{j_0 - j_1} \det B_{1j_0}.$$

Hence, by definition,

$$\det C = c_{11} \det C_{11} - c_{12} \det C_{12} + \cdots + (-1)^{n-1} c_{1n} \det C_{1n}$$
$$= -b_{11} \det B_{11} + b_{12} \det B_{12} + \cdots - (-1)^{n-1} b_{1n} \det B_{1n} = -\det B.$$

Thus i) holds for all $n \in \mathbf{N}$. Similar arguments establish parts ii) and iii). ∎

In the same way we can show that Theorem C.5 holds if "column" is replaced by "row." It follows that we can compute the determinant of a real matrix by expanding along any row or any column, with an appropriate adjustment of signs. For example, to expand along the $i$th row, interchange the $i$th row with the first row, expand along the new first row, and use Theorem C.5 to relate everything back to $B$. In particular, we see that

$$\det[b_{ij}]_{n \times n} = (-1)^{i+1} b_{i1} \det B_{i1} + (-1)^{i+2} b_{i2} \det B_{i2} + \cdots + (-1)^{i+n} b_{in} \det B_{in}.$$

The numbers $(-1)^{i+j} \det B_{ij}$ are called the *cofactors* of $b_{ij}$ in $\det B$

The operations in Theorem C.5 are called *elementary column operations*. They can be simulated by matrix multiplication. Indeed, an *elementary matrix* is a matrix obtained from the identity matrix by a single elementary column operation. Thus elementary matrices fall into three categories: $E(i \leftrightarrow j)$, the matrix obtained by interchanging the $i$th and $j$th columns of $I$; $E(\alpha i)$, the matrix obtained by multiplying the $i$th column of $I$ by $\alpha \neq 0$; and $E(\alpha i + j)$, the matrix obtained by multiplying the $i$th column of $I$ by $\alpha \neq 0$ and adding it to the $j$th column. Notice that an elementary column operation on $B$ can be obtained by multiplying $B$ by an elementary matrix; e.g., $E(i \leftrightarrow j)B$ is the matrix obtained by interchanging the $i$th and $j$th columns of $B$.

These observations can be used to show that the determinant is multiplicative.

**C.6 THEOREM.** *If $B, C$ are $n \times n$ real matrices, then*

$$\det(BC) = \det B \det C.$$

*Moreover, $B$ is invertible if and only if $\det(B) \neq 0$.*

PROOF. It is easy to check that

$$\det(E(i \leftrightarrow j)) = -1, \quad \det(E(\alpha i)) = \alpha, \quad \text{and} \quad \det(E(\alpha i + j)) = 1.$$

Hence, by Theorem C.5,

(7) $$\det(EA) = \det E \det A$$

holds for any $n \times n$ matrix $A$ and any $n \times n$ elementary matrix $E$.

The matrix $B$ can be reduced, by a sequence of elementary column operations, to a matrix $V$, where $V = I$ if $B$ is invertible, and $V$ has at least one zero column if $B$ is not invertible (see Noble and Daniel [9], p. 85). It follows that there exist elementary matrices $E_1, \ldots, E_p$ such that $A = E_1 \ldots E_p V$. Hence, by (7),

$$\det(B) = \det(E_1 \ldots E_p V)$$
$$= \det(E_1) \det(E_2 \ldots E_p V) = \cdots = \det(E_1 \ldots E_p) \det(V).$$

In particular, $B$ is invertible if and only if $\det B \neq 0$.

Suppose $B$ is invertible. Then $V = I$ and by (7),

$$\det(BC) = \det(E_1 \ldots E_p)\det(VC) = \det B \det C.$$

If $B$ is not invertible, then $BC$ is not invertible either (see Noble and Daniel [9], p. 204). Hence, $\det(BC) = 0$ and we have

$$\det(BC) = 0 = \det B \det C. \quad \blacksquare$$

The *transpose* of a matrix $B = [b_{ij}]$ is the matrix $B^T$ obtained from $B$ by making the $i$th row of $B$ the $i$th column of $B^T$, i.e., the $(i \times j)$th entry of $B^T$ is $b_{ji}$. The *adjoint* of an $n \times n$ matrix $B$ is the transpose of the matrix of cofactors of $B$, i.e.,

$$adj\,(B) = [(-1)^{i+j}\det B_{ij}]^T.$$

The adjoint can be used to give an explicit formula for the inverse of an invertible matrix.

**C.7 THEOREM.** *Suppose $B$ is a square real matrix. If $B$ is invertible, then*

$$(8) \qquad\qquad B^{-1} = \frac{1}{\det B}adj\,(B).$$

PROOF. Set $[c_{ij}] = B\,adj\,(B)$. By definition,

$$c_{ij} = (-1)^{1+j}b_{i1}B_{j1} + \cdots + (-1)^{n+j}b_{in}B_{jn}.$$

If $i = j$, then $c_{ij}$ is an expansion of the determinant of $B$ along the $i$th row of $B$, i.e., $c_{ii} = \det B$. If $i \neq j$, then $c_{ij}$ is a determinant of a matrix with two identical rows so $c_{ij}$ is zero. It follows that

$$B\,adj\,(B) = \det B \cdot I.$$

We conclude by Theorem C.4 that (8) holds. $\quad \blacksquare$

In particular,

$$(9) \qquad\qquad \begin{bmatrix} a & b \\ c & d \end{bmatrix}^{-1} = \frac{1}{ad - bc}\begin{bmatrix} d & -b \\ -c & a \end{bmatrix}.$$

The following result shows how the determinant can be used to solve systems of linear equations. (This result is of great theoretical interest but of little practical use because it requires lots of storage to use on a computer. Most packaged routines which solve systems of linear equations use methods more efficient than Cramer's Rule, e.g., Gaussian elimination.)

**C.8 THEOREM** [CRAMER'S RULE]. *Let $c_1, c_2, \ldots, c_n \in \mathbf{R}$ and $B = [b_{ij}]_{n \times n}$ be a square real matrix. The system*

(10)
$$b_{11}x_1 + b_{12}x_2 + \cdots + b_{1n}x_n = c_1$$
$$b_{21}x_1 + b_{22}x_2 + \cdots + b_{2n}x_n = c_2$$
$$\vdots \qquad\qquad \vdots$$
$$b_{n1}x_1 + b_{n2}x_2 + \cdots + b_{nn}x_n = c_n$$

*of $n$ linear equations in $n$ unknowns has a unique solution if and only if the matrix $B$ has a nonzero determinant, in which case*

$$x_j = \frac{\det C(j)}{\det B},$$

*where $C(j)$ is obtained from $B$ by replacing the $j$th column of $B$ by the column matrix $[c_1 \ldots c_n]^T$. In particular, if $c_j = 0$ for all $j$ and $\det B \neq 0$, then the system (10) has only the trivial solution $x_j = 0$ for $j = 1, 2, \ldots, n$.*

PROOF. The system (10) is equivalent to the matrix equation

$$BX = C,$$

where $B = [b_{ij}]$, $X = [x_1 \ldots x_n]^T$, and $C = [c_1 \ldots c_n]^T$. If $\det B \neq 0$, then by Theorem C.7,

$$X = B^{-1}C = \frac{1}{\det B} adj\,(B)C.$$

By definition, $adj\,(B)C$ is a column matrix whose $j$th "row" is the number

$$(-1)^{1+j}c_1 \det B_{1j} + (-1)^{2+j}c_2 \det B_{2j} + \cdots + (-1)^{n+j}c_n \det B_{nj} = \det C(j).$$

(We expanded the determinant of $C(j)$ along the $j$th column.) Thus $x_j = \det C(j)$.

Conversely, if $BX = C$ has a unique solution, $B$ can be row reduced to $I$. Thus $B$ is invertible, i.e., $\det B \neq 0$. ∎

## D.  QUADRIC SURFACES

A *quadric surface* is a surface which is the graph of a relation in $\mathbf{R}^3$ of the form

$$Ax^2 + By^2 + Cz^2 + Dx + Ey + Fz + Gxy + Hyz + Izx = J,$$

where $A, B, \ldots, J \in \mathbf{R}$ and not all $A, B, C, G, H, I$ are zero. We shall only consider the cases when $G = H = I = 0$. These include the following special types.

1. The *ellipsoid*, the graph of

$$Ax^2 + By^2 + Cz^2 = 1,$$

where $A, B, C$ are all positive.

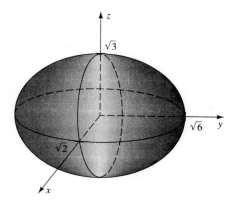

**Figure D.1**

2. The *hyperboloid of one sheet*, the graph of

$$Ax^2 + By^2 + Cz^2 = 1,$$

where two of $A, B, C$ are positive and the other is negative.

3. The *hyperboloid of two sheets*, the graph of

$$Ax^2 + By^2 + Cz^2 = 1,$$

where two of $A, B, C$ are negative and the other is positive.

4. The *cone*, the graph of

$$Ax^2 + By^2 + Cz^2 = 0,$$

where two of $A, B, C$ are positive and the other is negative.

5. The *paraboloid*, the graph of

$$z = Ax^2 + By^2,$$

where $A, B$ are both positive or both negative.

6. The *hyperbolic paraboloid*, the graph of

$$z = Ax^2 + By^2,$$

where one of $A, B$ is positive and the other is negative.

The *trace* of a surface $S$ in a plane $\Pi$ is defined to be the intersection of $S$ with $\Pi$. Graphs of many surfaces, including all quadrics, can be visualized by looking at their traces in various planes. We illustrate this technique with a typical example of each type of quadric.

**D.1 Example.** The ellipsoid $3x^2 + y^2 + 2z^2 = 6$.

SOLUTION. The trace of this surface in the $xy$ plane is the ellipse $3x^2 + y^2 = 6$. The trace of this surface in the $yz$ plane is the ellipse $y^2 + 2z^2 = 6$. And the trace of this surface in the $xz$ plane is the ellipse $3x^2 + 2z^2 = 6$. This surface is sketched in Figure D.1. ∎

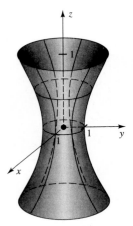

**Figure D.2**

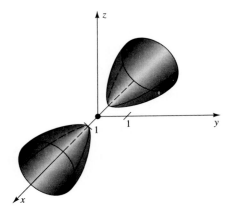

**Figure D.3**

**D.2 Example.** The hyperboloid of one sheet $x^2 + y^2 - z^2 = 1$.

SOLUTION. The trace of this surface in the plane $z = a$ is the circle $x^2 + y^2 = 1 + a^2$. The trace of this surface in $x = 0$ is the hyperbola $y^2 - z^2 = 1$. This surface is sketched in Figure D.2. ∎

**D.3 Example.** The hyperboloid of two sheets $x^2 - y^2 - z^2 = 1$.

SOLUTION. The trace of this surface in the plane $z = 0$ is the hyperbola $x^2 - y^2 = 1$. The trace of this surface in $y = 0$ is the hyperbola $x^2 - z^2 = 1$. This surface has no trace in $x = 0$. This surface is sketched in Figure D.3. ∎

**D.4 Example.** The cone $z^2 = x^2 + y^2$.

SOLUTION. The trace of this surface in the plane $z = a$ is the circle $x^2 + y^2 = a^2$. The trace of this surface in $y = 0$ is a pair of lines $z = \pm x$. This surface is sketched in

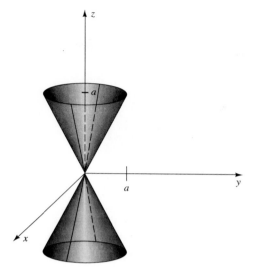

**Figure D.4**

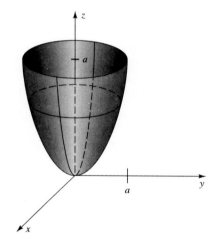

**Figure D.5**

Figure D.4. ∎

**D.5 Example.** The paraboloid $z = x^2 + y^2$.

SOLUTION. If $a > 0$, the trace of this surface in the plane $z = a$ is the circle $x^2 + y^2 = a$. The trace of this surface in $y = 0$ is the parabola $z = x^2$. This surface is sketched in Figure D.5. ∎

**D.6 Example.** The hyperbolic paraboloid $z = x^2 - y^2$.

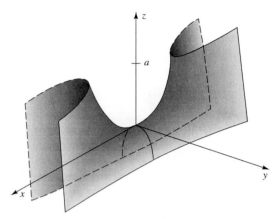

**Figure D.6**

SOLUTION. The trace of this surface in the plane $z = a$ is the hyperbola $a = x^2 - y^2$. (It opens up around the $xz$ plane when $a > 0$, and around the $yz$ plane when $a < 0$.) The trace of this surface in the plane $y = 0$ is the parabola $z = x^2$. This surface is sketched in Figure D.6. (Note: The scale along the $x$ axis has been exaggerated to enhance perspective, so the hyperbolas below the $z = 0$ plane are barely discernible.) ∎

## E.  VECTOR CALCULUS AND PHYSICS

Throughout this section $C = (\varphi, I)$ is a smooth arc in $\mathbf{R}^2$, $S = (\psi, E)$ is a smooth surface in $\mathbf{R}^3$, $\{t_0, \ldots, t_N\}$ is a partition of $I$, and $\{R_1, \ldots, R_N\}$ is a grid on $E$.

**E.1 Remark.** *The integral*

$$(10) \qquad \iint_S d\sigma = \int_E \|N_\psi(u, v)\| \, d(u, v)$$

*can be interpreted as the surface area of $S$.*

Let $(u_j, v_j)$ be the lower left-hand corner of $R_j$ and suppose $R_j$ has sides $\Delta u$, $\Delta v$ (see Figure E.1). If $R_j$ is small enough, the trace of each piece $S_j = (\psi, R_j)$ is approximately equal to the parallelogram determined by the vectors $\Delta u \, \psi_u$ and $\Delta v \, \psi_v$. Hence, by Exercise 7 in Section 8.1,

$$A(S_j) \approx \|(\Delta u \, \psi_u(u_j, v_j)) \times (\Delta v \, \psi_v(u_j, v_j))\|$$
$$= \|N_\psi(u_j, v_j)\| \, \Delta u \Delta v = \|N_\psi(u_j, v_j)\| \, |R_j|.$$

Summing over $j$, we obtain

$$A(S) \approx \sum_{j=1}^{N} \|N_\psi(u_j, v_j)\| \, |R_j|$$

which is a Riemann sum of the integral (10).

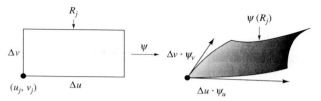

**Figure E.1**

**E.2 Remark.** *If $w$ is a thin wire lying along $C$, whose density (mass per unit length) at a point $(x, y)$ is given by $g(x, y)$, then*

$$\int_C g \, ds$$

*can be interpreted as the mass of $\omega$.*

Since mass is the product of density and length, an approximation to the mass of the piece of $w$ lying along $C_k = (\phi, [t_{k-1}, t_k])$ is given by

$$g(t_k) \cdot L(C_k) = \int_{t_{k-1}}^{t_k} g(t_k) \|\phi'(t)\| \, dt$$

(see Definition 13.4). Summing over $k$, an approximation to the mass of $w$ is

$$\sum_{j=1}^{N} \int_{t_{k-1}}^{t_k} g(t_k) \|\phi'(t)\| \, dt,$$

which is nearly a Riemann sum of the integral

$$\int_I g(\phi(t)) \|\phi'(t)\| \, dt = \int_C g \, ds.$$

The following remark has a similar justification.

**E.3 Remark.** *If $S$ is a thin sheet of metal whose density at a point $(x, y, z)$ is given by $g(x, y, z)$, then*

$$\iint_S g \, d\sigma$$

*can be interpreted as the mass of $S$.*

Work done by a force $F$ acting on an object as it moves a distance $d$ is defined to be $W = Fd$. There are many situations where the force changes from point to point. Examples include the force of gravity (which is weaker at higher altitudes), the velocity of a fluid flowing through a constricted tube (which gets faster at places where the tube narrows), the force on an electron moving through an electric field, and the force on a copper coil moving through a magnetic field.

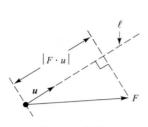

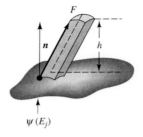

**Figure E.2a**                                    **Figure E.2b**

**E.4 Remark.** *If an object acted on by a force $F : \mathbf{R}^3 \to \mathbf{R}$ moves along the curve $C = (\phi, I)$, then the unoriented line integral*

$$\int_C F \, ds$$

*can be interpreted as the work done by $F$ along $C$.*

An approximation to the work done along $C_k = (\phi, [t_{k-1}, t_k])$ is

$$W_k \approx F(\phi(t_k)) \|\phi(t_k) - \phi(t_{k-1})\| = \int_{t_{k-1}}^{t_k} F(\phi(t_k)) \|\phi'(t)\| \, dt.$$

Summing over $k$, we find that an approximation to the total work along $C$ is given by

$$\sum_{k=1}^{N} \int_{t_{k-1}}^{t_k} F(\phi(t_k)) \|\phi'(t)\| \, dt,$$

which is nearly a Riemann sum of the integral

$$\int_I F(\phi(t)) \|\phi'(t)\| \, dt = \int_C F \, ds.$$

The following remark explains why $F \cdot T$ is called the *tangential* component of $F$ and $F \cdot \boldsymbol{n}$ is called the *normal* component of $F$.

**E.5 Remark.** *Let $\boldsymbol{u}$ be a unit vector in $\mathbf{R}^2$ (respectively, $\mathbf{R}^3$) and $F$ be a function whose range is a subset of $\mathbf{R}^2$ (respectively, $\mathbf{R}^3$). If $\ell$ is the line in the direction $\boldsymbol{u}$ passing through the origin, then $|F \cdot \boldsymbol{u}|$ is the length of the projection of $F$ onto $\ell$ (see Figure E.2a).*

Let $\theta$ represent the angle between $\boldsymbol{u}$ and $F$. By (2) in Section 8.1,

$$|F \cdot \boldsymbol{u}| = \cos\theta \|F\| \, \|\boldsymbol{u}\| = \cos\theta \|F\|.$$

Hence, by trigonometry, $|F \cdot \boldsymbol{u}|$ is the length of the projection of $F$ onto $\ell$. Notice that $F \cdot \boldsymbol{u}$ is positive when $\theta$ is acute and negative when $\theta$ is obtuse.

Combining Remarks E.4 and E.5, we see that $\int_C F \cdot T \, ds$ represents the work done by the tangential component of a force field $F : \mathbf{R}^3 \to \mathbf{R}^3$ along $C$.

**E.6 Remark.** *If $S = (\psi, E)$ is a thin membrane submerged in an incompressible fluid which passes through $S$, and $F(x, y, z)$ represents the velocity vector of the flow of that fluid at the point $(x, y, z)$, then the oriented integral of $F \cdot \boldsymbol{n}$ can be interpreted as the volume of fluid flowing through $S$ in unit time.*

Let $\{E_j\}$ be a grid which covers $E$, and let $h$ be the length of the line segment obtained by projecting $F$ onto the normal line to $S$ at a point $(x_j, y_j, z_j) \in \psi(E_j)$ (see Figure E.2b). If $E_j$ is so small that $F$ is essentially constant on the trace of $S_j = (\psi, E_j)$, then an approximation to the volume of fluid passing through $S_j$ per unit time is given by

$$V_j = A(S_j) \cdot h = A(S_j) \cdot F(x_j, y_j, z_j) \cdot \boldsymbol{n}$$
$$= A(S_j) F(\psi(u_j, v_j)) \cdot N_\psi(u_j, v_j) / \|N_\psi(u_j, v_j)\|.$$

Summing over $j$ and replacing $A(S_j)$ by $\|N_\psi\| \, |E_j|$ (see Remark E.1 above), we see that an approximation to the volume $V$ of fluid passing through $S$ per unit time is given by

$$\sum_{j=1}^{N} F(\psi(u_j, v_j)) \cdot N_\psi(u_j, v_j) |E_j|.$$

This is a Riemann sum of the oriented integral

$$\iint_E F(\psi(u, v)) \cdot N_\psi(u, v) \, dA = \iint_S F \cdot \boldsymbol{n} \, d\sigma.$$

## F.  EQUIVALENCE RELATIONS

A *partition* of a set $X$ is a family of nonempty sets $\{E_\alpha\}_{\alpha \in A}$ such that

$$X = \bigcup_{\alpha \in A} E_\alpha \quad \text{and} \quad E_\alpha \cap E_\beta = \emptyset$$

for $\alpha \neq \beta$. A *binary relation* $\sim$ on $X$ is a subset of $X \times X$. If $(x, y)$ belongs to $\sim$, we shall write $x \sim y$. Examples of binary relations include $=$ on $\mathbf{R}$, $\leq$ on $\mathbf{R}$, and "parallel to" on the class of straight lines in $\mathbf{R}^2$.

A binary relation is called an *equivalence relation* if it satisfies three additional properties.

[THE REFLEXIVE PROPERTY]   For every $x \in X$, $x \sim x$.

[THE SYMMETRIC PROPERTY]   If $x \sim y$, then $y \sim x$.

[THE TRANSITIVE PROPERTY]   If $x \sim y$ and $y \sim z$, then $x \sim z$.

Notice that $=$ is an equivalence relation on $\mathbf{R}$, "parallel to" is an equivalence relation on the class of straight lines in $\mathbf{R}^2$, but $\leq$ is not an equivalence relation on $\mathbf{R}$ (it fails to satisfy the Symmetric Property).

If $\sim$ is an equivalence relation on a set $X$, then

$$\overline{x} := \{y \in X : y \sim x\}$$

is called the *equivalence class* of $X$ which contains $x$.

**F.1 THEOREM.** *If $\sim$ is an equivalence relation on a set $X$, then the set of equivalence classes $\{\overline{x} : x \in X\}$ forms a partition of $X$.*

PROOF. Since $\sim$ is reflexive, each equivalence class $\overline{x}$ contains $x$, i.e., $\overline{x}$ is nonempty. Suppose $\overline{x} \cap \overline{y} \neq \emptyset$, i.e., some $z \in X$ belongs to both these equivalence classes. Then $z \sim x$ and $z \sim y$. By the Symmetric Property and the Transitive Property, we have $x \sim y$, i.e., $y \in \overline{x}$. By the Transitive Property, it follows that $\overline{y} \subseteq \overline{x}$. Reversing the roles of $x$ and $y$, we also have $\overline{x} \subseteq \overline{y}$. Thus $\overline{x} = \overline{y}$. ∎

# References

1. APOSTOL, TOM M., *Mathematical Analysis*. Reading, Massachusetts: Addison-Wesley Publishing Co., 1974.

2. BOAS, RALPH P., JR., *Primer of Real Functions*, Carus Monograph 13. New York: Mathematical Association of America and John Wiley and Sons, Inc., 1960.

3. GRIFFITHS, HUBERT B., *Surfaces*. London and New York: Cambridge University Press, 1976.

4. HOCKING, JOHN G. AND GAIL S. YOUNG, *Topology*. Reading, Massachusetts: Addison-Wesley Publishing Co., 1961.

5. KLINE, MORRIS, *Mathematical Thought from Ancient to Modern Times*. New York: Oxford University Press, 1972.

6. LOOMIS, LYNN H. AND SHLOMO STERNBERG, *Advanced Calculus*. Reading, Massachusetts: Addison-Wesley Publishing Co., 1968.

7. MARSDEN, JERROLD E., *Elementary Classical Analysis*. New York, N.Y.: W. H. Freeman & Co. Publishers, 1990.

8. MUNKRES, JAMES R., *Elementary Differential Topology*, Annals of Mathematical Studies. Princeton: Princeton University Press, 1963.

9. NOBLE, BEN AND JAMES W. DANIEL, *Applied Linear Algebra*, 2nd ed. Englewood Cliffs: Prentice-Hall, Inc., 1977.

10. PRICE, G. BALEY, *Multivariable Analysis*. New York: Springer-Verlag New York, Inc., 1984.

11. RUDIN, WALTER, *Principles of Mathematical Analysis*, 3rd ed. New York: McGraw-Hill Book Co., 1976.

12. SPIVAK, MICHAEL, *Calculus on Manifolds*. New York: W. A. Benjamin, Inc., 1965.

13. TAYLOR, ANGUS E., *Advanced Calculus*. Boston: Ginn and Company, 1955.

14. WIDDER, DAVID V. *Advanced Calculus*, 2nd ed. Englewood Cliffs: Prentice-Hall, Inc., 1961.

15. ZYGMUND, ANTONI, *Trigonometric Series*, Vol. I, 2nd ed. London and New York: Cambridge University Press, 1968.

# Answers and Hints
# to Selected Exercises

## CHAPTER 1

### 1.1 Ordered Field Axioms.

   2. a) $(-3, 7)$. b) $(-3, 5)$. c) $(-1, -1/2) \cup (1, \infty)$. d) $(-2, 1)$.
   3. b) Consider the cases $c = 0$ and $c \neq 0$.
   4. To prove (7), multiply the first inequality in (7) by $c$ and the second inequality in (7) by $b$. Prove (8) and (9) by contradiction.
   5. a) Apply (6) to $1 - a$. b) Apply (6) to $a - 1$. c) Observe that $(\sqrt{a} - \sqrt{b})^2 \geq 0$.
   6. a) Use uniqueness of multiplicative inverses to prove $(nq)^{-1} = n^{-1}q^{-1}$. b) Use part a). c) Use proof by contradiction for the sum. Use a similar argument for the product and identify all rationals $q$ such that $xq \in \mathbf{Q}$ for a given $x \in \mathbf{R} \setminus \mathbf{Q}$. d) Use the Multiplicative Properties.
   7. a) Prove $|x| \leq 1$ implies $|x + 1| \leq 2$. b) Prove $-1 \leq x \leq 2$ implies $|x + 2| \leq 4$.
   8. a) $n > 99$. b) $n \geq 20$. c) $n \geq 23$.
   9. Show first that the given inequality is equivalent to $2a_1 b_1 a_2 b_2 \leq a_2^2 b_1^2 + a_1^2 b_2^2$.
   10. a) Observe that $|xy - ab| = |xy - xb + xb - ab|$ and $|x| < |a| + \varepsilon$.
   11. a) The Trichotomy Property implies i), the Additive and Multiplicative Properties imply ii).

### 1.2 The Well-Ordering Principle.

   3. $\sum_{k=1}^{n} \binom{n}{k} x^{n-k} h^{k-1}$.
   4. See Exercise 5 in Section 1.1.
   5. Observe that $x^2 - x - 2 < 0$ for all $0 < x < 2$.
   6. b) First prove that $2n + 1 < 2^n$ for $n = 3, 4, \ldots$.
   8. a) Show that $n^2 + 3n$ cannot be the square of an integer when $n > 1$. b) The expression is rational if and only if $n = 9$.
   10. This recursion, discovered by P.W.Wade, generates all Pythagorean triples $a, b, c$ which satisfy $c - b = 1$.

### 1.3 The Completeness Axiom.

   1. a) $\inf E = 1$, $\sup E = 8$. b) $\inf E = (3 - \sqrt{29})/2$, $\sup E = (3 + \sqrt{29})/2$. c) $\inf E = a$, $\sup E = b$. d) $\inf E = 0$, $\sup E = \sqrt{2}$. e) $\inf E = 0$, $\sup E = 2$. f) $\inf E = -1$, $\sup E = 2$. g) $\inf E = 0$, $\sup E = 3/2$.

2. Prove $\sup E$ must be an integer.

3. Notice that $a - \sqrt{2} < b - \sqrt{2}$ and use Exercise 6c in Section 1.1.

5. b) Apply Theorem 1.20 to $-E$.

6. b) Apply the Completeness Axiom to $-E$.

8. $t_1 \leq t_2 \leq \ldots$.

9. Use the proof of Theorem 1.24 as a model.

10. After showing that $\sup A$ and $\sup B$ exist, prove $\max\{\sup A, \sup B\} \leq \sup E$.

## 1.4 Functions, Countability, and the Algebra of Sets.

1. a) $f^{-1}(x) = (x + 7)/3$. b) $f^{-1}(x) = 1/\log x$. c) $f^{-1}(x) = \arctan x$. d) $f^{-1}(x) = (-3 + \sqrt{33 + 4x})/2$. e) $f^{-1}(x) = (x - 2)/3$ when $x \leq 2$, $f^{-1}(x) = x - 2$ when $2 < x \leq 4$, and $f^{-1}(x) = (x + 2)/3$ when $x > 4$. f) $f^{-1}(x) = (1 - \sqrt{1 - 4x^2})/2x$ when $x \neq 0$, and $f^{-1}(0) = 0$.

2. If $\phi$ is 1–1 from a set $E$ into $A$, is $\psi(x) := f(\phi(x))$ 1–1 from $E$ into $B$?

3. a) $f(E) = (-4, 16)$, $f^{-1}(E) = (0, 4/5)$. b) $f(E) = [0, 16]$, $f^{-1}(E) = [-2, 2]$. c) $f(E) = [-1/4, 2]$, $f^{-1}(E) = ((-1 - \sqrt{5})/2), (-1 + \sqrt{5})/2)$.
   d) $f(E) = (\log(7/4), \log(31)]$, $f^{-1}(E) = [-1 - \sqrt{4e^5 - 3})/2, (-1 - \sqrt{4\sqrt{e} - 3})/2) \cup ((-1 + \sqrt{4\sqrt{e} - 3})/2, -1 + \sqrt{4e^5 - 3})/2]$.
   e) $f(E) = [-1, 1]$, $f^{-1}(E) = \cup_{k \in \mathbf{Z}}[2k\pi, (2k + 1)\pi]$.

5. a) $[-1, 2]$. b) $[0, 1]$. c) $[0, 1]$. d) $\{0\}$.

6. First prove that $f(A) \setminus f(B) \subseteq f(A \setminus B)$ and $A \subseteq f^{-1}(f(A))$ hold whether $f$ is 1–1 or not.

10. a) Prove it by induction on $n$. b) Use Exercise 9.

## CHAPTER 2

### 2.1 Limits of Sequences.

2. b) Apply Definition 2.1 with $\varepsilon/3$ in place of $\varepsilon$.

4. b) Definition 2.1 works for any $\varepsilon$ including $\varepsilon/C$.

### 2.2 Limit Theorems.

1. c) You may use Exercise 4.

2. a) $-3$. b) $1/5$. c) $\sqrt{2}$. (See Exercise 4.) d) 0.

4. You may wish to prove that $\sqrt{x_n} - \sqrt{x} = (x_n - x)/(\sqrt{x_n} + \sqrt{x})$.

5. Use Theorem 1.24.

7. If $x = \lim_{n \to \infty} x_n$ exists, what is $\lim_{n \to \infty} x_{n+1}$?

9. a) See Exercise 1c) in Section 1.2.

10. a) Modify the proof of Theorem 1.24.

### 2.3 The Bolzano–Weierstrass Theorem.

1. You only need to prove that $\{x_n\}$ *has* a convergent subsequence, not actually find it.

4. See Exercise 4a in Section 1.2

5. Prove $x \leq \sqrt{2x + 3}$ for $-3/2 \leq x \leq 3$.

6. See Exercise 4b in Section 1.2.

8. Prove that $\{x_n\}$ is monotone.

9. a) See Exercise 5c in Section 1.1.

## 2.4 Cauchy Sequences.

    2. You may use Theorem 2.29.

    6. You may use Exercise 4.

    7. Is it Cauchy? (See Exercise 1c in Section 1.2.)

    8. a) Use the Bolzano–Weierstrass Theorem.

## 2.5 Limits Supremum and Infimum.

    1. a) $2, 4$. b) $-1, 1$. c) $-1, 1$. d) $1/2, 1/2$. e) $0, 0$. f) $0, \infty$. g) $\infty, \infty$.

    4. a) First prove $\inf_{k \geq n} x_k + \inf_{k \geq n} y_k \leq \inf_{k \geq n}(x_k + y_k)$. c) By b), the first and final inequalities can only be strict if neither $\{x_n\}$ nor $\{y_n\}$ converges.

    7. Let $s = \inf_{n \in \mathbf{N}}(\sup_{k \geq n} x_k)$ and consider the cases $s = \infty$, $s = -\infty$, and $s \in \mathbf{R}$.

    8. Let $s = \liminf_{n \to \infty} x_n$ and consider the cases $s = \infty$, $s = 0$, and $0 < s < \infty$.

# CHAPTER 3

## 3.1 Two–sided Limits.

    1. See Example 3.3.

    2. b) The limit is zero. c) Does it exist as an extended real number? Why not?

    3. a) $1/2$. b) $3/2$. c) $0$. d) $n$. e) $0$.

    7. b) Use Exercise 8.

    9. b) Use Exercise 8.

## 3.2 One–sided Limits and Limits at Infinity.

    2. a) $-\infty$. b) $0$. c) $0$. d) $1$. e) $\infty$.

    3. a) $-3$. b) $0$. c) $-\infty$. d) $\pi/2$. e) $0$. f) It does not exist.

    4. b) Use Theorem 3.8.

    6. You may use $\cos x \to 1$ as $x \to 0$.

    8. Prove that if $f(x)$ does not converge to $L$ as $x \to \infty$, then there is a sequence $\{x_n\}$ such that $x_n \to \infty$ but $f(x_n)$ does not converge to $L$ as $n \to \infty$.

    9. See Exercise 5 in Section 2.2.

## 3.3 Continuity.

    1. c) Recall that $2^x = e^{x \log 2}$.

    2. c) Recall that $\sqrt{x}$ is continuous on $[0, \infty)$.

    8. b) Use part a) to show $f(x) \equiv f(mx/m) = mf(x/m)$ first. d) If the statement is true, then $m$ must equal $f(1)$.

    9. a) Begin by showing $f(0) = 1$.

## 3.4 Uniform Continuity.

    1. c) Recall that $\sin 2x - \sin 2a = 2 \sin(x - a) \cos(x + a)$.

    6. c) and e) Prove $f(x) = x$ and $g(x) = x^2$ are both uniformly continuous on $(0, 1)$ but only one of them is uniformly continuous on $[0, \infty)$.

    7. a) This is a function analogue of the Monotone Convergence Theorem.

    9. You may wish to prove that if $P(x) = a_n x^n + \cdots + a_0$ is a polynomial of degree $n \geq 1$ whose leading coefficient satisfies $a_n > 0$, then $P(x) \to \infty$ as $x \to \infty$.

# CHAPTER 4

## 4.1 The Derivative.

2. Use Definition 4.1 directly.
4. a) Use (ii) and (vi) to prove $\sin x \to 0$ as $x \to 0$. Use (iii) to prove $\cos x \to 1$ as $x \to 0$. b) First prove $\sin x = \sin(x - x_0) \cos x_0 + \cos(x - x_0) \sin x_0$ for any $x, x_0 \in \mathbf{R}$. c) Inequality (vi) and $0 \le 1 - \cos x \le 1 - \cos^2 x$ play a prominent role here. d) Use (iv) and part c).

## 4.2 Differentiability Theorems.

1. a) $(5x^2 - 6x + 3)/(2\sqrt{x})$, $x > 0$. b) $-(2x+1)/(x^2 + x - 1)^2$, $x \ne (-1 \pm \sqrt{5})/2$. c) $(1 + \log x)x^x$, $x > 0$. d) $f'(x) = (3x^2 + 4x - 1)(x^3 + 2x^2 - x - 2)/|x^3 + 2x^2 - x - 2|$ for $x \ne 1, -1, -2$.
2. a) $3a + c$. b) $(2b - d)/8$. c) $bc$. d) $bc$.
3. No, $f$ is not differentiable at 0.
8. a) Observe that $y^n = x^m$ and use Exercise 6 in Section 4.1 together with the Chain Rule. b) To handle the case $q < 0$, first prove that $(x^{-1})' = -x^{-2}$ for all $x \ne 0$.

## 4.3 The Mean Value Theorem.

1. a) 3. b) $-\infty$. c) $e^{1/6}$. d) 1. e) $-1/\pi$. f) $-1$.
3. b) First prove that if $g(x) = e^{-1/x^2}/x^k$ for some $k \in \mathbf{N}$, then $g(x) \to 0$ as $x \to 0$. Next, prove that given $n \in \mathbf{N}$, there are integers $N = N(n) \in \mathbf{N}$ and $a_k = a_k^{(n)} \in \mathbf{Z}$ such that

$$f^{(n)}(x) = \begin{cases} \sum_{k=0}^{N}(a_k/x^k)e^{-1/x^2} & x \ne 0 \\ 0 & x = 0. \end{cases}$$

(Note: Although for each $n \in \mathbf{N}$ many of the $a_k$'s are zero, this fact is not needed in this exercise.)
4. b) Find the maximum of $f(x) = \log x/x^\alpha$ for $x \in [1, \infty)$.
11. This is the only exercise in this section which has nothing to do with the Mean Value Theorem or L'Hôpital's Rule.
12. a) Compare with Exercise 3 in Section 4.1.

## 4.4 Monotone Functions and the Inverse Function Theorem.

1. a) $a > -3$. b) $a \ge -3/4$. c) $f$ is strictly decreasing on $(-\infty, 1]$ and strictly increasing on $[1, \infty)$.
2. a) $1/\pi$. b) $1/e$. c) $1/\pi$.
3. b) $1/4e$.
6. $f(x) = \pm\sqrt{\alpha}x + c$ for some $c \in \mathbf{R}$.
7. Observe that if $x = \sin y$, then $\cos y = \sqrt{1 - x^2}$.
8. Use Theorem 4.29.
9. Use Darboux's Theorem.
10. Use Darboux's Theorem and Lemma 4.28.

# CHAPTER 5

## 5.1 The Riemann Integral.

    4. a) Use the Sign Preserving Property.

    5. First show that $\int_I f(x)\,dx = 0$ for all subintervals $I$ of $[a,b]$.

    8. Notice that $|x_j - x_{j-1}| \le \|P\|$ for each $j = 1, 2, \ldots, n$.

## 5.2 Riemann Sums.

    1. a) $1/4$. b) $\pi a^2/4$. c) $9$. d) $(3/2)(b^2 - a^2) + (b - a)$. (Note: If $a \ge -1/3$ or $b \le -1/3$, the integral represents the area of a trapezoid; if $a < -1/3 < b$, the integral represents the difference of the areas of two triangles, one above the $x$ axis and the other below the $x$ axis.)

    3. Do not forget that $f$ is bounded.

    5. b) You may use the fact that $\int x^n\,dx = x^{n+1}/(n+1)$.

    8. a) If $|f(x_0)| > M - \varepsilon/2$ for some $x_0 \in [a,b]$, can you choose a nondegenerate interval $I$ such that $|f(x)| > M - \varepsilon$ for all $x \in I$? b) See Example 2.21.

## 5.3 The Fundamental Theorem of Calculus.

    1. a) $15$. b) $1$. c) $(4^{100} - 1)/300$. d) $(e^2 + 1)/4$. e) $(e^{\pi/2} + 1)/2$. f) $4\sqrt{3} - 2\sqrt{11}$.

    3. a) $2xf(x^2)$. b) $h(t)+\sin t\cdot h(\cos t)$. c) $g(-t)$. d) Integrate by parts with $u = f(x)$.

    8. Use the Fundamental Theorem of Calculus.

    10. a) See Exercise 4 in Section 5.1. b) See Exercise 3 in Section 4.3.

## 5.4 Improper Riemann Integration.

    1. a) $3/2$. b) $\pi$. c) $3/2$. d) $4$.

    2. a) $p > 1$. b) $p < 1$. c) $p > 1$. d) $p > 1$.

    3. Compare with Example 5.44.

    4. a) diverges. b) diverges. c) converges. d) converges. e) converges.

    9. Integrate by parts first.

    10. You might begin by verifying $\sin x \ge \sqrt{2}/2$ for $x \in [\pi/4, \pi/2]$ and $\sin x \ge 2x/\pi$ for $x \in [0, \pi/4]$.

## 5.5 Functions of Bounded Variation.

    4. Combine Lemma 4.28 and Theorem 3.39.

    9. For the bounded case, prove $(L) \int_a^b |f'(x)|\,dx \le \operatorname{Var} f \le (U) \int_a^b |f'(x)|\,dx$.

## 5.6 Convex Functions.

    5. Use Remark 5.60.

# CHAPTER 6

## 6.1 Introduction.

    2. a) $1/(1 + \pi)$. b) $5/6$. c) $21/4$. d) $e/(e - 2)$.

    3. a) $1$. b) $\log(2/3)$. c) $-1 + \pi/4$.

    4. $|x| \le 1$

    6. b) Consider the Geometric Series.

7. c) Notice that if the partial sums of $\sum_{k=1}^{\infty} b_k$ are bounded, then $b = 0$.

8. b) See Exercise 7b. d) First prove that if $a_k \geq 0$ and $\sum_{k=0}^{\infty} a_k$ diverges, then $\sum_{k=0}^{\infty} a_k = \infty$.

9. a) Is $na_{2n} \leq \sum_{k=n}^{\infty} a_k$?

10. Use Corollary 6.9.

## 6.2 Series With Nonnegative Terms.

1. c) If $p > 1$, are there constants $C > 0$ and $q > 1$ such that $\log k / k^p \leq C k^{-q}$?

2. a) No, you cannot apply the $p$-Series Test to $k^{1-1/k}$ because the exponent $p := (1 - 1/k)$ is NOT constant, but depends on $k$. d) Try the Integral Test.

3. It converges when $p > 1$ and diverges when $0 \leq p \leq 1$.

5. See Exercise 7 in Section 4.4.

9. It diverges when $0 < q \leq 1$ and converges when $q > 1$.

## 6.3 Absolute Convergence.

2. a) Convergent. b) Divergent. c) Convergent. d) Divergent. e) Convergent. f) Divergent. g) Convergent.

6. a) $(1, \infty)$. b) $\emptyset$. c) $(-\infty, -1) \cup (1, \infty)$. d) $(1/2, \infty)$. e) $(-\infty, \log_2(e))$ (use Stirling's formula when $p = \log_2(e)$). f) $(1, \infty)$.

9. See Exercise 8 in Section 6.1.

10. See Definition 2.32.

## 6.4 Alternating Series.

1. d) Use Example 6.32.

2. a) $[-1, 1)$. b) $(-\sqrt[3]{2}, \sqrt[3]{2})$. c) $(-1, 1]$. d) $[-3, -1]$.

3. a) Absolutely convergent. b) Absolutely convergent. c) Absolutely convergent. d) Conditionally convergent. e) Absolutely convergent.

5. See Exercise 6.32.

6. See Exercise 6.32.

8. Is it Cauchy?

9. Let $c_k = \sum_{j=k}^{\infty} a_j b_j$ and apply Abel's Formula to

$$\sum_{k=n}^{m} a_k \equiv \sum_{k=n}^{m} \frac{c_k - c_{k+1}}{b_k}.$$

## 6.5 Estimation of Series.

1. a) at most 100 terms. b) at most 15 terms. c) at most 10 terms. (To prove $\{a_k\}$ is monotone, show $a_{k+1}/a_k < 1$.)

2. a) $p > 1$.

3. a) $n = 5$. b) $n = 7$. c) $n = 10$. d) $n = 7$.

## 6.6 Additional Tests.

1. a) Divergent. b) Absolutely convergent. c) Divergent. d) Absolutely convergent.

2. a) Absolutely convergent for $p > 0$ and divergent for $p \leq 0$. b) Absolutely convergent for $p > 0$ and divergent for $p \leq 0$. c) Absolutely convergent for $|p| < 1/e$, conditionally convergent for $p = -1/e$, and divergent otherwise. (Use Stirling's formula when $p = \pm 1/e$.)

4. It actually converges absolutely.

# CHAPTER 7

## 7.1 Uniform Convergence of Sequences.

2. a) 2. b) $9/2$. c) $\pi/2$.
4. a) Use Exercise 3c).
6. a) This is different from Theorem 7.9 because $E$ is not necessarily an interval.
7. Modify the proof of Example 4.21 to show $(1 + x/n)^n \uparrow e^x$ as $n \to \infty$. To prove this is a uniform limit, choose $N$ so large that $[a, b] \subset [-N, N]$ and find the maximum of $e^x - (1 + x/N)^N$ on $[a, b]$.

## 7.2 Uniform Convergence of Series.

1. b) Recall that $|\sin \theta| \leq |\theta|$ for all $\theta \in \mathbf{R}$.
5. See Exercise 3a in Section 6.1.
6. Is there a connection between $\sum_{k=1}^{\infty} k^{-1} \sin(xk^{-1})$ and $\sum_{k=1}^{\infty} \cos(xk^{-1})$?
7. Use Abel's Formula.
9. See Example 6.32.

## 7.3 Power Series.

1. a) $(-2, 2)$. b) $(3/4, 5/4)$. c) $[-1, 1)$. d) $[-1/\sqrt{2}, 1/\sqrt{2}]$. (Use Raabe's Test for the endpoints.)
2. a) $f(x) = 3x^2/(1 - x^3)$ for $x \in (-1, 1)$. b) $f(x) = (2 - x)/(1 - x)^2$ for $x \in (-1, 1)$.
   c) $f(x) = 2(\log x + 1/x - 1)/(1 - x)$ for $x \in (0, 2)$, $x \neq 1$, and $f(1) = 0$. d) $f(x) = \log(1/(1 - x^3))/x^3$ for $x \in [-1, 1)$, $x \neq 0$, and $f(0) = 1$. ~ discontinuous at $x = 1$.
4. Use Exercise 8 in Section 2.5 to prove that if $\limsup |a_k/a_{k+1}| < R$ then there is an $r < R$ such that $\{|a_k r^k|\}$ is increasing for $k$ large, i.e., that $\sum_{k=1}^{\infty} a_k r^k$ diverges for some $r < R$.
8. Use the method of Example 7.36 to estimate $|f'(x)|$.
9. First prove that the radius of convergence of $\sum_{k=0}^{\infty} a_k x^k$ is $\geq 1$.
10. a) Use Theorem 6.35 to estimate $\log(n!) = \sum_{k=1}^{n} \log k$. b) $x \in (-1/e, 1/e)$.

## 7.4 Analytic Functions.

1. a) $\cos(3x) = \sum_{k=0}^{\infty} (-9)^k x^{2k}/(2k)!$. b) $2^x = \sum_{k=0}^{\infty} x^k \log^k 2/k!$.
   c) $\cos^2 x = 1 + \sum_{k=1}^{\infty} (-1)^k 2^{2k-1} x^{2k}/(2k)!$. d) $\sin^2 x + \cos^2 x = 1$. e) $x^3 e^{x^2} = \sum_{k=0}^{\infty} x^{2k+3}/k!$.
2. a) $\log(1 - x) = -\sum_{k=1}^{\infty} x^k/k$. b) $x^2/(1 - x^3) = \sum_{k=0}^{\infty} x^{3k+2}$.
   c) $e^x/(1 - x) = \sum_{k=0}^{\infty} (\sum_{j=0}^{k} 1/j!) x^k$. d) $x^3/(1 - x)^2 = \sum_{k=1}^{\infty} k x^{k+2}$.
   e) $\arcsin x = \sum_{k=0}^{\infty} \binom{-1/2}{k} (-1)^k x^{2k+1}/(2k + 1)$. (Use Theorem 7.54.)
3. a) $(x^2 - 1)e^x = -1 - x \sum_{k=2}^{\infty} (k^2 - k - 1) x^k/k!$.
   b) $e^x \cos x = \sum_{k=0}^{\infty} (\sum_{j \in A_k} (-1)^j/((2j)!(k - 2j)!)) \cdot x^k$, where $A_k := \{j \in \mathbf{N} : 0 \leq j \leq k/2\}$.
   c) $\sin x/e^x = \sum_{k=1}^{\infty} (\sum_{j \in A_k} (-1)^{k-j+1}/((2j + 1)!(k - 2j - 1)!)) \cdot x^k$, where $A_k := \{j \in \mathbf{N} : 0 \leq j \leq (k - 1)/2\}$.
   d) $f(x) = \sum_{k=0}^{\infty} x^k \log^{k+1} a/(k + 1)!$.
4. a) $\log_{10} x = \sum_{k=1}^{\infty} (-1)^{k+1} (x - 1)^k/(k \log 10)$, valid for $x \in (0, 2]$. b) $x^2 + 2x - 1 = 2 + 4(x - 1) + (x - 1)^2$, valid for $x \in \mathbf{R}$. c) $e^x = \sum_{k=0}^{\infty} e(x - 1)^k/k!$, valid for $x \in \mathbf{R}$.

10. See Exercise 4 in Section 5.1 and use analytic continuation.

11. First use the Binomial Series to verify that $(1+x)^\beta \geq 1 + x^\beta$ for any $0 < x < 1$.

### 7.5 Applications.

1. The first seven places of the only real root are given by $-0.3176721$.

6. Choose $r_0$ as in the proof of Theorem 7.60, define $\{x_n\}$ by (19), and find a $\delta$ so that $|f(x_0)| \leq \delta$ implies $|x_n - x_{n-1}| < r_0^{n+1}$.

## CHAPTER 8

### 8.1 Algebraic Structure.

1. a) $(a, a, a)$, $a \neq 0$. b) $(a, (20 - 8a)/7, (8+a)/7)$, $a \neq 0$. c) $x - 2y - z = 0$.

4. vi) Write $\|\boldsymbol{x} \times \boldsymbol{y}\|^2 = (\boldsymbol{x} \times \boldsymbol{y}) \cdot (\boldsymbol{x} \times \boldsymbol{y})$ and use parts iv) and v).

5. Use $\boldsymbol{a}, \boldsymbol{b}, \boldsymbol{c}$ to produce a normal to $\Pi$.

9. If $(x_0, y_0, z_0)$ does not lie on $\Pi$, let $(x_2, y_2, z_2)$ be a point on $\Pi$ different from $(x_1, y_1, z_1)$, let $\theta$ represent the angle between $\boldsymbol{w} := (x_0 - x_2, y_0 - y_2, z_0 - z_2)$ and the normal $(a, b, c)$, and compute $\cos \theta$ two different ways, once in terms of $\boldsymbol{w}$ and a second time in terms of the distance from $(x_0, y_0, z_0)$ to $\Pi$.

10. b) Diamonds and squares.

### 8.2 Limits of Sequences.

2. a) $(0, -3)$. b) $(1, 0, 1)$. c) $(-1/2, 1, 0)$.

9. b) Use the Bolzano–Weierstrass Theorem.

### 8.3 Limits of Functions.

1. a) Dom $f = \{(x, y) : x \neq 1, y \neq 1\}$ and the limit is $(0, 3)$. b) Dom $f = \{(x, y) : x \neq 0, y \neq 0\}$ and the limit is $(1, 0, 1)$. c) Dom $f = \{(x, y) : (x, y) \neq (0, 0)\}$ and the limit is $(0, 0)$. d) Dom $f = \{(x, y) : (x, y) \neq (1, 1)\}$ and the limit is $(0, 0)$.

2. a) $\lim_{y \to 0} \lim_{x \to 0} f(x, y) = \lim_{x \to 0} \lim_{y \to 0} f(x, y) = 0$, but $f(x, y)$ has no limit as $(x, y) \to (0, 0)$. b) $\lim_{y \to 0} \lim_{x \to 0} f(x, y) = 1/2$, $\lim_{x \to 0} \lim_{y \to 0} f(x, y) = 1$, so $f(x, y)$ has no limit as $(x, y) \to (0, 0)$. c) $\lim_{y \to 0} \lim_{x \to 0} f(x, y) = \lim_{x \to 0} \lim_{y \to 0} f(x, y) = 0$, and $f(x, y) \to 0$ as $(x, y) \to (0, 0)$.

### 8.4 The Total Derivative.

3. Note that $\cos(\theta + h) - \cos \theta \to 0$ as $h \to 0$ and $\sin(\theta + k) - \sin \theta \to 0$ as $k \to 0$.

## CHAPTER 9

### 9.1 Interior, Closure, and Boundary.

1. a) $E^o = (a, b)$, $\overline{E} = [a, b]$, $\partial E = \{a, b\}$. b) $E^o = \emptyset$, $\overline{E} = E \cup \{0\}$, $\partial E = \overline{E}$. c) $E^o = E$, $\overline{E} = [0, 1]$, $\partial E = \{1/n : n \in \mathbf{N}\} \cup \{0\}$. d) $E^o = \mathbf{R}$, $\overline{E} = \mathbf{R}$, $\partial E = \emptyset$.

2. a) Closed. $E^o = \{(x, y) : x^2 + 4y^2 < 1\}$ and $\partial E = \{(x, y) : x^2 + 4y^2 = 1\}$. b) Closed. $E^o = \emptyset$ and $\partial E = E$. c) Neither open nor closed. $E^o = \{(x, y) : y > x^2, 0 < y < 1\}$, $\overline{E} = \{(x, y) : y \geq x^2, 0 \leq y \leq 1\}$, and $\partial E = \{(x, y) : y = x^2, 0 \leq y \leq 1\} \cup \{(x, 1) : -1 \leq x \leq 1\}$. d) Open. $\overline{E} = \{x^2 - y^2 \leq 1, -1 \leq y \leq 1\}$ and $\partial E = \{(x, y) : x^2 - y^2 = 1, -\sqrt{2} \leq x \leq \sqrt{2}\} \cup \{(x, 1) : -\sqrt{2} \leq x \leq \sqrt{2}\} \cup \{(x, -1) : -\sqrt{2} \leq x \leq \sqrt{2}\}$.

5. Notice that $\boldsymbol{a} \in E^c$ and $E^c$ is open.

8. To show $f$ is continuous at $a$, consider the open interval $I = (f(a) - \varepsilon, f(a) + \varepsilon)$.

## 9.2 Compact Sets.

1. a) Compact. b) Compact. c) Not compact. $H = E \cup \{(0, y) : -1 \le y \le 1\}$. d) Not compact. There is no compact set $H$ which contains $E$.

5. See Exercise 9 in Section 9.1.

7. a) Use Theorem 9.22.

## 9.3 Connected Sets.

6. Try a proof by contradiction.

9. Use Exercise 8. You may assume that $\mathbf{R}^n$ is connected.

10. a) A polygonal path in $E$ can be described as the image of a continuous function $f : [0, 1] \to E$. Use this to prove every polygonal path is connected. c) Prove that if $E$ is not polygonally connected, then there are nonempty open sets $U, V \subset E$ such that $U \cap V = \emptyset$ and $U \cup V = E$.

## 9.4 Continuous Functions.

2. b) Note that $f^{-1}(-1, 1)$ is not open. Does this contradict Theorem 9.33?

4. You may wish to prove that $A$ is relatively closed in $E$ if and only if $E \setminus A$ is relatively open in $E$.

8. See Theorem 3.40.

## 9.5 Applications.

2. See Exercise 7 in Section 7.2.

4. a) $\omega_f(t) = 1$ for all $t$. b) $\omega_f(t) = 0$ if $t \ne 0$ and $\omega_f(0) = 1$. c) $\omega_f(t) = 0$ if $t \ne 0$ and $\omega_f(0) = 2$.

7. a) $\sqrt{1/2}$. b) $f(0)/3$. c) $1/4$. d) $(e^4 - 1)/(2e^2)$.

9. d) You may wish to use Lemma 4.28.


## CHAPTER 10

## 10.1 Introduction.

10. a) Show that if $E$ is not bounded, then there exist $x_n \in E$ and $a \in X$ such that $\rho(x_n, a) \to \infty$ as $n \to \infty$.

## 10.2 Limits of Functions.

1. a) $\mathbf{R}$. b) $[a, b]$. c) $\emptyset$. d) $\{x\}$ if $E$ is infinite, $\emptyset$ if $E$ is finite. e) $\emptyset$.

9. b) See the proof of Theorem 3.26.

## 10.3 Interior, Closure, and Boundary.

1. a) $E^o = (a, b)$, $\overline{E} = [a, b]$, $\partial E = \{a, b\}$. b) $E^o = \emptyset$, $\overline{E} = E \cup \{0\}$, $\partial E = \overline{E}$. c) $E^o = E$, $\overline{E} = [0, 1]$, $\partial E = \{1/n : n \in \mathbf{N}\} \cup \{0\}$. d) $E^o = \mathbf{R}$, $\overline{E} = \mathbf{R}$, $\partial E = \emptyset$.

2. a) Closed. $E^o = \{(x, y) : x^2 + 4y^2 < 1\}$ and $\partial E = \{(x, y) : x^2 + 4y^2 = 1\}$. b) Closed. $E^o = \emptyset$ and $\partial E = E$. c) Neither open nor closed. $E^o = \{(x, y) : y > x^2, 0 < y < 1\}$, $\overline{E} = \{(x, y) : y \ge x^2, 0 \le y \le 1\}$, and $\partial E = \{(x, y) : y = x^2, 0 \le y \le 1\} \cup \{(x, 1) : -1 \le x \le 1\}$. d) Open. $\overline{E} = \{x^2 - y^2 \le 1, -1 \le y \le 1\}$

and $\partial E = \{(x, y) : x^2 - y^2 = 1, -\sqrt{2} \le x \le \sqrt{2}\} \cup \{(x, 1) : -\sqrt{2} \le x \le \sqrt{2}\} \cup \{(x, -1) : -\sqrt{2} \le x \le \sqrt{2}\}$.

5. Notice that $a \in E^c$ and $E^c$ is open.

8. See the description of relative balls following Definition 10.22.

9. To show $f$ is continuous at $a$, consider the open interval $I = (f(a) - \varepsilon, f(a) + \varepsilon)$.

## 10.4 Compact Sets.

1. a) Compact. b) Compact. c) Not compact. $H = E \cup \{(0, y) : -1 \le y \le 1\}$. d) Not compact. There is no compact set $H$ which contains $E$.

5. See Exercise 10 in Section 10.3.

8. a) Notice that if $\cap H_k = \emptyset$, then $\{X \setminus H_k\}$ covers $X$.

10. a) Let $x_k \in E$. Does $E$ contain a point $a$ such that each $B_r(a)$, $r > 0$, contains $x_k$ for infinitely many $k$'s? b) See Exercise 10a) in Section 10.1.

## 10.5 Connected Sets.

6. Try a proof by contradiction.

9. Use Exercise 8.

10. a) A polygonal path in $E$ can be described as the image of a continuous function $f : [0, 1] \to E$. Use this to prove every polygonal path is connected. c) Prove that if $E$ is not polygonally connected, then there are nonempty open sets $U, V \subset E$ such that $U \cap V = \emptyset$ and $U \cup V = E$.

## 10.6 Continuous Functions.

2. b) Note that $f^{-1}(-1, 1)$ is not open. Does this contradict Theorem 10.58?

4. You may wish to prove that $A$ is relatively closed in $E$ if and only if $E \setminus A$ is relatively open in $E$.

9. See Theorem 3.40.

## CHAPTER 11

## 11.1 Partial Derivatives and Partial Integrals.

1. a) $f_{xy} = f_{yx} = e^y$. b) $f_{xy} = f_{yx} = -\sin(xy) - xy \cos(xy)$. c) $f_{xy} = f_{yx} = -2x/(x^2 + 1)^2$.

2. a) $f_x = 2x + y \cos(xy)$ and $f_y = x \cos(xy)$ are continuous everywhere on $\mathbf{R}^2$. b) $f_x = y/(1+z)$, $f_y = x/(1+z)$, and $f_z = -xy/(1+z)^2$ are continuous except when $z = -1$. c) $f_x = x/\sqrt{x^2 + y^2}$ and $f_y = y/\sqrt{x^2 + y^2}$ are continuous everywhere except at the origin.

3. a) $f_x = (2x^5 + 4x^3 y^2 - 2xy^4)/(x^2 + y^2)^2$ for $(x, y) \ne (0, 0)$ and $f_x(0, 0) = 0$. $f_x$ is continuous on $\mathbf{R}^2$. b) $f_x = (2x/3) \cdot (2x^2 + 4y^2)/(x^2 + y^2)^{4/3}$ for $(x, y) \ne (0, 0)$ and $f_x(0, 0) = 0$. $f_x$ is continuous on $\mathbf{R}^2$.

5. a) 1. b) $\sqrt{2}/2$.

6. a) 9/10. b) $e^{-\pi}/2$.

9. c) Choose $\delta > 0$ such that $|\phi(t)| < \varepsilon$ for $0 \le t < \delta$ and break the integral in part b) into two pieces, one corresponding to $0 \le t \le \delta$ and the other to $\delta \le t < \infty$. d) Combine part b) with Theorem 11.10.

10. a) $\mathcal{L}\{te^t\} = 1/(s - 1)^2$. b) $\mathcal{L}\{t \sin \pi t\} = 2s\pi/(s^2 + \pi^2)^2$. c) $\mathcal{L}\{t^2 \cos t\} = 2(s^3 - 3s)/(s^2 + 1)^3$.

## 11.2 The Definition of Differentiability.

1. a)
$$Df(x,y) = \begin{bmatrix} \cos x & 0 \\ y & x \\ 0 & -\sin y \end{bmatrix}.$$

b)
$$Df(s,t,u,v) = \begin{bmatrix} t & s & 2u & 0 \\ -2s & 0 & v & u \end{bmatrix}.$$

c)
$$Df(t) = \begin{bmatrix} 1/t \\ -1/(1+t)^2 \end{bmatrix}.$$

d)
$$Df(r,\theta) = \begin{bmatrix} \cos\theta & -r\sin\theta \\ \sin\theta & r\cos\theta \end{bmatrix}.$$

5. b) The function $f$ might not be differentiable when $\alpha = 1$.
8. a) $z = 0$. b) $2x - 2y - z = 0$.
9. $(-1/2, -1/2, 1/2)$, $2x + 2y + 2z = -1$.
10. b) $ax + by = 1$, where $a^2 + b^2 = 1$. c) $x + y - z = \pm 1$.

## 11.3 Differentiability Theorems.

1. a)
$$D(f+g)(\boldsymbol{a}) = 4 \quad \text{and} \quad D(3f - 2g)(\boldsymbol{a}) = 2.$$

b)
$$D(f+g)(\boldsymbol{a}) = \begin{bmatrix} 3 & -3 \end{bmatrix} \quad \text{and} \quad D(3f-2g)(\boldsymbol{a}) = \begin{bmatrix} -1 & 1 \end{bmatrix}.$$

c)
$$D(f+g)(\boldsymbol{a}) = \begin{bmatrix} 0 & \pi \end{bmatrix} \quad \text{and} \quad D(3f-2g)(\boldsymbol{a}) = \begin{bmatrix} 5\pi & 3\pi \end{bmatrix}.$$

d)
$$D(f+g)(\boldsymbol{a}) = \begin{bmatrix} 2 & 1 & 0 \\ 3 & -2 & 1 \end{bmatrix} \quad \text{and} \quad D(3f-2g)(\boldsymbol{a}) = \begin{bmatrix} 1 & -2 & -5 \\ -1 & 4 & 3 \end{bmatrix}.$$

e)
$$D(f+g)(\boldsymbol{a}) = \begin{bmatrix} 1 & 1 \\ 1 & 1 \\ -1 & 1 \end{bmatrix} \quad \text{and} \quad D(3f-2g)(\boldsymbol{a}) = \begin{bmatrix} 3 & -2 \\ -2 & 3 \\ 2 & -2 \end{bmatrix}.$$

2. c)
$$D(f \cdot g)(\boldsymbol{a})(1,1,1) = 10 \quad \text{and} \quad D(f \times g)(\boldsymbol{a})(1,1,1) = (-4,0,1).$$

3.
$$\frac{\partial w}{\partial p} = \frac{\partial F}{\partial x}\frac{\partial x}{\partial p} + \frac{\partial F}{\partial y}\frac{\partial y}{\partial p} + \frac{\partial F}{\partial z}\frac{\partial z}{\partial p}, \quad \frac{\partial w}{\partial q} = \frac{\partial F}{\partial x}\frac{\partial x}{\partial q} + \frac{\partial F}{\partial y}\frac{\partial y}{\partial q} + \frac{\partial F}{\partial z}\frac{\partial z}{\partial q},$$

$$\frac{\partial^2 w}{\partial p^2} = \frac{\partial}{\partial p}\left(\frac{\partial F}{\partial x}\right)\frac{\partial x}{\partial p} + \frac{\partial F}{\partial x}\frac{\partial^2 x}{\partial p^2} + \frac{\partial}{\partial p}\left(\frac{\partial F}{\partial y}\right)\frac{\partial y}{\partial p}$$

$$+ \frac{\partial F}{\partial y}\frac{\partial^2 y}{\partial p^2} + \frac{\partial}{\partial p}\left(\frac{\partial F}{\partial z}\right)\frac{\partial z}{\partial p} + \frac{\partial F}{\partial z}\frac{\partial^2 z}{\partial p^2}$$

$$= \frac{\partial F}{\partial x}\frac{\partial^2 x}{\partial p^2} + \frac{\partial F}{\partial y}\frac{\partial^2 y}{\partial p^2} + \frac{\partial F}{\partial z}\frac{\partial^2 z}{\partial p^2}$$

$$+ \frac{\partial^2 F}{\partial x^2}\left(\frac{\partial x}{\partial p}\right)^2 + \frac{\partial^2 F}{\partial y^2}\left(\frac{\partial y}{\partial p}\right)^2 + \frac{\partial^2 F}{\partial z^2}\left(\frac{\partial z}{\partial p}\right)^2$$

$$+ 2\frac{\partial^2 F}{\partial x\partial y}\left(\frac{\partial x}{\partial p}\right)\left(\frac{\partial y}{\partial p}\right) + 2\frac{\partial^2 F}{\partial x\partial z}\left(\frac{\partial x}{\partial p}\right)\left(\frac{\partial z}{\partial p}\right)$$

$$+ 2\frac{\partial^2 F}{\partial y\partial z}\left(\frac{\partial y}{\partial p}\right)\left(\frac{\partial z}{\partial p}\right).$$

    8. Compute the derivative of $f \cdot f$ using the Dot Product Rule.

  13. Notice that by Exercise 11 in Section 11.2, this result still holds if "$f$ is in $C^2$" is replaced by "the first-order partial derivatives of $f$ are differentiable."

## 11.4 The Mean Value Theorem and Taylor's Formula.

    3. a) $f(x,y) = 1-(x+1)+(y-1)+(x+1)^2+(x+1)(y-1)+(y-1)^2$. b) $\sqrt{x}+\sqrt{y} = 3+(x-1)/2+(y-4)/4-(x-1)^2/8-(y-4)^2/64 +(x-1)^3/16\sqrt{c^5}+(y-1)^3/16\sqrt{d^5}$ for some $(c,d) \in L((x,y);(1,4))$. c) $e^{xy} = 1+xy+((dx+cy)^4+12(dx+cy)^2xy+12x^2y^2)e^{cd}/4!$ for some $(c,d) \in L((x,y);(0,0))$.

    4. Notice that by Exercise 11 in Section 11.2, this result still holds if "$f$ is in $C^p$" is replaced by "the $(p-1)$-st order partial derivatives of $f$ are differentiable."

    8. Apply Taylor's Formula to $f(a+x,b+y)$ for $p=3$, $x = r\cos\theta$, and $y = r\sin\theta$, and prove that

$$\frac{4}{\pi r^2}\int_0^{2\pi} f(a+r\cos\theta, b+r\sin\theta)\cos(2\theta)\,d\theta = f_{xx}(a,b) - f_{yy}(a,b) + F(r),$$

where $F(r)$ is a function which converges to 0 as $r \to 0$.

    9. c) Let $(x_2, t_2)$ be the point identified in part b), and observe by one-dimensional theory that $u_t(x_2, t_2) = 0$. Use this observation and Taylor's Formula to obtain the contradiction $w_{xx}(x_2, t_2) - w_t(x_2, t_2) \geq 0$.

  10. a) $dz = 2x\,dx + 2y\,dy$. b) $dz = y\cos(xy)\,dx + x\cos(xy)\,dy$. c) $dz = (1 - x^2 + y^2)y/(1 + x^2 + y^2)^2\,dx + (1 + x^2 - y^2)x/(1 + x^2 + y^2)^2\,dy$.

  11. $dw = .05$ and $\Delta w \approx 0.049798$.

  12. $L$ must be measured with no more than 5% error.

## 11.5 The Inverse Function Theorem.

    1. a) Since $f(x,y) = (a,b)$ always has a solution by Cramer's Rule and $Df$ is constant,

$$Df^{-1}(a,b) = \begin{bmatrix} 5/17 & 1/17 \\ -2/17 & 3/17 \end{bmatrix} \quad \text{(see Theorem C.7).}$$

b) Since $f((2k+1)\pi/2, -(2k+1)\pi/2) = f(2k\pi, -2k\pi) = (0, 1)$,

$$Df^{-1}(0, 1) = \begin{bmatrix} 0 & 1 \\ 1 & -1 \end{bmatrix} \quad \text{or} \quad \begin{bmatrix} 1 & 1 \\ 0 & -1 \end{bmatrix} \quad \text{(see Theorem C.7)}$$

depending on which branch of $f^{-1}$ you choose.
c) Since $f(\pm 2, \pm 1) = f(\pm 1, \pm 2) = (2, 5)$,

$$Df^{-1}(2, 5) = \begin{bmatrix} \mp 1/3 & \pm 1/3 \\ \pm 2/3 & \mp 1/6 \end{bmatrix} \quad \text{or} \quad \begin{bmatrix} \pm 2/3 & \mp 1/6 \\ \mp 1/3 & \pm 1/3 \end{bmatrix} \quad \text{(see Theorem C.7)}$$

depending on which branch of $f^{-1}$ you choose.
d) Since $f(0, 1) = (-1, 0)$, one branch of $f^{-1}$ satisfies

$$D(f^{-1})(-1, 0) = \begin{bmatrix} -1/2 & 1 \\ -1/2 & 0 \end{bmatrix} \quad \text{(see Theorem C.7)}.$$

4. $F(x_0, y_0, u_0, v_0) = (0, 0)$, $x_0^2 \neq y_0^2$, and $u_0 \neq 0 \neq v_0$, where $F(x, y, u, v) = (xu^2 + yv^2 + xy - 9, xv^2 + yu^2 - xy - 7)$.

6. a)    $\quad\quad f^{-1}(s, t) = ((s + \sqrt{s^2 - 4t})/2, (s - \sqrt{s^2 - 4t})/2)$.

   b)    $\quad D(f^{-1})(f(x, y)) = \begin{bmatrix} x/(x-y) & 1/(y-x) \\ y/(y-x) & 1/(x-y) \end{bmatrix} \quad \text{(see Theorem C.7)}.$

8. See Remark 11.17.

## 11.6 Optimization.

1. a) $f(1/3, 2/3) = -13/27$ is a local minimum and $(-1/4, -1/2)$ is a saddle point.
   b) Let $k, j \in \mathbf{Z}$. $f((2k+1)\pi/2, j\pi) = 2$ is a local maximum if $k$ and $j$ are even, $f((2k+1)\pi/2, j\pi) = -2$ is a local minimum if $k$ and $j$ are odd, and $((2k+1)\pi/2, j\pi)$ is a saddle point if $k + j$ is odd. c) This function has no local extrema. d) $f(0, 0) = 0$ is a local minimum if $a > 0$ and $b^2 - 4ac < 0$, a local maximum if $a < 0$ and $b^2 - 4ac < 0$, and $(0, 0)$ is a saddle point if $b^2 - 4ac > 0$.
2. a) $f(2, 0) = 8$ is the maximum and $f(-4/5, \sqrt{21}/5) = -9/5$ is the minimum. b) $f(1, 2) = 17$ is the maximum and $f(1, 0) = 1$ is the minimum. c) $f(1, 1) = f(-1, -1) = 3$ is the maximum and $f(-1, 1) = -5$ is the minimum.
3. a) $f(2, 0) = 2$ is the minimum and $f(1/2, \pm\sqrt{15}/2) = 17/4$ is the maximum. b) $f(\pm 2/\sqrt{5}, \pm 1/\sqrt{5}) = 0$ is the minimum and $f(\pm 1/\sqrt{5}, \mp 2/\sqrt{5}) = 5$ is the maximum. c) $\lambda = xy$, $3\mu = x + y$, $f(\pm 1/\sqrt{2}, \mp 1/\sqrt{2}, 0) = -1/2$ is the minimum and $f(\pm 1/\sqrt{6}, \pm 1/\sqrt{6}, \mp 2/\sqrt{6}) = 1/6$ is the maximum. d) $f(1, -2, 0, 1) = 2$ is the minimum and $f(1, 2, -1, -2) = 3$ is the maximum.
7. b) If $DE < 0$, then $ax + by + cz$ has no extremum subject to the constraint $z = Dx^2 + Ey^2$.
8. b) $f(2, 2, 4) = 48$ is the minimum. There is no maximum.

# CHAPTER 12

## 12.1 Jordan Regions.

1. a) $V(E;\mathcal{G}_1) = 3/4$, $V(E;\mathcal{G}_2) = 7/16$, $V(E;\mathcal{G}_3) = 15/64$; $v(E;\mathcal{G}_m) = 0$ for all $m$.
   b) $V(E;\mathcal{G}_1) = 1$, $v(E;\mathcal{G}_1) = 0$; $V(E;\mathcal{G}_2) = 13/16$, $v(E;\mathcal{G}_2) = 3/16$; $V(E;\mathcal{G}_3) = 43/64$, $v(E;\mathcal{G}_3) = 21/64$.
   c) $V(E;\mathcal{G}_1) = 1$, $v(E;\mathcal{G}_1) = 0$; $V(E;\mathcal{G}_2) = 1$, $v(E;\mathcal{G}_2) = 1/4$; $V(E;\mathcal{G}_3) = 15/16$, $v(E;\mathcal{G}_3) = 1/2$.
2. c) First prove that $E$ is a Jordan region if and only if there exist grids $\mathcal{G}_m$ such that $V(\partial E;\mathcal{G}_m) \to 0$ as $m \to \infty$.
4. d) Apply part c) to $E_1 = (E_1 \setminus E_2) \cup E_2$. e) Apply parts c) and d) to $(E_1 \cup E_2) = (E_1 \setminus (E_1 \cap E_2)) \cup (E_2 \setminus (E_1 \cap E_2)) \cup (E_1 \cap E_2)$.
5. a) See Theorem 9.15 or 10.39. b) You may use Exercise 4a).
7. a) Use Corollary 11.30. c) See Exercise 6 in Section 1.4.

## 12.2 Riemann Integration on Jordan Regions.

2. a) 4.
3. a) $-1$. b) $1/2$.
5. Show that the difference converges to zero as $r \to 0+$.
6. b) Area$(E)$.
8. b) The hypothesis $H^o \neq \emptyset$ can be dropped when $\inf_{\boldsymbol{x} \in H} f(\boldsymbol{x}) < f(\boldsymbol{x}_0) < \sup_{\boldsymbol{x} \in H} f(\boldsymbol{x})$.

## 12.3 Iterated Integrals.

1. a) $5/6$. b) $8(2\sqrt{2} - 1)/9$. c) $(1 - \cos(\pi^2/4)) \cdot (2/\pi)$.
2. a) $E = \{(x,y) : 0 \le x \le 1, x \le y \le x^2 + 1\}$ and $\int_0^1 \int_x^{x^2+1} (x + y)\, dy\, dx = 71/60$. b) $E = \{(x,y,z) : 0 \le y \le 1, \sqrt{y} \le x \le 1, 0 \le z \le x^2 + y^2\}$ and $3\int_0^1 \int_{\sqrt{y}}^1 (x^2 + y^2)\, dx\, dy = 26/105$. c) $E = \{(x,y) : 0 \le x \le 1, 0 \le y \le x\}$ and $\int_0^1 \int_y^1 \sin(x^2)\, dx\, dy = (1 - \cos(1))/2$. d) $E = \{(x,y,z) : 0 \le x \le 1, 0 \le y \le x^2, x^3 \le z \le 1\}$ and $\int_0^1 \int_{\sqrt{y}}^1 \int_{x^3}^1 \sqrt{x^3 + z}\, dz\, dx\, dy = 4(2\sqrt{2} - 1)/45$.
3. a) $2/35$. b) 1. c) $492/5$. d) $1/8$.
4. a) $3\pi$. b) $91/30$. c) $88/105$. d) $1/18$.
7. a) See Exercise 6.

## 12.4 Change of Variables.

1. a) $\pi(1 - \cos 4)/4$. b) $3/10$. c) $(\sqrt{2} + \log(1 + \sqrt{2}))(b^3 - a^3)/6$. (Recall that the indefinite integral of $\sec\theta$ is $\int \sec\theta = \log|\sec\theta + \tan\theta| + C$.)
2. a) $(\pi\sqrt{3}/3)\sin 3$. b) $16^2/(3 \cdot 5 \cdot 7)$.
3. a) $(6\sqrt{6} - 7)4\pi/5$. b) $\pi(4e^3 - 2(\sqrt{8} - 1)e^{\sqrt{8}})$. c) $16\sqrt{2}/15$.
4. b) $\pi r^2 d/a$.
5. a) $4/27$. b) $9/112$. c) $3(e - 1)/e$. d) 5. (Use the change of variables $x = u + v$, $y = u - v$.)
6. See Exercise 5 in Section 12.2.
9. See Exercise 7 in Section 8.1.
10. d) $\pi^{n/2}$.

## 12.5 Partitions of Unity.

3. See Theorem 7.58.

## 12.6 The Gamma Function and Volume.

5. Let $\psi_n$ represent the spherical change of variables in $\mathbf{R}^n$ and observe that the cofactor $|A_1|$ of $-\rho \sin \varphi_1 \sin \varphi_2 \ldots \sin \varphi_{n-2} \sin \theta$ in the matrix $D\psi_n$ is identical to $\Delta_{\psi_{n-1}}$ if in $\Delta_{\psi_{n-1}}$, $\theta$ is replaced by $\varphi_{n-2}$ and each entry in the last row of $D\psi_{n-1}$ is multiplied by $\sin \theta$.

8. $r^2 \mathrm{Vol}(B_r(\mathbf{0}))/(n+2)$.

# CHAPTER 13

## 13.1 Curves.

5. a) This curve spirals up the cone $x^2 + y^2 = z^2$ from $(0, 1, 1)$ to $(0, e^{2\pi}, e^{2\pi})$ and has arc length $\sqrt{3}(e^{2\pi} - 1)$. b) This curve coincides with the graph of $x = \pm y^{3/2}$, $0 \le y \le 1$ (looking like a stylized gull in flight), and has arc length $2(\sqrt{13^3}-1)/27$. c) This curve is a straight line segment from $(0, 0, 0)$ to $(4, 4, 4)$ and has arc length $4\sqrt{3}$. d) The arc length of the astroid is 6.

6. a) 27. b) $ab(a^2 + ab + b^2)/(3(a+b))$. c) $12\pi$. d) $(5 + 3\sqrt{5})/2$.

7. b) Use Dini's Theorem.

9. Analyze what happens to $(x, y)$ and $dy/dx := (dy/dt)/(dx/dt)$ as $t \to -\infty$, $t \to -1-$, $t \to -t+$, $t \to 0$, and $t \to \infty$. For example, prove that, as $t \to -1-$, the trace of $\phi(t)$ lies in the fourth quadrant and is asymptotic to the line $y = -x$.

11. b) Take the derivative of $\nu' \cdot \nu'$ using the Dot Product Rule. d) Observe that $\phi(t) = \nu(\ell(t))$ and use the Chain Rule to compute $\phi'(t)$ and $\phi''(t)$. Then calculate $\phi' \times \phi''$ directly.

## 13.2 Oriented Curves.

1. a) A spiral on the elliptic cylinder $y^2 + 9z^2 = 9$ oriented clockwise when viewed from far out the $x$ axis. b) A cubical parabola (it looks like a stylized gull in flight) on the plane $z = x$ oriented from left to right when viewed from far out the plane $y = x$. c) A sine wave on the parabolic cylinder $y = x^2$ oriented from right to left when viewed from far out the $y$ axis. d) An ellipse sliced by the plane $x = z$ out of the cylinder $y^2 + z^2 = 1$ oriented clockwise when viewed from far out the $x$ axis. e) A sine wave traced vertically on the plane $y = x$ oriented from below to above when viewed from far out the $x$ axis.

2. a) 128/3. b) $-\pi\sqrt{2}/2$. c) 0.

3. a) 5. b) $\pi(-1 + \sqrt{5})/2$. c) $|R|(2 - a - b)/2$. d) $-\sin(1) + 1/3$.

4. c) There exist functions $\psi$ and $\tau$ on $[0, 1]$ which are $\mathcal{C}^1$ on $(0, 1) \setminus \{j/N : j = 1, \ldots, N\}$ such that $\tau' > 0$ and $\psi = \phi_j \circ \tau$ on $((j-1)/N, j/N)$ for each $j = 1, \ldots, N$.

7. c) If $F$ is conservative, consider the case when $C$ is smooth first. If $(*)$ holds, use parts a) and b) to prove that $F$ is conservative.

8. Use Jensen's Inequality.

## 13.3 Surfaces.

1. a) $\sqrt{2}\pi(b^2 - a^2)$. b) $4\pi a^2$. c) $4\pi^2 ab$.

2. a) $\phi(u, v) = (u, v, u^2 - v^2)$, $E = \{(u, v) : -1 \le u \le 1, -|u| \le v \le |u|\}$, $\psi_1(t) = (1, t, 1 - t^2)$, $\psi_2(t) = (-1, t, 1 - t^2)$, $\psi_3(t) = (t, t, 0)$, $\psi_4(t) = (t, -t, 0)$,

$I_1 = I_2 = I_3 = I_4 = [-1, 1]$, and $\iint_S g\,d\sigma = 22/3$. b) $\phi(u, v) = (u, u^3, v)$, $E = [0, 2] \times [0, 4]$, $\psi_1(t) = (t, t^3, 4)$, $\psi_2(t) = (t, t^3, 0)$, $\psi_3(t) = (0, 0, t)$, $\psi_4(t) = (2, 8, t)$, $I_1 = I_2 = [0, 2]$, $I_3 = I_4 = [0, 4]$, and $\iint_S g\,d\sigma = (4/27)(145^{3/2} - 1)$.
c) $\phi(u, v) = (3\cos u \cos v, 3\sin u \cos v, 3\sin v)$, $E = [0, 2\pi] \times [\pi/4, \pi/2]$, $\psi_1(t) = ((3/\sqrt{2})\cos t, (3/\sqrt{2})\sin t, 3/\sqrt{2})$, $\psi_2(t) = (3\cos t, 3\sin t, 0)$, $I = [0, 2\pi]$, and $\iint_S g\,d\sigma = 27\pi/2$.

5. b) Use Theorem 12.66.
6. If you got $52\pi$, you gave up too much when you replaced $\|(x, y)\|$ by 3.

### 13.4 Oriented Surfaces.

1. a) Since the $x$ axis lies to the left of the $yz$ plane when viewed from far out the positive $y$ axis, the boundary can be parametrized by $\phi(t) = (3\sin t, 0, 3\cos t)$, $I = [0, 2\pi]$, and $\int_{\partial S} F \cdot T\,ds = -9\pi$. b) The boundary can be parametrized by $\phi_1(t) = (0, -t, 1 + 2t)$, $I_1 = [-1/2, 0]$; $\phi_2(t) = (t, 0, 1 - t)$, $I_2 = [0, 1]$; and $\phi_3(t) = (-t, (1 + t)/2, 0)$, $I_3 = [-1, 0]$; and $\int_{\partial S} F \cdot T\,ds = -1/12$. c) The boundary can be parametrized by $\phi_1(t) = (2\sin t, 2\cos t, 4)$, $I_1 = [0, 2\pi]$, and $\phi_2(t) = (\cos t, \sin t, 1)$, $I_2 = [0, 2\pi]$, and $\int_{\partial S} F \cdot T\,ds = 3\pi$.
2. a) $\pi/2$. b) 16. c) $2\pi^2 ab^2$. d) $\pi/8$.
3. a) $-14/15$. b) $4\pi a^3/3$. c) $(3b^4 + 8a^3 - 8(a^2 - b^2)^{3/2})\pi/12$. d) $-2\pi/3$.
4. b) Use Theorem 12.66.

### 13.5 Theorems of Green and Gauss.

1. a) $8/3$. b) $3\log 3 + 2(1 - e^3)$. c) $-15\pi/4$.
2. a) $(b - a)(c - d)(c + d - 2)/2$. b) $-1/6$. c) 0.
3. a) $2(5 + e^3)$. b) $\pi$. c) 8. d) $\pi abc(a + b + c)/2$.
4. a) $224/3$. b) $2(8\sqrt{2} - 2)/15$. c) $24\pi$.
5. b) $3/2$. c) $\mathrm{Vol}(E) = (1/3)\int_{\partial E} x\,dy\,dz + y\,dz\,dx + z\,dx\,dy$. d) $2\pi^2 ab^2$.
9. c) Use Exercise 8 and Gauss's Theorem.
10. e) Use Green's Theorem and Exercise 5 in Section 12.2.

### 13.6 Stokes's Theorem.

1. a) $-\pi/4$. b) $27\pi/4$.
2. a) 0. b) $-3\pi$. c) $-10\pi$. d) $-1/12$.
3. a) $4\pi/5$. b) $-\pi/(8\sqrt{2})$. c) $28\pi$ (not $-28\pi$ because $\mathbf{i} \times \mathbf{k} = -\mathbf{j}$). d) $32\pi$. e) $-\pi$.
4. a) $18\pi$. b) $8\pi$. c) $3(1 - e) + 3\pi/2$. d) 0.
10. b) $2\pi$.

## CHAPTER 14

### 14.1 Introduction.

1. a) $a_0(x^2) = 2\pi^2/3$, $a_k(x^2) = 4(-1)^k/k^2$, and $b_k(x^2) = 0$ for $k = 1, 2, \ldots$ b) All Fourier coefficients of $\cos^2 x$ are zero except $a_0(\cos^2 x) = 1/2$ and $a_2(\cos^2 x) = 1/2$.
6. a) $a_k(f) = 0$ for $k = 0, 1, \ldots$, $b_k(f) = 4/(k\pi)$ when $k$ is odd and 0 when $k$ is even.
   c) You may wish to use Theorem 9.41.

## 14.2 Summability of Fourier Series.

    5. c) See Exercise 4b in Section 5.1.

    8. See Theorem 9.50.

## 14.3 Growth of Fourier Coefficients.

    4. See Exercise 4a in Section 14.2 and Theorem 7.12.

## 14.4 Convergence of Fourier Series.

    1. Note: It is not assumed that $f$ is periodic.

    2. c) $\pi^2/8$.

    4. a) Use Abel's Formula. For the first identity, you must show that $S_N \rho^N \to 0$ as $N \to \infty$ for all $\rho \in (0,1)$ if $\sum_{k=0}^{\infty} a_k r^k$ converges for all $r \in (0,1)$.

    5. a) Prove that for each fixed $h$, $a_k(f(x+h)) = a_k(f)\cos kh + b_k(f)\sin kh$.

## CHAPTER 15

## 15.1 Differential Forms on $\mathbf{R}^n$.

    1. a) $dy\,dz - 3\,dz\,dx + 2\,dx\,dy$. b) $x^2\,dy\,dz + xy\,dz\,dx + yz\,dx\,dy$. c) $x^2 \cos x\,dx\,dy\,dw - dy\,dz\,dw$.

    2. a) $(2x+2y)\,dx\,dy$. b) $y\cos(xy)\,dx\,dz\,dw + x\cos(xy)\,dy\,dz\,dw - w\sin(zw)\,dx\,dy\,dz - z\sin(zw)\,dx\,dy\,dw$. c) $((x+y)/\sqrt{x^2+y^2})\,dx\,dy\,dz$.

    d) $(y\sin xe^{yz} - y\sin xe^{xy} + x\cos ye^{xy} - \cos xe^{xy})\,dx\,dy\,dz$.

## 15.2 Differentiable Manifolds.

    5. a) Note: If $\mathcal{B}$ is another atlas of $M$, then $\mathcal{A}$ and $\mathcal{B}$ are compatible.

## 15.3 Stokes's Theorem on Manifolds.

    1. $\pi^2 a^6/4$.

    2. $\sum_{j=1}^{n}(-1)^{j-1} a_1 \ldots \widehat{a_j} \ldots a_n \cdot a_j^2$.

# Subject Index

# Notation Index